INDUSTRIAL REFRIGERATION HANDBOOK

Other McGraw-Hill Books of Interest

CHEN and DEMSTER • *Variable Air Volume Systems for Environmental Quality*
GLADSTONE and HUMPHREYS • *Mechanical Estimating Guidebook for HVAC*
GLADSTONE et al • *HVAC Testing, Adjusting, and Balancing Field Manual*
GRIMM and ROSALER • *HVAC Systems and Components Handbook*
HAINES and WILSON • *HVAC Systems Design Handbook*
HARTMAN • *Direct Digital Controls for HVAC Systems*
LEVENHAGEN and SPETHMANN • *HVAC Controls and Systems*
MULL • *HVAC Principles and Applications Manual*
PARMLEY • *HVAC Design Data Sourcebook*
RISHEL • *HVAC Pump Handbook*
ROSALER • *HVAC Maintenance and Operations Handbook*
SUN • *Air Handling Systems Design*
WANG • *Handbook of Air Conditioning and Refrigeration*

INDUSTRIAL REFRIGERATION HANDBOOK

Wilbert F. Stoecker

McGraw-Hill
New York San Francisco Washington, D.C. Auckland Bogota
Caracas Lisbon London Madrid Mexico City Milan
Montreal New Delhi San Juan Singapore
Sydney Tokyo Toronto

Library of Congress Cataloging-in-Publication Data

Stoecker, W. F. (Wilbert F.)
Industrial refrigeration handbook / Wilbert F. Stoecker.
p. cm.
Updated ed. of: Industrial refrigeration. c1988-c1995.
Includes index.
ISBN 0-07-061623-X (alk. paper)
1. Refrigeration and refrigerating machinery. I. Stoecker, W. F. (Wilbert F.) Industrial refrigeration. II. Title.
TP492.S788 1998
621.5′6—dc21 97-47344
CIP

McGraw-Hill

A Division of The ***McGraw-Hill*** *Companies*

22 23 24 25 26 27 QVS/QVS 23 22 21 20 19

ISBN-13: 9781265830991

ISBN-10: 1265830991

The sponsoring editor for this book was Harold B. Crawford, the editing supervisor was Paul R. Sobel, and the production supervisor was Tina Cameron. It was set in Times Roman by Wilbert F. Stoecker.

CONTENTS

PREFACE

This *Industrial Refrigeration Handbook* is a combination and update of Volume I (1988) and Volume II (1995) *Industrial Refrigeration*, first published by Business News Publishing Company. The agreement between the original publisher and the McGraw-Hill Book Company has allowed all of the material to be consolidated into one book and has offered yet another possibility to provide new editions, particularly of Volume I.

The response to Volumes I and II has been gratifying. At least a portion of this acceptance has been due to these books being the only comprehensive books devoted to industrial refrigeration. In writing the first volume, the author was forced to make a selection of chapter topics and stop at some point in order to maintain a manageable book size. There were a number of additional subjects that merited inclusion, so a second volume had always been anticipated. Now the entire work appears under one cover.

The *Industrial Refrigeration Handbook* continues to target as its audience industry professionals, whether they are designers, contractors, operators, manufacturers of components, or sales and application engineers. The book concentrates on practice, but also chooses the fundamentals that provide the skeleton onto which the flesh of applications hang. In the world of operating plants, *theory* is often dismissed as an unnecessary parlor game. This book attempts to counter this impression by showing that the combination of theory with field observations and practical experience enhances the ability to explain the behavior of systems and solve operating problems. Without the framework of theory a new specific answer must be found for each new problem rather than being able to propose an entire group of solutions based on fundamental principles.

The author cannot claim personal credit for most of the ideas and approaches in this book. Most of the facts presented in this book are known to someone else or even a number of people. The book attempts to bring a body of this dispersed knowledge together in an organized form. Much of the substance of this book flows from following the North American and European literature on industrial refrigeration and from close contact and perpetual questioning of some of the outstanding professionals in the field of industrial refrigeration.

Because the book embraces the experience of so many practicing professionals, any effort to acknowledge all whose ideas are incorporated in this book will omit many names. There are, however, members of four groups who have been rich sources of experience. These organizations are the International Institute of Refrigeration, with headquarters in Paris, whose meetings have always enjoyed a strong emphasis on topics in industrial refrigeration and bring worldwide results from research and field experience into print. The second group is represented in the refrigeration committees of the American Society of Heating, Refrigerating and Air-Conditioning Engineers (ASHRAE). Especially valuable has been my association with the International Institute of Ammonia Refrigeration (IIAR) as Scientific Advisor for twelve years and member of the Research Committee. The members, officers, and Board of Directors of IIAR are a collective authority on industrial refrigeration. The final association that has been valuable in exposing issues and topic that appear in many locations of the book is my role in the semiannual Industrial Refrigeration Workshop. As associate director and one of the presenters in this week-long course, the opportunity to learn from the research of the other university staff members, the current practice and experience of the industry instructors, and the field problems confronted by the participants in the workshop are a gold mine of knowledge.

Chapter 1, The Field of Industrial Refrigeration, presents a panorama of the wide, but still generally recognized, segments of industrial refrigeration. Chapter 2, Properties, Cycles and Units, first covers SI units, which are used by most of the world and slowly being adopted in the United States. The major portion of this chapter focuses on the thermodynamic principles surrounding refrigeration. Chapter 3 concentrates on multistage, expecially two-stage, systems because they are so commonly used in the low-temperature systems typical of industrial refrigeration.

The next several chapters concentrate on the major components found in all refrigeration systems, namely compressors, evaporators, and condensers. Two major types of compressors serve the field of industrial refrigeration, reciprocating compressors (Chapter 4) and screw compressors (Chapter 5). The two major types of evaporators are air coolers and liquid coolers, which are dealt with in Chapter 6. Low-temperature air-cooling evaporators becomes frosted and thus require periodic defrost, which is usually accomplished by hot gas in industrial systems. The most popular type of condenser in industrial refrigeration service is the evaporative condenser, which is the emphasis of Chapter 7.

The next group of chapters connect the major components. Chapter 8, Liquid Recirculation, explains and analyzes a popular method of distributing refrigerant to the various locations throughout the plant. Chapter 9 explores the unique piping requirements in industrial refrigeration, and Chapter 10 addresses vessels. Vessels are common in industrial refrigeration and not, for example, in large air conditioning systems. Chapter 11 shows the wide range of valves and refrigerant controls needed for effective operation of the plant.

The chapter on refrigerants, now Chapter 12, has been completely rewritten in this and the predecesor editions, because the greatest upheaval in the history of the refrigerant industry has been the need to convert from CFCs to non-ozone-depleting refrigerants. Chapter 13 concentrates on safety, which is a requirement in using all refrigerants, but especially in systems using ammonia, which is the most widely used refrigerant in industrial refrigeration. The development of the microprocessor has profoundly affected all industrial operations, and Chapter 14, Electrical Control and Instrumentation outlines the impact on industrial refrigeration systems. The special provisions needed in industrial refrigeration for lubricating the compressors and handling the oil are studied in Chapter 15, and the perennial objective of operating the plant in an energy-efficient manner is considered in Chapter 16.

The most significant application of industrial refrigeration is in the refrigeration and freezing of food and how the refrigeration plant serves these roles and the special equipment needed are examined in Chapter 17. Proper design of the industrial refrigeration plant demands an accurate calculation of the refrigeration load, and these procedures are shown in Chapter 18. Buildings and spaces operating at low temperature require special treatment, as described in Chapter 19. For many years the field of brines and antifreezes has been stable and even static, but in the last few years new secondary coolants have been developed and these may open new applications (Chapter 20). The final text chapter, Chapter 21, addresses the subsection of industrial refrigeration, namely the ultra-low-temperature systems which bring a new set of requirements not normally encountered in conventional industrial refrigeration.

The appendix is intended to be more complete than most refrigeration books to bring into one location the thermodynamic data of the most-used industrial refrigerants.

W. F. Stoecker
Professor Emeritus, University of Illinois
1997

About the Author

Wilbert F. (Will) Stoecker is Professor Emeritus of Mechanical Engineering at the University of Illinois at Urbana-Champaign, and the author of two McGraw-Hill books, *Refrigeration and Air Conditioning*, now in its second edition, and *Design of Thermal Systems*, in its third edition. For 12 years, he has been a science adviser for the International Institute of Ammonia Refrigeration as well as a member of its research committee. Mr. Stoecker is also an associate director of the semiannual Industrial Refrigeration Workshop, a week-long course in which he also serves as a presenter.

CHAPTER 1

THE FIELD OF INDUSTRIAL REFRIGERATION

1.1 INDUSTRIAL REFRIGERATION AS DISTINGUISHED FROM COMFORT AIR CONDITIONING

Both air conditioning and industrial refrigeration have a common objective, cooling some substance. Both types of systems are constructed of common hardware—compressors, heat exchangers, fans, pumps, pipe, duct, and controls. The predominant working fluids are air, water, and refrigerants. A refrigeration system is an integral part of both systems.

While much is shared by both fields, there are enough differences in the systems, components, design practices, and business methods to justify separate treatment of industrial refrigeration. By almost every standard, the size of the comfort air conditioning industry dominates over industrial refrigeration, including the number of units sold and installed, gross monetary sales, and the number of engineers and technicians employed. Nevertheless, industrial refrigeration is a lively business of significant size, provides many technical challenges to its practitioners, and serves a crucial role in industry and society.

While there is always a danger in isolating one's area of technical activity and thus losing the input from related fields, industrial refrigeration should not be considered a branch of comfort air conditioning. Industrial refrigeration is characterized by special rather than standard jobs, and the fraction of total cost devoted to engineering and design usually is higher than with air conditioning. Many problems can occur in the lower temperatures typical of industrial refrigeration plants that do not occur in the usual comfort air conditioning temperatures. The refrigeration plant for an air- conditioning system is usually a factory-assembled package ready for connection to electrical, water, and air services. In industrial refrigeration, built-up systems are more common than complete packages because of the great variety of installations.

Another distinguishing feature is that industrial refrigeration systems usually consist of parallel compressors, condensers, and evaporators in contrast to packages with one of each of these components serving a comfort air conditioning assignment. When an industrial refrigeration plant is expanded, it is usually done by installing an additional compressor, condenser, and/or evaporator.

Another comparison is that air conditioning systems are usually characterized by duct networks carrying air and piping networks carrying water. Industrial refrigeration systems only occasionally employ air duct networks, but commonly incorporate extensive refrigerant distribution networks.

This chapter continues with an exploration of the activities encompassed by industrial refrigeration, then briefly describes some of the refrigeration assignments that illustrate the rich diversity of applications.

1.2 WHAT IS INDUSTRIAL REFRIGERATION?

One characteristic of industrial refrigeration is the temperature range it embraces. While evaporating temperatures may be as high as 15°C (60°F), the range extends down to about −60° or −70°C (−76° or −94°F). At temperatures much lower than about −70°C (−94°F) another industry called *cryogenics* takes over, producing and using liquefied natural gas, liquid nitrogen, liquid oxygen, and other low-temperature substances. If industrial refrigeration were described as the refrigeration that occurs in the food, chemical, and process industries, then probably two- thirds of the applications would be covered. Figure 1.1 shows the machine room of a refrigerated warehouse, which is one of the many examples of industrial refrigeration in the food industry. Another significant application exists in the manufacturing industry and in laboratories where special conditions, especially low temperatures, must be maintained. While the low temperature range may be the best single means of characterizing industrial refrigeration, some industrial heat pumping applications that reject heat at much higher temperatures than ambient could also be called industrial refrigeration.

The remainder of this chapter briefly describes some of the major applications of industrial refrigeration in food processing and preservation and in the chemical, process, and construction industries.

FIGURE 1.1
The machine room of a refrigerated warehouse that maintains a temperature of −23.5°C (−10°F) in the spaces where frozen food is stored. (*The Haskell Company*)

1.3 REFRIGERATED STORAGE OF UNFROZEN FOODS

The shelf life of most foods is enhanced by storage at low temperatures. Figure 1.2 shows the shelf life of several food products as a function of temperature[1]. The life of all products shown increases as the storage temperature drops, but for certain food products, such as bananas and certain kinds of apples, the optimum storage temperature is higher than the freezing temperature. For the products shown in Figure 1.2 and for most meats, fish, fruit, and vegetables, the storage life can be prolonged by maintaining the product at low temperatures. Many foods are stored in the unfrozen state and rarely are frozen, for example, bananas, apples, tomatoes, lettuce, cabbage, potatoes, and onions. The optimum storage temperatures of several products, from the perspective of both preservation and economy, are shown in Table 1.1. Some fruits can be stored at temperatures slightly below 0°C (32°F) without freezing, because the water in the fruit is in solution with sugars and other substances providing a lower freezing point than 0°C (32°F). Certain meat and poultry products maintain their quality better when stored *superchilled* at temperatures about −1°C (30°F) at which temperatures they remain unfrozen.

Immediately following harvest, the fruits or vegetables are often warm. While the products would cool eventually if stored in a refrigerated room, the rate of such cooling might be too slow. Sometimes a *precooling* process[3] (distinguished from room cooling) is applied. Forced air cooling, application of ice,

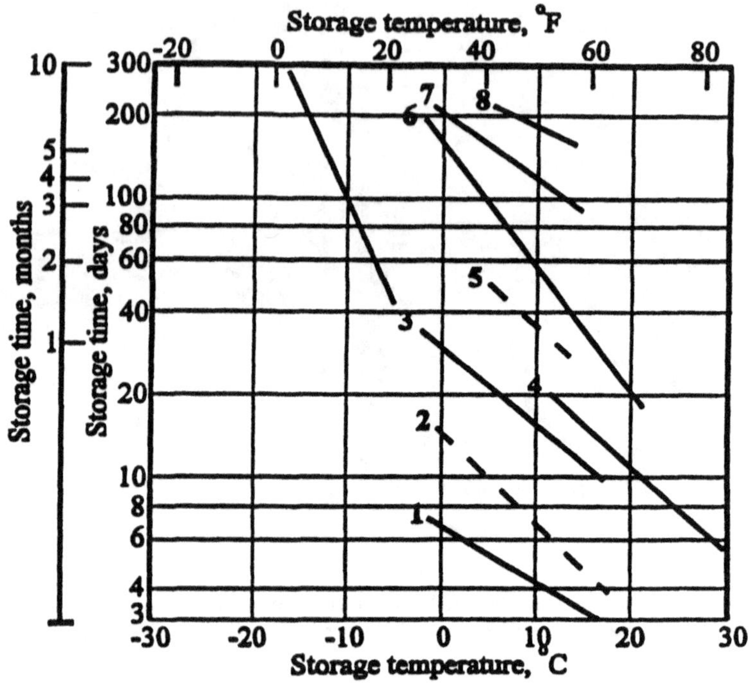

FIGURE 1.2
Rough indication of shelf life of several food products at various temperatures. The products are (1) chicken, (2) lean fish, (3) beef, (4) bananas, (5) oranges, (6) apples, (7) eggs, and (8) apples stored in controlled carbon dioxide atmosphere.

hydrocooling where the product is inundated with chilled water, and vacuum cooling are examples of methods of post-harvest cooling. These methods are examined further in Chapter 17, The Refrigeration and Freezing of Food. Vacuum precooling is one of the methods used wherein a load of product, such as lettuce, is placed into a chamber and a vacuum pulled until a small amount of liquid water on the leaves evaporates. The process provides rapid internal cooling of the lettuce head. Sometimes refrigeration plants are loaded on truck trailers (Figure 1.3) and moved to growing areas during the harvest season.

1.4 FROZEN FOOD

The history of frozen food goes back to middle 1800's. As early as 1861 fish was frozen with ice and salt and stored for 8 to 10 months. The first successful ocean shipment of frozen meat was on the French ship *Paraguay* in 1878 when it sailed from Buenos Aires to France. In 1880 the *S.S. Strathleven* transported a load of frozen meat from Australia to England. In North America during a scientific expedition to Labrador in the years 1912-1915 expedition member Clarence Bird-

TABLE 1.1
Recommended storage temperatures for several unfrozen food products[2].

Storage temperature		
Product	°C	(°F)
Apples	−1 to −0.5	(30 to 31)
(some varieties)	3 to 4	(38 to 40)
Avacados	4 to 13	(40 to 55)
Bananas	13 to 14	(56 to 58)
Cabbage	0	(32)
Cheese	0 to 1	(32 to 34)
Lettuce	0 to 1	(32 to 34)
Pears	−2 to 0	(29 to 31)
Poultry	−1 to 2	(30 to 35)
Strawberries	−0.5 to 0	(31 to 32)
Tomatoes	3 to 4	(38 to 40)

seye noticed that fish frozen at subzero temperatures maintained their quality for long periods of time.

The modern era for frozen foods began with the discovery of quick freezing. By freezing the product within several hours (or several minutes in the case of food in small particles) rather than days, the build-up of microscopic ice crystals in the interior of the product is avoided. The frozen food market in the U.S. has grown from 180,000 kg (400,000 lb) in 1942 to 6 billion kg (13 billion lb) in 1995.

Popular methods of freezing food include air-blast freezing (where low-temperature air passes at high velocity over food packages), contact freezing (packed or unpacked food is placed between refrigerated plates), immersion freezing (food is immersed in low-temperature brine), belt freezers which force low-temperature air at high velocity from underneath the belt, spiral freezers where the product is conveyed on a conveyer belt within a low-temperature chamber, and cryogenic freezing using liquid carbon dioxide or nitrogen.

In the distribution chain, frozen foods must be stored, perhaps first in large refrigerated warehouses, then later in small quantities near the grocery market. Storage temperatures for frozen foods are typically between −20°C (−5°F) and −23°C (−10°F), although some products such as fish and ice cream are usually stored at lower temperatures. Warehouses storing fish often maintain temperatures as low as −30°C (−22°F). A food like ice cream behaves similarly to brine or antifreeze. The water begins to freeze at a certain temperature which concentrates the remainder and lowers its freezing temperature. The range of temperatures through which ice cream passes as it freezes starts at about −2°C (28°F) and ultimately reaches −30°C (−22°F). Ice cream is packaged at about −5°C (23°F) at which temperature it is stiff, but will still flow. Then it moves to a hardening room kept at −30°C (−22°F) for complete freezing.

FIGURE 1.3
Refrigeration plant on a truck trailer for location near the field for post-harvest cooling of agricultural products. (*Post-Harvest Technologies*)

1.5 REFRIGERATION IN FOOD PROCESSING

The usual purpose for refrigerating and freezing foods is to preserve their quality, which technically is changing what would be natural processes of deterioration. Refrigeration may also play a role in facilitating a change in the character or chemical structure of food—called *processing* here. Some foods incorporating refrigeration in their processing are cheese, beverages (such as beer, wine, and juice concentrates), and instant coffee. The procedure for making cheese varies from one type to another, but all cheeses are made from milk to which is added a starting organism to cause the formation of curds and whey. Once the whey is drained, the curds form the basis of the cheese. Refrigeration serves a role in the manufacture of most cheeses during the curing step. Depending upon the type, the cheese is held at between 10°C (50°F) and 20°C (68°F) for a period ranging from several days to several months.

Although it might not be obvious to the general public, refrigeration is an important function in the bakery industry. Some applications include ingredient cooling and dough mixing; freezing of bread for later holding, thawing, and sale; refrigerating unfrozen dough products; freezing of dough for the food service industry and supermarkets; and the freezing of fried and baked products for sale to consumers for microwaving or toasting.

With beer, two main chemical reactions occur in the brewing operation: (1) converting grain starches into sugar; and (2) fermentation, which converts the sugar into alcohol and carbon dioxide. The fermentation process is exothermic, so if heat were not removed, the temperature would rise and ultimately terminate fermentation. Refrigeration maintains the fermenting mixture between 7°C and 13°C (45°F and 55°F), and also cools the storage facilities where the beer is kept for two or three months until it matures.

The production of wine also demands refrigeration. Following fermentation, the wine is held in refrigerated storage at 10°C (50°F) in insulated stainless steel tanks for six months to two years. In addition, wines require a cold stabilization process to precipitate potassium bitartrate (PB). PB is a harmless but unaesthetic crystal and would be particularly noticeable in bottles of white wine. PB will precipitate naturally at 10°C (50°F), but only over a lengthy period. To accelerate the process, the wine is chilled to −4°C (25°F) for ten days.

To concentrate some foods that are sensitive to high temperatures, water may be removed by evaporation or by freezing. One concept is to employ a refrigeration system as a heat pump with the condenser providing heat for vaporization of the water in the product and the evaporator liquefying this water vapor. Because of its high first cost this application is rare while the heat-pumping principle of thermocompression is preferred. In thermocompression the evaporated water vapor itself is pumped with a high-volume-flow compressor from the product fluid to a condenser. Of the concentrating techniques that apply refrigeration, freeze concentration[4] is the most popular. Some of the products that can be concentrated by freezing include citrus fruit juices, vinegar, beer and wines, coffee and tea extracts, sugar syrups, and aroma extracts. The most-used procedure is to bring the solution in contact with refrigerated surfaces on which a thin layer of ice freezes. The ice is continuously removed by scraping and then separated from the concentrated solution.

Another example of refrigeration being used for food processing is in the drying of fish.[5] Figure 1.4 schematically shows air at a low humidity and moderate temperature removing moisture from the product. Some water is then removed as the air passes through the evaporator of the refrigeration unit that provides thermal load for the refrigeration system. The discharge gas from the refrigerant compressor divides, some passing to the air-cooled condenser that warms the air and the other portion to an auxiliary condenser that essentially removes the energy introduced to the circuit by the compressor. This concept is particularly applicable in localities where the cost of electricity is moderate.

1.6 FREEZE DRYING

Freeze drying is a process that finds application in both the food industry as well as some process industries, such as pharmaceuticals. The processes include first freezing the product, then removing the water by sublimation in which the water in a solid state converts directly to vapor without passing through a liquid state. To carry out the sublimation process, heat is carefully added to the product being maintained under a vacuum in chambers such as one shown in Figure 1.5. To achieve the vacuum a combination of refrigeration and mechanical pumping is employed. The mechanical pump mostly removes air and other

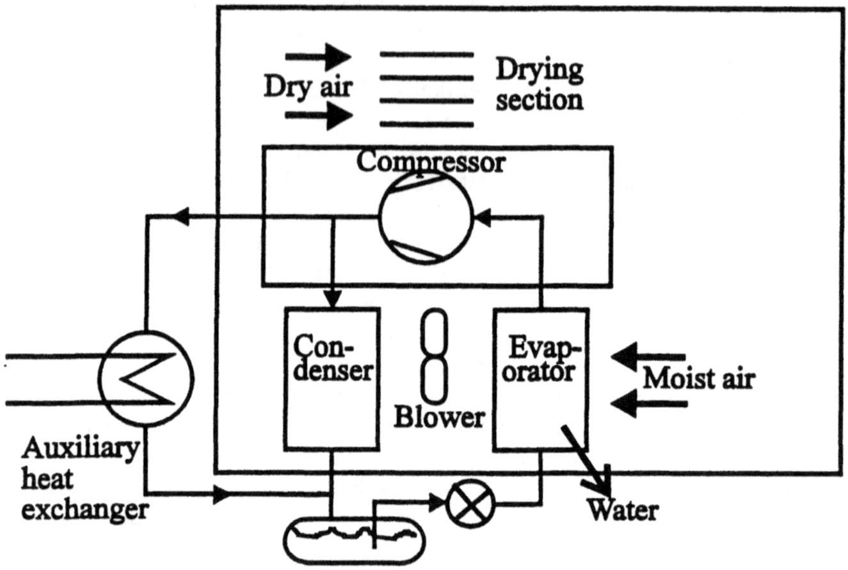

FIGURE 1.4
Heat pump for drying fish.

noncondensible gases, while the refrigeration system chills an evaporator to perhaps −50°C (−58°F). Water from the product deposits on the evaporator coils in the form of frost. Refrigeration thus serves the initial freezing process as well as the evaporator coils of the vacuum chamber.

The pharmaceutical industry uses freeze drying for certain products where the removal of water at higher temperatures would damage the product. The percentage of food that is preserved by freeze drying is very low, but the process finds a role for certain applications. Once a food product is freeze dried, it needs no further refrigeration and at the time of preparation of eating the addition of water restores the product to a taste remarkably close to the original quality.

1.7 INDUSTRIAL AIR CONDITIONING

If all air conditioning were to be classified into two groups—comfort air conditioning and industrial air conditioning—the basic difference would be in the objective. Comfort air conditioning is performed for people, while industrial air conditioning is directed toward processes. On the air side, industrial air conditioning differs in one or more features: level of temperature, need for precise humidity control, a higher standard of filtration, and the removal of contaminants. Industrial air conditioning appears in such applications as printing plants,

FIGURE 1.5
A vacuum chamber for freeze drying. (*VirTis Company*)

where humidity control is essential for proper registration of multicolor printing; textile plants, to help prevent fibers from breaking and to reduce static electricity; the manufacture of photographic materials; clean rooms for manufacturing of electronic equipment; and in a wide variety of laboratories. One difference between refrigeration for comfort and for industrial air conditioning might be in the higher precision of temperature control usually demanded for industrial air conditioning.

1.8 REFRIGERATION IN MANUFACTURING PLANTS

There are frequent and varied needs for refrigeration in the machining, forming, and fabrication of products made from metals, plastics, and other materials. An example is the cooling of the fluid sprayed on the point-of-contact between a machine tool and metal. In many factories, a network of pipes carries compressed air for air motors and other users of air power throughout the plant. When air, particularly humid air, is compressed and brought back to room temperature, water in the compressed air may condense and adversely affect the operation of equipment using the compressed air. To condense the water where it can be removed, some compressed air systems use a refrigerated heat exchanger located immediately after the air compressor.

Many manufacturing plants operate environmental test chambers to duplicate or even exaggerate extreme conditions to which the product might be subjected during normal use. These extremes almost always include high and low temperatures and high and low humidities. Environmental test chambers call for refrigeration to achieve their low temperatures. The task of the refrigeration plant in usually to provide a source of low-temperature air.

1.9 REFRIGERATION IN CONSTRUCTION

Two important uses of refrigeration in large construction projects are the cooling of large masses of concrete,[6] and the freezing of soil to facilitate excavation.[7] During the hardening of concrete, its chemical reaction is exothermic, and if heat were not removed, the high temperatures would stress the concrete. Concrete may be cooled by chilling the sand, gravel, water, and cement before mixing, or by imbedding chilled-water pipes in the concrete. Another application of refrigeration is the freezing of wet soil in the vicinity of excavations to prevent cave-ins. The typically pipes are driven into the soil surrounding the area where excavation is to occur. Then cold brine circulates through the pipe coils and freezes the soil in its vicinity. When the frozen zones overlap, excavation can proceed.

Another special application of refrigeration related to construction is when a refrigeration plant maintains permafrost in its frozen state beneath a building or other facility in cold climates. If the heat from the structure were permitted to thaw the permafrost, the structure would sink. An example of this application[8] is a hybrid thermosyphon serving a power plant in Gakona, Alaska. The permafrost is maintained frozen in winter by thermosyphon heat exchangers and in the summer by a mechanical refrigeration plant.

1.10 ICE RINKS AND FACILITIES FOR OTHER WINTER SPORTS

Some ice rinks serve specific purposes, such as for recreational or speed skating, hockey, and curling, while others are general purpose that at different times are used for different types of events. Desired ice temperatures[9] are typically −5°C (23°F) for hockey, −3.5°C (26°F) for figure skating, and −2°C (28°F) for recreational skating. The ice is maintained at a thickness of approximately 25 mm (1 in) by passing either cold brine or direct refrigerant (R-22 or ammonia) through a grid of pipes in the ice. The brine or refrigerant temperature is usually held about 5°C (9°F) below the desired ice temperature.

In addition to the widespread availability of ice rinks, some high-visibility facilities are constructed every four years for Olympic games for speed skating, figure skating, bobsledding, and tobogganing.

1.11 REFRIGERATION IN THE CHEMICAL INDUSTRY

The pharmaceutical industry has already been mentioned as a user of refrigeration in freeze-drying processes. An approach that seems popular in that industry is also to chill a fluid and distribute this cold fluid (glycol antifreeze or alcohol,

FIGURE 1.6
Turbocompressors compressing propane in a Texas LPG fractionation plant. (*York International Corporation*)

for example) to the location where refrigeration is needed. Low-temperature alcohol may chill the jacket of a vessel, or glycol may be sent to air-cooling coils located in a room where filling of powder is taking place that must be maintained at a low humidity.

The chemical, petrochemical, and oil-refining industries often need large-sized refrigeration plants. Some important operations often requiring refrigeration are:

- separation of one gas from another by liquefying more of one gas
- condensation of gases, such as capturing gas venting from a liquid storage tank
- solidification of one substance in a mixture to separate it from others
- maintenance of a stored liquid at low temperature to control pressure in the containing vessel
- removal of reaction heat
- humidity control for hygroscopic chemicals

Two major concepts in providing the refrigeration are direct and indirect. In direct refrigeration the product itself is the refrigerant and is compressed, condensed, and throttled to form liquid at a low temperature. The petrochemical industry frequently uses propane, ethane, and ethylene, for example, as refrigerants. Often these chemical plants are large and traditionally use high-capacity centrifugal compressors as shown in Figure 1.6. The facility shown is in a liquefied petroleum gas fractionation plant where 8 two-stage centrifugals compress

FIGURE 1.7
A packaged refrigeration plant to serve a process industry. (*Tomco Equipment Company*)

propane in the refrigeration cycle. An alternate approach is to provide indirect chilling through the use of a skid-mounted package that employs ammonia or a suitable halocarbon refrigerant. The ammonia or halocarbon refrigerant from the package serves the heat exchanger performing the cooling, or in some cases a pump delivers a low-temperature secondary refrigerant to the heat exchanger. The skid-mounted package in Figure 1.7 provides low-temperature refrigeration through a combination of two circuits, one using R-22 and the other carbon dioxide.

A list of just a few of the industries and the temperature range of their refrigeration needs is: precooling air prior to separation[10], −100°C (−148°F); condensation of hydrocarbon vapors[11] taking place at −55°C (−67°F); and in the manufacture of such chemicals as ammonia/urea, aniline dyes, butadiene, butyl rubber, chlorine, pesticides/herbicides, synthetic rubber, vinyl chloride, and xylenes, as well as in the recovery of solvents.

REFERENCES

1. Lorentzen, G., "The Role of Refrigeration in Solving the World Food Problem," Proceedings of the XIII International Congress of Refrigeration, Washington, 1971.
2. *Commodity Storage Manual*, The Refrigeration Research Foundation, Washington, DC, continually updated.
3. American Society of Heating, Refrigerating and Air-Conditioning Engineers, Atlanta, Georgia, from the 1994 *Handbook—Refrigeration*; Chapter 8, Commercial Freezing Methods; Chapter 10, Methods of Precooling Fruits, Vegetables, and Cut Flowers; and Chapter 18, Fruit Juice Concentrates.

4. Hartel, R. W., "Chapter 8, Evaporation and Freeze Concentration," *Handbook of Food Engineering*, Marcel Dekker, Inc., New York, 1992.
5. Strommen, I., "Incorporation of Heat Pumps in Drying Processes," Proceedings of the 3rd International Energy Agency Heat Pumps Conference, Tokyo, Pergamon Press, March 1990.
6. Nagataki S. and S. Ono, "Precooling in Concrete Construction," *Refrigeration*, (Japanese), 67(777): 717-731, 1992.
7. "Ground Stabilization—Review of Grouting and Freezing Techniques for Underground Openings," Report No. FRA-ORD and D-75-95, U. S. Department of Transportation, Washington, D.C., 1975.
8. Haynes, F. D., "Effects of Foundation Insulation and Thermosyphons on Permafrost," Proceedings of Fourth Internation Symposium on Thermal Engineering and Science for Cold Regions, U.S. Army Cold Regions Research and Engineering Laboratory, Hanover, New Hampshire, 1993.
9. American Society of Heating, Refrigerating and Air-Conditioning Engineers, Atlanta, Georgia, from the 1994 ASHRAE *Hand- book—Refrigeration*, Chapter 33, Ice Rinks.
10. du Toit, A. S., "The Application of Refrigeration in the Commerical Production of Fuels and Chemicals from Coal," *International Journal of Refrigeration*, 7(4): 233-236, 1984.
11. "A Cool Way to Reduce Emissions," *Chemical Engineering*, 95(8): 145-150, May 23, 1988.

CHAPTER 2

REFRIGERANT PROPERTIES, REFRIGERATION CYCLES AND SI UNITS

2.1 USING SI UNITS

The Metric Conversion Act of the United States, signed into law in December 1975, encouraged the adoption of the SI (Systeme International d'Unites) but permitted the conversion to be voluntary. Nations that had traditionally used the inch-pound (I-P) system, such as Canada, the United Kingdom, and Australia, made the change to SI mandatory and simply got on with the job. Nations that traditionally used the metric system also had to adjust to some new units. The new unit of pressure, for example, is pascal, which replaced kg/m^2, and watt replaced kcal/hr as the unit for refrigerating capacity. Because the process was voluntary in the United States, the conversion was herky-jerky and in some cases consisted of no conversion at all. Finally realizing the penalty in trying to conduct international business with an outdated set of units, President Reagan signed the Omnibus Trade and Competitiveness Act, mandating that all federal agencies adopt the metric system for procurements, grants, and other business activity by the end of 1992. In the private sector, ASHRAE President Damon Gowan (1990-1991) challenged ASHRAE to set a date by which time all ASHRAE publications

TABLE 2.1
Base SI units.

Quantity	Name	Symbol
length	meter	m
mass	kilogram	kg
time	second	s
electric current	ampere	A
thermodynamic temperature	kelvin	K

will be expressed in SI units. Gowan's leadership greatly influenced ASHRAE's Long Range Plan of 1991, which specifies that "ASHRAE shall implement a policy by 1997 that all publications use the international system of units (SI)." Other technical organizations in the United States are attempting to move on with the conversion process. For example, the American Society of Mechanical Engineers is attempting to convert all of their codes and standards to SI in the near future.

The practice of this book is to use *dual* units, with SI as the primary set and I-P in parentheses. Graphs will be shown with SI units on one axis and I-P on the opposite axis. In numerical examples the statements of the problems are expressed in dual units, the work is executed in SI, and the results presented in dual units. The conversion between SI and I-P units is a frequent requirement, so this chapter initially presents the multiplying factors for these conversions. The conversions are then grouped in consolidated form in the appendix, Tables A.1a to A.1k. Conversions of temperatures and enthalpies are the only situations where shifts as well as the application of multiplying factors are needed.

2.2 SOME BASE AND DERIVED SI UNITS

Table 2.1 shows the base SI units that are important in industrial refrigeration. Derived from these base units are some other frequently used units, shown in Table 2.2.

To avoid confusion, negative exponents are sometimes used for terms in the denominator. For example, the base units for pressure may be expressed as $kg/(m^1{\cdot}s^2)$ or $kg{\cdot}m^{-1}{\cdot}s^{-2}$.

The unit of force is the newton, defined through Newton's law as:

$$\text{force} = (\text{mass})(\text{acceleration})$$
$$1\ \text{N} = (1\ \text{kg})(1\ \text{m/s}^2)$$

The unit of energy is the joule, which is the application of a force through a distance,

TABLE 2.2
Some derived SI units with special names.

Quantity	Name	Symbol	Expression in terms of other units	Expression in terms of SI base units
frequency	hertz	Hz		1/s
force	newton	N		$m \cdot kg/s^2$
pressure	pascal	Pa	N/m^2	$kg/(m \cdot s^2)$
energy	joule	J	$N \cdot m$	$m^2 \cdot kg/s^2$
power	watt	W	J/s	$m^2 \cdot kg/s^3$
electric potential	volt	V	W/A	$m^2 \cdot kg/(s^3 \cdot A)$
electric resistance	ohm	Ω	V/A	$m^2 \cdot kg/(s^3 \cdot A^2)$
specific heat			$J/kg \cdot K$	$m^2/(s^2 \cdot K)$
thermal conductivity			$W/m \cdot K$	$m \cdot kg/(s^2 \cdot K)$

$$1 \text{ J} = (1 \text{ N})(1 \text{ m})$$

There are standard prefixes for multiples and submultiples of SI units, and these are shown in Table 2.3. For example, 1/1000 of an ampere is a milliampere, abbreviated as mA.

The orthodox use of SI permits only those units and multiples listed in Tables 2.1, 2.2, and 2.3, with the following two exceptions: degrees Celsius (°C) for temperature, and liter (L) for volume. The SI system bowed to the use of these exceptions in order for the numerical values of temperature and volume to appear in more convenient magnitudes. The relationship between degrees Celsius and Kelvin and between the liter and cubic meters are:

$$\begin{aligned} \text{Temperature, °C} &= \text{Temperature, K} - 273.15 \\ \text{Volume, L} &= 1000 \text{ (Volume, m}^3\text{)} \end{aligned}$$

2.3 CONVERSION BETWEEN SI AND I-P UNITS

Multiplication by an appropriate factor will convert a quantity expressed in one set of units into the other with two exceptions, temperature and enthalpy. For these variables, a shift is also required because of the arbitrary datum planes, as will be explained in Section 2.4. The quantities in both the SI and the I-P systems likely to be encountered in industrial refrigeration practice are listed in Table 2.4. The conversion factors are tabulated[1] in the appendix, Tables A.1a through A.1k.

TABLE 2.3
Multiples and submultiples of units.

Multiplying factor			Prefix	Symbol
1,000,000,000,000	=	10^{12}	tera	T
1,000,000,000	=	10^{9}	giga	G
1,000,000	=	10^{6}	mega	M
1,000	=	10^{3}	kilo	k
0.001	=	10^{-3}	milli	m
0.000 001	=	10^{-6}	micro	μ
0.000 000 001	=	10^{-9}	nano	n
0.000 000 000 001	=	10^{-12}	pico	p

Example 2.1. Convert the following:
(a) 920 kg/m^3 to lb/ft^3
(b) 2 in. of water to Pa
(c) 1,000,000 Btu to GJ
(d) 0.24 Btu/lb·°F to kJ/kg·K.

Solution. (a) From Table A.1b, (920 kg/m^3)(0.06243) = 57.43 lb/ft^3
(b) From Table A.1d, (2 in. of water)(249.08) = 498.1 Pa
(c) From Table A.1f, (1,000,000 Btu)(1055.06) = $1.005 \cdot 10^9$ J = 1.005 GJ
(d) From Table A.1g, (0.24 Btu/lb·°F)(4187) = 1005 J/kg·°F = 1.005 kJ/kg·°F

2.4 CONVERSIONS OF TEMPERATURE, ENTHALPY, AND ENTROPY

For all the quantitites listed in Table 2-4, the process of conversion between SI and I-P units requires only the multiplication by a constant, except for temperatures, which are based on arbitrary datum planes. There is thus a need to shift as well as proportion when converting between Celsius and Fahrenheit. On the other hand, since absolute temperatures have a base of zero, only a conversion factor need be applied when converting between Kelvin and Rankine, namely,

$$\text{temperature, K} = (\text{temperature, R})(5/9)$$
$$\text{temperature, R} = (\text{temperature, K})(9/5)$$

To convert between °C and °F,
temperature, °C = (temperature in °F − 32)(5/9)
temperature, °F = (temperature in °C(9/5) + 32

TABLE 2.4
Corresponding units in the SI and I-P systems.

Quantity	SI units	I-P Units
Length	m	inch, in; foot, ft
Area	m^2	in^2, ft^2
Volume	m^3	in^3, ft^3
Mass	kg	pound, lb
Density	kg/m^3	lb/ft^3
Specific volume	m^3/kg	ft^3/lb
Velocity	m/s	fps, fpm, fph
Acceleration	m/s^2	ft/s^2
Force	N	lb_f
Pressure	Pa = N/m^2	psi, psf, ft of water, in. of water, in of Hg
Volume flow rate	m^3/s, L/s	cfm, cfs, gpm
Mass flow rate	kg/s	lb/s, lb/m, lb/hr
Temperature	°C or K	°F or R
Energy	J = N·m	ft-lb, Btu
Power	W	Btu/hr, hp, tons of refrigeration
Enthalpy	J/kg, kJ/kg	Btu/lb
Specific heat	J/kg·°C	Btu/lb·°F
Entropy	J/kg·K	Btu/lb·R
Viscosity	Pa·s	lb_m/ft·s, lb_m/ft·hr, centipoise
Thermal conductivity	W/m·K	Btu/hr·ft·°F
Heat transfer coefficient	W/m^2·K	Btu/hr·ft^2·°F

Example 2.2. (a) What is 95°F in Celsius?
(b) What is −40°C in Fahrenheit?

Solution. (a) temperature = (95 − 32)(5/9) = 35°C.
(b) temperature = (−40)(9/5) + 32 = −40°F.

The two temperature scales coincide at −40°.

The symbol for enthalpy is h, and the standard datum plane for enthalpies in the I-P system is that the enthalpy of saturated liquid at −40°F is zero. The datum plane for enthalpies in SI units is 200 kJ/kg for saturated liquid at 0°C. These different datum planes result in a unique conversion for each refrigerant. The multiplying factor applies to differences of enthalpy as shown in Figure 2.1.

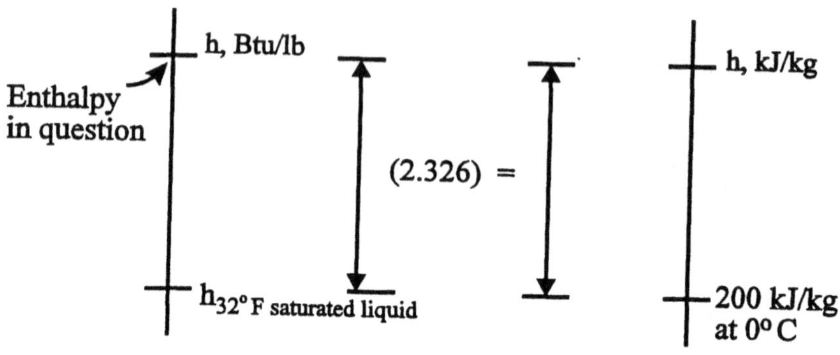

FIGURE 2.1
Conversion of refrigerant enthalpies.

In equation form,

$$(h,\ \text{Btu/lb} - h_{32F})2.326 = h,\ \text{kJ/kg} - 200\ \text{kJ/kg} \tag{2.1}$$

where $h_{32F} = h$ of saturated liquid at 32°F, Btu/lb

Equation 2.1 converts into one of two equations, depending upon whether the conversion is from SI to I-P or vice versa,

$$h,\ \text{kJ/kg} = 200 + (h,\ \text{Btu/lb} - h_{32F},\ \text{Btu/lb})(2.36) \tag{2.2}$$

$$h,\ \text{Btu/lb} = h_{32F},\ \text{Btu/lb} + (h, \text{kJ/kg} - 200)(0.4299) \tag{2.3}$$

The values of the enthalpy of saturated liquid at 32°F in Btu/lb for some important refrigerants are available from the tables in the appendix and are also listed in Table 2.5.

Example 2.3. (a) An R-22 property table shows the enthalpy at a certain condition to be 111.4 Btu/lb. What is the enthalpy in kJ/kg?
(b) Ammonia at a certain condition has an enthalpy of 341.8 kJ/kg. What is the enthalpy in Btu/lb?

Solution. (a) Using Equation 2.2,

$$h = 200 + (111.4 - 19.24)(2.326) = 414.4\ \text{kJ/kg}$$

(b) Using Equation 2.3,

$$h = 77.5 + (341.8 - 200)(0.4299) = 138.5\ \text{Btu/lb}$$

TABLE 2.5
Enthalpies in Btu/lb of saturated liquid at 32°F for use in Equations 2.2 and 2.3.

Refrigerant	h_f at 32 F, Btu/lb
R-22	19.24
R-23	25.18
R-507	23.59
R-125	18.32
R-134a	21.50
R-717, ammonia	77.60

Entropy in the two systems of units also has different bases. In the SI system the datum plane is an entropy value of 1.0 kJ/kg·K of saturated liquid at 0°C (32°F), while the base in the I-P system is 0 Btu/lb·R for saturated liquid at −40°F.

The conversion procedure corresponds to that used for enthalpy, but will not be pursued further since the need for converting entropies between the two systems of units in industrial refrigeration work is rare.

2.5 SOME IMPORTANT SI CONSTANTS

Several frequently used constants expressed in the SI system are:

standard gravitational acceleration = 9.80665 m/s^2
standard atmospheric pressure = 101,325 Pa
universal gas constant, $\mathcal{R}$ = 8314 J/(mol· K)
gas constant for air, R_{air} = 287 J/kg·K

In addition to fundamental constants, several important properties are applicable to air and water. Nominal values are:

density of water = 1,000 kg/m^3
density of air = 1.2 kg/m^3
specific heat of liquid water = 4.19 kJ/kg·K
specific heat of air at constant pressure = 1.004 kJ/kg·K
specific heat of air at constant volume = 0.717 kJ/kg·K

Example 2.4. What is the mass of air in a room of dimensions 6 x 9 x 3 m (19 ft, 8 in. x 29 ft, 6 in. x 9 ft, 10 in.) if the air temperature is 25°C (77°F) and the barometric pressure is 101.3 kPa (14.7 psi)?

Solution. The perfect gas equation which is applicable to air is

$$pv = RT$$

where p = pressure, Pa
v = specific volume, m^3/kg
R = gas constant for air = 287 J/kg·K
T = absolute temperature, K

Then the specific volume is:

$$v = \frac{(287)(25 + 273.15)}{101,300} = 0.845 \text{ m}^3/\text{kg } (13.5 \text{ ft}^3/\text{lb})$$

The room has a volume of (6)(9)(3) = 162 m^3 (5735 ft^3), so the mass of air contained in the room is:

$$162/0.845 = 191.7 \text{ kg } (423 \text{ lb})$$

2.6 REFRIGERANT PROPERTIES

In the next several sections the thermodynamic properties of refrigerants that are important in designing and analyzing systems will be explained. These properties include temperature, pressure, specific volume, density, enthalpy, and entropy. Numerical values of these important thermodynamic properties are presented in either tabular or graphical form. The most widely used graph is the pressure-enthalpy diagram, which will be examined. A state in which refrigerants often exist within refrigeration systems is called *saturated*, and in this state the temperature and pressure have definite relationships.

Throughout the explanation and discussion of properties we will be conscious that the chief importance of the properties is their effect on the refrigeration cycle. The refrigeration cycle will be explored later in the chapter, but to keep it in mind a schematic diagram of the familiar basic cycle is shown in Figure 2.2. It consists of four components—the compressor, condenser, expansion device, and evaporator. Also indicated on Figure 2.2 are the states (liquid or vapor) of the refrigerant at various points in the system.

2.7 TWO KINDS OF PRESSURE: GAUGE AND ABSOLUTE

Gauge pressure is the magnitude indicated by a precise gauge. Most pressure gauges, including the bourdon-tube type (Figure 2.3) commonly used in refrigeration practice, actually indicate the difference between the pressure of the fluid and atmospheric pressure. The bourdon-tube gauge incorporates a flexible tube that tends to straighten as the pressure increases. For a given fluid pressure the gauge will indicate a different value if the barometric pressure changes. The actual fluid pressure is called the ***absolute pressure*** and is related to the gauge pressure by the equation:

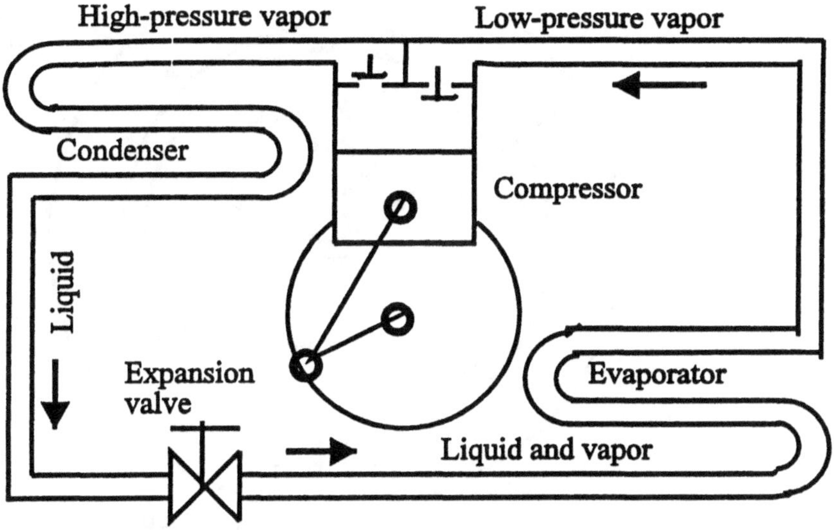

FIGURE 2.2
The basic refrigeration cycle.

$$\text{absolute pressure} = \text{gauge pressure} + \text{atmospheric pressure}$$

Tables and graphs are based on absolute pressures, so that the multitude of different atmospheric pressures that might exist at the site of the refrigeration plant do not have to be known. Many tables do, however, show gauge pressures, but these are derived from the absolute pressure with the assumption of standard atmospheric pressure of 101.325 kPa (14.7 psi, where psi is the abbreviation for pounds of force per square inch). To identify which type of pressure is being designated, we shall use the following designations:

absolute pressure in SI units:	kPa
gauge pressure in SI units:	kPa gauge
absolute pressure in I-P units:	psia
gauge pressure in I-P units:	psig

Example 2.5. Jim in New Orleans reads 125 kPa (18 psi) on his pressure gauge, while Ann in Denver reads the identical value on hers. In New Orleans, at the moment, the atmospheric pressure is 100 kPa (14.5 psia), while in Denver it is currently 90 kPa (13 psia). What are the absolute pressures of the refrigerant in the systems at these two locations?

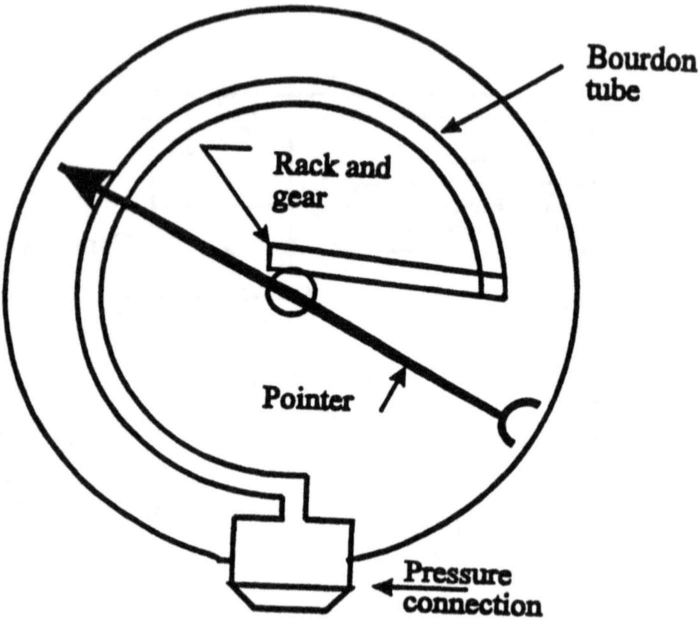

FIGURE 2.3
A bourdon-tube pressure gauge.

Solution. In New Orleans the refrigerant exists at an absolute pressure of 125 + 100 = 225 kPa (18 + 14.5 = 32.5 psia), while in Denver the refrigerant pressure is 125 + 90 = 215 kPa (18 + 13 = 31 psia).

2.8 SATURATION TEMPERATURES AND PRESSURES

When liquid and vapor coexist in a vessel and there is an equilibrium of temperature and pressure, the condition is called *saturated*. The liquid is called saturated liquid and the vapor is called saturated vapor. Figure 2.4 shows several vessels, two containing water and two ammonia. In its saturated state the pressure has a definite relationship to the temperature. As the temperature increases, the pressure also increases.

The tables of thermodynamic properties at saturated conditions of refrigerants dealt with in this book are presented in the Appendix, Tables A.13 through A.27. Table 2.6 is an excerpt of the temperatures and corresponding saturation pressures of ammonia extracted from Tables A.25 and A.27. At saturated conditions, if the pressure is known, the temperature of the refrigerant can be found from the table, and similarly, if the temperature is known, the pressure can be found.

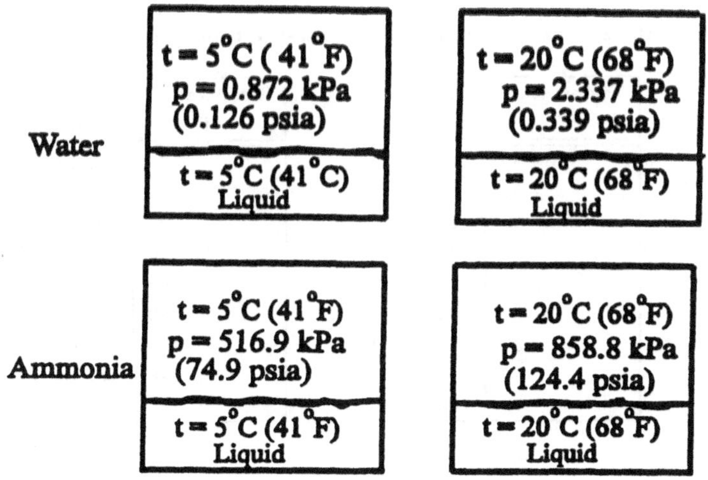

FIGURE 2.4
Pressure-temperature relationships for water and ammonia at saturated conditions.

Example 2.6. What is the temperature of R-507 if the refrigerant is saturated and at a pressure of 1931 kPa (280 psia)?

Solution. Interpolate between pressures of 1925.5 kPa corresponding to 41°C and 1972.4 kPa corresponding to 42°C to find the saturation temperature of 41.1°C (106°F).

In what situations in a refrigeration plant do saturated conditions occur? Certainly in vessels containing both liquid and vapor. But saturated conditions prevail in condensers and evaporators as well, as illustrated in Figure 2.5. Actually saturated conditions do not precisely occur in these heat exchangers, because equilibrium does not exist between liquid and vapor. In order to achieve heat transfer there must be temperature differences between the refrigerant vapor, refrigerant liquid, and the tube wall. For practical purposes, however, measuring the pressure in a condenser or evaporator is a means of determining the temperature of, respectively, the condensing or boiling refrigerant.

Example 2.7. The reading of a pressure gauge on an ammonia evaporator is 131 kPa gauge (19 psig). What is the evaporating temperature?

Solution. Assuming that the atmospheric pressure is 101.3 kPa (14.7 psia), the absolute pressure is 232.3 kPa (33.7 psia). From Table A.25 or A.27 by interpolation, the evaporating temperature is −15.3°C (4.5°F).

TABLE 2.6
Excerpt of the saturation temperature-pressure table for ammonia.

Temperature °C	Pressure kPa
−30	119.15
−20	189.59
−10	289.93
0	428.18
10	613.29
20	855.01
30	1163.8
40	1550.9
50	2028.1

Temperature °F	Pressure psia
−20	18.24
0	30.32
20	48.06
40	73.11
60	107.32
80	152.69
100	211.40
120	285.76
140	378.23

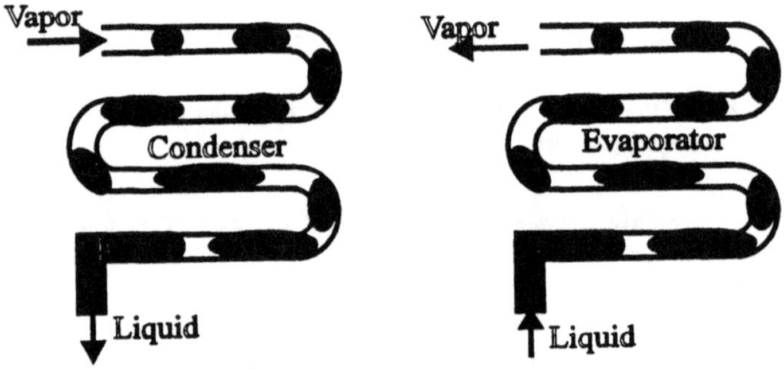

FIGURE 2.5
The refrigerant is essentially saturated in condensers and evaporators.

2.9 DENSITY AND SPECIFIC VOLUME

Two additional properties that are valuable are the density and the specific volume of the refrigerant, as illustrated in Figure 2.6. The density is the mass of refrigerant occupying a unit volume and expressed in units of kg/m^3 (lb/ft^3). The specific volume is the inverse of density and is thus the volume occupied by unit mass and expressed in units of m^3/kg (ft^3/lb). Usually density is used for liquids and specific volume for vapor, which results in numerical values in a convenient magnitude.

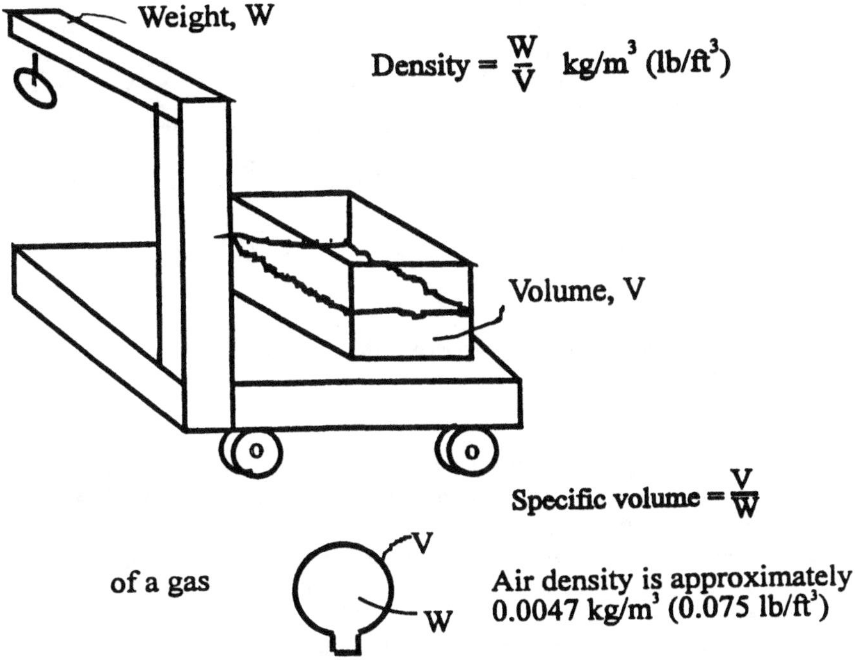

FIGURE 2.6
Density and specific volume

Example 2.8. A flow rate of 0.3 kg/s (40 lb/min) of ammonia circulates through a refrigeration system operating at a condensing temperature of 30°C (86°F) and an evaporating temperature of 0°C (32°F). Liquid leaves the condenser saturated and vapor leaves the evaporator saturated. What are the volume flow rates of liquid leaving the condenser and of vapor leaving the evaporator?

Solution. From Table A.25 or Table A.27, Properties of Ammonia at Saturated Conditions, the density of saturated liquid at 30°C (86°F) is 595.4 kg/m^3 (37.20 lb/ft^3) and the specific volume of saturated vapor at 0°C (32°F) is 290.2 L/m^3 or 0.290 m^3/kg (4.648 ft^3/lb). The volume flow rate of liquid leaving the condenser is

$$(0.3\ \text{kg/s})/(595.4\ \text{kg/m}^3) = 0.0005039\ \text{m}^3/\text{s or } 0.504\ \text{L/s}\ (1.07\ \text{cfm})$$

and leaving the evaporator

$$(0.3\ \text{kg/s})(0.290\ \text{m}^3/\text{kg}) = 0.087\ \text{m}^3/\text{s}\ (184.3\ \text{cfm})$$

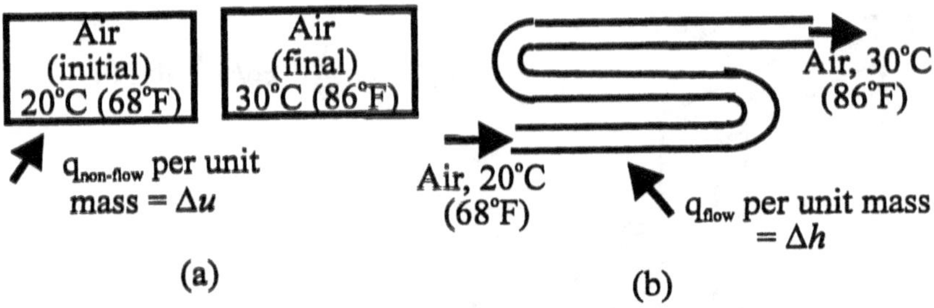

FIGURE 2.7
(a) Heat added to a gas in a nonflow process, and (b) in a flow process.

2.10 ENTHALPY

To analyze the energy characteristics of a refrigeration system the rates of heat flow and compressor power must be quantified. A key term in facilitating such calculations is *enthalpy*. For almost all situations we will encounter, we could think of enthalpy as *heat content*. The precise definition of enthalpy is

$h = u + pv$ or in I-P units: $h = u + pv/J$

where h = enthalpy, kJ/kg (Btu/lb)
u = internal energy, kJ/kg (Btu/lb)
p = pressure, kPa (lb_f/ft^2)
v = specific volume, m^3/kg (ft^3/lb)
J = conversion factor = 778 ft-lb_f/Btu

To illustrate the difference between internal energy and enthalpy, consider two processes of heating air from one temperature to another, as shown in Figure 2.7. In Figure 2.7a the process occurs in a closed vessel, while in Figure 2.7b the air flows through a heat exchanger. To increase the temperature of air from 20°C (68°F) to 30°C (86°F), for example, the required quantity of heat that must be transferred differs in the two situations. In the process of Figure 2.7a the quantity of the heat that must be added per unit mass in the vessel is Δu, while in the flow process of Figure 2.7b the quantity of heat added to each unit mass flowing through the heat exchanger Δh. The point is that the quantity of heat added in each of the processes of Figure 2.7 could be considered as a change in *heat content*, and because these quantities differ, the term heat content is ambiguous.

Almost all processes occurring in refrigeration plants are of the flow type, like in Figure 2.7b, so the change in enthalpy applies, and this is why we almost exclusively use enthalpy in making energy calculations. To summarize, the rate of heat transfer in kW (Btu/min) in a heat exchanger, such as the evaporator in Figure 2.8a, is the product of the mass rate of flow $\dot{m}$ kg/s (lb/min) and the difference of enthalpies in kJ/kg (Btu/lb) leaving and entering the evaporator.

$$q_{evap} = \dot{m}(h_2 - h_1) \qquad (2.4)$$

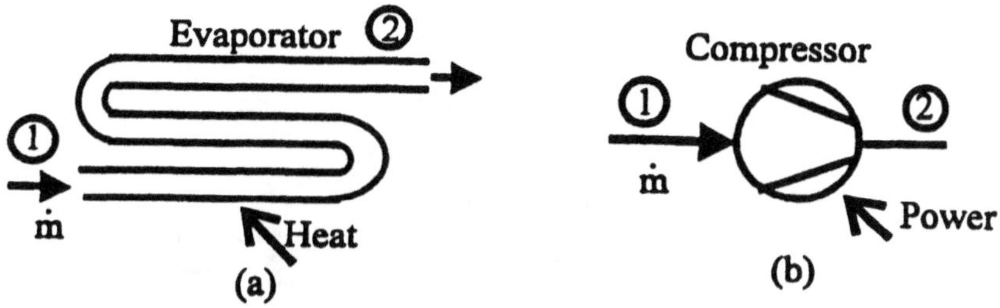

FIGURE 2.8
(a) Rate of heat transferred in a heat exchanger, (b) Power delivered to a compressor.

Likewise, the power in kW (Btu/min) supplied to the compressor in Figure 2.8b, assuming no heat losses or gains from the environment, equals the mass rate of refrigerant flow multiplied by the increase in enthalpy of the refrigerant in passing through the compressor,

$$\text{Power}_{compr} = \dot{m}(h_2 - h_1) \tag{2.5}$$

A process that illustrates a frequently used term is shown in Figure 2.9, where saturated liquid enters a heat exchanger and leaves as saturated vapor. The enthalpy of saturated liquid is designated by the term h_f and that of the saturated vapor is h_g. The change in enthalpy in converting saturated liquid to saturated vapor is called the *latent heat* and given the symbol h_{fg}. The tables of properties of refrigerants at saturated conditions in the appendix list the values of h_f and h_g. All the terms in these saturation tables have now been introduced with the exception of the entropy, whose significance will be addressed during the development of the pressure-enthalpy diagram starting in Section 2.12.

Example 2.9. A condenser receives saturated R-22 vapor at a flow rate of 1.5 kg/s (463 lb/min) and condenses it to saturated liquid. (a) What is the temperature of the refrigerant leaving the condenser, and (b) what is the rate of heat extraction from the refrigerant?

Solution. (a) Neglecting pressure drop of the refrigerant as it flows through the condenser, the pressure at the inlet and outlet are the same, and since the refrigerant is saturated, the temperature at the outlet of the condenser is also 30°C (86°F).

(b) From Table A.2 or A.3, the values of enthalpy of saturated liquid h_f and of saturated vapor h_g at 30°C (86°F) are 236.6 kJ/kg (34.9 Btu/lb) and 413.6 kJ/kg (111.1 Btu/lb), respectively. The rate of heat transfer q is thus:
$q = \dot{m}(h_2 - h_1) = (1.5 \text{ kg/s})(236.6 - 413.6 \text{ kJ/kg})$, or
$q = -265.5$ kW (− 15,100 Btu/min).

The negative sign on the thermal quantities only indicates that heat is flowing from the fluid.

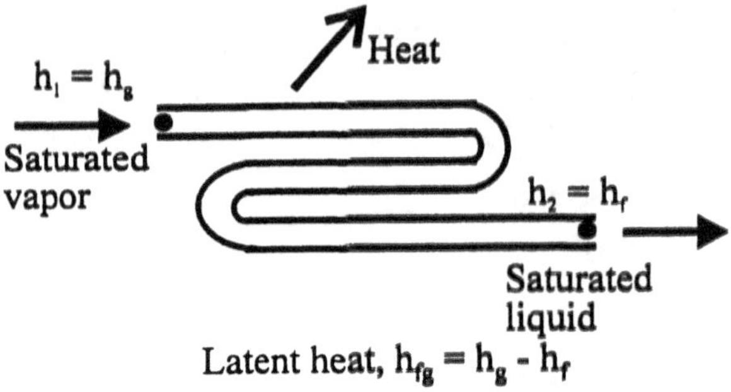

FIGURE 2.9
Latent heat in converting saturated vapor to saturated liquid.

2.11 SUPERHEATED VAPOR AND SUBCOOLED LIQUID

The state of refrigerant that we have focused on so far is the saturated condition, where liquid and vapor coexist and are in equilibrium—thus have the same temperature and the same pressure. We progress now to two additional states: superheated vapor and subcooled liquid. The starting point to reach these conditions can be chosen as in Figure 2.10. In Figure 2-10a some of the saturated vapor from the vessel is removed and heated to raise its temperature. The state of the vapor is changed to superheated, because the temperature of the vapor is above the saturation temperature at the existing pressure. No liquid can exist simultaneously with superheated vapor, except temporarily until the liquid is vaporized by the higher-temperature vapor. Superheated vapor exists at various locations throughout refrigeration systems. One place is at the entrance to the compressor, where vapor from the evaporator, even if it leaves the evaporator saturated, probably receives some heat in the suction line to the compressor. Refrigerant leaving the compressor is almost certainly superheated.

An opposite process can be performed to obtain subcooled liquid, as in Figure 2.10b. Here some saturated liquid is removed from the vessel and cooled to drop its temperature below that of the saturation temperature. No vapor can exist, except temporarily in nonequilibrium, with subcooled liquid. Subcooled liquid can exist at numerous points throughout the refrigeration system, such as at the exit of the condenser. Subcooled liquid can also be generated by increasing the pressure of saturated liquid by sending it through a pump. If there is little or no change in temperature through the pump, the outlet liquid possesses a temperature lower than the saturation temperature corresponding to the new higher pressure.

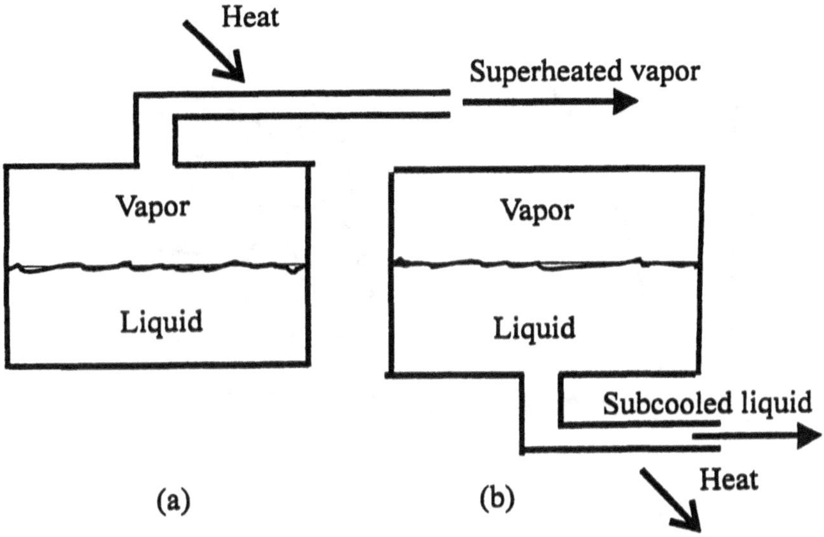

FIGURE 2.10
(a) Superheated vapor formed by adding heat to saturated vapor at constant pressure, and (b) subcooled liquid formed by removing heat from saturated liquid at constant pressure.

2.12 THE PRESSURE-ENTHALPY DIAGRAM

A valuable tool for those analyzing refrigeration systems is the pressure-enthalpy diagram. This diagram provides the information to determine the magnitudes of various properties and is also useful in displaying processes that are taking place in the plant. While there is some duplication of information provided by the saturation tables and the pressure-enthalpy diagram, some data are found in one source and not in the other. Other coordinates could have been chosen for a chart on which refrigerant properties are presented. Why pressure and enthalpy were chosen for the coordinates is not known for certain, but pressure is an easily-measured value in a refrigeration system and enthalpy is a key variable in analyzing the energy effectiveness of a plant. The pressure-enthalpy diagram is particularly valuable for finding properties in the superheated vapor region. The general form of the pressure-enthalpy diagram is shown in Figure 2.11, and the sections that follow will build up the chart, one property at a time.

2.13 SATURATED LIQUID AND VAPOR LINES

The major regions on the pressure-enthalpy diagram are defined by the lines of saturated liquid and saturated vapor. Considering first the saturated liquid line,

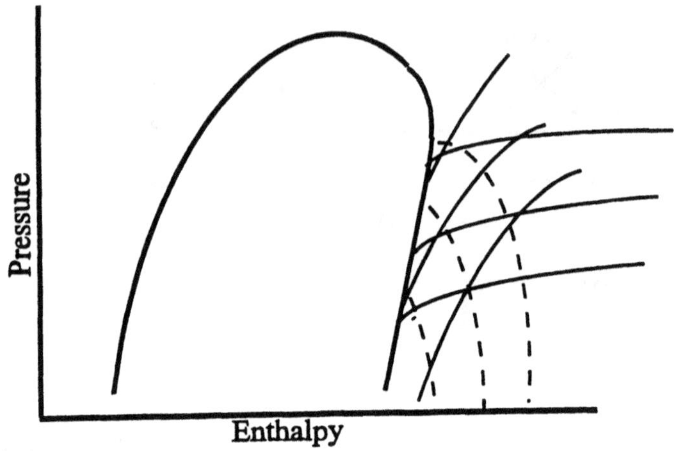

FIGURE 2.11
The pressure-enthalpy diagram.

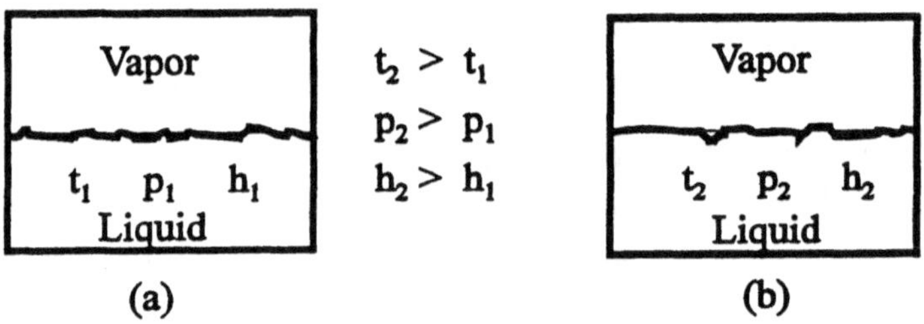

FIGURE 2.12
A vessel with saturated liquid at (a) a low temperature, pressure, and enthalpy, and (b) at a high temperature, pressure, and enthalpy.

the orientation of this line can be predicted by picturing a vessel containing both liquid and vapor as in Figure 2.12. The refrigerant in Figure 2.12a at state 1 is saturated, because both liquid and vapor are present. If this liquid is heated, the temperature rises, and the pressure rises too. The conversion from state 1 to state 2 is furthermore one where the heat content or enthalpy increases. The only purpose of showing the process is to illustrate that on a pressure-enthalpy diagram, such as in Figure 2.13, the line representing saturated liquid moves upward to the right.

The next step in building the pressure-enthalpy diagram is to show the saturated vapor line, which also appears on Figure 2.13. Figure 2.9 presented a conversion from saturated vapor to saturated liquid in which the change of enthalpy is called the latent heat, h_{fg}. For each of the temperatures in Figure 2.13

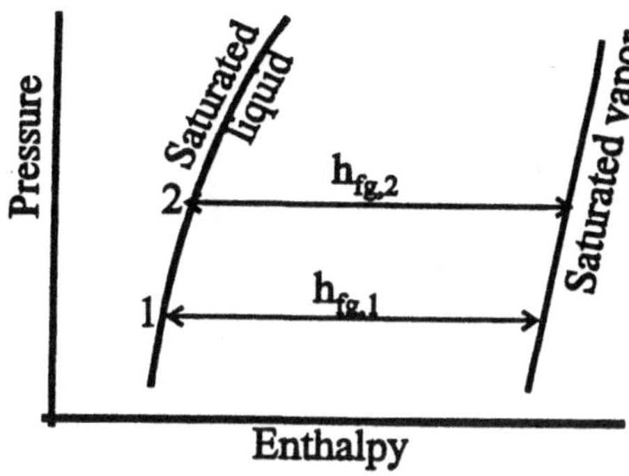

FIGURE 2.13
Saturated liquid and saturated vapor lines on a pressure-enthalpy diagram.

there is a unique value of h_{fg} that permits locating the point on the saturated vapor line opposite the corresponding point on the saturated liquid line.

The complete outline of the pressure-enthalpy diagram is shown in Figure 2.14, which defines the various states of existence of the refrigerant: subcooled liquid, liquid/vapor mixture, and superheated vapor. The region to the left of the saturated liquid line is subcooled liquid, which would be formed by the process in Figure 2.10b. The region enclosed within the saturated liquid and saturated vapor lines represents a mixture of liquid and vapor. Close to the saturated liquid line the fraction of liquid in the mixture is high and close to the saturated vapor line the fraction of vapor is high. To the right of the saturated vapor line is superheated vapor, which would be generated in the process illustrated in Figure 2-10a. A special condition is the point at the very top of the saturation curves, called the *critical point*. At this point liquid and vapor are indistinguishable. The magnitudes of the critical temperature and pressure have practical significance, because refrigerating cycles that operate near the critical point are, in general, not very efficient.

If the position of a point can be located on the diagram, the value of the pressure can be determined by moving horizontally to the pressure scale, and the enthalpy can be read by moving vertically to the enthalpy scale.

2.14 LINES OF CONSTANT TEMPERATURE

Several lines of different constant temperatures are shown on the pressure-enthalpy diagram of Figure 2.15. The lines are shown in the three regions—subcooled liquid, liquid/vapor mixture, and superheated vapor. In the mixture region the lines are horizontal, because saturated conditions exist here. Thus, a given pressure, which is described by a horizontal line, establishes a certain temperature. In the subcooled liquid region the lines of constant temperature are vertical. This

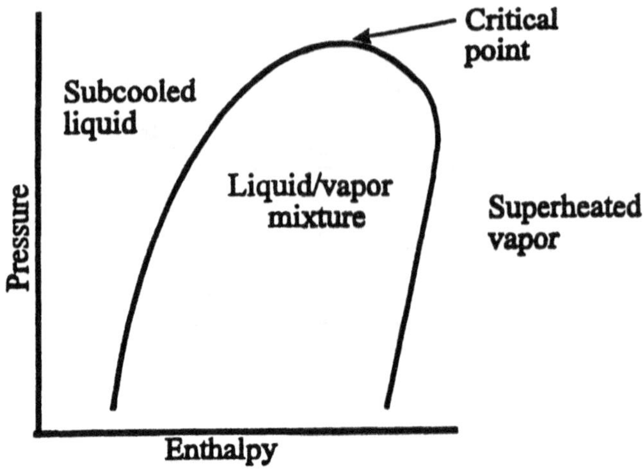

FIGURE 2.14
The complete outline of the pressure-enthalpy diagram.

orientation may be explained by acknowledging that the liquid is essentially incompressible. Thus if the pressure of liquid at a given temperature in a cylinder is increased by applying a force through a piston, the piston doesn't move. Since in order to exert work on the liquid the force must move through a distance, but the piston remains motionless. So increasing the pressure changes neither the temperature nor the enthalpy, and the constant-temperature line is vertical. The fact that lines of constant temperature are vertical in the subcooled liquid region has some implications that will be illustrated in Example 2-10.

Example 2.10. Ammonia exists at a temperature of 15°C (59°F) and a pressure of 855 kPa (124 psia). (a) What is the state of the refrigerant, and (b) what is its enthalpy?

Solution. (a) At a pressure of 855 kPa (124 psia) the saturation temperature is 20°C (68°F). Since the prevailing temperature of 15°C (59°F) is lower than this saturation temperature, the liquid is subcooled.
(b) The saturation tables can be used to find the enthalpy and the question is whether to look up the value of enthalpy of liquid at the existing temperature or the existing pressure. The temperature determines the enthalpy of subcooled liquid, so corresponding to 15°C (59°F) the enthalpy is found to be 270.2 kJ/kg (107.7 Btu/lb).

While Figure 2.15 might suggest that lines of constant temperature will appear in all the regions of the pressure-enthalpy diagram, there is no need to show the lines in the subcooled liquid region,. because we know that the lines there are vertical. Similarly, the lines in the mixture region are horizontal, so there is no need to show them. The lines are useful in the superheat region, and it is here that they are shown on the working pressure-enthalpy diagrams.

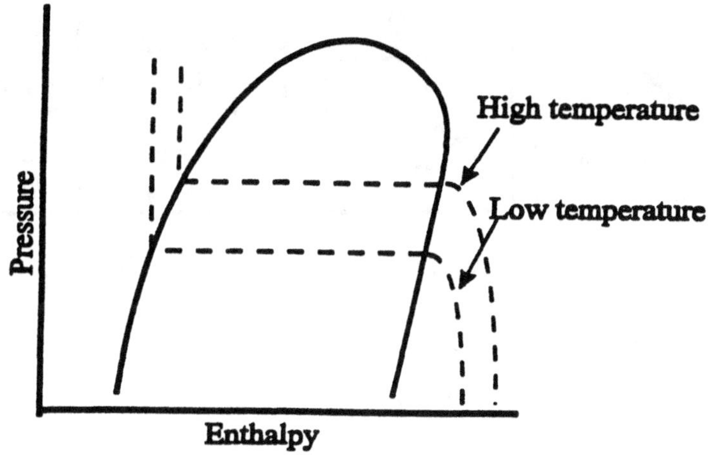

FIGURE 2.15
Lines of constant temperature on the pressure-enthalpy diagram.

Example 2.11. What is the enthalpy of ammonia existing at a temperature of 30°C (86°F) and a pressure of 500 kPa (72.5 psia)?

Solution. At a pressure of 500 kPa (72.5 psia) the saturation temperature (from Table A.25 or A.27) is 4.2°C (39.6°F). The actual temperature is higher than the saturation temperature, so the refrigerant is superheated. The saturation tables do not apply, but the condition can be located in the superheat region of the pressure-enthalpy diagram, Figure A.12 or Figure A.13. The point is located at the intersection of the horizontal pressure line at 500 kPa (72.5 psia) and the line of constant temperature of 30°C (86°F). From this point move either up or down to the horizontal enthalpy scale to read 1532 kJ/kg (650 Btu/lb).

2.15 SPECIFIC VOLUME

As explained in Section 2.9, the specific volume is the volume occupied by unit mass of the substance, thus possesses units of m^3/kg (ft^3/lb). To provide numbers of convenient magnitudes, liters per kilogram (L/kg) is sometimes chosen as the unit in SI. Typically, specific volumes are used in reference to gases while the reciprocal of specific volume, density, is applied to liquids. Two situations where knowledge of the specific volume of a refrigerant is important are in computing vapor velocities in pipes and in computing the volume flow rate in a compressor.

The influences of the temperature and the pressure on specific volume of a vapor are illustrated in Figure 2.16. In Figure 2.16a a given mass of refrigerant contained in a cylinder maintained at a constant pressure experiences a rise in temperature from state A to state B. The gas expands, resulting in an increase in volume, and because the mass remains constant the specific volume increases as well. In Figure 2.16b the process from state C to state D takes place at

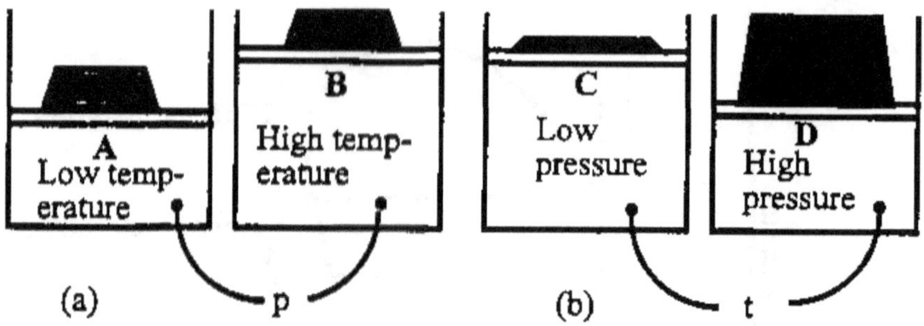

FIGURE 2.16
(a) An increase in specific volume as the temperature increases at constant pressure, and (b) a decrease in specific volume as the pressure increases at constant temperature.

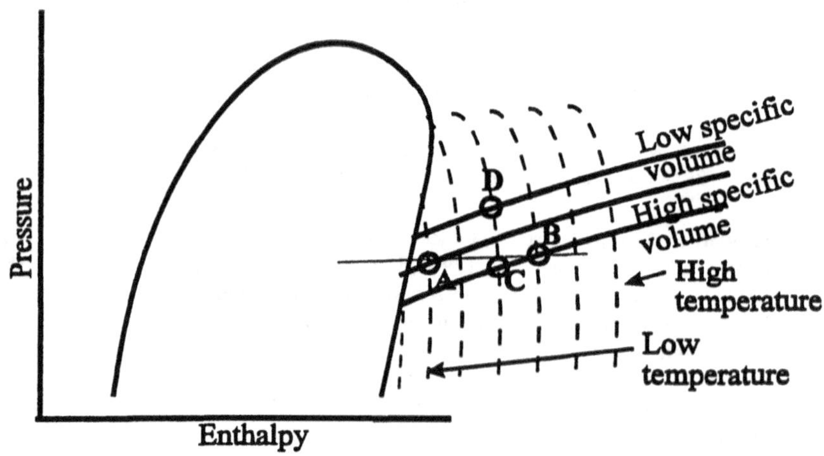

FIGURE 2.17
Constant specific volume lines on the pressure-enthalpy diagram.

constant temperature that is maintained by appropriate cooling. The light weight is replaced by a heavy weight which compresses the gas, decreasing its specific volume.

On the pressure-enthalpy diagram, which is our focus at the moment, lines of constant specific volume appear as shown in Figure 2.17. The paths of the two processes illustrated in Figure 2.16, A-B and C-D, indeed verify the increase in specific volume in process A-B and the decrease of specific volume in process C-D.

Example 2.12. What is the specific volume of R-22 at a temperature of 30°C (86°F) and a pressure of 280 kPa (40.6 psia)?

Solution. From the pressure-enthalpy diagram for R-22, Figure A.1 or A.2, at the intersection of the pressure and temperature lines, the specific volume is found to be 0.100 m^3/kg (1.6 ft^3/lb).

2.16 ENTROPY

Entropy is a property that in the study of thermodynamics provides useful insight on such tasks as evaluating the the effectiveness of processes. Observation of any of the saturation tables of refrigerants in the appendix shows values of entropy given for saturated liquid and saturated vapor. While the concept of entropy is necessary for a thorough study of thermodynamics, we will not delve deeply into the nature of the property. Our uses of entropy will be limited to two applications (1) the compression of a refrigerant vapor which ideally is represented by the constant entropy lines, and (2) use of entropy to help describe an ideal refrigeration cycle, as will be applied in Section 2.17.

It is not necessary to understand entropy completely for our limited purposes, but for the inquisitive person the definition of entropy expressed in terms of a heat-transfer process is:

$$ds = \left(\frac{dq}{T}\right)_{rev} \tag{2.6}$$

where s = entropy, kJ/kg·K (Btu/lb·R)
q = magnitude of heat transfer, kJ/kg (Btu/lb)
T = absolute temperature, kelvin (rankine)

The letter d preceding s and q indicates a very small or differential quantity as illustrated in the graph of a heat-transfer process as shown on temperature-entropy coordinates in Figure 2.18. Revision of Equation 2.6 gives

$$(dq)_{rev} = T(ds) \tag{2.7}$$

which in Figure 2.18 shows that the area under the curve represents the quantity of heat transferred, dq.

The subscript *rev* has important thermodynamic implications and requires that the process take place with no losses such as friction. In the reversible compression of a spring, for example, the spring must be able to repay precisely the same force at each position of compression that was exerted to compress the spring. In the thermal process, heat must be transferred while the source of heat and the object are at the same temperature. Such a condition would permit the process to reverse itself by giving up heat from the object to the original source with no temperature differences. Of course, we know that there must be a temperature difference for heat to flow, which indicates that the reversible process is the ideal and is the basis of comparison of our actual process.

We will return to graphs displayed on the temperature-entropy coordinates in the next section, which introduces refrigeration cycles, but first we continue

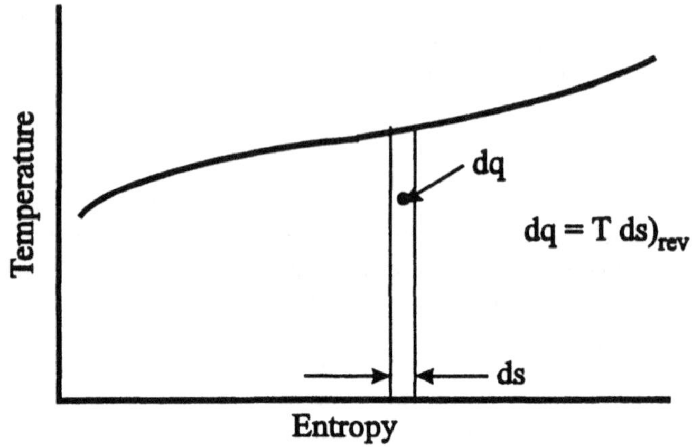

FIGURE 2.18
The area under a curve of a heat-transfer process when shown on temperature-entropy coordinates represents the magnitude of heat transfer.

the development of the pressure-enthalpy diagram and, in particular, predict the path of lines of constant entropy in the superheat region. A key property of lines of constant entropy in the superheat region on the pressure-enthalpy diagram is that they represent an ideal compression:

lines of constant entropy on p-h diagram = compression lines

An ideal compression is considered one with no heat transfer in or out of the compressor (called an adiabatic process) and is reversible (as described above). If we accept the fact that the lines of constant entropy represent an ideal compression, we can predict their orientation on the pressure-enthalpy diagram, as in Figure 2.19. In a compression process we expect the pressure to increase and also the enthalpy to increase, as was proposed in Figure 2.8a. The lines of constant entropy thus trend upward to the right in the superheat region.

Example 2.13. R-22 saturated vapor at a temperature of −15°C (5°F) is compressed in a process with neither friction nor heat transfer to a pressure of 1400 kPa (203 psia). How much work is exerted on each kg (lb) of refrigerant compressed?

Solution. Referring to the pressure-enthalpy diagram for R-22, Figure A.1 or Figure A.2, locate saturated vapor at a temperature of −15°C (5°F) where the entropy is 1.775 kJ/kg·K (0.226 Btu/lb·R). From the saturation tables or from the pressure-enthalpy diagram, the enthalpy of the vapor at the beginning of compression is 399 kJ/kg (105 Btu/lb). Move upward to the right along the constant-entropy line until reaching the pressure of 1400 kPa (203 psia), which is the condition at the end of compression. Here the enthalpy is 438 kJ/kg (121.6

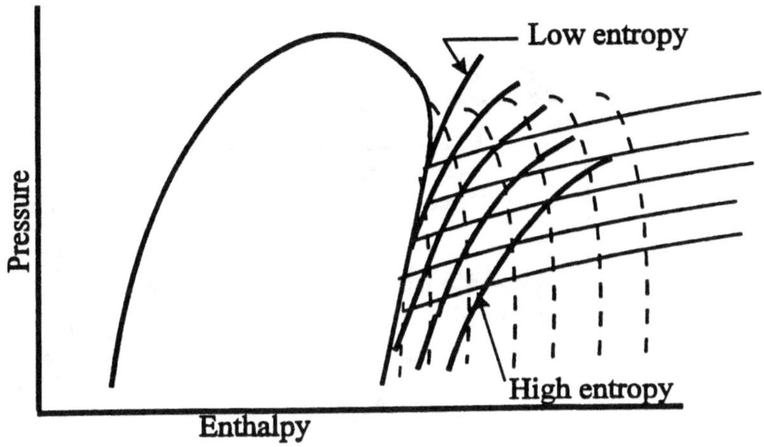

FIGURE 2.19
Constant entropy lines on the pressure-enthalpy diagram.

Btu/lb), and, incidentally, the temperature is 62°C (144°F). The work of compression is the difference in enthalpy during the process:

$$\text{Work of compression} = \Delta h = 438 - 399 = 39 \text{ kJ/kg } (16.7 \text{ Btu/lb})$$

2.17 THE IDEAL REFRIGERATION CYCLE–THE CARNOT CYCLE

We move now from the study of refrigerant properties to the examination of cycles. A method of achieving continuous refrigeration is to pass a working substance, called a refrigerant, through a series of processes, one of which can extract heat at low temperature. One such cycle is the Carnot refrigeration cycle, which is an ideal one and the most efficient possible between given temperature levels.

Why even consider a cycle that is unattainable? There are several reasons. One is that the Carnot cycle offers a convenient means of showing the influence of operating temperatures. Another is that the Carnot cycle gives a quick check on the effectiveness and ratios of heat-transfer rates of a real system. If a calculation during a design of a plant shows an efficiency superior to that of the Carnot cycle, that calculation is suspect. The components that comprise the Carnot refrigeration cycle are a compressor, two heat exchangers, and an engine, arranged as shown in Figure 2.20.

Furthermore, the processes must be conducted in specific ways:

1-2. Compression with no friction and no transfer of heat.

2-3. Heat rejection while the refrigerant maintains a constant temperature.

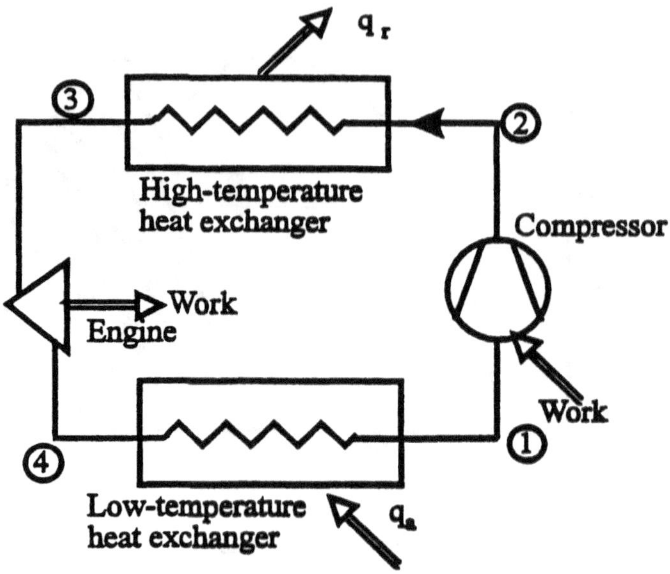

FIGURE 2.20
A Carnot refrigeration cycle.

3-4. Expansion in the engine with no friction and no transfer of temperature.

4-1. Heat absorption while the refrigerant maintains a constant temperature.

Because processes 1-2 and 3-4 are frictionless, they are thermodynamically reversible, a term introduced in Section 2.16. Furthermore, because no heat is transferred to the surroundings, the processes are adiabatic. Reversible adiabatic processes take place at constant entropy, as indicated in Section 2.16, so the Carnot cycle consists of two constant-entropy and two constant-temperature processes, as shown in Figure 2.21 on the coordinates of absolute temperature and entropy.

As Figure 2.18 showed, the area under a line on the T-s diagram representing a reversible process indicates the amount of heat transferred. Thus, the area on Figure 2.21 under the process 2-3 represents the amount of heat rejected in that process, and the area under line 4-1 indicates the amount of heat absorbed in the refrigeration process.

Example 2.14. In a Carnot refrigeration cycle the following temperatures and entropies prevail:

T_1 = 250 K (450 Rankine = 450 R)
T_2 = 300 K (540 Rankine = 540 R)
s_2 = 1.2 kJ/kg·K
s_3 = 0.9 kJ/kg·K.

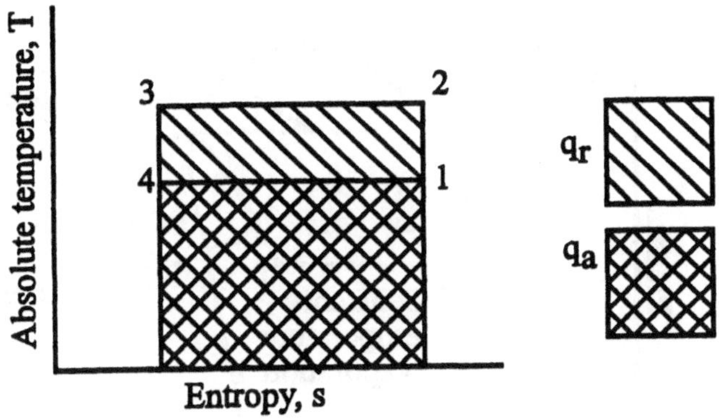

FIGURE 2.21
Carnot cycle displayed on temperature-entropy coordinates.

What are the magnitudes of refrigeration q_a and the heat rejected q_r per kg of refrigerant circulated in the cycle?

Solution. Because of the rectangular shape of the Carnot cycle, the remaining temperatures and entropies can be established:

$T_3 = T_2 = 300$ K (540 R)
$T_4 = T_1 = 250$ K (450 R)
$s_1 = s_2 = 1.2$ kJ/kg·K
$s_4 = s_3 = 0.9$ kJ/kg·K

The magnitude of refrigeration q_a is represented by the area under line 4-1:

$$q_a = T_1(s_1 - s_4) = 250(1.2 - 0.9) = 75 \text{ kJ/kg (32.24 Btu/lb)}$$

and the area under line 2-3 represents the heat rejected q_r

$$q_r = T_2(s_2 - s_3) = 300(1.2 - 0.9) = 90 \text{ kJ/kg (38.69 Btu/lb)}$$

Another important piece of information that can be calculated about a Carnot cycle is *net work per unit of mass circulated.* Figure 2.20 shows the compressor receiving and the engine providing work. An energy balance states that the energy into the cycle equals that leaving:

$$q_a + \text{net work into cycle} = q_r$$

The difference between q_r and q_a is net work,

$$\text{Net work into cycle} = q_r - q_a = (T_2 - T_1)(s_2 - s_3) \qquad (2.8)$$

In Example 2.14 the net work would be 90 − 75 = 15 kJ/kg (6.45 Btu/lb).

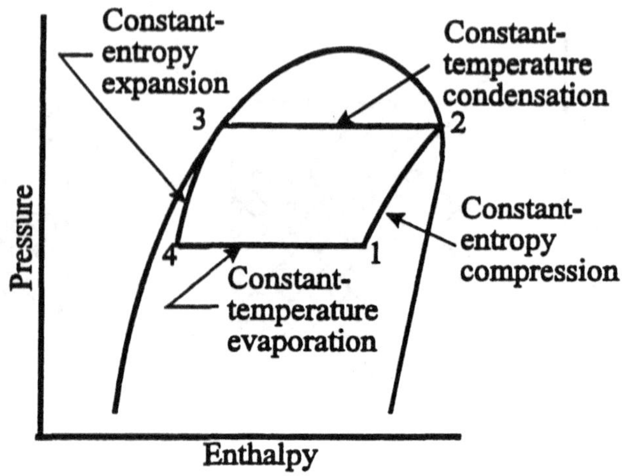

FIGURE 2.22
The Carnot cycle shown on the pressure-enthalpy diagram.

2.18 ACHIEVING THE CARNOT CYCLE WITH A REAL REFRIGERANT

If the Carnot cycle is the most efficient possible between two temperature levels, the goal should be to reproduce it in practice. We acknowledge that frictionless compressions and expansions are not possible, but the goal of constant-temperature heat transfer is worthy of pursuit. Such a process is achievable if a real refrigerant is condensed or evaporated at constant pressure, during which the state is saturated. A Carnot cycle fitting within the saturation lines of the p-h diagram is shown in Figure 2.22. Alternate axes on which the cycle can be displayed are the temperature-entropy coordinates of Figure 2.23 with the rectangular diagram of the Carnot cycle placed within the envelope.

Pick up the description of the cycle in process 4-1 where the refrigerant evaporates at constant pressure. The evaporation terminates at point 1 with some liquid still mixed with the vapor. Point 1 is placed such that, after an isentropic compression to the high pressure, the position will be point 2 on the saturated vapor line, or perhaps even within the liquid-vapor mixture region. The next process, 2-3, is a constant-pressure condensation, which is a constant-temperature process as well. The condensation terminates at saturated liquid, point 3. The isentropic expansion in the engine carries the process into the mixture region back to point 4.

Several equipment problems would likely result from attempting to operate the cycle of Figures 2.22 and 2.23. The revisions of this idealized cycle into a practical one will be described in Sections 2.25 and 2.26. A degree of practicality already has emerged in the use of a *condenser* for high-temperature heat rejection

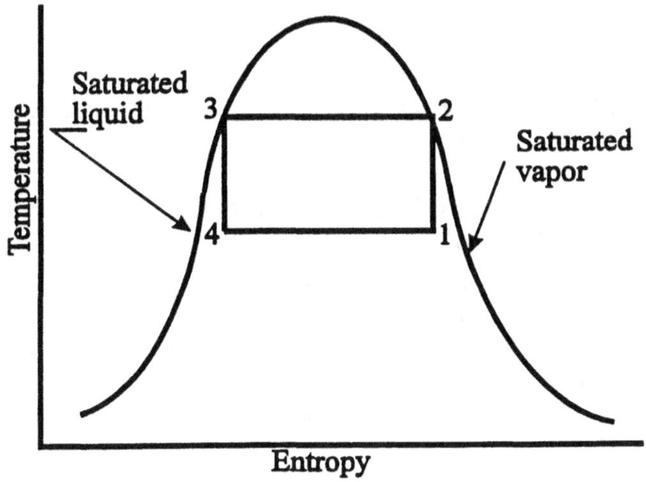

FIGURE 2.23
The Carnot cycle using a real refrigerant confined within the saturated liquid and vapor lines on temperature-entropy coordinates.

and an *evaporator* for low-temperature heat absorption. Before addressing the changes needed to convert the Carnot cycle to a practical refrigeration cycle, several conclusions can be drawn from the Carnot cycle which carry through to actual plant operation. These conclusions apply to the efficiency of a refrigeration cycle that is explained in the next section.

2.19 EFFICIENCY OF A REFRIGERATION CYCLE—THE COEFFICIENT OF PERFORMANCE

The efficiency of power-generating cycles is defined as the ratio of the energy output to the energy input. This definition works well for power cycles, but would be ambiguous for a refrigeration cycle. The thrust of the efficiency of a power cycle may be adapted to the refrigeration cycle by defining the efficiency as *the desired quantity divided by what must be paid to achieve the desired quantity.* The desired quantity is the refrigeration, and the expenditure is the net work. To avoid confusion this term is not called an efficiency, but instead called *the coefficient of performance*, and abbreviated COP.

$$\text{Coefficient of performance} = \text{COP} = \frac{\text{desired quantity}}{\text{expenditure}}$$

For the Carnot cycle, since from Equation 2.8 the net work is $q_r - q_a$, the COP is:

$$COP_{Carnot} = \frac{q_a}{q_r - q_a} = \frac{T_1(s_2 - s_3)}{(T_2 - T_1)(s_2 - s_3)} = \frac{T_1}{T_2 - T_1} \tag{2.9}$$

Since we would like to expend the least power for a given refrigeration rate, we seek a high value of the COP.

2.20 CONDITIONS FOR HIGH COP OF THE CARNOT CYCLE

A high COP is desirable since it indicates that the work necessary to motivate the cycle will be small for a given amount of refrigeration. Examination of Equation 2.9 shows that decreasing the condensing temperature T_2 increases the COP. Also, an increase in the evaporating temperature T_1 increases the COP, from its influence on both the numerator and the denominator. A one-degree increase in evaporating temperature exerts greater influence than a one-degree decrease in condensing temperature. The influence of these temperatures is one of the useful insights provided by the Carnot cycle and carries through to actual refrigeration plants. These objectives permeate the endeavor to operate refrigeration plants efficiently, so for minimum power requirements for a given refrigeration capacity:

- maintain the condensing temperature as low as possible, and
- maintain the evaporating temperature as high as possible.

Knowing which way to change T_1 and T_2 does not imply that we have complete liberty to set their values. If we did, we would simply operate with $T_1 = T_2$, which results in a COP of infinity. Then we could achieve all the refrigeration we needed without an energy expenditure. The refrigeration cycle pumps heat from one temperature level to a higher one. In Figure 2.24, for example, the Carnot cycle pumps heat from a cold room at −20°C (−4°F, 253.15 K, 456 R) and discharges heat to atmosphere at a temperature of 35°C (95°F, 308.15 K, 555 R). The temperature T_1 can be no higher than the cold room temperature of −20°C (−4°F), and T_2 can be no lower than the atmospheric temperature of 35°C (95°F). The coefficient of performance of the cycle operating between the absolute temperatures of 253.15 and 308.15 K (455.67 and 554.67 R) is:

$$COP = \frac{253.15}{308.15 - 253.15} = \frac{455.67}{554.67 - 455.67} = 4.60$$

The suggestion that T_1 could be the same as the temperature of the cold room and that T_2 could be the same as the temperature of the atmosphere continues the concept of reversibility. A reversible heat-transfer process is one where heat is transferred even though there is no difference in temperature. Such a condition is an idealization and is physically impossible. In order to transfer heat in a real system, T_1 must be *lower* than the cold room temperature and T_2 must be *higher* than the atmospheric temperature. These temperature differences, ΔT_1

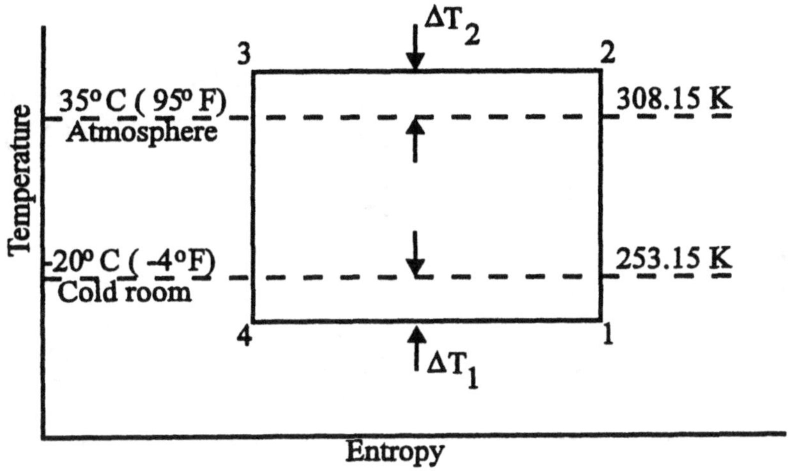

FIGURE 2.24
Temperature limitation imposed on the refrigeration cycle.

and ΔT_2, respectively, as shown in Figure 2.24, lower T_1 and raise T_2, decreasing the COP. To achieve a high COP, then, the focus shifts to the ΔTs and the attempt to keep the ΔTs as small as possible. In actual refrigeration plants, the magnitudes of the ΔTs are largely controlled by the amount of area chosen for the condenser and evaporator. Reduction of ΔT to zero requires infinite heat transfer area and thus infinite cost. Chapter 6 on evaporators and Chapter 7 on condensers treat the choice of size of these components in economic terms.

Example 2.15. A refrigeration system in a food processing plant operates with an evaporating temperature of −20°C (−4°F) and discharges heat to the atmosphere at 25°C (77°F). If the temperature at which heat is discharged were raised to 50°C (122°F), the energy could be used for process heating. If the system operates on the Carnot cycle and the cost of electric energy is $0.07 per kWh, what would be the additional cost of compressor energy to raise the condensing temperature to a level that would provide 1 GJ (948,000 Btu) of process heat?

Solution. To make a fair comparison, both cycles should be analyzed using the same magnitude of refrigeration, as shown in Figure 2.25. The amount of refrigeration, as established from Figure 2.25b is:

$$\text{Refrigeration} = (1 \cdot 10^9 \text{ J}) \left(\frac{253.2}{323.2}\right) = 783 \text{ MJ } (742,000 \text{ Btu})$$

The work $\mathcal{W}$ required for the refrigeration-only cycle is:

$$\mathcal{W} = (783 \text{ MJ}) \left(\frac{298.2 - 253.2}{253.2}\right) = 139 \text{ MJ } (131,900 \text{ Btu})$$

The work required for the heat-pump cycle is:

$$\mathcal{W} = (783 \text{ MJ}) \left(\frac{323.2 - 253.2}{253.2}\right) = 216 \text{ MJ } (205,200 \text{ Btu})$$

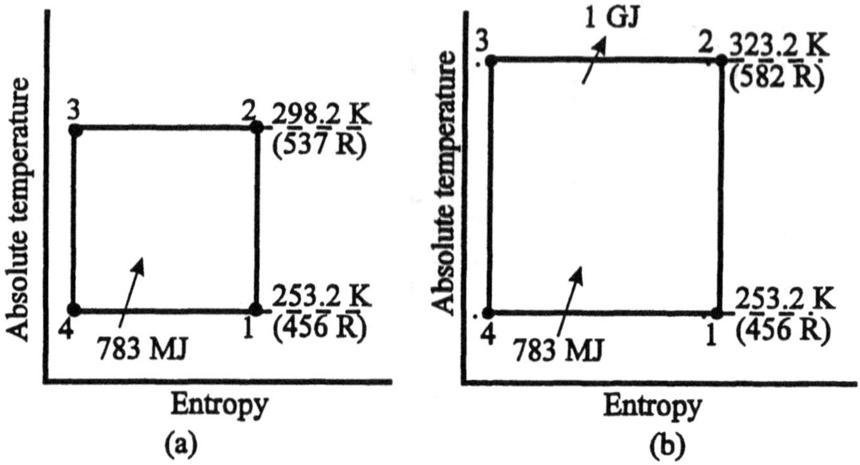

FIGURE 2.25
(a) Refrigeration-only, and (b) heat-pumping cycle in Example 2.15.

To obtain 1 GJ (948,000 Btu) of process heat the additional cost of compressor energy is 216 − 139 = 77 MJ (73,300 Btu). The cost of electric energy is expressed on a kWh basis. One MJ = 1,000 kJ, and 1 kJ = 1 kWs or 1/3600 kWh. So 1 MJ = 1,000/3600 = 0.278 kWh. The additional cost of electricity to recover 1 GJ of process heat is:

Cost = (77 MJ per GJ)(0.278 kWh)($0.07 per kWh)

Cost = $1.50 per GJ (948,000 Btu)

which needs to be compared with the cost of heat from a fossil-fueled boiler.

2.21 STEADY-FLOW ENERGY EQUATION

An important tool of thermodynamics applicable to refrigeration practice is the steady-flow energy equation. We have already used, perhaps unknowingly, special cases of the the steady-flow energy equation in Section 2.10 when explaining enthalpy and its use. We stated in Equation 2.4 that the rate of heat transfer in a heat exchanger such as a condenser or evaporator equals the product of the mass rate of flow and the difference of enthalpies in and out of the exchanger. We also indicated in Equation 2.5 that the power required by a compressor (in the absence of heat gain or loss) was the product of the mass rate of flow and the enthalpy difference. The steady-flow energy equation relates the energy terms of the process shown schematically in Figure 2.26. The form of the equation is slightly different in SI and I-P units, so both will be shown.

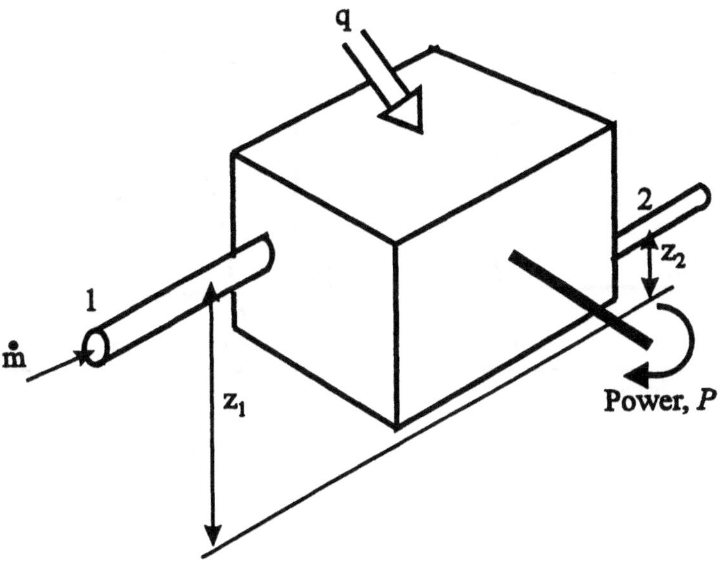

FIGURE 2.26
Initial and final positions, (states 1 and 2) in a steady-flow stream undergoing energy changes.

In SI, the equation is:

$$\dot{m}(h_1 + \frac{V_1^2}{2} + z_1 g) + q = \dot{m}(h_2 + \frac{V_2^2}{2} + z_2 g) + \mathcal{P} \quad (2.10)$$

and in I-P units:

$$\dot{m}(h_1 + \frac{V_1^2}{2g_cJ} + \frac{gz_1}{g_cJ}) + q = \dot{m}(h_2 + \frac{V_2^2}{2g_cJ} + \frac{gz_2}{g_cJ}) + \mathcal{P} \quad (2.11)$$

where

$\dot{m}$	=	mass rate of flow of fluid, kg/s (lb/s)
h	=	enthalpy, J/kg (Btu/lb)
V	=	velocity, m/s (ft/s)
z	=	elevation above arbitrary datum plane, m (ft)
q	=	rate of heat transfer into fluid, W (Btu/s)
$\mathcal{P}$	=	energy rate (shaft power) performed by fluid, W (Btu/s)
J	=	conversion constant, 778 ft·lb/Btu
g	=	local gravitational acceleration, m/s^2 (ft/s^2)
g_c	=	standard gravitational acceleration, 32.2 ft/s^2

To arrive at Equation 2.4, $q = \dot{m}(h_2 - h_1)$, Equations 2.10 and 2.11 indicate the requirement that there was no work done ($\mathcal{P} = 0$) and that there are negligible

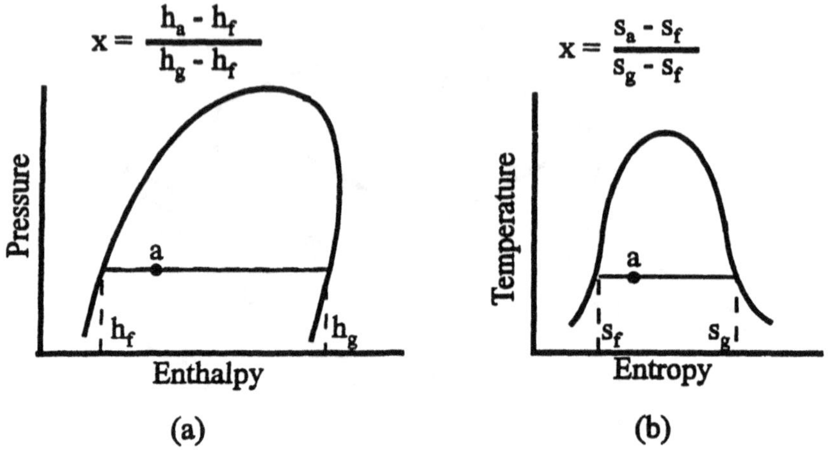

FIGURE 2.27
Quality expressed as the ratio of lengths (a) on a pressure-enthalpy diagram, and (b) on a temperature-entropy diagram.

changes in velocity and elevation. Equation 2.5, $\mathcal{P} = \dot{m}(h_1 - h_2)$, implies that in addition to negligible changes in velocity and elevation, the process is adiabatic, thus $q = 0$.

The steady-flow energy equation is a useful tool in other situations experienced in refrigeration systems. An application that will be used soon will be to explain what occurs in an expansion valve.

2.22 STATE OF REFRIGERANT EXPRESSED BY QUALITY

A thermodynamic property that turns out to be a poor choice of terms is *quality*, because it has nothing to do with the purity or desirability of the fluid. Instead it defines the fraction of vapor in a sample of the liquid/vapor mixture:

$$\text{Quality, } x = \frac{\text{mass of vapor}}{\text{total mass of liquid and vapor}} \tag{2.12}$$

In the diagrams of Figure 2.27 the quality x is represented by the ratio of distances on the enthalpy and entropy scales.

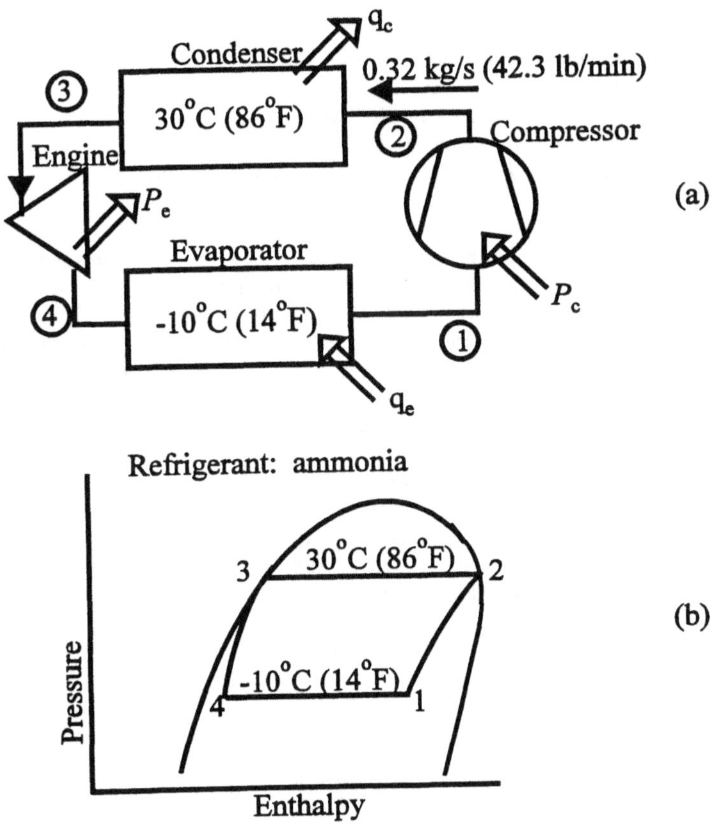

FIGURE 2.28
(a) Schematic diagram of the refrigeration plant, and (b) the pressure-enthalpy diagram of the system in Example 2.16.

2.23 ANALYSIS OF A CARNOT CYCLE USING REFRIGERANT ENTHALPIES

An important reason for analyzing refrigeration cycles is to design and/or troubleshoot problems in the capacities of evaporators and condensers and in the power requirements of compressors. Before leaving the Carnot cycle, the procedure for calculating energy flow rates using refrigerant enthalpies will be illustrated. The same concepts apply to actual operating cycles.

Example 2.16. A refrigeration plant shown in Figure 2.28a operates on the Carnot cycle shown in Figure 2.28b. The refrigerant is ammonia and circulates with a flow rate of 0.32 kg/s (42.3 lb/min). The condensing temperature is 30°C (86°F), and the evaporating temperature is −10°C (14°F).

Compute:

(a) the enthalpies at all points in the cycle,
(b) the refrigeration rate,
(c) the power required by the compressor,
(d) the power delivered by the engine,
(e) the rate of heat rejected at the condenser, and
(f) the COP.

Solution. (a) Because the cycle is a Carnot cycle, the compression, 1-2, and the expansion through the engine, 3-4, are isentropic. Thus, $s_1 = s_2$, and $s_3 = s_4$. Processes 2-3 and 4-1 take place at constant temperature. Values of enthalpy and entropy for points 2 and 3 can be read directly from the saturation property table for ammonia, Table A.12 (Table A.14). Point 2 is saturated vapor and 3 is saturated liquid, both at 30°C (86°F).

$h_2 = 1486$ kJ/kg (630.5 Btu/lb)
$s_2 = 5.264$ kJ/kg·K (1.188 Btu/lb·R)
$h_3 = 342$ kJ/kg (138.6 Btu/lb)
$s_3 = 1.489$ kJ/kg·K (0.287 Btu/lb·R)

Point 1 is a mixture of liquid and vapor at −10°C (14°F) where the entropy $s_1 = s_2$, and s_2 is the entropy of saturated vapor at 30°C (86°F). The quality at point 1, x_1, is given by the equation,

$$x_1 = \frac{s_1 - s_f}{s_g - s_f}$$

where s_f and s_g are the entropies of saturated liquid and vapor, respectively, at −10°C (14°F). Thus,

$$x_1 = \frac{5.264 - 0.829}{5.757 - 0.829} = 0.900$$

The expression for x_1 in terms of enthalpies is

$$x_1 = 0.900 = \frac{h_1 - h_f}{h_g - h_f} = \frac{h_1 - 154.0}{1450.5 - 154.0}$$

from which h_1 can be computed as $h_1 = 1320.9$ kJ/kg (559.4 Btu/lb).

In a similar manner h_4 can be calculated,

$$x_4 = \frac{1.489 - 0.829}{5.757 - 0.829} = 0.134 = \frac{h_4 - 154.0}{1450.5 - 154.0}$$

then $h_4 = 327.7$ kJ/kg (132.4 Btu/lb).

Summarizing,

$h_1 = 1320.9$	kJ/kg	(559.4 Btu/lb)
$h_2 = 1486$	"	(630.5 ")
$h_3 = 342$	"	(138.6 ")
$h_4 = 327.7$	"	(132.4 ")

(b) The refrigeration developed by each unit mass of refrigerant is called the *refrigerating effect*, which is:

$$h_1 - h_4 = 1320.9 - 327.7 = 993.2 \text{ kJ/kg (427 Btu/lb)}$$

The refrigerating capacity, q_e is the product of the mass rate of flow and the refrigerating effect,

$$q_e = (0.32 \text{ kg/s})(993.2 \text{ kJ/kg}) = 317.8 \text{ kW (18,070 Btu/min)}$$

(c) The power required by the compressor $\mathcal{P}_c$ is:
$\mathcal{P}_c$ = (flow rate)$(h_2 - h_1)$ = (0.32 kg/s)(1486 − 1321)
$\mathcal{P}_c$ = 52.8 kJ/s = 52.8 kW (70.8 hp)
(d) The power delivered by the engine $\mathcal{P}_e$ is:
$\mathcal{P}_e$ = (flow rate)$(h_3 - h_4)$ = (0.32 kg/s)(342 − 328)
$\mathcal{P}_e$ = 4.5 kJ/s = 4.5 kW (6.0 hp)
(e) The rate of heat rejected at the condenser, q_c is:
q_c = (flow rate)$(h_2 - h_3)$ = (0.32 kg/s)(1486 − 342)
q_c = 366.1 kJ/s = 366.1 kW (20,820 Btu/min)
(f) The coefficient of performance of the cycle is the refrigerating rate divided by the net power. The net power is the compressor power less the power provided by the engine, 52.8 − 4.5 = 48.3 kW (2747 Btu/min). Then the COP is

$$\text{COP} = \frac{317.8 \text{ kW}}{48.3 \text{ kW}} = \frac{18,070 \text{ Btu/min}}{2747 \text{ Btu/min}} = 6.58$$

An alternate method of computing the COP for the Carnot cycle is through the use of Equation 2.9,

$$\text{COP} = \frac{T_1}{T_2 - T_1} = \frac{-10 + 273.15}{(30 + 273.15) - (-10 + 273.15)} = 6.58, \text{ or}$$

$$\text{COP} = \frac{14 + 459.7}{(86 + 459.7) - (14 + 459.7)} = 6.58$$

2.24 REFRIGERATION CAPACITY EXPRESSED IN I-P UNITS—TONS OF REFRIGERATION

The standard unit of refrigerating capacity in the I-P system of units is the ton of refrigeration, where

1 ton of refrigeration = 12,000 Btu/hr = 200 Btu/min

The origin of the term suggests the early history of refrigeration in ice plants. A one-ton ice plant could freeze 2,000 lb of ice per day, which corresponds to

$$\frac{(2000 \text{ lb})(144 \text{ Btu/lb, latent heat})}{24 \text{ hr}} = 12,000 \text{ Btu/hr}$$

The refrigerating capacity in Example 2.16 is 18,070 Btu/min, or

(18,078)/(200 Btu/min per ton) = 90.4 tons.

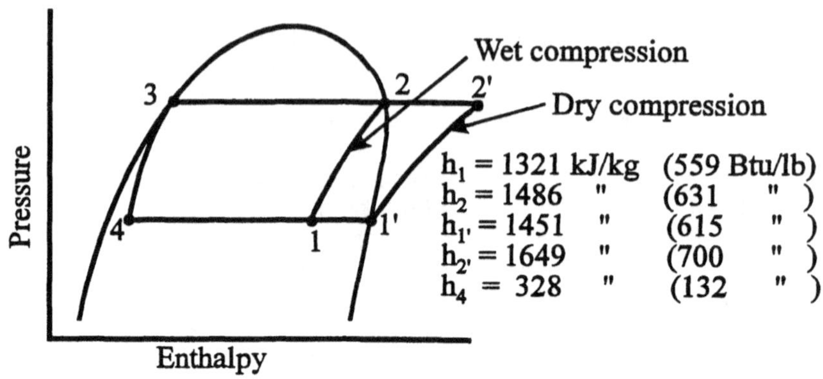

FIGURE 2.29
Wet versus dry compression.

2.25 DRY VERSUS WET COMPRESSION

In this section and the next, we progress from the Carnot cycle toward a more realistic cycle called the *standard vapor-compression cycle.* Two processes in the cycle shown in Figure 2.28b are revised—compression and expansion. We examine first the compression process, 1-2, displayed in Figure 2.29. The compression in the Carnot cycle is called *wet compression,* because it occurs completely in the mixture region with some liquid present.

Some operating problems are associated with wet compression. With reciprocating compressors, liquid on the cylinder walls diminishes lubrication effectiveness. Also, if liquid droplets remain unvaporized at the end of compression, the incompressible liquid may be forced into limited clearance volume and valves could be damaged. A further difficulty is that of controlling the fraction of liquid at point 1 to provide saturated vapor at point 2 after compression.

As will be discussed in Chapter 5, oil is injected to seal the space between the rotors of a screw compressor, and this oil can be cooled by direct injection of refrigerant. Laboratory tests show, however, that the expected improvement of performance with wet compression is not realized.

Instead of the wet compression of path 1-2, the standard vapor compression cycle assumes dry compression wherein the vapor entering the compressor is saturated vapor at point 1′. The compression then proceeds to point 2′ in the superheated vapor region. There are two influences on the efficiency of the cycle by the change from wet to dry compression—one an advantage and the other a disadvantage. The advantage is that the refrigerating effect increases from $(h_1 - h_4)$ to $(h_{1'} - h_4)$, thus from 993 kJ/kg (427 Btu/lb) to 1123 kJ/kg (483 Btu/lb), in the conditions of Example 2-16. On the other hand the work of compression also increases from $(h_2 - h_1)$ of 165 kJ/kg (71 Btu/lb) to $(h_{2'} - h_{1'})$ of 198 kJ/kg (85 Btu/lb). Thus the improvement in capacity is 13.1% while the additional work expended is 20.0%. The change to dry compression therefore reduces the COP of the cycle.

2.26 THROTTLING VALVE VERSUS AN EXPANSION ENGINE

The Carnot cycle envisions the liquid at point 3 in the cycle dropping in pressure through an engine, and the power developed by the engine helping to drive the compressor, thereby reducing the net power that must be supplied to the compressor. Building the hardware to accomplish this task has been an illusive goal, because the challenges include a reliable engine that handles a mixture of liquid and vapor, a simple way of transferring the power developed in the engine to the compressor, and regulating the flow rate of refrigerant through the engine so that the evaporator is supplied with the correct flow rate of refrigerant. A flow rate supplied to the evaporator that is too low penalizes the rate of heat transfer, and a flow rate too high could flood liquid out of the evaporator into the compressor.

The difficulties of manufacturing engines that work reliably with the mixture of liquid and vapor are formidable, so most refrigeration systems accomplish the necessary drop in pressure from point 3 to point 4 in a controlled throttling valve, called an *expansion valve.* Currently at least one commercial refrigeration system uses an expansion engine[3], which is applied to a water chiller using R-134a as the refrigerant. The engine is a turbine that delivers its power to the compressor through a direct shaft connection. Across the expansion valve, shown schematically in Figure 2-30, the refrigerant property relations prevailing derive from the steady-flow energy equation, Equations 2.10 and 2.11.

There is no shaft taking out work, so $\mathcal{P} = 0$, and even if the line is oriented vertically, the change in energy due to the change in elevation is negligible. While there will be a velocity change because the refrigerant at point 4 contains some vapor and has a higher specific volume, the magnitude of the change in kinetic energy, $V^2/2$ $(V^2/2g)$ is negligible. The final assumption is that the rate of heat transferred to or from ambient is zero. After all such simplifications are made, the steady-flow energy equation for the expansion valve reduces to:

$$h_3 = h_4 \tag{2.13}$$

The substitution of an expansion valve for the expansion engine results in a pressure-enthalpy diagram as in Figure 2.31. This revision from the Carnot cycle imposes a double penalty: the net power requirement is increased because of the loss of the contribution of the engine, and the refrigerating effect is decreased. In Example 2.16 the work of engine was $h_3 - h_4$ or $342 - 328$ kJ/kg = 14 kJ/kg (6.0 Btu/lb), and when the expansion valve is substituted this useful work is eliminated resulting in an 8.5% increase in power. The refrigerating effect changes from 993 kJ/kg (427 Btu/lb) to 979 kJ/kg (421 Btu/lb) which is a reduction of

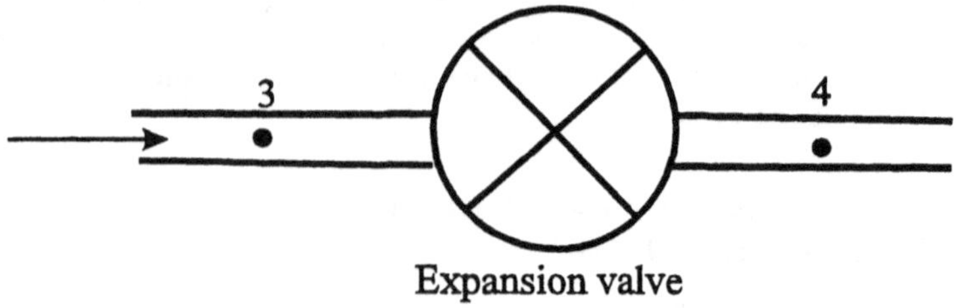

where:

$$z_3 = z_4$$
$$V_3 = V_4$$
$$P = 0$$
$$q = 0$$

FIGURE 2.30
Refrigerant flow through an expansion valve.

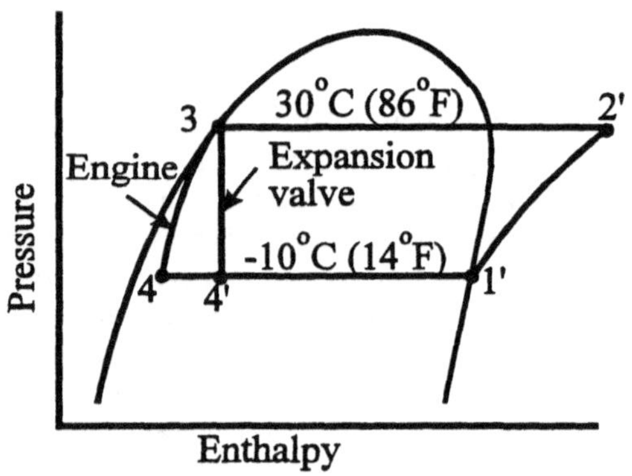

FIGURE 2.31
Pressure-enthalpy diagram showing the substitution of an expansion valve for an expansion engine.

1.4%. The conversion to the expansion valve is even more penalizing than the change to dry compression.

2.27 THE STANDARD VAPOR-COMPRESSION CYCLE

From the revisions to the Carnot cycle we are led to a popular convention for analyzing refrigeration cycles—the standard vapor- compression cycle. The

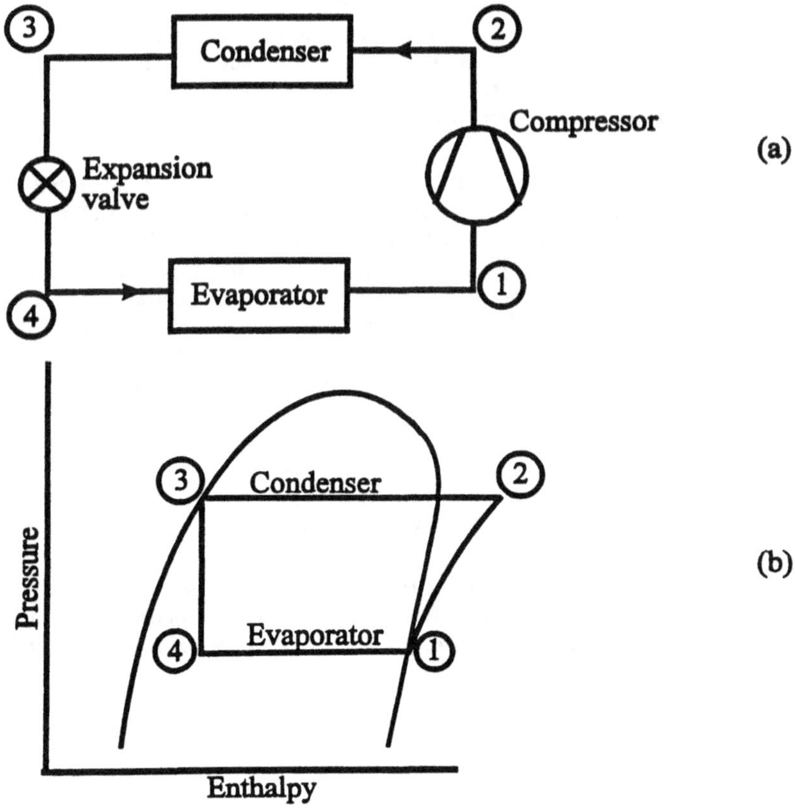

FIGURE 2.32
The standard vapor compression cycle with (a) the flow diagram, and (b) the pressure-enthalpy diagram.

schematic diagram of the equipment and the points on the pressure-enthalpy diagram that correspond to key positions in the cycle are shown in Figure 2.32. The following are the specifications of the four processes:

1-2, constant-entropy compression to the condensing pressure
2-3, constant-pressure desuperheating and condensation to saturated liquid
3-4, constant-enthalpy expansion through the expansion valve to the low pressure
4-1, constant-pressure evaporation to saturated vapor

Example 2.17. An R-22 system that develops a refrigerating capacity of 150 kW (42.6 tons of refrigeration) operates on the standard vapor-compression cycle with an evaporating temperature of $-5°C$ (23°F) and a condensing temperature of 35°C (95°F). Compute the following:

(a) evaporating and condensing pressures
(b) enthalpies at all points in the cycle
(c) mass flow rate of R-22 circulating through the system
(d) volume flow rate measured at the compressor suction
(e) power required by the compressor
(f) coefficient of performance

Solution. (a) From Table A.13 (A.14), properties of R-22 at saturated conditions, the evaporating pressure is 421.7 kPa (61.18 psia) at −5°C (23°F) and the condensing pressure is 1355.1 kPa (196.6 psia) at 35°C (95°F).

(b) The enthalpy at point 1 is that of saturated vapor at the evaporating temperature, and the enthalpy at point 3 is that of saturated liquid at the condensing temperature. The enthalpy at point 2 is found by moving along a constant-entropy line from point 1 to the condensing pressure. The enthalpy leaving the expansion valve, h_4, is the same as that entering, h_3. The values are:

$$\begin{aligned} h_1 &= 402.7 \text{ kJ/kg (106.4 Btu/lb)} \\ h_2 &= 431.8 \text{ kJ/kg (118.9 Btu/lb)} \\ h_3 &= 242.8 \text{ kJ/kg (37.64 Btu/lb)} \\ h_4 &= 242.8 \text{ kJ/kg (37.64 Btu/lb)} \end{aligned}$$

(c) Each kg (lb) of refrigerant that circulates through the system performs $402.7 - 242.8 = 159.9$ kJ/kg (68.7 Btu/lb) of refrigeration. In order to accomplish 150 kW (42.6 tons) of refrigeration, the mass rate of flow of R-22 $\dot{m}$ must be

$$\dot{m} = \frac{150 \text{ kW}}{159.9 \text{ kJ/kg}} = 0.938 \text{ kg/s}$$

or in I-P units

$$\dot{m} = \frac{(42.6 \text{ tons})(200 \text{ Btu/min per ton})}{68.7 \text{ Btu/lb}} = 124 \text{ lb/min}$$

(d) The volume rate of flow Q at the compressor suction (point 1) is the mass rate of flow multiplied by the specific volume at that position which is 55.2 L/kg or 0.0552 m^3/kg (0.884 ft^3/lb).

$$Q = (0.938 \text{ kg/s})(0.0552 \text{ m}^3\text{/s}) = 0.0518 \text{ m}^3\text{/s (109.6 ft}^3\text{/min)}$$

(e) The compressor power is the product of the work of compression in kJ/kg (Btu/lb) and the mass rate of flow. The work of compression is $h_2 - h_1$ or $431.8 - 402.7 = 29.1$ kJ/kg (12.5 Btu/lb). So the power requirement for an ideal compressor is

$$\begin{aligned} \text{Compressor power} &= (0.938 \text{ kg/s})(29.1 \text{ kJ/kg}) \\ \text{Compressor power} &= 27.3 \text{ kW (36.6 hp)} \end{aligned}$$

(f) The coefficient of performance is the ratio of useful refrigeration to the power required, thus

$$\text{COP} = \frac{150 \text{ kW}}{27.3 \text{ kW}} = 5.50$$

2.28 HORSEPOWER PER TON

The COP has been used frequently throughout this chapter as a measure of the effectiveness of the system and is a term that is widely used in the design and analysis of systems. In the United States another measure of the efficiency of the system that is sometimes used is the term ***horsepower per ton of refrigeration***. These two terms can be readily converted from one to another by the equation

$$\text{Horsepower per ton} = \frac{4.715}{\text{COP}} \tag{2.14}$$

In Example 2.17, for example, the effectiveness of the system in horsepower per ton is (36.6 hp)/(42.6 tons) = 0.86. This value could have been derived from the COP of 5.50 using Equation 2.14,

$$\text{Horsepower per ton} = 4.715/5.50 = 0.86$$

2.29 VARIATIONS IN THE STANDARD VAPOR-COMPRESSION CYCLE

The standard vapor-compression cycle is a useful concept, and the quantitative values that it provides are useful in the design and analysis of refrigeration systems. Certain idealizations are inherent in the standard cycle and we look ahead to the performance characteristics of real systems. For this reason, translation of the results to operating systems will be considered in the next section. We need to know, however, how real systems deviate from the standard vapor-compression cycle, and the first modification is in the flow diagram of the cycle as shown in Figure 2-33. Most industrial refrigeration systems incorporate a *receiver*, which accumulates the liquid coming from the condenser and feeds it to the evaporator as needed. The level of liquid in the receiver rises and falls as liquid shifts in and out of the evaporator and condenser at various operating conditions. The presence of the receiver does not change the cycle diagram as appears in Figure 2.32b.

Several deviations from the standard vapor-compression cycle that are typical of operating systems are shown in Figure 2.34. These variations include drops in pressure in the condenser and evaporator as well as in the connecting piping. Also there may be a small amount of subcooling of the liquid leaving the condenser and superheating of the vapor leaving the evaporator. If the type of condenser is one where refrigerant condenses in the tubes (in contrast to condensing in the shell of a shell-and-tube condenser), the pressure will drop as the refrigerant passes through the heat exchanger. Similarly, the pressure of refrigerant drops in an evaporator when boiling in the tubes. The refrigerant leaving industrial refrigeration condensers may be slightly subcooled, but subcooling is not normally desired since it indicates that some of the heat-transfer surface that should be used for condensation is used for subcooling. At the outlet of the evaporator it is crucial for protection of the compressor that there be no liquid, so to be safe it is preferable for the vapor to be slightly superheated.

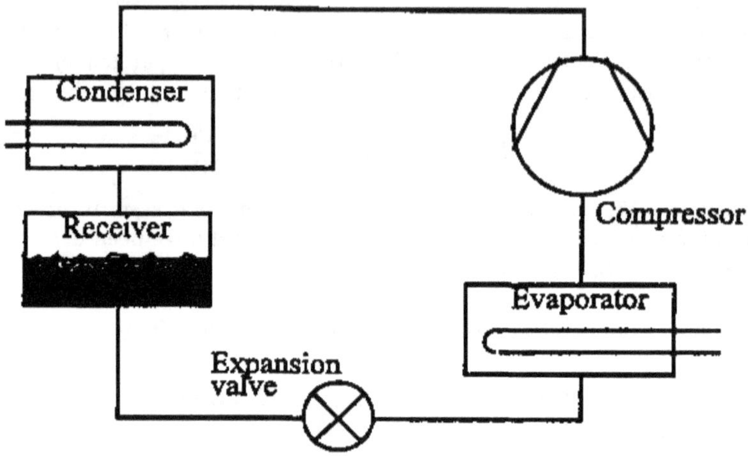

FIGURE 2.33
Addition of a liquid receiver to the refrigeration cycle.

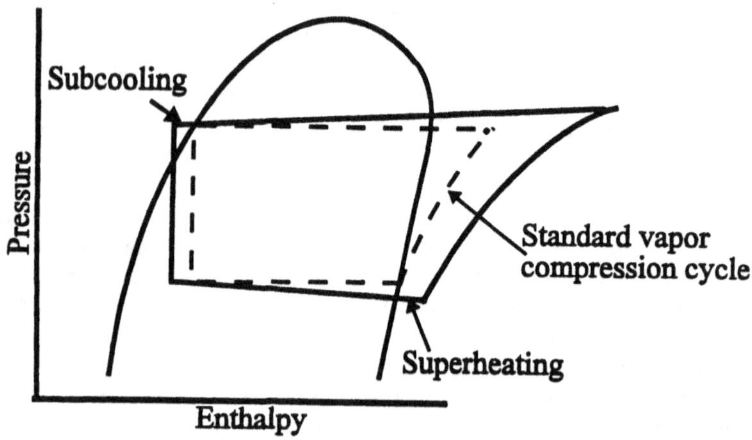

FIGURE 2.34
Deviations of the real cycle from the standard vapor-compression cycle.

2.30 USEFULNESS OF THE THERMODYNAMIC FOUNDATION

In the study of cycles, this chapter has proceeded toward the goal of real cycles starting with the very most efficient one, the Carnot cycle, and then to the standard vapor-compression cycle. The standard vapor-compression cycle lies somewhere between the Carnot cycle and actual system operation in its expression of reality. The Carnot and standard vapor-compression cycle provide insight into the character of refrigeration, and these cycles have additional value as tools to numerically estimate the performance of actual plants.

First consider the relationship of the Carnot and the standard vapor compression cycles. Converting the Carnot cycle to the standard vapor-compression cycle (Sections 2.25 and 2.26) introduced some inefficiencies into the cycle. If $COP_{std.vc}$ designates the COP of the standard vapor-compression cycle and COP_{Carnot} that of the Carnot cycle, the ratio of COPs, $COP_{std.vc}/COP_{Carnot}$, for most industrial refrigerants is an approximate function of the temperature difference between condensing and evaporating temperatures, as shown in Figure 2.35. For quick estimates of the COP of a standard vapor-compression cycle, data of Figure 2.35 can be applied to the simply-computed Carnot COP, $T_{evap}/(T_{cond} - T_{evap})$. In Example 2.17, the temperature difference between condensing and evaporating is 40°C (72°F) which from Figure 2.35 suggests a ratio of COPs of about 0.83. When applied to the COP_{Carnot} of 268.15/40 = 6.7 the prediction of the $COP_{std.vc} = (0.83)(6.7) = 5.56$ which agrees well with the value calculated in Example 2.17.

The next comparison is between the performance indicated by the standard vapor-compression cycle and the performance of actual hardware. To translate the COP of the standard vapor-compression cycle to actual equipment, simply multiply by the compressor efficiency. Usual compressor efficiencies range from 80% for low pressure ratios to 70% for high pressure ratios. Test this procedure with the operating conditions of Example 2.17. If a compressor efficiency of 80% is assumed, the predicted COP would be (5.50)(0.80) = 4.40 corresponding to 1.07 hp/ton. Catalog data of a six-cylinder ammonia compressor manufactured by Vilter Manufacturing Corporation, operating at 35°C (95°F) and −5°C (23°F) condensing and evaporating temperatures, respectively, quotes a refrigerating capacity of 368 kW (104.6 tons) and a power requirement of 81.2 kW (108.9 hp). Thus, the actual horsepower per ton is 1.04 corresponding to a COP of 4.53.

Another occasion where the procedures explained in this chapter are used is in computing flow rates at various points in the refrigeration cycle. The mass flow rates, the volume flow rates, and the velocities are necessary, for example, in computing the pressure drop of the refrigerant flowing in pipes.

Plant managers and operators strive constantly to conserve energy and Section 2.20 explained that a fundamental approach to energy conservation is to provide the compressor with as high a suction pressure and as low a condensing pressure as possible. It is useful to know what magnitudes of improvement are possible by raising the evaporating temperature and reducing the condensing

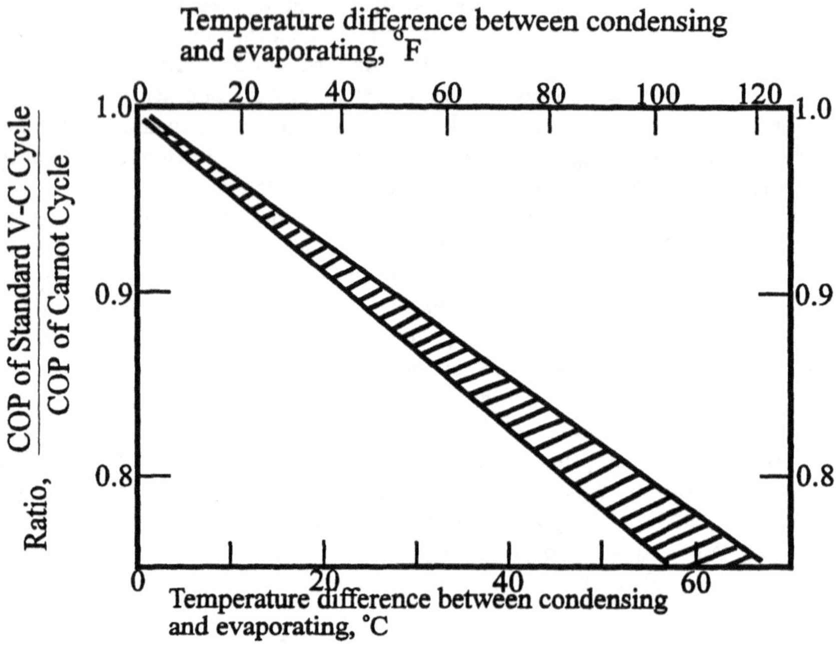

FIGURE 2.35
Approximate ratio of the coefficients of performance of the standard vapor-compression cycle to the Carnot cycle for most industrial refrigerants.

temperature. The expression for the Carnot COP provides a good estimate of possible improvements.

Example 2.18. What is the percentage improvement in the COP (horsepower per ton) of elevating the evaporating temperature 1°C and reducing the condensing temperature 1°C at the following two operating conditions: (a) 5°C (40°F) evaporating and 30°C (86°F) condensing temperatures, respectively, and (b) −10°C (14°F) evaporating and 35°C (95°F) condensing temperatures, respectively.

Solution. (a) The base COP is $(5 + 273.15)/(30 - 5) = 11.126$. The COP that results when the evaporating temperature is raised 1° is $(6 + 273.15)/(30 - 6) = 11.631$ or a 4.5% increase in COP per °C (2.5% per °F). Reducing the condensing temperature by 1°C raises the COP to $(5 + 273.15)/(29 - 5) = 11.59$ or a 4.2% increase in COP per °C (2.3% per °F).

(b) The base COP is $(-10 + 273.15)/(35 - (-10)) = 5.848$ and a computation similar to that of part (a) yields an improvement in COP of 2.4% per °C (1.33% per °F) for a 1°C increase in evaporating temperature and a COP improvement of 2.3% per °C (1.26% per °F) for a reduction in condensing temperature.

A few comments and interpretations applicable to Example 2.18 are as follows. The Carnot COP was used directly for the percentage improvement calculations because the corrections of Figure 2.35 and the compressor efficiency would simply cancel out. However, the corrections for the translation of Carnot to the standard vapor-compression cycle (Figure 2.35) are also affected by the change in evaporating or condensing temperature and contributes a secondary effect on the calculation. Each °C increase in evaporating or decrease in condensing temperature improves the COP ratio by about 0.04 (0.022 per °F). Thus the calculation using only the Carnot COP slightly underestimates the potential improvement.

A further observation from Example 2.18 is that the percentage improvement in the COP in part (a) with its small lift of temperatures is greater than in part (b) with its large lift of temperatures. A superficial and erroneous conclusion is that a change in temperature of a given amount is not as important with high temperatures lifts as it is with low temperature lifts. The opposite conclusion is correct, however, when the analysis is extended to the energy requirement which ultimately appears on the electric bill. For a given kW of refrigeration the power requirement in the plant (now including the data of Figure 2.35 and the compressor efficiency) for the operating conditions of part (a) would be

$$\text{Power, kW} = \frac{(1\text{ kW refrigeration})}{(\text{COP}_{Carnot})(0.9\text{ from Figure 2.35})(0.80\text{ compressor effic.})}$$

Power = 0.125 kW per kW of refrigeration

A similar calculation for the conditions in part (b) shows a power requirement of (1 kW)/[(5.848)(0.81)(0.78 efficiency)] = 0.27 kW. Application of the percentage improvement of a 1°C higher evaporating temperature calculated in Example 2. 18 shows a power reduction of 0.0056 kW with the conditions of part (a) and 0.0065 kW under the operating conditions of part (b). The effect on compressor power per unit refrigerating capacity is thus more prounouced at high temperature lifts.

Knowledge of the fundamentals of refrigerant properties and cycles has equal importance to the refrigeration engineer that the ability to tackle and block has to a football player and possession of a sense of rhythm has to a musician. The fundamentals are the foundation on which to construct the practice. Good refrigeration engineers frequently return to these fundamentals in their design of new plants and in the analysis of problems in existing systems.

REFERENCES

1. Stoecker, W. F., *SI Units for the HVAC/R Professional,* Business News Publishing Company, Troy, Michigan, 1992.
2. American Society of Heating, Refrigerating and Air-Conditioning Engineers, Inc., Atlanta, Georgia, from the 1993 ASHRAE *Handbook— Fundamentals.*
3. Brasz, J. J, "Improving the Refrigeration Cycle with Turbo-Expanders," International Congress of Refrigeration, The Hague, International Institute of Refrigeration, 1995.

CHAPTER 3

MULTISTAGE SYSTEMS

3.1 MULTISTAGE COMPRESSION IN INDUSTRIAL REFRIGERATION

A significant fraction of industrial refrigeration plants operate with a large difference between evaporating and condensing temperatures—perhaps 50°to 80°C (100°to 150°F). This large temperature lift imposes both problems and opportunities for the system. An opportunity is to use multistage compression, which although increasing the first cost over single-stage compression, also alleviates some problems and can save on total compressor power. In multistage compression, the refrigerant basically flows in series through two or more compressors with special processes performed on the refrigerant between stages. This chapter concentrates on two-stage compression, but the principles could be extended to three-stage compression which is an arrangement sometimes used in extremely low-temperature installations.

The use of two-stage compression opens up the opportunity to use two key processes which characterize multistage systems—flash gas removal and intercooling. It is instructive to examine these processes individually before combining them, as they exist in real systems. The number and temperature level of evaporators in the system imposes another variation on the configuration of multistage systems. The chapter concludes with an examination of cascade circuits, which are a form of multistage systems.

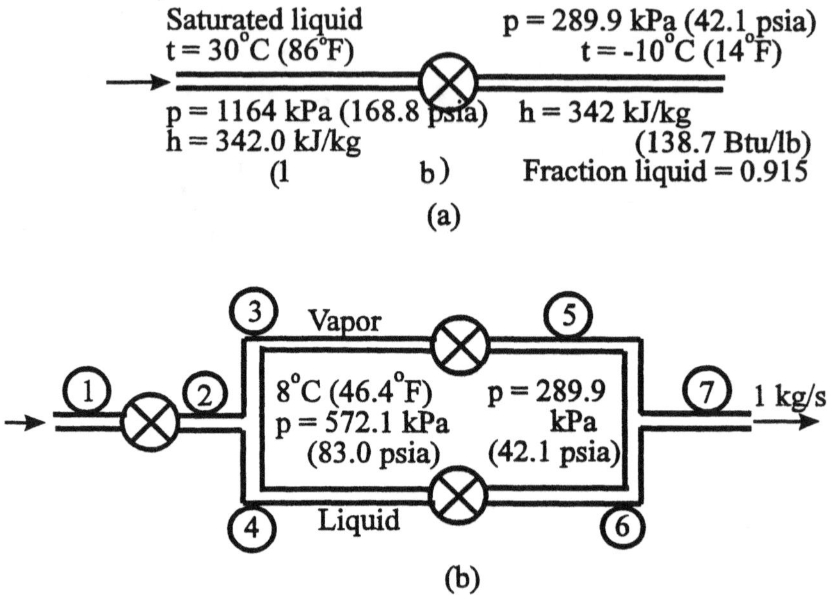

FIGURE 3.1
Expansion of saturated liquid (a) in a one-step process and (b) in a partial expansion followed by separation then continued expansion separately of liquid and vapor.

3.2 LOSSES IN THE EXPANSION VALVE PROCESS

First focus on the process that occurs in the expansion valve, where liquid enters and a mixture of liquid and vapor leaves. The familiar vertical line on the pressure-enthalpy diagram, such as process 3-4 in the standard vapor compression cycle of Figure 2.32, indicates a progressive increase in the fraction of vapor. This vapor is unable to absorb heat from the evaporator, but requires compressor power to return it to the condenser pressure. To seek more insight into this process, explore the expansion of ammonia-saturated liquid at 30°C (86°F) and a pressure of 1163.8 kPa (168.8 psia) to a pressure of 289.9 kPa (42.1 psia) corresponding to a saturation temperature of −10°C (14°F). Figure 3.1a shows the process taking place in one step. Figure 3.1b shows the equivalent process but with the expansion occurring as follows: (a) a drop in pressure to 572.1 kPa (83.0 psia), (b) a separation of the liquid and vapor, (c) separate expansion of liquid and vapor to the final pressure of 289.9 kPa (42.1 psia), and finally (d) combining the mixture of liquid and vapor at point 6 with the vapor at 5 to achieve at point 7 the same conditions that occurred after the single-stage expansion. Table 3.1

TABLE 3.1
Conditions at the statepoints in Figure 3.1b.

Point	Temperature		Pressure		Flow rate, kg/s		Enthalpy	
	°C	°F	kPa	psia	liquid	vapor	kJ/kg	Btu/lb
1	30	86.0	1164	168.8	1.000	—	342.0	138.7
2	8	46.4	572.1	83.0	0.915	0.085	342.0	138.7
3	8	46.4	572.1	83.0	—	0.085	1469.9	623.5
4	8	46.4	572.1	83.0	0.915	—	237.2	114.8
5	−2.15	28.1	289.9	42.1	—	0.085	1469.9	623.5
6	−10	14.0	289.9	42.1	0.856	0.059	237.2	114.8
7	−10	14.0	289.9	42.1	0.855	0.145	342.0	138.7

shows conditions at the statepoints of Figure 3.1b with the flow rates that would occur for a total flow rate of 1 kg/s.

The purpose of comparing the two techniques for conducting the expansion is to focus on the flash gas that progressively develops as the expansion proceeds. In particular, when the vapor at point 3 is expanded to the low pressure this vapor is useless for refrigeration. Its enthalpy is too high to achieve any refrigeration in the evaporator, and its temperature of −2.15°C (28.1°F) is actually higher than the −10°C (14°F) that will be the evaporating temperature. In order to drop the temperature of the vapor at point 5 to −10°C (14°F), a small amount of liquid at 6 must vaporize to match the state at point 7 to that which occurs in the single-stage expansion. Not only is the vapor at point 3 useless for refrigeration, the continued expansion to the low pressure only entails the expenditure of power to recompress it. The conclusion is that the vapor at point 3 should be removed and compressed back to the condensing pressure without expanding it further to the low pressure. This objective defines the function of flash-gas removal.

3.3 FLASH-GAS REMOVAL

Suitable equipment to partially expand the refrigerant and then remove the flash gas is shown schematically in Figure 3.2. Liquid refrigerant from the condenser or high-pressure receiver passes through a level-control valve. The liquid, being more dense than the vapor, separates and flows on to the expansion valve of the evaporator. A flash-gas compressor draws off the vapor from the separating vessel or flash tank and compresses it to the condensing pressure where it joins the vapor from the main compressor.

Because the level-control valve in Figure 3.2 is appearing for the first time in this book, it is important to realize that it performs the function of an expansion valve. In controlling the level of liquid in the vessel, the valve opens wider if the liquid level begins to drop. Similarly, if the liquid level begins to rise, the valve closes more. The pressure in the flash tank is controlled by the pumping capacity

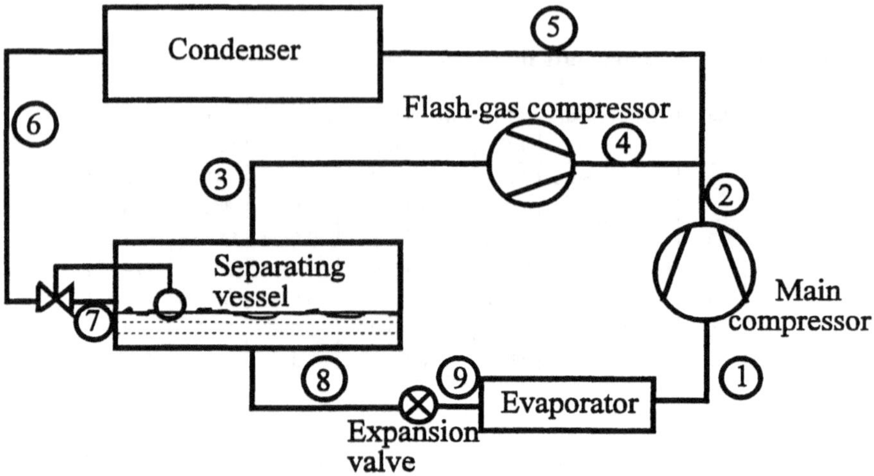

FIGURE 3.2
A two-stage compression system with flash-gas removal.

of the flash gas compressor in relation to the flow rate of liquid passing on to the evaporator. For a given flow rate of liquid leaving the vessel, a low pumping capacity of the flash-gas compressor results in a high intermediate pressure. This is true, because at steady-state operation the flow rate of vapor that forms due to flashing must equal the rate of removal by the compressor. If this pumping rate is low, the process adjusts by generating only a low flow rate of vapor, which is associated with a small drop in pressure through the valve. The pressure-enthalpy diagram displaying the statepoints of the refrigerant in the equipment of Figure 3.2 is shown in Figure 3.3.

Example 3.1. An ammonia system operating with an evaporating temperature of −30°C (−22°F) and a condensing temperature of 35°C (95°F) separates flash gas at a temperature of −2°C (28.4°F) and delivers it to the condenser through a separate compressor. If the refrigerating capacity is 200 kW (56.9 tons of refrigeration), what are the power requirements if the system operates (a) single stage, and (b) with flash-gas removal?

Solution. *a. Single-stage.* In single-stage compression the enthalpies are as follows:

leaving the evaporator	=	1423.6 kJ/kg (603.6 Btu/lb)
leaving the compressor	=	1796.9 kJ/kg (764.1 Btu/lb)
leaving the condenser	=	366.4 kJ/kg (149.1 Btu/lb)

The mass rate of flow through the system is:

$$\dot{m} = \frac{200 \text{ kW}}{1423.6 - 366.4} = 0.189 \text{ kg/s} \quad (25.0 \text{ lb/min})$$

and the power requirement is:

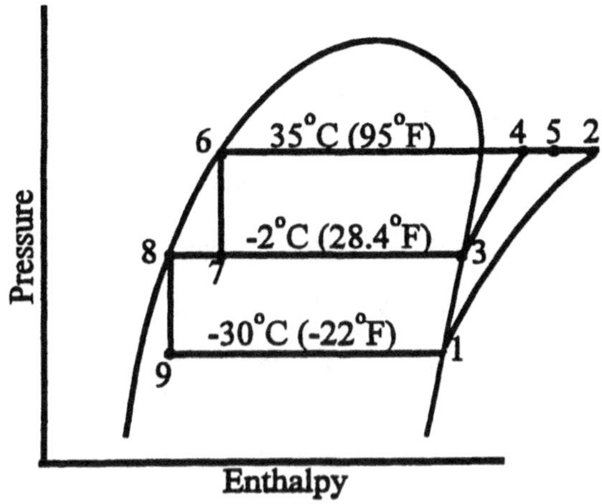

FIGURE 3.3
Pressure-enthalpy diagram of flash-gas removal process in Figure 3.2.

$$\text{Compressor power} = (0.189 \text{ kg/s})(1796.9 - 1423.6 \text{ kJ/kg}) = 70.55 \text{ kW } (94.61 \text{ hp})$$

b. With flash-gas removal. In the system of Figure 3.3 the following enthalpies apply:

$$h_1 = 1423.6 \text{ kJ/kg } (603.6 \text{ Btu/lb})$$
$$h_2 = 1796.9 \text{ kJ/kg } (764.1 \text{ Btu/lb})$$
$$h_3 = 1459.7 \text{ kJ/kg } (619.1 \text{ Btu/lb})$$
$$h_4 = 1633.5 \text{ kJ/kg } (693.9 \text{ Btu/lb})$$
$$h_6 = 366.4 \text{ kJ/kg } (149.1 \text{ Btu/lb})$$
$$h_7 = 366.4 \text{ kJ/kg } (149.1 \text{ Btu/lb})$$
$$h_8 = 190.7 \text{ kJ/kg } (73.6 \text{ Btu/lb})$$
$$h_9 = 190.7 \text{ kJ/kg } (73.6 \text{ Btu/lb})$$

The mass rate of flow through the evaporator, $\dot{m}_1$, is:
$\dot{m}_1 = (200 \text{ kW})/(h_1 - h_9) = 0.162$ kg/s (21.45 lb/min),
and
$\dot{m}_8 = \dot{m}_9 = \dot{m}_2 = \dot{m}_1 = 0.162$ kg/s (21.45 lb/min).

The mass rate of flow of flash gas, $\dot{m}_3$, can be calculated from a mass and energy balance about the separating vessel:

$$\dot{m}_7 = \dot{m}_3 + \dot{m}_8 \quad \text{or} \quad \dot{m}_7 = \dot{m}_3 + 0.162 \text{ kg/s} \tag{3.1}$$

and

$$\dot{m}_7 h_7 = \dot{m}_3 h_3 + \dot{m}_8 h_8 \tag{3.2}$$

or:

$$\text{or: } (\dot{m}_3 + 0.162)366.4 = \dot{m}_3(1459.7) + (0.162)(190.7)$$

Then:

$$\dot{m}_3 = 0.162\left(\frac{366.4 - 190.7}{1459.7 - 366.4}\right) = 0.0260 \text{ kg/s } (3.44 \text{ lb/min})$$

The power requirements of the compressors are:

Main compressor:

$\dot{m}_1(h_2 - h_1) = 0.162(1796.9 - 1423.6) = 60.47$ kW (81.1 hp)

Flash-gas compressor:

$\dot{m}_3(h_4 - h_3) = 0.0260(1633.5 - 1459.7) = 4.52$ kW (6.06 hp)

$$\text{The total compressor power} = 60.47 + 4.52 = 64.99 \text{ kW } (87.2 \text{ hp}).$$

The percentage saving resulting from flash-gas removal is:

$$\left(\frac{70.55 - 64.99}{70.55}\right)100 = 7.9\%$$

The magnitude of saving from flash-gas removal depends on the thermodynamic properties of the refrigerant and the magnitude of the temperature lift from the evaporator to the condenser. Figure 3.4 shows the percent saving in total compressor power for several refrigerants and for a range of evaporating temperatures. Flash gas is removed at an optimum intermediate temperature, and this temperature will be explained further in Section 3.10. The percent saving increases as the evaporating temperature drops, which supports the contention that flash-gas removal and, in general, two-stage compression are most effective at low evaporating temperatures.

The halocarbons exhibit higher percent savings than does ammonia, but it should not be assumed that two-stage systems provide greater improvement in efficiency when using halocarbons. The other process associated with two-stage compression (intercooling) must yet be incorporated.

3.4 LIQUID SUBCOOLING

Some of the same benefits of direct flashing of liquid into vapor at the intermediate pressure can be achieved by liquid subcooling. A popular method of subcooling, shown in Figure 3.5, immerses a pipe coil in the liquid of the intermediate-pressure vessel. Warm liquid from the condenser enters the heat-exchanger coil and transfers heat to the lower-temperature liquid. Liquid subcooling is a form of flash-gas removal, because some liquid in the vessel vaporizes and is drawn off at the intermediate pressure.

Compared to direct flashing of Figure 3.2, the liquid subcooler has the advantage of maintaining the liquid at a high pressure. Therefore, the subcooled liquid can travel long distances and endure some drops in pressure without flashing into vapor. The liquid leaving a direct flash tank is saturated and flashes into

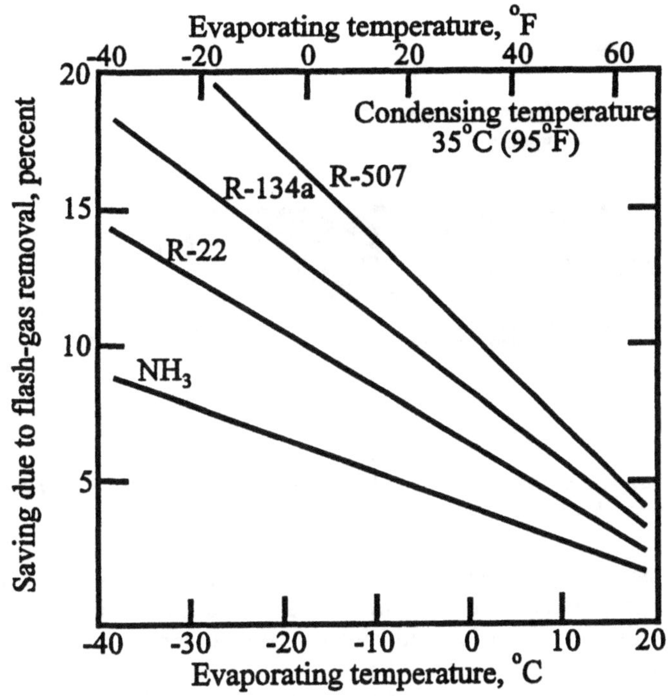

FIGURE 3.4
Percent saving in total compressor power resulting from flash-gas removal at the optimum intermediate temperature.

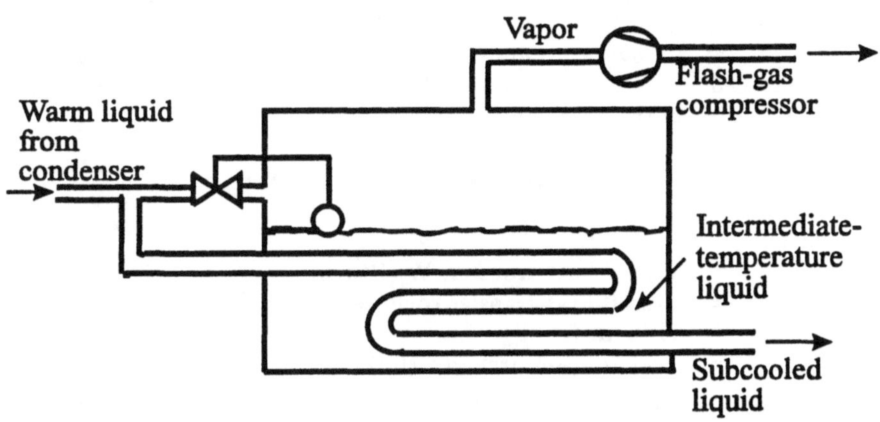

FIGURE 3.5
A liquid subcooler using a coil immersed in the liquid of an intermediate-pressure vessel.

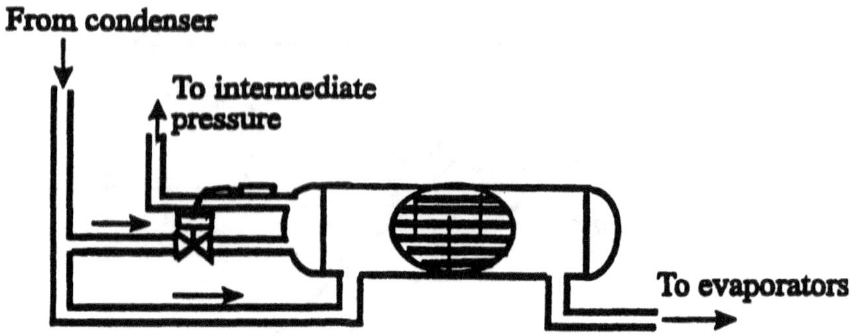

FIGURE 3.6
Liquid subcooling with an external shell-and-tube heat exchanger with boiling refrigerant controlled by an expansion valve.

vapor when the pressure drops due to friction or a rise in elevation. Also, since the liquid is cool, it will absorb heat from the warm ambient, which may also cause flashing. The disadvantage of the subcooler in comparison to the direct flash tank is that liquid cannot be cooled all the way down to the saturation temperature of the liquid because the heat exchanger must operate with a temperature difference between the leaving subcooled liquid and the intermediate-temperature liquid.

The selection of the length of tube of the immersed subcooler is not usually the result of a detailed heat-transfer calculation. Instead, the fabricator often inserts as much tube as convenient, and the system lives with the result. In certain cases it may be profitable to design the coil more carefully to balance the installed cost against the saving. The overall-heat-transfer coefficient, which is the U-value in the equation:

$$q = UA(\text{Mean temperature difference}) \tag{3.3}$$

where:

q = rate of heat transfer, W (Btu/hr)
U = overall-heat-transfer coefficient, $W/m^2 \cdot K$ ($Btu/hr \cdot ft^2 \cdot {}^\circ F$)
A = heat transfer area, m^2 (ft^2)

The U-value is a function of the boiling heat-transfer coefficient at the outside of the tube and the convection coefficient of the flowing liquid refrigerant inside the tube. Using some typical values of boiling heat-transfer coefficients suggested by Ayub[1] for R-22 and ammonia, approximate U-values can be calculated, as shown in Table 3.2.

TABLE 3.2
U-values of immersed coil subcoolers,
W/m²·K (Btu/hr·ft²·°F)

Refrigerant	Tube-side velocity, m/s (ft/s)	
	0.75 (2.46)	2.0 (6.56)
R-22	75 (13)	125 (22)
Ammonia	180 (32)	275 (49)

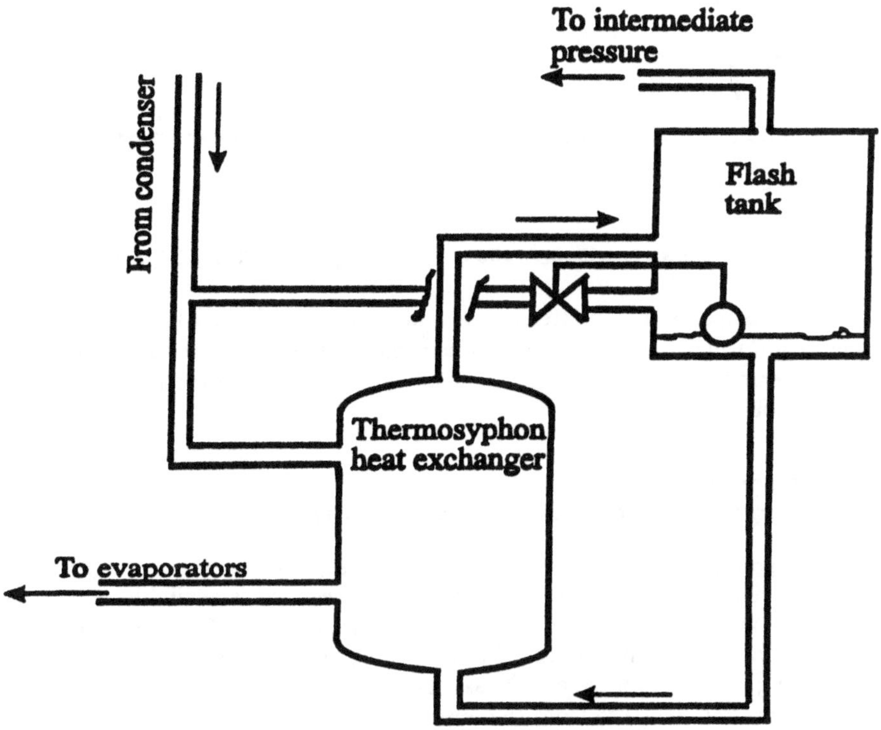

FIGURE 3.7
Liquid subcooling with an external heat exchanger of the thermosyphon type.

A guideline sometimes used by one designer[2] is to install a heat-transfer area of immersed coil of 2.5 m² for every 100 kW (1 ft² per ton) of refrigeration capacity at the evaporator. A heat exchanger of this size conforms with the data in the table to provide a reasonable temperature drop of liquid.

Another class of heat exchangers for subcooling liquid is shown in Figure 3.6, which depicts a shell-and-tube heat exchanger in which the boiling liquid is regulated by a thermostatic (superheat-controlled) expansion valve. The heat exchanger in Figure 3.7 is of the thermosyphon type, which is provided with liquid from the flash tank and this liquid partially vaporizes in the vertical tubes

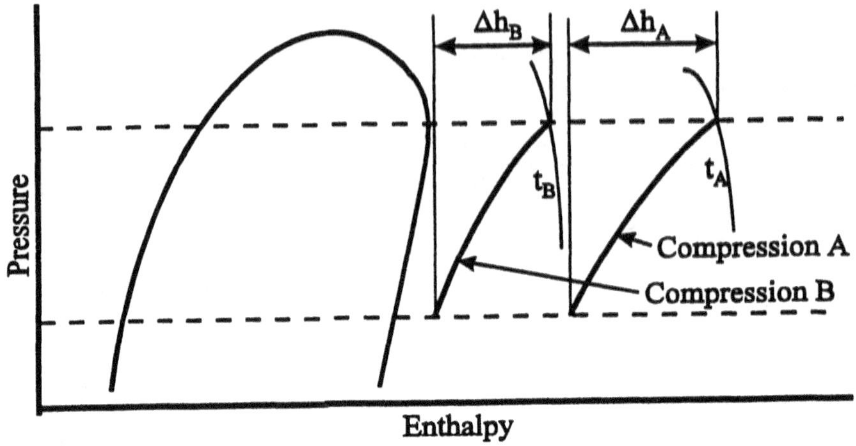

FIGURE 3.8
Comparison of compressions between two given pressures with differing initial temperatures.

of the exchanger and circulates by natural convection. Higher heat-transfer coefficients are usually possible with the external heat exchangers in comparison to the immersed type of Figure 3.5, and they are much more practical for a retrofit installation.

3.5 INTERSTAGE DESUPERHEATING—INTERCOOLING

The foregoing two sections have isolated for separate analysis the process of flash-gas removal—one of the two major processes that is available in two-stage compression. The other major process is desuperheating of the discharge vapor from the low-stage compressor. The two principal advantages of interstage desuperheating, or intercooling, are the saving in compressor power and the reduction in discharge temperature from the low-stage compressor.

Both of these advantages can be demonstrated on the pressure-enthalpy diagram, as in Figure 3.8, for a compression between two given pressures. The initial temperature of the refrigerant in compression A is high compared to that of compression B. A general observation of the compression lines (lines of constant entropy) is that they become flatter as they move to regions of greater superheat. The consequence of this change of slope is that the increase in enthalpy during the compression, which indicates the power required in the compression, is greater with high inlet temperatures. In Figure 3.8, Δh_A is greater than Δh_B. Another observation from Figure 3.8 is that the discharge temperature after Compression B is lower than that after Compression A. This feature is particularly important in reciprocating compressors, but is also significant in screw compressors, despite the internal cooling inherently provided during compression by the sealing oil.

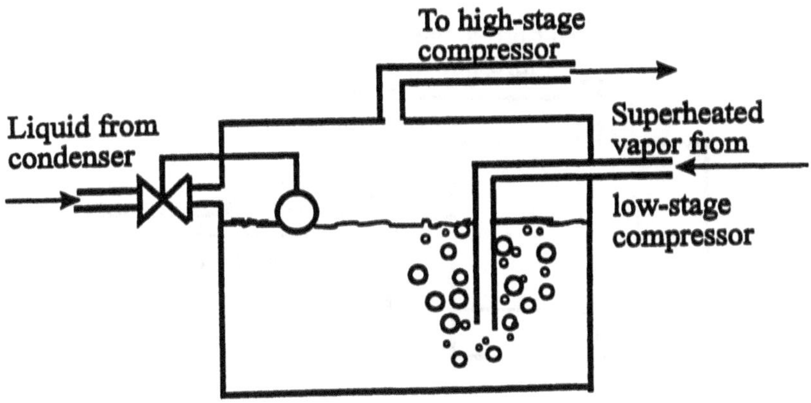

FIGURE 3.9
Desuperheating vapor by bubbling it through liquid in the intercooler.

Two of the methods of achieving intercooling are shown in Figures 3.9 and 3.10. The traditional method, Figure 3.9, is to immerse the outlet of the discharge line of the low-stage compressor below the liquid level in the intercooler providing bubbling of vapor through the liquid. Some of the liquid in the vessel evaporates to provide desuperheating of the vapor. The method will usually achieve a close approach of the temperature of the discharge vapor to the liquid temperature. The second approach[3] is to spray liquid into the discharge line from the low-stage compressor, as in Figure 3.10, and in so doing vaporize some of the liquid to desuperheat the discharge vapor.

The concept of bubbling superheated vapor through the liquid in the vessel, as shown in Figure 3.9, is effective as a heat-transfer process, but has several disadvantages. One is that in order to be effective, the outlet of the low-stage discharge line should be between 0.6 to 1.2 m (2 to 4 ft) below the surface of the liquid. Due to the static head of the liquid, the point of discharge will be at a slightly higher pressure than at the surface of the liquid. The compressor must, therefore, expend more energy to overcome this additional pressure. A second disadvantage of the bubbler desuperheater is that the process churns the liquid and vapor in the vessel. Because one of the functions of the vessel is to separate liquid from vapor so that only dry vapor passes on to the compressor, the turbulence makes this task more difficult.

While usually not able to achieve the degree of desuperheating possible in the bubbler, the spray method of Figure 3.10 causes less disturbance in the vessel. In some cases the supply of liquid comes directly from the condenser through an expansion valve. The superheat-control (thermostatic) expansion valve, which will be discussed further in Chapter 11, mounts a sensor on the vapor line to the high-stage compressor, and if the vapor temperature is too high, the valve opens to admit more refrigerant. A failure of this valve would permit liquid to continue

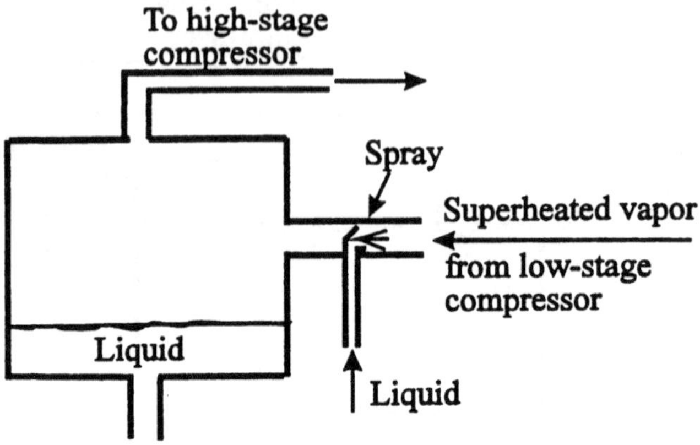

FIGURE 3.10
Desuperheating vapor by spraying liquid into the discharge line.

to pour into the vessel. An approach that avoids this danger is possible when liquid is pumped from the vessel in the liquid recirculation concept (Chapter 8). Centrifugal pumps handling liquid close to its saturated condition require a small flow rate even when the demand drops to zero. This bypass flow can simply serve as the desuperheating spray.

3.6 ANALYSIS OF THE INTERCOOLING PROCESS

Just as flash-gas removal was analyzed separately, the intercooling process will also be explored independently. The flow diagram of a two-stage system using only intercooling is shown in Figure 3.11a and the corresponding pressure-enthalpy diagram in Figure 3.11b.

Intercooling is often used in two-stage air compression, and in this application water from a cooling tower or the tap is the cooling medium. The cooling water is assumed to be free-of-charge, so all of the reduction in power in the high-stage compressor represents a net saving in power. When intercooling refrigerant, however, the cooling water that is normally available is not cold enough to desuperheat the refrigerant to the saturation temperature. The vaporization of some refrigerant, which must then be compressed by the high-stage compressor, is an expenditure of power that must be deducted from the reduction in power attributable to the lower inlet temperature.

Example 3.2. The intercooling system shown in Figure 3.11a operates with ammonia at the following saturation temperatures: evaporating, −35°C (−31°F); intermediate, 0°C (32°F); and condensing, 35°C (95°F). What is the saving in power of the intercooled cycle, expressed in percent, compared to single-stage operation?

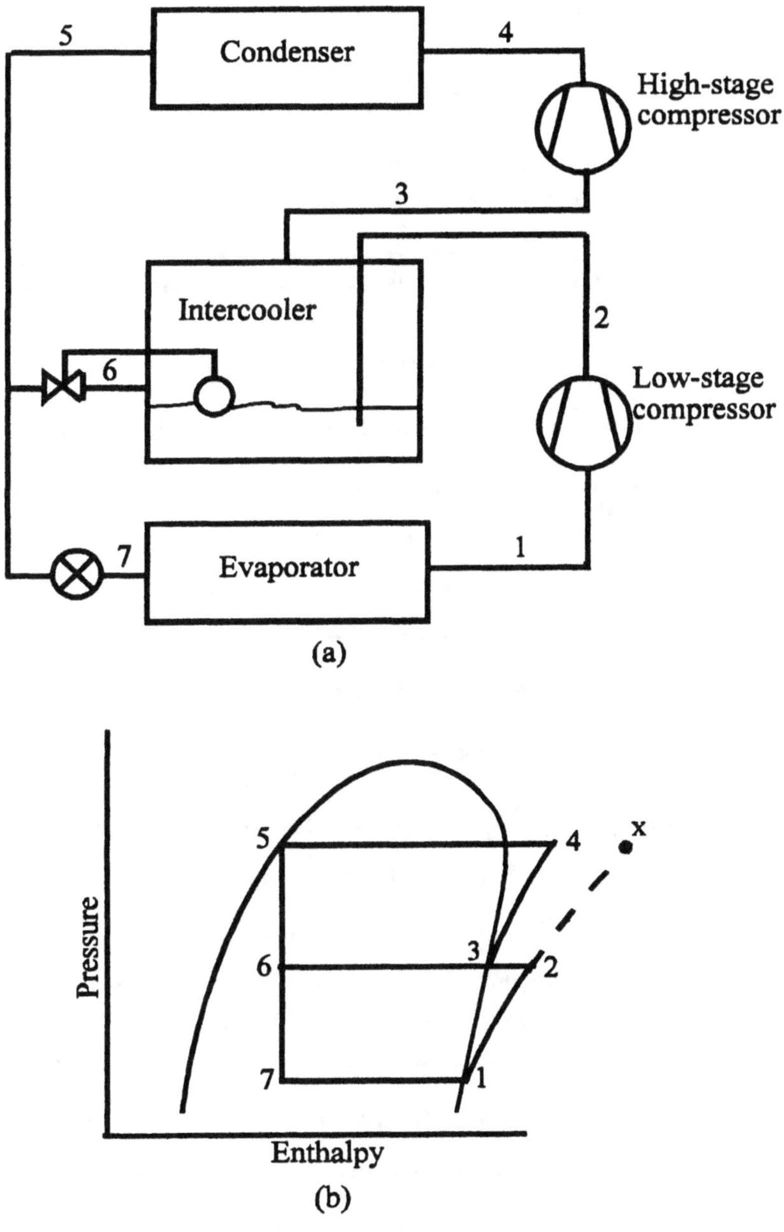

FIGURE 3.11
(a) Two-stage system with intercooling only, and (b) the corresponding pressure-enthalpy diagram.

Solution. The following enthalpies apply to the system and pressure-enthalpy diagram in Figure 3.11:

$$\begin{aligned} h_1 &= 1416.2 \text{ kJ/kg} \quad (600.4 \text{ Btu/lb}) \\ h_2 &= 1621.4 \quad (688.7) \\ h_3 &= 1462.7 \quad (620.4) \\ h_4 &= 1628.2 \quad (691.6) \\ h_5 = h_6 = h_7 &= 366.4 \quad (149.1) \end{aligned}$$

Assuming an arbitrary refrigeration capacity at the evaporator of 100 kW (28.4 tons of refrigeration), the flow rate through the evaporator and low-stage compressor is:

$$\dot{m}_1 = \dot{m}_2 = (100 \text{ kW})/(h_1 - h_7)$$

$$\dot{m}_1 = \dot{m}_2 = (100)/(1416.2 - 366.4) = 0.09526 \text{ kg/s } (12.6 \text{ lb/min})$$

The flow rate through the high-stage compressor, $\dot{m}_3$, can be calculated from a mass balance:

$$\dot{m}_6 + \dot{m}_2 = \dot{m}_3. \tag{3.4}$$

combined with an energy balance:

$$\dot{m}_6 h_6 + \dot{m}_2 h_2 = \dot{m}_3 h_3 \tag{3.5}$$

Combining Equations 3.4 and 3.5 yields:

$$\dot{m}_3 = \dot{m}_2 \left(\frac{h_2 - h_6}{h_3 - h_6}\right)$$

$$\dot{m}_3 = 0.09526 \left(\frac{1621.4 - 366.4}{1462.7 - 366.4}\right) = 0.1090 \text{ kg/s } (14.4 \text{ lb/min})$$

The total power, P_{int}, of the two compressors is:

$$P_{int} = \dot{m}_2(h_2 - h_1) + \dot{m}_3(h_4 - h_3)$$

$$P_{int} = 0.09526(1621.4 - 1416.2) + 0.1090(1628.2 - 1462.7)$$

$$P_{int} = 19.55 + 18.03 = 37.58 \text{ kW } (50.4 \text{ hp})$$

In single-stage compression the power requirement, P_{ss}, is:

$$P_{ss} = \dot{m}_1(h_x - h_1) = 0.09526(1834.4 - 1416.2) = 39.84 \text{ kW } (53.4 \text{ hp})$$

The percentage saving is power due to intercooling is thus:

$$\text{Saving} = \left(\frac{39.84 - 37.58}{39.84}\right) 100 = 5.7\%$$

In Example 3.2 the difference in power requirement attributable to intercooling appears in the high-stage compressor. When intercooling is applied, the

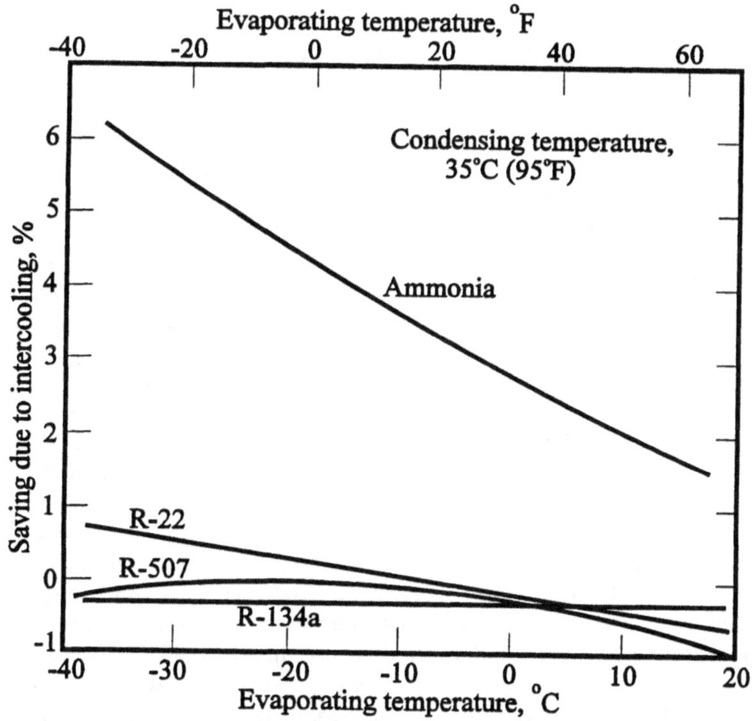

FIGURE 3.12
Percent saving in total compressor power resulting from intercooling at the optimum intermediate temperature.

flow rate of refrigerant handled by the high-stage compressor increases, but the work of compression for each kg (lb) of refrigerant decreases. With the statepoints shown in Figure 3.11b, the comparison of high-stage power is as follows:

Without intercooling	With intercooling
$\dot{m}_1(h_x - h_2)$	$\dot{m}_3(h_4 - h_3)$
0.09526(1834.4 − 1621.4)	0.1090(1628.2 − 1462.7)
20.29 kW (27.2 hp)	18.03 kW (24.2 hp)

As was true of flash-gas removal only, the intercooling process with ammonia becomes more effective as the evaporating temperature drops, as Figure 3.12 shows. For the three halocarbon refrigerants shown, however, intercooling provides little or no improvement in the power requirements in contrast to the results indicated in Figure 3.4 where the halocarbons react more positively to flash-gas removal than does ammonia. Since flash-gas removal and intercooling are so frequently used in conjunction with one another, the combination of the influences will be shown later for the standard two-stage system.

3.7 INTERCOOLING IN AN OIL-COOLED SCREW COMPRESSOR

Although Chapter 5, Screw Compressors, explains more thoroughly the features and operation of screw compressors, an intriguing situation surrounding intercooling of these machines will be addressed here. Screw compressors operate with oil injected into the refrigerant during compression. This oil absorbs some of the heat of compression and leaves the compressor with the refrigerant at a moderately high temperature. After separation from the refrigerant, as in Figure 3.13, the oil is cooled in a heat exchanger before it returns for injection into the compressor. In one concept of oil cooling, the heat from the heat exchanger flows directly to the condenser. Figure 3.13 applies to the low-stage compressor whose discharge gas is to be intercooled.

Figure 3.13 shows that the refrigerant leaves the compressor at a temperature of approximately 60°C (140°F) instead of the 110 to 120°C (230 to 250°F) typical of a reciprocating ammonia compressor. Furthermore, when the popular thermosiphon type of oil cooler is used, the heat removed from the oil passes directly to the condenser without requiring any refrigerant to vaporize and cool it in the intercooler. The efficiency of the intercooling process is thereby improved. Continued desuperheating with refrigerant to a temperature approaching the saturation temperature at the intermediate pressure is, however, still desirable. This continued desuperheating is advisable because directing 60°C (140°F) vapor to the high-stage compressor would result in unacceptably high oil and refrigerant temperatures leaving the high-stage compressor, particularly for ammonia. Also, the benefits of intercooling from 60°C (140°F) are still waiting to be achieved.

3.8 TWO EVAPORATOR-TEMPERATURE LEVELS AND ONE COMPRESSOR

Multipressure systems are usually understood to be those with two stages of compression, but a frequently-encountered configuration consists of only one compressor and two or more evaporators operating at different temperatures. The usual approach to accommodate this arrangement is shown in Figure 3.14, where the compressor operates with the same saturated suction temperature as the lowest-temperature evaporator. A pressure-regulating valve, also called an evaporator-pressure regulator or a back-pressure valve, regulates the pressure in the higher-temperature evaporator.

This arrangement is used where a low temperature in an evaporator may be undesirable, or even damaging. Examples include refrigerated spaces storing unfrozen food where air temperatures below freezing may injure the product. Another example is a liquid-chilling evaporator, where if the liquid freezes the tubes could burst. The throttling of the suction vapor in the line from the high-temperature evaporator introduces a thermodynamic loss. The application of the evaporator-pressure regulator of Figure 3.14 finds application mostly in small- and medium-sized systems where the additional investment cost and complexity

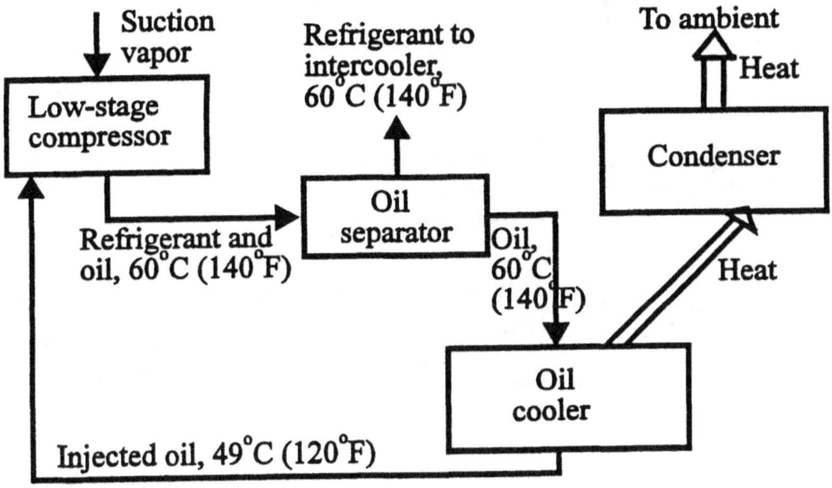

FIGURE 3.13
A partial intercooling provided by oil cooling in a low-stage screw compressor.

of a two-stage system with flash-gas removal and intercooling cannot be justified by the saving in power.

3.9 STANDARD TWO-STAGE SYSTEM WITH ONE EVAPORATING TEMPERATURE

The two features made available by two stages of compression—flash- gas removal and intercooling—have been analyzed separately. A system[4] with two stages of compression and one level of evaporating temperature usually incorporates both processes, as shown in Figure 3.15. The corresponding pressure-enthalpy diagram is shown in Figure 3.16. The flash-gas removal may be achieved either by direct flashing of liquid at the interstage pressure (Figure 3.2) or with a liquid subcooler (Figure 3.5). One of the variations of intercooling, Figures 3.9 or 3.10, may also be chosen.

Example 3.3. An R-507 two-stage system with flash-gas removal and intercooling provides 200 kW of refrigeration at an evaporating temperature of −40°C (−40°F) when operating with a condensing temperature of 35°C (95°F). The intermediate temperature is −5°C (23°F).

(a) What are the enthalpies of the refrigerant at all points in the system?

(b) Compute the flow rates through each compressor.

(c) What are the power requirements of the compressors?

(d) What would be the power required in a single-stage R-507 system with these evaporating and condensing temperatures, and the percentage saving in

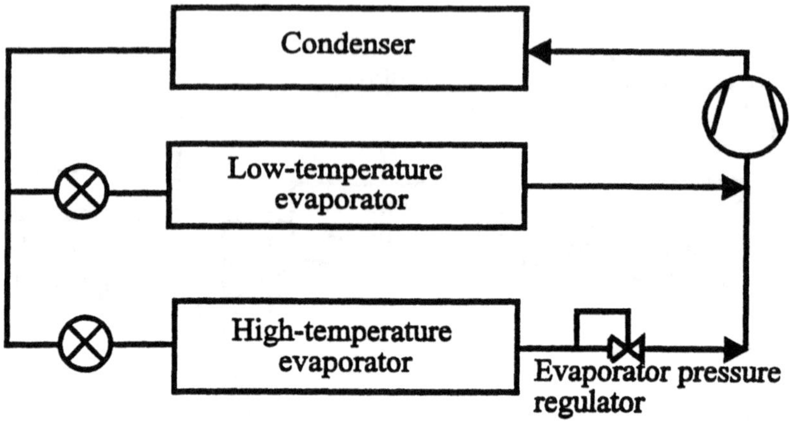

FIGURE 3.14
Using an evaporator-pressure regulator to elevate the temperature of one evaporator in a system with one compressor.

power through the use of a two-stage system?

Solution. (a) The enthalpies of R-507 at the positions designated in Figures 3.15 and 3.16 are:

Position	Enthalpy, kJ/kg	Enthalpy Btu/lb
1	340.5	84.0
2	366.8	95.3
3	360.5	92.6
4	382.6	102.1
5 and 6	249.5	44.9
7 and 8	193.2	20.7

(b) The flow rate through the low-stage compressor, which is the same as the flow rate through the evaporator is:

$$\dot{m}_1 = \frac{200\text{ kW}}{h_1 - h_8} = \frac{200}{340.5 - 193.2}$$

$$\dot{m}_1 = 1.358\text{ kg/s } (179.6\text{ lb/min})$$

The flow rate through the high-stage compressor, $\dot{m}_3$, can be found through a combined mass and heat balance about the flash-tank/intercooler. The mass balance is:

$$\dot{m}_6 + \dot{m}_2 = \dot{m}_7 + \dot{m}_3$$

The flow rate $\dot{m}_1$ has already been calculated as 1.358 kg/s (179.6 lb/min), and, from examination of the flow diagram, equals $\dot{m}_2$ and $\dot{m}_7$. The energy balance is represented by the equation:

$$\dot{m}_6 h_6 + \dot{m}_2 h_2 = \dot{m}_7 h_7 + \dot{m}_3 h_3$$

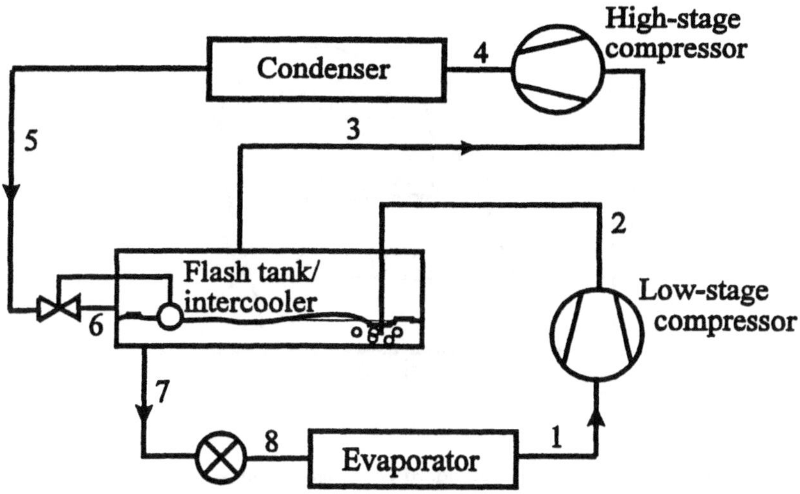

FIGURE 3.15
A standard two-stage system with flash-gas removal and intercooling.

The flow diagram shows that $\dot{m}_6 = \dot{m}_3$, thus the energy balance equation becomes:

$$249.5\dot{m}_3 + (1.358)((366.8) = (1.358)(360.5) + 360.5\dot{m}_3$$

so the flow rate through the high-stage compressor, $\dot{m}_3$, is:

$$\dot{m}_3 = 2.124 \text{ kg/s } (281.0 \text{ lb/min}).$$

(c) The power requirements of the compressors are, respectively:

low-stage:	1.358(366.8 − 340.5)	=	35.71 kW (47.9 hp)
high-stage:	2.124(382.6 −360.5)	=	46.93 kW (62.9 hp)
total power:	35.71 + 46.93	=	82.64 kW (110.8 hp)

(d) A single-stage R-507 system operating between −40°C (−40°F) evaporating and 35°C (95°F) condensing temperatures would require 108.7 kW (145.8 hp) compressor power, so the percentage saving through the use of the two-stage system is:

$$\frac{108.7 - 82.6}{108.7} = 24.0\%$$

3.10 REFRIGERANT FLOWS IN A FLASH TANK/INTERCOOLER

Figure 3.17 shows another way to visualize the processes that take place in the flash-tank/intercooler of a two-stage system. A stream that is 100% liquid enters at the left from the condenser. A large portion of this liquid is cooled and passes

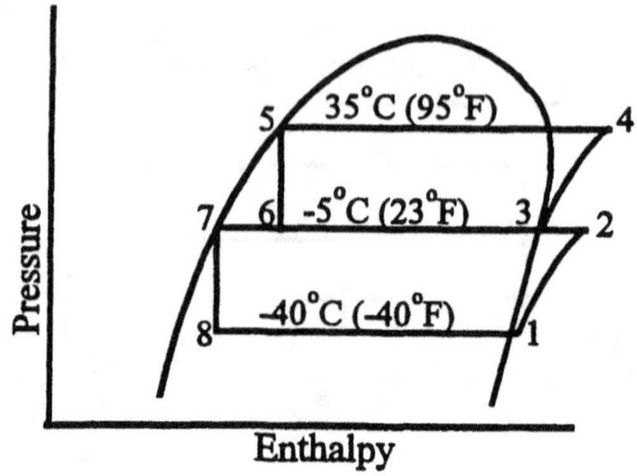

FIGURE 3.16
Pressure-enthalpy diagram for the standard two-stage system of Figure 3.15.

on to the evaporators, and the cooling of this liquid is accomplished by the vaporization of some of the incoming liquid. Another portion of the incoming liquid stream vaporizes to desuperheat the vapor received from the low-stage compressor.

3.11 OPTIMUM INTERMEDIATE PRESSURE

When the intermediate pressure in a two-stage system is selected at a very high value or a very low value, the benefits are reduced. In fact, if the intermediate pressure is the same as either the condensing or the evaporating pressure, the system reverts to single stage. We are led to the conclusion, then, that there must be some optimum intermediate pressure that results in minimum total power requirements. In the two-stage compression of air with ideal compression in both compressors, the optimum intermediate pressure is the geometric mean of the suction and discharge pressure,

$$p_{intermediate} = \sqrt{(p_{suction})(p_{discharge})} \tag{3.6}$$

The optimum intermediate pressure in a two-stage refrigeration system is quite close, but not equal, to that shown in Eq. 3.6. The reason for the deviation is that the cooling process in air compression is without cost, but in the refrigeration system there is a power cost associated with compressing the refrigerant vapor that performs the cooling. Figure 3.18 shows a graph of the total power in a two-stage R-22 system operating with an evaporating temperature of −30°C (−22°F) and a condensing temperature of 35°C (95°F). The optimum intermediate pressure in two-stage refrigeration systems is slightly higher than that

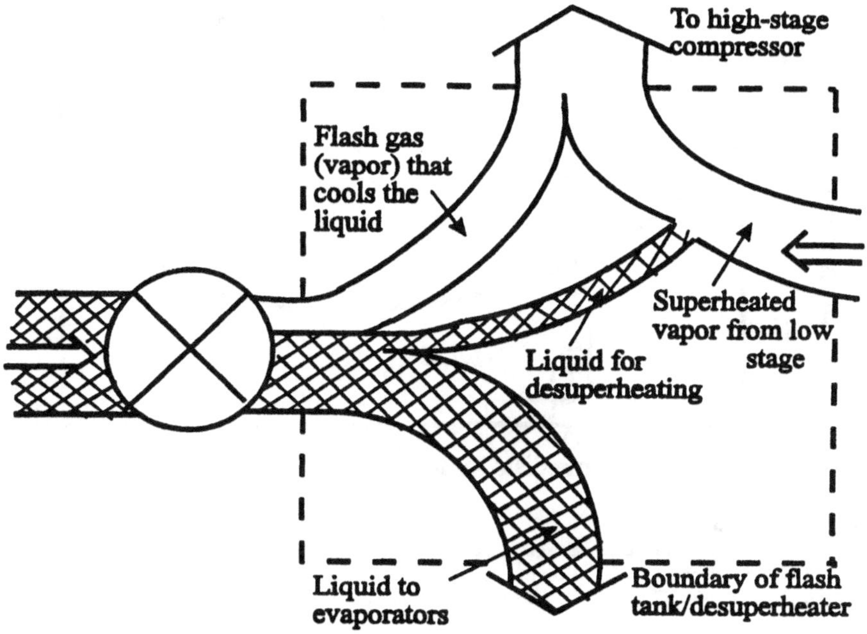

FIGURE 3.17
Refrigerant streams in a flash-tank/desuperheater.

indicated by Eq. 3.6, because of the bias toward a higher intermediate pressure to reduce the flow rate of refrigerant performing the desuperheating.

A conclusion that can be drawn from Fig. 3.18 is that the curve showing the power is quite flat, so a few degrees difference in intermediate temperature has little practical influence on power required by the plant. The relative insensitivity of the power to the intermediate temperature is particularly apparent when comparing the values shown in Fig. 3.18 to the power required in single-stage compression, which is 36.1 kW (48.4 hp), for these evaporating and condensing temperatures. The noncritical nature of the intermediate temperature is fortunate, because in those systems with evaporators at the intermediate temperature, (see Sec. 3.12), the temperature requirements of these evaporators dictate the intermediate saturation temperature.

3.12 TWO-STAGE COMPRESSION WITH TWO TEMPERATURE LEVELS OF EVAPORATION

Many plants serve evaporators at several different temperature levels. Examples are particularly frequent in the food refrigeration industry where a facility maintains some space at, for example, −23°C (−10°F) for storing frozen food and

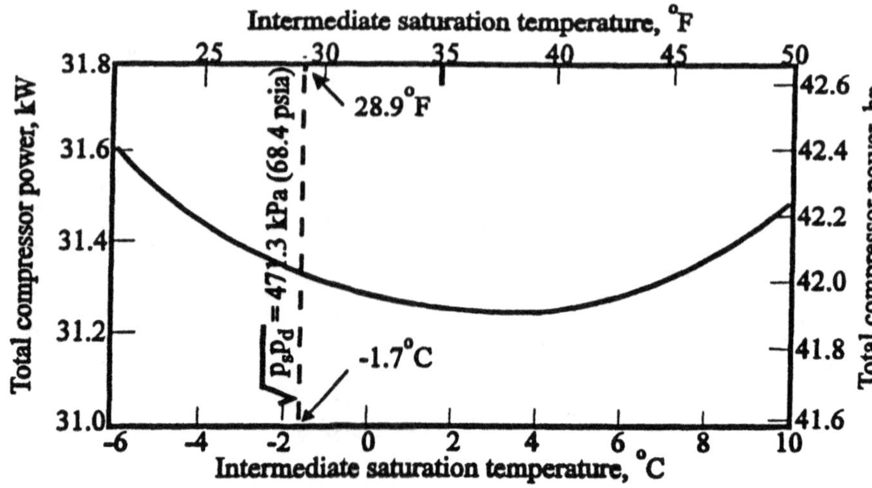

FIGURE 3.18
Total power required in a two-stage R-22 system with a refrigerating capacity of 100 kW (28.4 tons of refrigeration) as a function of the saturated intermediate temperature. The evaporating temperature is −30°C (−22°F), and the condensing temperature is 35°C (95°F).

another space at 2°C (36°F) for unfrozen fruits and vegetables. The flow diagram of a refrigeration system to meet such varied temperature needs is shown in Figure 3.19. This system uses a subcooler for the liquid passing to the low-temperature evaporator. The intermediate- temperature evaporator is fed with liquid from the condenser or in some cases liquid mechanically pumped from the subcooler/intercooler. While it would be possible to direct the refrigerant leaving the medium-temperature evaporator directly to the suction line of the high-stage compressor, this stream is normally passed to the vessel. This arrangement is mandatory when the medium-temperature evaporator receives pumped liquid and operates with liquid overfeed. Return to the vessel is also a good precaution as well when expansion valves control the liquid flow to the evaporator and may fail, which would allow liquid to flood out of the evaporator.

When the plant operates with two temperature levels of the evaporation, the intermediate pressure is regulated to meet the temperature needs of the medium-temperature evaporator. The intermediate pressure thus may not match the optimum for minimum total power, but as Figure 3.18 shows, a moderate variation in this pressure results in only a small penalty in increased power. In designing a system with two levels of evaporating temperature, the flow rate of refrigerant through the low-stage compressor is unaffected by the existence of the medium-temperature evaporator. The high-stage compressor, on the other hand, must accommodate the flow attributable to the medium-temperature evaporator plus that normally associated with the low-stage compressor, flash-gas removal, and desuperheating.

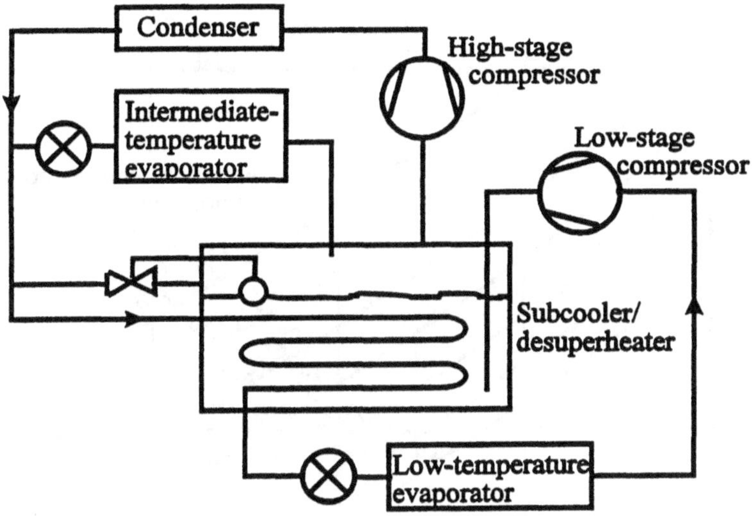

FIGURE 3.19
A two-stage system with both low- and medium-temperature evaporators.

3.13 SHORTCUTS IN MULTISTAGE SYSTEM DESIGN CALCULATIONS

This chapter has explained how to compute capacity requirements of the low- and high-stage compressors in two-stage systems. The precise method requires the application of mass and energy balances around the intermediate-pressure devices. Repeated two-stage calculations are tedious, although design and sales organizations that perform frequent analyses can readily computerize these calculations. Even for the manual process, the calculations can be simplified through the use of a graph such as Figure 3.20.

A chart such as Figure 3.20 can accurately represent the mass and energy balance equations, because it is only the evaporating and the intermediate temperatures that influence this calculation. The application of Figure 3.20 is simply to multiply the kW (tons of refrigeration) capacity of the low-temperature evaporator by the multiplying factor to find the required refrigeration capacity of the high-stage compressor. The values of the ratios from Figure 3.20 show higher quantities than obtained with the procedures explained in this chapter. The reason is that Figure 3.20 acknowledges the inefficiencies of the booster compressor, which causes a greater load on the desuperheater than would be true of an ideal compression. The calculation procedures of this chapter can be made more realistic by using the actual power of the low-stage compressor rather than the ideal isentropic power.

Figure 3.20 applies when the low-stage compressor is of the reciprocating type and to some configurations of screw compressor installations. The features

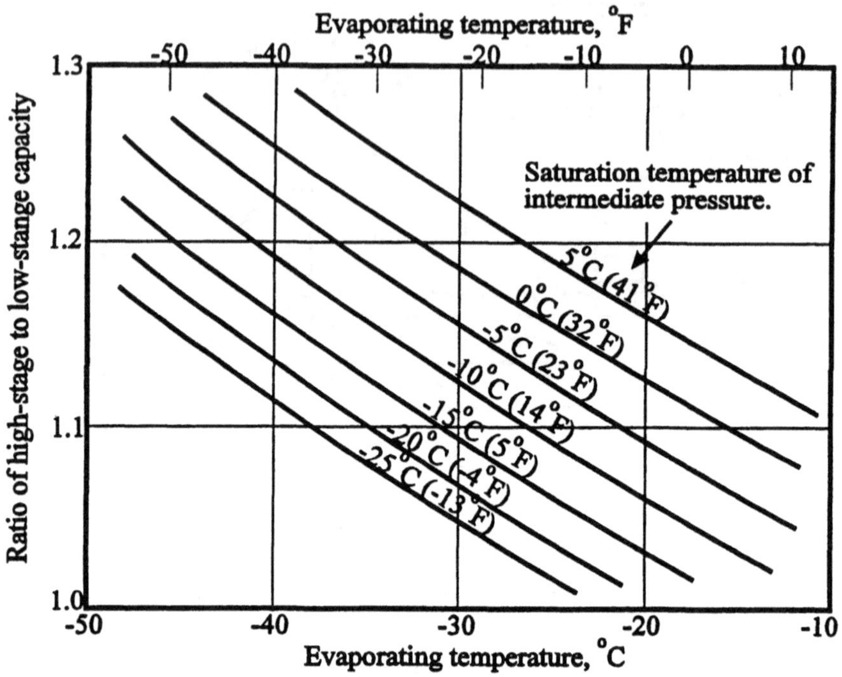

FIGURE 3.20
Ratio of required capacities of the high-stage to the low-stage compressor in a single-evaporator ammonia two-stage system. (*Courtesy of Vilter Manufacturing Corporation*)

of screw compressor systems will be explained in greater depth in Chapter 5, but at the moment only those equipment characteristics will be discussed that are necessary to understand the analysis of two-stage operation with a low-stage screw compressor. To seal the spaces between the rotors of a screw compressor, oil is injected which intermingles with the refrigerant being compressed. The oil absorbs some of the compression heat and must be cooled before reinjection. Two popular methods of oil cooling—direct injection of liquid refrigerant and cooling with a thermosiphon heat exchanger—are shown in Figures 3.21 and 3.22, respectively.

In Figure 3.21, liquid refrigerant is injected directly into the compressor, and its vaporization neutralizes the heat added by compression. The discharge vapor leaving the compressor is not completely desuperheated, and the remainder of the desuperheating process takes place in the flash tank/intercooler with the vaporization of more liquid refrigerant. From a thermodynamic standpoint, it makes no difference whether refrigerant vaporizes within the compressor or within the flash tank/intercooler to perform the superheat. Consequently, the factors shown in Figure 3-20 apply to systems where the oil in the low-stage compressor is cooled by direct admission of refrigerant.

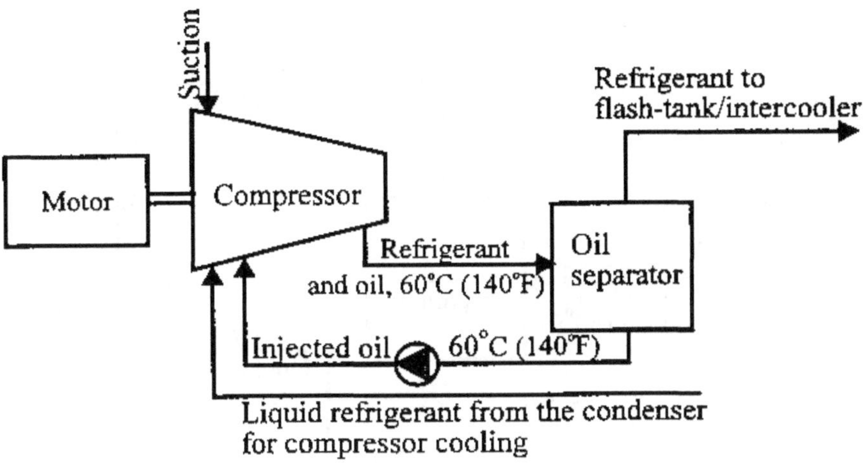

FIGURE 3.21
Cooling the oil injected into a low-stage screw compressor by direct admission of liquid refrigerant.

A different, and usually more favorable, situation results with the arrangement in Figure 3.22 where the oil is first separated from the the refrigerant vapor that is discharged from the compressor, then cooled in a thermosiphon heat exchanger. The thermosiphon heat exchanger receives liquid refrigerant, boils a portion of it in cooling the oil, and thus maintains natural convection circulation. The vapor generated in the thermosiphon heat exchanger passes directly to the condenser with the result that some of the heat of compression in the low-stage compressor is rejected directly to atmosphere. The amount of liquid that would otherwise be allocated to desuperheating and which must be compressed by the high-stage compressor is reduced by 15 to 30%. More precise data of the percentage will be presented in Chapter 5, Screw Compressors, but it is sufficient here to call attention to the fact that the multiplying factor of Figure 3.20 is reduced for the compressor cooling arrangement in Figure 3.22.

3.14 COMPRESSOR SELECTION AND CONTROL

For a given temperature of the low-temperature evaporator and a given condensing temperature, the intermediate pressure is controlled by the relative pumping rates of the low- and high-stage compressors. The basic rules for controlling pressures, and thus the saturation temperatures, in a two-stage system are:

- Regulate the pressure of the low-temperature evaporator by the pumping capacity of the low-stage compressor

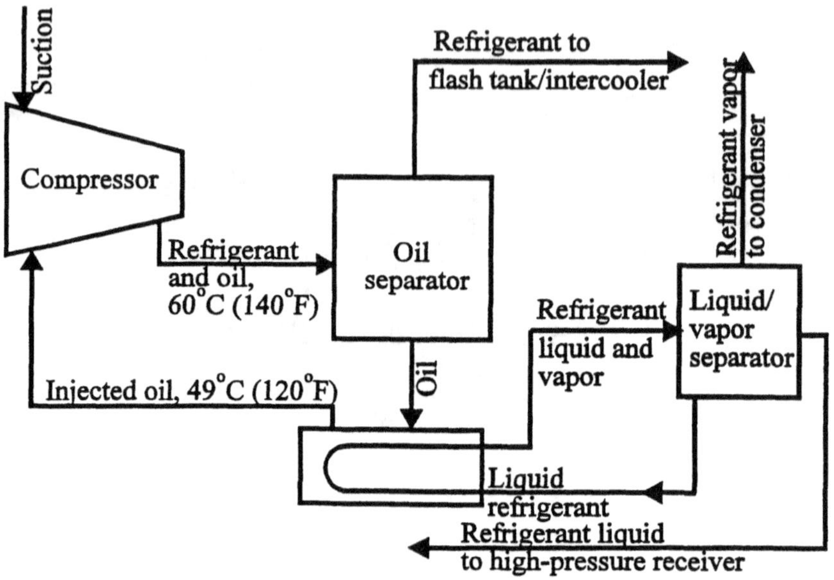

FIGURE 3.22
Cooling the oil injected into a low-stage screw compressor by an external thermosiphon heat exchanger which in turn is cooled by boiling refrigerant.

- Regulate the intermediate pressure by the pumping capacity of the high-stage compressor

Following these rules for control of the pressures is not an unreasonable assignment, because suction pressure control of both reciprocating and screw industrial compressors is quite standard.

The two stages of compression must work in conjunction, because the high-stage compressor must pump the same mass flow rate as the low-stage compressor plus the extra flow rate attributable to flash-gas removal and desuperheating, as well as that of an intermediate-temperature evaporator. The responses of mass flow rates and saturation temperatures to an increase in the refrigeration load in the low-temperature evaporator are shown in Figure 3.23. The system is assumed to have been in a steady-state operating condition, when at time T the refrigeration load on the low-temperature evaporator suddenly increases. This increase in load may appear as an abrupt increase in air temperature on the evaporator coil. The first response is for the evaporating temperature and pressure to increase. With a higher evaporating temperature the low-stage compressor is able to pump a higher flow rate which raises the intermediate pressure. Quickly the suction pressure controllers on both compressors sense values that are too high,

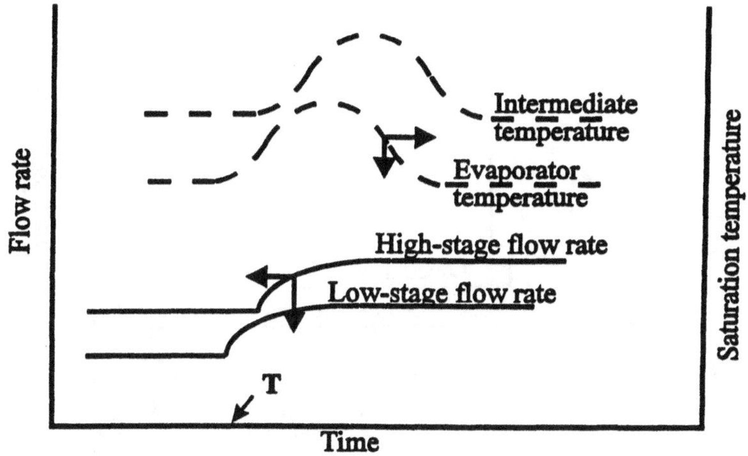

FIGURE 3.23
Response of flow rates and saturation temperatures of a two-stage system to an increase in refrigeration load.

so the controllers increase the flow rates of both compressors. The new steady-state conditions will settle at higher flow rates than original. A portion of the increase in flow rate is due to the initial increase in evaporating and intermediate pressures, and the other part is due to the controllers increasing the compressor capacities to restore the pressures to their settings.

3.15 SIDE PORT OF A SCREW COMPRESSOR FOR LIQUID SUBCOOLING

One of the options available on a screw compressor is a *side port* which provides an entry into the gas spaces during the compression process. This feature allows gas to be compressed from the intermediate to the discharge pressure, avoiding the power requirement of the compression from the suction to the intermediate pressure. One of the arrangements that uses the side port forms what is often called an economizer, as in Figure 3.24. This configuration is similar to the subcoolers in Figures 3.5 and 3.6, except that the separate compressor for the flash gas is not necessary.

The economizer is a low cost method of achieving some of the benefits of two-stage compression, but requiring only a single compressor. The efficiency is not as high, however, as two-stage compression with two separate compressors. Also there are intricacies in the performance characteristics when using the side port of a screw compressor. These characteristics will be explained in greater detail in Chapter 5, after the construction and principles of operation of the screw compressor have been explored.

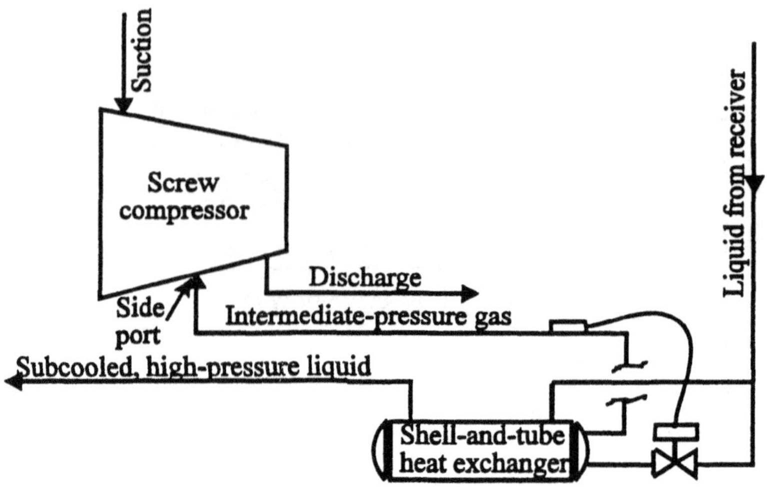

FIGURE 3.24
Using the side port of a screw compressor as an economizer to subcool liquid.

3.16 DECIDING BETWEEN SINGLE- AND TWO-STAGE COMPRESSION

The designer of a low-temperature refrigeration system must decide whether to choose a single- or a two-stage arrangement. It is hard to imagine any facility where the first cost of the two-stage system is not greater than a single stage plant. The savings in energy and the other advantages of the two-stage design must, therefore, compensate for the higher first cost.

Figure 3.25 shows the savings of two-stage operation in comparison to single stage, incorporating flash-gas removal and intercooling, and asssuming ideal compression. The savings increase as the evaporating temperature drops, which is one reason why two-stage systems are so popular for low-temperature refrigeration. Designers usually begin considering two-stage systems when the evaporator temperature drops below about −10°C (15°F).

The extent of savings achieved by two-stage operation, as shown in Figure 3.25, is a function of the properties of the refrigerant. It is interesting that the least efficient refrigerants in single-stage systems experience the greatest percent savings when converted to two-stage. In fact, the two-stage operation with an evaporating temperature of −40°C (−40°F) of ammonia, R-22 and R-134a are all equally efficient, even though in single-stage operation ammonia is most efficient, followed by R-22 and R-134a.

If the plant uses reciprocating compressors, a two-stage arrangement may be necessary to limit the pressure ratio of the compressor to the typical value of about 8. If the condensing temperature is 35°C (95°F), the pressure ratio of 8 is exceeded at evaporating temperatures below −6°C (21°F). This limitation

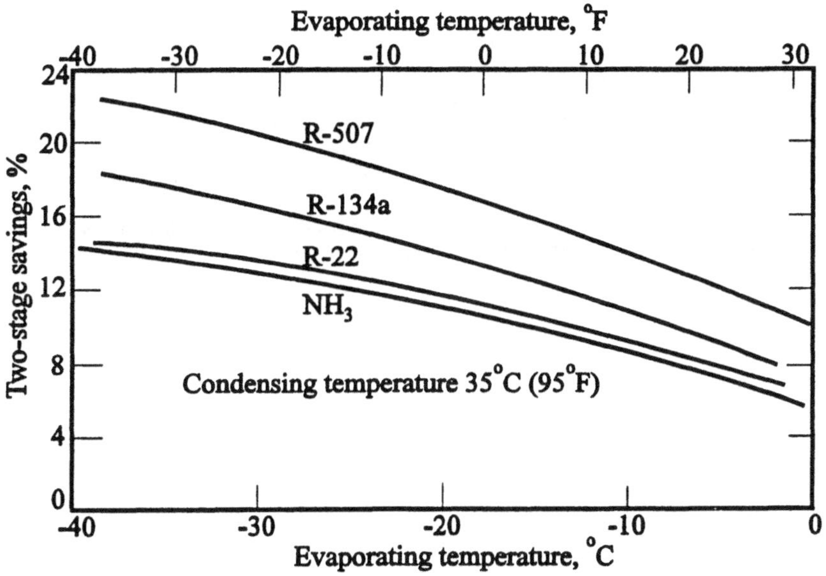

FIGURE 3.25
Percent savings in power of two-stage systems employing flash-gas removal and desuperheating in comparison to single-stage operation.

on the pressure ratio does not apply to screw-type compressors which are capable of operating with ratios as high as 20. But another influence is that of compression efficiency, since Figure 3.25 is based on 100%-efficient compressors. The efficiency of a compressor deteriorates as the pressure ratio increases, so the compressors in a two-stage system operate more efficiently. Use as an example the case of a two-stage system where the evaporator operates at −40°C (−40°F), at which condition Figure 3.25 indicates for ammonia a saving of 14.5%. If actual compressors are considered, the efficiency under single-stage operation might be 68%, while at the lower pressure ratios of two-stage operation the efficiency of each compressor might be 75%. The saving when considering actual compressors and their inefficiencies would climb to 22.5%.

3.17 GROUNDWORK NOW LAID

This chapter and the previous one has provided the thermodynamic basis for the work to follow. With this background, the study can now proceed to the components of the system, including the compressor, evaporator, and condenser. The chapters on these components, as well as those on piping, vessels, control and operation call on the three fundamental fields: thermodynamics, fluid mechanics and heat transfer. These three fundamental subjects are among the most important bases needed in technical work in refrigeration.

REFERENCES

1. Ayub, Z. H., "Shell-Side Two-Phase Heat Transfer Coefficient for Coils in R-22 and Ammonia," *ASHRAE Transactions*, 97 (Part I): 814-815, 1991.
2. Richards, W. V., Personal communication.
3. Cole, R. A., "Energy Efficient Gas Cooling for Two-Stage Ammonia Refrigerating Plants," Proceedings of 1992 International Refrigeration Conference—Energy Efficiency and New Refrigerants, International Institute of Refrigeration, Purdue University, 1992.
4. American Society of Heating, Refrigerating and Air-Conditioning Engineers, Atlanta, Georgia, from the 1994 ASHRAE *Handbook—Refrigeration*.

CHAPTER 4

RECIPROCATING COMPRESSORS

4.1 TYPES OF COMPRESSORS

The main types of compressors used in industrial refrigeration are screw, reciprocating, centrifugal, and rotary vane. Another type that is manufactured in large numbers is the scroll compressor, but so far these compressors are not available in the size normally encountered in industrial refrigeration. Rotary-vane compressors are still in use as low-stage compressors, but very few new ones are installed.

Centrifugal compressors have been standard in large-capacity chemical and process industry plants where they are driven by electric motors or by steam or gas turbines. Centrifugal compressors are also widely used for chilling water in air-conditioning applications. Manufacturers use such refrigerants as R-123 and R-134a in these packages. After condensing the longer list of compressor types, the two that remain, and the ones that will be addressed in this book, are reciprocating and screw compressors. This chapter focuses on reciprocating compressors, while Chapter 5 concentrates on screw compressors.

Two styles of construction of compressors are open and hermetically sealed. In the open-type compressor, as in Figure 4.1, the shaft extends out of the compressor and is connected externally to the electric motor which drives the compressor. In the hermetically-sealed motor-compressor unit, the entire assembly is encapsulated and only the refrigerant lines and electrical connections, and no rotating shaft, penetrate the housing. These hermetically sealed units are always

FIGURE 4.1
A 16-cylinder open-type reciprocating compressor for industrial refrigeration. (*Vilter Manufacturing Corporation*).

used for domestic refrigerators, window-type air conditioners, and other small appliances. They are expected to run for decades with no leaks of refrigerant.

Some larger-size compressors are semi-hermetic, where the heads of the compressors can be removed to gain access to the pistons and valves for servicing. Semi-hermetic compressors available on the market are generally limited to a maximum refrigeration capacity of about 150 kW (40 tons of refrigeration).

In either a hermetic or semi-hermetic unit, the refrigerant is in contact with the windings of the motor, so halocarbon refrigerants which do not attack copper are used in this type of compressor. Since ammonia will react with copper, ammonia compressors are of the open type. At one time the shaft seal was a troublesome cause of ammonia leaks, but the quality of shaft seals has progressively improved. Nevertheless, the development of a hermetic ammonia compressor remains an attractive goal, and such approaches as using aluminum windings for the motor, encapsulating the rotor so that ammonia does not contact the motor windings[1], or connecting the motor and compressor through a magnetic drive[2] are all approaches being considered. On the other hand, an open-type compressor is usually more efficient than the hermetic type, because the suction vapor in a hermetic compressor passes over the motor to cool it, and in so doing the vapor is superheated, thereby requiring more power for compression.

A central emphasis of this chapter is the effect of suction and discharge pressures on the refrigeration capacity and power requirement of the compressor. The suction and discharge pressures are the primary influences of the rest of the system on the compressor. The nature of the refrigeration load at the evaporator strongly influences the suction pressure, and the ambient conditions translate through the condenser to control the discharge pressure. The performance characteristics that are most important to the user of the compressor are the refrigerating capacity and the power requirement. The compressor has a dominating influence on system performance, so the ability to predict influences of changes in conditions imposed on the system, such as fluid temperatures at the evaporator or air temperatures at the condenser, requires an understanding of compressor performance. Certain idealizations will initially be the basis of explaining the performance characteristics, but then the chapter shows that these idealized trends prevail in real compressors. The latter part of the chapter explains some capabilities and limitations of the reciprocating compressor and also describes auxiliaries necessary for reliable operation.

4.2 THE ROLE OF RECIPROCATING COMPRESSORS

In the early 1970s the builder of a large refrigeration facility would install a row of reciprocating compressors in the machine room to provide the compression requirements. Today, the same owner would choose one or two screw compressors to achieve the same capacity. At one time, reciprocating compressors were the standard choice for industrial refrigeration plants—sometimes combined with rotary-vane compressors. The high-volume capacity of the rotary-vane machine suited it for the low stage in a two-stage system. The screw compressor has made inroads into the reciprocating compressor market to such an extent that it is appropriate to evaluate the role of the reciprocating compressor for industrial refrigeration.

Here are some observations regarding the current role and future prospects of the reciprocating compressor in industrial refrigeration. During the several decades of its shrinking market, some manufacturers of reciprocating compressors disbanded their production. The result is that only the strong producers of reciprocating compressor remain today. Sales of reciprocating compressors seem to have stabilized at this lower level in North America, while in Europe and Asia the reciprocating compressors has maintained a strong position. The service and repair requirements remain less sophisticated for the reciprocating compressor.

Customers often prefer the screw compressor because of lower maintenance costs. The screw compressor might be run for 30,000 to 60,000 hours before overhaul, while plant operators have traditionally serviced reciprocating compressors annually. However, many manufacturers of reciprocating compressors are recommending that while the cylinder heads should be removed annually for a routine inspection that the machines can be operated between 10,000 and to 20,000 hours before overhaul. What has happened to the requirement of annual overhaul? To a considerable extent it had been a tradition, and in some cases was encouraged by the manufacturer to further the profitable sale of repair parts. When the operating conditions are not abusive, annual overhauls of compressors may not be necessary due to the improved quality of materials and manufacturing processes available today.

The efficiencies of the reciprocating and screw compressors are of the same order of magnitude at full load. In refrigeration capacities less than about 350 kW (100 tons of refrigeration) the reciprocating compressor is slightly more efficient. At part load, the reciprocating compressor with cylinder unloaders almost always enjoys a higher efficiency than the screw compressor. The first cost of a single reciprocating compressor is normally less than that of a screw compressor of the same pumping capacity. The first cost comparison switches in favor of the screw compressor in large sizes when one screw compressor can provide the capacity of multiple reciprocating compressors. The screw compressor can work against higher compression ratios than can the reciprocating compressor. Thus, a screw compressor is capable of operating single stage from a low-temperature evaporator in a two-stage system during periods of low capacity demands.

The maximum refrigeration capacity of the largest reciprocating compressor available on the market in a typical high-stage application is approximately 900 kW (250 tons of refrigeration), while one large screw compressor can deliver over 4000 kW (1200 tons). The owner of a large-capacity plant will achieve lower capital costs by choosing screw compressors, and accommodate the machines in much less space. If the refrigeration demand on the plant is greater than that capable of being provided by one reciprocating compressor, two units would be needed, which is more costly than one screw compressor, but the reciprocating compressor installation has the advantage of having one unit as a backup if the other is out of service.

It must be accepted that the reciprocating compressor will provide only a portion of the compressor capacity for the industry of the future. Its place is likely to be in plants of less than 1000 kW (300 tons), and many such plants exist. In many packaged refrigeration applications, such as ice makers, reciprocating compressors are usually favored. The treatment of reciprocating compressors in this book reflects the evaluation that reciprocating compressors, although providing much less capacity for the industry than the screw compressor, are still extremely important. Furthermore, some of the analyses of the performance of reciprocating compressors translate directly to screw compressors as well.

4.3 AN OVERVIEW OF THE PERFORMANCE OF RECIPROCATING COMPRESSORS

The refrigeration professional who selects and operates compressors should have a clear understanding of how the evaporating and condensing temperatures affect the refrigerating capacity and power requirements of the compressor. The compressor is often called the heart of the refrigeration system, but it must interact with the other components of the system. Each component affects the performance of the total system, and the two operating variables to which the compressor is most sensitive are the evaporating and condensing temperatures. The two performance characteristics of most importance to the designer and operator are the refrigerating capacity and the power requirement.

Throughout this book and throughout the industry the compressor is characterized as having *refrigerating capacity*. Of course, refrigeration takes place at the evaporator, not the compressor. Indicating that the compressor possesses a certain refrigerating capacity in kW (tons of refrigeration) means that the compressor is capable of pumping the flow rate of refrigerant that will provide the stated refrigeration capacity at the evaporator.

A portion of this chapter is devoted to showing and explaining the influences of evaporating and condensing temperatures on the refrigerating capacity and power requirement of the compressor. The designer and operator of refrigeration systems must constantly reckon with how the conditions imposed on the compressor—the saturated suction temperature and the condensing temperature—influence the two key performance factors, the refrigeration capacity and the power requirement. Beyond knowing the trends, the implications of these trends are crucial. In addition to a citing these performance trends, the next several sections will explain from thermodynamics and the pumping characteristics of reciprocating compressors why the trends occur.

4.4 EFFECT OF THE EVAPORATING TEMPERATURE ON VOLUMETRIC EFFICIENCY

The volumetric efficiency is a key term in explaining trends in the refrigerating capacity and power requirement of reciprocating compressors. The volumetric efficiency of a compressor, η_v in percent, is defined by the equation:

$$\eta_v = \frac{\text{volume rate entering compressor, m}^3\text{/s (ft}^3\text{/min)}}{\text{displacement rate, m}^3\text{/s (ft}^3\text{/min)}}(100) \qquad (4.1)$$

The displacement rate is the volume rate swept through by the pistons during their suction strokes. The volumetric efficiency is less than 100% because of such factors as leakage past the piston rings, pressure drop through the suction and discharge valves, heating of the suction gas when it enters the cylinder by the warm cylinder walls, and the reexpansion of gas remaining in the cylinder following discharge. The volumetric efficiency can be calculated from catalog data of the compressor, as in Example 4.1.

Example 4.1. What is the volumetric efficiency of an eight-cylinder Vilter 458XL ammonia compressor operating at 1200 rpm when the saturated suction temperature is −1°C (30.2°F) and the condensing temperature is 30°C (86°F)? The bore and stroke of the compressor are 114.3 by 114.3 mm (4-1/2 by 4-1/2 in). The catalog lists the refrigerating capacity at this condition as 603.1 kW (171.5 tons).

Solution. The volume swept by one piston during a stroke is:

$$\text{swept volume} = (\pi\ 0.1143^2/4\ \text{m}^2)(0.1143\ \text{m}) = 0.001173\ \text{m}^3\ (0.0414\ \text{ft}^3)$$

The displacement rate is the displacement volume of one cylinder multiplied by the number of cylinders and the number of strokes per second:

$$\text{displacement rate} = (0.001173\ \text{m}^3)(20\ \text{rev/s})(8\ \text{cylinders}) = 0.1877\ \text{m}^3/\text{s}\ (397.7\ \text{cfm})$$

The mass flow rate can be computed by dividing the refrigerating capacity by the refrigerating effect. The enthalpy of ammonia leaving the condenser and entering the evaporator is 342.0 kJ/kg (138.7 Btu/lb) and the enthalpy leaving the evaporator is 1460.8 kJ/kg (619.6 Btu/lb). The mass flow rate $\dot{m}$ is:

$$\dot{m} = \frac{603.1\ \text{kW}}{1460.8 - 342\ \text{kJ/kg}} = 0.539\ \text{kg/s}\ (71.3\ \text{lb/min})$$

The specific volume of the refrigerant entering the compressor is 0.3007 m^3/kg (4.82 ft^3/lb), so the actual volume flow rate is:

$$\text{volume flow rate} = (0.539\ \text{kg/s})(0.3007\ \text{m}^3/\text{kg}) = 0.1621\ \text{m}^3/\text{s}\ (343.4\ \text{cfm})$$

Equation 4.1 can now be applied to compute η_v:

$$\eta_v = \text{volumetric efficiency} = \frac{0.1621\ \text{m}^3/s}{0.1877\ \text{m}^3/s}(100) = 86.4\%$$

The variable that most directly controls the volumetric efficiency is the pressure ratio—the ratio of the absolute discharge pressure to the absolute suction pressure. Figure 4.2 shows η_v, calculated in the manner of Example 4.1, using catalog data for an 8-cylinder ammonia compressor operating at 1170 rpm.

Of the factors that influence the volumetric efficiency, a significant one is the reexpansion of refrigerant that remains in the cylinder when the piston completes its discharge stroke. The gas remaining at the end of the discharge stroke is called *clearance gas* and is the amount retained in the *clearance volume.* The effect of the clearance gas and clearance volume is shown in Figure 4.3 on a piston/cylinder laid on its side. The volumes in the pressure-volume graph correspond to those shown in the cylinder below. The clearance volume is designated as V_c, and as the piston starts its suction stroke the gas trapped in the clearance volume must first expand down to the suction pressure before refrigerant can be drawn on the cylinder.

If at a particular time the suction pressure is p_{s1}, the piston moves until the pressure in the cylinder is p_{s1} at which point the volume in the cylinder is

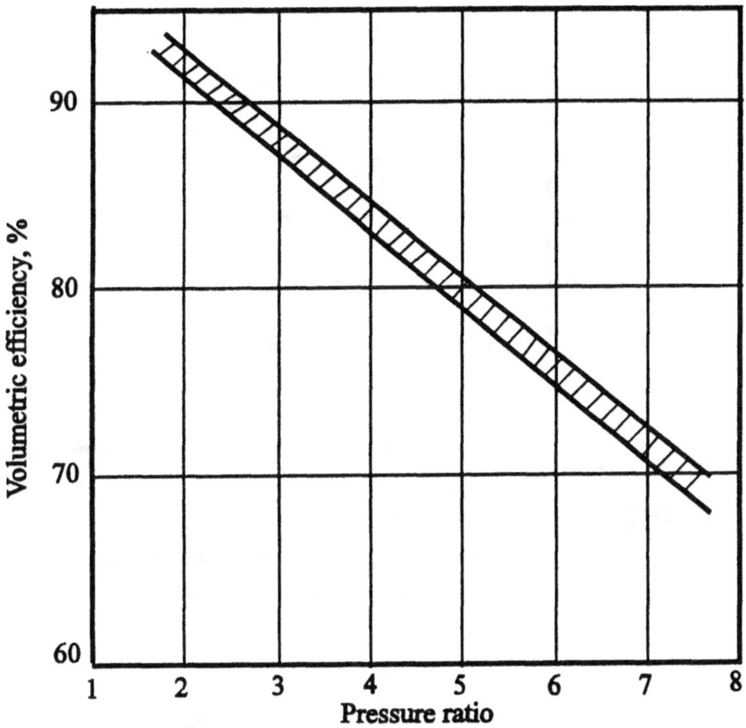

FIGURE 4.2
Band of volumetric efficiencies of an 8-cylinder Sabroe 108L ammonia compressor operating at 1170 rpm.

V_1. It is at this position that refrigerant can be drawn into the cylinder, so the intake volume would be $(V_3 - V_1)$. Define a new type of volumetric efficiency and call it the *clearance volumetric efficiency*, η_{vc}. Keeping with the philosophy of volumetric efficiency as expressed in Eq. 4-1, the clearance volumetric efficiency is:

$$\eta_{vc} = \left(\frac{V_3 - V_1}{V_3 - V_c}\right) 100 \tag{4.2}$$

The magnitude of V_c is a function of the design and construction of the compressor, and most manufacturers try to keep this volume to a minimum. V_c is often expressed as a percentage of the swept volume in a term called the percent clearance, m

$$\text{Percent clearance} = m = \left(\frac{V_c}{V_3 - V_c}\right) 100 \tag{4.3}$$

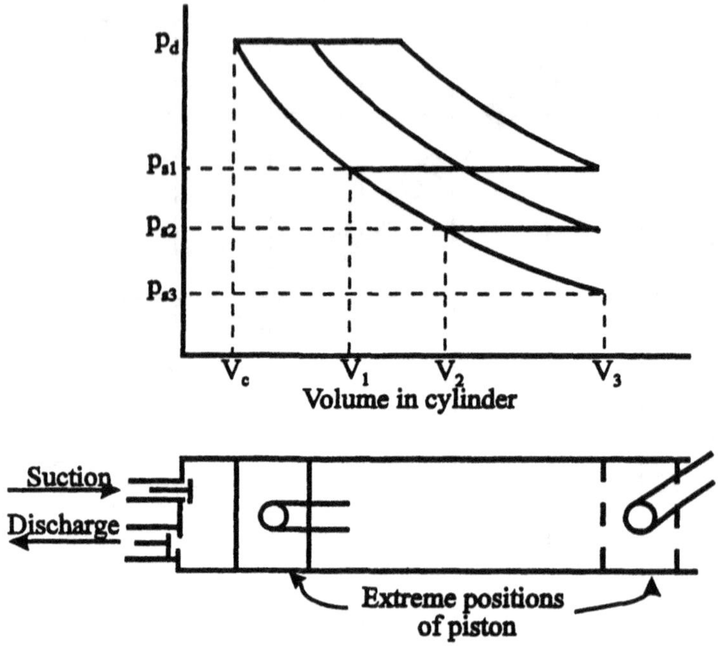

FIGURE 4.3
Pressure-volume diagram corresponding to several piston positions.

Add zero in the form of $V_c - V_c$ to the numerator of Eq. 4-2. Then:

$$\eta_{vc} = \left(\frac{V_3 - V_c + V_c - V_1}{V_3 - V_c}\right) 100 = 100 + \left(\frac{V_c - V_1}{V_3 - V_c}\right) 100$$

$$\eta_{vc} = 100 - 100\left(\frac{V_c}{V_3 - V_c}\right)\left(\frac{V_1}{V_c} - 1\right) = 100 - m\left(\frac{V_1}{V_c} - 1\right) \qquad (4.4)$$

Finally, the volume V_1 can be related to V_c by assuming that the expansion of the clearance gas to the suction pressure is an isentropic expansion—the reverse of an isentropic compression. The relationship of pressures and specific volumes in an isentropic process between point a and point b can be approximated by the relationship:

$$p_a v_a^n = p_b v_b^n \quad \text{or} \quad \left(\frac{v_b}{v_a}\right) = \left(\frac{p_a}{p_b}\right)^{1/n} \qquad (4.5)$$

where the exponent n is unique for each substance. For ammonia the value of n is about 1.28 and for R-22 its value is approximately 1.11.

The term V_1/V_c in Eq. 4.4 can be expressed in terms of the pressure ratio, $(p_c/p_1)^{1/n}$. The expression for the clearance volumetric efficiency then becomes:

$$\eta_{vc} = 100 - m[(\text{pressure ratio})^{1/n} - 1] \tag{4.6}$$

Example 4.2. The percent clearance of the ammonia compressor whose actual volumetric efficiency is shown in Fig. 4.2 is 3%. At a pressure ratio of 5.0, what is the clearance volumetric efficiency?

Solution. Applying Eq. 4.6,

$$\eta_{vc} = 100 - 3(5^{1/1.28} - 1) = 92.5\%$$

At a pressure ratio of 5.0, Figure 4.2 showed the actual volumetric efficiency η_v of that compressor to be 80%. The ideal volumetric efficiency would be 100%, and this ideal value is diminished by the various real processes (reexpansion of clearance gas, fluid friction, cylinder heating, etc.). The clearance volumetric efficiency of 92.5% in Example 4.2 explains approximately one-third of the drop of volumetric efficiency from 100% to the actual value of 80%.

4.5 INFLUENCE OF EVAPORATING TEMPERATURE ON REFRIGERATING CAPACITY.

Along with the power requirement, the refrigerating capacity is a key characteristic of a compressor. The meaning of refrigerating capacity of a compressor should first be explained, because no refrigeration takes place at the compressor. For a compressor to possess a certain refrigerating capacity means that the compressor is capable of compressing the flow rate of refrigerant from its suction pressure to its discharge pressure that will provide the specified heat-transfer rate at the evaporator.

The overall equation that expresses the refrigeration rate is:

$$q_r = \dot{V}_d\,(\eta_v/100)\,(1/v_s)\,\Delta h_{ev} \tag{4.7}$$

where:

q_r = refrigeration rate, kW [(tons of refrigeration)200]
$\dot{V}_d$ = displacement rate, m^3/s (ft^3/min)
η_v = actual volumetric efficiency, percent
v_s = specific volume of gas entering the compressor, m^3/kg (ft^3/lb)
Δh_{ev} = refrigerating effect, kJ/kg (Btu/lb)

Equation 4.7 expresses in compact form what was executed in stepwise fashion in Example 4.1, and will now be used to show the trend in performance variables as influenced by the evaporating temperature with a constant condensing temperature. We arbitrarily choose a condensing temperature of 30°C (86°F)

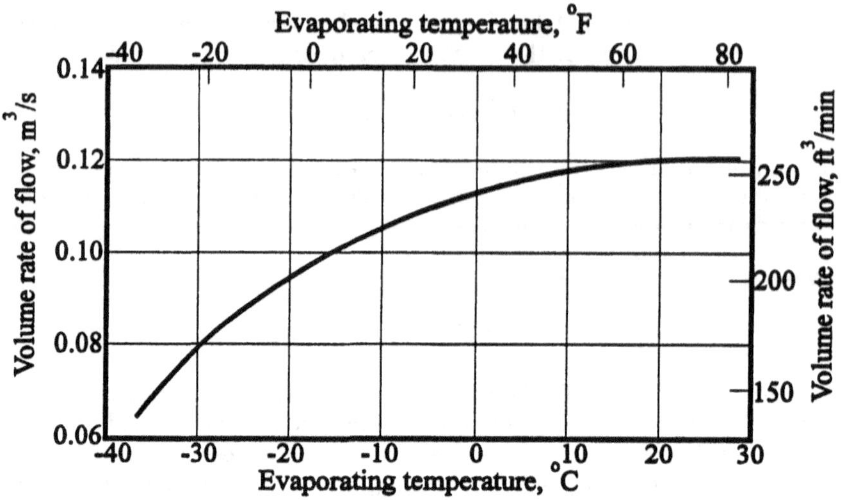

FIGURE 4.4
Effect of evaporating temperature on volume rate of flow measured at the compressor suction of an 8-cylinder compressor with a displacement rate of 0.123 m^3/s (260 cfm) operating with a condensing temperature of 30°C (86°F).

and predict the refrigeration capacity over a range of evaporating temperatures. Furthermore, calculations will be performed for the compressor whose volumetric efficiency is shown in Figure 4.2 and which has a displacement rate $\dot{V}_d$ of 0.123 m^3/s (260 cfm).

The volume rate of flow is available from a part of Eq. 4.7:

$$\dot{V} = \dot{V}_d(\eta_v/100) \tag{4.8}$$

where $\dot{V}$ = volume rate of flow measured at the compressor suction, m^3/s (cfm)

This trend is shown in Figure 4.4, where as the evaporating temperature drops the pressure ratio increases and the volumetric efficiency drops off.

The next objective will be to show the trend in the mass rate of flow $\dot{m}$, which can be done by introducing the specific volume of the suction gas v_s:

$$\dot{m} = \frac{\dot{V}}{v_s} \tag{4.9}$$

where $\dot{m}$ = mass rate of flow, kg/s (lb/min)

The magnitudes of specific volumes are unique for each refrigerant, so we will arbitrarily illustrate the principles using ammonia from this point on. Figure 4.5 shows the variation in specific volume of saturated vapor leaving the evaporator and entering the compressor. As the evaporating temperature drops, the

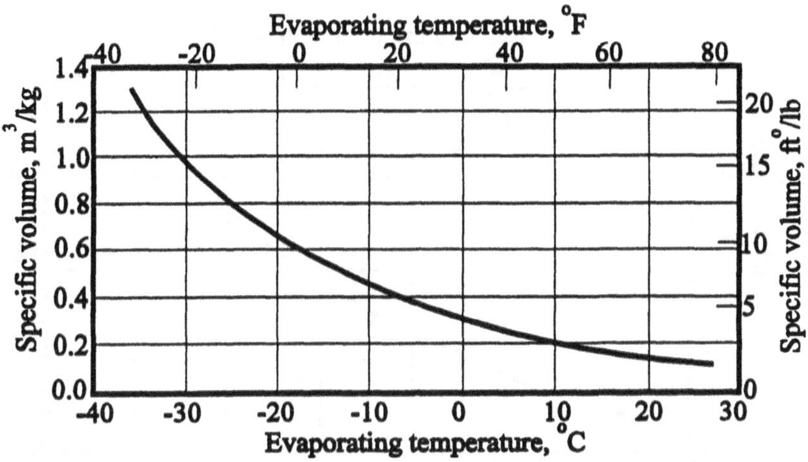

FIGURE 4.5
Variation in the specific volume of ammonia suction vapor with evaporating temperature.

volume flow rate decreases and the specific volume increases, both effects causing a decline in the mass flow rate. This trend is shown in Figure 4.6.

The final term to be brought into Equation 4.7 is the refrigerating effect, Δh_{ev}, which is the change in enthalpy experienced by the refrigerant as it passes through the evaporator. Because the catalog data on which the volumetric efficiencies of Figure 4.2 are based specifies 5.6°C (10°F) subcooling of the liquid entering the expansion device and 5.6°C (10°F) of useful superheat leaving the evaporator those additions to the refrigerating effect are incorporated in the graph of Figure 4.7. The refrigerating effect is only mildly influenced by the evaporating temperature.

This foregoing sequence of graphs culminates in the goal of showing the effect of evaporating temperature on the refrigeration capacity, Figure 4.8. The values shown in the figure are realistic and match catalog data. This fact is not surprising, because the volumetric efficiency (Figure 4.2) used to develop Figure 4.8 was actually calculated from a revised form of Equation 4.7 using catalog data for the refrigerating capacity.

An immediate observation from Figure 4.8 is that the refrigerating capacity always decreases as the evaporating temperature drops. At high evaporating temperatures the decrease in refrigeration capacity is approximately 4% per °C (2.2% per °F) and at low evaporating temperatures, near the maximum pressure ratios of reciprocating compressors, the decrease in refrigerating capacity is approximately 9% per °C (5% per °F). Since the usual objective is to achieve high refrigerating capacities, what can be done to maintain a high evaporating temperature? The temperature of the product or process is usually imposed on the designer and operator, which fixes the maximum ideal evaporating temperature. But the evaporating temperature must be lower than the temperature of the

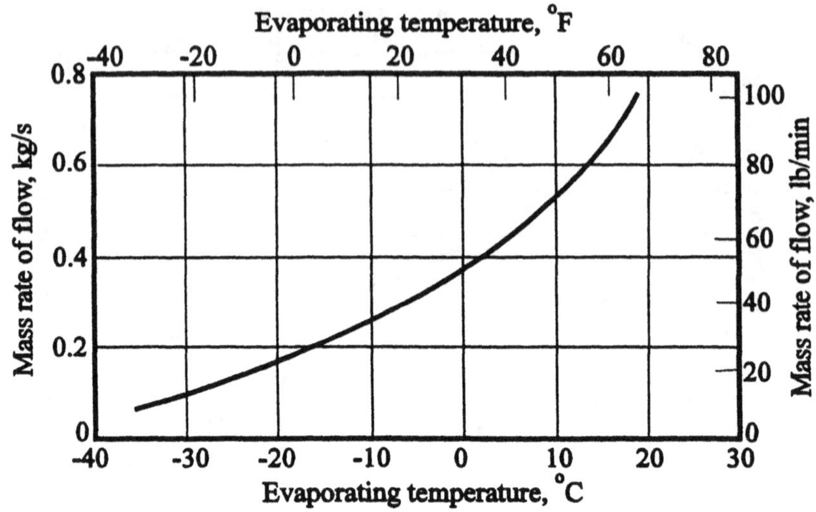

FIGURE 4.6
Influence of the evaporating temperature on the mass rate of flow of ammonia for the reciprocating compressor introduced in Figure 4.4.

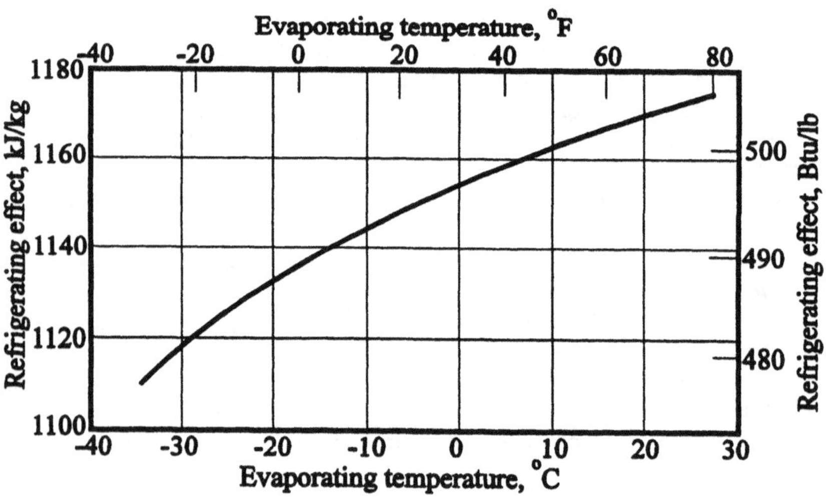

FIGURE 4.7
Refrigerating effect for ammonia as a function of the evaporating temperature, with a condensing temperature of 30°C (10°F) assuming 5.6°C (10°F) of liquid subcooling and 5.6°C (10°F) suction superheat.

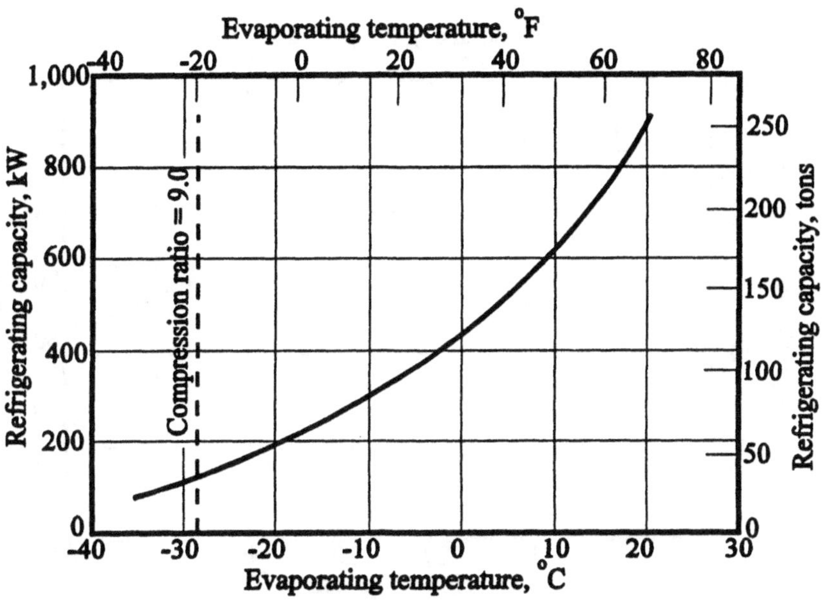

FIGURE 4.8
Effect of the evaporating temperature on refrigerating capacity of the ammonia compressor in Figure 4.4. The condensing temperature is constant at 30°C (86°F).

product or process in order to provide the temperature difference that motivates the flow of heat. It is in the potential reduction of this temperature difference that some latitude might exist to elevate the evaporating temperatures. Choosing a larger heat-transfer area for the evaporator or improving the heat-transfer coefficients by such measures as higher fluid velocities are possibilities under the control of the designer and operator.

4.6 INFLUENCE OF CONDENSING TEMPERATURE ON REFRIGERATING CAPACITY

In a manner similar to that used to explore the effect of evaporating temperatures on the refrigeration capacity, the influence of condensing temperatures can be examined. Equation 4.7 once again becomes the tool, and all the terms change as the condensing temperature varies with the exception of the specific volume entering the compressor, v_s, which is a function of the evaporating temperature only. For the compressor whose volumetric efficiency is shown in Figure 4.2, the effect of condensing temperature on the refrigerating capacity for an evaporating temperature of −10°C (14°F) is shown in Figure 4.9.

The refrigerating capacity always decreases as the condensing temperature increases, so the measures that are discussed in Chapter 7 on condensers are potential tools to keep the condensing temperature low. Compared to the influence

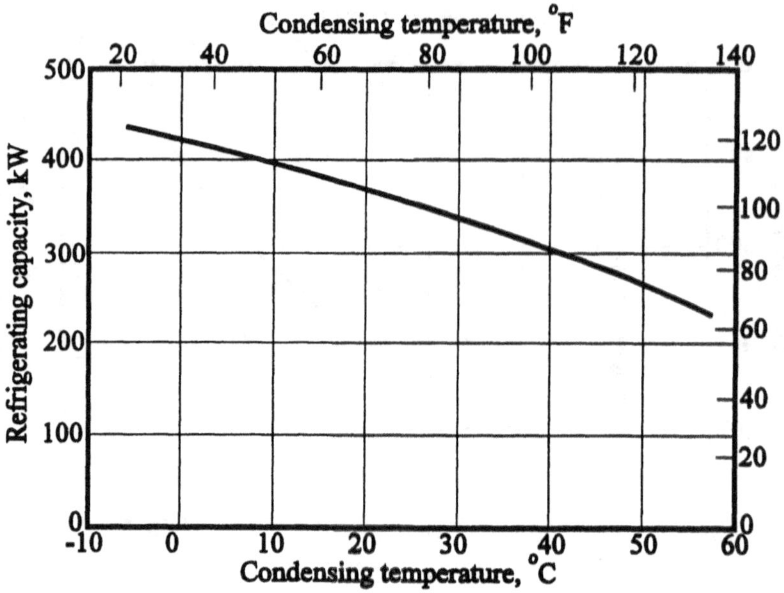

FIGURE 4.9
Effect of condensing temperature on the refrigerating capacity of an 8-cylinder ammonia compressor with a displacement rate of 0.123 m^3/s (260 cfm) operating with an evaporating temperature of −10°C (14°F).

of the evaporating temperature, each degree change in the condensing temperature affects the refrigerating capacity to a lesser extent than a degree change in evaporating temperature. The reason for this difference is that changes in the evaporating temperature also exert a considerable effect on the specific volume entering the compressor, while the condensing temperature does not.

The comparison of the influences of evaporating and condensing temperatures on the refrigerating capacity also appears on the complete map of refrigeration capacity, as shown in Figure 4.10, as controlled by the evaporating and condensing temperatures.

4.7 POWER REQUIRED BY A RECIPROCATING COMPRESSOR

The previous several sections have concentrated on the refrigerating capacity that a reciprocating compressor develops and how the evaporating and condensing temperatures governs it. The other major characteristic of any compressor is its power requirement. One way to approach the question of how the evaporating and condensing temperatures affect the power requirement is to apply the equation:

$$P' = \dot{m}\,\Delta h_{ideal} \tag{4.10}$$

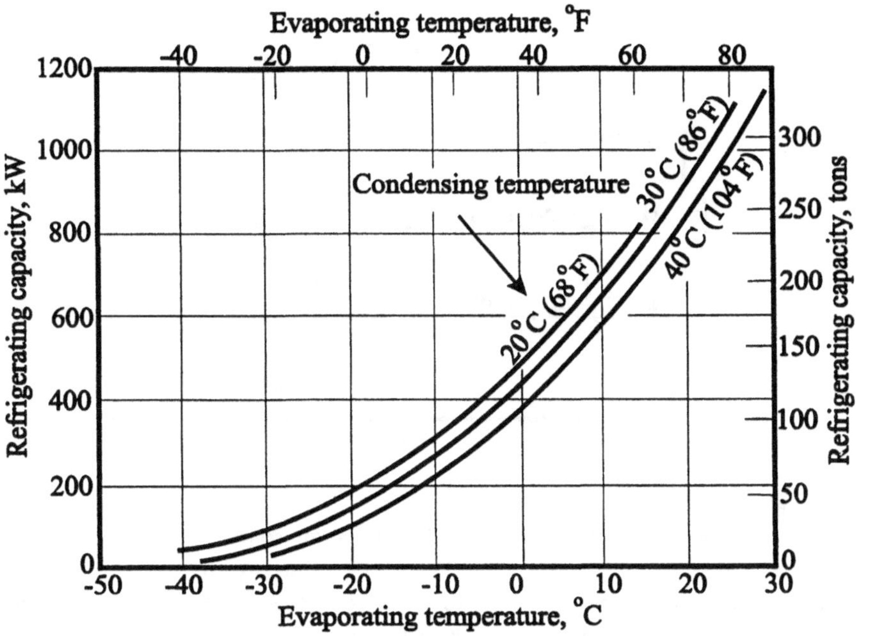

FIGURE 4.10
Effect of both the evaporating and condensing temperatures on refrigeration capacity for the 8-cylinder ammonia compressor.

where:

P' = Power required if the compression is adiabatic and frictionless, kW (Btu/min)

Δh_{ideal} = ideal work of compression, kJ/kg (Btu/lb)

Equation 4.10 is semi-realistic, because the actual mass rate of flow, $\dot{m}$, is available from Figure 4.6. On the other hand, the change in enthalpy during the compression, Δh_{ideal}, used in Equation 4.10 is ideal. This combination is effective, however, in showing the trend in the power requirement.

Explore first the effect of evaporating temperature on the work of compression, as shown in Figure 4.11. With a given condensing temperature, the ideal work of compression decreases as the evaporating temperature increases, until the work of compression shrinks to zero when the evaporating temperature reaches the same value as the condensing temperature.

Figure 4.12 shows the trends of $\dot{m}$, Δh_{ideal}, and the power as the evaporating temperature changes while the condensing temperature remains constant. Starting at very low evaporating temperatures an increase in this temperature provides a progressive increase in the mass rate of flow. Figure 4.11 suggests that the Δh_{ideal} progresses from its high value at low evaporating temperatures to zero when the evaporating temperature reaches the condensing temperature. Equation 4.10 specifies that the compressor power is the product of these two

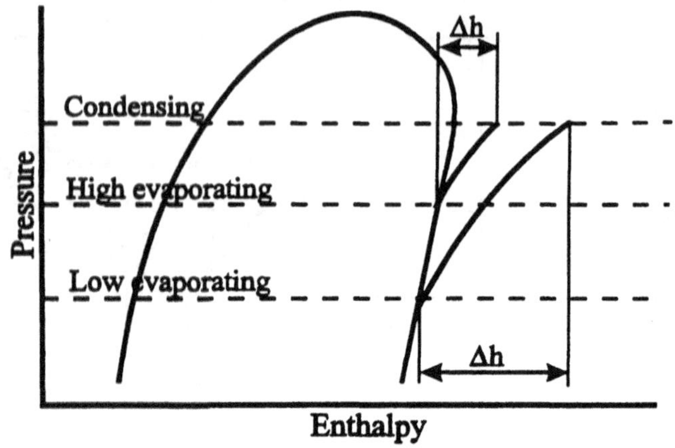

FIGURE 4.11
Effect of evaporating temperature on the ideal work of compression.

terms, so the power would be expected to be low at both very low and very high evaporating temperatures. between those extremes the power requirement reaches a peak. The trends indicated by the power curve in Figure 4.12 are significant and may be surprising. Someone analyzing the power requirement of a reciprocating compressor for the first time may expect that raising the suction pressure will lighten the load on the compressor and lower the draw of power.

The range of pressure ratios against which reciprocating compressors operate is typically between about 2.5 and 8 or 9. As Figure 4.12 shows, in this range of pressure ratios the power required by the compressor increases as the suction pressure and temperature increase. This trend appears in the industrial refrigeration plant, for example, if the refrigeration load on the evaporator increases. The increase in refrigeration load almost certainly is precipitated by an increase in temperature of the product being cooled, which in turn raises the evaporating temperature. As a result, the power requirement of the compressor increases, often resulting in overload of the motor that drives the compressor.

Another motor that might be subject to overload is the one driving a high-stage compressor in a two-stage system. If for some reason the intermediate pressure increases, the high-stage compressor feels the additional power requirement. Still another situation where the power curve of Figure 4.12 explains a potential motor overload is during a pull-down of temperature. If facility has been idle and then is brought into service, the evaporating temperature starts high and progressively drops. The power requirement of the compressor passes through the peak, and unless it does so quickly, the compressor motor might overload. Most reciprocating compressors in industrial applications are equipped with cylinder unloaders, as will be discussed in Sec. 4.18, which may need to be activated by special intervention during this peak power requirement. Normal activation of

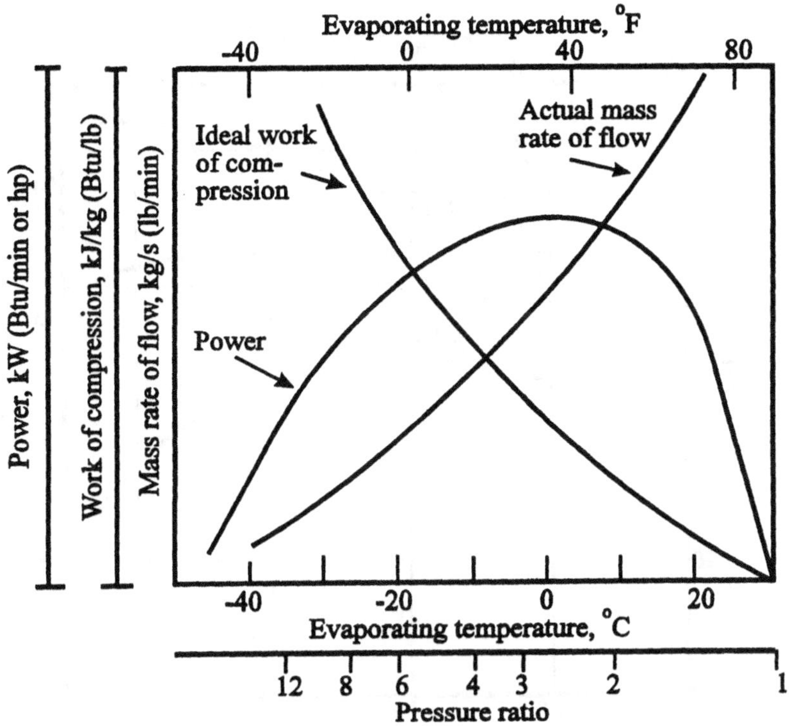

FIGURE 4.12
Effect of evaporating temperature on the mass rate of flow, the ideal work of compression and the compressor power requirement. The condensing temperature is 30°C (86°F).

these cylinder unloaders occurs only when the evaporating temperature drops below the normal operating setting.

Figure 4.12 shows the power requirement of a compressor if the compression were ideal. Figure 4.13, on the other hand, presents the power requirements of the actual compressor that has been the subject of analysis in this chapter. These trends derive directly from catalog data. One of the conclusions from an examination of Figure 4.13 is that the power increases toward a peak as the evaporating temperature increases. Curves for three different condensing temperatures appear on Figure 4.13, so the influence of the condensing temperature can be determined. The power requirement always increases with an increase in condensing temperature, at least within the normal range of operation.

Of possible interest, but of no practical value in industrial refrigeration, is the fact that the power requirement with a constant evaporating temperature reaches a peak as the condensing temperature increases. One of the safety tests of small window air conditioners is to stop the condenser fan, which results

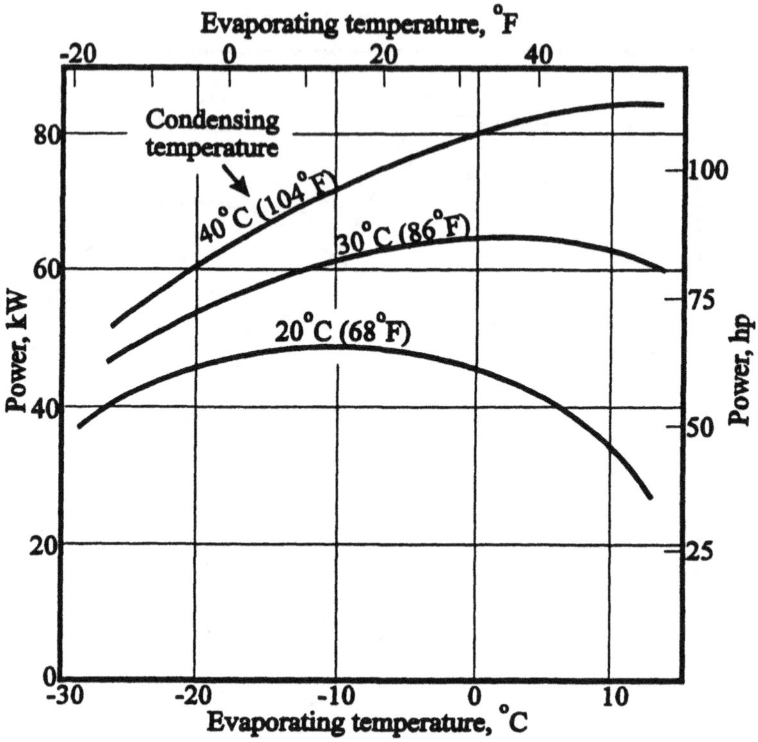

FIGURE 4.13
Actual power requirement of an 8-cylinder Sabroe 108L ammonia compressor operating at 1170 rpm.

in extreme discharge pressures. The compressor continues to operate, although there is no additional compression of refrigerant because the volumetric efficiency has dropped to zero.

The influences of both evaporating temperature and condensing temperature on the refrigerating capacity and the power requirement have now been analyzed. The next two sections address byproducts of these trends.

4.8 ADIABATIC COMPRESSION EFFICIENCY

Equation 4.10 presented a specially-defined compressor power requirement using the ideal work of compression Δh_{ideal}. This ideal work of compression applies to a process which is both adiabatic (no transfer of heat) and frictionless. The actual power requirement for the compressor that has been the subject of study is shown in Figure 4.13, and the actual work of compression, Δh_{comp}, can be

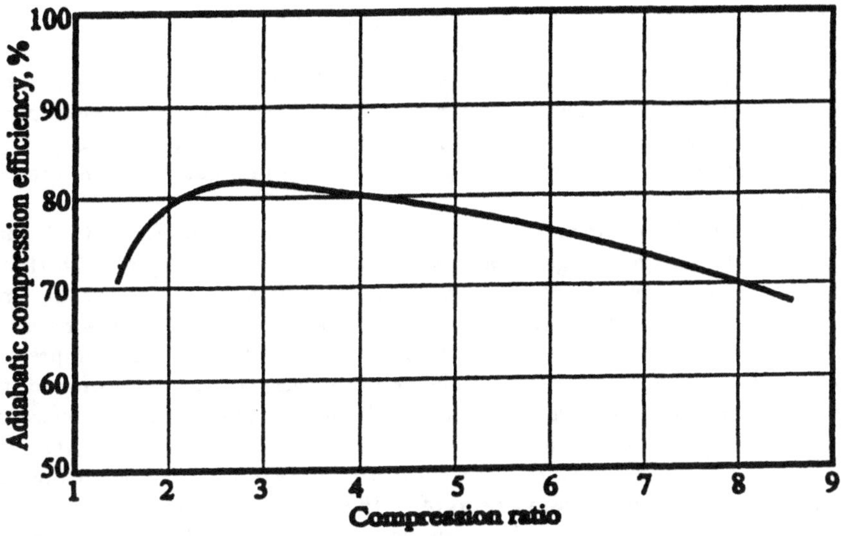

FIGURE 4.14
Adiabatic compression efficiency as a function of the compression ratio.

calculated from the equation:

$$P = \dot{m}\,\Delta h_{comp} \tag{4.11}$$

where P is the power required by the actual compressor.

The ratio of the ideal to the actual work of compression is defined as the adiabatic compression efficiency, η_c:

$$\eta_c = \frac{\Delta h_{ideal}}{\Delta h_{comp}} \tag{4.12}$$

Using the actual power requirements of Figure 4.13 to determine Δh_{comp} and the values of Δh_{ideal}, the adiabatic compression efficiency can best be correlated by the compression ratio, as demonstrated by Figure 4.14. Such factors as the friction due to the mechanical rubbing of metal parts and the friction of the flow of refrigerant are losses that reduce the compression efficiency. The value of η_c drops at higher compression ratios because of increased forces of the rubbing parts, such as shafts on bearing and piston rings on cylinders. There is also a dropoff of η_c at low compression ratios and this reduced efficiency is probably due to flow friction. In fact, at a compression ratio of 1.0 the value of Δh_{ideal} is zero, so any actual work, even though small, drives η_c to zero.

Knowledge of the adiabatic compression efficiency has several important uses. In the first place, the value of η_c is a tool in comparing the effectiveness of two different compressors. Another use is to quickly estimate the work of compression for an operating compressor by determining Δh_{ideal} and dividing by

η_c, expressed as a fraction. The trend shown in Figure 4.14 that is applicable to a specific compressor is fairly typical of most good compressors, namely η_c is about 70% at high compression ratios and 80% at low compression ratios. Chapter 2 in Sections 2.27 and 2.28 proposed and analyzed a ***standard vapor-compression cycle*** in which the compression is assumed to be ideal. The compressor power calculated in that analysis would also be ideal, but can be converted easily to the actual power requirement by dividing by an appropriate adiabatic compression efficiency.

The trend in η_c shown in Figure 4.14 reveals another useful fact. Chapter 3 analyzed two-stage refrigeration systems and showed in Figure 3.25 that a reduction in total power is possible through the application of flash-gas removal and desuperheating. The compressor power of a single-stage compressor used as the basis of comparison and the power of the low- and high-stage compressors were based on ideal compression. If the actual power were to be estimated, the ideal work of compression must be divided by η_c. If the system were operating single-stage with a compression ratio of 9, for example, the adiabatic compression efficiency would be less than 70%. By converting to two stage and operating each compressor with a compression ratio of 3, the adiabatic compression efficiency would be elevated to 80%. This comparison shows that with an actual plant, the reduction in total power achieved to shifting to two-stage operation is even greater than that shown in Figure 3.25.

4.9 EFFECT OF EVAPORATING AND CONDENSING TEMPERATURES ON SYSTEM EFFICIENCY

The influence of both the evaporating and condensing temperatures on the refrigerating capacity and the power requirement of reciprocating compressors has now been studied. Figure 4.10 is a summary of the influence of these temperatures on the refrigerating capacity and Figure 4.13 the influence on the power requirement. The designer and operator of a plant might be interested as well in the efficiency, as expressed by the coefficient of performance, COP, or the horsepower per ton. The importance of these variables lies in the fact that the plant has a certain refrigeration duty to perform, and the desire is to accomplish that duty with the minimum power requirement. The data in Figures 4.10 and 4.13 are the bases for computing the COP:

$$\text{COP} = \frac{\text{refrigerating capacity, kW (Btu/min)}}{\text{compression power, kW (Btu/min)}} \tag{4.13}$$

Figure 4.15 also shows values of horsepower per ton (hp/ton), which varies inversely as the COP through the equation:

$$\text{hp/ton} = \frac{4.715}{\text{COP}}$$

The COP always increases with an increase in evaporating temperature and decreases with an increase in condensing temperature. Designers and operators

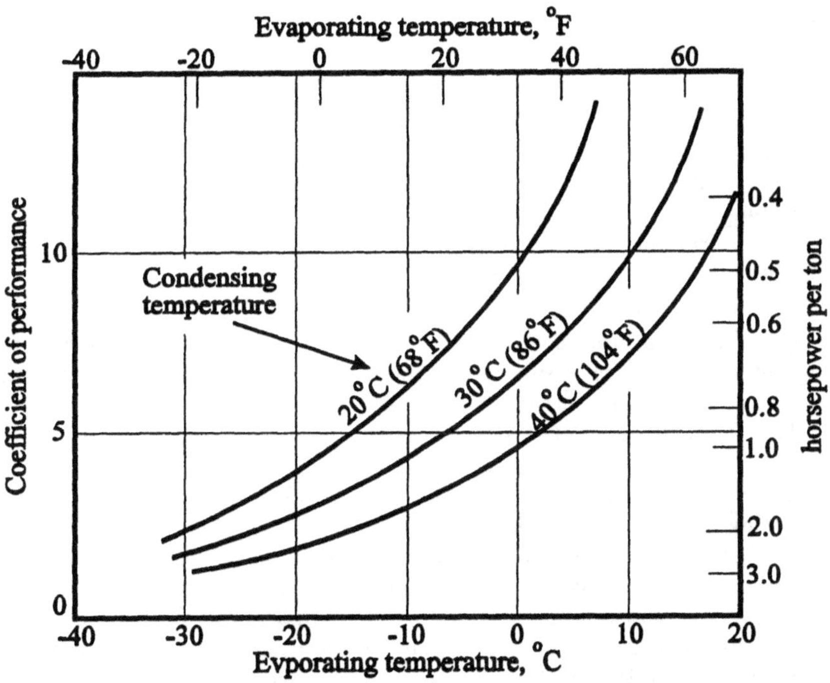

FIGURE 4.15
Coefficient of performance and horsepower per ton as a function of the evaporating and condensing temperatures.

frequently want to know the possible reduction in power for a given refrigerating capacity through an increase in the evaporating temperature or a decrease in the condensing temperature of one degree. Table 4.1 shows some approximate values, based on the data of Figure 4.15, that could serve as guidlines for potential savings. As expected, the improvement per degree in temperature expressed as a percentage will be greatest when the temperature lift from the evaporating to condensing temperatures is small and least when one degree change is a small fraction of the total temperature lift.

4.10 EFFECT OF LIQUID SUBCOOLING ON CATALOG RATINGS

Compressor manufacturers express refrigerating capacity in their catalogs based on a certain number of degrees of subcooling of the liquid entering the expansion valve and a certain number of degrees of superheat of vapor leaving the evaporator. These two conditions affect the refrigerating effect, Δh_{ev}, in Eq. 4.7, and thus affect both the catalog rating as well as the actual refrigeration capacity

TABLE 4.1
Percent reduction in power per °C (°F).

	Small lift of temperature	Large lift of temperature
Increase of evaporating temperature	5^{+}% per °C 2.8^{+}% per °F	3^{+}% per °C 1.7^{+}% per °F
Decrease of condensing temperature	5^{-}% per °C 2.8^{-}% per °F	3^{-}% per °C 1.7^{-}% per °F

developed by the compressor. Furthermore, the purchaser of a compressor deciding between offerings of two competing suppliers should know how to adjust the ratings to put them on the same basis.

The temperature of the liquid arriving at the expansion valve must be equal or lower than the condensing temperature in order for the expansion valve to function properly. Every additional degree of subcooling of this liquid lowers the entering enthalpy and adds to the refrigerating effect and the refrigeration capacity. Some manufacturers rate their compressors on the basis of saturated liquid entering the expansion valve, and others on the basis of a specified number of degrees of subcooling. To establish a common basis for comparison, the capacities listed in different catalogs can be multiplied by a correction factor, illustrated in Fig. 4.16, which is the ratio of the refrigerating effects:

$$F_{sc} = \frac{\Delta h_{ev,s}}{\Delta h_{ev,c}} \tag{4.14}$$

where:

F_{sc} = factor that corrects the refrigeration capacity to that for liquid entering saturated
$\Delta h_{ev,s}$ = refrigerating effect if liquid enters saturated
$\Delta h_{ev,c}$ = refrigerating effect when subcooling is used in catalog

An important question is, "What is a reasonable amount of subcooling to expect?" The answer is, "Not much," because most condensers in industrial refrigeration plants operate with very little subcooling. But there are situations where the liquid subcooling is appreciable, and these are in two-stage systems, as studied in Chapter 3, where subcooled liquid supplies the low-stage evaporators. Realizing that the low-stage subsystem is fed with liquid that may be appreciably subcooled emphasizes that the selection of the low-stage compressor should not be made directly from catalog data. Instead, the required evaporator capacity should be multiplied by the ratio of refrigerating effects $\Delta h_{ev,c}/\Delta h_{ev,a}$, where $\Delta h_{ev,a}$ is the refrigerating effect with the actual entering enthalpy. The resulting capacity is the value to be sought of a compressor listed in the catalog.

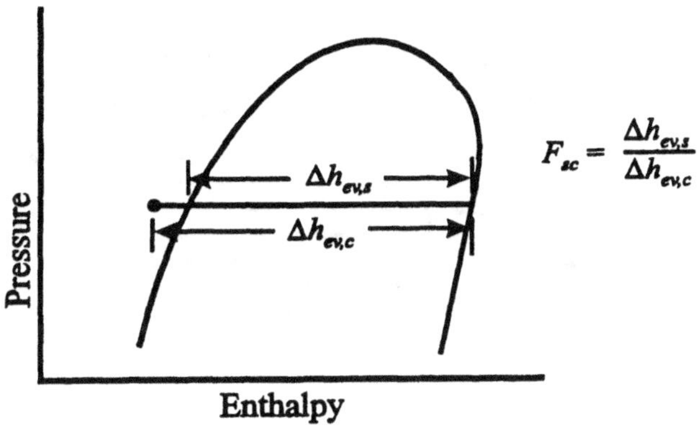

FIGURE 4.16
Correction factor F_{sc} to translate the refrigerating capacity to that if liquid enters the expansion valve saturated.

4.11 EFFECT OF SUPERHEATING OF SUCTION VAPOR ON CATALOG RATINGS

Another specification that often varies from the compressor catalog of one manufacturer to another is the number of degrees of superheat of the vapor entering the compressor. If a catalog specifies some superheat of the vapor, it is almost always understood that the superheat can be counted as useful refrigeration. Whether this assumption is true depends on the configuration of the evaporator. If the superheating occurs within the evaporator, as shown in Figure 4.17a, the refrigerating effect is useful. On the other hand, if the vapor that enters the compressor is superheated because of heat absorbed in the suction line, as in Figure 4.17b, the amount of this superheating is not productive, and the refrigerating capacity shown in the catalog must be reduced by the amount of this superheating. Whether the superheating occurs within the evaporator is usually a function of the type of expansion valve, and applies only when a superheat-controlled expansion valve is used (See Chapter 11 for further description).

A perspective[3] on the question of how liquid subcooling and vapor superheating are specified is that equipment selected without reading the fine print may be short of capacity by about 5% when actually installed and set into service. Because load calculations can hardly ever be precise, and because plants are usually installed with more capacity than absolutely necessary, the difference in behavior of the plant may not be noticed. On the other hand, in the bidding stage, a price difference of 5% may shift the choice from one supplier to another. The subcooling and superheating issue may, therefore, be more important from a marketing than from an engineering standpoint.

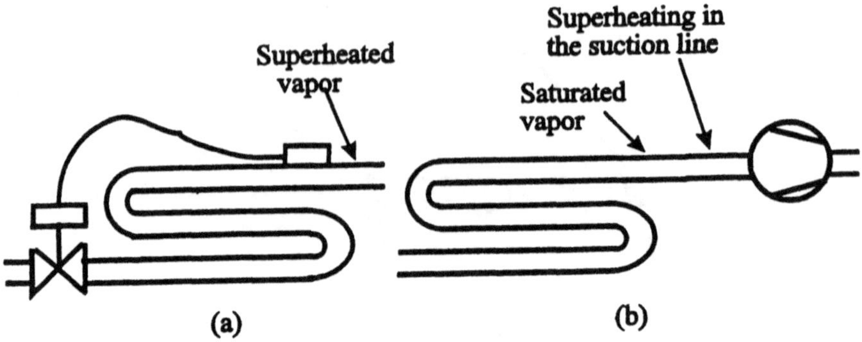

FIGURE 4.17
Superheated vapor entering the compressor where (a) the superheat provides useful refrigeration, and (b) where the superheat does not add to the refrigeration.

4.12 PRESSURE DROP BETWEEN EVAPORATOR AND COMPRESSOR

The life of a manufacturer of any type of compressor is complicated when a designer or contractor fails to account for the pressure drop of refrigerant in the suction line between the evaporator and compressor. The person ordering the compressor knows, for example, that the evaporating temperature is to be −30°C (−22°F) and orders a compressor to operate at this same saturated suction temperature of −30°C (−22 °F). After the compressor is installed and in operation it proves to be short of capacity. The reason is that the pressure drop in the suction line increases the specific volume of the suction gas, so the mass rate of flow drops. This phenomenon is particularly penalizing at low evaporating temperatures. Table 4.2 shows the percentage reduction in refrigerating capacity for ammonia and R-22 at two different evaporating temperatures and at drops in saturation temperature of 0.5 and 1.0°C (0.9 and 1.8°F). While it is true that at the lower saturated suction temperature caused by pressure drop in the suction line, the power requirement also decreases, as was illustrated in Figure 4.13. However, if the compressor is already attempting to deliver full capacity, it cannot take advantage of this reduction in power to increase the refrigerating capacity.

4.13 CONSTRUCTION DETAILS

Most industrial reciprocating compressors are built with two, three, or four banks of cylinders, each bank incorporating either two or four compressors side-by-side in the axial orientation. This arrangement allows a choice of compressor with 4, 6, 8, 12, or 16 cylinders. As the cutaway view in Figure 4.18 shows, the pistons are equipped with rings that are lubricated with oil forced through passages

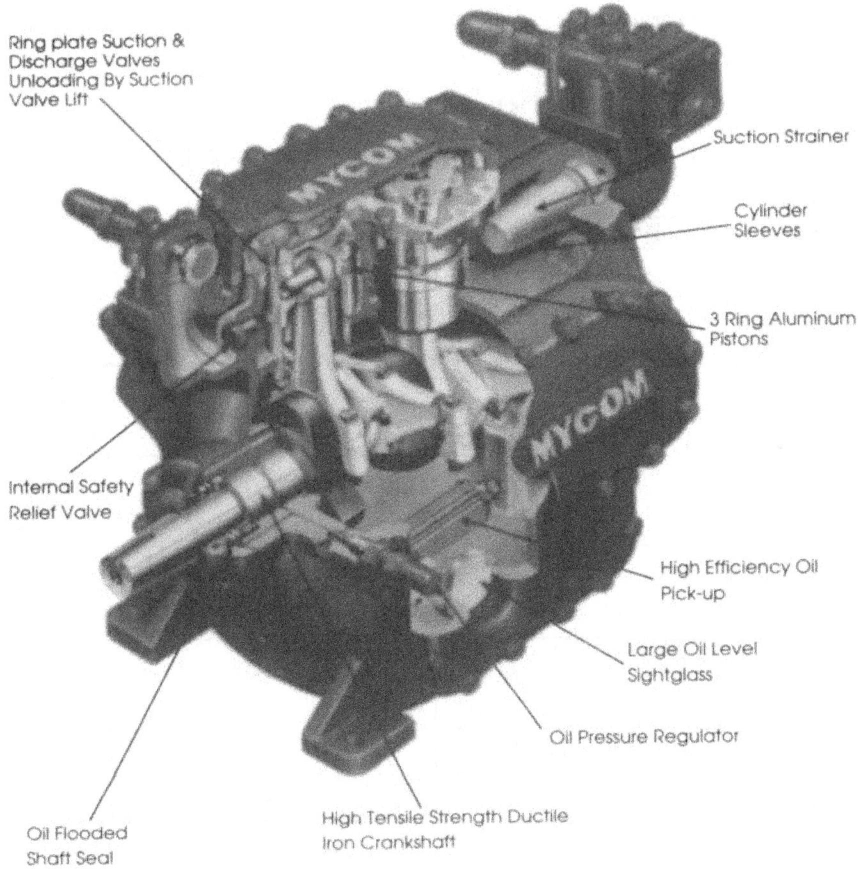

FIGURE 4.18
Cutaway view of a reciprocating compressor. (*Mycom Corporation*)

in the connecting rods. Spring-loaded safety heads permit the heads to relieve the extreme pressure that might be caused by liquid becoming trapped at the top of the piston. Two bearings support the crankshaft of the compressor of Figure 4.18, while 12- and 16-cylinder compressors are provided with a center bearing as well. The oil pump is driven off the shaft and delivers oil through a filter and oil cooler that is usually water-cooled, but the option is available on some compressors to cool the oil with refrigerant. Both the suction and discharge valves are of the spring-loaded plate type.

TABLE 4.2
Reduction in compressor capacity caused by a drop in saturation temperature in the suction line.

	Evaporating temperature, °C (°F)			
	−20°C (−4°F)		−40°C (−40°F)	
Refrigerant	0.5°C drop	1.0°C drop	0.5°C drop	1.0°C drop
Ammonia	2.1%	4.2%	2.6%	5.2%
R-22	2.0%	3.9%	2.3%	4.6%

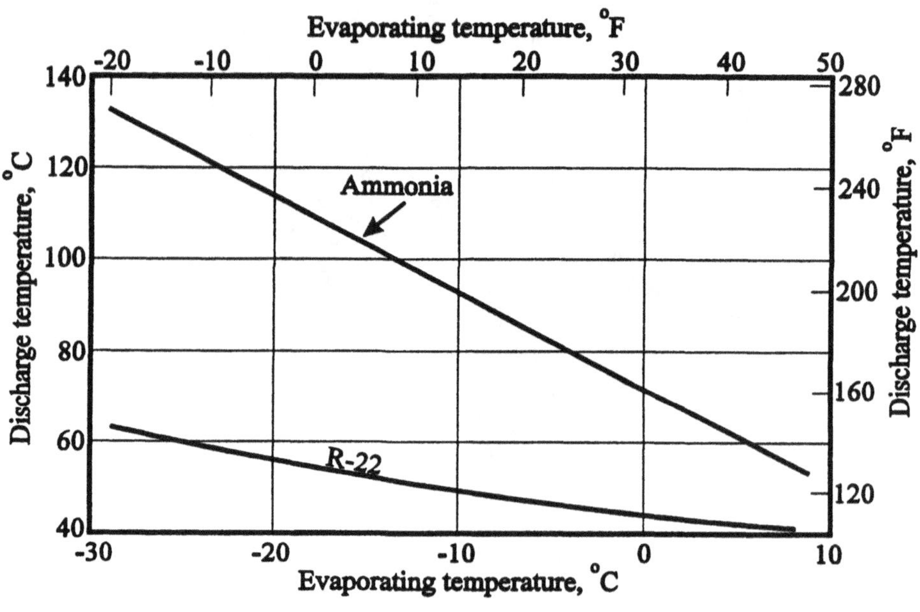

FIGURE 4.19
Discharge temperatures for ideal, adiabatic compressions from saturated vapor to a pressure corresponding to a condensing temperature of 30°C (86°F).

4.14 MAXIMUM PRESSURE RATIOS AND PRESSURE DIFFERENTIALS

In catalogs showing refrigerating capacity and power requirements of reciprocating compressors, manufacturers do not show performance for all combinations of evaporating and condensing temperatures. Also, some manufacturers show data for conditions that are flagged and the user is asked to consult the manufacturer before selecting a compressor at those conditions. An important reason for avoiding some of these combinations is because of pressures—pressure ratio and/or the pressure difference. Ratios of discharge-to-suction pressures higher than 8 or

9 normally are not recommended because the discharge temperature, particularly with ammonia, may rise too high. Another benefit of alerting a potential user to the prospect of high compression ratios is that two-stage compression should be considered for this application.

The reason for limiting the pressure differential is to restrict the load on the bearings and crankshaft. The maximum pressure difference varies between 1000 and 2000 kPa (150 to 300 psi) depending upon the bore, stroke, and other construction features of the compressor. The pressure difference restriction should be kept in mind, particularly in the range of high evaporating and condensing temperatures.

Example 4.3. What is the maximum lift of temperature permitted for an R-22 compressor that is restricted to a 1500 kPa (218 psi) pressure difference if

(*a*) The evaporating temperature is −10°C (14°F),

(*b*) The evaporating temperature is 20°C (68°F)?

Solution. (*a*) The evaporating pressure at −10°C (14°F) is 354.3 kPa, so the maximum permitted condensing pressure is 1854.3 kPa (209 psia), which corresponds to a condensing temperature of 48°C (118.4 °F). Under this condition the temperature lift is 48 − (−10°C) = 58°C (104.4°F).

(*b*) When the evaporating temperature is 20°C (68°F), which corresponds to 909 kPa (131.8 psia), the maximum condensing pressure is 2409 kPa (349 psia). This condensing pressure corresponds to a saturation temperature of 58.8°C (137.8°F). The temperature lift at this higher pressure range is 58.8 − 20 = 38.8°C (69.8°F), which is more restrictive than for the low evaporating temperature.

4.15 DISCHARGE TEMPERATURES AND WATER-COOLED HEADS

Figure 4.19 shows some adiabatic discharge temperatures that would occur with ammonia and R-22 were the compressions ideal (frictionless) and with no transfer of heat. The actual discharge temperatures would be higher than those shown because of inefficiencies of the compressor if negligible heat is lost to the ambient. Because the cylinders and heads are hot, there is natural convection of heat to air, but particularly in the case of ammonia, more intensive cooling is needed. It is standard, then, for ammonia compressors to be equipped with water-cooled heads, thereby keeping valves cooler to prolong their life and preventing the breakdown of oil at high temperatures. Sometimes R- 22 compressors are equipped with water-cooled heads. Manufacturers recommend that discharge temperatures not exceed a temperature of approximately 135°C (275°F).

A modest provision that must be planned by the designer and provided by the builder of the plant is a small supply of cooling water to the compressor. The flow rate of water is usually regulated by a control valve that maintains an outlet cooling water temperature of approximately 45°C (113°F). Typical figures for the flow rate of cooling water are 0.7 kg/s of cooling water for each

100 kW of refrigeration (1 gpm per 25 tons of refrigeration). With an assumed temperature rise of 15°C (29°F) with those cooling water flow rates, ammonia leaves the compressor at a temperature 16°C (29°F) lower than it would have if there had been no cooling of the head.

Two precautions to follow to avoid condensation of refrigerant in the head of the cylinder are: (1) ensure that the temperature of the cooling water never falls below the condensing temperature, and (2) terminate the supply of cooling water when the compressor is not in operation.

4.16 LUBRICATION AND OIL COOLING

Although small compressors may be able to achieve adequate lubrication of the moving parts by splash lubrication, virtually all reciprocating compressors used in industrial refrigeration practice are provided with forced lubrication. A positive-displacement pump draws oil from the crankcase and delivers the oil to bearings, cylinder walls, and to the shaft seal on many compressors. Most pumps are driven off the compressor shaft and some are nonreversible, which fixes the required direction of compressor rotation.

Several auxiliaries to the compressor lubrication system are the oil cooler, a crankcase heater, a separator to remove oil from the discharge gas, and safety cutouts. Particularly on large compressors, the oil is passed through a water-cooled heat exchanger that cools the oil. The rate of water flow required for cooling is of the order of 10 liters/min (several gallons per minute). Another guideline is to set the water-flow rate such that the leaving water temperature is about 45°C (113°F), and then rely on the compressor manufacturer to have provided a cooler large enough to maintain a satisfactory oil temperature with this flow rate. A typical oil temperature during operation is 50°C (122°F).

Crankcase heaters automatically come into service during compressor shutdown. If the oil is permitted to become cool during shutdown, the refrigerant—particularly the halocarbons—will dissolve in the oil. Upon startup, the refrigerant boils off, causing oil foaming and possible oil carryout from the compressor.

The type of oil separator traditionally found in the discharge line of reciprocating compressor is a small vessel using abrupt changes of direction of the oil-laden refrigerant to separate oil droplets that then periodically are returned to the compressor crankcase. The oil concentration leaving this type of separator is in the range of 50 to 80 ppm. When oil-injected screw compressors appeared they required a much more efficient separator, so the coalescing type, which is discussed more thoroughly in Chapter 5, was developed. The coalescing separator reduces the oil concentration to approximately 5 ppm. This type of separator is also now available for reciprocating compressors and is quite widely used on new installations.

Typical safety cutouts associated with the oil system are those that shut off the compressor if a high oil temperature or a low oil pressure occur. The oil pressure cutout usually senses the pressure differential across the pump, which

TABLE 4.3
Effect of rotative speed on the volumetric and compression efficiencies for a compressor operating at 35°C (95°F) condensing and −10°C (14°F) evaporating temperatures.

	800 rpm	1200 rpm	1600 rpm
Volumetric efficiency	80%	80%	80%
Compression efficiency	80%	76%	72%

typically must be higher than about 100 kPa (15 psi). The cutout could be set to shut down the compressor after a 90-second duration of low pressure. This time delay permits the compressor a time interval to build up the oil pressure on startup.

4.17 COMPRESSOR SPEED

The usual range of rotative speed of reciprocating compressors is between 800 and 1800 rpm. The minimum rotative speed is generally dictated by the requirement of the oil pump, which because it runs off the compressor shaft needs a minimum speed in order to develop an acceptable pressure rise of the oil. There is no sharp value of the maximum speed, but the life expectancy of the compressor is influenced by the speed. The impact on suction and discharge valves increases with speed, and friction considerations usually suggest desirable piston speeds at approximately 4 m/s (800 fpm). With the typical length of piston stroke, this piston speed translates to 1200 rpm.

The compressor speed also influences the volumetric and compression efficiencies[4], as Table 4.3 shows.

The volumetric efficiency is fairly constant because the degradation of efficiency due to the pressure drop of the refrigerant vapor at high flow rates is compensated by the relative reduction in leakage past the piston rings. The compression efficiency drops as the speed increases, because certain losses, such as pressure drop through valve ports, increase as the square of the flow rate.

4.18 CAPACITY REGULATION

Almost all refrigeration plants are subject to varying refrigeration loads and, in general, plants operate at part load. The capacity must be reduced, otherwise the evaporating temperature would drop until the plant capacity matched the load. An evaporating temperature that is too low may damage the product. Small plants simply cycle the compressor on and off to provide a time-average match of capacity to load. In larger plants the capability to modulate the capacity down to at least 25% of full capacity is desirable. As the cost of variable-frequency drives decreases and their reliability improves, this method of capacity modulation may become more prominent. For the time being, however, cylinder unloading is the standard technique for adjusting the pumping capacity of reciprocating compressors.

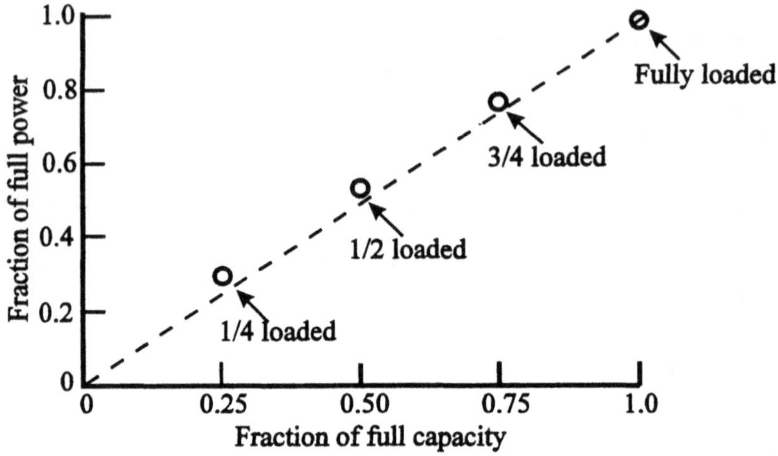

FIGURE 4.20
Typical catalog data of power-capacity relationships for a compressor during cylinder unloading.

Multicylinder compressors can be unloaded by holding open the suction valve on a cylinder. During its intake stroke, the piston draws suction gas into the compressor, but then on its return stroke, instead of compressing the vapor, the piston pushes the refrigerant back into the suction manifold. The suction valve can be held open by valve-lifting pins which in turn are actuated by oil from the lubricating oil pump or discharge gas that is controlled by solenoid valves. In some designs, the normal position of the unloaders is such that oil pressure is required to activate compression. This arrangement automatically starts the compressor with the cylinders unloaded, because oil pressure is not available until the compressor is at least partially up to speed. The two variables most frequently sensed to regulate the operation of the cylinder unloaders are the suction pressure or the outlet temperature of the liquid being chilled in the evaporator.

To indicate part-load efficiency, compressor manufacturers typically suggest a power-capacity relationship during cylinder unloading, like that shown in Figure 4-20. This graph could apply to an 8-cylinder compressor that can unload 2, 4, and 6 cylinders. In order to operate at 60% capacity, for example, the compressor would switch back and forth between the half-unloaded and the one-fourth-unloaded conditions. The percent of full capacity closely matches the percent of total cylinders pumping, but the power requirement runs several percent higher than the linear relationship. One report[5] on field behavior of compressors during unloading indicates, as in Figure 4-21, a slightly lower efficiency than that indicated in Figure 4-20.

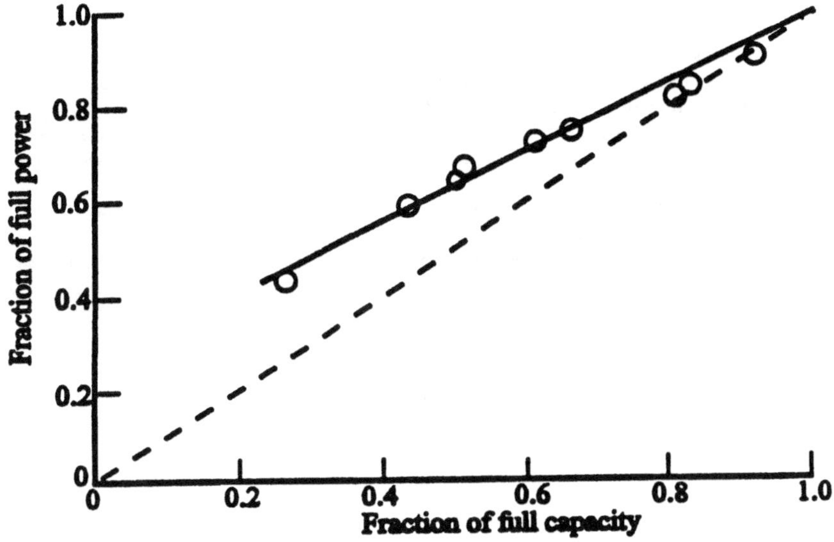

FIGURE 4.21
Power-capacity relationship of a 70-kW (20-ton) water chiller during cylinder unloading.

Power to overcome friction in pumping an idle piston is to be expected, but there is a compensating positive influence. At part-load operation the condenser and evaporator heat transfer rates drop so the condensing temperature decreases and the evaporating temperature increases. Both effects reduce power and compensate for the losses of the idle cylinders.

REFERENCES

1. Lampugnano, V. and N. Kawamura, "New Approaches to Ammonia System Designs," Proceedings of the Annual Meeting of the International Institute of Ammonia Refrigeration, Washington, DC, pp. 328-341, March 1993.
2. "Air Cooled Liquid Chillers," N. R. Koeling bv, AD Krimpen a.d. Ijssel, Netherlands.
3. Cole, R. A. and J. R. Cole, "Compressor Ratings: What Do They Mean?" *Heating/Piping/Air Conditioning*, 66(2): 41-50, February 1993.
4. "Capacities and Equipment Manual, Technical Data," Sabroe Refrigeration AS, Aarhus, Denmark, 1993.
5. Leverenz, D. J. and N. E. Bergan, "Development and Validation of a Reciprocating Chiller Model for Hourly Energy Analysis Programs," *ASHRAE Transactions*, 89(Part 1A): 156-175, 1983.

CHAPTER 5

SCREW COMPRESSORS

5.1 TYPES OF SCREW COMPRESSORS

The two major categories of screw compressors are *twin screw* and *single screw*. The twin-screw compressor is widely used and has many years of operating experience. It occupies a position alongside reciprocating and centrifugal types as a standard choice of refrigeration compressors. The single-screw type, described in Section 5.22, is becoming well established because of the efforts of several manufacturers. The twin- screw will simply be referred to in this chapter as the *screw compressor*.

The invention and evolution of the screw compressor bears a heavy Swedish imprint through a succession of firms beginning about the turn of the century with the company of the Ljungstrom brothers—a name that became associated later with the Ljungstrom air preheater for power plants. In 1913 the brothers organized a subsidiary, Svenska Turbinfabriks Aktiebolaget Ljungstrom, also known by the acronym STAL. Following some successes and reverses, the Ljungstrom brothers resigned from the company in the 1920s and a new chief engineer, Alf Lysholm, was appointed, who provided the firm with several inventions, including that of the screw compressor.

The early screw compressors were fraught with many deficiencies in design and operation which had to be solved one by one. In 1951 the name of AB Ljungstroms Angturbin was changed to Svenska Rotor Maskiner AB (SRM). Up until this time screw compressors were equipped with synchronizing gears and operated dry, but during the 1950s the practice of injecting oil began and this development gave the screw compressor new impetus. Used primarily for air compressors initially, development work on the application of screw compressors to refrigerants began in the 1950s. The improvement of the rotor profiles to provide ease of manufacture and efficient performance has been an ongoing emphasis in the screw compressor development. By now about a million air compressors and nearly 100,000 refrigerant screw compressors have been manufactured.

Section 4.2 in the previous chapter on reciprocating compressors presented a picture of the competitive situation between reciprocating and screw compressors. The conclusion of that discussion was that during the past several decades the screw compressor has gained much of the compressor market in industrial refrigeration, particularly in large-capacity units. Of the refrigeration capacity installed each year the screw compressor serves more of this capacity than does the reciprocating type, so the principles, applications, and procedures described in this chapter are especially important. Screw compressors are available in volume capacity ranges from about 0.05 to 1.5 m^3/s (100 to 3300 cfm), driven by motors ranging in output from 25 to 1250 kW, and operating at usual speeds of 3550 rpm (2950 rpm with 50-Hz power).

This chapter describes the screw compressor and explains how it works. The performance of the basic compressor is first explored, particularly as it encounters changes in evaporating and condensing temperatures. Capacity regulation of a screw compressor is typically achieved through the use of a special valve which provides continuous-capacity modulation over a wide range. Screw compressors are basically constant-volume-ratio machines, the implications of which will be explored. Oil is injected in screw compressors for sealing the spaces between the lobes, and this oil must subsequently be separated and cooled. End users of refrigeration plants usually buy screw compressors incorporated in packages that include the necessary auxiliaries, which will be described. The chapter concludes with an explanation of the single-screw compressor.

5.2 HOW THE SCREW COMPRESSOR WORKS

A cross-sectional view of two pairs of rotational elements, called rotors, of the screw compressor with two different profiles is shown in Figure 5.1. The male rotor here has four lobes and the female rotor.six gullies, and this combination of numbers of lobe/gullies is most common. Other combinations, such 3/5 and 5/7 are sometin.es available. Another view of the rotors presenting the third dimension is shown in Figure 5.2.

Some of the popular nominal diameters[1] of the rotors are 125, 160, 200, 250, and 320 mm. Manufacturers often offer two or three rotor lengths for each rotor diameter and the length-to-diameter ratios usually fall in a range of 1.12 to 1.70. The rotors slip into a housing as indicated by the exploded view, as shown in Figure 5.3, that also shows some of the main elements of the compressor.

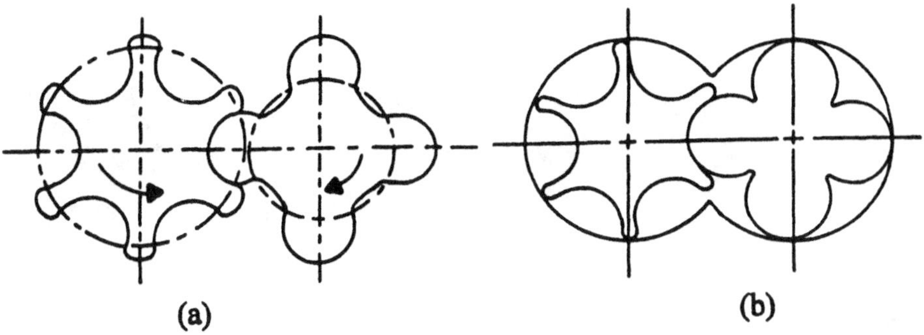

FIGURE 5.1
Screw compressor rotors with *(a)* symmetric profile, and *(b)* asymmetric profile.

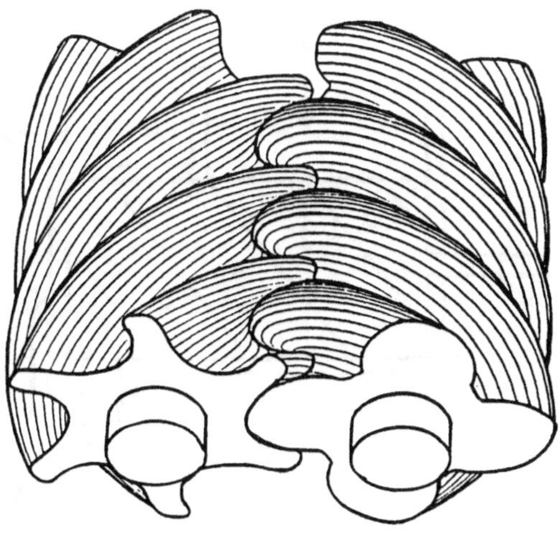

FIGURE 5.2
Screw compressor rotors.

The separate processes experienced by the vapor in passing through the compressor are (1) filling of a cavity with suction gas, (2) sealing of gas between the rotors and housing, (3) reducing the volume of the cavity to perform the compression, and (4) uncovering the discharge opening to expel the compressed gas to the discharge line. One way to picture these processes is by observing a side view of the screws in Figure 5.4 whose threads move to the right as the rotors turn. The suction vapor enters the top of the rotors, and as the rotors turn a cavity appears at 1. Cavity 2 is continuing to fill, and cavity 3 is completely filled. Cavity 4 has now trapped gas between its threads and the housing. Cavity 5 is in the compression process with the volume shrinking as the cavity bears against the end of the housing. When the thread of the rotor reaches the discharge

FIGURE 5.3
Exploded view of main elements of a screw compressor. *(Courtesy Sullair Refrigeration)*

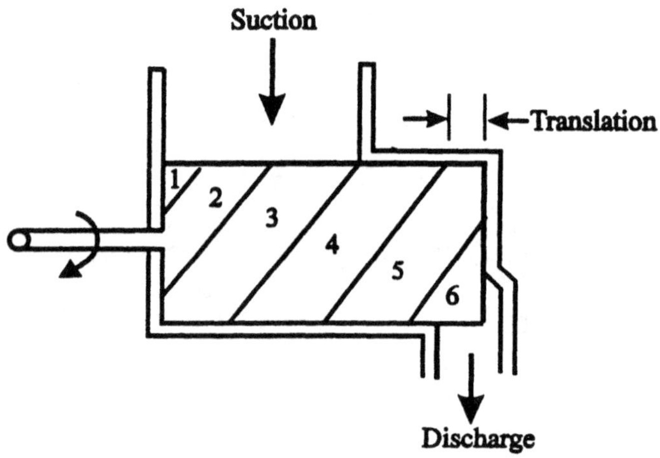

FIGURE 5.4
Visualization of the intake, compression and discharge processes of a screw compressor.

port, the compressed gas flows into the discharge line. A translation process is indicated in Figure 5.4, which is an interval occurring between the time the cavity is sealed until compression begins. This translation process takes up about 30° of the rotation of the rotor.

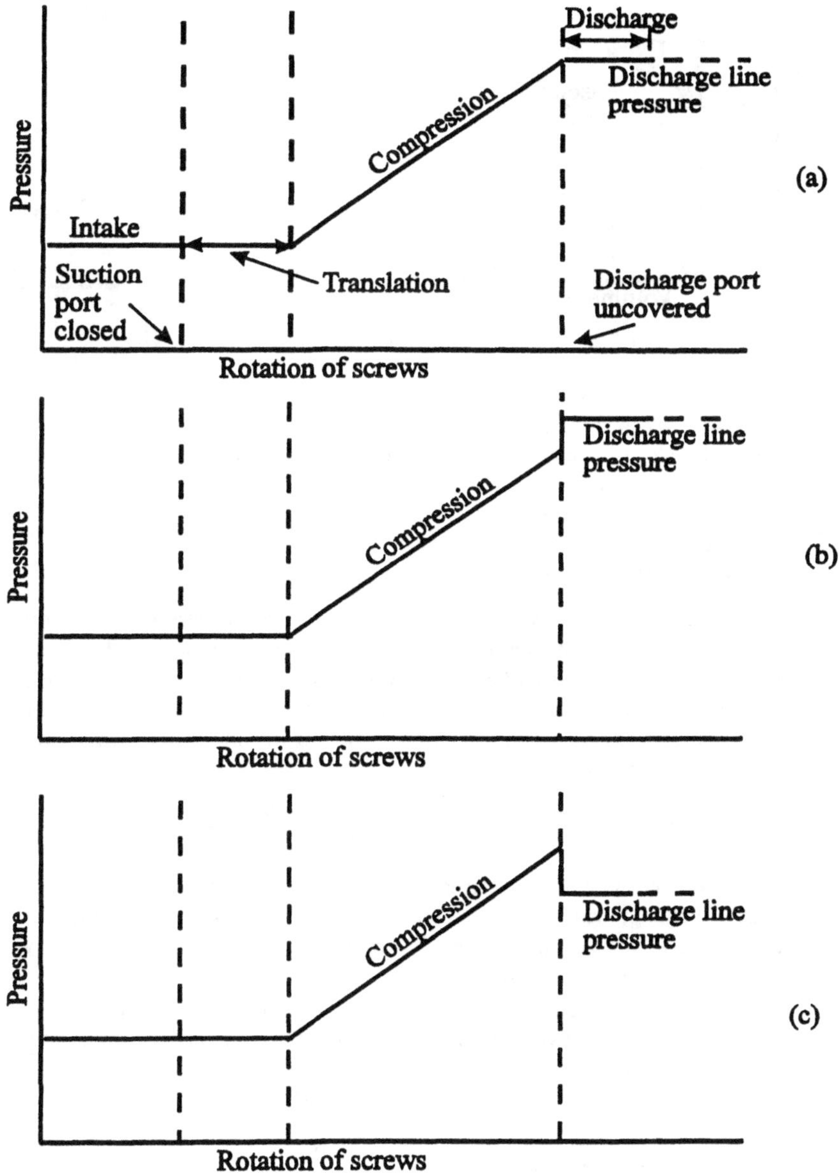

FIGURE 5.5
Pressures during intake, translation, compression, and discharge when (a) the discharge-line pressure equals, (b) when the discharge-line pressure is higher, and (c) the discharge-line pressure is lower than the built-in discharge pressure.

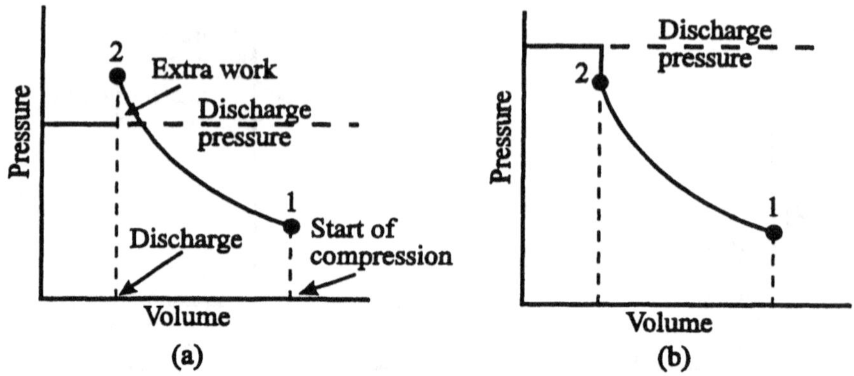

FIGURE 5.6
(a) Over-compression and (b) under-compression shown on a pressure-volume diagram where the area under the curves indicate work applied to the refrigerant.

5.3 PERFORMANCE CHARACTERISTICS OF A BASIC SCREW COMPRESSOR

To begin the explanation of why the screw compressor possesses its unique capacity, power, and efficiency characteristics, a basic machine will be analyzed. This compressor is assumed to be operating at constant speed and without the capacity control capabilities that will be introduced in Section 5.8. A fundamental characteristic of the basic screw compressor is its built-in volume ratio, v_i, which is defined as follows:

$$v_i = \frac{\text{volume in cavity when suction port closes}}{\text{volume in cavity when discharge port uncovers}}$$

In contrast to the reciprocating compressor, the screw compressor has no suction and discharge valves but accepts a certain volume of suction gas in a cavity and reduces this volume a specific amount before discharge. Some typical values of v_i used by manufacturers are 2.6, 3.6, 4.2, and 5.0. For a given rotor diameter each different v_i is associated with a different rotor length. Each v_i corresponds to a certain pressure ratio that varies from one refrigerant to another. Table 5.1 presents estimates of pressure ratios using the following equation that is applicable to an isentropic compression of a perfect gas:

$$\text{Pressure ratio} = \left(\frac{\text{suction volume}}{\text{discharge volume}}\right)^k = v_i^k$$

where k = ratio of specific heats, c_p/c_v, which is approximately 1.29 for ammonia and 1.18 for R-22.

If the pressure ratio against which the compressor pumps is precisely equal to that developed within the compressor, then the discharge port is uncovered at

TABLE 5.1
Pressure ratios corresponding to built-in volume ratios for ideal compression.

Built-in volume ratio	Ammonia	R-22
2.6	3.4	3.1
3.6	5.3	4.5
4.2	6.4	8.9
5.0	8.0	6.7

the instant that the pressure of the refrigerant in the cavity has been raised to that of the discharge line, and the compressed gas is expelled into the discharge line by the continued rotation of the screws. This situation is represented by Figure 5.5a, which shows the pressure changes in one cavity between the screws as rotation progresses. It is rare, however, that the developed pressure within the compressor precisely matches that prevailing in the discharge line. Figures 5.5b and 5.5c demonstrate what happens when the developed pressure is lower or higher, respectively, than the discharge-line pressure.

In Figure 5.5b the compressed refrigerant has not yet reached the discharge-line pressure when the discharge port is uncovered, so there is a sudden rush of gas from the discharge line into the compressor that almost instantaneously increases the pressure. Thereafter, the continued rotation of the screws expels this gas as well as the refrigerant ready to be discharged.

The third situation, as shown in Figure 5.5c, occurs when the discharge-line pressure is lower than that achieved within the compressor. At the instant the discharge port is uncovered there is a sudden rush of gas out of the compressor into the discharge line.

Another picture of the drawbacks of the mismatch of the internally developed pressure and that in the discharge line is shown in the pressure-volume diagrams of Figure 5.6 which concentrate only on the compression and discharge processes. Since the area under a compression or expansion curve on the pressure-volume diagram indicates work done in the process, the horn in Figure 5.6a indicates nonproductive work in the process. Losses are caused by *unrestrained expansions* that are the result of gas under pressure venting freely from a high pressure to a low pressure. Unrestrained expansions occur in both overcompression and undercompression, but the most penalizing of the two is in overcompression where a limited volume of gas in the compressor vents to the extensive volume of the discharge line. The losses are less in under- compression where the gas expands unrestrained from the extensive volume of the discharge line to fill the compressor cavity.

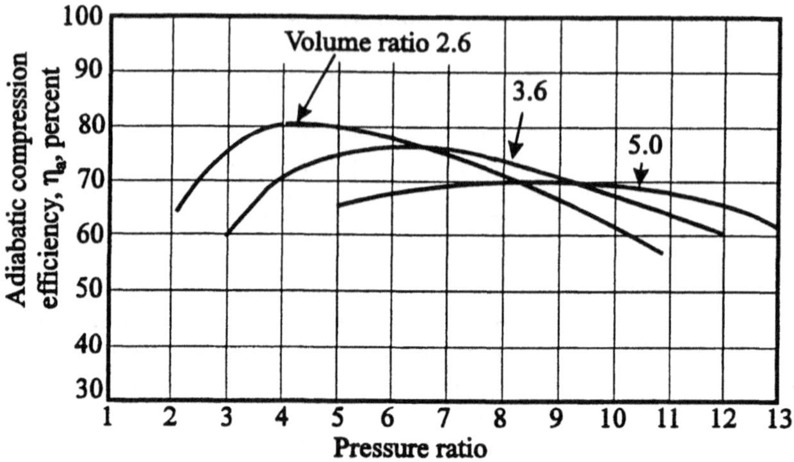

FIGURE 5.7
Adiabatic compression efficiency of ammonia screw compressors.

5.4 ADIABATIC COMPRESSION EFFICIENCY OF A SCREW COMPRESSOR

The adiabatic compression efficiency η_a was defined for the reciprocating compressor in Equation 4.12:

$$\eta_a = \frac{\text{isentropic work of compression, kJ/kg (Btu/lb)}}{\text{actual work of compression, kJ/kg (Btu/lb)}} \times 100 \qquad (5.1)$$

The compression efficiency of reciprocating compressors is strongly influenced by the pressure ratio, and for a screw compressor the additional influence is that of the built-in volume ratio, as shown in Figure 5.7.

The pattern is that the efficiency reaches a peak at a certain discharge-to-suction pressure ratio, and the pressure ratio for optimum η_a is a function of the built-in volume ratio. The mechanisms of how the screw compressor works now are used to explain some of the trends appearing in Figure 5.7.

The ideal situation is where the pressure in the cavity during compression builds up at the instant the discharge port is uncovered to precisely the discharge-line pressure, as in Figure 5.5a. Table 5.1 predicted pressure ratios corresponding to several built-in volume ratios for ammonia and R-22 and shows that the pressure ratio is higher than the volume ratio. The efficiency curves in Figure 5.7 reach their peaks at pressure ratios slightly higher than those shown in Table 5.1 for a given volume ratio.

Two reasons for the shifts from expectations are:

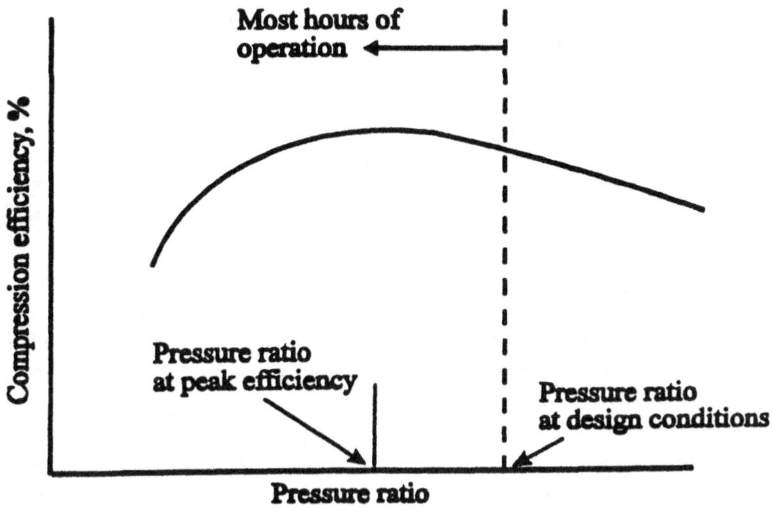

FIGURE 5.8
Selecting a compressor with its peak efficiency occurring lower than the design pressure ratio.

1. Some cooling is performed during compression rather than taking place adiabatically, and
2. There is some leakage of refrgierant so that the ideal pressure ratio is not achieved.

Overcompression prevails to the left of the peak efficiency, and undercompression to its right. The previous section commented that overcompression results in greater losses than does undercompression, and this fact is borne out by the rapid dropoff of η_a to the left of the peak efficiency.

The trends demonstrated in Figure 5.7 provide guidance for choosing the built-in volume ratio when a compressor is being selected. It is not the best strategy to select a compressor with its peak efficiency occurring at the design pressure ratio. It is preferable to choose a compressor that exhibits its peak efficiency at a pressure ratio lower than design, as shown in Figure 5.8. In industrial refrigeration systems the suction pressure usually remains nearly constant, but the condensing pressure will likely drift lower than design during most of the hours of operation. Thus, in Figure 5.8 the pressure ratio shifts downward from the design condition toward the value where the peak efficiency occurs. Furthermore, the low efficiencies experienced during overcompression are avoided.

5.5 EFFECT OF EVAPORATING AND CONDENSING TEMPERATURES ON REFRIGERATING CAPACITY

It is important for the designer and operator of a system to know how the suction and discharge pressures influence the capacity and power, because few plants operate with these pressures constant. An awareness of the influence on power requirements is needed both for analyses of energy requirements and for selection and operation of the motor without overloading it.

For the reciprocating compressor, Figure 4.10 showed the effect of evaporating and condensing temperatures on the refrigerating capacity, and Figure 4.13 the effect of these temperatures on the power requirement. Figure 5.9 shows the influence of the evaporating and condensing temperatures on the refrigerating capacity, with the evaporating temperature exerting the major influence. The power required by the screw compressor appears in Fig. 5.10, exhibiting the characteristic peak in the curve at a given condensing temperature. These trends for the screw compressor are similar to those of the reciprocating compressor, but there are also some differences worth noting. For both types of compressors the refrigerating capacity is influenced by the volumetric efficiency, the specific volume of the suction vapor, and the pressure ratio. The power is affected by the volumetric efficiency and work of compression.

Because the screw compressor completes its expulsion of gas with virtually no volume remaining, there is no clearance volume to reexpand, as is the case with the reciprocating compressor. It would be expected, then, that the volumetric efficiency and refrigerating capacity drop off less as the pressure ratio increases. Table 5.2 shows the comparison of refrigerating capacity and power of a screw and reciprocating compressor as the evaporating temperature changes. Indeed at the higher condensing temperature of 35°C (95°F) there is a greater dropoff in capacity of the reciprocating compressor as the evaporating temperature decreases. But at the lower condensing temperature of 20°C (68°F), the percentage reduction in capacity is about the same for the two compressors.

The modest drop in refrigerating capacity of the screw compressor at the higher condensing temperature is accompanied by very little reduction in the power requirement as the evaporating temperature drops. In contrast to the reciprocating compressor, the screw compressor operates more favorably in a temperture pull-down situation. The capacity of the reciprocating compressor drops off, even though the motor would have the capability of providing additional power.

5.6 PRESSURE DROP BETWEEN EVAPORATOR AND COMPRESSOR.

Section 4.12 addressed for the reciprocating compressor the precaution that there will always be some pressure drop in the suction line between the evaporator and compressor. If an evaporator is operating at a temperature of −30°C (−22°F) a compressor selected for this saturated suction temperature will be about 5%

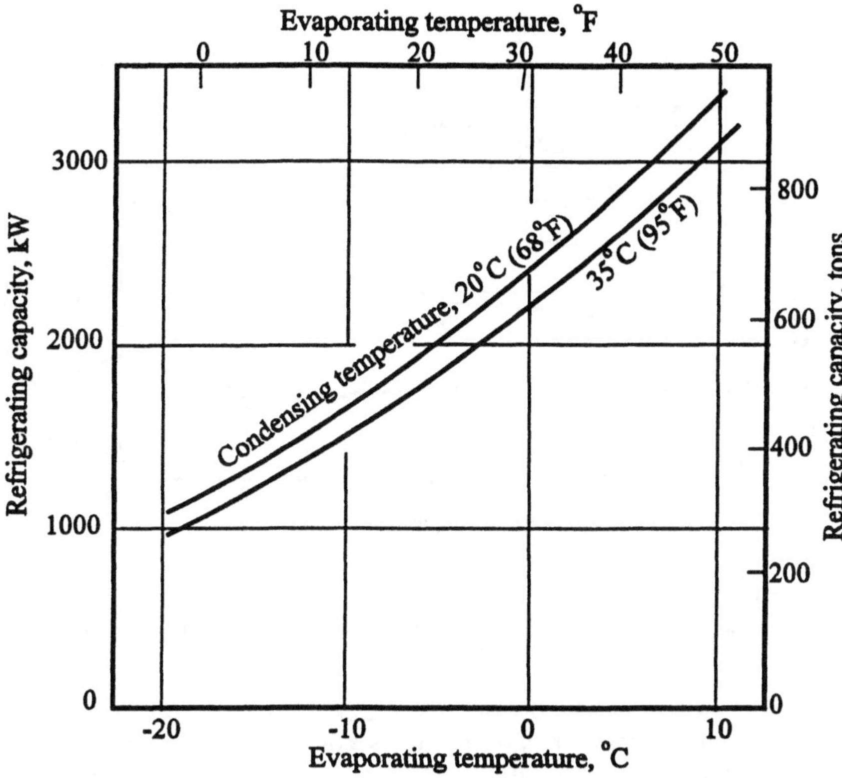

FIGURE 5.9
Effect of evaporating and condensing temperatures on the refrigerating capacity of an ammonia screw compressor. (*Model RWB-II 222, Frick Company*)

TABLE 5.2
Comparison of refrigerating capacity and power of screw and reciprocating compressor with changes in evaporating and condensing temperatures. Values shown are percentages referred to the base evaporating temperature of 5°C (41°F).

Evaporating temperature, °C (°F)	Capacity, kW (tons)				Power, kW (hp)			
	Condensing temperature				Condensing temperature			
	20°C (68°F)		35°C (95°F)		20°C (68°F)		35°C (95°F)	
	Recip.	Screw	Recip.	Screw	Recip.	Screw	Recip.	Screw
5 (41)	1.00	1.00	1.00	1.00	1.00	1.00	1.00	1.00
0 (32)	0.838	0.834	0.714	0.835	1.02	1.05	0.972	1.01
−10 (14)	0.572	0.564	0.481	0.565	0.986	1.09	0.872	0.980
−20 (−4)	0.38	0.375	0.321	0.371	0.848	1.09	0.710	0.877

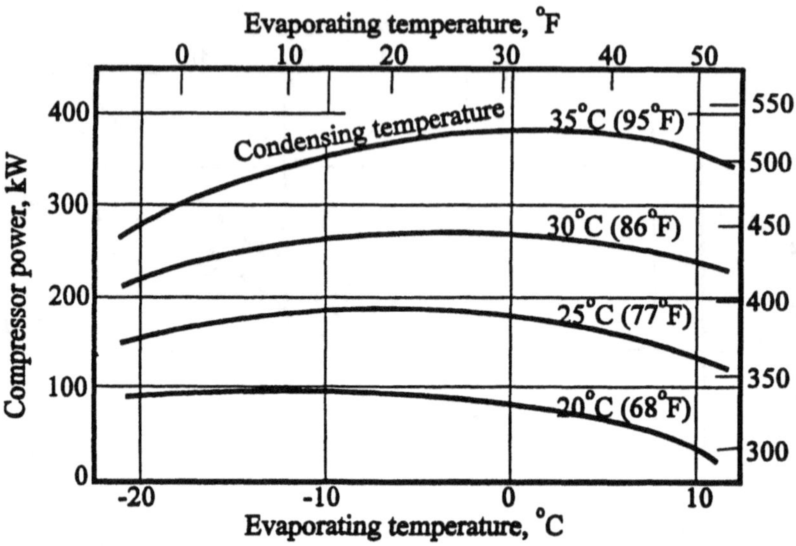

FIGURE 5.10
Effect of evaporating and condensing temperatures on the power requirement of an ammonia screw compressor. (*Model RWB-II 222, Frick Company*)

short of capacity if its actual saturated suction temperature is −31°C (−23.8°F). When ordering a compressor, then, it is essential that the compressor be selected for a lower saturated suction temperature than the evaporating temperature. This requirement applies equally to screw and reciprocating compressors.

5.7 CATALOG SPECIFICATIONS OF LIQUID SUBCOOLING AND SUCTION SUPERHEATING

Sections 4.10 and 4.11 in Chapter 4 on reciprocating compressors described how to compare the ratings of several compressor manaufacturers when they are based on different amounts of liquid subcooling and/or suction superheat. The procedures are the same for the screw compressor.

5.8 CAPACITY CONTROL AND PART-LOAD PERFORMANCE

The most common device for achieving a variation in refrigerating capacity with a screw compressor is the slide valve, as illustrated in Figure 5.11. The slide valve is cradled between the rotors and consists of two members, one fixed and the other movable. The compressor develops full capacity when the movable portion bears on the fixed member.

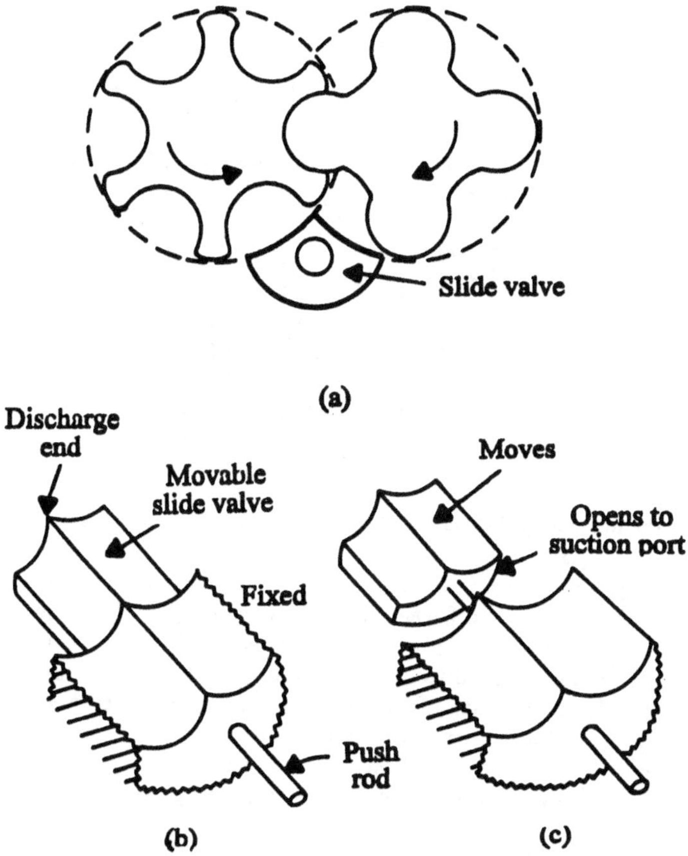

FIGURE 5.11
A slide valve for capacity control of a screw compressor: *(a)* its position relative to the rotors, *(b)* slide at full-capacity position, and *(c)* slide at reduced-capacity position.

For capacity reduction the movable portion of the slide separates from the fixed portion so that some of the gas that has filled the cavity during the suction process is not compressed. Instead, as Figure 5.12 illustrates, at the beginning of volume reduction in the cavity, the gap in the slide valve permits some of the gas to vent back to the suction. The slide valve permits a smooth, continuous modulation of capacity from full to 10% of full capacity.

Even though the slide valve can provide smooth changes of capacity the method results in reduced efficiency at part load. This reduction in efficiency is shown in Figure 5.13 where the percent of full power is related to percent of full capacity[2]. The 45° line shown in Figure 5.13 represents the ideal where a given percentage of full capacity requires the same percentage of full power. The curves show, however, that the percentage of full power always exceeds the percentage

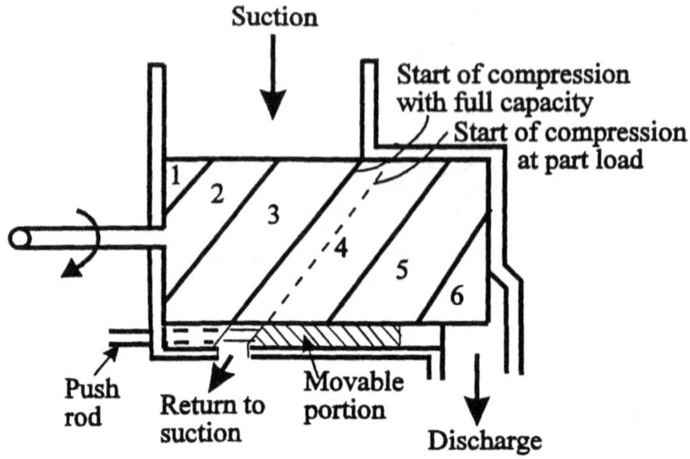

FIGURE 5.12
Side view of the function of the slide valve at *(a)* full capacity, and *(b)* partial capacity.

of full capacity. Two reasons for the drop in efficiency associated with opening of the slide valve are (1) the friction of the gas venting back to the suction, and (2) the changing of the v_i of the compressor which is assumed to be properly matched to the external conditions at full load. The recommendation that plant operators draw from the data of Figure 5.13 is to operate screw compressors as close to full load as the mix of compressors allow.

The percent capacity reduction does not vary linearly with the motion of the slide valve. The precise relation varies from compressor to compressor, but the general curve is as shown in Figure 5.14. The relationship shows that small changes of position of the slide valve at high capacity have a dominant influence on the capacity.

5.9 VARIABLE-SPEED DRIVE OF SCREW COMPRESSOR

If reducing the capacity using a slide valve results in a decline of compression efficiency, it is reasonable to explore other means of capacity control. The most attractive alternative is the use of variable-speed drive, provided either by a two-speed motor or by a frequency inverter that furnishes infinite variations of speed. The speed boundaries within which either of these concepts must operate is generally between 800 and perhaps 5000 or 6000 rpm. At low speeds the ratio of leakage gas to that pumped increases, so both the volumetric and the compression efficiency decrease. At high speeds the high pressure drop through the passages of the compressor reduces the compression efficiency. Furthermore, the noise level increases as the speed increases, and mechanical limits of the moving parts come into play. The rotor tip speed is the best criterion to guide

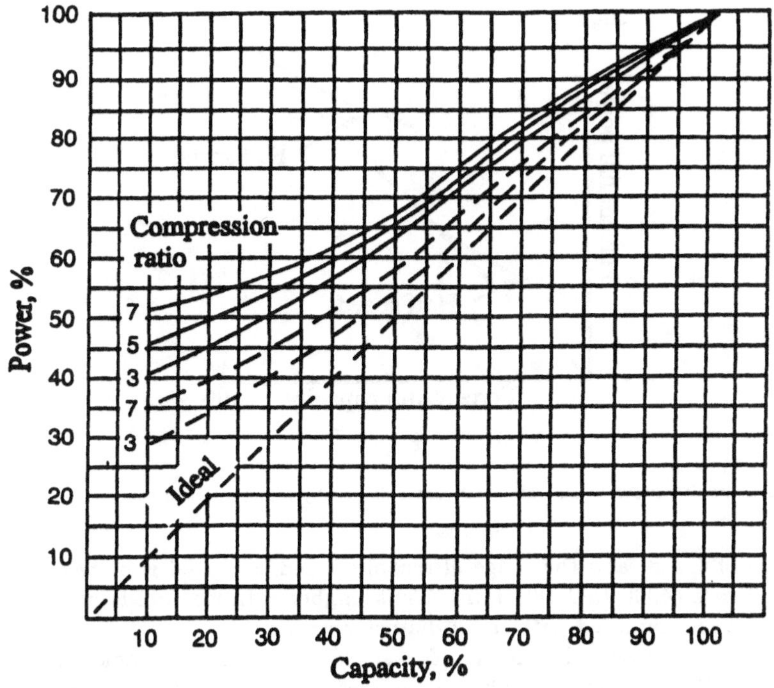

FIGURE 5.13
Part-load power requirements of a screw compressor. The solid lines apply to constant condensing and evaporating temperatures, while the dashed lines reflect a drop in condensing temperature and increase in evaporating temperature at part load.

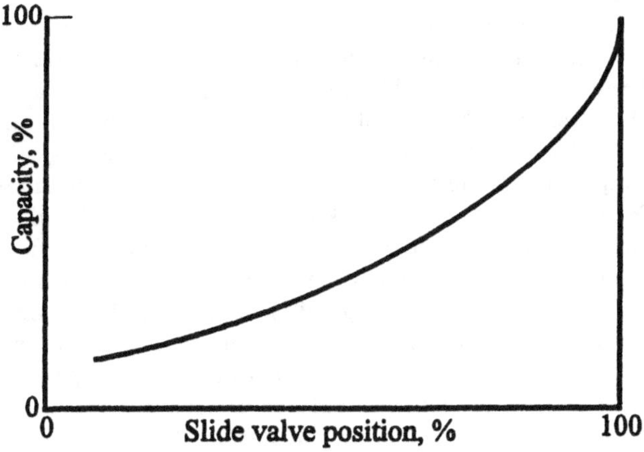

FIGURE 5.14
Variation in the compressor capacity as a function of the slide-valve position.

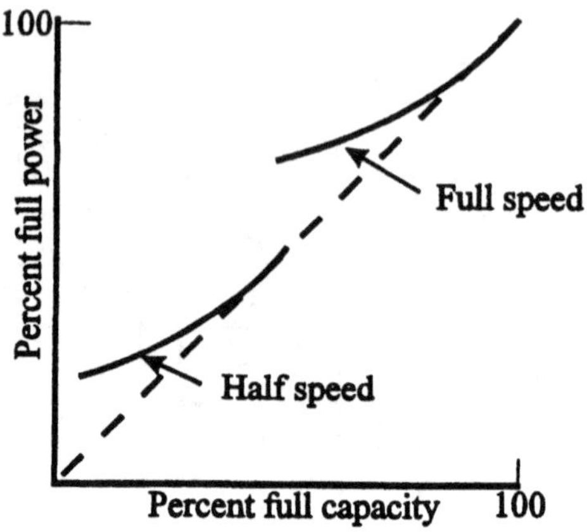

FIGURE 5.15
The power-capacity curve of a screw compressor driven by a two-speed motor.

the choice of speed, and the range of recommended tip speed is between about 20 and 50 m/s (4000 to 10000 fpm). Thus, higher rotative speeds are possible for compressors equipped with small-diameter rotors.

Screw compressors driven by a two-speed motors are available for certain applications. Such a compressor has the potential of a percent power versus percent capacity relationship as shown in Figure 5.15. The characteristics shown in Figure 5.15 assume that the compressor is equipped with a slide valve. Dropping the compressor speed from 3600 to 1800 rpm will normally reduce the volumetric efficiency only several percent. A further consideration is that two-speed motors are slightly less efficient than single-speed ones.

The variable-frequency inverter receives power at 60 Hz and converts the frequency to a different value which drives the motor with the corresponding speed change. With a variable-frequency drive no slide valve whatsoever is needed, which reduces the cost of the unit slightly and eliminates the occasional replacement of the slide valve due to its wear. The critical speed of the compressor may lie in the desired range of operation, but a standard capability of variable-frequency inverters is to skip over the narrow band of frequencies associated with critical speed of the motor. An additional benefit of a variable-frequency inverters is that a frequency higher than the power-line frequency of 60 Hz can also be developed. This capability permits the handling of peak loads with a compressor slightly smaller than would normally have been necessary.

Even though the listing of the advantages of variable-speed drive suggests that the method of capacity control would be in wide use, such is not the case in industrial refrigeration. The economic calculations for justifying the additional cost of the inverter are complex because of the various efficiencies that must be considered, including those of the inverter, the motor when operating at off-design speed and the efficiency of the compressor equipped with the slide valve.

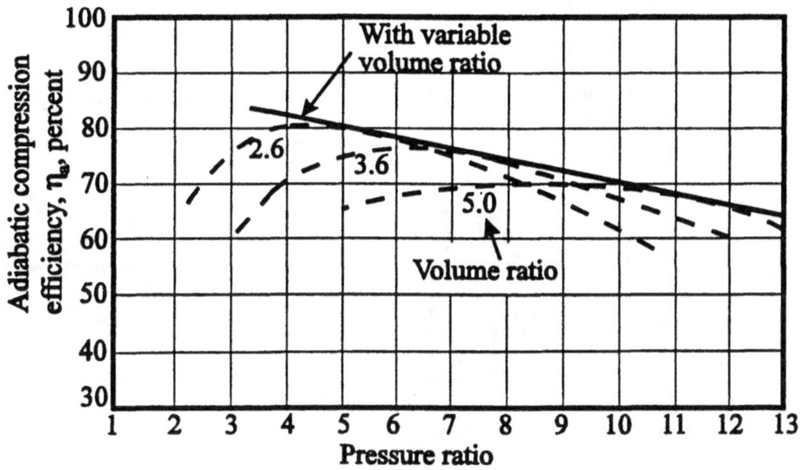

FIGURE 5.16
Maintaining peak compression efficiency with a variable-volume ratio device during changes in the pressure ratio.

Furthermore, the load profile expected for the compressor must be evaluated. A compressor that operates nearly fully loaded most of the time is probably more efficient with a slide valve for capacity control. On the other hand, if the first cost of inverters continues to drop, marginal cases will fall toward the choice of the variable-speed inverter.

5.10 VARIABLE VOLUME RATIO COMPRESSORS

Figure 5.7 showed that a given compressor with a fixed volume ratio is most efficient at one ratio of discharge-to-suction pressures. In a typical two-stage industrial refrigeration plant the high-stage compressor receives suction vapor from the intermediate pressure and discharges to the condensing pressure. The low-stage compressor pumps between the pressure of the low-temperature evaporators and the intermediate pressure. The only pressure of these three that is likely to vary is the condensing pressure due to changes in ambient conditions. The low-stage compressor can be selected with a built-in volume ratio as close as possible to that imposed by the combination of evaporating and intermediate pressures. The built-in volume ratio is determined at the time the compressor is manufactured by positioning the point at which discharge begins. For a high built-in volume ratio, for example, the discharge port is uncovered late in the compression in contrast to a machine with a low built-in volume ratio.

The high-stage compressor in a two-stage system or the compressor in a single-stage system experiences variations in the compression ratio. A goal is to

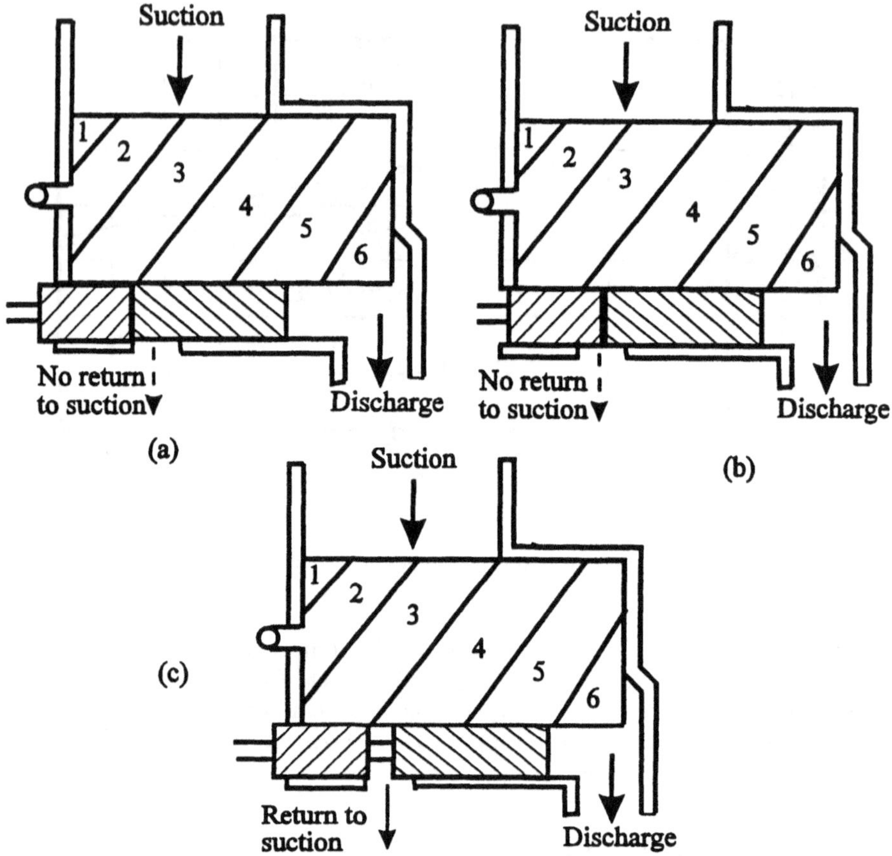

FIGURE 5.17
A variable v_i device at the following operating conditions: (a) full load and low v_i, (b) full load and high v_i, and (c) part load and high v_i.

apply some device that permits the volume ratio of the compressor to change as the imposed conditions change. The ideal performance would be that shown in Figure 5.16 where the compression efficiency developed rides along the peaks of the various volume ratios. This device and process is called *variable built-in volume ratio* or simply *variable* v_i.[3,4,5] Variable v_i devices function in conjunction with the slide valve that controls the capacity, as illustrated in Figure 5.17.

The variable v_i device of Figure 5.17 consists of two parts which can move independently. In Figure 5.17a the two parts have no gap between them, so no refrigerant vapor vents back to the suction and the compressor operates at full capacity. The discharge port is uncovered when the cavities of the rotors move past the right end of the right member. If the v_i is to be increased but full capacity maintained, both parts move to the right, as in Figure 5.17b. At this position the discharge is delayed so that the pressure in the cavities builds up more before discharging. If the high value of v_i is to be maintained, but the capacity reduced, the left member backs off which vents some vapor back to the suction, as in Figure 5.17c.

The motion of the two members requires a complex control, and there are limitations in achieving the desired v_i when the capacity must also be reduced[6]. If the capacity has been reduced by as much as 50%, the variable v_i portion of the control may no longer be able to meet its requirements.

5.11 OIL INJECTION AND SEPARATION

The screw compressor is provided with oil to serve three purposes: (1) sealing of internal clearances between the two rotors and between the rotors and housing, (2) lubrication of bearings, and (3) actuation of the slide valve. The circulation and distribution of oil is illustrated in Figure 5.18. All the oil supplied to the compressor leaves with the refrigerant and flows to the oil separator. The separator removes the oil from the refrigerant and the refrigerant passes on to the intermediate stage (in the case of a low-stage compressor) or to the condenser (in the case of a high-stage compressor). The oil from the separator is warm because it has absorbed some of the heat of compression and must pass through an oil-cooling heat exchanger before distribution to the three streams serving the compressor. The means of rejecting heat from this exchanger will be discussed in the next several sections.

The type of oil separator used in screw compressor packages is called the coalescing type which will be discussed in more detail in Chapter 15, Lubrication and Oil Handling. This type of separator is much more efficient than the older-style inertia separators that have been traditionally used for reciprocating compressors. A strong trend prevails now to use coalescing separators for reciprocating compressors as well. Coalescing separators are expected to pass no liquid oil, so the only oil that escapes the separator is in vapor form, which means that the oil concentration in the refrigerant leaving the separator will be of the order of 5 ppm. Dirty oil will plug the coalescing elements requiring that the element be replaced, a procedure possible by access through the handhole. Normally the coalescing element lasts for several years.

The end user of the refrigeration facility usually has neither the need nor opportunity to choose the size of the oil separator. This decision is made by the assembler of the compressor package. The size must be such that velocities are not so high that they carry liquid oil out of the separator. The critical operating condition is at low discharge temperatures when the discharge gas experiences a high mass flow rate and high specific volume, both contributing to high refrigerant velocities. Limiting velocities are in the neighborhood of 0.76 m/s (150 fpm) for ammonia and 0.38 m/s (75 fpm) for R-22. Sudden drops in pressure in the separator which would generate foam should be avoided, because coalescing separators are not capable of preventing carryover of foam.

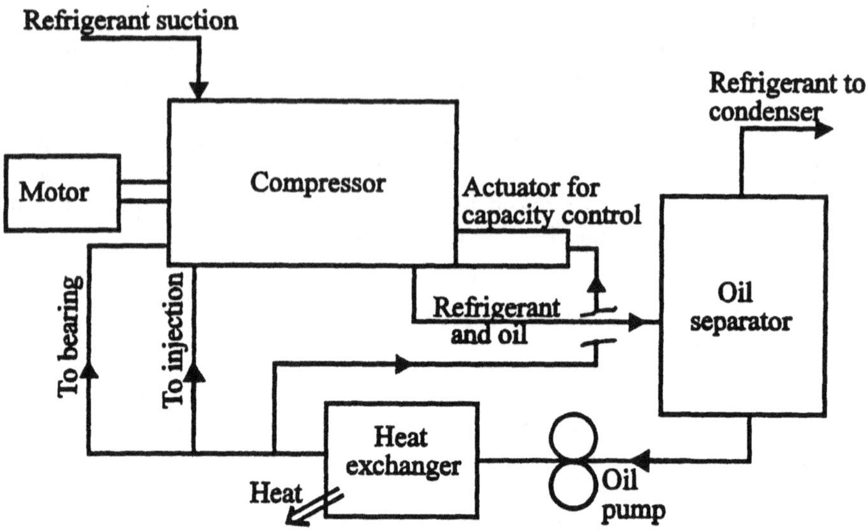

FIGURE 5.18
Flow and distribution of oil serving a screw compressor.

The injection oil flow rate should be adequate to seal the internal clearances, as well as to lubricate the moving parts and to cool the gas being compressed. On the other hand, excessive oil quantities will result in undesirable hydraulic hammer. Knowledge of the rate of flow of injection oil is necessary to properly design the oil-cooling heat exchanger. Often, however, when the oil cooler is part of the package it is selected by the assembler, and the system designer and operator may not need to know the injection oil flow rate. Some compressor manufacturers publish the information[7] in their catalogs, and the order of magnitude of oil flow rates range from about 0.065 to 0.11 L/min per kW of refrigeration (0.06 to 0.1 gpm/ton of refrigeration) for high-stage machines.

5.12 OIL COOLING METHODS

The injected oil that seals the clearances in the compressor is intimately mixed with the refrigerant undergoing compression. The refrigerant vapor becomes hot during compression and transfers some heat to the oil as it passes through the compressor. The oil must be cooled before reinjection, and four of the important methods of oil cooling are: (1) direct injection of liquid refrigerant, (2) external cooling with a thermosyphon heat exchanger, (3) external cooling with cooling water or antifreeze, and (4) pumping of liquid refrigerant into the refrigerant/oil mixture as it leaves the compressor. The first two methods are the most popular and will be addressed in more detail in Sections 5.13 and 5.14, respectively. This section describes external cooling using a cooling liquid or antifreeze and the introduction of liquid refrigerant into the discharge stream.

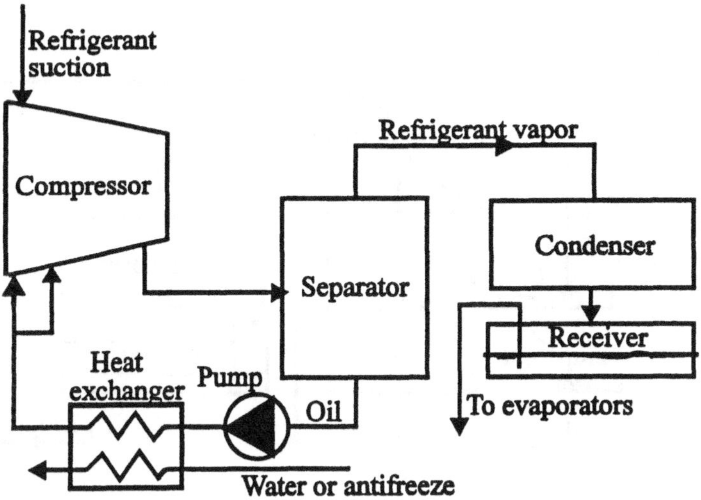

FIGURE 5.19
Oil cooling using an external heat exchanger rejecting heat to water or antifreeze.

The method of cooling oil with an external heat exchanger that rejects heat to cooling water or antifreeze, as illustrated in Figure 5.19, was the earliest, widely-used concept. In warm climates cooling water could serve the heat exchanger, and this cooling water could come from a cooling tower or a closed-circuit cooler. In the closed-circuit cooler the fluid being cooled flows through pipe coils with water spraying from the top and ambient air flowing up through the sprays and coils. Some plants simply draw water from the pan of the evaporative condenser, but this practice is not recommended. In the first place, this water is likely to collect foreign matter that could deposit on the tubes of the heat exchanger, reducing its heat-transfer effectiveness. Also, in cold climates the water in the evaporative condenser must sometimes be drained to avoid freezing, and in that situation no means of oil cooling would be available. Even with the closed-circuit cooler, in cold climates antifreeze must be circulated to avoid freezing water.

The second method[8] of oil cooling, as illustrated in Figure 5.20, extracts liquid refrigerant from the receiver and delivers it into the discharge of the compressor. Evaporation of this pumped liquid cools the refrigerant/oil vapor mixture to the desired 49°C (120°F) temperature before the mixture enters the separator. A temperature sensor at some point following the injection of liquid regulates a variable-speed pump to adjust the flow rate of liquid. In this method and the method of direct injection the temperature of the oil separator operates at 49°C (120°F) in contrast to the external heat exchanger and thermosyphon

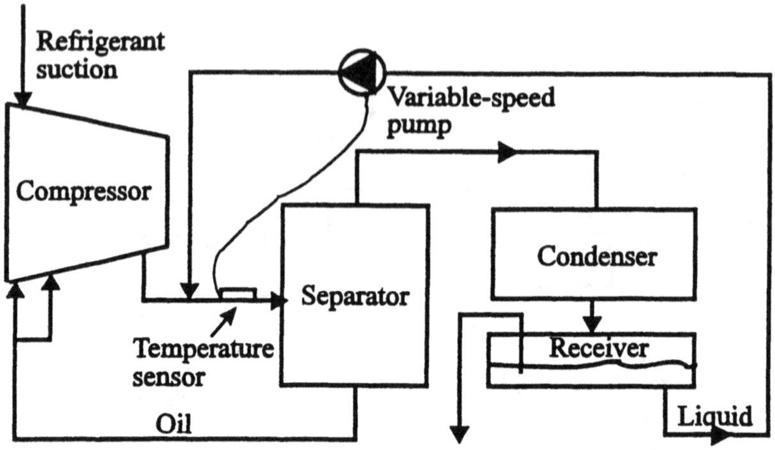

FIGURE 5.20
Cooling oil by injection of liquid into the compressor discharge line.

systems where the operating temperature is that of typical discharge, namely, 60°C (140°F). Oil separators function slightly better when the temperature is low, because the viscosity as well as the vapor presssure of the oil are lowered.

5.13 OIL COOLING BY REFRIGERANT INJECTION

Several decades ago the oil-cooling system using the external heat exchanger cooled by water or antifreeze lost favor compared to direct injection of refrigerant. While this concept in turn has been supplanted in many cases by the thermosyphon system, many new systems use direct refrigerant injection for oil cooling. Liquid refrigerant from the receiver is injected into the compressor at an early stage of compression in a manner illustrated by Figure 5.21. A valve regulates the flow of liquid to maintain the desired discharge temperature of the refrigerant/oil mixture. A conventional superheat-controlled expansion valve (sometimes called a TXV or a thermo-valve), as will be described in Chapter 11, Valves and Refrigerant Controls, can be adapted for this service. The usual valve controls the amount of superheat at the position of its sensor, but the objective of the valve in Figure 5.21 is to control the temperature at 49°C (120°F). The adaptation consists of providing a pressure reference to the valve, which it tries to match by the fluid pressure, and thus the desired temperature, in its sensing bulb.

Cooling the oil with direct injection of refrigerant imposes penalties both in the reduction in refrigeration capacity as well as an increase in power. Figure 5.22 shows the magnitude[9] of these penalties for an ammonia compressor.

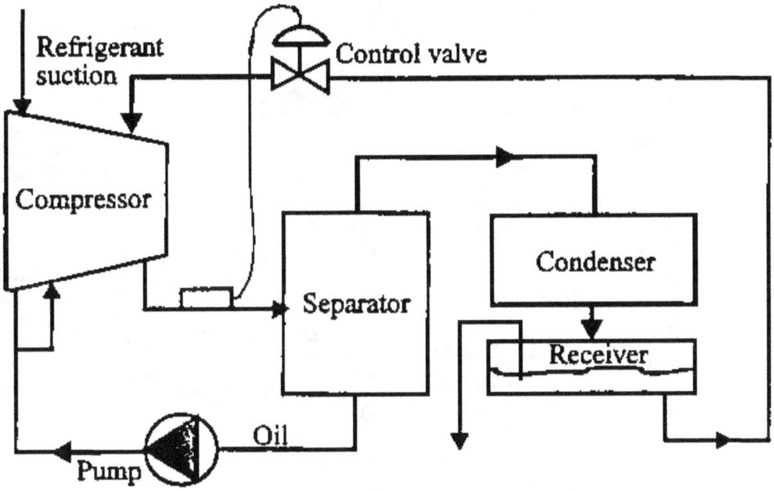

FIGURE 5.21
Cooling oil by direct injection of liquid refrigerant at an early stage of the compression process.

Not all manufacturers agree on the magnitude of power penalties[10], reporting instead that the power penalty is about 1% for all conditions. Both penalties increase with an increase in pressure ratio against which the compressor operates. The losses are thus significant when the compressor operates single-stage with a low evaporating temperature. The losses are moderate for high- and low-stage compressors in a two-stage system.

A limitation in the system operating with direct-injected compressors is how low the condensing temperature is permitted to fall. To conserve energy, most plants operate with as low a condensing temperature as the ambient temperatures can provide. One of the plant characteristics that may limit the degree to which the condensing pressure is allowed to fall is the need for adequate pressure to force liquid through the control valve in Fig. 5.21. A typical minimum condensing pressure recommended by manufacturers of screw compressors is 860 kPa (125 psia).

The decision of whether to choose direct injection as the method of oil cooling is primarily an economic one. A plant cooling oil with a thermosyphon system will be more efficient, but at the expense of a higher first cost in comparison to direct injection. Direct injection boasts the lowest first cost of the oil-cooling methods, and this feature is a major contributor to its selection.

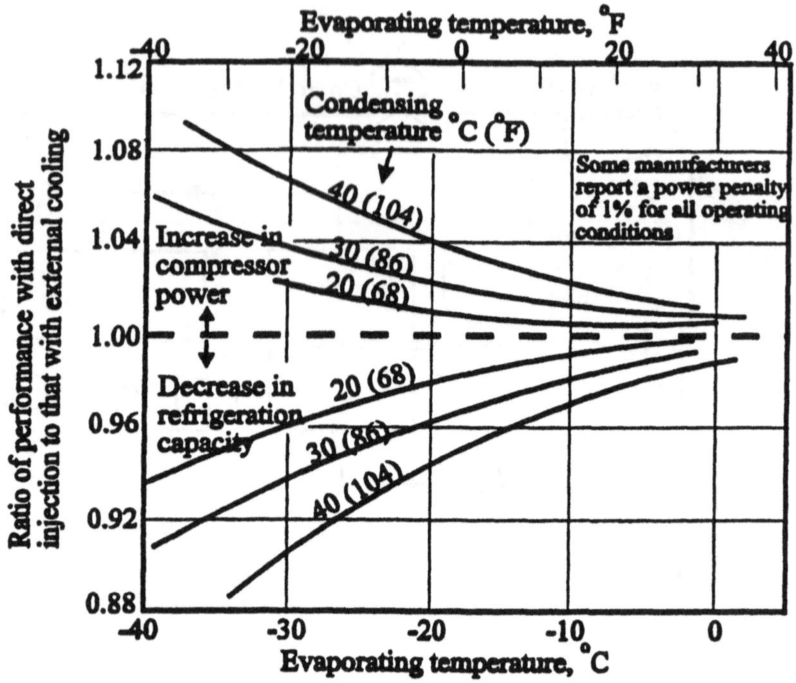

FIGURE 5.22
Penalties in refrigeration capacity and power requirement for ammonia screw compressors provided with direct-injection oil cooling.

5.14 OIL COOLING WITH A THERMOSYPHON HEAT EXCHANGER

The thermosyphon concept in oil cooling achieves heat transfer by boiling liquid refrigerant at the condensing pressure. Furthermore, the boiling refrigerant flows by natural convection (thermosyphon effect) through the heat exchanger. Flow diagrams of two thermosyphon systems are illustrated in Figures 5.23 and 5.24. Liquid refrigerant flows down from a receiver and some of this liquid boils in the heat exchanger. The mixture of liquid and vapor rises from the heat exchanger and returns to the receiver. Circulation of the refrigerant in the loop takes place because of the greater desity in the liquid leg in contrast to the return line where the mixture of liquid and vapor flows.

The difference of the systems in the two diagrams is that in Figure 5.23 that the liquid level in the system receiver is above that of the heat exchanger, a requirement for the natural circulation to take place. If the level in the system receiver is at or below that of the heat exchanger, as in Figure 5.24, an additional receiver, called the thermosyphon or pilot receiver, must be installed.

In the receivers of both thermosyphon systems shown in Figs. 5.23 and

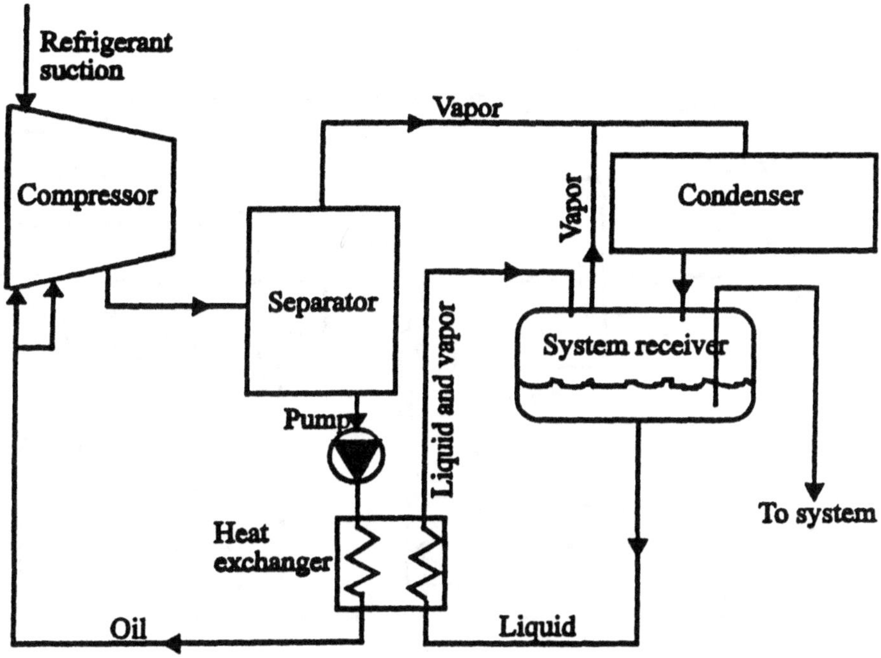

FIGURE 5.23
A thermosyphon oil cooling installation where the level of the system receiver is above the level of the heat exchanger.

5.24, the entrance to the liquid line that drains refrigerant to the oil cooler is positioned lower than the entrance to the line feeding liquid to the evaporators. The strategy is that the oil cooler has the first call on liquid refrigerant. If the high side of the system should run short of refrigerant, the liquid supply for cooling the oil has first priority in order to protect the compressors.

The designer of the refrigeration system is normally not called on to design the heat exchanger that cools oil. In fact, manufacturers usually guard jealously their right to select the heat exchanger. In the first place, the heat exchanger is an integral part of the package that is assembled at the factory. Of equal importance is that the heat exchanger is a minor contributor to the total cost of the package, and an improper heat exchanger could result in damage to the expensive compressor.

Each compressor is equipped with its own heat exchanger, and in multiple-compressor installations one thermosyphon system can supply several heat exchangers. The designer of the refrigeration system is responsible for designing the thermosyphon system which includes selecting the size of the thermosyphon receiver and the sizes of three main lines. These pipes include: (1) the liquid/vapor line from the heat exchanger to the receiver, (2) the liquid line from the receiver

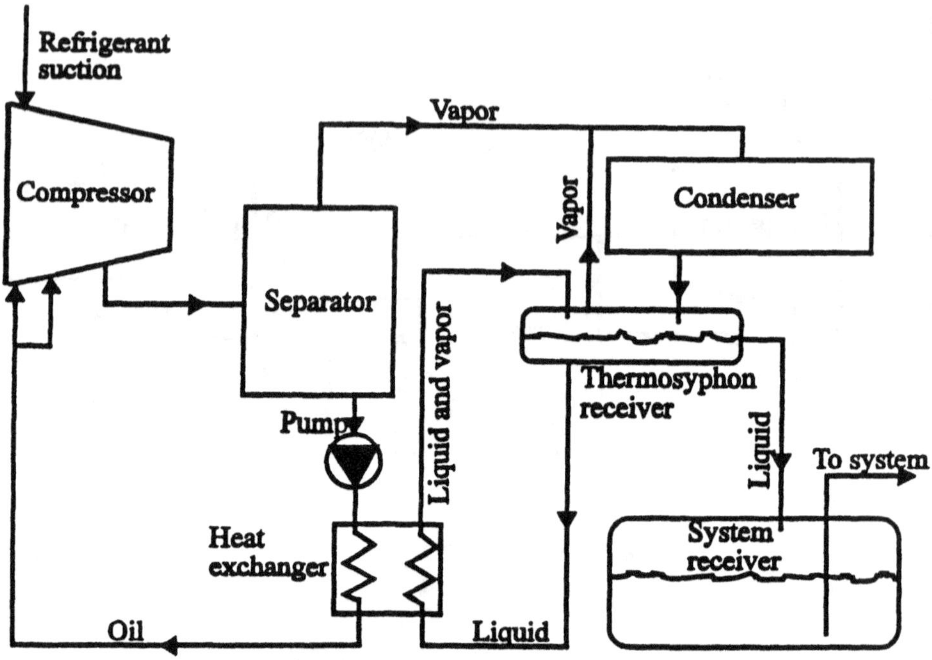

FIGURE 5.24
A thermosyphon oil cooling installation where the level of the system receiver is at or below the level of the heat exchanger, requiring an additional receiver.

to the heat exchanger, and (3) the vapor line from the receiver to the header carrying discharge vapor to the condenser(s). The diagrams of the thermosyphon systems shown in Figures 5.23 and 5.24 are schematic, and there are additional piping details that must be provided for proper drainage of the condenser, as will be discussed in Chapter 7 on condensers.

A basic piece of data is the fraction of the heat equivalent of the total of the compressor power and refrigeration load that is absorbed by the injected oil and thus must be removed in the oil cooler. This percentage is presented in Figure 5.25 and is needed for the thermal analysis of a two-stage system. It would also be used if the designer is making detailed calculation of the thermosyphon system. To simplify the task of designing the thermosyphon system, Reference 11 bypasses some of these calculation details by providing recommendations on the size of these components, and those guidelines will be quoted here.

The preliminary steps in the basic procedure of selecting the components in the system are to determine the flow rates.

(1) Determine the heat rejection rate at the oil cooler, q_{oc},

$$q_{oc} = (q_{tot})(\text{percent from Figure 5.25})/100 \qquad (5.2)$$

where q_{tot} = refrigeration capacity + heat equivalent of compressor power

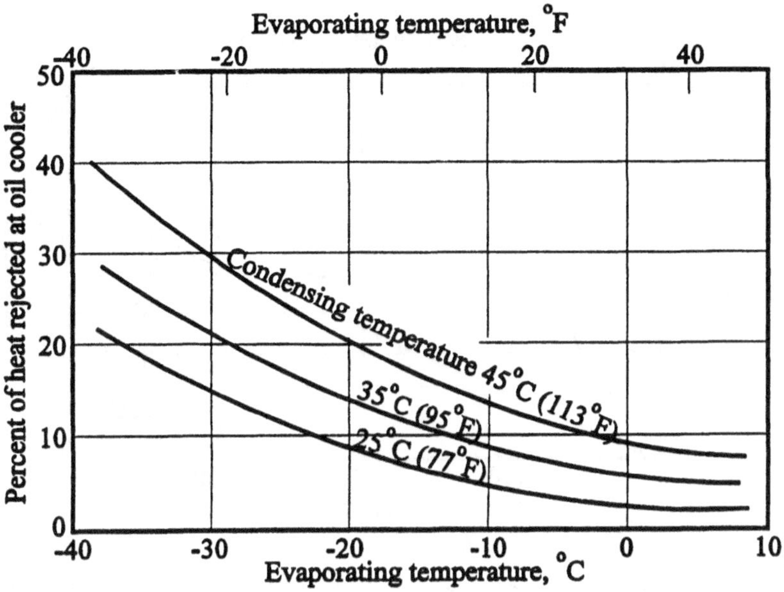

FIGURE 5.25
Percentage of heat input (total of refrigeration load and compressor power) that is absorbed by the injected oil in a screw compressor.

(2) Compute the evaporation rate, $\dot{m}_{ev}$, where

$$\dot{m}_{ev} = \frac{q_{oc}}{\text{enthalpy of evaporation}} = \frac{q_{oc}}{h_g - h_f} \tag{5.3}$$

(3) Calculate the flow rate through the oil cooler, $\dot{m}_{oc}$, assuming a recirculation ratio of 2:1 for R-22 and 3:1 for ammonia,

$$\dot{m}_{oc} = 2\dot{m}_{ev} \quad \text{for R-22} \quad \text{and} \quad \dot{m}_{oc} = 3\dot{m}_{ev} \quad \text{for ammonia}$$

Thermosyphon receiver. The maximum heat rejection rate of the heat exchanger controls the size of the receiver. Specifically, the size of the receiver is chosen so that a reserve for five minutes of operation, thus $\dot{m}_{ev}$, is available if the supply of liquid from the condenser is interrupted. It is expected that the outlet to the system receiver is at about the midpoint in the thermosiphon receiver. Thus, the thermosiphon receiver should be twice the size of the volume of five minutes of refrigerant evporation.

$$\text{Volume of receiver} = \frac{(5\text{ min})(\dot{m}_{ev})}{\text{density of liquid refrigerant in mass per unit volume}}$$

Figure 5.26 shows recommended volumes for R-22 and ammonia on this basis. Usually these receivers are about 1.7 m long (5 or 6 feet). Figure 5.26 shows

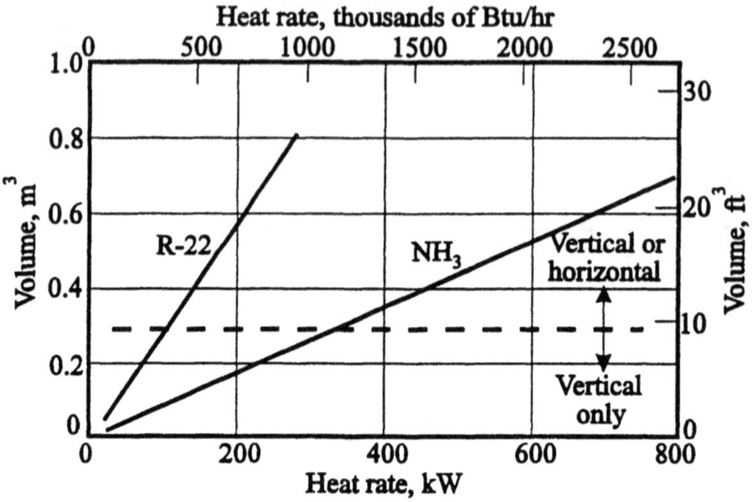

FIGURE 5.26
Volume of the thermosiphon receiver as a function of the heat rate of the oil cooling heat exchanger.

that a smaller receiver is adequate for ammonia compared to R-22, because the high latent heat of ammonia permits it to transfer a given amount of heat with a lower flow rate than for R-22.

Liquid line from receiver to the heat exchanger. This section of line carries a flow rate greater than the rate evaporated, because a properly operating thermosiphon system circulates unevaporated liquid back to the receiver. Designers of thermosyphon systems strive for a circulation ratio of 3:1 for ammonia and 2:1 for R-22, where the circulation ratio means the rate supplied to the heat exchanger divided by the rate evaporated.

The recommended pressure gradients[11] for this pipe are 22.6 Pa/m (0.1 psi per 100 ft) for ammonia and 113 Pa/m (0.5 psi per 100 ft) for R-22. With the assumption of a condensing temperature of 35°C (95°F) at which temperature the enthalpy of evaporation for ammonia is 1124 kJ/kg (483.2 Btu/lb) and for R-22 it is 172.6 kJ/kg (74.2 Btu/lb), the following equations may be used to compute the required pipe size, D, in inches to abide by the pressure gradients and circulation ratios specified above.

For ammonia

$$D, \text{ in.} = 0.472(\dot{m}_{oc}, \text{ lb/min})^{0.37} = 2.88(\dot{m}_{oc}, \text{ kg/s})^{0.37} \tag{5.4}$$

For R-22

$$D, \text{ in.} = 0.350(\dot{m}_{oc}, \text{ lb/min})^{0.37} = 2.13(\dot{m}_{oc}, \text{ kg/s})^{0.37} \tag{5.5}$$

Liquid/vapor line from heat exchanger to thermosiphon receiver. The recommended pressure gradients[11] for the liquid/vapor return line are 9.04 Pa/m

TABLE 5.3
Flow-rate carrying capacities of various line sizes in the vent pipe between the receiver and the condenser.

Pipe size, inch	Ammonia		R- 22	
	kg/s	lb/min	kg/s	lb/min
1-1/2	0.0529	7	0.121	16
2	0.0907	12	0.219	29
2-1/2	0.166	22	0.378	50
3	0.295	39	0.680	90
4	0.529	70	1.17	155
5	0.907	120	2.12	280
6	1.66	220	3.63	480
8	5.29	700	7.56	1000

(0.04 psi per 100 ft) for ammonia and 45.2 Pa/m (0.2 psi per 100 ft) for R-22. To abide by these pressure gradients, the required pipe sizes are given by the following equations:
For ammonia

$$D, \text{in.} = 0.572(\dot{m}_{oc}, \text{lb/min})^{0.37} = 3.49(\dot{m}_{oc}, \text{kg/s})^{0.37} \tag{5.6}$$

For R-22

$$D, \text{in.} = 0.423(\dot{m}_{oc}, \text{lb/min})^{0.37} = 2.58(\dot{m}_{oc}, \text{kg/s})^{0.37} \tag{5.7}$$

Vapor line from the receiver to the condenser header. This line may at first have the appearance of an equalizer line through which there is flow only when the pressures at the terminal points of the pipe need to be balanced. If that were the case, only a small pipe, perhaps of 1-inch size, would suffice. On the contrary, a flow of refrigerant equal to $\dot{m}_{ev}$ passes through this line. To motivate this flow, the pressure in the thermosiphon receiver must be higher than the entrance to the condenser. In other words, the refrigerant must gain in pressure as it passes through the condenser and the drain line to the thermosyphon receiver. A way in which this can be done is to trap the liquid drain line from the condenser and provide a liquid leg in this line to compensate for the pressure drop. The details of this arrangement will be provided in Chapter 7, but for now it is important to realize that the size of this line should be generous to keep the pressure drop low.

The recommended minimum pipe sizes[11] for various flow rates with ammonia and R-22 are given in Table 5.3.

Difference in elevation from the receiver to the heat exchanger. The thermosyphon concept operates because of the higher pressure developed down the liquid leg in comparison to the magnitude of pressure reduction of the less-dense mixture of liquid and vapor flowing upward in the line between the heat exchanger and the receiver. Since the pressure difference is proportional to the vertical distance over which this difference in density prevails, a certain minimum vertical distance should be provided between the liquid level in the thermosiphon receiver and the heat exchanger. Reference 11 recommends a minimum elevation difference of 1.8 m (6 ft).

Example 5.1. Design the thermosiphon oil-cooling system serving an ammonia screw compressor operating with an evaporating temperature of −20°C (−4°F) and a condensing temperature of 35°C (95°F). The full-load refrigerating capacity and power requirement at these conditions are 1025 kW (291.4 tons of refrigeration) and 342 kW (458.5 hp), respectively.

Solution. The combined refrigeration and power input is:

$$1025\ \text{kW} + 342\ \text{kW} = 1367\ \text{kW}$$

or

$$(291.4\ \text{tons})(12{,}000) + (458.5\ \text{hp})(2545) = 4{,}664{,}000\ \text{Btu/hr}$$

Figure 5.25 indicates that at the prevailing evaporating and condensing temperatures, 14% of the total energy input is absorbed by the oil:

$$q_{oc} = (1367\ \text{kW})(0.14) = 191.4\ \text{kW}\ (653{,}000\ \text{Btu/hr})$$

As a preliminary step, compute the evaporation rate of ammonia,

$$\dot{m}_{ev} = \frac{q_{oc}}{h_g - h_f} = \frac{191.4\ \text{kW}}{1124\ \text{kJ/kg}} = 0.1703\ \text{kg/s}$$

$$\text{or}\quad \dot{m}_{ev} = \frac{653{,}000\ \text{Btu/hr}}{(483.2\ \text{Btu/lb})(60\ \text{min/hr})} = 22.52\ \text{lb/min}$$

Designing for a circulation ratio of 3, which is typical for ammonia,

$$\dot{m}_{oc} = 3\dot{m}_{ev} = 0.511\ \text{kg/s}\ (67.6\ \text{lb/min})$$

Thermosiphon receiver. If one-half the receiver should be able to contain a five-minute evaporation rate, the volume of the receiver, V_{rec} is

$$V_{rec} = \frac{2(5\ \text{min})(60\ \text{sec/min})(0.1703\ \text{kg/s})}{(\rho_{liquid} = 587.6\ \text{kg/m}^3)} = 0.174\ \text{m}^3\ (6.31\ \text{ft}^3)$$

a size which corresponds to Fig. 5.26.

Choosing a length L of 1.83 m (6 ft) in the equation for volume $\pi(D^2/4)L$,

$$D = \sqrt{\frac{(0.174\ \text{m}^3)(4)}{\pi(1.83)}} = 0.348\ \text{m}\ (13.7\ \text{in})$$

Choose the next largest diameter which is 16 in (0.4064 m).

Line size from receiver to heat exchanger. Applying Eq. 5.4 to the flow rate of $\dot{m}_{oc}$,

$$D = 0.472(67.6\ \text{lb/min})^{0.37} = 2.88(0.511\ \text{kg/s})^{0.37} = 2.24\text{in}$$

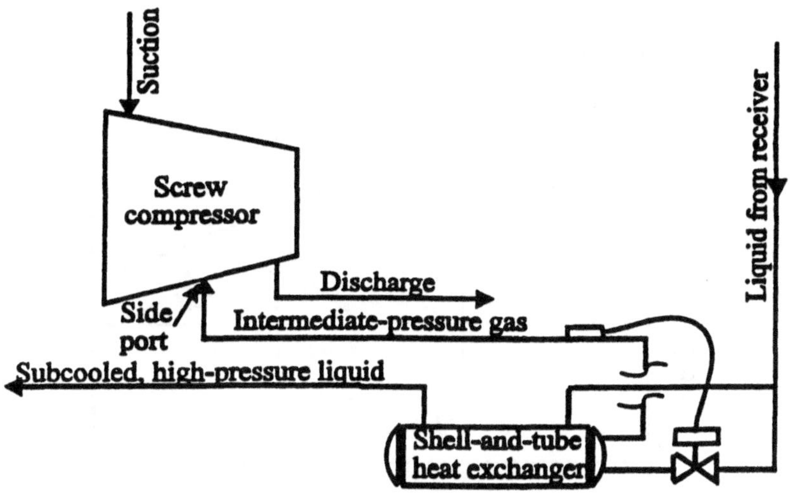

FIGURE 5.27
Using the side port of a screw compressor to provide a two-stage benefit.

Choose a 2-1/2 in size.

Liquid/vapor line size between the heat exchanger and receiver. Using Eq. 5.6,

$$D = 0.572(67.6\ \text{lb/min})^{0.37} = 3.49(0.511\ \text{kg/s})^{0.37} = 2.72\ \text{in}$$

Choose a 3-inch line.

Vapor line from receiver to condenser. Entering Table 5.3 with the evaporation rate of 0.1703 kg/s (22.52 lb/min), we find that a 2-1/2-inch size would almost be adequate, but choose a 3-inch line to minimze the pressure drop.

5.15 ECONOMIZER CIRCUIT USING A SIDE PORT

Examination of Figure 5.4 suggests that it would be possible to provide an opening in the housing of the compressor to tap into a cavity during compression. The refrigerant in Cavity 5, for example, is at a pressure somewhere between suction and discharge. Refrigerant can be supplied through this opening at an intermediate pressure, and the compressor continues the compression of all the refrigerant. This feature opens the possibility of a liquid subcooler as was first illustrated in Figure 3.24 which is reproduced as Figure 5.27. This opening, often called the side port, offers within one compressor some of the advantages of a multiple-compressor, two-stage installation. A subsystem as shown in Figure 5.27 is called an economizer.

Manufacturers of screw compressors are usually able to choose the position of the side port so that the desired intermediate pressure can be provided.

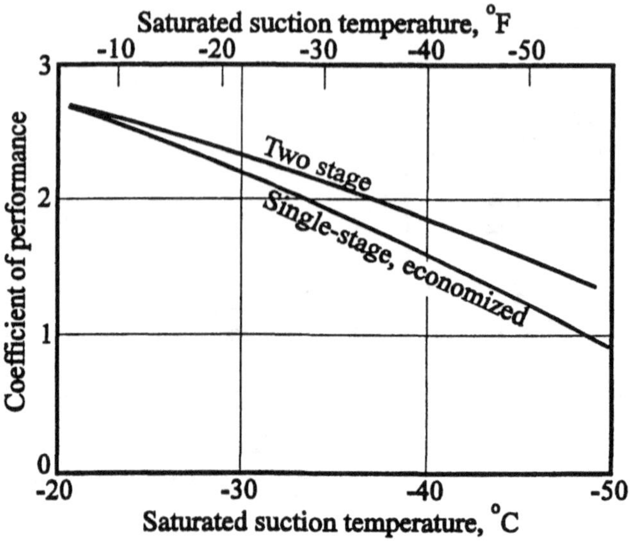

FIGURE 5.28
Comparison of the coefficients of performance of a two-stage ammonia system with an economized single-stage compressor equipped with a flash-type subcooler.

Once the position has been established, however, it is fixed and the compressor has no flexibility to maintain a constant intermediate pressure as the suction and discharge pressures change. When operating at its optimum conditions the economizer cycle provides a significant benefit with a low first-cost investment. For example, the shell-and-tube heat exchanger, including its control, shown in Figure 5.27 is a low-cost addition. When the side port is placed into service, the volume flow rate drawn in at the compressor suction is not affected, because the gas is already trapped by the time the side port is uncovered. Additional refrigeration capacity is provided, however, because the liquid flowing to the evaporators has been chilled and its enthalpy reduced. The power reqirement of the compressor will increase because of the additional gas to be compressed from the side-port pressure to the condensing pressure.

The exhileration of discovering a low-cost replacement for a two-stage system that requires an extra compressor must be tempered somewhat by the realization that the economizer cycle in its best operation is not quite as efficient as two stage. Stegmann[12] analyzed the performance of the side port and several of his findings will be reported in the next several pages. Figure 5.28 shows the comparative coefficients of performance (COP) at various suction temperatures when operating with a condensing temperature of 35°C (95°F).

One reason for the inability of the economized system using a side port to attain the efficiency of a two-stage system is illustrated in Figure 5.29 which shows that the pressure within the cavity changes during the time that the side port is uncovered. In the early stage of admission there is an unrestrained

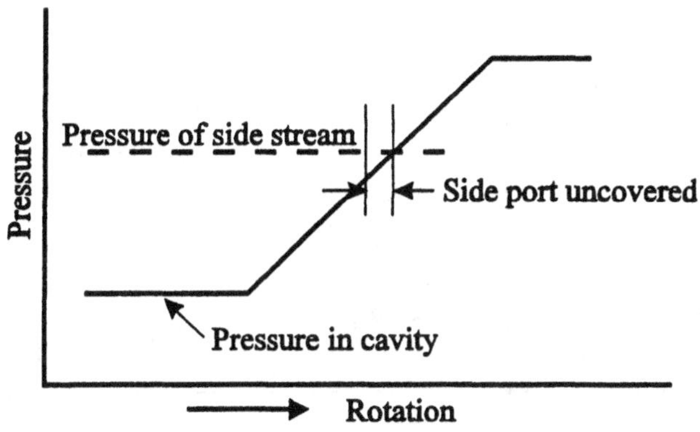

FIGURE 5.29
Unrestrained expansion of side-port gas during the admission into the compressor.

expansion, as discussed in Section 5.3, of the side-port gas as it flows into the compressor. This unrestrained expansion consitutes a thermodynamic loss.

From an understanding of the process associated with an economizer operation using the side port, it can be inferred that the capacity of the system will increase. This increase occurs, because the enthalpy of liquid reaching the expansion valve is reduced, even though the volume flow rate at the inlet to the compressor remains unchanged. Due to the admission of additional gas during the compression process, the power requirement increases. Both of these effects[13] are shown in Figure 5.30, which presents multiplying factors for both refrigerating capacity and power with respect to the noneconomized system. Since the factors for the increase of refrigerating capacity exceed those for the increase in power, the economized cycle shows an improvement over the noneconomized cycle.

While the use of the side port with an economizer shows performance advantages with only a moderate additional first cost, there are some limitations. The economizer cycle is most effective when the compressor is operating at full refrigeration capacity. With compressors equipped with slide valves for capacity control, the opening of the slide valve changes the pressure within the compressor at the side port. As Figure 5.31 shows, when the slide valve is in a partial capacity position, the point at which the gas is trapped in the cavity moves further to the right. Because the start of compression is delayed, the pressure in the cavity is low when the side port is first uncovered. Thus, the pressure at the side port progressively drops as the slide valve opens. One consequence of this pressure change is that the optimum intermediate pressure no longer prevails, and the improvement of the flash-gas removal by liquid subcooling diminishes. The slide valve can even move to the point where the side port uncovers at the instant the gas is trapped in the cavity. At this point the side-port pressure has dropped to the suction pressure and the economizer is completely ineffective.

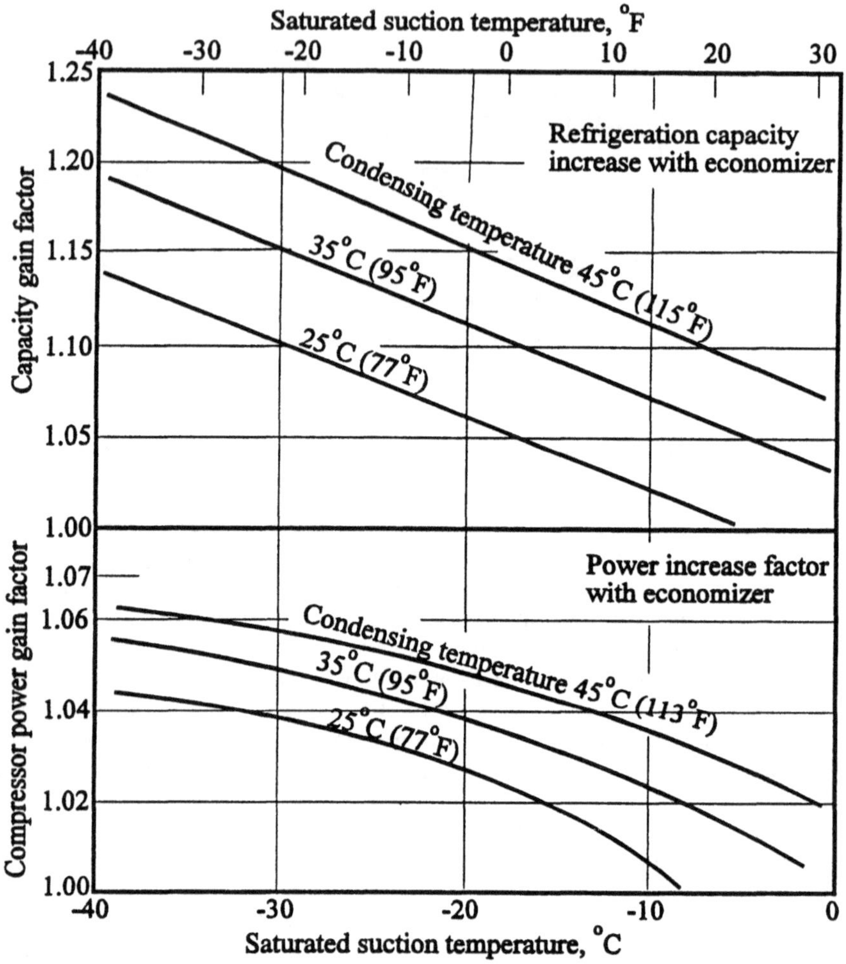

FIGURE 5.30
Multiplying factors for the refrigerating capacity and the power requirement in an ammonia system when operating with an economized cycle.

The equipment shown in Figure 5.27 with the shell-and-tube liquid subcooler turns out to be one of the best means of exploiting the side port. Other potential applications include using a flash tank with the vapor drawn off by the side port. As discussed in Section 3.4 in the chapter on multistage systems, the flash tank drops the temperature of liquid more than the liquid subcooler which must operate with a temperature difference between the leaving subcooled liquid and the saturation temperature of the side-port pressure. But the drop in side-port pressure as the slide valve opens has consequences over and above the

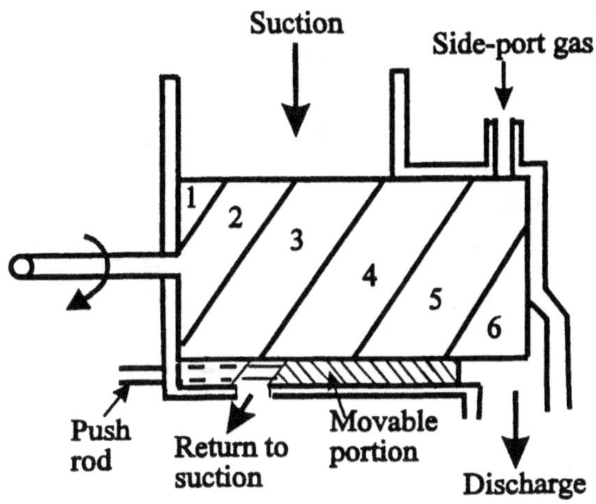

FIGURE 5.31
Opening of the slide valve drops the side-port pressure.

reduction in efficiency. The pressure of the liquid leaving the flash tank also drops and may not provide enough pressure to force the liquid through downstream valves.

Another potential application of the side port is to provide the suction for an intermediate-temperature evaporator. Here again there are limitations imposed by the prospect of the drop in side-port pressure. In the food industry the intermediate-temperature evaporator is often serving spaces storing unfrozen food where the drop in evaporating temperatures much below freezing could damage products. A conclusion is that the side port offers attractive possibilities, but it also has limitations.

5.16 THE SCREW COMPRESSOR AS THE LOW-STAGE COMPRESSOR

When applied as a low-stage (booster) compressor, several economies can be achieved in the design of the screw compressor. A majority of low-stage compressors serve low-temperature applications in the food industry, including the freezing of food and in refrigerating spaces storing frozen food. There are three operating pressures in two-stage systems, namely at the low-temperature evaporators, at the intermediate-temperature evaporators or the subcooler/desuperheater, and at the condensers. Of these three pressures, the only one expected to vary appreciably during operation is the condensing temperature because of changes in the ambient conditions. The suction and discharge temperatures of the low-stage compressor are likely to remain fairly constant, so the low-stage compressor can be selected to be optimum for the given pressure ratio. The capacity control of the high-stage compressor, for example, normally uses the suction pressure of

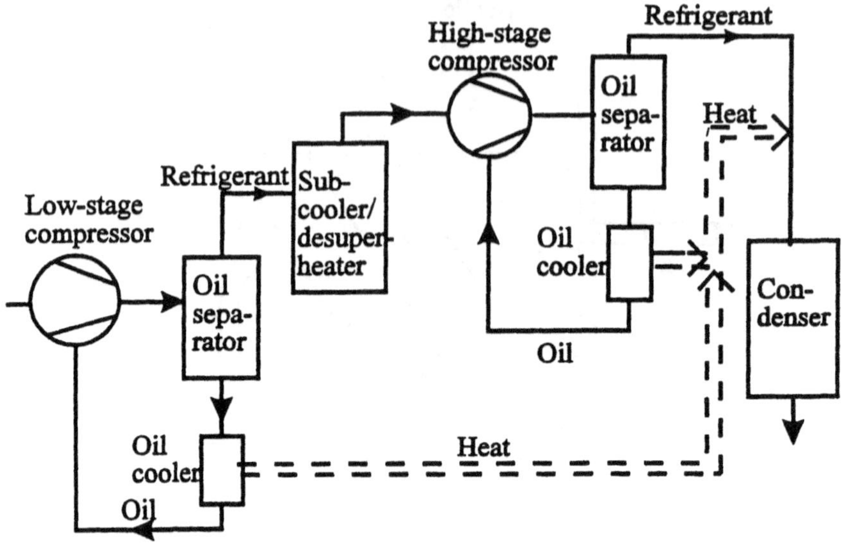

FIGURE 5.32
Transfer of heat from the low-stage oil cooler directly to the condenser when using a thermosyphon or external liquid oil cooler.

that compressor to regulate the slide valve. Because its discharge and suction pressures do not vary much, the variable v_i feature is unnecessary, so the cost of this device and its controlling mechanism can be eliminated.

For capacity control, the slide valve offers a means of quite precise control—usually activated by the suction pressure. When the booster compressor experiences a reduction in the refrigeration capacity, it is not so crucial to reduce the compressor capacity. When serving a food freezer or a frozen food refrigerated space, it is not usually detrimental that the evaporating temperature drops when the load falls off. In fact, the freezing time is shortened in that process, or in the case of a refrigerated storage space, the thermal capacity of the products is so great that little change in temperature results.

Several low-cost approaches to capacity control of low-stage compressors are capacity reduction with simple plug valves that open as needed to vent gas that has been trapped in the cavities back to the suction chamber. The usual application of these valves is to provide a choice of 100%, 75% or 50% capacity. The control of these valves is done with low-cost electromechanical relays, and microprocessor control is not needed. Two methods of still lower first cost, although less efficient, are the installation of a throttling valve in the suction line or the controlled return of discharge gas to the suction line in what is called *hot-gas bypass*. A further use of hot-gas bypass is to be able to start the compressor unloaded, a capability offered to the high-stage compressor through the use of its slide-valve. Still another option to address low loads, for example, during off duty in a food production facility is to arrange the piping in such a way that one compressor operates single-stage during low-production periods. Since screw compressors can operate, although with reduced efficiency, against high pressure ratios, they can function with no damage from −40°C (−40°F) evaporating to 35°C (95°F) discharge in single-stage operation, for example.

Screw booster compressors offer a further economy if the oil is cooled externally in contrast to internal cooling as with the direct injection of refrigerant. Figure 5.32 shows that the heat liberated by the oil in a heat exchanger cooled by the thermosyphon process or by an external liquid cooler passes directly to the condenser. When directly injected refrigerant cools the oil, this heat passes on to the subcooler/desuperheater where the high-stage compressor must pump additional refrigerant in order to transfer this heat to the condenser.

5.17 VIBRATION AND NOISE

It is not reasonable to expect an assembly the size of a screw compressor package that converts a large rate of energy into mechanical and fluid motion to be whisper quiet. The responsibility of designing and building a quiet package rests with the manufacturer and assembler, but the responsibility of an adequate foundation or base is usually under control of the user. The major contributor to noise is usually not directly attributable to the compressor, and most of the noise is likely to be generated by the oil separator, the discharge line, and the motor. These components generate noise primarily because of the pulsations of flow which are characteristic of the screw compressor. Noise generation increases when there is under- or overcompression[14].

The designer, installer, and user have responsibility for providing an adequate foundation or base for the compressor package. The predominant vibration frequency[15] produced in a screw compressor package is the lobe-passing frequency of between 200 and 600 Hz. These frequencies are high enough that an excessive mass of the foundation is not necessary, but the foundation must be able to support the full weight of the compressor package. The floor should be a minimum of 250 mm (6 in) thick in which are imbedded L- or J-type anchor bolts. A 250 mm (6 in) thick housekeeping pad of concrete on top of the floor is recommended.

The mounting of the suction and discharge refrigerant lines is also important. These lines should not be fastened rigidly to the structure in order to avoid transmitting vibrations. Rubber or spring supports may be required, and these supports must be capable of supporting the piping to minimize loading on the compressor flanges and to support the piping when it is disconnected from the compressor. When the compressor is driven by an engine, the engine manufacturer should be consulted for foundation recommendations.

FIGURE 5.33
A screw compressor package. (*Mycom Corporation*)

5.18 SCREW COMPRESSOR PACKAGES

Rather than buying a bare compressor and assembling the associated components in the field, screw compressor packages are almost universally factory-assembled. In the first place, the packager selects properly sized components for the application. Furthermore, assembly in a factory environment will usually provide better control than a field assembly.

A screw compressor package is shown in Figure 5.33. The major components in a package, some of which are noted in Figure 5.33, are the compressor, its motor, coupling, the oil separator, and an oil pump and filter. The microprocessor controller is also installed at the factory so that all the control wiring can be installed by experienced technicians. The oil separator is classified as a vessel, so it is equipped with safety relief valves. The suction line before the compressor is equipped in sequence with the suction stop valve, a check valve, and, just before the compressor, a suction-line filter. At the refrigerant exit from the oil separator are found another check valve and the discharge stop valve.

The purpose of the suction and discharge check valves is to prevent backflow of refrigerant when the compressor shuts down. Backflow could cause rotors to spin backward and damage them. Also backflow may drive oil out of the compressor into the suction line. The engineer selecting the compressor package should be aware that the pressure drop through the suction and discharge stop valves is not usually acknowledged in the catalog ratings of the compressor.

5.19 MOTOR SELECTION

Certain steps are standard when selecting an electric motor[16] for a screw compressor. The motor must accommodate the design load imposed by the compressor on a continuous basis. Even a potential motor overload (when the suction pressure rises above the design value, for example) can be handled by the automatic unloading of the compressor using the slide valve.

In selecting the motor to drive a screw compressor, the power and torque demanded of the motor during start-up must be evaluated carefully. Even though the slide valve is to be in its fully unloaded position, the moments of inertia of the rotors are relatively high. The torque delivered by the motor should be 20% higher than required by the compressor throughout the entire speed range experienced during startup[13]. The draw of current during startup may be five to seven times that occurring during full-speed operation, depending upon the type of starter. Since the heating of the windings is proportional to the square of the electrical current, the rate of motor heating is 25- to 50-times that of normal operation. The duration of this high-current period must be kept short by choosing a drive motor with adequate starting torque.

5.20 THE PLACE OF THE SCREW COMPRESSOR IN THE MARKET

Section 4.2 in the chapter on reciprocating compressors addressed the shift in the market in the past several decades from reciprocating to screw compressors. Since these two types of compressors are virtually the only types used in industrial refrigeration, the roles of the two compressors are interrelated. From the time in the early 1970s when the screw compressor appeared in the refrigeration market, the screw compressor has captured the major share of the industrial refrigeration practice, at least from the standpoint of refrigeration capacity. The screw compressor has proven to be a reliable machine usually capable of 50,000 hours or more of operation between overhauls.

The limitation of pressure ratio of 8 or 9 confronted by the reciprocating compressor does not apply to the screw type, which is capable of performing adequately up to pressure ratios of 20 or more. While the capability of working against a high pressure ratio is available, the benefits of multistage compression usually lead to the choice of multiple compressors and multistage compression. Even in a multistage plant, however, there may be periods of light refrigeration load where it is advantageous to abandon the savings of multistage compression and operate one screw compressor in a single-stage mode from the low evaporating pressure to the condensing pressure.

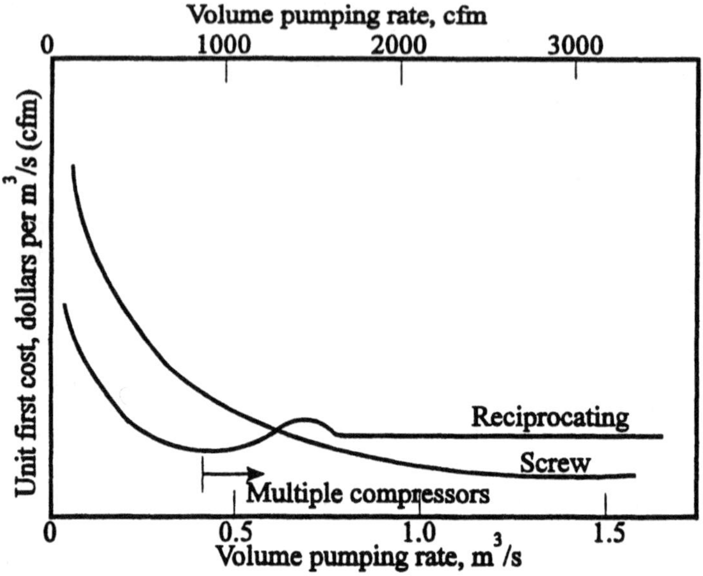

FIGURE 5.34
Comparative first costs of reciprocating and screw compressors.

The maximum pressure sustainable by screw compressors is about 2400 kPa (350 psia), and the compressor can operate against a pressure difference of approximately 1700 kPa (250 psi). The minimum practical suction pressure is of the order of 17 kPa (2.5 psia).

The typical first-cost comparison between the screw and reciprocating compressor is shown in Figure 5.34. A single reciprocating compressor will normally cost less than a single screw compressor. The maximum pumping rate of reciprocating compressors is of the order of 0.33 to 0.47 m^3/s (700 to 1000 cfm), so for higher refrigeration loads two reciprocating compressors are required. At that point and at higher refrigeration capacities the screw compressor is lower in first cost.

A controversial subject is the tolerance of screw compressors to slugging with liquid refrigerant. An enthusiastic promoter of screw compressors might assert that the screw compressor is unaffected by liquid slugging. It may be true that the screw compressor is less sensitive to slugging in comparison to the reciprocating type, but a slug of liquid refrigerant may vaporize and interrupt the flow of injected oil resulting in scored rotors. One engineer recommends an inspection of the machine following any known instance of slugging.

The size and capacity range of the screw compressor extends from the small mini-screw machines to large units driven by motors of 1000 kW or more. The mini-screw unit is available in refrigerating capacities as low as about 100 kW (28 tons of refrigeration). Due to its small-diameter rotors the mini-screw is usually not as efficient as the larger-size units. The mini-screw of one manufacturer operating with -6.7°C (20°F) evaporating temperature and 35°C (95°F) condensing temperature develops a coefficient of performance of 3.94. At corresponding evaporating and condensing temperature, a large compressor exhibits a COP of 4.10, thus, about 4% better efficiency.

5.21 MAINTENANCE AND SERVICE OF SCREW COMPRESSORS

Many users of screw compressors hire service and contracting organizations to perform nonroutine maintenance and repairs on the machines. Major overhauls and service require the attention of specialists, but some important monitoring and service functions can be the responsibility of operating personnel. Valuable data[17] to be recorded and analyzed include: suction pressure, suction temperature, suction superheat, discharge pressure, discharge temperature, oil pressure at the compressor, pressure drop across the oil filter, oil temperature, and oil level. Several of these items, such as the suction pressure and temperature and the discharge pressure are conditions imposed by the system, but they should be normal in order for the compressor to function properly.

A routine maintenance task is to periodically clean the oil filter, for which the shutdown of the compressor is required. The service valves on either side of the filter are first closed, then the purge valve opened to relieve refrigerant pressure. The filter canister is then opened and the cartridge removed for flushing with clean oil. After replacing the cartridge the canister is filled with clean oil, closed, and is then ready for resumption of service.

Another maintenance procedure usually handled by the operating staff is the periodic replacement of the elements in the coalescing separator. The compressor is shut down, the discharge valve closed, and the suction check valve opened in order to vent the refrigerant to the low-pressure region of the system. The coalescing elements are accessible following removal of the manway. A new gasket is provided for the manway upon reassembling.

High quality oil is essential because of the crucial role the oil performs in operation of the compressor. When initially charging oil into the compressor, two samples should be taken and one analyzed in the laboratory as the control reference. The other sample should be kept in a clear glass container and used as the reference basis for clarity of the oil extracted every 2000 hours of operation. A systematic oil analysis program should determine the viscosity of the oil at two temperatures, 38°C (100°F) and 100°C (212°F). In addition the moisture content should be measured as well as mineral content (iron, antimony, phosphorus, copper, and magnesium). The presence of minerals may indicate a problem in the refrigeration system or the compressor itself. All the data from the periodic oil analysis are used as an indication of when to change oil.

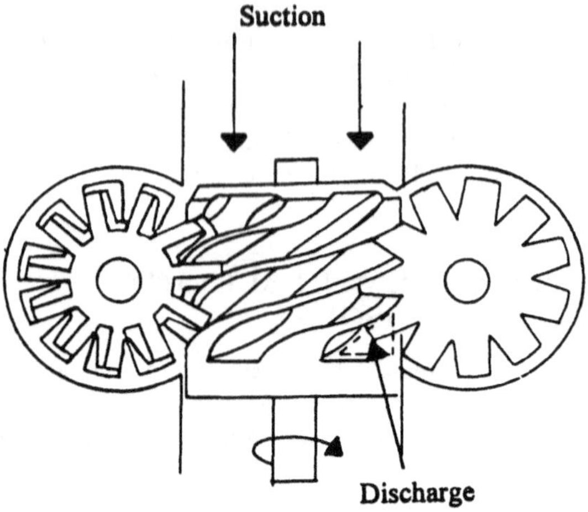

FIGURE 5.35
Screw of the single-screw compressor in the center with the two star wheels on either side.

5.22 SINGLE-SCREW COMPRESSORS

The single-screw compressor was developed by a French physicist/engineer, Bernard Zimmern[18], who began his work on the compressor in the early 1960s. French and U.S. patents were issued in 1964-1965. Initial sales in the U.S. were of machines compressing air, but refrigeration models followed.[19,20]

The screw of the single-screw compressor is a cylindrical member with helical grooves, shown in the center of Figure 5.35. Mated in the screw are two flat star wheels on either side of the screw that rotate in opposite directions from one another.

These star wheels rotate in the plane of the center of the screw shaft. A gas-tight housing encloses the screw and star wheels, Figure 5.36, and the screw rotates with a slight clearance in a cylindrical mantel that forms part of the housing.

The mantel contains two slots in which the star wheels run. Only the screw is driven from the outside, and this screw then drives the two star wheels. Capacity control is provided by a variable-return port controlled by a sliding vane that regulates the position where compression begins.

Compression occurs simultaneously in the upper and lower halves of the compressor. This combined action results in negligible net radial loads on the screw bearings. The only bearing loads in the machine, other than from the weight of the parts, are small loads on the star wheel shafts due to high-pressure gas acting on one side of each tooth during meshing.

The machine, like the twin-screw compressor, has few moving parts—one screw and two star wheels. Manufacurers seek to extend favorable compression efficiencies to smaller sizes than are now appropriate for the twin-screw compressor. In the development of single-screw compressors one of the difficulties

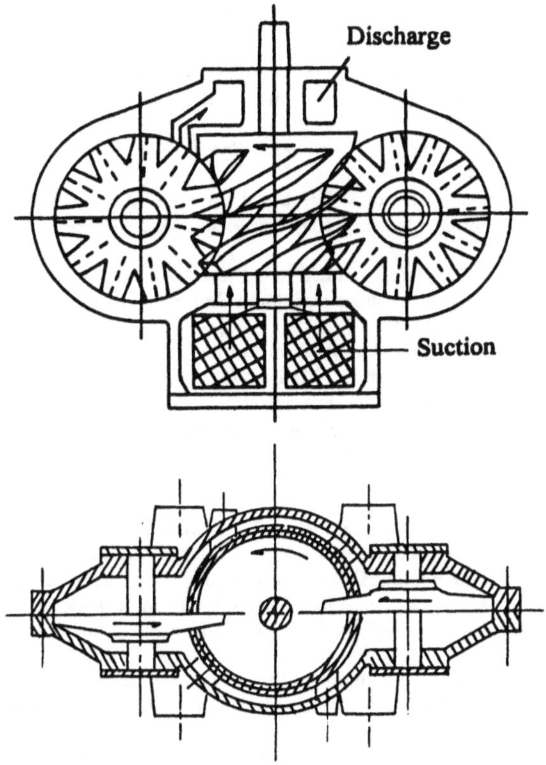

FIGURE 5.36
Top and end sectional views of the single-screw compressor and housing.

has been in discovering materials for the star wheels that resist wear. Currently manufacturers of single-screw compressors seem to prefer a composite of steel and glass-fiber reinforced plastic.

When viewing a screw compressor package, such as the one shown in Fig. 5.33, the oil separator stands out as the largest-size component. Were some concept developed to eliminate this component, the size of the package could shrink. One such method[18,20] is to seal the gaps between the star wheels and the rotor by liquid refrigerant—the same refrigerant that the compressor is pumping. Oil is still required for lubrication, but none is injected for sealing. Compressors designed for sealing by liquid refrigerant are built with smaller clearances between the star wheel and rotor than is true of compressors that are sealed with oil. The concept can be applied to halocarbon refrigerants, such as R-134a and R-22, and for air conditioning applications where the pressure ratio is moderate. The use of liquid refrigerant sealing has not so far been successful in ammonia compressors nor for industrial refrigeration applications.

REFERENCES

1. Dreksler, M. Y, "Compound Screw Compressors Low Temperature Applications", Proceedings of Annual Meeting, International Institute of Ammonia Refrigeration, Washington, DC, 1991.
2. "Screw Compressor Technical Manual," The Stal Refrigeration Group, Bensalem, Pennsylvania, 1983.
3. Pillis, J. W., "Development of a Variable Volume Ratio Screw Compressor," Proceedings of the 8th Annual Meeting of the International Institute of Ammonia Refrigeration, 1983.
4. Sjoholm, L. "Variable Volume-Ratio and Capacity Control in Twin-screw Compressors," Proceedings of the 1986 International Compressor Engineering Conference, Vol. II, Purdue University, 1986.
5. Jansson, L. "Modern Screw Compressors Applied to Liquid Chillers Using Ammonia as a Refrigerant," Proceedings of the 17th Annual Meeting of the International Institute of Ammonia Refrigeration, San Diego, 1995.
6. Shaw, D. N., "Screw Compressors Control of v_i and Capacity—the Conflict," Proceedings of the International Compressor Conference," Purdue University, 1988.
7. TM-VRS-Sec II, Page 8, Vilter Manufacturing Corporation, Milwuakee, Wisconsin, 1989.
8. Vilter V-PLUS^R, Vilter Manufacturing Corporation, Milwaukee, Wisconsin.
9. Data provided by Svenska Rotor Maskiner, AB, Stockholm, Sweden, 1990.
10. "Specifications—Engineering Data—Dimensions, RWBII-Plus Rotary Screw Compressor Units," Frick/Frigid Coil/Imeco Division of York International, 1991.
11. "Thermosyphon Oil Cooling," Frick/Frigid Coil/Imeco Division of York International, 1991.
12. Stegmann, R., "Taking Advantage of the Screw Compressor Side Port," Proceedings of the 16th Annual Meeting of the International Institute of Ammonia Refrigeration, Washington, DC, 1994.
13. "RWB II Plus Rotary Screw Compressor Units—Engineering Data", Frick/Frigid Coil/Imeco Division of York International, 1991.
14. Koai, K and W. Soedel, "Contributions to the Understanding of Flow Pulsations Levels and Performance of a Twin ScrewCompressor Equipped with a Slide Valve and a Stopper for Capacity Control," Proceedings of the International Compressor Engineering Conference, Purdue University, 1990.
15. Pillis, J. W., "How Firm a Foundation for Screw Compressors?", *Air Conditioning, Heating and Refrigeration News,* April 12, 1993.
16. "Vilter VRS Rotary Screw Compressor Technical Manual," Vilter Manufacturing Corporation, Milwaukee, WI, 1989.
17. "Frick RWB II Rotary Screw Compressor Units—Installation, Operation and Maintenance," Frick/Frigid Coil/Imeco Division of York International, 1983.
18. Zimmern, B. "Single Screw Compressor With and Without Oil Injection: A Comparison in the Field of Heat Pumps," Pager B2-125, International Congress of Refrigeration, Paris. 1983.
19. "The Hallscrew Compressor—Selection Guide,: APV Hall Products, Ltd., Dartford, Kent England, 1985.
20. "VSS Single Screw Compressor Units," Vilter Manufacturing Corporation, Milwaukee, Wisconsin, 1994.
21. "McQuay SeasonPak^R ALS Air Cooled Packaged Water Chillers," McQuay—Snyger General, 1994.

CHAPTER 6

EVAPORATORS—AIR COILS AND LIQUID CHILLERS

6.1 WHERE THE REFRIGERATION LOAD ENTERS THE SYSTEM

The evaporator is the component of the refrigeration system where a fluid stream or a product is cooled. As Fig. 6.1 shows, the evaporator is the interface between the process and the refrigeration system. With the exception of direct cooling a product, such as in a plate freezer, most evaporators chill air or such liquids as water, brine, or antifreeze.

6.2 TYPES OF EVAPORATORS

Two major categories of evaporators used in industrial refrigeration practice are air coils and liquid chillers. A typical air coil is shown in Fig. 6.2 and several types of liquid chillers are shown in Fig. 6.3. In the air coil, refrigerant flows through the tubes and air passes over the outside of the tubes. For effective heat transfer, fins are fastened to the outside of the tubes and the air flows between the fins.

The liquid-chilling evaporators in Figs. 6.3a and 6.3b are of the shell-and-tube design, while Figs. 6.3c is a plate-type chiller. In Fig. 6.3a the refrigerant boils in the shell while the liquid flows through the tubes. In Fig. 6.3b the roles of

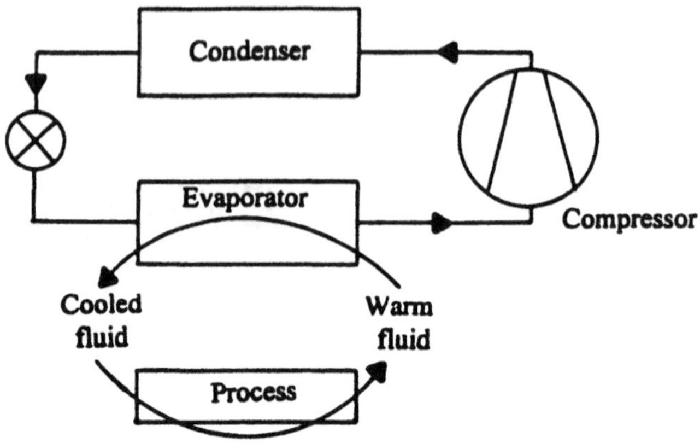

FIGURE 6.1
Cooling of a fluid stream at the evaporator.

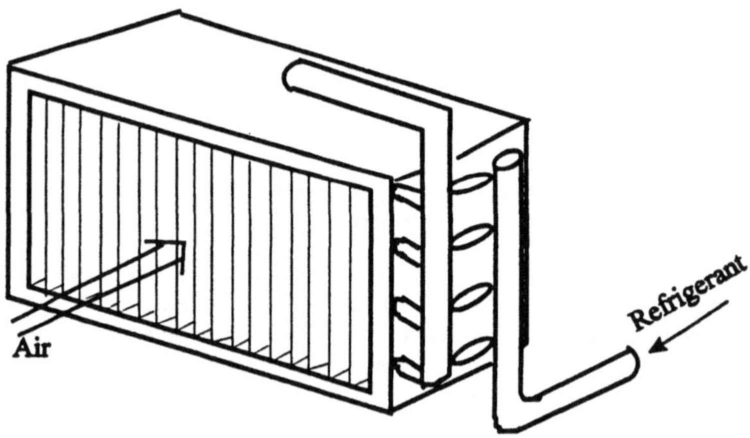

FIGURE 6.2
An air cooling coil.

the tubes and shell are reversed. The plate-type evaporator in Fig. 6.3c is growing in market share and is an adaptation of the plate-type heat exchanger used for many years in the food industry. Some of its popularity is attributable to its compactness and also that the refrigerant charge is less than in a shell-and-tube evaporator. The reduced charge characteristic is attractive both when ammonia and the new chlorine-free refrigerants are used. A low-charge ammonia system is desirable for safety reasons, and a low-charge chlorine-free system minimizes the cost of these expensive refrigerants.

This chapter is divided into three parts: Part I, Evaporator Performance, addresses the principles applicable to all types of evaporators; Part II, Air Coils,

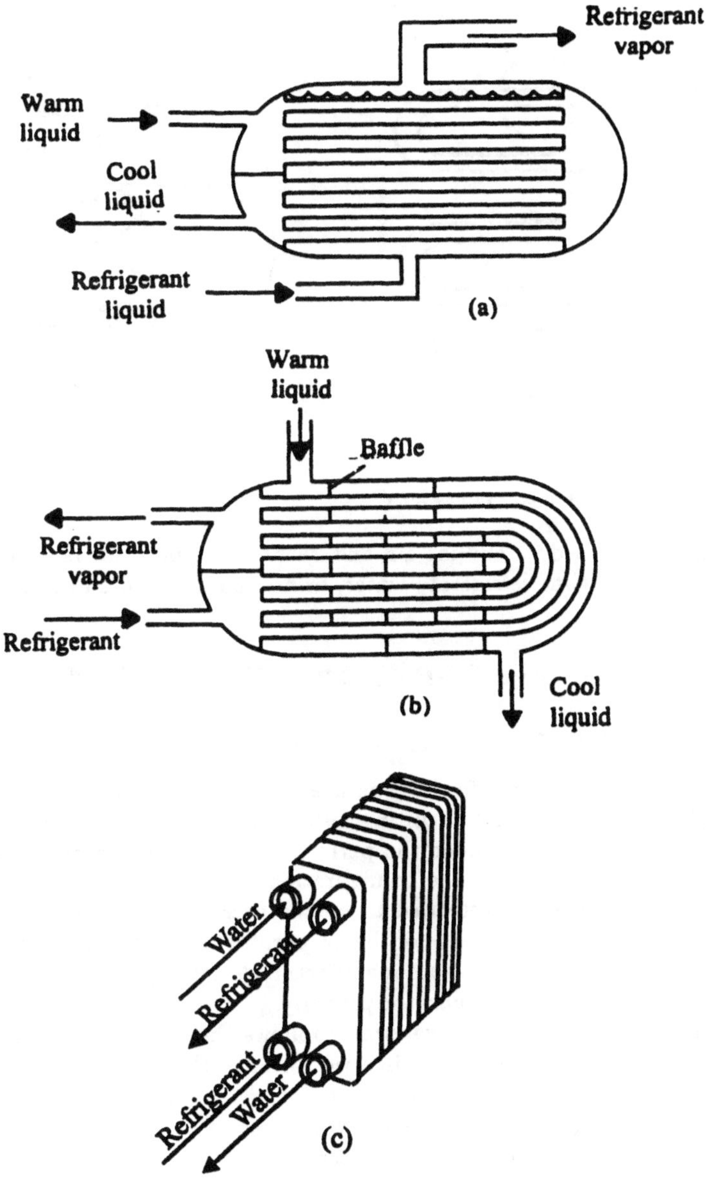

FIGURE 6.3
Liquid-chilling evaporators: (a) shell-and-tube with the refrigerant boiling in the shell, (b) shell-and-tube with the refrigerant boiling in the tubes, and (c) a plate-type evaporator.

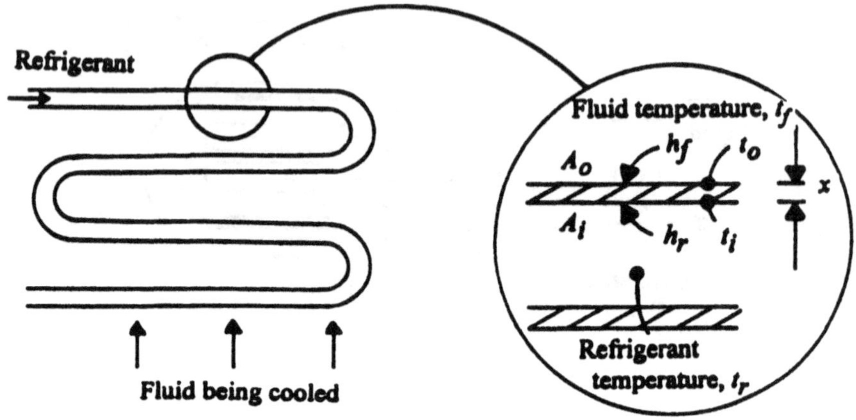

FIGURE 6.4
Heat-transfer coefficients in an evaporator.

treats evaporators used for chilling air, and Part III, Liquid Chillers, concentrates on the evaporators that refrigerate water, brines, antifreezes, and other liquids.

PART I. EVAPORATOR PERFORMANCE

6.3 EVAPORATOR HEAT TRANSFER

This chapter is not intended for the engineers of evaporator manufacturers who must decide on the refrigerant circuiting, tube arrangement, and other important details. These designers consider the performance, cost of materials, and ease of manufacturing in deciding the configuration of the evaporator. Instead the audience of this chapter is intended to be the user of evaporators who should understand the basic performance of the evaporator as a heat exchanger, how to properly select the evaporator from manufacturer's catalogs, and how to install, operate, and maintain the evaporator properly.

As a heat exchanger, the evaporator follows the rules of heat transfer. In the evaporator of Fig. 6.4 heat flows in series through three resistances—the fluid side, the metal of the tube, and the refrigerant.

Many engineers visualize a heat-flow process as an analogy to the electric flow process with the correspondence of terms shown in Table 6.1.

Ohm's law for electricity states that

$$I = \frac{V}{R}$$

The corresponding heat-flow equations expressed in terms used in Fig. 6.4 are:

TABLE 6.1
Electrical and heat-flow analogy.

Quantity	Symbol and units	
	Electrical	Heat flow
Flow rate	I, ampere	q, W (Btu/hr)
Potential	V, volts	Δt, °C (°F)
Resistance	R, ohm	x/kA or $1/hA$, °C/W (°F/Btu per hr)

fluid to outside surface of the tube,

$$q = \frac{t_f - t_o}{\frac{1}{h_f A_o}} \tag{6.1}$$

outside of tube to inside of tube

$$q = \frac{t_o - t_i}{\frac{x}{kA_{mean}}} \tag{6.2}$$

inside of tube to refrigerant

$$q = \frac{t_i - t_r}{\frac{1}{h_r A_i}} \tag{6.3}$$

where A_o = external area of evaporator tube, m^2 (ft^2)
A_{mean} = mean cylindrical area of evaporator tube, m^2 (ft^2)
A_i = inside area of tube, m^2 (ft^2)
h_f = heat-transfer coefficient on fluid side, $W/m^2 \cdot °C$ ($Btu/hr \cdot ft^2 \cdot °F$)
h_r = heat-transfer coefficient on refrigerant side, $W/m^2 \cdot °C$ ($Btu/hr \cdot ft^2 \cdot °F$)
k = conductivity of the tube, $W/m \cdot °C$ ($Btu/hr \cdot ft \cdot °F$)
q = rate of heat transfer, W (Btu/hr)
t_f = fluid temperature, °C (°F)
t_o = temperature of outside surface of tube, °C (°F)
t_i = temperature of inside surface of tube, °C (°F)
t_r = temperature of refrigerant °C (°F)
x = thickness of tube, m (ft)

Application of the electrical analogy provides a simple means of deriving an expression for the overall U-value, where U is a term which when multiplied by the overall temperature difference, $t_f - t_r$, and the area yields the rate of heat transfer q, W (Btu/hr).

$$q = UA(t_f - t_r) \tag{6.4}$$

where U = overall heat-transfer coefficient, $W/m^2 \cdot °C$ ($Btu/hr \cdot ft^2 \cdot °F$).

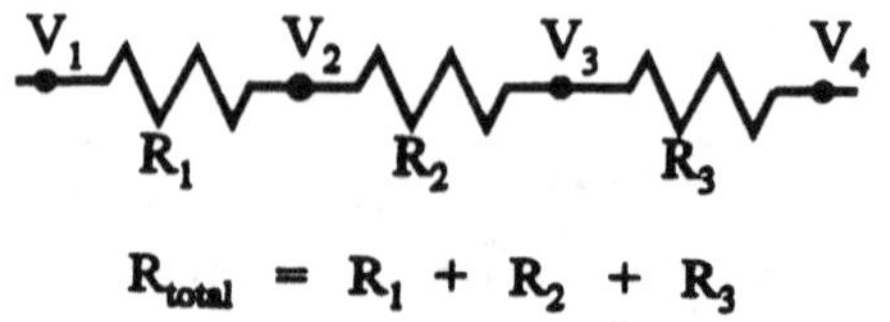

FIGURE 6.5
Total electrical resistance is the sum of the individual resistances.

When electrical resistances are connected in series, as in Fig. 6.5, the resistance of the combined circuit is the sum of the individual resistances,

$$R_{total} = R_1 + R_2 + R_3$$

so the current flow through the circuit I is the driving force, $V_1 - V_4$, divided by the total resistance,

$$I = \frac{V_1 - V_4}{R_{total}} = \frac{V_1 - V_4}{R_1 + R_2 + R_3}$$

When translating the resistance concept of electricity to heat flow, the resistances to the rate of heat flow in each of the stages are the denominators of Eqs. 6.1, 6.2, and 6.3, respectively. The total heat-transfer resistance is the sum of the individual resistances, each of which incorporates its own unique area.

$$R_{total} = \frac{1}{h_f A_o} + \frac{x}{k A_{mean}} + \frac{1}{h_r A_i}$$

then

$$q = \frac{t_f - t_r}{R_{total}} = \frac{t_f - t_r}{\frac{1}{h_f A_o} + \frac{x}{k A_{mean}} + \frac{1}{h_r A_i}} \tag{6.5}$$

From a comparison of Eqs. 6.4 and 6.5, the expression for the U-value is found to be:

$$\frac{1}{UA} = \frac{1}{h_f A_o} + \frac{x}{k A_{mean}} + \frac{1}{h_r A_i} \tag{6.6}$$

When heat flows through stages where different magnitudes of area exist, a unique U-value is associated with each area:

$$\frac{1}{UA} = \frac{1}{U_o A_o} = \frac{1}{U_i A_i} \tag{6.7}$$

Example 6.1. What is the U-value of an evaporator formed of steel pipe when the fluid-side (air) heat-transfer coefficient is 60 W/m^2·°C (10.6 Btu/hr·ft^2·°F) and the heat-transfer coefficient on the refrigerant side is 1200 W/m^2·°C (211

Btu/hr·ft^2·°F). The pipe has an outside diameter of 26.7 mm (1.05 in) and an inside diameter of 20.9 mm (0.824 in). The thermal conductivity of steel is 45 W/m·°C (26 Btu/hr·ft·°F).

Solution. Arbitrarily decide to compute the U-value based on the inside area, so a combination of Eqs. 6.6 and 6.7 provides an expression for U_i:

$$\frac{1}{U_i} = \frac{A_i}{h_f A_o} + \frac{x A_i}{k \frac{A_o + A_i}{2}} + \frac{1}{h_r} \tag{6.8}$$

The ratios of the cylindrical areas through which heat flows are the same as the diameter ratios,

$$\frac{A_i}{A_o} = \frac{20.9}{26.7} = 0.783 \quad \text{and} \quad \frac{A_i}{(A_o + A_i)/2} = \frac{20.9}{(26.7 + 20.9)/2} = 0.878$$

The tube thickness is

$$x = \frac{26.7 - 20.9}{2} = 2.9 \text{ mm} = 0.0029 \text{ m } (0.114 \text{ in})$$

The values of the resistances in Eq. 6.8 are

$$\frac{1}{U_i} = \frac{0.783}{60} + \frac{(0.0029)(0.878)}{45} + \frac{1}{1200}$$

Then

$$U_i = \frac{1}{0.01305 + 0.000057 + 0.000833}$$

$$U_i = 71.9 \text{ W/m}^2\cdot°\text{C } (12.7 \text{ Btu/hr}\cdot\text{ft}^2\cdot°\text{F})$$

The U-value based on the outside area is:

$$U_o = U_i \left(\frac{A_i}{A_o}\right)$$

$$U_o = 56.3 \text{ W/m}^2\cdot°\text{C } (9.9 \text{ Btu/hr}\cdot\text{ft}^2\cdot°\text{F})$$

6.4 EXTENDED SURFACE (FINS)

A comparison of the magnitudes of the various heat-transfer resistances in Example 6.1 shows that the major resistance is on the fluid side, where the fluid in this case is air. The heat-transfer resistance on the air side is 0.01305 m^2·°C/W (0.0741 hr·ft^2·°F/Btu) while on the refrigerant side the resistance is 0.000833 m^2·°C/W (0.00473 hr·ft^2·°F/Btu), so the air-side resistance is 20 times that of the refrigerant side. This comparison suggests immediately where attention should be directed if an increase in the U-value were desired. Certainly not on the refrigerant side, because even if the coefficient could somehow be doubled, the U-value would increase by only 3%. No, the resistance to attack is that on the

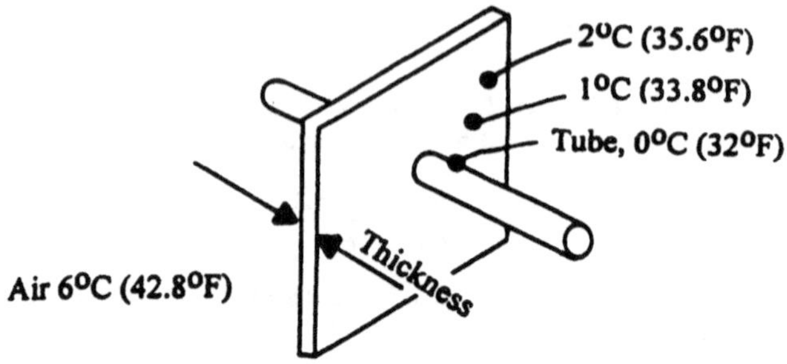

FIGURE 6.6
A fin on a tube of an air coil illustrating progressive reduction in effectiveness of the fin surface further from the tube.

air side, and this resistance can be decreased only by increasing h_f or increasing the area ratio.

Several methods of increasing h_f are to increase the air velocity and to increase the turbulence by introducing irregularities in the heat-transfer surface. Increasing the air velocity will increase h_f, but at the expense of additional fan power. The coil manufacturer seeks the optimum air velocity which provides a respectable h_f, but requires a reasonably sized fan and motor. A further consideration affected by the fan power is that the power to the fan motor ultimately appears as refrigeration load, and perhaps 10 to 20% of the heat removed by the air coil was ultimately introduced by the fan and its motor.

The standard approach to reducing the air-side resistance is to increase the area ratio A_o/A_i by the application of extended surface or fins. The air coil shown in Fig. 6.2 is equipped with fins which are formed from flat metal plates that are then punched and the tubes inserted in the holes. When the tubes are in position, they are expanded either hydraulically or mechanically to provide good thermal contact between the tube and fin. The section of fin associated with one tube is shown in Fig. 6.6 in a situation where the tube temperature is 0°C (32°F) and the air temperature is 6°C (42.8°F). If the entire fin were at the same temperature as the tube, namely 0°C (32°F), the resistance on the air side would be as stated in Eq. 6.8, $A_i/h_f A_o$. Figure 6.6 indicates, however, that the fin temperature increases at positions progressively further removed from the tube.

So all of the air-side area A_o is not 100% effective. The effectiveness of the fin is usually given the symbol η and the equation for the U-value of the finned coil is:

$$\frac{1}{U_i} = \frac{A_i}{h_f A_o \eta} + \frac{x A_i}{k A_{mean\ tube}} + \frac{1}{h_r} \qquad (6.9)$$

The values of η for commercial coils generally range between 0.3 and 0.7, and the effectiveness is a function of such factors as the choice of fin material

TABLE 6.2
Influence of some choices of fin dimensions and materials on the fin effectiveness η and the overall heat-transfer capacity of the coil.

Increase of variable	Effect on fin effectiveness η	Effect on overall heat transfer of coil	Effect on cost
Distance between tubes	Decrease	Increase	Increases
Air-side heat-transfer coefficient	Decrease	Increase	Increase of fan power
Thermal conductivity	Increase $k_{aluminum} = 4k_{steel}$	Increase	See Sec. 6.26
Fin thickness	Increase	Increase	Increase because of more metal

(usually steel or aluminum), fin thickness, and distance from the tube. The effects of several choices on fin effectiveness and on the overall heat-transfer capacity of the coil are shown in Table. 6.2. The designer for the coil manufacturer spends nights trying to juggle these various decisions to provide the maximum heat-transfer rate for a given cost of the coil.

Most of the foregoing discussion has implied that the evaporator is an air-cooling coil, and indeed fins are almost universal on air coils, but not always. In some food plant applications the evaporator is composed of unfinned tubes. The reason is that these coils are easier to clean for hygienic purposes, even though the heat-transfer rate suffers. It should be pointed out that while the bare-tube coil is easier to clean, the coils must often be deep with a large number of rows of tubes in the air-flow direction which complicates cleaning.

A further comment to complete the discussion of extended surface is that fins are sometimes used on the refrigerant side in liquid-chilling evaporators. In water-chilling evaporators where refrigerant flows in the tubes, the tubes are often equipped with inner fins that are sometimes rifled. This style is especially adaptable to copper tubes which can be used in halocarbon refrigeration systems, but not ammonia. For ammonia systems, aluminum tubes can be provided with internal and/or external integral fins, and even for steel tubes the possibility of external integral fins exists.

6.5 TEMPERATURE DIFFERENCE: BETWEEN ENTERING FLUID AND REFRIGERANT

Engineers who expect the appearance of the logarithmic-mean-temperature difference or complex heat-exchanger effectiveness factors to be encountered in selecting and analyzing the performance of evaporators will be pleased to know of a simplification. The streamlined process is made possible by the fact that the

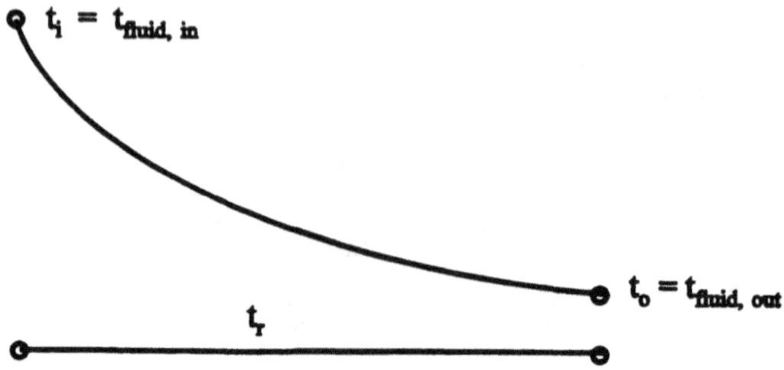

FIGURE 6.7
Distributions of the temperatures of the refrigerant and the fluid being chilled in an evaporator.

temperature of one of the fluids (the refrigerant) remains constant through the evaporator. The temperature distributions with a constant refrigerant temperature are shown in Fig. 6.7.

The rate of heat transfer is the product of UA and the log-mean-temperature difference,

$$q,\ \text{kW (Btu/hr)} = UA\left[\frac{(t_i - t_r) - (t_o - t_r)}{\ln[(t_i - t_r)/(t_o - t_r)]}\right] \tag{6.10}$$

$$q = UA\left[\frac{t_i - t_o}{\ln[(t_i - t_r)/(t_o - t_r)]}\right] \tag{6.11}$$

Since the fluid-flow rate $\dot{m}$ and the specific heat c_p remain constant, and $q = \dot{m}c_p(t_i - t_o)$, Eq. 6.11 becomes

$$\frac{q}{UA(t_i - t_o)} = \frac{\dot{m}c_p}{UA} = \frac{1}{\ln[(t_i - t_r)/(t_o - t_r)]}$$

Also

$$\ln\left(\frac{t_i - t_r}{t_o - t_r}\right) = \frac{UA}{\dot{m}c_p} \quad \text{and} \quad \frac{t_i - t_r}{t_o - t_r} = e^{UA/\dot{m}c_p}$$

Rearranging,

$$(t_i - t_r)[e^{-(UA/\dot{m}c_p)}] = t_o - t_r = -[(t_i - t_o) - (t_i - t_r)] = -[\frac{q}{\dot{m}c_p} - (t_i - t_r)]$$

and

$$q = \dot{m}c_p[1 - e^{(-UA/\dot{m}c_p)}](t_i - t_r) = (Factor)(t_i - t_r) \tag{6.12}$$

Equation 6.12 shows that for a given evaporator operating with a constant flow rate that the term *Factor* is constant and the rate of heat transfer q is proportional to the temperature difference between the temperatures of the entering fluid and the refrigerant. In their catalogs, then, manufacturers only need to present the term *Factor* for each evaporator and each fluid flow rate.

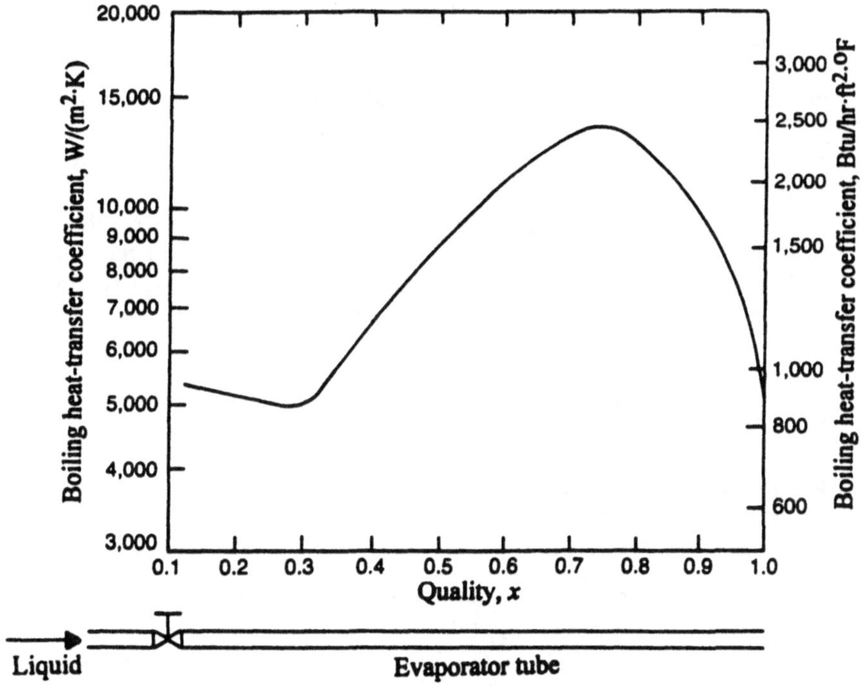

FIGURE 6.8
Heat-transfer coefficient as refrigerant boils in an evaporator tube[1].

6.6 REFRIGERANT BOILING INSIDE TUBES

The mechanism of refrigerant boiling inside a tube of an evaporator is complex. There are more than 4000 technical papers on the subject, so a person could make a career of studying boiling heat-transfer. Even the prediction of heat-transfer coefficients for a given refrigerant in a certain size tube with a specified flow rate is difficult. Fortunately, the designer and manufacturer of the evaporator take over that task, and the engineer who selects or uses the evaporator normally does not need to. However, the application engineer should understand what occurs during the boiling process.

One evaporator concept is *direct expansion*, discussed further in Sec. 6.27. The direct-expansion evaporator receives refrigerant from the expansion valve with a small fraction of vapor as shown in Fig. 6.8.

As the warm tube adds heat to the refrigerant, progressively more refrigerant evaporates, and the velocity increases until the refrigerant leaves the evaporator saturated or superheated. Figure 6.8 also shows typical boiling heat-transfer coefficients corresponding to the position along the evaporator tube. The changes in the heat-transfer coefficient are associated with differing patterns of flow[2] as the fraction of vapor and the velocity change along the tube. At the entering section of the evaporator, bubbles and plugs of vapor flow along with the liquid. Further along the tube, the flow becomes annular with high-velocity vapor rushing through the center and the liquid clinging to the inside surface of the tube. Still later in the evaporator, the flow converts to a mist and eventually there could be a nonequilibrium mixture of superheated vapor and liquid until all the liquid finally evaporates.

The benefit of being aware of a distribution of heat-transfer coefficients as in Fig. 6.8 is to understand why such concepts as liquid recirculation (Sec. 6.7) have some heat-transfer advantages and to be able to diagnose operating problems attributable to the refrigerant-side heat transfer.

6.7 METHODS OF SUPPLYING REFRIGERANT TO EVAPORATORS—DIRECT EXPANSION, FLOODED EVAPORATORS AND LIQUID RECIRCULATION

The techniques for feeding the evaporators with refrigerant provide one form of categorizing the types of evaporators.

Direct expansion. In what is referred to as direct expansion, liquid refrigerant enters the expansion valve and only vapor leaves the evaporator, as shown in Fig. 6.9. One of the most popular types of expansion valves that facilitates this control is the *superheat-controlled valve*, which is also called a thermo-valve, thermostatic valve, or a TXV. More on this type of control will be covered in Chapter 11 on valves and refrigerant controls, but it is sufficient at this point to explain that the valve controls the flow rate of refrigerant such that the vapor leaves the evaporator superheated by from 4 to 7°C (7 to 12 °F). Direct expansion is limited to evaporators where the refrigerant evaporates in the tubes.

If the sensing bulb detects a higher-than-setpoint superheat at the evaporator outlet, the valve opens further. The evaporator fed by a superheat-controlled expansion valve is probably the lowest in first cost of the three methods described here. It is used widely with halocarbon refrigerants at moderate refrigerating temperatures, but its use is limited in low-temperature applications and for ammonia. More on the possibilities of direct-expansion coils with ammonia will be found in Sec. 6.27.

Flooded evaporator. The flooded evaporator, shown in Fig. 6.9, relies on natural convection to circulate more refrigerant through the evaporator than what evaporates. All inside surfaces of the evaporator are thus wetted with liquid refrigerant. The vapor formed in the evaporator is separated in the surge drum and flows to the suction line.

A level-control valve admits liquid refrigerant to replace the amount vaporized. The difference in static pressure in the liquid leg is greater than that of the mixture of vapor and liquid in the evaporator tubes, and this difference in

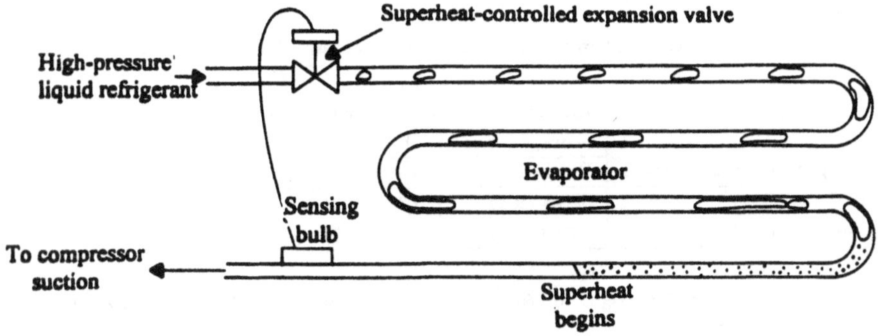

FIGURE 6.9
Superheat-controlled (thermovalve) expansion valve in a direct-expansion valve/coil arrangement.

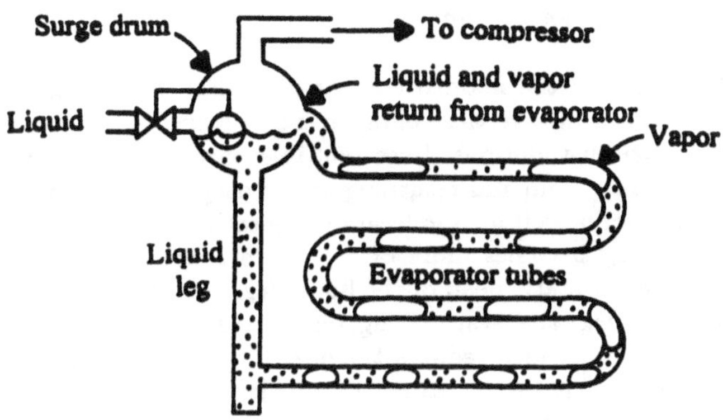

FIGURE 6.10
A flooded evaporator.

pressure is the motivation for the flow of refrigerant.

Several advantages of flooded evaporators in comparison to direct expansion are:

- the evaporator surfaces are used more effectively because they are completely wetted
- problems in distributing refrigerant in parallel-circuit evaporators are less severe
- saturated vapor rather than superheated vapor enters the suction line, so the temperature of suction gas entering the compressor is likely to be lower, which also reduces the discharge temperature from the compressor.

TABLE 6.3
Penalties in evaporating temperature due to static head of liquid in the evaporator.

	Increase in evaporating temperature, °C per m (°F per ft)	
Evaporating temperature	Refrigerant-22	Ammonia
0°C (32°F)	0.774 (0.425)	0.392 (0.215)
-40°C (-40°F)	2.81 (1.54)	1.77 (0.97)

Several disadvantages of the flooded evaporator in comparison to the direct-expansion evaporator are:

- the first cost is higher
- more refrigerant is needed to fill the evaporator and surge drum
- oil is likely to accumulate in the surge drum and evaporator and must be periodically or continuously removed.

The configuration of the flooded coil shown in Fig. 6.10 where the tubes of a circuit are arranged in a vertical plane exists in some applications, such as in ice thermal storage installations. Here multiple parallel circuits are located side by side each feeding from and returning to the surge drum or header pipes. The vertical circuit works well for moderate evaporating temperatures, but the height of the liquid leg becomes more of a consideration for low-temperature evaporators. Because the static head of the liquid leg results in a higher pressure at the bottom of the leg, the evaporating temperature is higher at the lower section of the evaporator, which reduces the heat-transfer rate. Table 6.3 shows penalties in the evaporating temperature per unit length of static head for R-22 and ammonia at two different evaporating temperatures. Because liquid R-22 is more dense than liquid ammonia, the penalty for a given magnitude of liquid head is greater for R-22.

In low-temperature evaporators it becomes more crucial to keep the static head at a minimum. For this reason, evaporator coils are often constructed as shown in Fig. 6.11, where the evaporating sections of the circuits rise, as necessary, but on an incline. The compromise that emerges is that the Δp that motivates the flow is also proportional to the difference in elevation between the liquid in the surge drum and that of the bottom of the tubes.

Associated with this compromise is how much elevation of the surge drum above the coil to provide. Often the coil is installed beneath the ceiling and a large vertical distance between the surge drum and the coil poses headroom problems. A survey[3] of some coil manufacturers revealed that typical distances from the bottom of the surge drum to coil to be between 0.15 and 0.25 m (6 to 10 in). When physical restrictions force other arrangements, such as placing the surge drum above the roof in order to conserve headroom in the refrigerated space, this information should be communicated with the coil manufacturer.

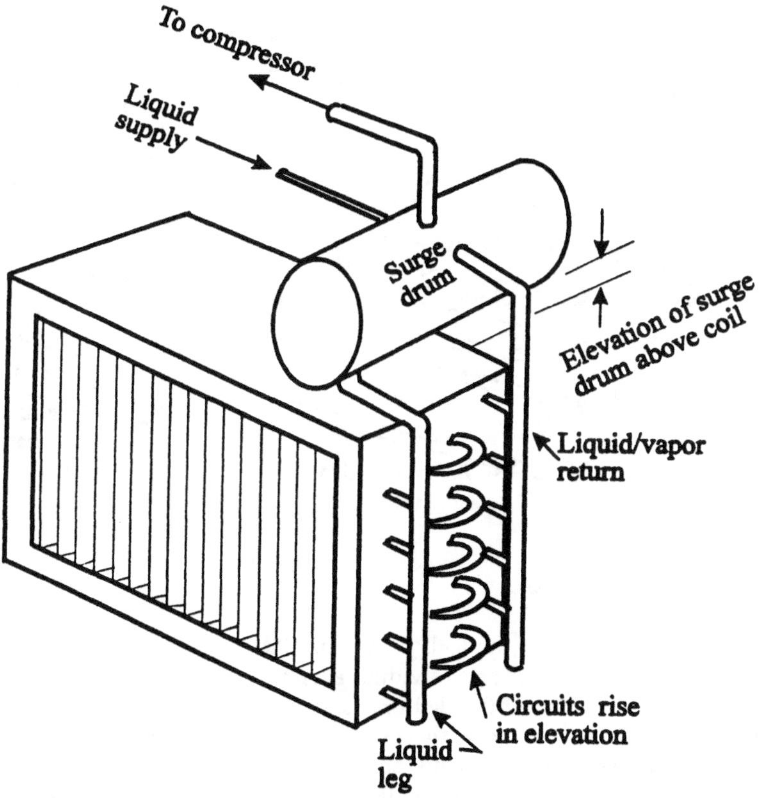

FIGURE 6.11
A flooded air-cooling coil for low-temperature application.

One coil manufacturer[4] bases its design on circulating twice as much liquid as evaporates and on a coil pressure drop of approximately 1.4 kPa (0.2 psi) and arrives at a height from the liquid level in the surge drum to the top of the coil of approximately 0.46 m (1.5 ft).

Liquid recirculation. The third major method of feeding evaporators is forced liquid recirculation, as shown in Fig. 6.12, where liquid from either a mechanical pump or a gas-pressure pumping arrangement supplied the evaporators.

A greater liquid flow rate enters the evaporator than is evaporated, so a mixture of liquid and vapor leaves the evaporator and passes back to a vessel that separates vapor from liquid. The separated liquid, along with the makeup liquid admitted through the level-control valve, drops to the pump for recirculation to the evaporators. Both forced-recirculation systems and flooded coils boast the advantages of good heat transfer and saturated vapor passing to the compressor. The assurance of overfeeding under varying loads is probably stronger with forced

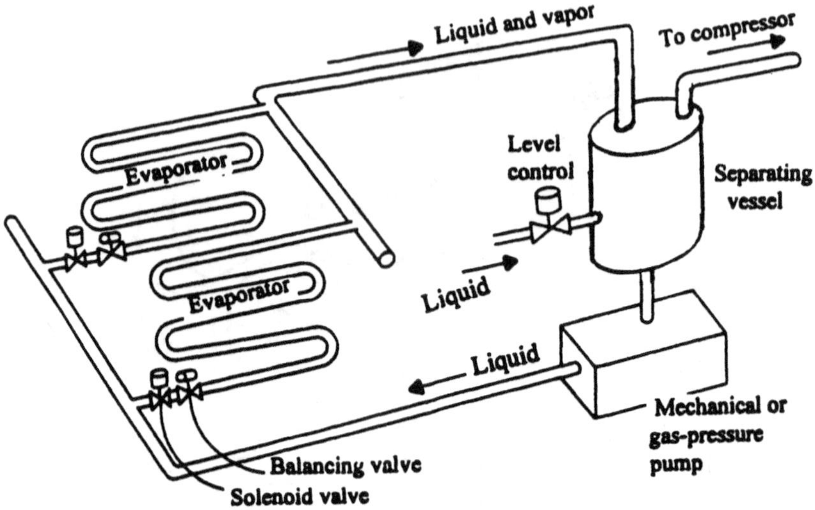

FIGURE 6.12
Forced liquid recirculation.

recirculation, however, and all oil accumulates at one location—the separating vessel—rather than in the individual evaporators.

Properly designed flooded evaporators and evaporators operating with liquid recirculation operate with equal effectiveness. What, then, is the basis for choosing between flooded coils and a liquid recirculation system serving multiple coils? With a liquid recirculation system the pump and separating vessel add to the first cost, which is not true of flooded coils. On the other hand the maintenance cost of the system using flooded coils is usually high because of the multiple locations where oil must be removed and because of the maintenance cost on the controls of a large number of evaporators. These considerations lead to favoring flooded coils for small systems (perhaps three or four evaporators) and pumped liquid recirculation for larger installations.

6.8 OPTIMUM EVAPORATING TEMPERATURES

The size of the evaporator selected is dictated by the design temperature difference, namely the temperature difference between the entering fluid and the refrigerant. Typical values for this temperature difference are discussed in Sec. 6.16 on the selection of air-cooling coils. In many situations the choice of this temperature difference, and thus the size of the evaporator, is purely economic, with trends as shown in Fig. 6.13. An increase in evaporator area results in a lower temperature difference (high evaporating temperature) with the consequent saving in compressor power over the life of the facility.

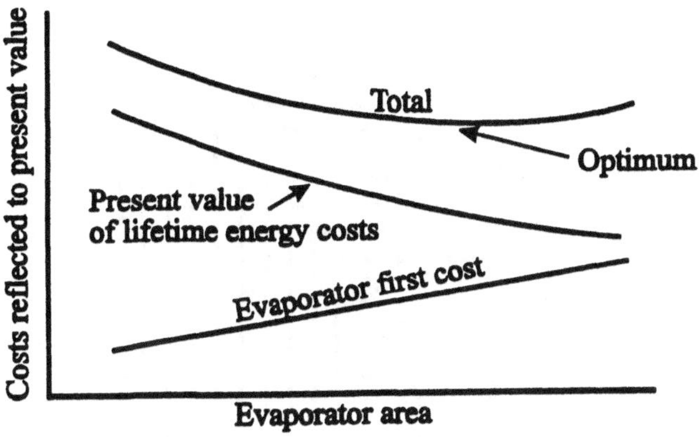

FIGURE 6.13
Optimum evaporator area for minimum total of the first cost of the evaporator and the present worth of the lifetime compressor energy cost.

The compressor energy costs which occur throughout the life of the system are reflected back to a present value using an applicable interest rate.

PART II. AIR COILS

6.9 CONSTRUCTION OF AIR COILS

The number of air-cooling coils in operation in industrial refrigeration plants far exceeds the number of liquid-chilling evaporators installed. Before explaining the performance of air coils, the physical features of the several types of industrial refrigeration coils will be presented. The major components of air coils are tubes, tube sheets, fins, and drain pan.

Tubes. The tubes are pipes that enclose the refrigerant. The most common materials used for tubes are carbon steel, copper, aluminum, and stainless steel. If ammonia is the refrigerant any of the four materials except copper may be used, and most halocarbon systems apply coils with copper tubes. The most common sizes of steel-tube coils for ammonia service are 3/4, 7/8 and 1 in, although 5/8 inch tubes are also used. For smaller halocarbon coils, 1/2-in copper tubes are sometimes used.

Tube sheets. At each end of the coil a heavy plate supports the tubes by having holes through which the tubes pass. The pattern of these holes defines whether the tubes are *in-line* or *staggered*. The coil elements in Fig. 6.14 depict a staggered pattern for the tubes. A coil with a staggered-fin pattern enjoys slightly improved heat transfer, but at the expense of a slight increase in air pressure drop.

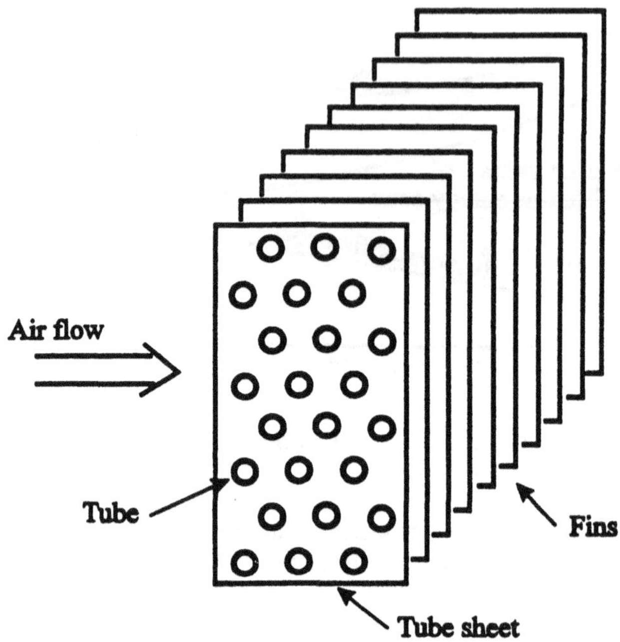

FIGURE 6.14
Tube sheet and fins of a coil with the tubes in a staggered pattern.

Fins. Section 6.4 explained from a heat-transfer standpoint why extended surface, or fins, are a logical feature of air-cooling coils. These fins may be appllied by wrapping a strip of metal in a helical fashion around the tube and then bonding it to the tube. Much more common, however, is the use of plate or flat fins, as appear in Fig. 6.14. The materials available for these fins are the same as for tubes, and typical combinations of tube/fin materials are:

- copper tube/aluminum fin for halocarbon air-cooling coils
- aluminum tube/aluminum fin for halocarbon or ammonia air-cooling coils
- carbon steel tube/carbon steel fin for air-cooling coils using ammonia, halocarbons, antifreezes or water in the tubes
- stainless steel tube/stainless steel fin when special cleaning provisions are required on the air side

Stainless steel is usually used only for extremely low temperature, where there is a corrosive atmosphere, or whenever periodic cleaning is necessary. The thermal conductivity is less than that of carbon steel which itself is about one-fourth that of aluminum. The cost of a stainless steel coil may be five or more times that of a steel coil of comparable size.

The application of the coil, particularly whether it will become frosted, determines to a large extent the spacing of fins. In air conditioning coils with thin aluminum fins, the spacing may be 470 per m (12 fins per inch, FPI), while industrial coils are usually built with 118 or 158 fins per m (3 or 4 FPI). Coils serving spaces where the air temperature is below freezing usually have a fin spacing of 118 per m (3 per inch).

Bonding of the fins to the tubes. The fin must form a good bond to the tube, otherwise there will be additional heat-transfer resistance through air gaps. A steel tube/steel fin coil will be galvanized, which is a process whereby the entire coil is dipped in molten zinc. The zinc provides a protective surface against corrosion and also gives an effective bond between the tube and fin. For non-galvanized coils the tubes are usually expanded against the collar of the fin to yield a tight fit. The tube is usually expanded by forcing a hardened ball at the end of a rod through the tube after the fin plates are stacked on the tubes.

Circuiting of the coil. In direct-expansion coils using halocarbon refrigerants, the general direction of the flow of refrigerant through the circuits is downward, while in order for flooded coils to function properly the general direction of the refrigerant flow must be upward. Forced liquid recirculation coils may be circuited either as upward flow (bottom feed) or as downward flow (top feed), as illustrated in Fig. 6.15. The coil designer chooses the length of the circuit such that the refrigerant when flowing with appropriate velocity receives enough heat in passing through the circuit to vaporize the desired fraction of refrigerant.

Orifices in liquid recirculation coils. One refrigerant circuit in the coils of Fig. 6.15 consists of six passes back and forth through the coil. There are a number of parallel passes, and the upper circuits in a coil are prone to receive an inadequate flow of liquid. In order to strive toward an equal distribution of refrigerant flow, orifices are placed at the entrance of each circuit, as shown by the cutaway in Fig. 6.15a. These orifices, as shown in Fig. 6.16, are thin metal discs with a hole. The holes are usually located eccentrically near the bottom so that oil which accumulates in the coil during refrigeration operation may more easily flow out of the coil during hot-gas defrost. Often the diameter of the orifices is greater for the upper circuits to achieve uniform distribution of refrigerant.

Drain pan for a low-temperature coil. All coils are equipped with drain pans, because in the normal operation of a cooling coil that operates above freezing temperature, some water vapor will be condensed from the air. This condensate is collected by the drain plan and drained to some convenient destination. Coils that operate below freezing temperatures must be defrosted periodically, and once again a drain pan is necessary to collect the melted frost to drain it away. The drain pan must be kept warm so that the melted frost does not refreeze, and when the method of defrost is by hot gas (see Sec. 6.22) a source of heat is available for the drain pan. Figure 6.17 shows that hot gas that reaches the coil for purposes of defrost is first passed through pipes embedded in the drain pan. Hot gas first comes to the drain pan and then flows on to the coil to defrost it.

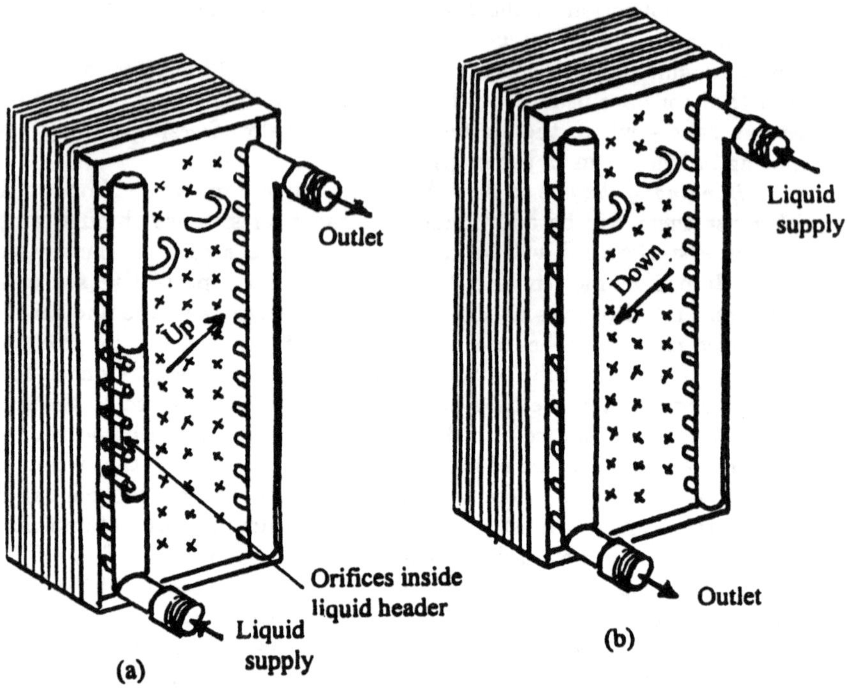

FIGURE 6.15
Forced liquid recirculation coils arranged for (a) bottom feed and upflow, and (b) top feed and downflow.

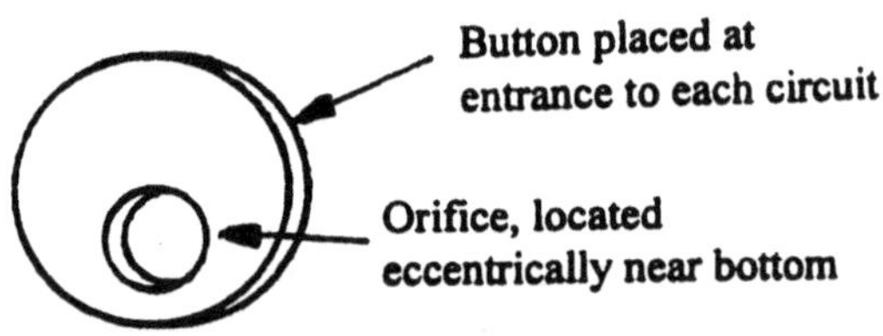

FIGURE 6.16
An orifice placed at the entrance to each circuit in a liquid recirculation coil.

6.10 DEFINITION OF TERMS APPLIED TO AIR COILS

The previous section on coil construction necessarily explained some of the terminology associated with air-cooling coils. Several other terms that also appear frequently are:

- *Circuit:* the route of the refrigerant from the time it enters a tube until it leaves.

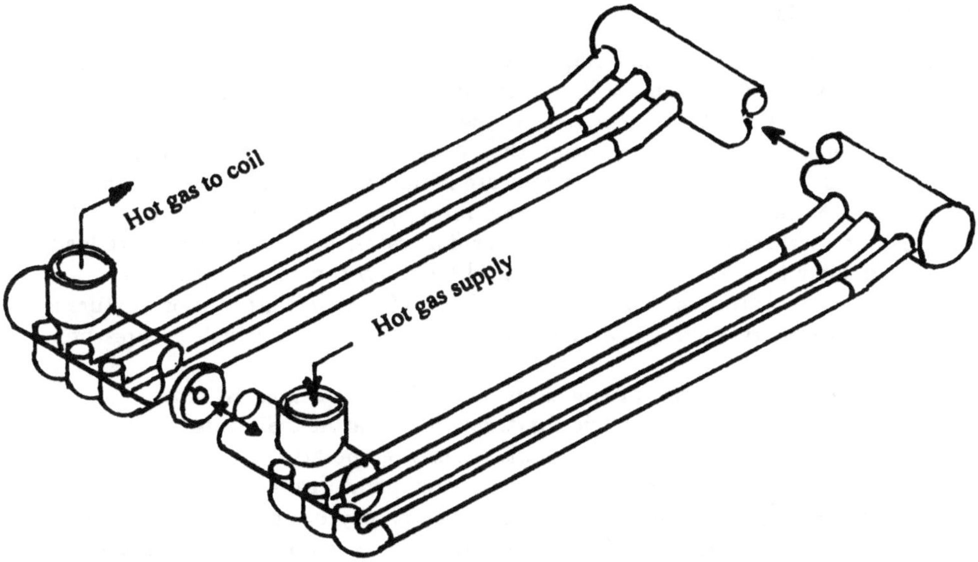

FIGURE 6.17
Pipe coils embedded in the drain pan in order to facilitate drainage of the melted frost in a low-temperature coil.

- *Coil depth:* defines the number of rows of tubes the air passes from the entrance to the exit of the coil.
- *Face area:* the cross-sectional area through which the air passes as it enters the coil, or in other words the finned length of the coil multiplied by the finned width.
- *Face velocity:* the volume rate of air flow divided by the face area.
- *Header:* A common pipe, as shown in the coils of Fig. 6.15, from which all circuits are supplied refrigerant (liquid or supply header) or a common pipe which gathers refrigerant leaving all the tube circuits (return or suction header).
- *Pass:* The flow of refrigerant through one straight section of the circuit.
- *Prime surface:* the air-side area of tubes that is in contact with air.
- *Return bends:* or U-bends are short sections of curved pipe to direct the refrigerant leaving one pass to the entrance of the next pass.
- *Secondary surface:* the surface area of both sides of the fins in contact with air.
- *Temperature difference:* in the jargon of industrial refrigeration two temperature differences may be encountered—the *TD*, which is the temperature difference between incoming air and the refrigerant and the ΔT, which is the change in air temperature through the coil.

6.11 PROPERTIES OF AIR—THE PSYCHROMETRIC CHART

Most industrial refrigeration evaporators cool air in refrigerated spaces. Even in some evaporators that cool a liquid, the liquid (water or brine, for example) is often sent to a coil that cools air. In most cases, removal of moisture from air is a companion process to reducing the air temperature, so a knowledge of the properties of the air/water-vapor mixture is fundamental to understanding the behavior of air passing through an evaporator coil. This background equips the engineer to choose coil surface areas, air-flow rates, performance at part load, and to know how to accommodate special humidity control requirements.

A valuable tool in relating the properties of the air/water-vapor mixture is the psychrometric chart. The four psychrometric charts that will be presented first are:

Figure 6.18. Psychrometric chart, normal temperatures, SI units
Figure 6.19. Psychrometric chart, low temperatures, SI units
Figure 6.20. Psychrometric chart, normal temperatures, I-P units
Figure 6.21. Psychrometric chart, low temperatures, I-P units

To help identify the properties displayed on psychrometric charts, a skeleton chart is shown in Fig. 6.22. The horizontal scale is the ***dry-bulb temperature*** in °C (°F), which is different from the ***wet-bulb temperature*** to be described later. The vertical scale is the moisture content, W, in units of kg of water per kg of dry air (lb of water per lb of dry air).

Immediately, an inconsistency arises. How could any water be contained in dry air? A preferable designation is kg (lb) of water associated with a kg (lb) of dry air. The psychrometric chart is based on a unit mass of dry air which normally remains constant in processes, even when the amount of water increases or decreases. The curved line that forms the upper-left border of the chart is the saturation line, which relates the moisture content of saturated air to temperature. The region below and to the right of the saturation line is unsaturated air capable of holding additional water vapor.

Another indication of the moisture content is the ***relative humidity***, ϕ, expressed as a percentage. One curve of constant relative humidity is shown in Fig. 6.22. The various relative humidity curves simply proportion off the vertical distance from the base of the diagram, where the moisture content is zero, to the saturation curve. Thus, the 40% relative humidity line lies 40% of the vertical distance upward from the base to the saturation line.

The next property to be highlighted in Fig. 6.22 is the wet-bulb temperature, shown by lines that run downward to the right from the saturation line. The wet-bulb temperature is that measured by a thermometer whose bulb is wrapped with a wet wick. Moisture evaporates from the wet bulb if the air in contact with the bulb is unsaturated. The difference between dry- and wet-bulb temperatures is a measure of the extent that the air is unsaturated. At temperatures below 0°C (32°F), the water on the bulb freezes, but even then the sublimation of the ice to water vapor cools the wet bulb relative to the dry bulb.

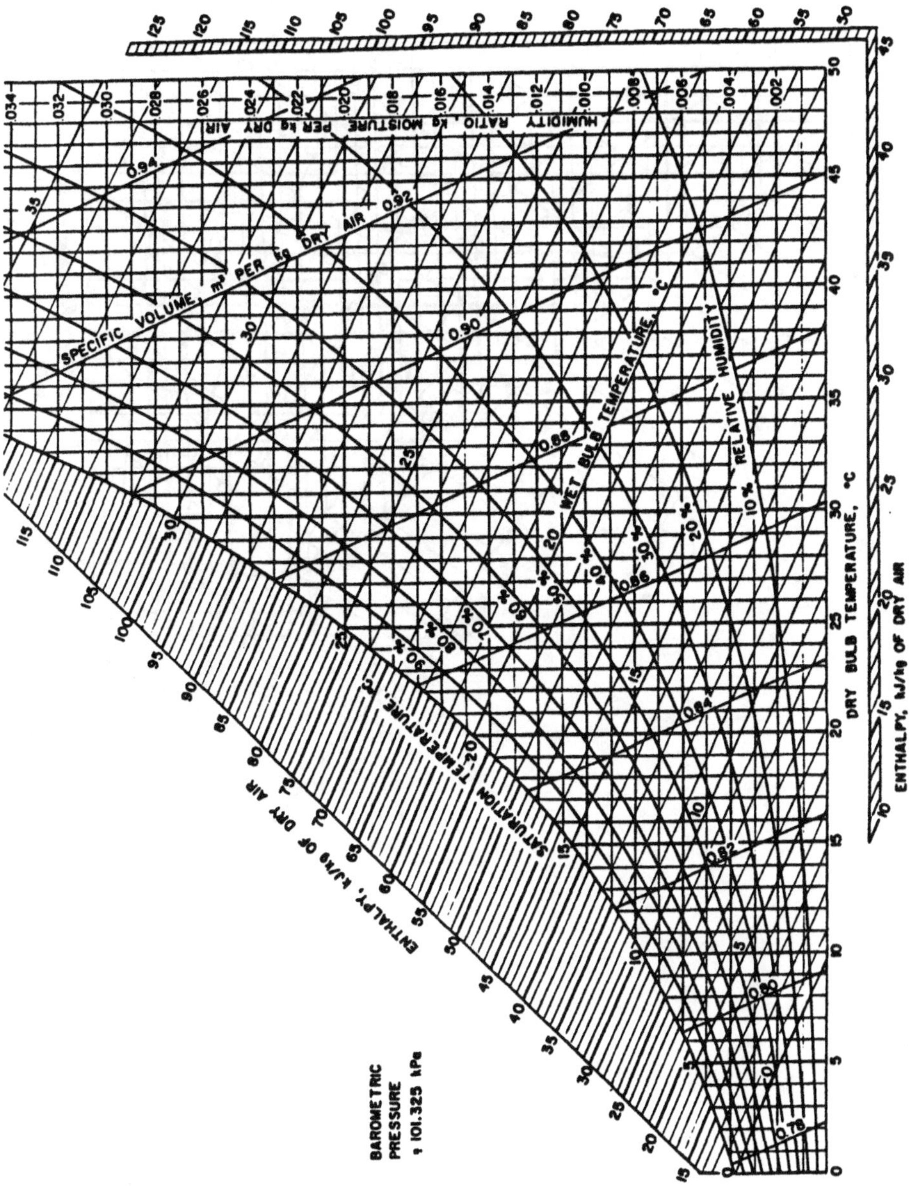

FIGURE 6.18
Psychrometric chart for air at normal temperatures. The barometric pressure is 101.325 kPa.

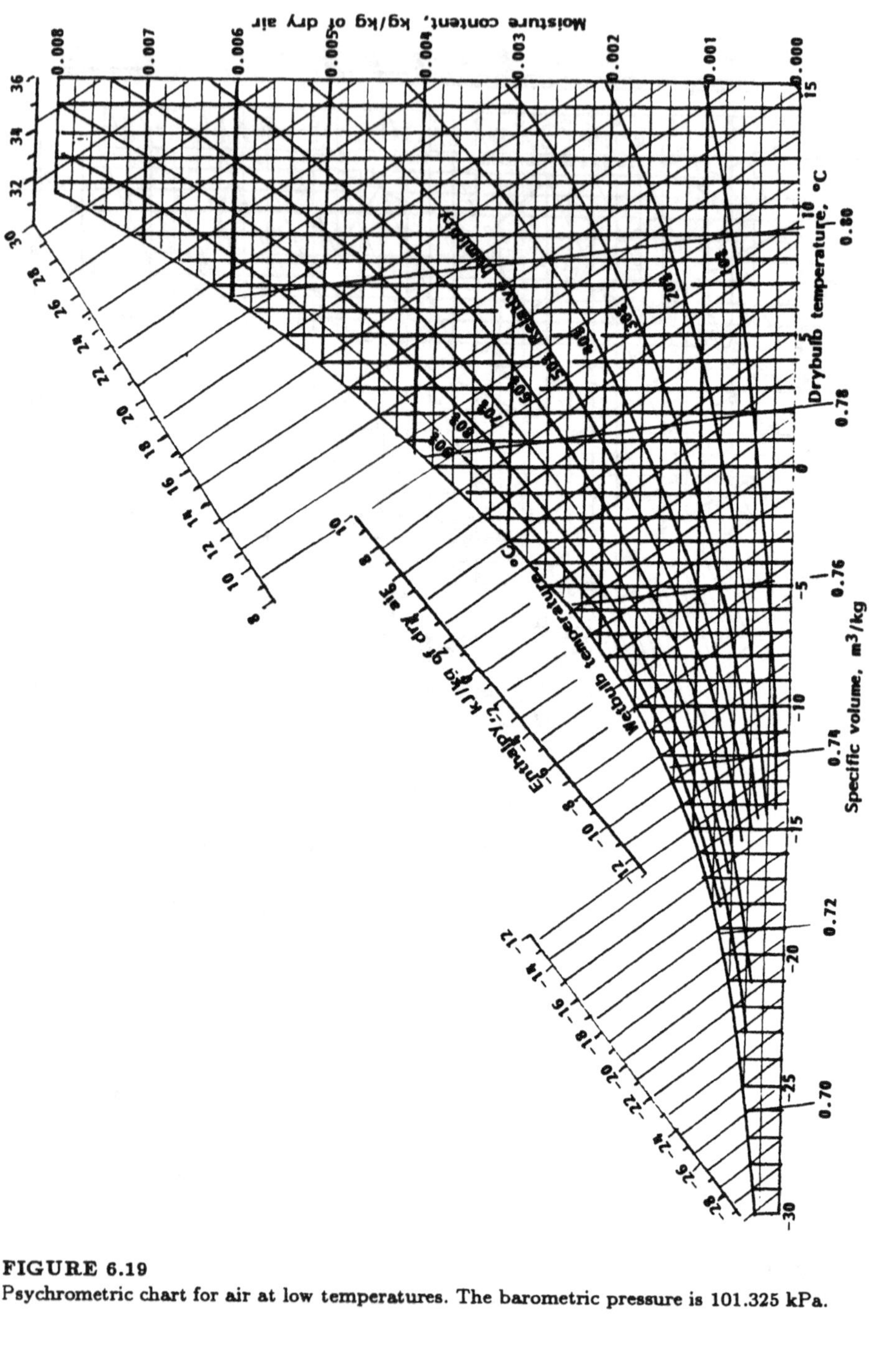

FIGURE 6.19
Psychrometric chart for air at low temperatures. The barometric pressure is 101.325 kPa.

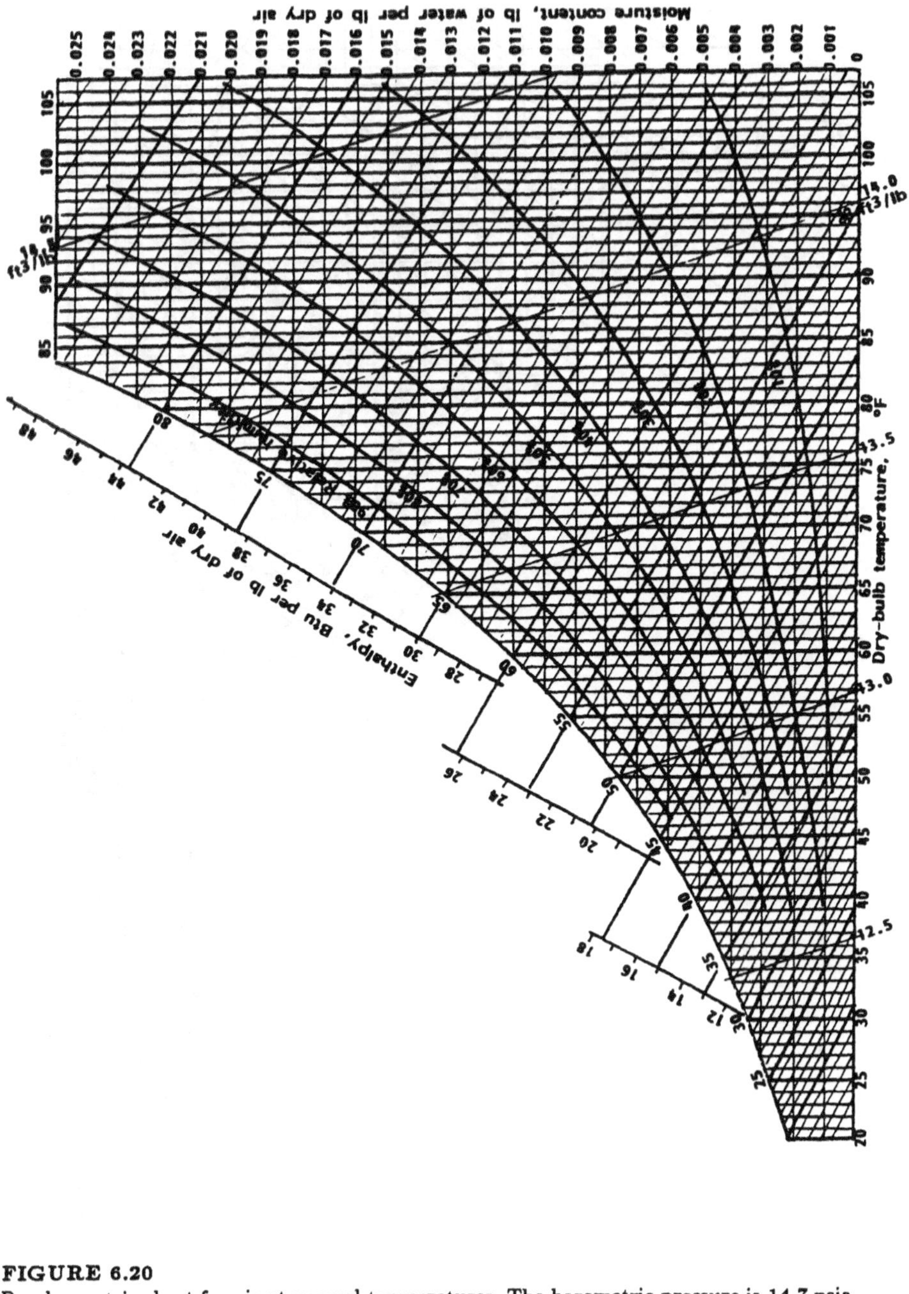

FIGURE 6.20
Psychrometric chart for air at normal temperatures. The barometric pressure is 14.7 psia.

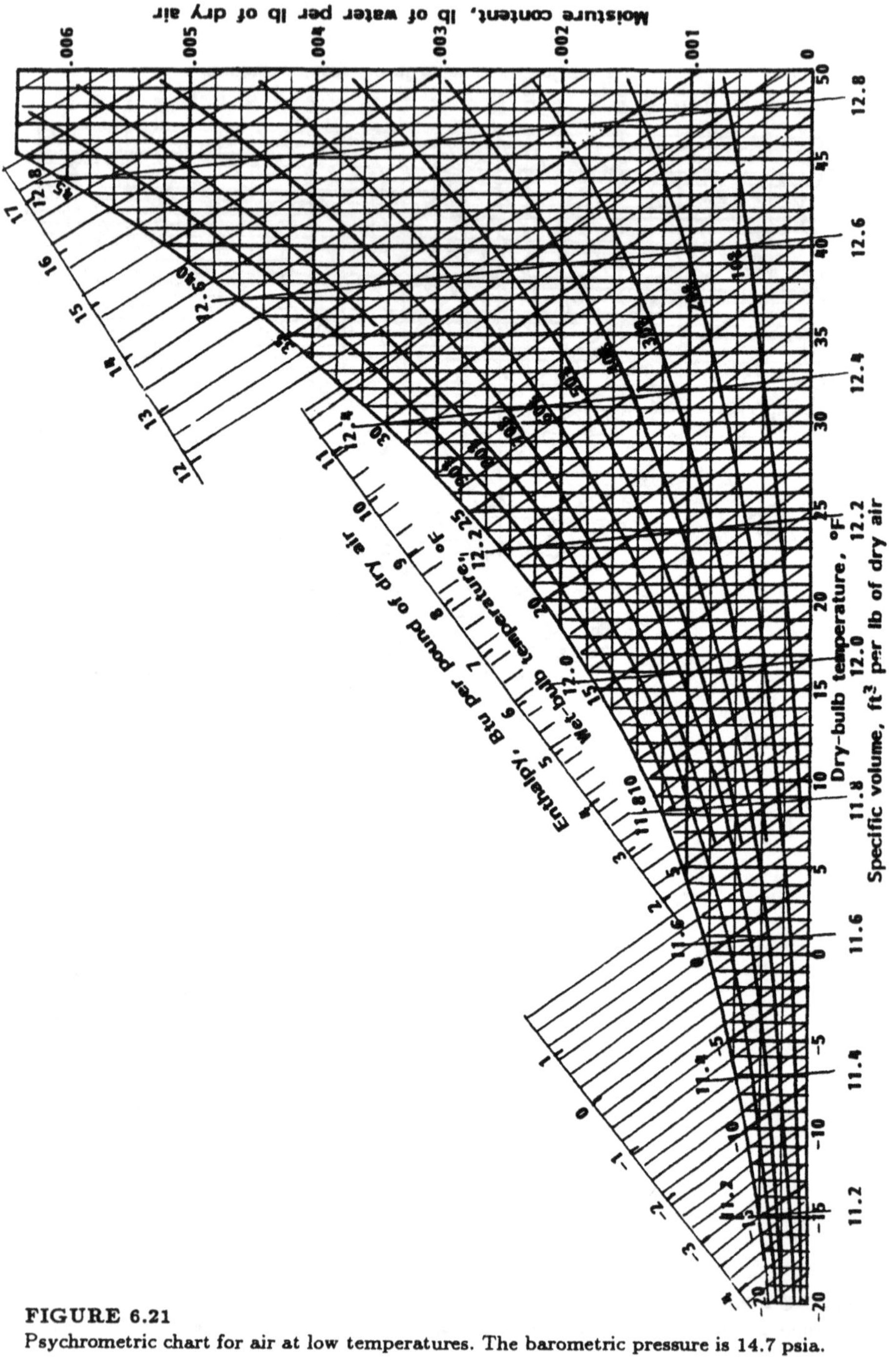

FIGURE 6.21
Psychrometric chart for air at low temperatures. The barometric pressure is 14.7 psia.

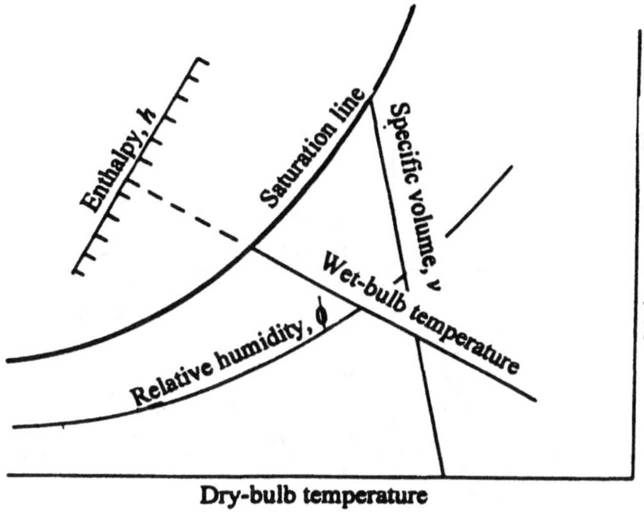

FIGURE 6.22
A skeleton psychrometric chart.

The enthalpy scale is shown in Fig. 6.22 to the left of the saturation line. To determine enthalpy in kJ per kg of dry air (Btu per lb of dry air), locate the point on the psychrometric chart and follow the line of constant wet bulb to the enthalpy scale.

Lines of constant specific volume in m^3/kg of dry air (ft^3/lb of dry air) are nearly vertical on the chart.

Example 6.2. Based on the properties given, determine the other psychrometric properties from Fig. 6.18, 6.19, 6.20, or 6.21 for the following cases:

(a) 30°C (86°F) dry-bulb temperature and a moisture content of 0.018 kg/kg (0.018 lb/lb).

(b) Saturated air at 20°C (68°F).

(c) Dry-bulb temperature of −10°C (14°F) and a wet-bulb temperature of −12°C (10.4°F).

Solution. The solutions are shown in Table 6.4.

Example 6.3. An air-flow rate of 2.2 m^3/s (4660 cfm) enters an evaporator coil at 4°C (39.2°F) and a relative humidity of 90%. Air leaves the coil at 0.5°C (32.9°F) and a relative humidity of 98%.

(a) What is the refrigeration capacity of the coil?

(b) What is the rate of water removal from the air?

Solution. From either Fig. 6.18 (Fig. 6.20) or Fig. 6.19 (Fig. 6.21), the specific volume of the entering air is found to be 0.791 m^3/kg (12.65 ft^3/lb). The mass

TABLE 6.4
Solutions to Example 6.2

	Part 6.2a	Part 6.2b	Part 6.2c, using the low-temperature chart
Dry-bulb temperature	30°C (86°F)	20°C (68°F)	−10°C (14°F)
Wet-bulb temperature	25°C (77°F)	20°C (68°F)	−12°C (10.4°F)
Moisture content, W	0.018 kg/kg (0.018 lb/lb)	0.0147 kg/kg 0.0147 lb/lb	0.00066 kg/kg 0.00066 lb/lb
Enthalpy, h	75.4 kJ/kg (40.6 Btu/lb)	57.5 kJ/kg (32.4 Btu/lb)	−8.8 kJ/kg (3.9 Btu/lb)
Specific volume, v	0.88 m^3/kg (14.15 ft^3/lb)	0.85 m^3/kg (13.6 ft^3/lb)	0.745 m^3/kg (11.9 ft^3/lb)
Relative humidity, ϕ	67%	100%	40%

rate of flow of dry air which carries along its water vapor is:

$$\text{Mass flow rate} = \frac{2.2\ \text{m}^3/\text{s}}{0.791\ \text{m}^3/\text{kg}} = 2.78\ \text{kg/s}\ (368\ \text{lb/min})$$

The entering and leaving states of the air are shown schematically in Fig. 6.23.

The rate of heat transferred to the coil, q, is:

$$q = (0.368\ \text{lb/min})(14.3 - 12.1\ \text{Btu/lb}) = 809.6\ \text{Btu/min} = 4.05\ \text{tons}$$

The rate of water removal from the air is:

$$\text{Water rate} = (2.78\ \text{kg/s})(0.0045 - 0.0039\ \text{kg/kg}) = 0.00167\ \text{kg/s}\ (0.22\ \text{lb/min})$$

We have implied that the lines of constant wet-bulb temperature and lines of enthalpy are coincident on the psychrometric chart. This statement is not correct, because there is a slight deviation of slope of the two lines. The charts shown in this chapter are accurate enough for most engineering work, but if greater precision is desired, other charts may be obtained that recognize the difference of the slope of the two lines. The *ASHRAE Handbook*[5] presents a psychrometric chart where both the lines of constant wet-bulb temperature and constant enthalpy are shown on the chart. Another approach[6] taken on charts prepared by this firm is to show lines of *enthalpy deviation*, which provide values to correct the enthalpy values when using the wet-bulb lines to locate the condition of the air.

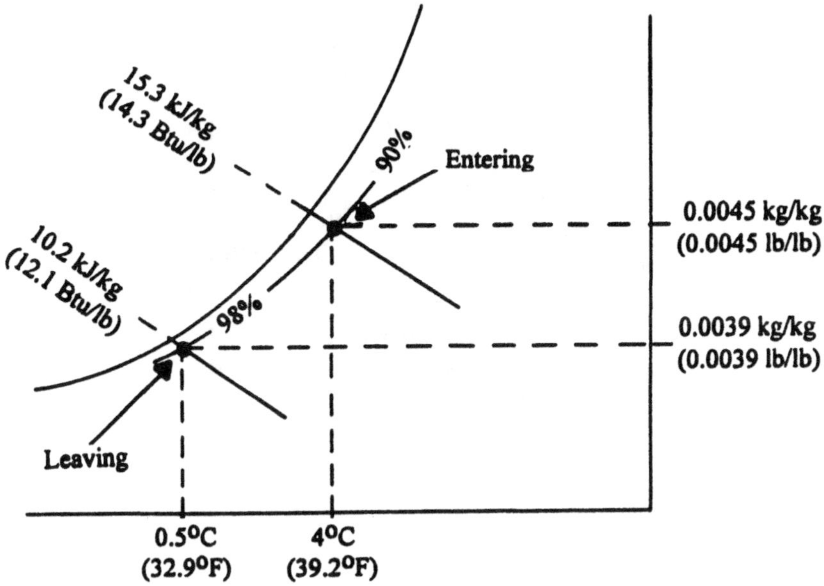

FIGURE 6.23
Inlet and outlet conditions of air in the coil of Example 6.3.

6.12 DEW-POINT TEMPERATURE

There is no mention on most psychrometric charts of a property that is important in its own right, but also provides understanding for events that occur at many locations throughout the refrigeration facility. The dew-point temperature of air is defined as the temperature that results if the air is cooled with no change in its moisture content until the air reaches the saturation conditions. In Fig. 6.24, the temperature of air at condition *a* is reduced until the air is saturated. The temperature at which saturation is reached, t_{dp}, is the dew-point temperature. One significance of the dew-point temperature is that if air at condition X is in contact with a surface having a temperature of t_{dp}, condensation of moisture will occur on that surface.

Example 6.4. The condition of air in a meat-processing room is 12°C (53.6°F) and the relative humidity is 70%. The surface temperature of the support metal for the cooling coils is 4°C (39.2°F). What is the potential operating problem that exists?

Solution. The dew-point temperature of the air in the process room, found by moving horizontally on the psychrometric chart from the condition of the air to the saturation line is 6.8°C (44.2°F). Since some metal surfaces exist at the lower temperature of 4°C (39.2°F), some moisture will condense on these surfaces. The drip of moisture on the food products is prohibited.

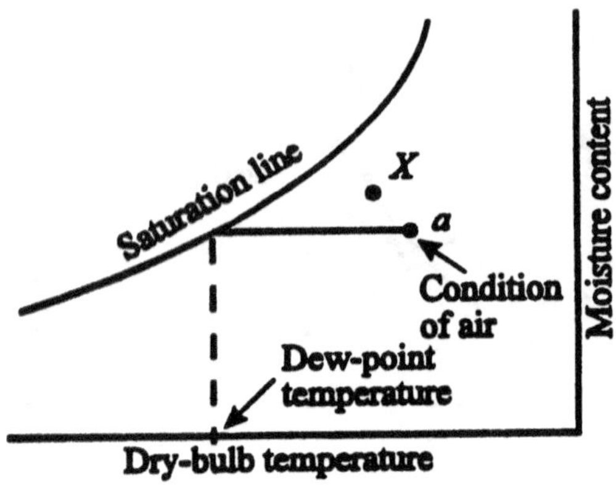

FIGURE 6.24
Dew-point temperature of air.

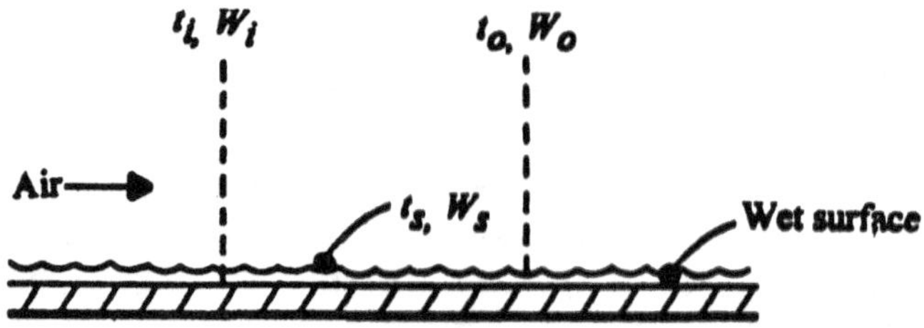

FIGURE 6.25
Heat and mass transfer between air and a wetted surface.

6.13 THE STRAIGHT-LINE LAW

A process that occurs frequently in refrigeration practice is the transfer of heat and mass (water) between air and a wetted surface, as shown in Fig. 6.25. The dry-bulb temperature of the air entering the section is t_i and its moisture content is W_i.

At the water surface, the air is in equilibrium with the water, so the air is saturated and at the same temperature as the water, t_s, and the moisture content of the air at the surface is the same as the saturated air, W_s. The temperature difference between the air and the wet surface is the driving force for the transfer of heat between them. The water vapor pressure is linearly proportional to the moisture content, W, so the difference between W_i and W_s is the driving force for the transmission of water vapor at the entrance to the section.

The handy straight-line law explains numerous processes by predicting the path of the air conditions on the psychrometric chart. The straight-line law applies to the situation shown in Fig. 6.25, where air contacts a wetted surface. The statement of the straight-line law is

The straight-line law:

The path on the psychrometric chart drives toward the saturation line at the temperature of the wetted surface.

The straight-line law is well verified by fundamental laws of heat and mass transfer.[7,8] It shows that from the change in properties of the air shown on the psychrometric chart, the directions of heat transfer and water-vapor transfer are indicated by the changes in enthalpy and moisture content, respectively.

Figure 6.26 shows several different situations where entering air at condition i contacts a wet surface of temperature t_s.

Figure 6.26a shows the situation prevailing in a cooling and dehumidifying coil where, even if the coil is dry when starting operation, moisture from the air condenses on the cold surface of the coil, such that thereafter the surface is wetted and the straight-line law applies. The air flowing across the surface changes state from i to o, and during that process the enthalpy indicates that it gives up heat to the coil and the refrigerant carries it away.

In Fig. 6.26b, the situation portrayed is that of evaporative cooling, as shown schematically in Fig. 6.27. Because no heat is added externally to the water, the energy balance requires that the enthalpy of entering air equals that leaving, thus $h_i = h_o$.

The process of constant enthalpy in Fig. 6.26b results in a drop in air temperature, but an increase in moisture content, W. Evaporative cooling is rare in industrial refrigeration practice, but Fig. 6.26b is a transition from the cooling and dehumidification process to the heating and humidification process of Fig. 6.26c.

Our interest in Fig. 6.26c lies not in the heating and humidification of the air, but in the removal of heat from the wetted surface. This is the process that occurs in a cooling tower and an evaporative condenser—a topic encountered again in Chapter 7 in the study of condensers.

6.14 CONDITION OF AIR PASSING THROUGH A COIL

The principles presented so far in this chapter have provided the foundation for making useful judgments about actual plant operation. The straight-line law and a few heat-transfer principles equip the refrigeration professional to predict qualitative trends and to know how to correct certain evaporator problems.

The predominant criterion for evaluating evaporator coil performance is the condition of air leaving the coil. The assignment of the evaporator is to hold a certain temperature in the space, which it does by removing heat at the proper rate from the air passing through the coil. The outlet air temperature

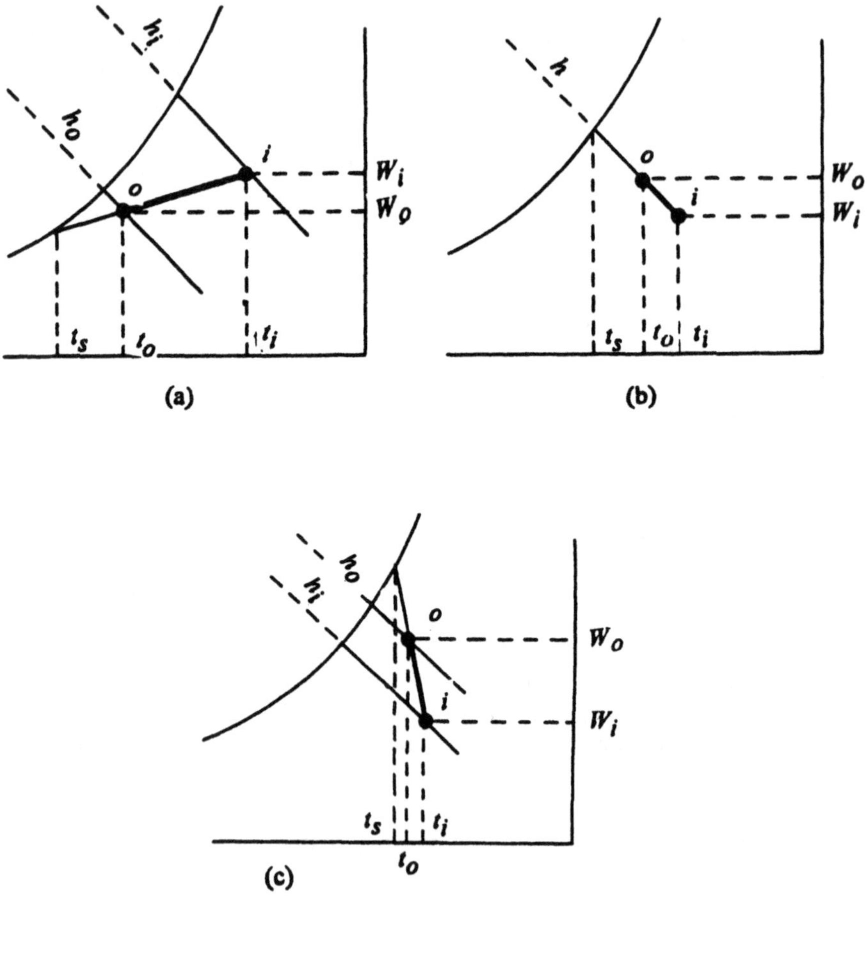

FIGURE 6.26
Several applications of the straight-line law: (a) cooling and dehumidification, (b) evaporative cooling, and (c) heating and humidification.

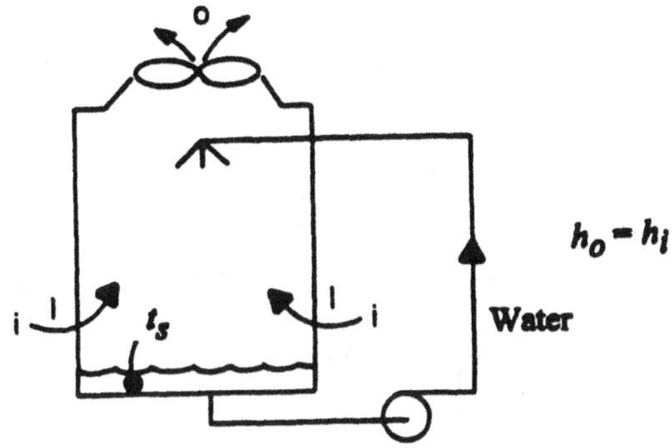

FIGURE 6.27
Evaporative cooler represented by Fig. 6.26b.

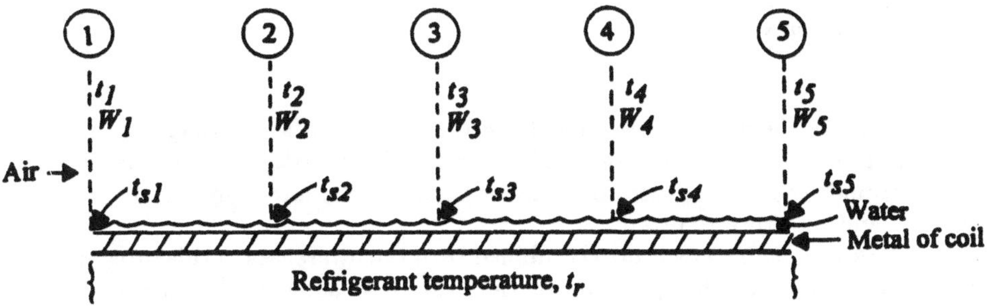

FIGURE 6.28
Cooling and dehumidifying of air in a coil with constant refrigerant temperature.

from the coil is often sensed and used as the basis of controlling the evaporator. The outlet dry-bulb temperature is not the only concern because the amount of moisture removed from the air is often crucial too. In some situations, such as a high-humidity produce storage room, as little moisture as possible should be removed. But the coils at a loading dock for a frozen-food storage room should remove as much moisture as possible to reduce the amount of water vapor carried into the low- temperature space with the infiltration air.

Predicting the condition of outlet air can be approached by analyzing the state of the air as it passes through the coil. Figure 6.28 imagines the coil to be a flat surface with refrigerant boiling at a temperature of t_r on one side of the metal. Air enters at a temperature of t_1 and a moisture content of W_1.

A key realization is that the wetted surface temperatures, t_{s1}, t_{s2}, etc., progressively decrease in the direction of air flow through the coil. The reason for this progression is that t_s adjusts itself, so that the same rate of heat flows

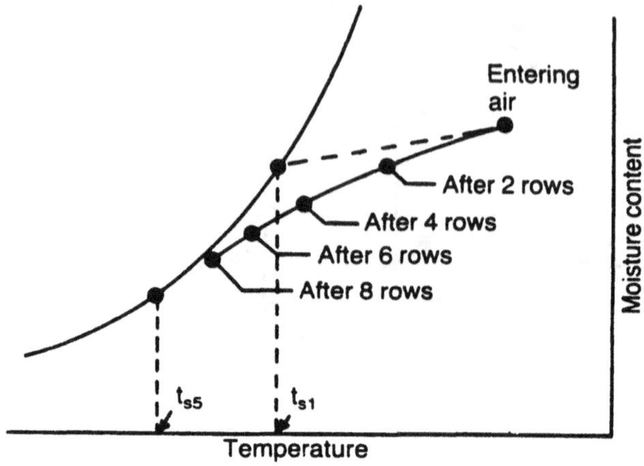

FIGURE 6.29
A coil condition curve.

from the air to the wetted surface as from the wetted surface through the metal to the refrigerant. With the drop in temperature and moisture content as the air flows through the coil, t_s also drops in response to the lower potential for heat transfer and moisture removal.

The straight-line law is now combined with the awareness of this progressive drop in wetted surface temperature to yield the curve showing the air conditions in the coil. Fig. 6.29.

Suppose that the flat surface of Fig. 6.28 was an idealization of a coil having 8 rows of tubes in the direction of air-flow, and each of the sections in Fig. 6.28 represents two rows of tubes. In Fig. 6.29, the entering air, in conformity to the straight-line law, drives toward t_{s1} on the saturation line. As the air moves through the coil, it drives toward progressively lower temperatures on the saturation line. The state of the air, called the coil condition curve, in Fig. 6.29 is the result.

A few observations from Fig. 6.29 are confirmed by catalog data from coil manufacturers. Each succeeding row of tubes does less work toward lowering the temperature or removing moisture from the air. The greatest rate of heat transfer is where the air enters the coil, because the air temperature and moisture content are the highest here. The curve becomes steeper as air passes through the coil, which indicates that the ratio of moisture removal to the temperature drop is greatest at the coil's air outlet.

TABLE 6.5
Influence of design or operating parameters on outlet air conditions from an evaporator coil.

Parameter, increase of:	Effect on outlet air conditions		Refrigerating capacity	Typical range
	temperature	moisture content		
Face area	lower	lower	higher	depends on refrigerating capacity
Number of rows of tubes deep	lower	lower	higher	four to eight
Number of fins per unit length	lower	lower	higher	115 to 300 fins per m (3 to 8 fins per in)
Air flow rate	higher	higher	higher	Face velocity 2 to 4 m/s (400 to 800 fpm)
Refrigerant temperature	higher	higher	lower	3 to 8°C (5 to 15°F) below entering air

6.15 HOW SELECTION AND OPERATION OF A COIL AFFECT ITS PERFORMANCE

Some of the most important factors over which the system designer and operator have some control are the:

- face area of the coil, the cross-sectional area of the air stream entering the coil
- number of rows of tubes deep
- fin spacing
- air-flow rate (and air velocity)
- refrigerant temperature.

Table 6.5 shows the influence of an individual increase in these parameters.

Face area. An increase provides more heat transfer surface, and the surface also assumes a lower temperature (Fig. 6.30a).

Number of rows of tubes deep. Each additional row of tubes further reduces the air temperature and moisture content, limited by the saturation conditions at the refrigerant temperature (Fig. 6.30b).

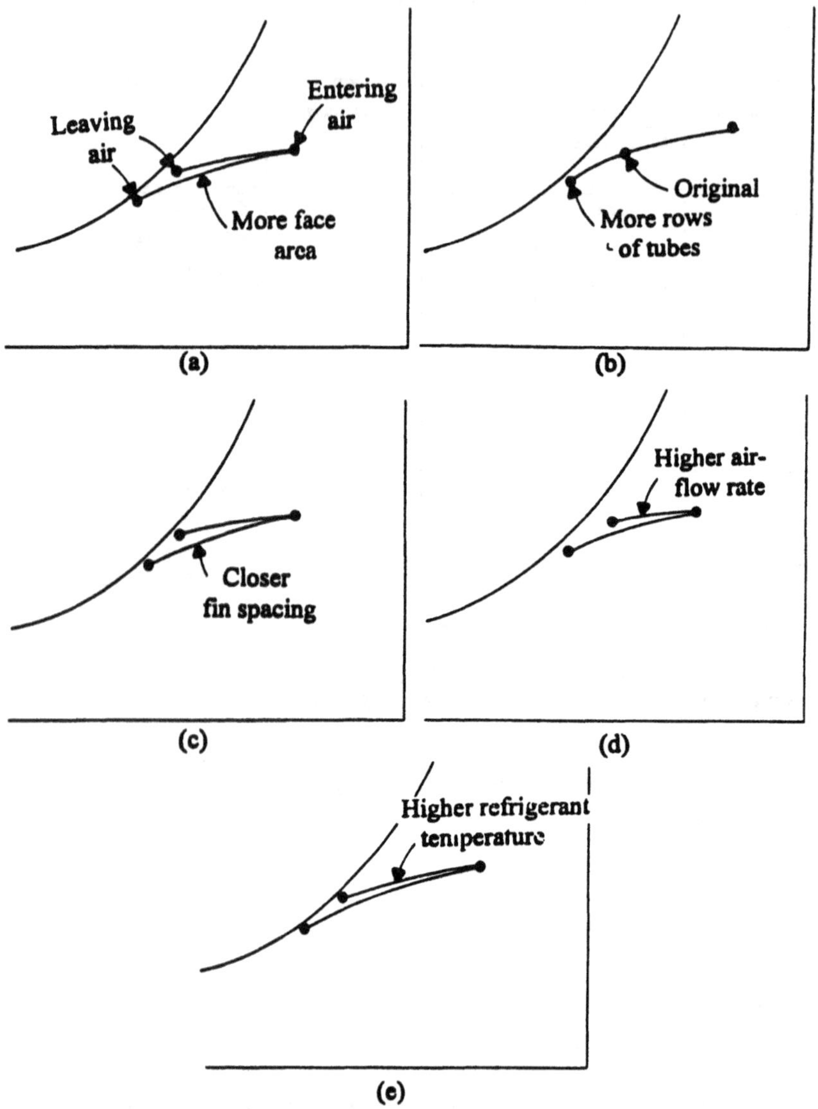

FIGURE 6.30
Influence of design and operating parameters on outlet air conditions from a coil.

Closer fin spacing. Closer spacing provides more heat-transfer area, resulting in lower outlet temperature and moisture content (Fig. 6.30c). Also see Section 6.20 on frosting of coils.

Higher air-flow rate With a higher flow rate, the outlet temperature and moisture content increase, but in the equation for the rate of heat transfer,

$$q = (\text{flow rate})(h_{entering} - h_{leaving})$$

the flow rate increases by a greater percentage than the percent decrease in enthalpy difference. The refrigeration capacity q thus increases (Fig. 6.30d).

Higher refrigerant temperature. A higher temperature raises the coil surface temperature all along the coil. Thus the temperature and moisture content are not reduced by as much. Also, as seen from Fig. 6.30e, the ratio of moisture removal to temperature drop is not as great—a characteristic desired in high-humidity rooms.

The preceding discussion concentrated on changing only one condition at a time. System designers who select coils may adjust two or more conditions to obtain desired outlet air properties. For example, suppose that the amount of available space does not permit the coil face area desired. Instead, a deeper coil could be selected and combined with a slightly higher air-flow rate to closely match the original outlet air conditions.

6.16 SELECTING AN AIR COIL FROM A CATALOG

A critical piece of data that manufacturers of air-cooling evaporators provide for their coils is the rate of heat transfer per degree temperature difference, which appears in the equation:

$$q \text{ kW (Btu/hr)} = R(t_{air,in} - t_{refrig}) \tag{6.13}$$

where R = rating, kW/°C (Btu/hr per °F)

Table 6.6 shows an exerpt from a manufacturer's catalog which specifies both the R-value and the flow rate delivered by the fans. The table applies to two-fan models, which is one of the choices among one-, two- and three-fan units. Two fin spacings are offered, 118 fins per m (3 fins per inch) and 158 fins per m (4 fins per inch). The coil with wider fin spacing having less heat-transfer area than the 158 fpm (4 fpi) coil shows lower R-values. The next line in Table 6.6 indicates the external static pressure (ESP) in inches of water against which the fans must operate. If the coil is hung in the refrigerated space with no attached ductwork to direct the air flow, there is no external static pressure. The R-values shown in the table are expressed in units of Btu/hr per °F temperature difference between the entering air and refrigerant. The air flow rate flowing through the coil is expressed in ft^3/min (cfm). The power requirement of each of the two fans

TABLE 6.6
Exerpt from a rating table of a two-fan evaporator coil. (*Courtesy Krack Corporation.*)

		4 FINS/INCH				3 FINS/INCH			
		NO ESP		1/4 INCH ESP		NO ESP		1/4 INCH ESP	
MODEL	HP EACH	BTUH/°TD	CFM	BTUH/°TD	CFM	BTUH/°TD	CFM	BTUH/°TD	CFM
2S-536	1-1/2	18980	31000			17550	32000		
2S-538		22620	30000			20920	31500		
2S-5310		26000	29200			24700	30000		
2S-536	2	20540	36200	19240	31800	19000	38000	17870	32800
2S-538		24960	35400	23140	30800	23320	37600	21400	32000
2S-5310		28860	34600	26520	30000	27420	36200	25200	31200
2L-536	3	20880	37000	19760	34200	19310	39000	18280	36000
2L-538		25480	36900	24400	33400	23560	38400	22900	35200
2L-5310		29380	35800	28000	32800	27920	37600	26600	34800
2L-536	5	23920	47600	22800	44400	22130	50000	29090	46800
2L-538		29380	46800	28000	43400	27980	48800	25900	45600
2L-5310		33800	45400	32240	42400	32120	47400	30620	44600
2L-536	7-1/2	26260	56000	25740	54000	24290	57000	23810	55200
2L-538		33000	53200	31720	51200	30860	54200	29340	52400
2L-5310		36800	51800	35360	49600	34960	53400	33600	50800

is specified, and this information serves two purposes: to indicate what electric current is required, and to provide data to compute the internal heat load for the refrigeration load calculation.

Predicting the rate of heat transfer using Eq. 6.13 is simple, and may seem too simple when realizing the complexity often encountered with logarithmic mean temperature differences that apply to heat exchangers. Section 6.5 has, however, already shown that for heat exchangers where one fluid remains at an essentially constant temperature, such as in evaporators, the simplification is valid.

Manufacturers often use model numbers to provide some information on the features of the coil, and such is the case of the model numbers in Table 6.6. The 536 coils are 6 rows of tubes deep, 538 eight rows, and 5310 ten rows. Further information provided by the manufacturer includes the volume of the coil and its weight, which is a function of the materials. For the 536 coils, for example, the approximate weights are 2130 kg (4700 lb), 1270 kg (2800 lb) and 1090 kg (2400 lb), respectively for steel, copper/aluminum, and aluminum materials, respectively.

The face area of the coil is 4.9 m^2 (52.8 ft^2) which is information needed for one routine check that should be made when selecting coils, particularly for above freezing operation. The concern is that if the face velocity if greater than

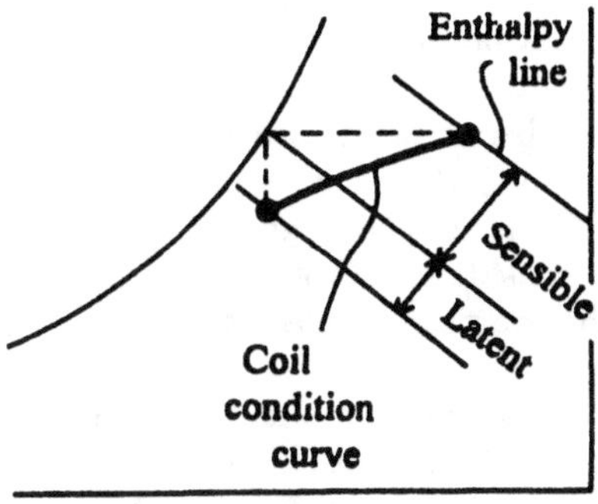

FIGURE 6.31
Sensible and latent heat removal in a coil.

TABLE 6.7
Typical temperature differences—entering air to refrigerant—for several applications.

Application		$t_{air,in} - t_{refrig}$
Below freezing	Storage and blast freezer	5.5 to 6.5°C (10 to 12°F)
Above freezing	Low humidity	11 to 17°C (20 to 30°F)
	High humidity	2.2 to 4.4°C (4 to 8°F)

about 3.0 m/s (600 fpm) the water that condenses on the coil surfaces can be blown off into the refrigerated space. For coils operating with surfaces below 0°C (32°F), the moisture deposits as frost, and adheres to the coil. Face velocities of 4 m/s (800 fpm) or more are typical of low-temperature coils. The highest velocity indicated in Table 6.6 is 5.5 m/s (1080 fpm) when the flow rate is 27 m^3/s (57,000 cfm). To limit the face velocity to 3.0 m/s (600 fpm) a set of fans delivering 15 m^3/s (31,700 cfm) or less should be chosen.

Part of the selection process is to decide what temperature difference, $t_{air,in} - t_{refrig}$, to use. This choice is embodied in the optimization process discussed in Sec. 6.8, because the choice of the temperature difference directly influences the heat-transfer area that is selected, and for a given air temperature in the space a higher refrigerant temperature offers the potential of energy saving at the compressor. Realistically, the optimization of only one or several evaporators in a multiple-evaporator installation may not permit elevating the saturated suction temperature if there are other evaporators that require a low suction temperature. Table 6.7 shows some typical temperature differences used in various applications.

The heat-transfer rates shown in Table 6.6 are called *dry* ratings and refer only to sensible heat. If the coil condition curve of Fig. 6.29 is subdivided into sensible and latent heat, as in Fig. 6.31, the sensible heat is that portion associated with a drop in dry-bulb temperature, while the latent heat removal occurs because of the condensation of water vapor.

When the air conditions and coil temperature are such that moisture condenses, this latent heat becomes an additional capacity for the coil. For produce storage rooms in the 0-5°C (32-40°F) temperature range, the coil may develop an additional 20%-40% capacity for a given $t_{air,in} - t_{refrig}$. While this additional coil capacity may be considered a bonus, remember that the refrigerant side of the evaporator and the compressor must still provide the total capacity—sensible plus latent.

The catalogs of most coil manufacturers do not reflect the thermal load imposed by fan motor(s). All the electric power delivered to the fan motor eventually appears in the refrigerated space as heat load. The fan-motor load could be deducted from the coil capacity that the manufacturer presents, but the addition of the electrical input to the total refrigeration load is left to the designer of the system as part of the refrigeration load calculation.

Example 6.5. One of the air-cooling coils is to be selected for a space storing frozen food where the air temperature is −23.3°C (−10°F). The refrigeration plant provides a suction pressure corresponding to −28.9°C (−20°F). The coil is ceiling-hung with no external static pressure air resistance. The coil is to provide a refrigeration capacity of 70 kW (20 tons of refrigeration). Select a coil from Table 6.6.

Solution. Since the performance factors of the table are presented in inch-pound units, use the I-P system for the selection. The required coil capacity is 20 tons or 240,000 Btu/hr, and the temperature difference available between the air entering the coil and the refrigerant is $[(-10°F) - (-20°F)] = 10°F$. The coil must therefore have a performance factor of $240{,}000/10 = 24{,}000$ (Btu per hr)/°F or higher. Assume that the space does not experience a heavy humdification and frost load due to infiltration, so decide that a 4 fin-per-inch choice will be satisfactory. Several options from Table 6.6 are:

Model 2S-5310 with 1-1/2 hp fans providing a performance factor of 26,000
Model 2S-538 with 2-hp fans providing a performance factor of 24,960
Model 2L-536 with 5-hp fans with a performance factor of 23,920 (close).

Although the 2L-536 would be the smallest and lowest-price coil, the jump to the 5-hp fans with its significantly higher heat input might not favor this choice. On the other hand, stepping up from 1-1/2 to 2 hp fan motors permits reduction from a 10-row coil to an 8-row coil, so the 2S-538 would be an appropriate choice.

While there were several possible choices of coil offered by Table 6.6, all of the coils represented in the table had a face area of 4.9 m^2 (52.8 ft^2), and the complete manufacturer's catalog would offer additional choices. Perhaps a 6-row coil with a larger face area would offer a preferred combination of coil cost and fan-power requirements.

A standard design practice is to install greater coil refrigerating capacity in a space than that needed to meet the maximum refrigeration load when defrosting of coils is expected. The basis for this practice is that for a period of time during the day certain coils will be out of service for defrost. If, for example, the coils in a space are expected to undergo defrost 2 hours out of 24 hours, or 8-1/3% of the time, the installed capacity of the coils is increased by that percentage. The compressor capacity need not be increased, however, because the actual refrigeration capacity must only meet the design load, which is simply shifted among the active coils.

Communication with the coil manufacturer at the time of the order is highly recommended. Typical information that can be useful to the coil manufacturer is:

- the space temperature
- the saturated suction temperature
- the refrigeration capacity
- which refrigerant is to be used
- the type of feed, for example, whether liquid recirculation, flooded coils, or direct expansion. If liquid recirculation is chosen whether top or bottom feed should be specified as well as the recirculation rate.

If at all possible in the case of liquid recirculation, the temperature of the supply liquid should be close to the evaporating temperature. If exceptionally cold liquid is supplied to the coil, the section where this liquid enters will frost at a more rapid rate than the rest of the coil. The heavily frosted section will absorb less heat from air and propagate the frosted region. A different problem occurs when the liquid is unusually warm, such as sometimes occurs when a gas-pumping system provides the recirculation. When the warm liquid passes through the throttling valve at the coil entrance, a high vapor fraction of refrigerant develops. With the bottom-feed coil equipped with orifices, this vapor may choke the flow at these orifices unless the manufacturer knows that they should be enlarged.

6.17 HUMIDITY CONTROL IN REFRIGERATED ROOMS

In many applications, control of the relative humidity in the refrigerated space is important. When storing many varieties of fresh produce, the humidity should be kept high to maintain product quality. In certain other applications, such as for rooms chilling hot red meat carcasses, humidity should be kept low to avoid fog formation and to prevent moisture from condensing on surfaces and dripping on product[9].

To maintain high humidities, choose large coils operating with small air-to-refrigerant temperature differences. The air-flow rate will be high to provide the refrigeration duty with the small drop in air temperature. For low humidities, select small coils with low air-flow rates and large air-to-refrigerant temperature differences.

This section points out the limitations of coils in meeting extreme conditions and explains why humidifiers (for high humidity) or reheat (for low humidity) may be necessary. Most applications where the control of humidity is important are in spaces operating in the temperature ranges of 0-10°C (32-50°F), so Figs. 6.32 and 6.33 provide psychrometric charts with this range of temperature enlarged.

To dramatize the challenge of selecting coils for high-humidity spaces, consider a room to be matintained at 1°C (34°F) and 95% relative humidity. Other refrigeration load data and characteristics of the space are:

floor area	=	1425 m^2 (15,340 ft^2)
room volume	=	10,000 m^3 (353,000 ft^3)
outdoor design conditions	=	35°C (95°F) dry-bulb temperature 25.5°C (78°F) wet-bulb temperature
sensible refrigeration load due to heat conduction through the structure, lights and motors, product cooling	=	150 kW (512,000 Btu/hr)
infiltration estimated at 0.15 air changes per hour	=	(10,000 m^3)(0.15)
	=	1500 m^3/hr (52,970 ft^3/hr)
the mass flow rate of infiltration air	=	

$$\frac{1500 \text{ m}^3/\text{hr}}{(3600 \text{ s/hr})(0.895 \text{ m}^3/\text{kg at outdoor conditions})} = 0.466 \text{ kg/s } (3700 \text{ lb/hr})$$

The sensible load associated with the infiltrated air is:

$$(0.466 \text{ kg/s})(35 - 1°\text{C})(1.0 \text{ kJ/kg}\cdot°\text{C}) = 15.84 \text{ kW } (54,000 \text{ Btu/hr})$$

In the exchange of outdoor and room air in the infiltration process, air with a moisture content of 0.017 kg/kg (0.017 lb/lb) enters to displace room air having a humidity ratio of 0.0038 kg/kg (0.0038 lb/lb). The rate of moisture introduced is, then:

$$(0.466 \text{ kg/s})(0.0170 - 0.0038 \text{ kg/kg}) = 0.00615 \text{ kg/s } (54,000 \text{ Btu/hr})$$

The required refrigeration rate at the coil to condense this water vapor is:

$$(0.00615 \text{ kg/s})(2500 \text{ kJ/kg}) = 15.4 \text{ kW } (52,500 \text{ Btu/hr})$$

The total of the sensible and latent loads from the infiltrated air is 31.2 kW (106,600 Btu/hr) which may be checked by computing the product of the air-flow rate and the differences in air enthalpy of the outdoor and interior air,

$$\text{total infiltration load} = (0.466 \text{ kg/s})(78.0 - 10.6 \text{ kJ/kg}) = 31.4 \text{ kW } (107,100 \text{ Btu/hr})$$

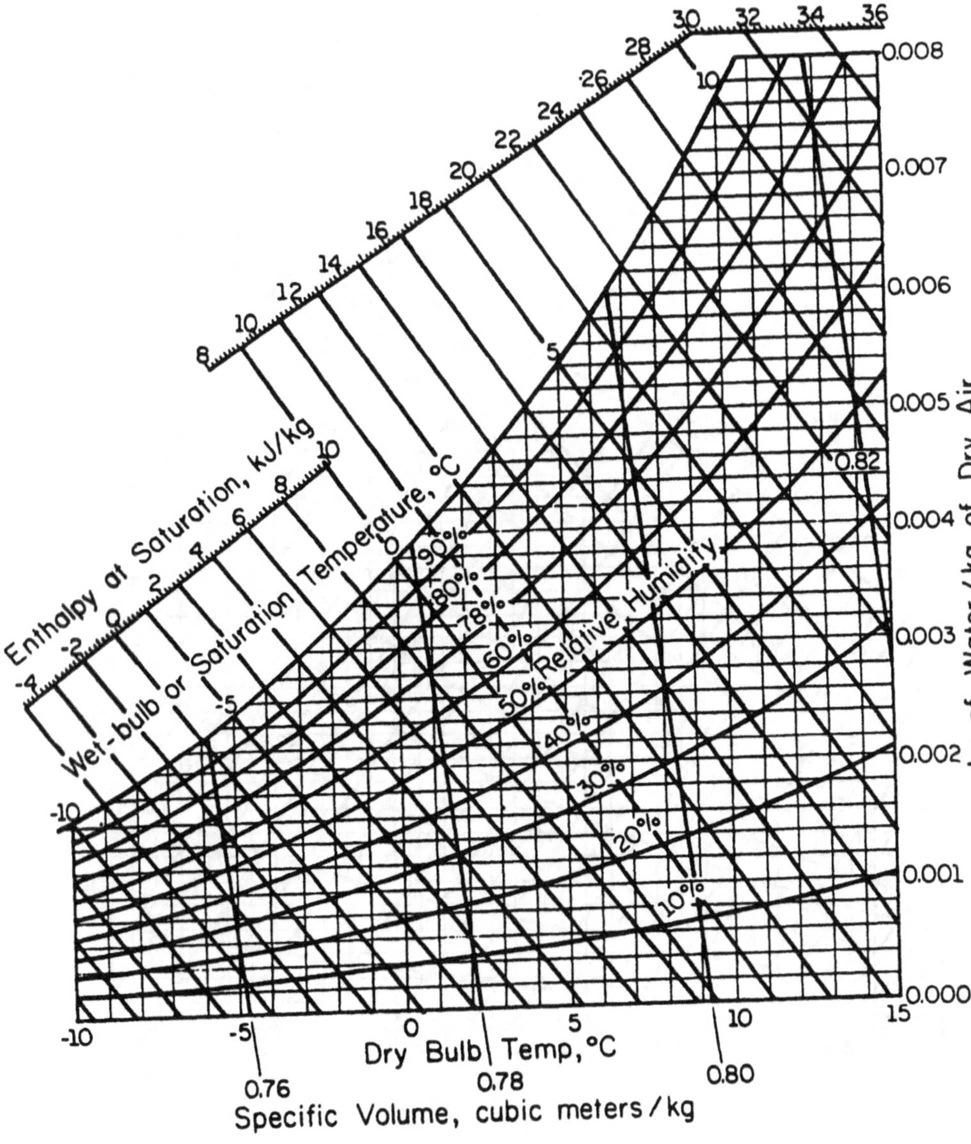

FIGURE 6.32
Psychrometric chart for air near 0°C (in SI units).

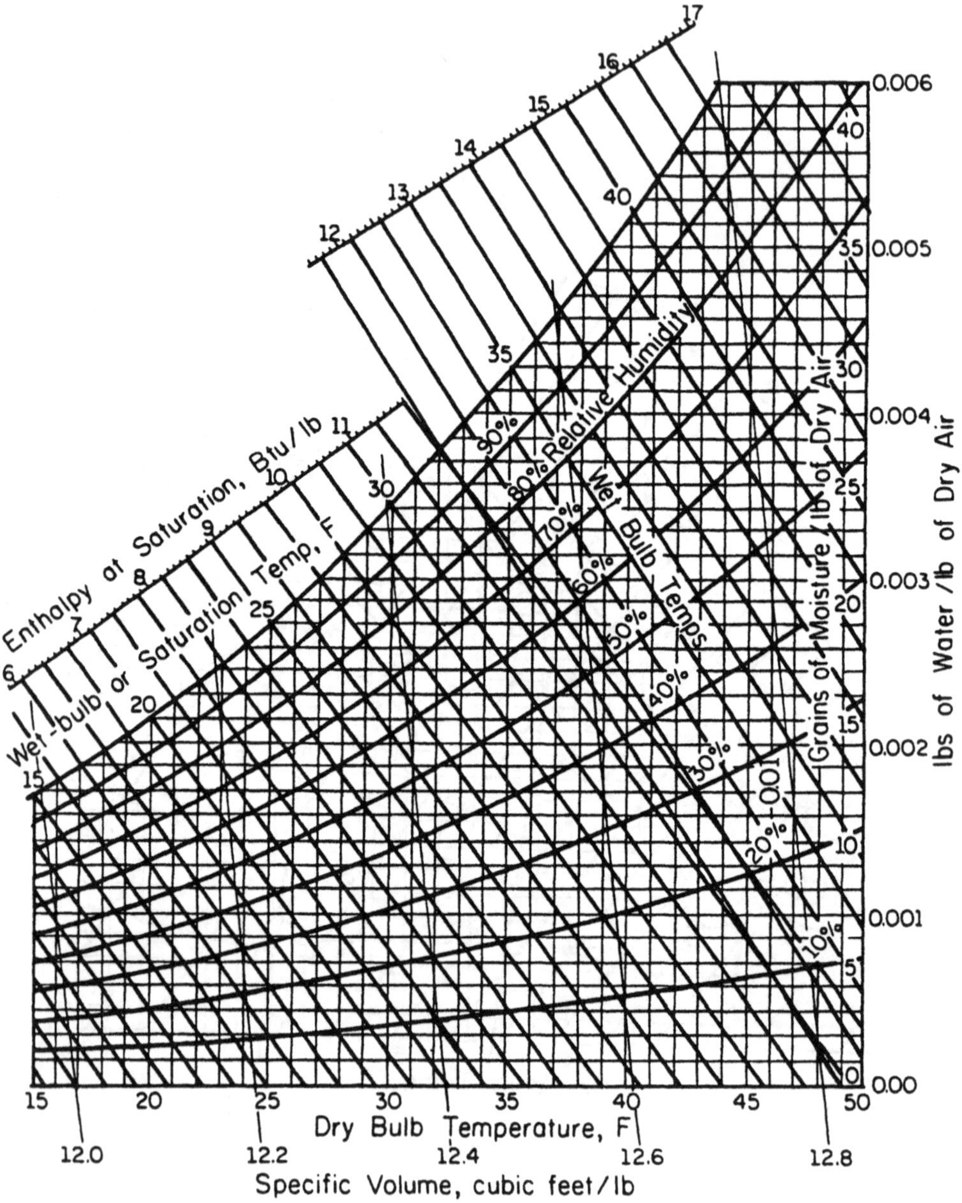

FIGURE 6.33
Psychrometric chart for air near 32°F (in I-P units.

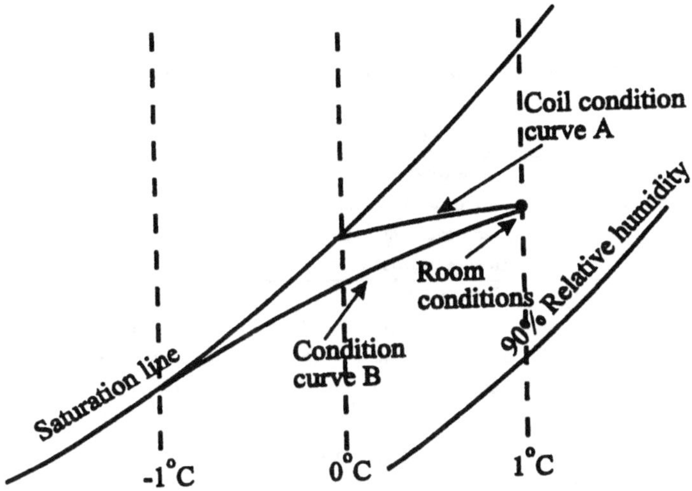

FIGURE 6.34
Expanded sketch of psychrometric chart displaying conditions at a coil serving a high-humidity room.

To summarize, the space loads are:

sensible:	150 + 15.8 =	165.8 kW (565,700 Btu/hr)
latent:		15.4 kW (52,500 Btu/hr)

Presenting the numerical values (which, incidentally, came from an actual load calculation) shows how the designer is squeezed to maintain high humidity in this room. The slope of the coil condition curve to meet the combination of sensible and latent loads, as shown in Fig. 6.29, will be very flat. An expanded sketch of the pertinent portion of the psychrometric chart is shown in Fig. 6.34, indicating that coil-condition-curve A intersects the saturation line at approximately 0°C (32°F).

When correlating coil condition curve A to the straight-line law, the average wetted surface temperature of the coil should be slightly lower than 0°C (32°F), and the refrigerant temperature would be perhaps, −1°C (30°F). Thus, the temperature difference—entering air to evaporating refrigerant—would be approximately 2°C (3.6°F), rather than the typical commercial Δt mentioned in the previous section. An alternate approach is to operate with a larger air-to-refrigerant temperature difference and combine with humidification. Coil condition curve B shows a steep slope resulting from operating at a higher air-to-refrigerant temperature difference. Without auxiliary humidification, the design relative humidity of 95% cannot be maintained, because the rate of moisture removal at the coil is higher than that entering with the infiltration air. Table 6.8 summarizes the two different approaches to maintaining high humidities.

TABLE 6.8
Two strategies for maintaining high humidities in refrigerated spaces.

Strategy	Implications
Operating with low air-to-refrigerant temperature differences	Large total area of coils, thus, large-size coils and/or a large number of them. Additional coils mean more fans and the sensible loads that their motors impose on the refrigerated space.
Higher air-to-refrigerant temperature difference in combination with humidifiers	Moderate total coil area, thus typical size and number of coils. Additional latent load imposed on coils, because the water vapor introduced by the humidifiers must constantly be removed by the coils.

The recommendation of operating with high air-flow rates to maintain a high humidity should be qualified when refrigerating floral products or leafy vegetables. Experience indicates that when these products are not covered or protected in some way from air currents, that increased velocities actually dry out the product, even though the humidity of the air has been increased. The conclusion is that if elevation of the humidity incorporates increasing the air-flow rate over the coil, it is imperative to protect the product from air currents in the refrigerated space.

An additional consideration is what happens at part load—the condition when outdoor air is cool and dry. In one sense, the moisture brought in with the high-humidity infiltrated air was a benefit because it helped maintain high humidity in the space. When the outdoor air has low humidity, this source of moisture is no longer available. The saving feature, however, is that the sensible load will also have dropped when the outdoor conditions are mild, which permits the coils to be operated with a low air-to-refrigerant temperature difference. Many designers anticipate that, when the design conditions are accommodated, the part-load operation will not require special provisions.

At the opposite end of the humidity spectrum is the requirement to maintain low humidity in the space. The following example illustrates a typical difficulty and how it might be solved.

A space storing seeds is to be maintained at 5°C (41°F) and 50% relative humidity and has design refrigeration loads of 110 kW (375,300 Btu/hr) sensible and 20 kW (68,200 Btu/hr) latent. To satisfy the ratio of sensible and latent loads, the air leaving the coil must be along the load-ratio line, shown in Fig. 6.35a.

The load-ratio line is positioned by proportioning Δh_s and Δh_l such that they match the ratio of the sensible-to-latent loads, namely, 110 kW to 20 kW. Figure 6.35a shows that the load-ratio line does not intersect the saturation line until an unreasonably low temperature of −19°C (−2°F). The remedy, as shown in Fig. 6.35b, is to perform cooling and dehumidification with a moderate refrigerant temperature, and then reheat the air back to the load-ratio line. The

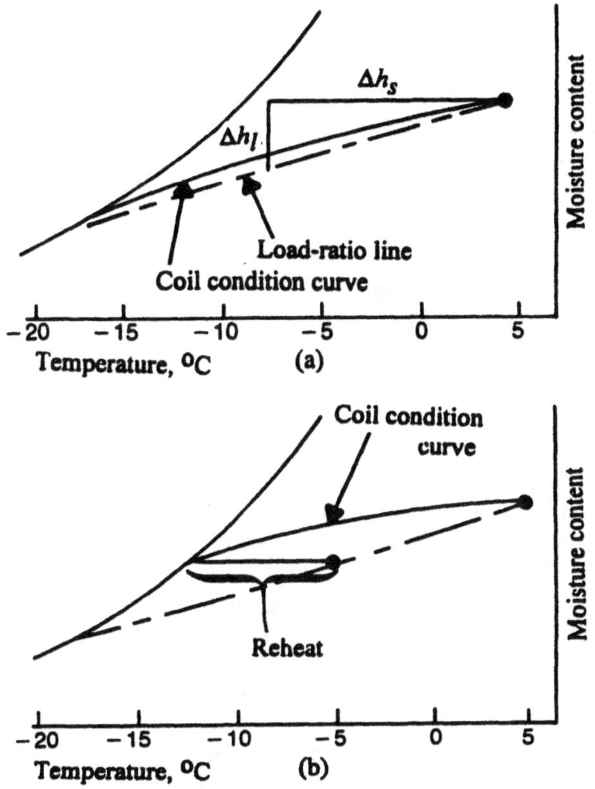

FIGURE 6.35
Meeting the requirements of a low-humidity room by (a) operating with an extremely low temperature off the coil, or by (b) utilizing reheat.

reheat energy need not necessarily be applied to the outlet air of the coil, but can be introduced anywhere in the space. The net result of the supplementary heat is to change the load-ratio line of the space. When the layout of the refrigeration system is appropriate, the heat for reheat can be supplied by discharge vapor from the compressor, which results in energy effectiveness of the operation.

Another low-humidity application where reheat may be beneficial is on loading docks adjacent to frozen-food storage areas. The temperature setting for the dock may be approximately 7°C (45°F) and when the outdoor temperature drops to this temperature range there may be little call for cooling by the dock coils. The relative humidity could rise in the dock area such that this high-humidity air infiltrates to the low- temperature space resulting in the deposit of considerable frost on the coils. Many operators equip their docks with reheat coils supplied with compressor discharge gas to force the refrigeration coils to cool and especially dehumidify the air when the outdoor temperatures drop to near the temperature settings on the dock.

6.18 FAN AND MOTOR PERFORMANCE AND SELECTION

Section 6.16 investigated the selection of coils from a manufacturer's catalog, and that operation constituted the major decision with respect to the coil. There are several auxiliary choices often available to the designer, namely, whether to select—

- draw-through or blow-through arrangement
- a propeller or centrifugal fan
- a single- or two-speed motor

The manufacturer is responsible for offering coil, fan, and motor combinations that provide the rate of air-flow and the velocities that develop adequate heat-transfer coefficients to transfer the specified rate of heat transfer. The manufacturer is also responsible for determining the motor capacity that will drive the fan under the specific application conditions, for example, conveying high-density air through a frosted coil in a low-temperature space.

One characteristic of a fan-coil combination is the *throw*, for which there should be a definition. Unfortunately, there doesn't seem to be one. In air conditioning practice, throw is defined as the distance from an air outlet to the point where the velocity has dropped to 0.5 m/s (100 fpm). In industrial refrigeration practice, the common usage of the term is the distance from the outlet of the coil to a surface, such as a wall, or to the position where another fan-coil exerts an influence. When designers refer to *throwing* air 30, 60, or 90 m (100, 200, or 300 ft), they imply that there is enough air circulation at that distance to avoid objectional pockets of high temperature. Despite the vague use of the term, throw is an important concept and is affected by several designer choices.

One option available to the designer is to select either a draw-through or a blow-through arrangement of the fan and coil (Fig. 6.36). The blow-through arrangement has the thermal advantage that the heat introduced by the fan motor is absorbed by the air before entering the coil, while in the draw-through configuration, the motor heat warms the air upon leaving the coil. Thus, the coil operates somewhat more effectively in the blow-through arrangement because of the higher mean air temperatures. The advantage of draw-through is that the throw is greater than with the blow-through. Outlet air velocities from the fan in a draw- through arrangement might be as high as 20 m/s (4000 fpm) to achieve 60 m (200 ft) throw.

Coils are usually available equipped with either a propeller fan or a centrifugal fan. These two types of fans have somewhat different pressure-flow and power-flow characteristics, as shown in Figs. 6.37a and 6.37b. If the resistance to air-flow remains constant under all operating conditions, so will the air-flow rate and power. In this case, knowledge of the static pressure and fan power is required only for the design conditions. Resistance does change when frost forms on a low-temperature coil. In Figs. 6.37a and 6.37b, two pressure-flow curves are

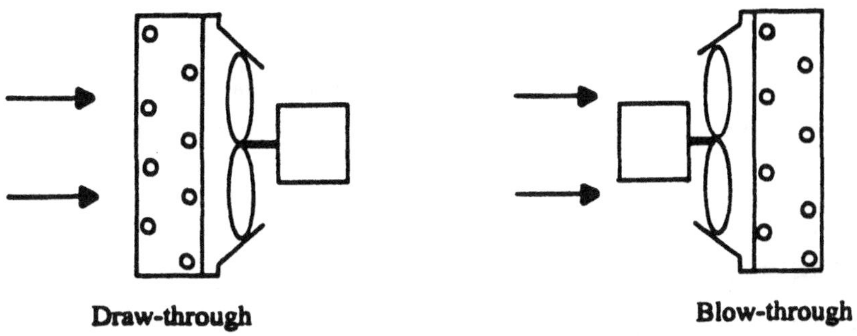

FIGURE 6.36
Draw-through and blow-through arrangements of the fan and coil.

shown—one for a clear coil and the other for a frosted coil. One difference in the response of the fans is that the power required by the motor driving the propeller fan increases, while that of the centrifugal fan drops as frost builds. The propeller fan is used much more frequently than the centrifugal fan in conjunction with refrigeration coils. This is because the shape of the propeller fan curve generally permits a more compact package, and because it is more efficient than the centrifugal fan at the low static pressure that is typical of the coil application. A disadvantage of the centrifugal fan in low-temperature applications is that belt life may be short. A centrifugal fan might be chosen when a very long throw is needed, because the centrifugal fan in a draw-through arrangement can direct a jet for a long distance. An installation requiring long duct work attached to the coil is another situation where the centrifugal can be considered.

Several options are available to reduce the refrigeration capacity of the coil at light refrigeration load:

- interrupting the supply of refrigerant to the coil
- elevating the evaporating temperature
- shutting off the fan or shutting off one fan in a multiple-fan unit
- reducing the air flow over the coil by shifting a two-speed motor to low speed

The last two techniques in the above list are associated with the motor and its control. An attractive feature of two-speed operation is that when the flow rate drops to half its full value, the static pressure drops to 1/4 and the fan power to 1/8 their rated values. The electric power demanded by the motor may not drop fully by 1/8 because the motor efficiency at the low speed drops off abruptly at low loads, as shown in Fig. 6.38. Nevertheless, the power saving is usually significant, which provides a bonus by also reducing the heat load introduced to the space by the fan motor.

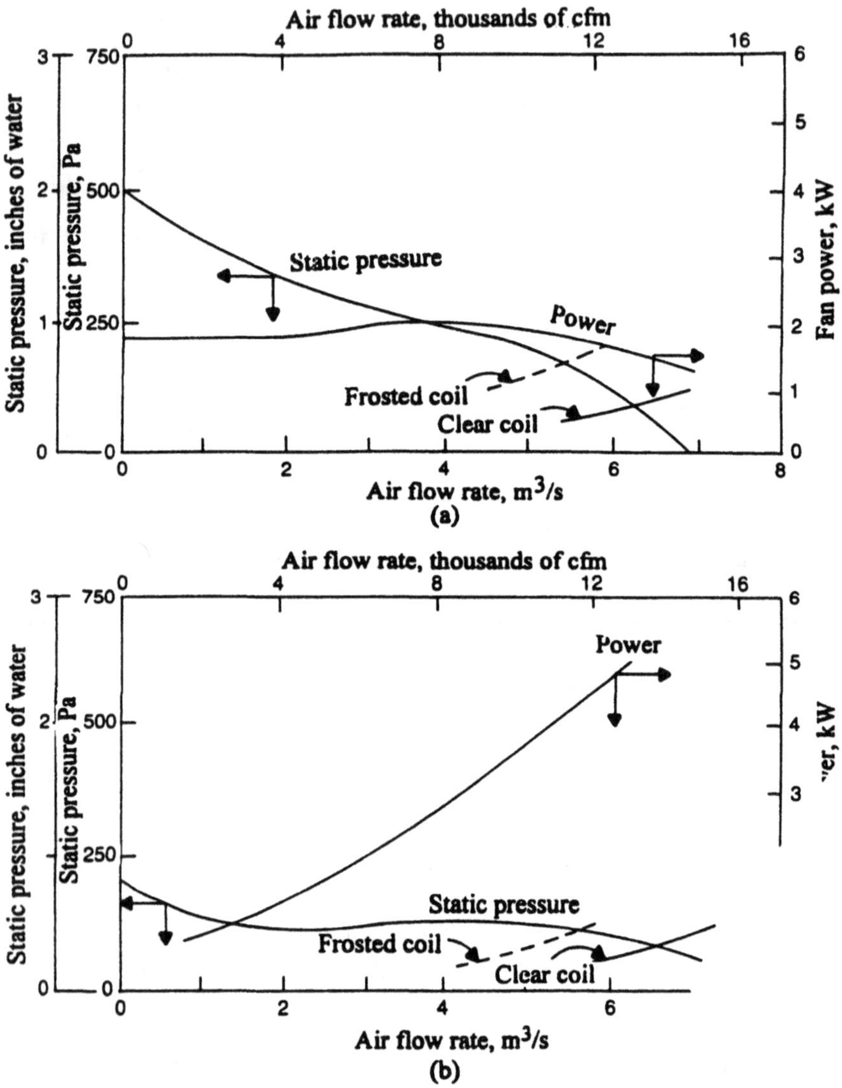

FIGURE 6.37
Static pressure and power for (a) a propeller fan, and (b) a centrifugal fan—both operating at constant speeds.

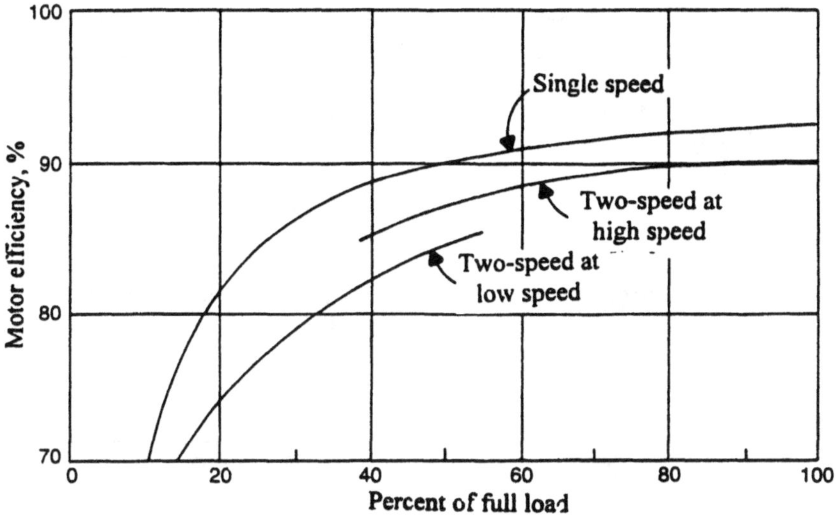

FIGURE 6.38
Efficiencies of single-speed and two-speed electric motors[10].

In a multiple-fan unit one or two fans can be cycled off to reduce the capacity and internal heat load from motors. When restarting an idle motor, the operating fan(s) should be stopped first. The reason for this procedure is that when one fan is operating, the idle one is likely to be spinning in reverse, and when power is applied to it for startup, the motor may overload. Another version of two-speed fan operation that is adaptable to a centrifugal fan installation having several scrolls is to apply pony motors. Here a motor is connected to each end of the shaft, with the speed of one motor twice that of the other. The fan is then driven by one motor while the other idles.

One of the fan laws states that the power required by a fan operating at constant speed varies as the density of air that it delivers. Thus if the motor is sized for a fan operating with air at room temperature and a density of 1.14 kg/m^3 (0.0714 lb/ft^3) and actually operates in a space at −21°C (−6°F) where the air density is 1.39 kg/m^3 (0.0870 lb/ft^3), the actual load on the motor will be about 20% higher than rated. This potential misapplication is inherently compensated by the fact that the temperature of the motor windings is a principal criterion influencing the rating. Since the motor is operating in a low-temperature space, it usually can be loaded higher than its rating.

6.19 THE NUMBER AND PLACEMENT OF COILS

Decisions about the number of coils, their locations in the refrigerated space, and the directions of air discharge are interrelated and influenced by room geometry, ceiling arrangement, and product placement. The sum of the refrigeration

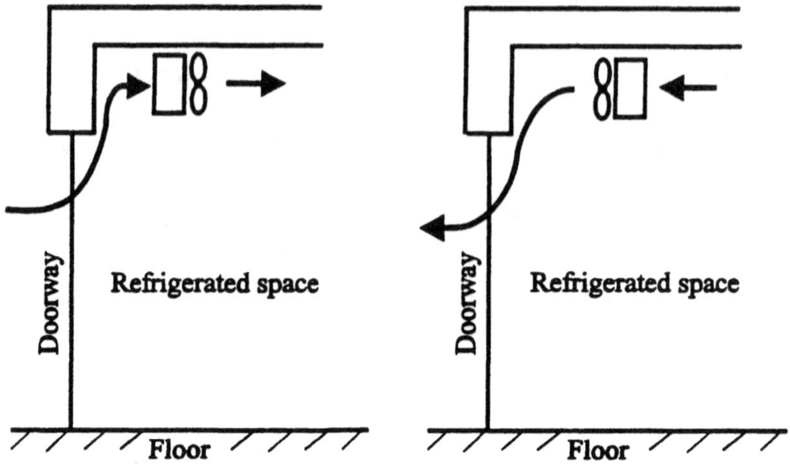

FIGURE 6.39
Do not mount coils above a doorway such that they draw warm air in through or discharge cold air out through the doorway. Instead, arrange for the air to flow past the doorway.

capacities of the coils must equal or exceed the design refrigeration load, but the number of coils into which the total capacity is divided is a compromise. A small number of large coils usually results in the lowest first cost (and lowest maintenance cost) because of the reduction in piping, valves, and controls for the coils. On the other hand, distribution of the total capacity into a large number of coils helps avoid pockets of high temperature in the space.

The goal of achieving uniform temperatures is dominant in the selection of the number of coils and their placement. Several guidelines frequently used by system designers include:

- select the coils and place them on the basis of a throw of 30 to 60 m (100 to 200 ft)
- blow the discharge air in the same orientation as the beams
- blow across doors and not away or toward them (Fig. 6.39)
- direct the discharge air down aisles
- direct the discharge air downward from ceiling coils in high-rise storage facilities

Sometimes, discharge air is ducted from the coil to convey it further, but most coils probably discharge directly into the space. There is a danger of ducts in low-temperature spaces becoming frosted without any automatic means of defrost. Coils are occasionally mounted near the floor for easy servicing (Fig. 6.40), and the air ducted to the ceiling. Dampers should be installed in the duct, and should close during defrost. Otherwise, the warm coil creates a chimney effect that retards the defrost rate. One of the situations where the floor-mounted unit

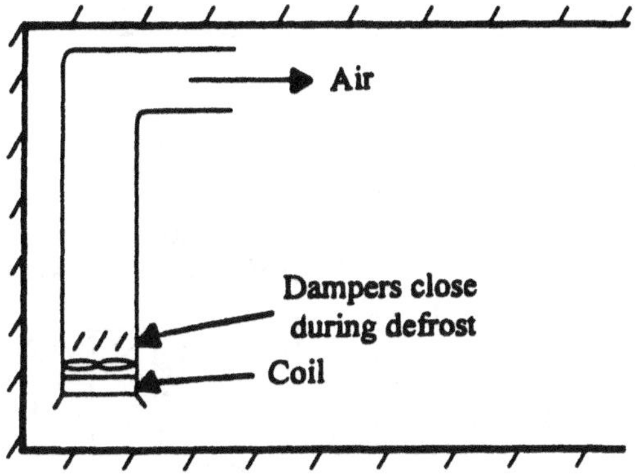

FIGURE 6.40
A floor-mounted coil.

would be considered is when coils are being installed in an existing facility where the roof structure may not be able to accommodate the weight of ceiling-hung or rooftop units.

A concept that is gaining in popularity is the penthouse placement of coils, as illustrated in Fig. 6.41. The coils are clustered in a penthouse and the air directed downward and then in various horizontal directions using short-elbow ducts. The penthouse installation of coils is likely to be the arrangement of highest first cost, but there are some operating advantages of this concept[11]. Space that is otherwise available for the storage of product or processing operations is not occupied by coils. The arrangement yields a safer installation than ceiling-mounted coils where the refrigerant piping is located in the refrigerated space where it is in danger of damage from collisions by fork-lift trucks. The valve groups needed for hot-gas defrost are located outside the penthouse where they can be serviced by technicians more comfortably, safely, and effectively. The refrigerant piping is outdoors above the roof. If there is a need for service personnel to work on the coils or the fans, they can stand on the metal grating of the floor of the penthouse. For such occasions it is possible to manually actuate the defrost to warm the penthouse above the space temperature for a short period of time. Perhaps 15% of industrial refrigeration air coils are now installed in penthouses[11].

6.20 FROSTING OF LOW-TEMPERATURE COILS

When the surfaces of evaporator coils operate at temperatures below 0°C (32°F) and also below the dewpoint temperature of the air, frost will form on the coil. In certain special situations, *icing* can occur where moisture from the air first condenses to liquid water, then freezes to ice. The far more common situation is where the water vapor transforms directly into a solid state, frost. No redeeming merit has been discovered for frost, and no general means of preventing its

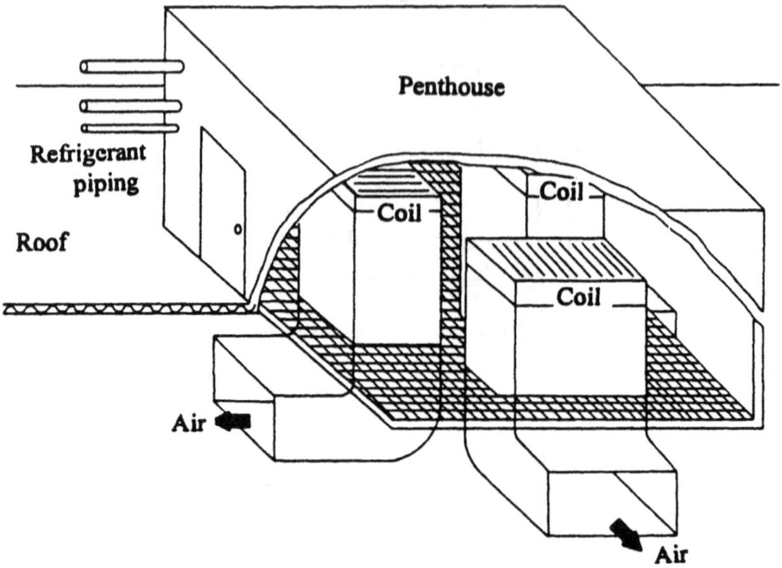

FIGURE 6.41
A penthouse installation of coils.

formation are known.

After accepting that we must live with frost when cooling air to subfreezing temperatures, the approach is to lessen its penalty and to remove it periodically. The two detrimental effects of frost cited most often are: (1) resistance to heat transfer; and (2) restriction of air- flow.

Of these two penalties, restricting the air-flow rate is the most serious. Some laboratory tests[12] were conducted where the rate of air-flow was maintained constant by progressively increasing the fan capacity while frost accumulated (Fig. 6.42). These tests indicated that the U-value was not severely reduced as long as the flow rate and velocity remained constant. The drop in air pressure, on the other hand, increases markedly with additional frost, as Fig. 6.43 shows.

Fan-coil units operating in the field are unable to maintain constant air-flow rates as frost accumulates, in contrast to the way the tests in Figs. 6.42 and 6.43 were conducted. Instead, because of its pressure-flow characteristics, the fan will deliver a lower air-flow rate as the pressure drop increases (Fig. 6.37). This reduction in air-flow rate and associated drop in air velocity reduces the U-value, as Fig. 6.42 shows, and is the major reason for the reduction in heat-transfer rate. This combined behavior of the fan and coil suggests criteria to sense the need for defrost: air-pressure drop across the coil or air-flow rate.

When selecting coils to operate under frosting conditions, the designer should lean toward coils with wide fin spacing and large heat-transfer area. Figure 6.44 shows some experimental data[13] demonstrating that coils with wide fin

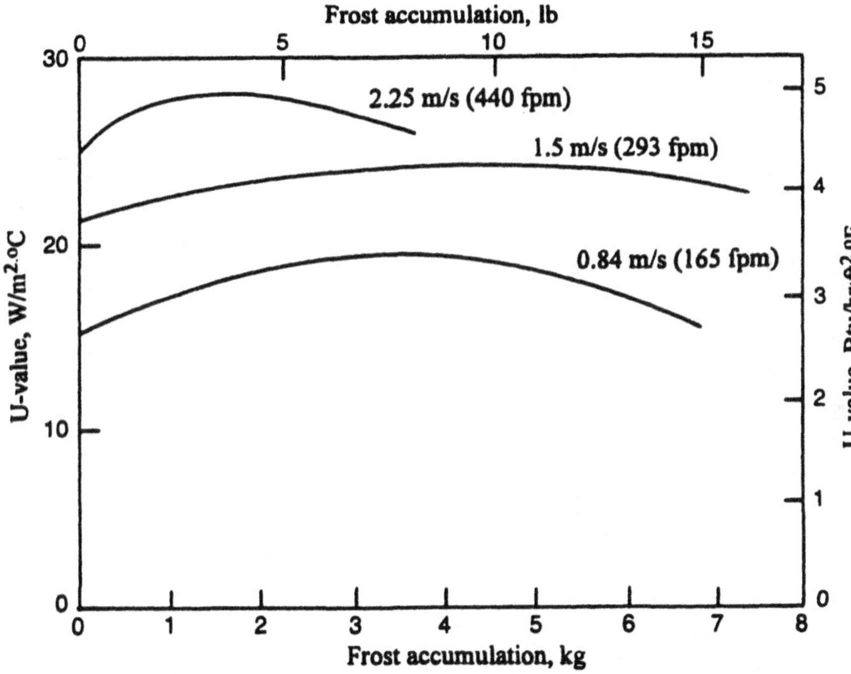

FIGURE 6.42
Overall U-value of a 5-row-deep coil under frosting conditions at three different air velocities[12]. The fin spacing of the coil was 6.3 mm (4 fins/in), and the entering air conditions were 0°C (32°F) and 72% relative humidity.

spacing are not subject to the rapid rise in air-pressure drop experienced by coils with closely spaced fins. Another goal is to select large coils with a low temperature difference between air and refrigerant. The benefit of this choice (see Section 6.17) is a lower rate of moisture removal and thus less rapid frost formation.

6.21 METHODS OF DEFROST

The most widely used methods of defrosting industrial refrigeration coils[14] are using: air, electric, water, and hot gas.

Air defrost may take several different forms. In refrigerated spaces operating at temperatures above about 2°C (36°F), it is possible to shut off the supply of refrigerant to the coil and allow heat from the space to melt the frost. An extremely long period is required unless the fan continues to operate, forcing air through the coil. Even so, the process may be slow, so the designer must insure that the capacity of operating coils is adequate to meet the refrigeration load. Another form of air defrost that was used in the past isolates the coil and the chamber, which is then supplied with warm air from an external source that blows over the coil to defrost it.

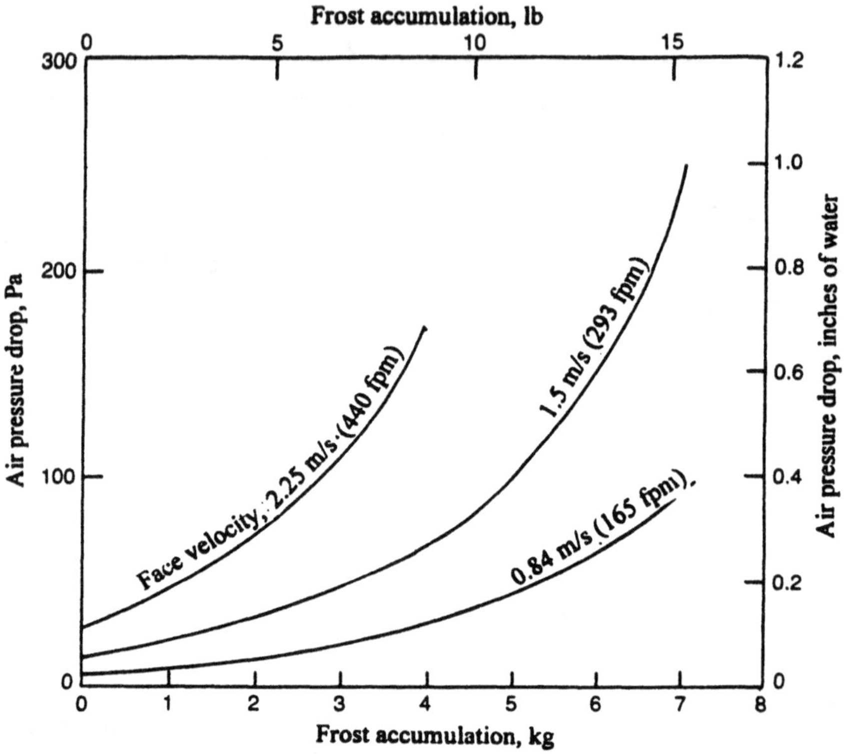

FIGURE 6.43
Air-pressure drop as influenced by frost accumulation for the coil described in Fig. 6.42.

For defrosting electrically, an electric resistance heater is mounted in good thermal contact to the coil. One design that achieves penetration of the heat into the inner sections of the coil is through the insertion during coil assembly of cylindrical heating elements that become dummy tubes. The first cost of electric defrost is probably the lowest of any method, but the operating costs may be the highest. The reason for the high cost is the expense of the electricity in comparison, for example, with hot-gas defrost where the defrost gas is a byproduct of the system operation.

Electric defrost is more common in smaller commercial coils than in large industrial coils, but electric defrost is often adaptable to flooded coils built of aluminum. The sequence of defrost is to first close a valve located between the surge drum and liquid leg, and also close the solenoid valve in the liquid supply line to the surge drum. The fan continues to operate, which supplies heat to the evaporator and evaporates or pumps down the coil. When most of the liquid ammonia is removed, the suction valve is closed, which isolates the evaporator. The heater is then turned on to defrost the coil. When the defrost has been

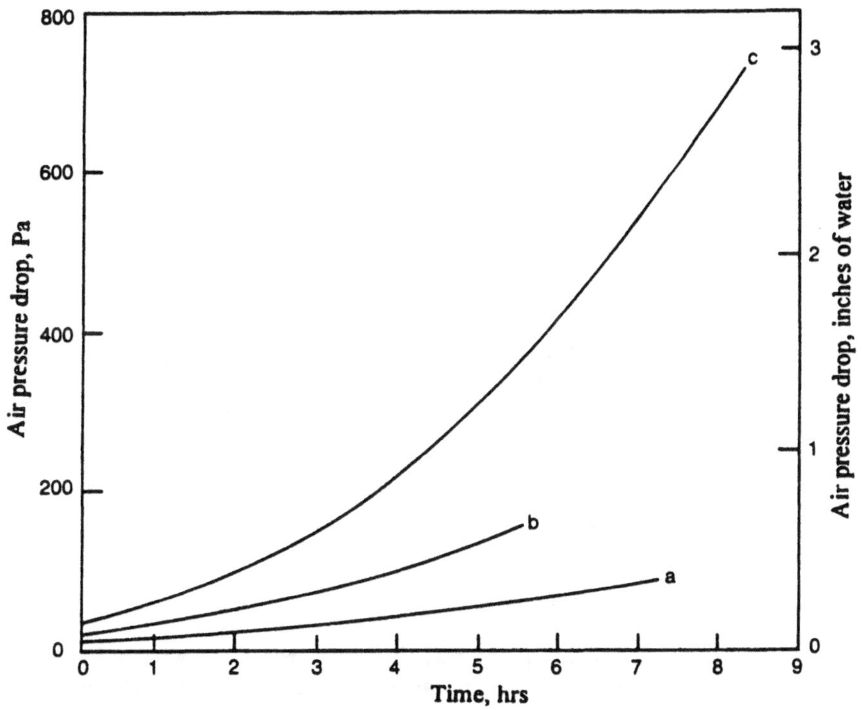

FIGURE 6.44
Increase in air-pressure drop with time for coils of different fin spacing. The face velocity was in the range of 3.2 to 3.4 m/s (625 to 675 fpm), and the entering relative humidity was 82%. The fin spacings are: Curve *a*, 15 mm (1.7 fins/in); Curve *b*, 10 mm (2.5 fins/in); and Curve *c*, 7.5 mm (3.4 fins/in)[13].

completed, the valve in the liquid leg and the suction valve are opened. The coil then becomes cold and freezes any drops of water that have not drained off. Thus, water is not blown off the coil when the fan resumes operation. Complete refrigeration operation resumes when the fans start and the liquid line solenoid valve opens.

Water defrost and hot-gas defrost are the most important methods, and each will be discussed later in their own sections.

Whatever defrost method is used, the drain line carrying defrost water from the coil to outside the space should be trapped, as in Fig. 6.45. Were no trap installed, cold air from the space could be blown to the outside (in the case of a blow-through coil), or warm, moist air from the ambient could be drawn into the space in the case of a draw-through coil. Each defrost water drain should be trapped individually.

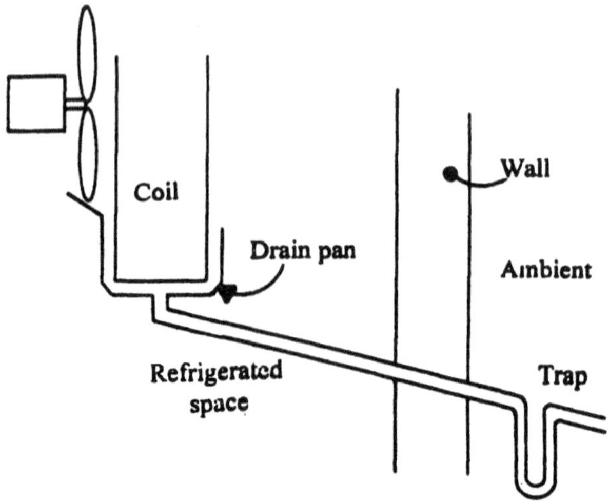

FIGURE 6.45
A trap in the drain line that carries defrost water to outside the refrigerated space.

6.22 HOT-GAS DEFROST

When defrosting an air coil using hot gas, the basic procedure is to interrupt the supply of liquid refrigerant to the evaporator, restrict the outlet of the coil, then supply high-pressure vapor. The interior of the coil then achieves a pressure such that the saturation temperature is high enough to melt the frost on the exterior of the coil. During defrost, then, the evaporator temporarily becomes a condenser.

The next several pages will be organized as follows: (1) standard piping and control sequence for a bottom-feed liquid recirculation coil, (2) defrosting a flooded coil, (3) recommended piping of the defrost gas supplying the coil and of the condensed refrigerant leaving the coil, and (4) recommended sequence for a safe defrost.

The elements[15] of a bottom-feed liquid recirculation coil equipped with hot-gas defrost are shown in Fig. 6.46. The coil has an odd number of rows of tubes with the liquid entering the header on the near side and the mixture of liquid and vapor leaving through the header on the far side and passing on through Suction Valve V1 to the liquid/vapor return line. Admission of liquid refrigerant to the coil is controlled by a solenoid valve that may serve several purposes. It may be connected to a thermostat that closes the valve when the air temperature in the space is satisfied. The liquid supply valve V2 also acts during the defrost cycle. The balancing valve that follows the solenoid valve in the line is set in conjunction with other coils in the system to insure adequate liquid refrigerant supply to all coils.

Table 6.9 shows the status of the four valves in Fig. 6.46 during refrigeration operation and during defrost. The two valves that figure into the defrost opera-

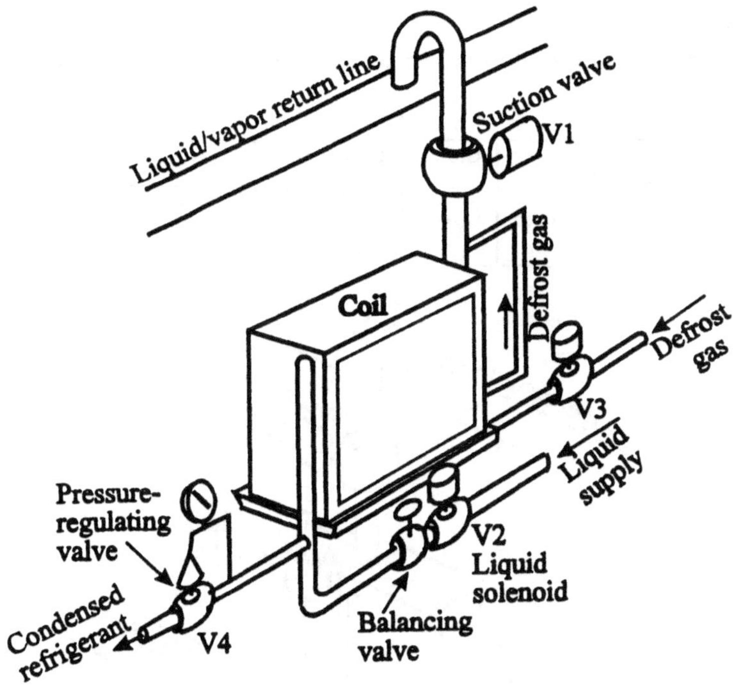

FIGURE 6.46
Elements of a bottom-feed liquid circulation coil equipped with hot-gas defrost.

tion are V3, the solenoid that controls the admission of defrost gas (hot gas) into the coil, and valve V4, a pressure-regulating valve that opens or closes as necessary in an attempt to maintain the pressure upstream of the valve. Thus, Valve V4 attempts to maintain the coil pressure constant. As Table 6.9 indicates, during refrigeration operation the refrigerant supply solenoid and the suction valve, valves V2 and V1, respectively, are open. The defrost gas valve, V3, is closed as is the pressure regulator, V4. Valve V4 is closed because during refrigeration operation the upstream pressure is low—much lower than the pressure setting of the valve. The regulating valve does what it can to try to increase the upstream pressure, which is to close.

During defrost, the liquid supply valve V2 and the suction valve V1 close, which isolates the coil from the normal segments of the refrigeration system. The defrost gas valve V3 opens, allowing vapor from the high-pressure receiver or the compressor discharge line to flow first into the tubes of the drain pan and then into the coil. The drain pan must be heated during defrost, otherwise water and frost would slide down from the coil as it defrosts, but then refreeze in the drain pan. At the initiation of defrost the pressure-regulating valve V4 is still closed. Defrost gas flows into the coil, bringing up its pressure. Since the coil is cold, the incoming vapor condenses in the coil, and during the defrost process

TABLE 6.9
Status of valves in Fig. 6.46 during refrigeration and during defrost operation.

Operation	V1	V2	V3	V4
Refrigeration	Open	Open	Closed	Closed, because valve is attempting to raise pressure
Defrost	Closed	Closed	Open	Opens when coil pressure rise rises above pressure setting

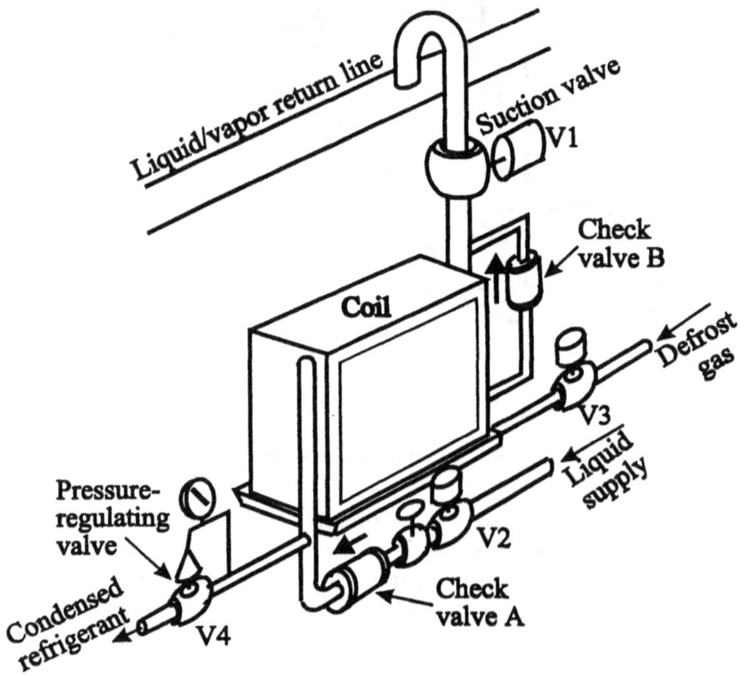

FIGURE 6.47
Addition of two check valves.

the evaporator coil temporarily acts as a condenser. The pressure in the coil continues to rise until the pressure setting of V4 is reached, which is typically about 620 kPa (90 psia or 75 psig) for ammonia or about 680 kPa (99 psia or 84 psig) for R-22. These pressure settings correspond to approximately 10°C (50°F) saturation temperatures of the refrigerants, and represent a high enough temperature to warm the coil and melt the frost from its outside surfaces. The disposition of the condensed refrigerant will be discussed in Sec. 6.23.

Several refinements must be made on the basic coil equipped with hot-gas defrost shown in Fig. 6.46. Figure 6.47 shows the installation of two check valves. The purpose of Check Valve A is to prevent high-pressure refrigerant

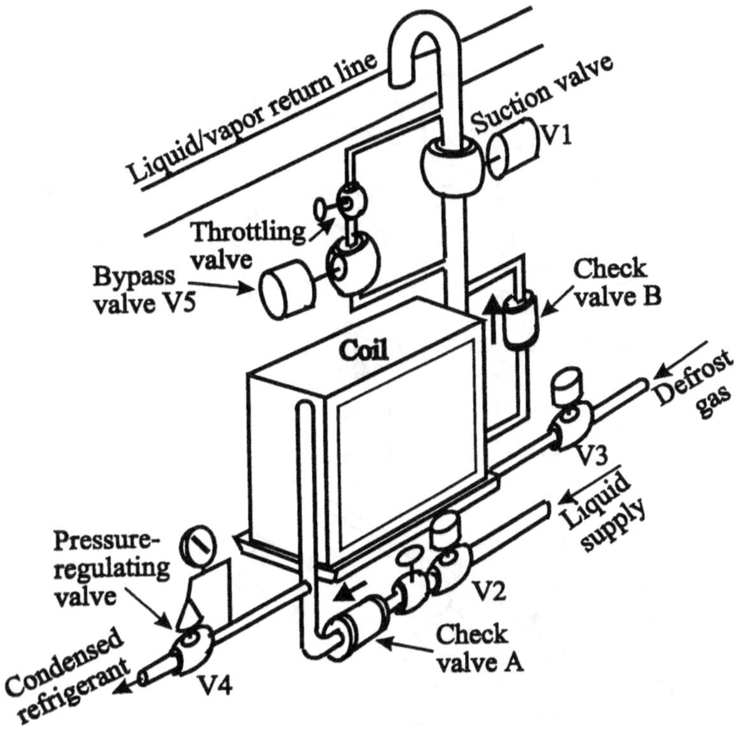

FIGURE 6.48
Bypass around the suction valve to slowly relieve the pressure in the coil at the termination of defrost.

from the coil from backing up through Solenoid Valve V2 during the defrost process. The pressure in the coil may reach 620 to 680 kPa (90 to 99 psia) during defrost and the liquid ahead of Valve V2 is only slightly higher than the operating evaporator pressure. The second check valve, Valve B, prevents liquid refrigerant from entering the drain pan during refrigeration operation. If the drain pan became as cold as the coil, it would collect frost which would drop off during the defrost process on product or whatever is beneath the coil.

The defrost process seems in recent years to have been the cause of occasional damage to equipment and in some cases the rupture of refrigerant lines, resulting in excessive cost and in several cases personal injury. The critical periods during a hot-gas defrost are at its initiation and at its termination. In both situations high pressure vapor that may be moving at a high velocity is brought into contact with cold liquid causing pressure shock waves. The conditions that occur at the initiation of defrost will be discussed in Sec. 6.23. The critical nature of the termination of defrost results when the mixture of liquid and vapor at a pressure of 600 to 700 kPa (85 to 100 psia) rushes into a low-pressure liquid/vapor return line, perhaps at a pressure below atmospheric. One or both of

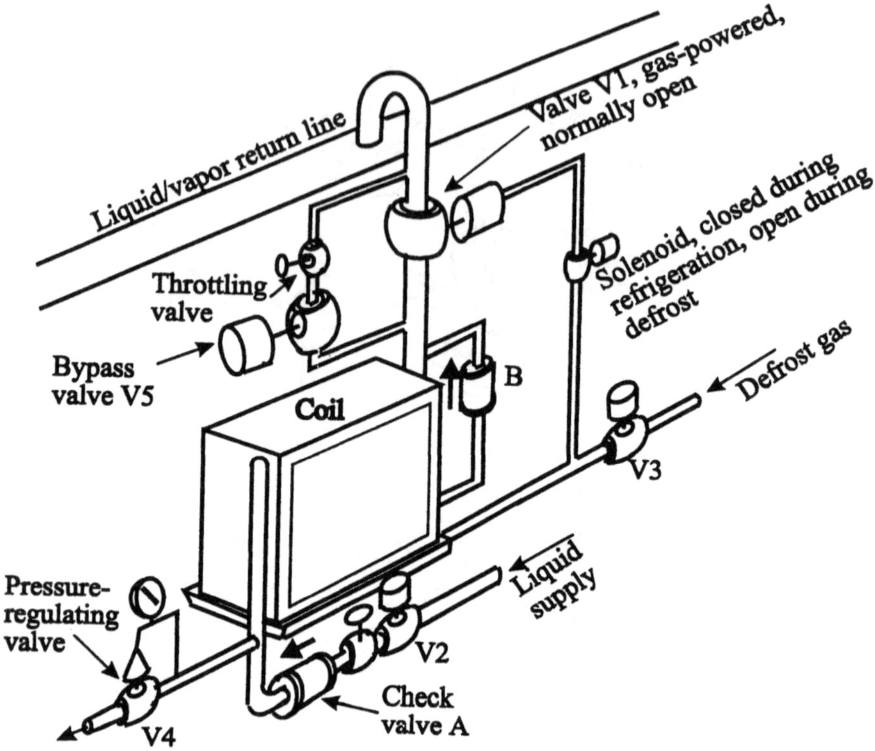

FIGURE 6.49
Defrost control group with gas-powered suction valve, bottom-feed liquid recirculation coil.

two phenomena may occur. The high-pressure vapor may drive the liquid in the return line at high velocity to the end of the suction pipe or to an elbow with such force that the pipe ruptures. Another event may be *condensation shock*,[16] wherein the high-temperature vapor condenses so rapidly on the cold liquid and collapses with such force that the resulting shock wave ruptures a pipe.

To prevent the extreme stresses at the end of defrost, Valve V1 should be equipped with a bypass that slowly drops the pressure in the coil before returning to refrigeration operation. In Fig. 6.48, this bypass consists of a solenoid valve V5 in series with a throttling valve, which opens before V1 opens to slowly relieve the coil pressure. The throttling valve is set to provide the desired rate of pressure decline. There remains the potential hazard that the bypass line, particularly the throttling valve may become plugged, in which case the bypass does not function. However, many hot-gas defrost processes are managed by a microprocessor-based controller which can sense the pressure in the coil and not open V1 until the pressure is low enough to be beyond the danger point.

Because Valve V1 is usually large it is often gas-powered, as shown in Fig. 6.49, drawing its power pressure from the defrost gas.

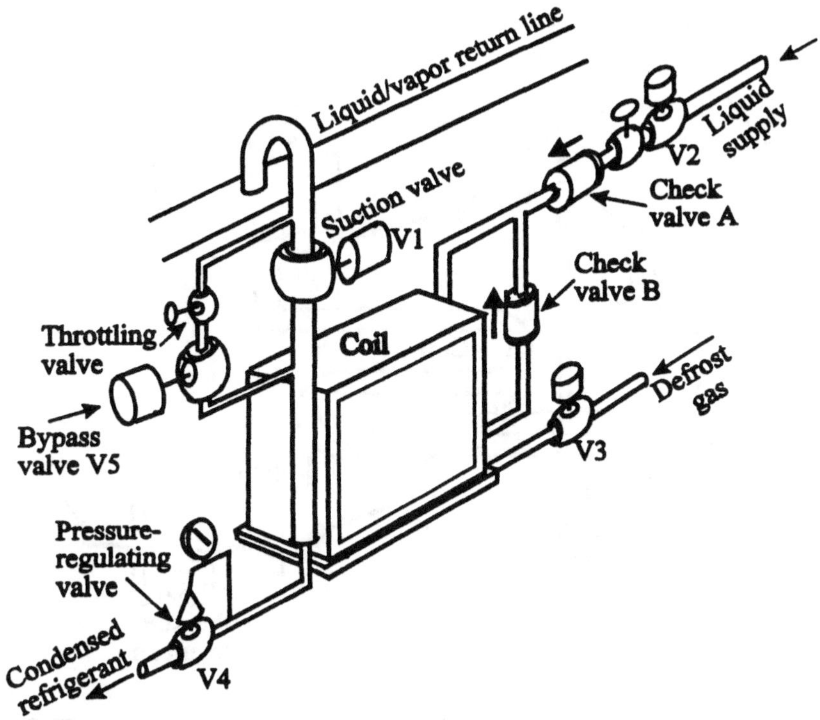

FIGURE 6.50
Defrost control group for a top-feed liquid-recirculation coil.

Operators should be cautioned that if there is excessive noise and shaking of pipes during a defrost, that this is not a normal condition, and the causes should be corrected.

When the liquid overfeed coil is fed from the top, some changes in the piping compared to the bottom-feed coil are required. Figure 6.50 shows the liquid entering the top of a header on the back side of the coil and the mixture of liquid and vapor leaving at the opposite end of the coil. During defrost the liquid valve V2 and the suction valve V1 are closed, but valve V3 admits defrost gas, first to the drain pan and then to the normal entrance of refrigerant to the coil. When the pressure rises sufficiently in the coil, the pressure-regulating valve V4 opens to allow condensed refrigerant to return to a low-pressure suction line.

The elements of the defrost arrangement for a flooded coil are shown in Fig. 6.51. As a preliminary step to defrost, the valve in the liquid supply line to the surge drum should be closed and the operation of the coil continued. This step reduces the amount of liquid in the coil and surge drum. To defrost the coil, isolate it by means of the two valves—one in the liquid leg and the other in the return line from the evaporator to the surge drum. The valve in the defrost gas line is then opened allowing high-pressure gas into the evaporator. Condensation of refrigerant vapor begins immediately, which begins warming the coil.

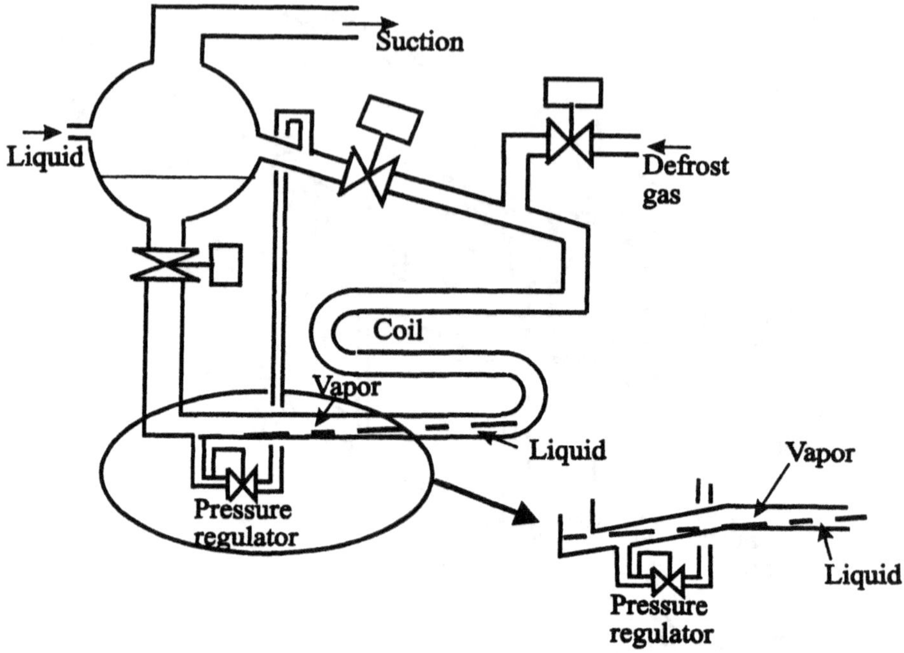

FIGURE 6.51
Valves and piping for defrosting a flooded coil, with the line from the liquid leg to the coil sloped upward to achieve a liquid seal during defrost.

The pressure inside the coil gradually increases until the setting of the pressure regulator is reached, at which time refrigerant is relieved from the evaporator to a pipe or vessel at low pressure, or to the surge drum as shown in Fig. 6.51. When venting to a suction line, the line should be equipped with a vessel to remove liquid before the refrigerant reaches the compressor, because liquid refrigerant issues from the pressure-regulating valve. Since flooded evaporators are often not associated with liquid-recirculation systems which always incorporate liquid/vapor separation vessels, the discharge from the pressure regulator to the surge drum has advantages. It may be necessary, however, to provide an enlarged surge drum to accommodate the liquid condensed in the coil during defrost.

The main diagram in Fig. 6.51 shows a horizontal line between the liquid leg and the lower tube of the evaporator. A problem sometimes occurs because of vapor reaching the pressure regulator restricting the discharge of liquid from the evaporator. In this case the lower row of tubes may not defrost properly. One approach[17] to prevent this problem is to slope the line from the liquid leg up to the tube, as shown in the inset of Fig. 6.51, which provides a liquid seal at the entrance to the pressure regulator and permits only liquid to flow through the regulator.

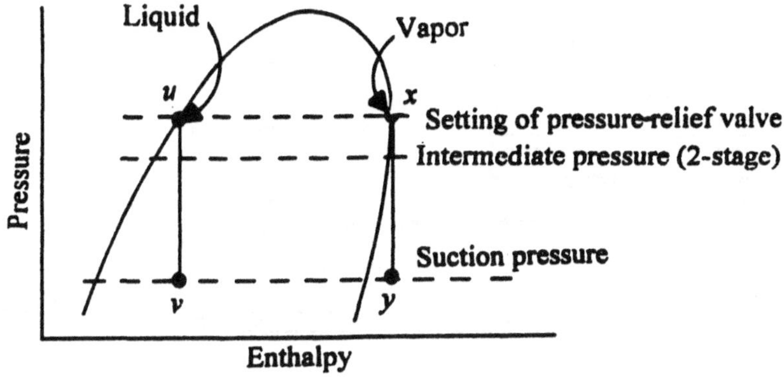

FIGURE 6.52
Pressure-enthalpy diagram of expansion of liquid and vapor being throttled through the pressure-regulating valve.

6.23 PROVISIONS EXTERNAL TO THE DEFROST CONTROL GROUP

The defrost process impacts the system beyond the confines of the coil itself and its associated defrost piping. For example, many plant operators maintain an artificially high condensing pressure, even when ambient comditions would permit the condensing temperature to drop, to provide sufficiently high defrost gas pressure. Since operating with low condensing pressure saves energy, the maintenance of artificially high condensing temperatures is challenged.[18] Some tests conducted with R- 22 in the laboratory and with ammonia in the field[19] indicated that a hot-gas supply pressure of 100 kPa (15 psi) above the setting of the pressure-relief valve could still achieve a defrost. Operators would want to be safe and not push this limit, but the values do suggest that some experimentation be conducted in the plant to determine how low the condensing temperatures can be safely operated.

Another issue is where to direct the discharge from the pressure-regulating valve during a defrost. It may be most convenient from the standpoint of simplicity of piping to discharge into the low-pressure liquid/vapor return line. This arrangement functions reliably, but results in some loss of efficiency. During the early stages of hot-gas defrost, the refrigerant passing through the pressure-regulating valve is likely to be in liquid form. In later stages, when the rate of refrigerant condensation has dropped, vapor is also likely to pass through the pressure-regulating valve. Figure 6.52 is a pressure enthalpy diagram showing throttling processes of both liquid and vapor.

The saturated liquid at u drops in pressure at constant enthalpy to point v, which is mostly liquid but also contains some flash gas. When the vapor at x throttles down to y, however, the process is wasteful because no refrigeration can be performed with vapor at y, and the compression from y to the condensing pressure incurs a cost of energy.

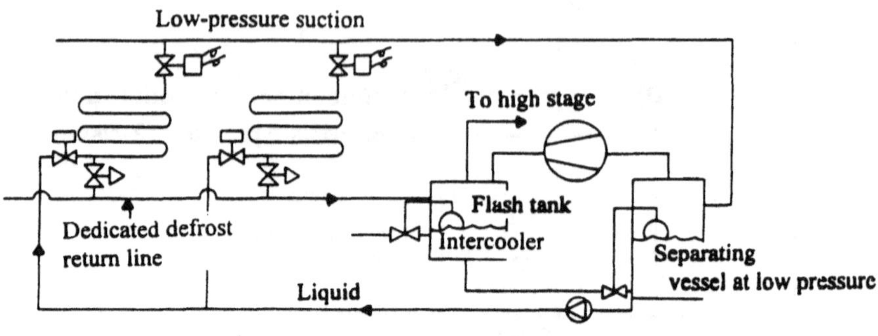

FIGURE 6.53
Returning discharge liquid and vapor from defrosting coils to the intermediate pressure in a two-stage system.

In two-stage compression systems, the intermediate pressure is likely to be lower than the setting of the pressure-regulating valve. When those relative pressures prevail, many designers provide a separate pipe to return the discharged refrigerant from defrosting coils to the intermediate-pressure intercooler/flash tank, as in Fig. 6.53, rather than to the low-pressure suction line. This arrangement improves the efficiency in several ways. Figure 6.52 shows the vapor at x only throttling down to the intermediate pressure, so the recompression from point y is avoided. The liquid at point u is much warmer than the liquid in the separating vessel of Fig. 6.53, and would cancel some potential refrigeration. Instead, the liquid is dumped into the intercooler/flash tank where the liquid temperature is the saturation temperature corresponding to the intermediate pressure, so less refrigeration is lost.

If the refrigerant leaving the pressure-regulating valve is discharged to the intermediate pressure, it is recommended that a check valve be placed in series with the regulating valve, as shown in Fig. 6.54. This check valve prevents refrigerant vapor from leaking back through the pressure- regulating valve when the evaporator is on the refrigeration cycle and operating at a lower pressure than the intermediate pressure.

A traditional question associated with the hot-gas defrost process is whether to take the defrost gas from the compressor discharge where it is superheated (as in Fig. 6.55a), or from the top of the liquid receiver where the gas is saturated (Fig. 6.55b).

The superheated vapor has a higher temperature than saturated vapor, but this benefit may be canceled by the lower heat-transfer coefficient of the gas in comparison to the condensation coefficient. In a plant with screw compressors, the discharge gas already has been cooled. When the defrost gas is taken off the top of the receiver, gas is drawn through the condenser or through its equalizer

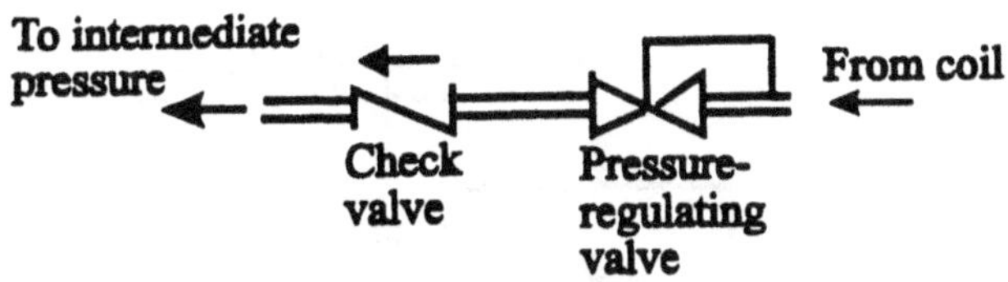

FIGURE 6.54
Placing a check valve in series with the pressure-regulating valve when discharging to intermediate pressure.

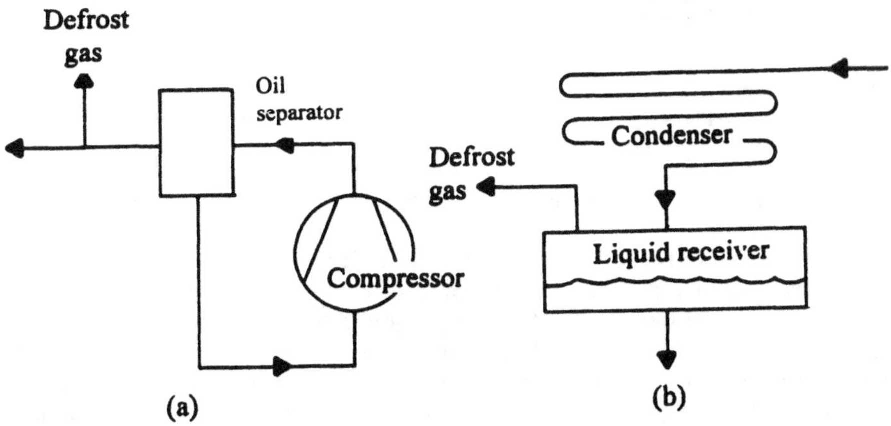

FIGURE 6.55
Source of defrost gas from (a) compressor discharge line, and (b) from the high-pressure liquid receiver.

line uncondensed and supplemented with vapor flashed from the liquid in the receiver. When the liquid flashes into vapor, the liquid temperature drops. But a potential benefit of extracting gas at this location is that the warm liquid provides some thermal storage to aid in supplying defrost gas if the amount being compressed is in short supply.

In developing the defrost plan, provision must be made for an adequate supply of defrost gas. An adequate supply is assured if several evaporators are refrigerating while one is being defrosted. A frequently used rule-of-thumb is that two evaporators should be in refrigeration service while one is defrosted. Some measurements of the flow of hot gas during defrost[19] indicate that this rule is safe, and even conservative when the condensing temperature is maintained low, as discussed above. Another estimate of the heat required for defrosting is in the range of 63 to 126 watts per m^2 of heat-transfer area (20 to 40 Btu/hr per ft^2). The defrost of a large industrial coil may require 10 to 15 minutes, but the defrost must continue longer because the frost that has melted and slipped down

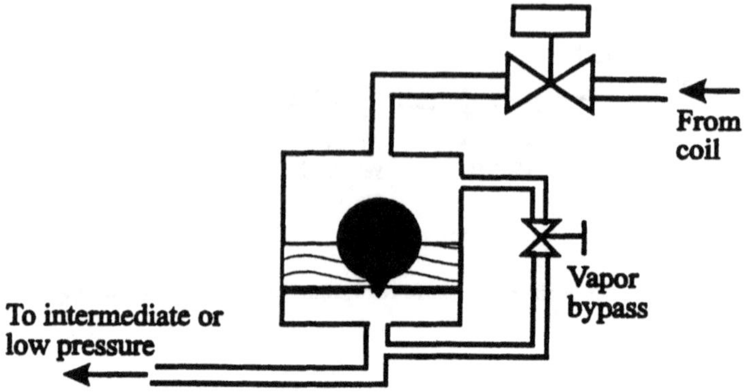

FIGURE 6.56
A liquid drainer replacing the pressure-regulating valve to control the flow of refrigerant condensate from the coil during defrost.

to the drain pan must be disposed of. This latter process may require another 10 to 15 minutes.

The pressure-regulating valve may introduce inefficiencies, as suggested by process *x*-*y* in Fig. 6.52. The sequence of events occurring in the pressure-regulating valve is likely to be that only liquid refrigerant arrives at this valve in the initial minutes of defrost. As the coil becomes warm, the condensation rate decreases and vapor along with liquid enters the regulating valve. Because passing vapor through the valve is inefficient, another approach to controlling the flow of condensed refrigerant is to use a liquid drainer, as in Fig. 6.56. The liquid drainer is a float valve that opens to allow liquid to pass, but closes when the liquid level drops. Also shown in Fig. 6.56 is a small vapor bypass to prevent vapor binding of the valve. The liquid drainer controls a different behavior of the pressure in the coil during defrost in comparison to the pressure-regulating valve. The regulating valve allows the pressure in the coil to rise to its pressure setting, where it remains. With the drainer, the pressure in the coil will continue to rise until it nearly reaches the pressure of the incoming defrost gas. Equipping the coil with the bypass valve of Fig. 6.48 is particularly critical so that high coil pressure is not suddenly discharged into the suction line when the defrost terminates.

6.24 PROCEDURES FOR A SAFE HOT-GAS DEFROST

Because of the high incidence of damage in recent years to industrial refrigeration plants attributable to hot-gas defrosts, the recommended procedures are now more tightly prescribed[20]. The critical times are at the initiation and at the termination of the defrost. The piping, valves and controls should be provided to achieve the following sequence.

Keep defrost gas mains free of liquid. Even before any defrost steps associated with the coil begin, the condition of the refrigerant in the defrost gas main from the machine room to the various coils should be addressed. These lines may run through low-temperature spaces or even above the roof and may be cold at times of the year. The moderately high-pressure defrost gas is likely to continuously condense, and the resulting liquid can be troublesome when the defrost begins. These mains should be equipped with liquid traps (like the float drainer of Fig. 6.56) which continuously pass liquid to a low-pressure pipe or vessel.

Pump out the coil. Close the solenoid valve in the liquid refrigerant supply line to the coil. Keep the suction valve open and continue fan operation, which will boil liquid refrigerant out of the coil.

Start defrost. Stop the fan, close the suction valve, and open the valve in the defrost line, preferably in stages. This is one of the crucial moments, because any retained liquid in the defrost mains that has not been purged and liquid remaining the coil after the pumpout can be propelled around the coil, resulting in possible damage. Two parallel valves, one small and the other capable of handling the total defrost gas flow, may be controlled so that the small one allows a low flow to build up the pressure in the coil before opening the main valve.

Complete the defrost. Once the defrost is in progress, allow it to continue until all the defrost water has melted and as much as possible drained. Failure to allow the pan to drain adequately will cause a progressive problem with the undrained water freezing in the pan when refrigeration resumes. On the next defrost the thickness of ice increases until it surrounds the lower tubes and possibly crushes them. Close the defrost gas valve.

Slowly relieve the pressure in the coil. The bypass valve around the main suction valve, as shown in Fig. 6.48, is first opened to slowly bleed down the pressure so that high-velocity vapor does not rush into the liquid/vapor return line. A high flow rate of vapor could propel the liquid in the line against end caps, elbows or valves. Furthermore, high-temperature vapor contacting cold liquid could cause condensation shock which reverberates throughout the piping. The combination of the bypass and main valves is available[21] in one body with the valve first opening to 10% of its capacity, then when the pressure differential has dropped to 150 kPa (22 psi) the main valve opens.

Open the solenoid in the liquid refrigerant supply line. When this valve is opened, the coil is once again refrigerating, but the fan is not yet started. The reason for the delay is to permit the coil to become cold and freeze droplets of water that cling to the surfaces during defrost. Were the fan started at this time, this water would be blown off the coil and into the space or on product where it would quickly freeze.

Restore fan operation. This final step completes the defrost sequence.

6.25 WATER DEFROST

The second-most-popular method of defrosting industrial refrigeration coils is by spraying water on the coil, although the method is a distant second to the popularity of hot-gas defrost. Water is sprayed over the coil and the mixture of water and melted frost is collected by the drain pan and passed outside the refrigerated space. There are some situations where water defrost is advantageous, and some where its use is not recommended. Spraying water on the coil when the objective is to remove water from the surface may seem incongruous, but the real goal is to attack the frost which is on the outside surface of the coil, and water defrost meets this objective.

Several advantages of water defrost in comparison to hot-gas defrost are: (1) inexpensive source of defrost medium, (2) short defrost time, and (3) provides a cleaning action on the coil. Water defrost may be appropriate when the refrigeration system is dedicated to only the coil(s) in question and the source of refrigerant vapor is limited for hot-gas defrost. Water defrost finds application in a variety of installations, including large low-temperature spaces, but one of the ideal applications is in a spiral freezer chamber operating in a production mode. In this application, the defrost must be rapid in order to return to production quickly. Furthermore, during the defrost and cleanup interval the temperature in the chamber usually rises above freezing temperatures, so the the presence of excessive water is not a problem. In a refrigerated warehouse, on the other hand, an increase of temperature in the refrigerated space is to be avoided.

Some design specifications and precautions surrounding water defrost[22] are:

- the rate of defrost water should be between 1 to 1.36 L/s per m^2 of coil face area (1-1/2 to 2 gpm per ft^2).
- a water temperature of about 16°C (61°F) is an acceptable compromise. The higher the water temperature the more rapid will be the defrost, but a high water temperature also results in excessive fogging in the neighborhood of the coil. The rate of vaporization of water into fog is controlled by the water vapor pressure which in turn is a function of the water temperature. Furthermore the water vapor pressure is 3-1/2 times higher at 30°C (86°F) than it is at 10°C (50°F).
- the rate of water to be discharged is that of the melted frost plus the defrost water, so the quantity that the drain pan and drain lines must handle will be considerable more than with a coil defrosted by hot gas.
- the solenoid valve controlling the defrost water should be in a warm environment so that the water line will not freeze. Also, from the position of this valve, the piping should be sloped so that negligible water is retained in the line between the valve and the sprayheads at the coil.
- the pumpout phase to evacuate the coil of refrigerant first specified for hot-gas defrost is equally important for water defrost.

6.26 ALUMINUM VERSUS STEEL COILS FOR AMMONIA

Section 6.9 has already listed the most popular choices of materials of construction for air coils, including copper tubes and aluminum fins for halocarbon coils and steel or aluminum for ammonia coils. Both all-steel galvanized coils and all-aluminum are widely used in industrial refrigeration and each has advantages and disadvantages. The early aluminum coils suffered bad publicity because of tubes and tube-bends rupturing, but aluminum coils manufactured today are more rugged than those of the past. The density of aluminum is one-third that of steel, which reflects directly on the weight of the coil and the strength of the required supporting members. The thermal conductivity of aluminum is 4-1/2 to 5 times that of steel, so it is expected that an aluminum coil of given construction will transfer a greater rate of heat than a steel coil. The advantages and disadvantages of aluminum for air coils can be summarized as follows:

Advantages of aluminum:

- lighter weight
- better heat transfer. While aluminum may possess five times the conductivity of steel, the influence of this factor applies only to heat transfer through the fins and through the tubes. The air-side and refrigerant-side heat-transfer coefficients dominate, so the favorable conductivity of aluminum results in a 10-12% improvement in heat-transfer rate for a coil of given construction.
- less corrosive in acidic and dry SO_2 atmospheres[23] (but see also disadvantages of aluminum)
- achieves more rapid defrost

Disadvantages of aluminum:

- inability to handle stress and physical blows
- higher cost than galvanized steel
- more difficult to repair in the field
- more corrosive when subjected to chlorine in cleaning solutions and when in contact with calcium chloride brine

6.27 DIRECT-EXPANSION AMMONIA COILS

Some veteran industrial refrigeration practicioners will state flatly that direct expansion should never be used with ammonia. This absolute condemnation of direct expansion for ammonia air coils should be challenged, because some of the unsuccessful attempts of the past can be prevented by proper coil design. Furthermore, a direct-expansion coil operates with a relatively low refrigerant quantity in the coil, and low-charge ammonia systems are the goal of safety-conscious designers.

The term direct expansion is synonomous with the use of a superheat-controlled expansion valve, as shown in Fig. 6.9. Typical expansion valves of this type require as much as 7°C (12.6°F) to open completely. For low-temperature coils the temperature difference between entering air (which is the same temperature of the space) and the refrigerant as listed in Table 6.7 is usually a maximum of 5.5°C (10°F). Consequently, there is not a temperature available that is high enough to open the valve completely. This situation is different than for spaces above the freezing temperature, and certainly for air-conditioning where there are large air-to-refrigerant temperature differences. A blanket statement, then, that direct expansion should not be used for low-temperature coils is normally good advice.

But even for the potential application of direct expansion in above-freezing applications, the coil must be properly designed[24]. Because of the high latent heat of ammonia, the flow rate to achieve a given refrigerating capacity will be low such that with the typical 3/4- or 1-in tubes the refrigerant velocity will be too low. The result of low velocity is that wavy and stratified flow results in which the heat-transfer coefficient is lower than when high velocities cause annular flow. A further requirement is to construct the circuits long enough that a sufficiently high circuit loading results. Reference 24 recommends no larger than 5/8-in tubes and each circuit carrying a refrigeration load of at least 7 kW (2 tons of refrigeration).

Another development that should be watched because of its pertinence to ammonia direct expansion is the availability of oils mutually soluble in ammonia. The coil could then operate in a manner similar to halocarbon installations where a sufficiently high velocity of the vapor leaving the evaporator carries oil back to the compressor.

6.28 ANTIFREEZE-SPRAYED COILS

An approach to controlling frost on low-temperature coils is to spray the coil with an antifreeze, such as ethylene glycol or propylene glycol. A characteristic of this equipment is that it generally provides a higher ratio of latent-to-sensible heat removal from the air than is true of a nonsprayed coil. The assembly[25,26,27] which can maintain a continuous operation is shown schematically in Fig. 6.57.

The glycol solution is sprayed on the surfaces of the evaporator to keep the coil free of frost. The air passing in contact with both the sprays and the evaporator surfaces is cooled and dehumidified. The glycol solution absorbs water, so a fraction of it is circulated to the regenerator to drive off the moisture. An alternate is to operate the regenerator intermittently. The regeneration of the glycol is accomplished by spraying the solution over a heating coil through which scavenging air, usually from the outdoors, passes. A property of glycol/water solutions is that the water-vapor pressure at a given liquid temperature is less than that of water alone[28]. The consequence of this property can be demonstrated on

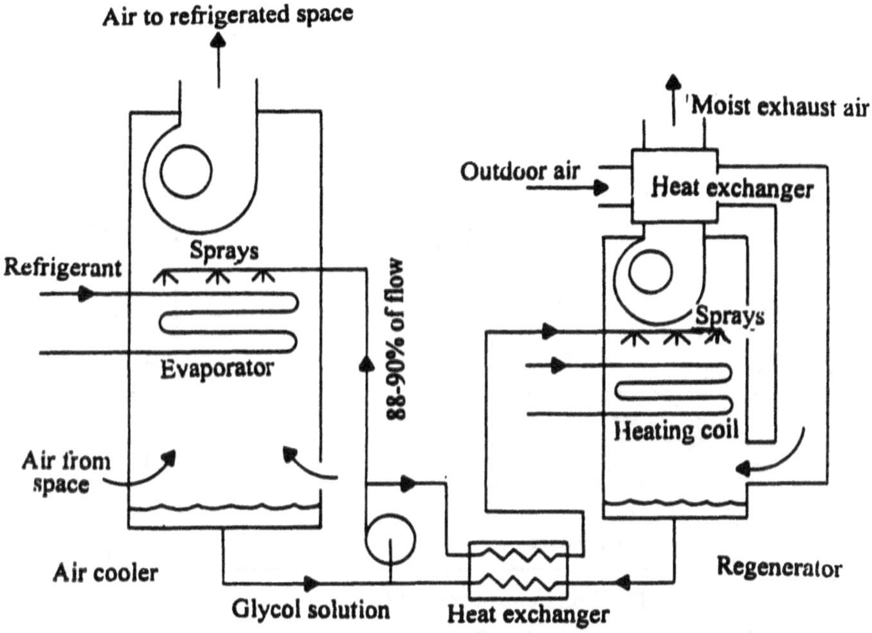

FIGURE 6.57
Schematic diagram of equipment that sprays the evaporator coil with a glycol solution while simultaneously regenerating the solution.

the psychrometric chart, Fig. 6.58, where the water-vapor pressure is chosen as the vertical scale.

A curve for the vapor pressure versus the liquid temperature of a 50%-by-mass solution of ethylene-glycol is superimposed on this psychrometric chart. The straight-line law holds for the air in contact with the glycol solution in the same manner as for an unsprayed coil. For the unsprayed coil, the path of air is A, while the air passing in contact with the glycol solution follows path B. Because the vapor pressure of the glycol solution at a given temperature is lower than that of the water or frost on the unsprayed coil, path B is steeper than path A, which results in a greater amount of dehumidification. The freezing temperature of this solution is −36°C (−32°F), so no frost will form on the coil as long as the coil surfaces are above this temperature.

The advantages of the sprayed coil over the unsprayed coil in subfreezing temperatures are:

- no need to shut down for defrost
- energy expenditure of hot-gas or water defrost is eliminated
- high latent heat removal when excessive humidity is a problem
- glycol provides a germicidal benefit, eliminating many bacteria, molds, and

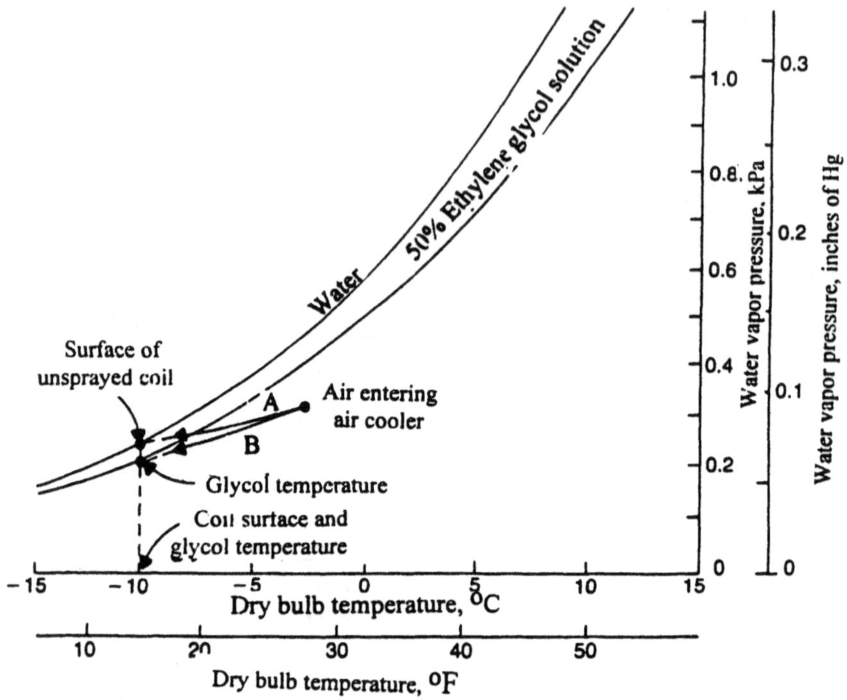

FIGURE 6.58
Condition of air driving according to the straight-line law to the surface temperature of an unsprayed coil (path A), and to the saturation curve of a glycol solution (path B).

other microorganisms.

The disadvantages are:

- higher first cost
- parasitic energy effects of the warm solution returning from the regenerator
- the additional operating cost of the pump and extra fan.

PART III. LIQUID CHILLERS

6.29 SHELL-AND-TUBE EVAPORATORS

In shell-and-tube evaporators the boiling refrigerant chills a liquid such as water or a brine/antifreeze. Some of the major forms of construction include: (1) refrigerant boiling in the tubes chilling in the shell the liquid flowing across the tube bundle with its direction repeatedly reversed by baffles, and (2) refrigerant boiling in the shell surrounding the tubes through which the liquid passes.

Refrigerant in tubes. Most water chillers using halocarbon refrigerants vaporize the refrigerant in the tubes using a superheat-control expansion valve and are thus direct expansion. An exception in the air-conditioning industry are the water chilling packages using centrifugal compressor. The refrigerant boils in the shell in these evaporators because excessive drops in pressure would result were the low-density refrigerant vapor to be sent through the tubes. In industrial refrigeration systems the circuits could be arranged for forced liquid overfeed or could be operated flooded using a surge drum.

Refrigerant in shell. Probably the most popular type of liquid-chilling evaporator in industrial refrigeration practice directs the liquid to be chilled through the tubes which are surrounded by the boiling refrigerant in the shell. These heat exchangers are usually called flooded chillers. Whenever the refrigerant boils in the shell there must be some way of separating the liquid from the vapor before the vapor leaves the evaporator and passes on to the compressor. Two approaches to achieving the liquid-vapor separation are shown in Fig. 6.59 and Fig. 6.60. In the liquid chiller of Fig. 6.59 the tubes do not fill the shell volume, but there is separation space left at the top of the shell. The level-control valve regulates the flow of liquid refrigerant to the bottom of the vessel. When ammonia is the refrigerant an oil pot is provided which is periodically drained of the oil that collects there. Another approach to separation is shown in Fig. 6.60 where a small vessel is mounted above the shell of the main evaporator, which now has all of its internal volume occupied by tubes. This design is used for brine chillers serving ice rinks as well as for many other brine-chilling applications.

The design of liquid chillers is a technical specialty not addressed here. Instead, this section emphasizes the concerns of one who selects and operates liquid chillers. As the required temperature of the liquid being chilled drops, a number of changes occur—all of them bad. The regrettable result is often the inability to chill the liquid to the required temperature. Such a failure is caused by the degradation of the heat-transfer coefficients, both of the refrigerant and the fluid being chilled, especially in the case of brines and antifreezes.

The total resistance to heat transfer, R_{total}, of the chiller is the sum of the individual resistances,

$$\frac{1}{U_o A_o} = R_{total} = \frac{1}{h_{refrig} A_o} + R_{metal} + \frac{1}{h_{liquid} A_i} \tag{6.14}$$

where h = heat-transfer coefficient, W/m^2·K (Btu/hr·ft^2·°F)
A = heat-transfer area, m^2 (ft^2), with subscript o indicating the outside and i the inside area

Equation 6.14 implies that the refrigerant is in the shell and liquid flows in the tubes. For the tube-side coefficient a standard equation is

$$\frac{hD}{k} = 0.023 \left(\frac{VD\rho}{\mu}\right)^{0.8} \left(\frac{\mu c_p}{k}\right)^{0.4} \tag{6.15}$$

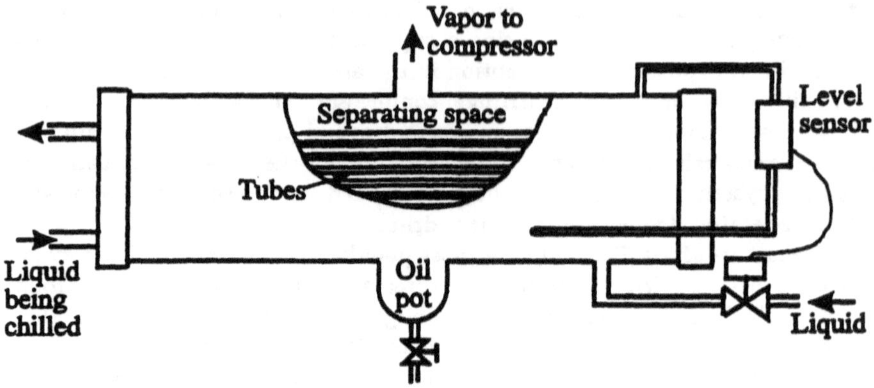

FIGURE 6.59
A shell-and-tube evaporator with refrigerant in the shell and in which separation space is provided above the tubes.

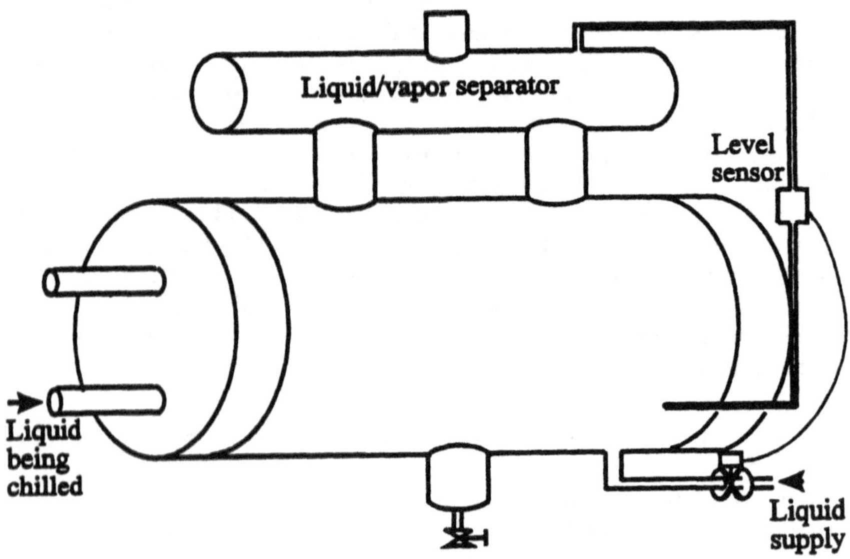

FIGURE 6.60
A shell-and-tube evaporator with refrigerant in the shell and an auxiliary vessel above the main evaporator to facilitate separation of liquid and vapor.

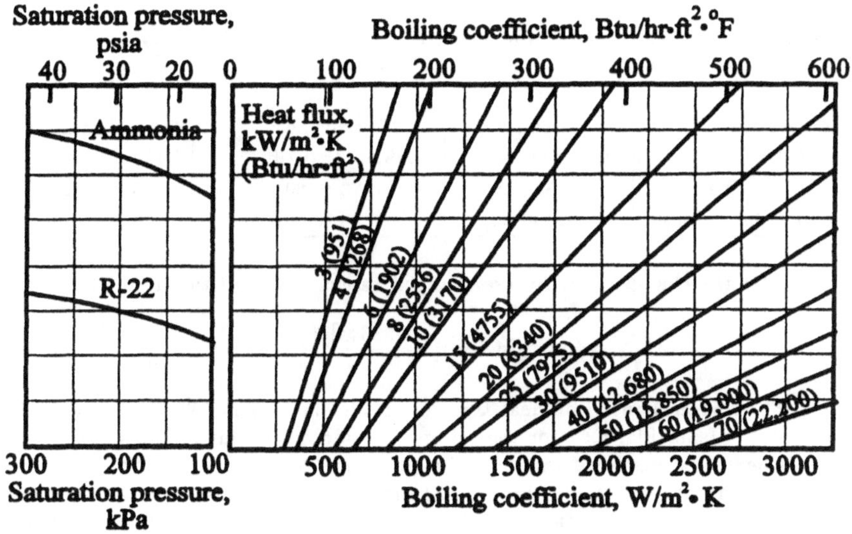

FIGURE 6.61
Heat-transfer coefficient for pool boiling from a horizontal cylinder.

where D = diameter of tube, m (ft)
k = thermal conductivity, W/m·°C (Btu/hr·ft·°F)
V = velocity, m/s (ft/s)
ρ = density, kg/m^3 (lb/ft^3)
μ = viscosity, Pa·s (lb/ft·s)
c_p = specific heat, J/kg·°C (Btu/lb·°F)

An estimate of the boiling coefficient of the refrigerant on the outside of the tubes is provided by the graph[29] in Fig. 6.61. Both for the liquid heat-transfer coefficient inside the tubes and for the boiling heat-transfer coefficient of the refrigerant outside the tubes the heat-transfer coefficients drop as the temperature drops.

Example 6.6. An evaporator chilling ethylene glycol to −5°C (23°F) using a 20% solution (freezing temperature of −10°C (14°F)) is to be revised to chill the glycol to −15°C (5°F) for which it is necessary to increase the strength of the solution to 40% (freezing temperature of −26°C (−15°F)). Meanwhile the evaporating temperature of the ammonia refrigerant is to be dropped from −10°C (14°F) to −25°C (−13°F). What is the effect of these changes on the heat-transfer coefficients inside and outside the tubes?

Solution. Table 6.10 shows property data from Chapter 20 on secondary coolants for the temperatures and concentrations of the ethylene glycol solution before and

TABLE 6.10
Effect on the viscosity, specific heat, and thermal conductivity of changing the temperature and concentration of ethylene glycol solution in Example 6.6.

Temperature and strength	Viscosity Pa·s (centipoise)	Specific heat kJ/kg·K (Btu/lb·°F)	Thermal conductivity W/m·K (Btu/hr·ft·°F)
−5°C (23°F), 20%	0.0036 (3.6)	3.73 (0.89)	0.46 (0.266)
−15°C (5°F), 40%	0.010 (10.0)	3.33 (0.79)	0.38 (0.220)

after the change. When these changed values are substituted into Eq. 6.15,

$$\text{Ratio of new } h_{liquid} \text{ to original} = \left(\frac{0.0036}{0.010}\right)^{0.4}\left(\frac{3.33}{3.73}\right)^{0.4}\left(\frac{0.38}{0.46}\right)^{0.6} = 0.593$$

The revised liquid-side heat-transfer coefficient is thus 59% of the original.

Using Fig. 6.61 for the refrigerant-side coefficient requires specification of a heat flux which we assume to be 10 kW/m^2 (3170 Btu/hr·ft^2). The saturation pressure drops from 290 kPa to 151 kPa (42.1 to 21.9 psia), so the ammonia-side heat transfer coefficient drops from 1870 W/m^2·K to 1700 W/m^2·K (329 to 300 Btu/hr·ft^2·hr).

Equations 6.14 and 6.15 along with Fig. 6.61 can be used by the designer or purchaser to estimate the size of a liquid chiller, although it is ultimately the manufacturer of the heat exchanger who is responsible, since the manufacturer is versed in the specialties of heat-exchanger design.

6.30 SPRAYED TUBE LIQUID CHILLERS

In the concept of spray chillers, as shown schematically in Fig. 6.62, rather than immersing the tubes in liquid as is done in the conventional shell-and-tube evaporator, liquid refrigerant from a circulating pump is sprayed over the tubes. The operating level of the liquid refrigerant in the evaporator is maintained below the bottom tubes. The idea of sprayed-tube evaporators is not new, and even though the overall heat-transfer coefficient of the sprayed-tube evaporator usually exceeds that of the flooded type, the additional complexity of the pump and spray assembly usually dissuaded designers from choosing them. In recent years, however, another strength of the spray-tube evaporator has emerged in importance, and that feature is the low refrigerant charge required.

The improvement in heat-transfer coefficient in comparison to a flooded evaporator occurs because the sprayed tube permits easy escape of the vapor bubbles, which once they form, insulate the heat-transfer surface. In fact, there is an optimum circulation rate in the sprayed-tube evaporator that yields the maximum boiling coefficient. When the flow rate is too high, the liquid film becomes thick and insulates the tubes. When the flow rate is too low some of the tube surfaces do not become wetted. A typical circulation rate[30] is 5 times the

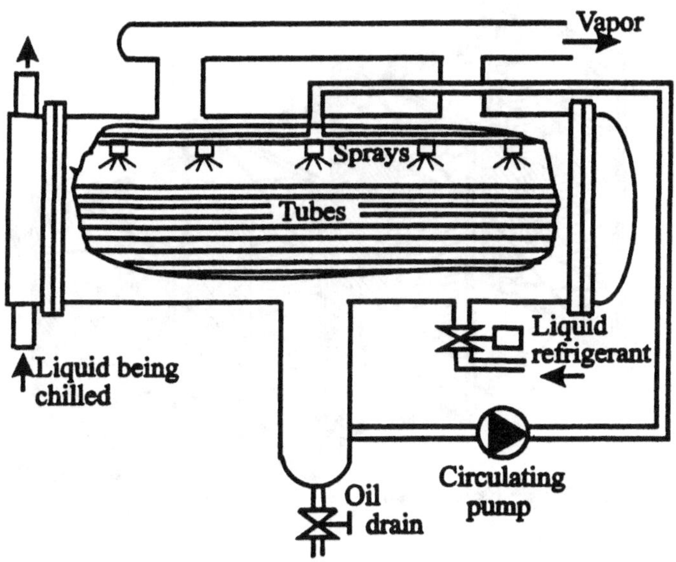

FIGURE 6.62
A sprayed-tube liquid chiller.

rate evaporated. The pump adds a small amount to the power cost, and if the pump fails, the evaporator is essentially out of operation. Often a reserve pump becomes part of the unit.

Reducing the refrigerant charge is an objective either from the standpoint of cost in the case of halocarbon refrigerants or for purposes of safety in the case of ammonia. As a comparison[30], a flooded chiller with a refrigerating capacity of 1407 kW (400 tons of refrigeration) would require an ammonia charge of 5900 kg (13,000 lb), while the sprayed-tube evaporator would require only 159 kg (350 lb). The liquid leg that provides an adequate static head to the pump inlet contains much of the refrigerant charge.

6.31 PLATE-TYPE EVAPORATORS

The plate-type of evaporator is the version of liquid chiller gaining most rapidly in popularity. It is an outgrowth of the *plate-and-frame heat exchanger* which has been applied in the food industry for many decades. The plate-and-frame heat exchanger consists of numerous plates that are gasketed in such a way that when these plates are bolted together one of the fluids flows between two of the plates and the other fluid between the pairs of adjacent plates. The plates are corrugated with a herringbone pattern that physically strengthens the plates and also promotes turbulence of the fluids, providing excellent convection heat-transfer coefficients. This type of heat exchanger is appealing to such food industries as dairies because at the end of a work shift the bolts holding the plates in position

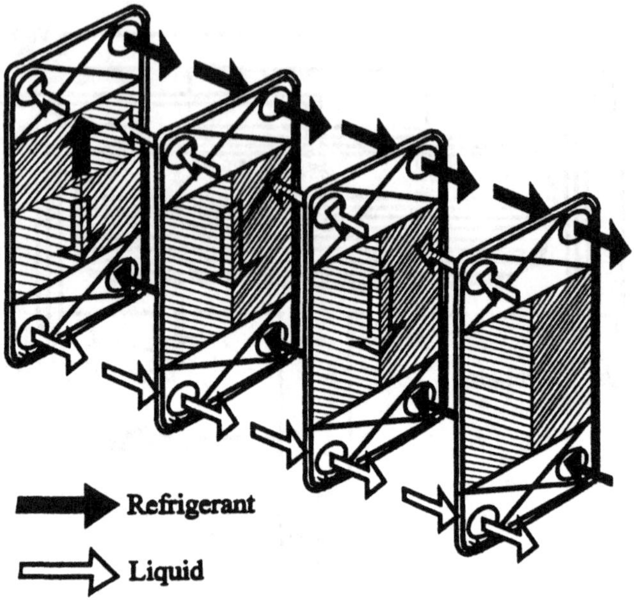

FIGURE 6.63
A plate-type liquid-chilling evaporator.

can be loosened, permitting access to all surfaces for cleaning.

The conversion of the plate-and-frame heat exchanger transferring heat between two liquids poses the challenge of how to seal the refrigerant passages. This goal is achieved by a construction shown in Fig. 6.63 where instead of each plate capable of being separated, pairs of plates forming the refrigerant passages are brazed or welded. For halocarbon refrigerants, normal brazing of the edges suffices, but for ammonia either nickel brazing or welding is necessary. The liquid flows downward between its two boundary plates, while the refrigerant flows upward, thus, counterflow to the liquid. The refrigerant enters the evaporator at the lower right, with a portion of the refrigerant flowing upward through the first pair of refrigerant plates and the remainder passing on to succeeding pairs. At the end refrigerant pair on the left of the diagram, the near plate is shown in an exploded view to illustrate the refrigerant flowing upward. Each stream of refrigerant leaves its pair at the upper right and joins the streams from the other pairs, finally leaving the evaporator in the upper right corner. The construction illustrated in Fig. 6.63 has a confined gasketed passage between one refrigerant pair and another, so while the edges of the plates are welded, there is a small gasketed joint.

The outside boundary of the liquid passages are the outside of the refrigerant pair. Some small plate-type evaporators for halocarbon refrigerants are manufacturered without gaskets by brazing all the connections. Industrial evaporators, which are usually larger, are made by bolting together the refrigerant pairs. It is possible, then, to dismantle the evaporator to clean the liquid-side surfaces should they become fouled. Such dismantling means that the refrigerant passages are also opened to air, so the refrigerant must first be evacuated.

The major strengths of the plate-type evaporator are: (1) high heat-transfer coefficients, (2) low refrigerant charge, and (3) small size. These advantages are related, because whenever the heat-transfer coefficient can be improved, the heat exchanger can be smaller for a given refrigerating capacity. A small exchanger automatically results in lower refrigerant charge. Ranges of overall heat-transfer coefficients are reported by one source[31] as 2500 to 4500 $W/m^2{\cdot}K$ (440 to 790 $Btu/hr{\cdot}ft^2{\cdot}°F$) for water/ammonia and 1500 to 3000 $W/m^2{\cdot}K$ (265 to 530 $Btu/hr{\cdot}ft^2{\cdot}°F$) for water/R-22. Another source[32] suggests values for a flooded or recirculated water/ammonia evaporator of 2840 to 3975 $W/^2{\cdot}K$ (500 to 700 $Btu/hr{\cdot}ft^2{\cdot}°F$). For direct-expansion ammonia, an overall heat-transfer coefficient can be expected in the range of 2275 to 3400 $W/m^2{\cdot}K$ (400 to 600 $Btu/hr{\cdot}ft^2{\cdot}°F$).

The three major types of refrigerant feed—direct expansion, flooded with surge drum, and forced liquid overfeed—are all used with this type of evaporator. Flooded and liquid overfeed usually function more effectively, but the direct-expansion arrangement is simplest. A difficulty with direct expansion is achieving uniform flow distribution to each of the refrigerant passages.

REFERENCES

1. Chawla, J. M., "A Refrigerant System with Auxiliary Liquid and Vapor Circuits," Proceedings of Meeting of Commissions II and III, International Institute of Refrigeration, London, 1970.
2. Baker, O., "Simultaneous Flow of Oil and Gas," *Oil and Gas Journal,* 53(12): 185-195, July 16, 1954.
3. "What is the Ideal Height of the Surge Drum Above a Flooded Evaporator Coil?" *NH_3 News*, International Institute of Ammonia Refrigeration, Sep./Oct., 1995.
4. Personal Communication, Edward Ames, August 28, 1996.
5. *ASHRAE Handbook of Fundamentals,* American Society of Heating, Refrigerating and Air-Conditioning Engineers, Atlanta, Georgia, 1997.
6. Carrier Corporation, a Subsidiary of United Technologies Corporation.
7. Stoecker, W. F., *Principles for Air Conditioning Practice,* Industrial Press, 1968.
8. Lewis, W. K., "The Evaporation of a Liquid into a Gas," *ASME Trans.*, 44:325, 1922.
9. Anderson, R. W., "How to Eliminate Condensate and Save Energy in Meat and Poultry Plants,: *ASHRAE Journal,* 24(8): 30-34, August 1982.
10. Pisati, A., "Single Circuit Multiple Compressor Water Chiller," International Conference on Components of Air Conditioning and Refrigerating Systems, sponsored by AICARR and REHVA, Milan, Italy, February 1984.
11. "Unit Cooler Location in Cold Storage Warehouses," Seminar at Annual Meeting of ASHRAE, New York, January 1987.
12. Stoecker, W. F. "How Frost Formation on Coils Affects Refrigeration Systems," *Refrigerating Engineering,* 65(2):42-46, February 1957.

13. Ivanova, V. S., "Aerodynamic Characteristics of Finned Air-Cooling Coils Under Frosting Conditions," *Kholodilnaia Technika*, 58(1):56-59, 1980.
14. *ASHRAE Handbook of Refrigeration*, American Society of Heating, Refrigerating and Air-Conditioning Engineers, Atlanta, Georgia, 1997.
15. "Hot Gas Defrost," Bulletin 90-10C, Refrigerating Specialties Company—Parker Hannifin, Broadview, Illinois, 1984.
16. Loyko, L., "Condensation-Induced Hydraulic Shock," Proceedings of 14th Annual Meeting of the International Institute of Ammonia Refrigeration, Washington, D.C., 1992.
17. Johnson, R., "Mother Nature's Legacy—Coil Frost," Krack Company, Addision, Illinois, 1992.
18. Cole, R. A., "Low Head Pressure Operation in Ammonia Systems," Seminary presentation, Annual Meeting of ASHRAE, June 1984.
19. Stoecker, W. F., J. J. Lux, and R. J. Kooy, "Energy Considerations in Hot-Gas Defrosting of Industrial Refrigeration Coils," *ASHRAE Transactions*, 89(Part II):549-573, 1983.
20. "Avoiding Component Failure in Industrial Refrigeration Systems Caused by Abnormal Pressure or Shock," Bulletin No. 116, International Institute of Ammonia Refrigeration, Washington, DC, 1992.
21. "Two-step solenoid valve," H. A. Phillips/Danfoss, St. Charles, Illinois, 1996.
22. Briley, G. C., and T. A. Lyons, "Defrosting Evaporators with Water," Proceedings of 16th Annual Meeting of the International Institute of Ammonia Refrigeration, Washington, DC., 1994.
23. Stencel, M. C., "Aluminum and Galvanized Steel Evaporators: Effects of the Operating Environment," Proceedings of the 15th Annual Meeting of the International Institute of Ammonia Refrigeration, Washington, DC, 1993.
24. Nelson, B. I., "Tech Tips: Direct Expansion (DX) of Ammonia," *The Coil Connection*, Colmac Coil Manufacturing Co., Colville, Washington, 1996.
25. "Desiccant Dehumidification and Pressure Drying Equipment" Chapter 22, *ASHRAE Handbook, HVAC Systems and Equipment*, American Society of Heating, Refrigerating, and Air-Conditioning Engineers, Atlanta, Georgia, 1996.
26. Devendorf., D. B., "The Control of Moisture in Food Plants," Proceedings of the 15th Annual Meeting of the International Institute of Ammonia Refrigeration, Washington, DC, 1993.
27. Griffith, W., "Enhancing Ammonia Refrigeration," *Engineered Systems*, 14(8): 34-44, August 1996.
28. "Glycols," Bulletin F-41515B, Union Carbide Corporation, New York, 1978.
29. Reference to Stephan, K., "The Computation of Heat Transfer to Boiling Coefficients," *Kaeltetechnik*, 15:23, 1963 in *ASHRAE Handbook—Fundamentals*, American Society of Heating, Refrigerating, and Air-Conditioning Engineers, Atlanta, Georgia, 1993.
30. Ayub, Z., M. C. Chyu and X. Zeng, "The Design and Operational Characteristics of Ammonia Spray Evaporators," Proceedings of 18th Annual Meeting of the International Institute of Ammonia Refrigeration, Washington, DC, 1996.
31. Koelet, P. C. "Industrial Refrigeration, Principles, Design and Applications," Marcel Decker, New York, 1992.
32. Villemain, D. K K. Gifford, J. Mitchell and M. Stromblad, "Technical Aspects of Semi-Welded Plate Heat Exchangers When Used as Evaporators and Condensers," Proceedings of 18th Annual Meeting of the International Institute of Ammonia Refrigeration, Washington, DC. 1996.

CHAPTER 7

CONDENSERS

7.1 TYPES OF CONDENSERS USED IN INDUSTRIAL REFRIGERATION

The three main types of condensers used in general refrigeration systems are:

- air-cooled
- water-cooled
- evaporative

All of these serve the industrial refrigeration field as well. In comparison to the air-conditioning industry, however, a lower percentage of air-cooled condensers and a higher percentage of evaporative condensers are operating in industrial refrigeration plants. In industrial refrigeration practice, it is common to connect the evaporative condensers in parallel—a concept not normally used in air conditioning.

The three types of condensers are shown schematically in Fig. 7.1a, 7.1b, and 7.1c. The air-cooled condenser in Fig. 7.1a condenses refrigerant vapor by rejecting heat to ambient air blown over the finned condenser coil with the aid of a fan, usually a propeller type.

Most all water-cooled condensers (Fig. 7.1b) condense refrigerant in the shell and on the outside of tubes through which water passes. The condenser cooling water picks up heat in passing through the condenser and this warm water is cooled by circulating through a cooling tower (Section 7.6). While

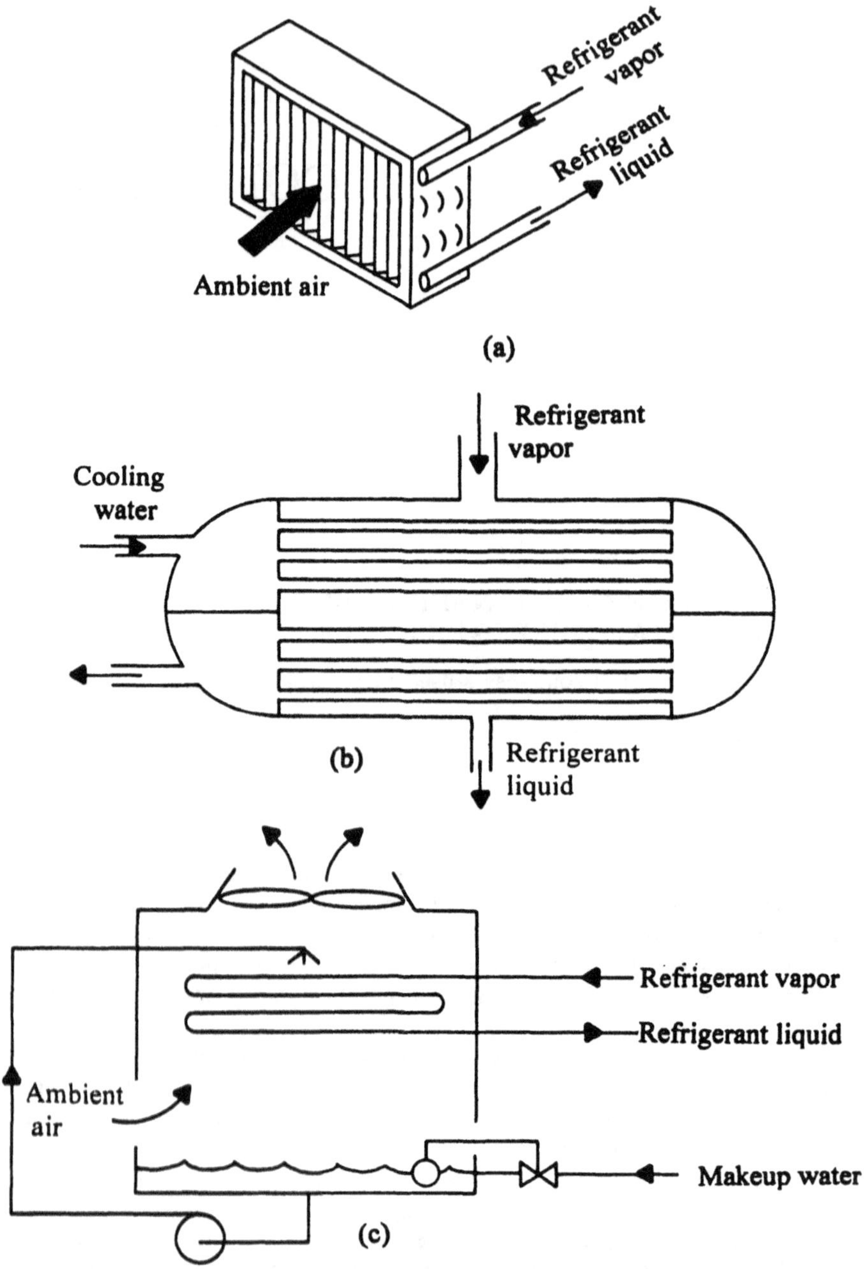

FIGURE 7.1
Types of condensers: (a) air-cooled, (b) water-cooled and (c) evaporative.

the shell-and-tube construction predominates for water-cooled condensers, plate-type condensers, sister of the plate-type evaporator explained in Sec. 6.31, are now appearing. The evaporative condenser of Fig. 7.1c might be considered a cooling tower, with the condenser tubes washed by the water spray. Ultimately, the heat rejected from the refrigeration plant is discharged to ambient air, except where the condenser is cooled by water from a well, lake, or stream.

This chapter first explores the condensing process outside and inside tubes. Next, the overall performance of water-cooled condensers and the translation of performance to noncatalog ratings is examined. An explanation of the performance of cooling towers, the constant companions of water-cooled condensers, is given. Because of their prevalence in industrial refrigeration plants, the emphasis of this chapter is on the performance, selection, application, and operation of evaporative condensers.

7.2 THE CONDENSING PROCESS

Nearly a century ago, heat-transfer pioneer, Willhelm Nusselt, proposed a model to predict the magnitude of a condensing coefficient for a special geometric situation[1]. Nusselt envisioned the condensation of vapor on a cold vertical plate, Fig. 7.2, as a process where vapor condenses on the plate and the condensate drains downward, with the condensate film becoming progressively thicker as it descends. The local condensing coefficient is taken to be the conductance through the condensate film—the conductivity of the liquid divided by the film thickness at that point. Nusselt developed the expression for the mean condensing coefficient as

$$h_c = 0.943 \left(\frac{g\rho^2 h_{fg} k^3}{\mu \Delta t L} \right)^{1/4} \tag{7.1}$$

where h_c = mean condensing coefficient, W/m²·°C (Btu/hr·ft²·°F)
g = acceleration due to gravity = 9.81 m/s² (4.17×10⁸ ft/hr²)
ρ = density of condensate, kg/m³ (lb/ft³)
h_{fg} = latent heat of vaporization of the refrigerant, kJ/kg (Btu/lb)
k = conductivity of condensate, W/m·°C (Btu/hr·ft·°F)
μ = viscosity of condensate, Pa·s (lb/ft·hr)
Δt = temperature difference, vapor to the plate, °C (°F)
L = vertical length of plate, m (ft)

The immediate question is where, if at all, does condensation occur on a vertical plate in industrial practice? Actually, a very old condenser design oriented the tubes vertically and water flowed by gravity down the inside of the tubes to ease their cleaning. The refrigerant in the shell condensed on the outside of the vertical tubes.

A slight modification of Eq. 7.1 applies to the widely used horizontal shell-and-tube condenser, Fig. 7.1b. The product of the number of tubes in a vertical

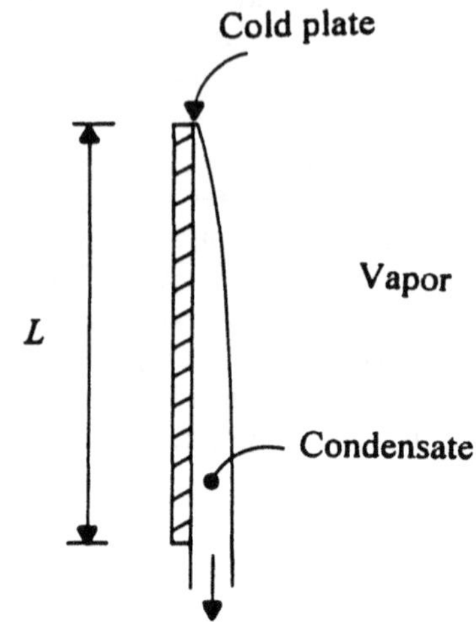

FIGURE 7.2
Condensation of a vapor on a cold vertical surface.

row multiplied by the diameter of the tubes replaces the vertical length of the plane L. White[2] found by experimental tests that the coefficient is 0.63 and Goto[3] measured 0.65, so the equation for N tubes of diameter D in a vertical row is:

$$h_c = 0.64\left(\frac{g\rho^2 h_{fg} k^3}{\mu \Delta t N D}\right)^{1/4} \tag{7.2}$$

Before leaving the condensing equations, an interesting comparison of the condensing coefficients of various refrigerants can be made. As Table 7.1 shows, the condensing coefficients of ammonia condensing on the outside of tubes far surpasses the coefficients of the other refrigerants shown. Experimental tests also show ammonia to have a higher condensing coefficient—five times that of the halocarbons in one study[4].

7.3 CONDENSATION INSIDE TUBES

In air-cooled and evaporative condensers, the refrigerant condenses inside tubes. The mechanism of condensation is complex and the flow regimes continue to change as the refrigerant passes through the tube[5]. Even though the state of the refrigerant is superheated vapor on entering the tube, condensation begins immediately and a spray regime develops. Later on the flow converts to annular then stratified with the liquid flowing along the bottom of the tube. Near the end of the condenser tube the flow regime is characterized as slug or plug. Fig-

TABLE 7.1
Condensing coefficients on the outside of tubes for several refrigerants. The condensing temperature is 30°C (86°F) and there are six 25-mm (1-in) tubes in a vertical row.

	Condensing coefficient	
Refrigerant	$W/m^2 \cdot °C$	$Btu/hr \cdot ft^2 \cdot °F$
R-22	1142	201
R-134a	1046	184
Ammonia	5096	897

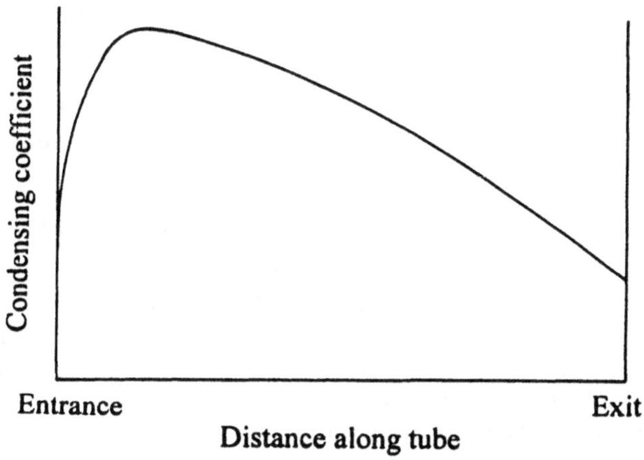

FIGURE 7.3
Variation in the condensing heat-transfer coefficient inside a tube.

ure 7.3 shows relative values[6] of the condensing coefficient throughout a tube. At the entrance to the tube with its superheated vapor content the coefficient is low, which is typical of convection heat transfer with a gas. The coefficient increases once surface condensation begins and is usually at its highest value during annular flow. As more and more condensed liquid flows with the vapor, the surface available for condensation decreases. Near the end of the condenser tube the coefficient drops quite low, because the process has approached that of convection heat transfer to a liquid.

The low heat-transfer coefficient near the end of the condenser tube when all or most of the vapor has condensed is pertinent to the plant operator. The reason is that backing liquid into an air-cooled or evaporative condenser shifts some heat-transfer area into the liquid subcooling mode which exhibits low heat-transfer coefficients.

7.4 HEAT-REJECTION RATIO

The *heat-rejection ratio* (HRR) is defined as the ratio of the rate of heat rejected at the condenser to that absorbed at the evaporator.

$$\text{HRR} = \frac{\text{rate of heat rejection at condenser}}{\text{rate of heat absorbed at evaporator}}$$

The designer and operator of the refrigeration system will usually characterize plant size by the refrigeration capacity. This capacity can be translated to a condenser capacity through the condenser-to-evaporator heat rejection ratio (HRR). The HRR is a function of the evaporating and condensing temperatures, but is also influenced by the compressor type and any supplementary cooling arrangements. The standard procedure for computing the HRR from catalog data of the compressor is to propose that the heat rejected at the condenser is composed of two contributions—the refrigerating capacity and the thermal equivalent of the power supplied to the compressor. The standard equation for computing the HRR is, therefore,

$$\text{HRR} = \frac{\text{refrigerating capacity} + \text{compressor power}}{\text{refrigerating capacity}} \tag{7.3}$$

where all the energy flow rates are expressed in the same units.

Figure 7.4 shows HRRs as functions of the evaporating and condensing temperatures. Changes of either of these temperatures affect both the refrigerating capacity and the power requirement of the compressor. The ideal HRR can be derived from knowledge of the Carnot cycle (Section 2.17), in which the ratio of area under the condensing line to that under the refrigeration line represents the HRR,

$$\text{HRR} = \frac{T_{cond}}{T_{refrig}} \tag{7.4}$$

where the temperatures T are in absolute, thus °C + 273.1 (°F + 459.7). Equation 7.4 assumes a 100% efficiency of the cycle and the compressor, and an improved expression that can be used when compressor catalog data are not readily available is

$$\text{HRR} = \left(\frac{T_{cond}}{T_{refrig}}\right)^{1.7} \tag{7.5}$$

Example 7.1. Estimate the HRR when the condensing temperature is 35°C (95°F) and the evaporating temperature is −10°C (14°F).

Solution. On the absolute scale the evaporating temperature is 263.1 K (473.6 R) and the condensing temperature is 308.1 K (554.6 R). The estimated HRR is

$$\text{HRR} = \left(\frac{308.1}{263.1}\right)^{1.7} = 1.31$$

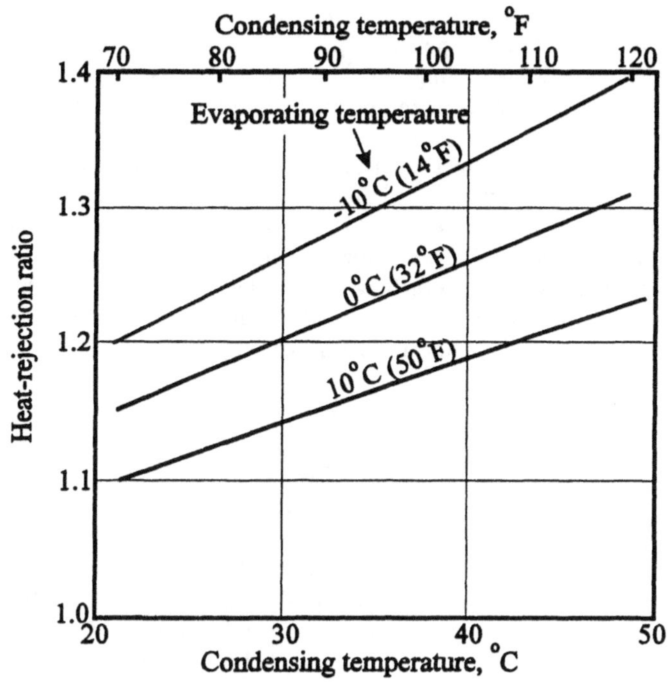

FIGURE 7.4
Typical values of the ratio of the heat rejected at the condenser to the refrigerating capacity, HRR, for ammonia and halocarbon refrigerants.

The value from Fig. 7.4 is 1.30.

Equation 7.3 is correct except for heat losses to the ambient or supplementary transfers of heat to other devices. The curves in Fig. 7.4 apply to open-type compressors, and the HRR will be higher for hermetic compressors servicing small halocarbon systems, because some of the motor heat enters the refrigerant stream. Also the HRR will be lowered if a reciprocating compressor uses water-cooled heads where the heat is rejected to a separate cooler or in a screw compressor where oil is cooled by a separate water or antifreeze circuit.

7.5 PERFORMANCE OF AIR AND WATER-COOLED CONDENSERS

Manufacturers of condensers provide performance data directed toward selecting equipment. By applying some fundamentals of heat transfer, a user can frequently translate catalog data to nondesign conditions. The strategy in extending catalog data to nondesign conditions is usually to compute the UA value (the product of the overall heat transfer coefficient and the heat-transfer area) and for situations where the UA remains essentially constant, apply this UA value to the new set of operating conditions. The temperature profiles are somewhat complex because of desuperheating and subcooling, as shown in Fig. 7.5a, but to approximate, assume the condensing temperature prevails throughout the condenser, as shown in Fig. 7.5b.

In the desuperheating section, the actual temperature difference between the refrigerant and cooling water is higher than the ideal, but this error is at least partially compensated for by the fact that the actual heat-transfer coefficient for the convection process is less than during condensation. Real condensers are rarely circuited strictly for counterflow or parallel flow. When one fluid is at a constant temperature, however, the flow pattern is immaterial, and an equation comparable to the one for evaporators, Eq. 6.11, applies:

$$q = UA\left[\frac{t_o - t_i}{\ln\left(\frac{t_c - t_i}{t_c - t_o}\right)}\right] \tag{7.6}$$

where q = rate of heat transfer, kW (Btu/hr)
UA = product of overall heat-transfer coefficient and area to which it applies, kW/°C (Btu/hr per °F)
t_c = temperature of condensing refrigerant, °C (°F)
t_i = temperature of entering cooling water, °C (°F)
t_o = temperature of leaving cooling water, °C (°F)

Example 7.2. The catalog for a Vilter 0.2 m × 2.13 m (8 in × 7 ft) R-22 condenser specifies a condensing capacity that accommodates a refrigeration load of 204 kW (58.1 tons) at the evaporator when the evaporating temperature is 4.4°C (40°F), the condensing temperature is 40.6°C (105°F), and a 9.8 L/s (156 gpm) flow rate of cooling water enters at 29.4°C (85°F).

What condensing temperature would prevail if the cooling water flow rate and its entering temperature remain constant, but the refrigeration capacity is half of the catalog value?

Solution. The rate of heat transfer q at the condenser with the original refrigeration load was:

$$q = (204 \text{ kW})(\text{heat rejection ratio}),$$

and at an evaporating temperature of 4.4°C (40°F) and a condensing temperature of 40.6°C (105°F), Fig. 7.4 shows a heat rejection ratio of 1.24, so q equals 253 kW (863,000 Btu/hr). The mass flow rate of cooling water is 9.8 kg/s (1300 lb/min), so the outlet water temperature t_o is:

$$t_o = 29.4 + \frac{253 \text{ kW}}{(9.8 \text{ kg/s})(4.19 \text{ kW/kg}\cdot{}^\circ\text{C})} = 29.4 + 6.2 = 35.6^\circ\text{C } (96.1^\circ\text{F})$$

The log-mean temperature difference is:

$$\text{LMTD} = \left[\frac{35.6 - 29.4}{\ln\left(\frac{40.6-29.4}{40.6-35.6}\right)}\right] = 7.69^\circ\text{C } (13.8^\circ\text{F})$$

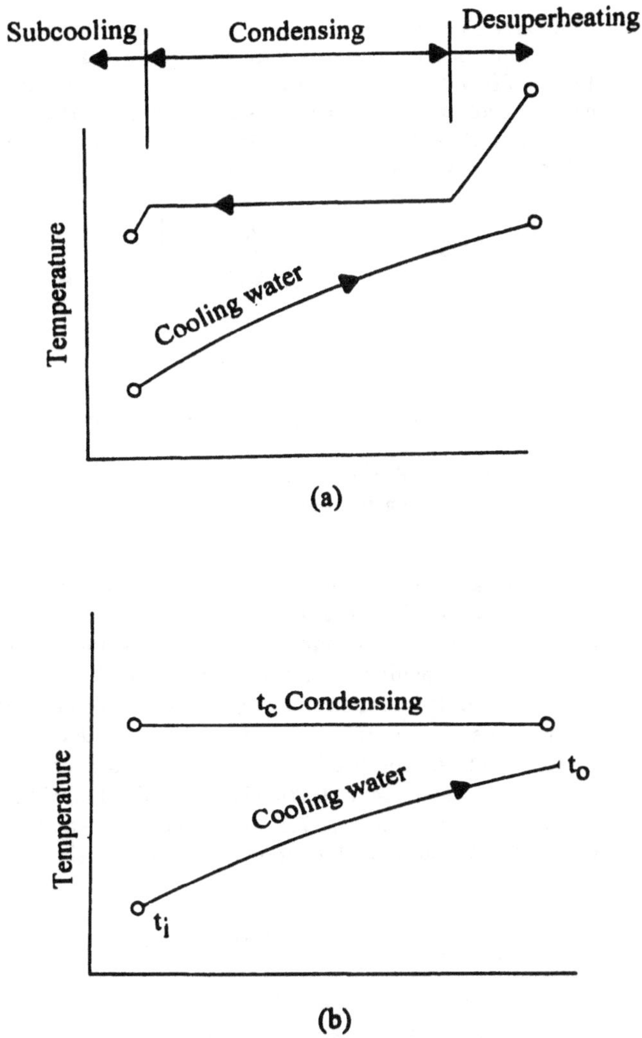

FIGURE 7.5
(a) Actual, and (b) idealized temperature profiles in a water-cooled condenser.

and the UA value is:

$$UA = \frac{253 \text{ kW}}{7.69°\text{C}} = 32.9 \text{ kW/°C } (62,400 \text{ Btu/hr per °F})$$

This UA value should remain essentially unchanged as the condensing capacity varies, so long as the water flow rate remains constant. At the half-load condition, the LMTD will be half the original value, 7.69 ÷ 2 or 3.85°C (6.92°F), and the condenser cooling water experiences half its original rise in temperature, so its new value is 29.4 + 3.1 = 32.5°C (90.5°F). The equation for the LMTD can be solved for the new condensing temperature,

$$\text{LMTD} = 3.85 = \left[\frac{32.5 - 29.4}{\ln\left(\frac{t_c - 29.4}{t_c - 32.5}\right)} \right]$$

$$\frac{t_c - 29.4}{t_c - 32.5} = e^{3.1/3.85} = 2.237$$

so the new t_c = 35.0°C (95°F), which is a reduction from the original t_c of 40.6°(105°F).

If greater precision is desired, perform an iteration which uses a revised HRR based on the new condensing temperature.

Example 7.1 illustrates a situation where the U-value remains constant from one condition to another. If the rate of water flow changes, the heat-transfer coefficient on the water side will also change so that the U-value no longer remains constant. It is advisable to consult the manufacturer in such a case.

The tubes of water-cooled condensers are subject to fouling caused by impurities in the water. Some measurements[7] made with suspended solids in cooling tower water indicated that the fouling factor (which is additional heat transfer resistance) can easily be on the order of 0.00004 $m^2 \cdot °C/W$ (0.0002 $hr \cdot ft^2 \cdot °F/Btu$) or higher. In the condenser of Example 7.1 the water-side heat transfer area is 3.78 m^2 (40.7 ft^2), so the U-value of the condenser based on the water-side area is:

$$\frac{32,900 \text{ W/°C}}{3.78 \text{ m}^2} = 8701 \text{ W/m}^2 \cdot °\text{C}$$

or 1533 Btu/($hr \cdot ft^2 \cdot °F$). The resistance is the reciprocal of this U-value or 0.000114 $m^2 \cdot °C/W$ (0.000645 $hr \cdot ft^2 \cdot °F/Btu$). The resistance when the fouling factor is included is 0.000154 $m^2 \cdot °C/W$ corresponding to a U-value of 6500 $W/m^2 \cdot °C$ (1143.6 $Btu/hr \cdot ft^2 \cdot °F$). Tube fouling in this case reduces the condensing capacity 25% compared with the clean condition. The user has some protection because condensers leave the factory with a higher U-value than indicated by catalog data. They are derated using a fouling factor specified in the catalog. The user, however, should be aware that if the cooling tower water is fouling the condenser, frequent tube cleaning can improve system performance.

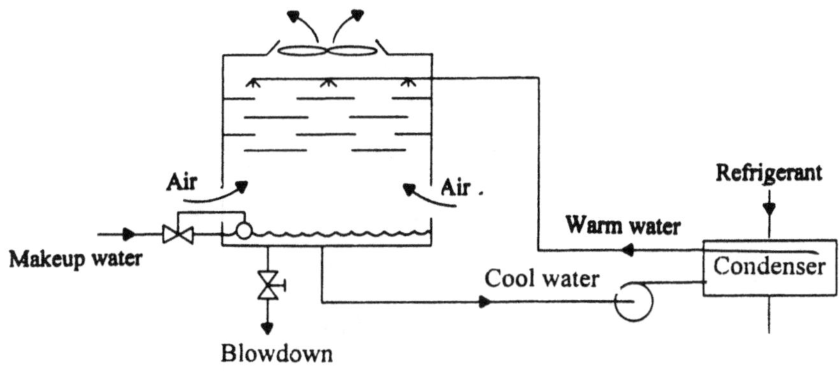

FIGURE 7.6
A cooling tower.

7.6 COOLING TOWERS

A cooling tower cools water by spraying it through a stream of ambient air. A schematic diagram of a cooling tower and the manner in which it serves the refrigeration condenser are shown in Fig. 7.6. The air- and water-flow patterns suggested in Fig. 7.6 are counterflow of air and water, a frequently used configuration. Another popular geometry is crossflow, in which the air is blown horizontally through the falling stream of water. Because some water evaporates into the air, a supply of makeup water must be provided. Also, because the makeup water contains some dissolved minerals, the concentration of these minerals in the sump water would progressively increase were a blowdown not provided.

The explanation of the heat- and mass-transfer process in a cooling tower starts with the recollection of the straight-line law first introduced in Sec. 6.13. The straight-line law states that when air is in contact with water, the change in air conditions is a straight line on the psychrometric chart directed toward the saturation line at the water temperature. This information is used to examine what happens to the enthalpy (heat content) of the air. If the enthalpy of air increases in the process, the enthalpy and temperature of the water must decrease. Consider first the special case shown in Fig. 7.7, where the wet-bulb temperature of the air equals the water temperature.

The path of the air moves toward the saturation line at the water temperature, which is along the wet-bulb temperature line. The wet-bulb temperature lines and the enthalpy lines are essentially parallel, so there is no change in the enthalpy of air, and the temperature of water does not change either. This is the process that takes place in evaporative coolers that reduce the air temperature in homes in arid regions.

If the temperature of the water is higher than the wet-bulb temperature of the air, as in Fig. 7.8, the enthalpy of the air increases from point 1 to point 2, so an energy balance requires that this heat must come from the water by cooling it from point 1′ to point 2′.

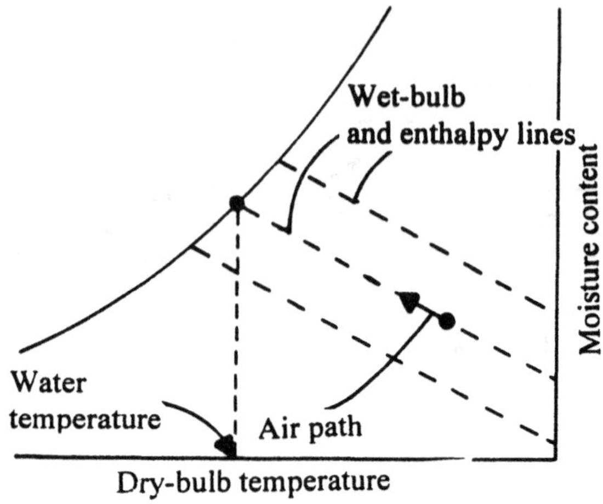

FIGURE 7.7
When the water temperature is the same as the wet-bulb temperature of the air, there is no change in water temperature.

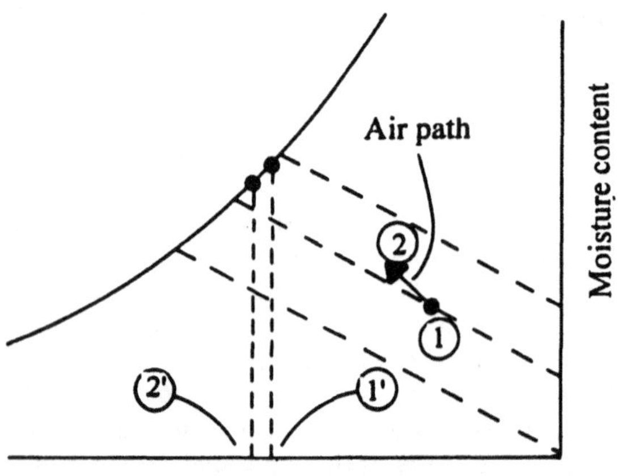

FIGURE 7.8
The enthalpy of air rises and the temperature of water drops when the water temperature is higher than the wet-bulb temperature of the air.

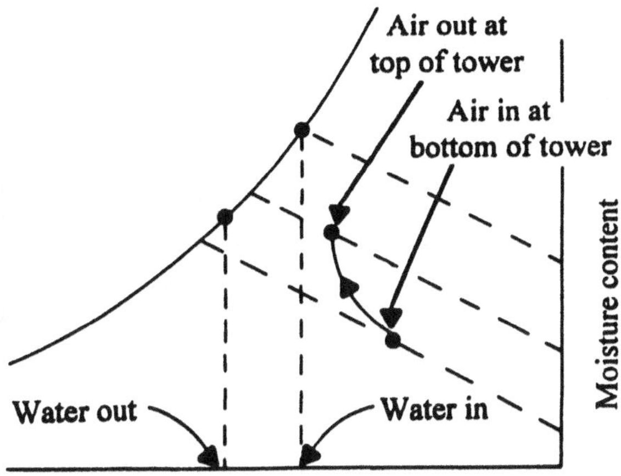

FIGURE 7.9
Conditions of air and water in a counterflow cooling tower.

When these elementary processes are expanded to a complete counterflow cooling tower, they show the pattern of air and water conditions as in Fig. 7.9. The air progressively increases in enthalpy, and while its dry-bulb temperature is shown decreasing in Fig. 7.9 as it rises through the tower, there could be situations where the temperature increases in passing through the tower.

The key concept implicit in Figs. 7.7 through 7.9 is that the leaving water temperature can approach the wet-bulb temperature of entering air. For this reason, catalog data for cooling towers show the ambient condition that affects cooling tower performance as the wet-bulb temperature, and dry-bulb temperatures may not even be indicated. When a constant heat load is imposed on the condenser and its cooling water, the leaving water temperature rides up as the ambient wet-bulb temperature increases in a trend as shown in Fig. 7.10. Because the heat load and the water-flow rate are constant, a fixed drop in water temperature (5°C or 9°F in this case) prevails over the entire range shown in the graph.

7.7 EVAPORATIVE CONDENSERS

The schematic diagram of the evaporative condenser shown in Fig. 7.1c illustrates that the evaporative condenser combines the functions of an air-cooled condenser and cooling tower. Refrigerant condenses within the tubes, and these tubes are sprayed with water through which an air stream passes. The evaporation of some water into the air is the dominant process of rejecting heat to the atmosphere.

To provide a comparison of the three forms of condensing equipment, namely the air-cooled condenser, the water-cooled condenser/cooling tower combination, and the evaporative condenser—some of the characteristics of each are

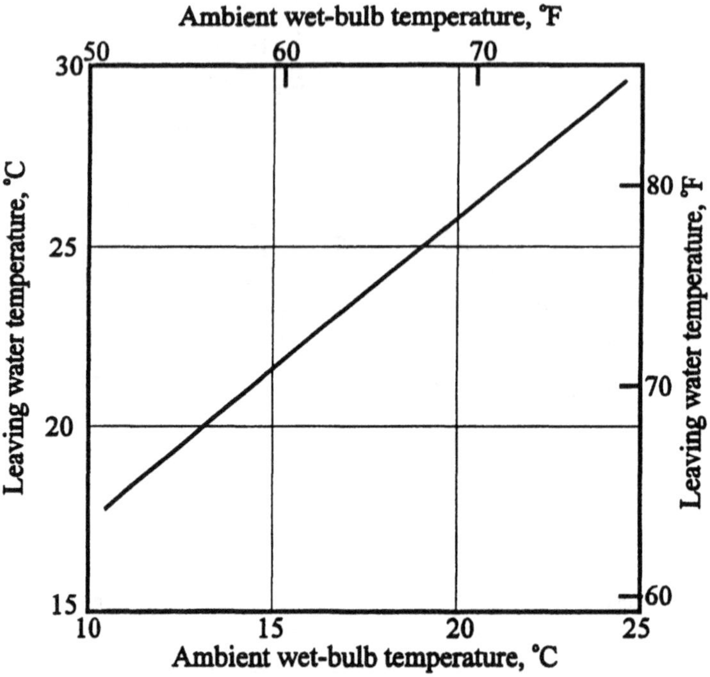

FIGURE 7.10
Leaving water temperature from a cooling tower as the ambient wet-bulb temperature changes. The heat load and water-flow rate are constant.

enumerated:

Air-cooled condenser. Usually lowest first cost of the three, and least maintenace cost as well, because no water circulates or evaporates.

Water-cooled condenser with cooling tower. Lower condensing temperature than with an air-cooled condenser, because the wet-bulb rather than the dry-bulb temperature of the air is the sink toward which the condensing temperature drives. When the distance between the compressor and the point of heat rejection is long, water can be piped to the cooling tower, rather than sending refrigerant, as must be done with the evaporative or air-cooled condenser.

Evaporative condenser. Compact and provides lower condensing temperatures than the air-cooled condenser and also lower than the water-cooled condenser/cooling tower combination. Figure 7.11 shows an evaporative condenser with a bit more detail than was presented in Fig. 7.1c.

The evaporative condenser is widely used in industrial refrigeration practice because it provides relatively low condensing temperatures. These temperatures conserve power and result in moderate compressor discharge temperatures

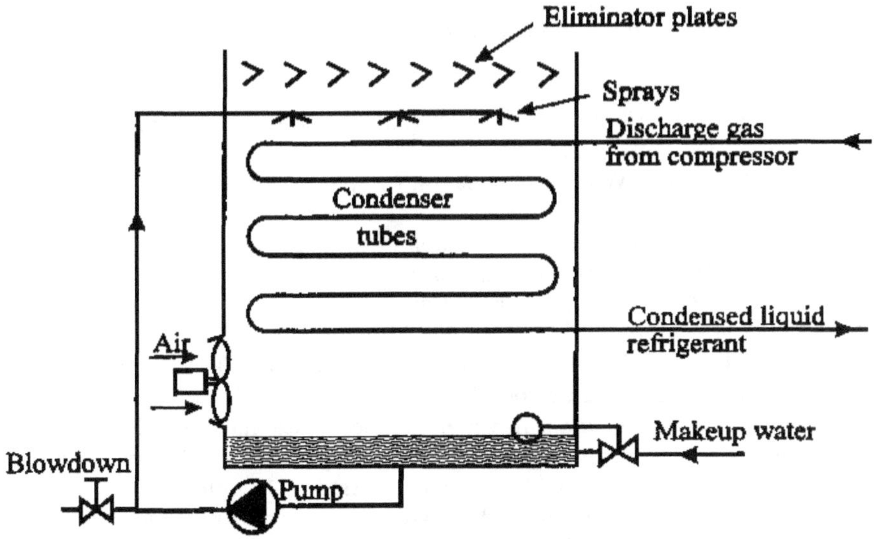

FIGURE 7.11
An evaporative condenser.

which may be important, especially for ammonia applications. The drawback of its maintenance requirements may not be prohibitive. Most industrial refrigeration plants employ service personnel, in contrast to smaller plants where the refrigeration facility is expected to operate without regular maintenance.

Most evaporative condensers are of the blow-through type using an axial-flow fan in preference to a centrifugal type. Eliminator plates are provided to avoid blowing water droplets out of the condenser. All condensers are equipped with blowdown to limit the buildup of minerals in the spray water, and this blowdown is usually installed in the pump discharge line, so that when the condenser is not in operation, the blowdown ceases.

7.8 NOMINAL SIZES AND RATES FOR EVAPORATIVE CONDENSERS

The design of an efficient evaporative condenser requires optimizing a number of factors, including tube size, tube length, tube spacing, refrigerant circuiting, air-flow rate, casing size, and spray-water flow rate. The condenser manufacturer/designer must draw on knowledge of refrigerant heat transfer, wetted-surface heat transfer (such as described in References 8 and 9), and a thorough understanding of fabrication economics and of the end-user operation. The condenser user is spared the responsibility of making most of these decisions and is best served with an understanding of how three variables affect the performance of an evaporative condenser: the wet-bulb temperature, air-flow rate, and spray-water flow rate. The next several sections will focus on those factors, but first a few nominal magnitudes applicable to many commercial evaporative condensers will be presented:

Heat-transfer area:

0.25 m^2 per kW of heat rejection (0.8 ft^2 per 1000 Btu/hr)

Spray water circulating rate:

0.018 L/s per kW of heat rejection (5 gph per 1000 Btu/hr)

Air volume flow rate:

0.03 m^3/s per kW of heat rejection (18 cfm per 1000 Btu/hr)

Air pressure drop through the condenser:

250-375 Pa (1 to 1-1/2 inches of water)

Rate of water evaporated:[10]

1.5 L/hr per kW of heat rejection (0.12 gph per 1000 Btu/hr)

Total rate of water consumption:[10]

with good quality makeup water the bleed rate may be as low as 50% of the evaporation rate, so the total rate evaporated and blown down may be about 2.2 L/hr per kW of heat rejection (0.18 gph per 1000 Btu/hr).

Many years ago the typical flow rate of spray water was quite low, perhaps 0.68 L/s per m^2 (1 gpm/ft^2), but this rate has gradually climbed so that it may run as high as 4.1 L/s per m^2 (6 gpm/ft^2) to achieve favorable capacity. The practical limit is reached when the spray water flow rate is so high that it restricts the air flow rate.

7.9 COMPARISON OF EVAPORATIVE CONDENSER WITH THE WATER-COOLED CONDENSER AND COOLING TOWER COMBINATION

Practically all of the remainder of this chapter will concentrate on evaporative condensers which are the predominant type used in industrial refrigeration. One of the reasons for the preference of evaporative condensers over the water-cooled condenser with cooling tower is the ability to achieve lower condensing temperatures. Figure 7.12 shows a heat rejection of 548 kW (1,870,000 Btu/hr) by means of an evaporative condenser (Fig. 7.12a) and the rejection of the same magnitude with a water-cooled condenser. The capacity of both the evaporative condenser and the cooling tower are controlled by the ambient wet-bulb temperature, which in this case is 25.6°C (78°F).

The comparison shows the ability to achieve a condensing temperature of 35°C (95°F) with the evaporative condenser, while with the water-cooled condenser the condensing temperature is 40.6°C (105°F), thus 5.6°C (10°F) higher. The superior performance of the evaporative condenser is explained by the avoidance of the intermediate fluid (the cooling-tower water) in the heat-transfer processes. The temperature of water leaving the cooling tower is 28.9°C (84°F) and can only approach the ambient wet-bulb temperature, and the condensing temperature can only approach the temperature of water returning to the tower

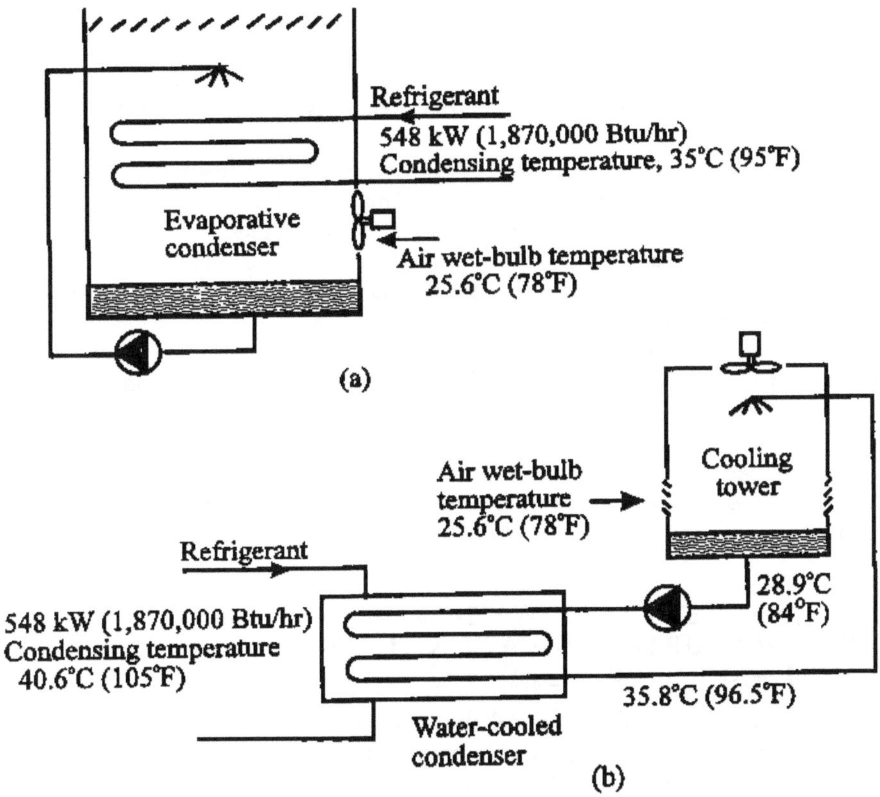

FIGURE 7.12
Achieving a lower condensing temperature with an evaporative condenser in comparison to the combination of a water-cooled condenser and cooling tower.

which is 35.8°C (96.5°F). The comparison could be accused of being biased in that it would be possible to lower the condensing temperature in Fig. 7.12b by enlarging either or both the cooling tower or the water-cooled condenser. This observation is correct, but the sizes of all the components in the comparison of Fig. 7.12 are based on sizes typically chosen for this application. Nevertheless, the comparison is not complete until the comparative first costs of the condensing subsystem are evaluated. Some designers contend that the evaporative condenser in Fig. 7.12 might even cost less than the sum of the water-cooled condenser and cooling tower.

Another factor in favor of the evaporative condenser is the lower water-pumping costs. The spray-water flow rate in the evaporative condenser is typically about one-third that of the flow rate circulated between the water-cooled condenser and the cooling tower. Furthermore, the length of water line between the water-cooled condenser and cooling tower will likely be much longer. This saving in water-pumping power must be balanced against the higher pressure drop in the refrigerant lines, particularly the vapor line between the compressor and condenser. Normally the compressor and water-cooled condenser will be close-coupled. The need for water treatment exists in both concepts, so this feature is not a factor.

These advantages of the evaporative condenser influence the industrial refrigeration industry to predominately favor this type of condensing system. The air-conditioning industry, on the other hand, usually chooses water-cooled condensers, so it is reasonable to ask whether the air-conditioning industry is unaware of the secret. Such is not the case, because many air-conditioning systems experience long distances between the compressor and the ultimate heat rejector. The compressor and condenser may be in the basement and the cooling tower on the roof of a multistory building. In many industrial refrigeration plants the evaporative condenser is on the roof of the machine room that houses the compressors, and the distance separating them may be only 6 to 12 m (20 to 40 ft). Also, when a centrifugal compressor serves a water-chilling system, the refrigerant chosen has a high specific volume, making condensing in the tubes less practical.

7.10 INFLUENCE OF WET-BULB TEMPERATURE ON EVAPORATIVE CONDENSER CAPACITY

It was the conclusion in Section 7.6 that the leaving water temperature from a cooling tower is controlled by the wet-bulb temperature of ambient air. Because the same process of heat and mass transfer occurs in both devices, the wet-bulb temperature also has a dominant influence on the capacity of evaporative condensers. Figure 7.13 shows relative capacities of an ammonia evaporative condensers to changes in wet-bulb temperature and condensing temperatures. The capacities are relative to a condenser operating with a condensing temperature of 35°C (95°F) and a wet-bulb temperature of 25°C (77°F). The trends are as expected, namely the capacity increases with a given wet-bulb temperature as the condensing temperature increases. Furthermore, at a given condensing temperature the capacity increases with a reduction in wet-bulb temperature.

Even though Fig. 7.13 indicates that the temperature difference between the condensing refrigerant and the entering wet-bulb influences the capacity, it is not to be assumed that the heat-rejection capacity is proportional to this difference in temperature. For an air-cooled condenser the heat-transfer rate is proportional to the temperature difference between the condensing refrigerant and the dry-bulb temperature of the entering air. For a water-cooled condenser the capacity is also proportional to the temperature difference between the refrigerant and entering water. For an evaporative condenser, as Fig. 7.14 shows, the level of temperatures as well as the temperature difference controls the capacity. This trend indicates that if an evaporative condenser develops a certain

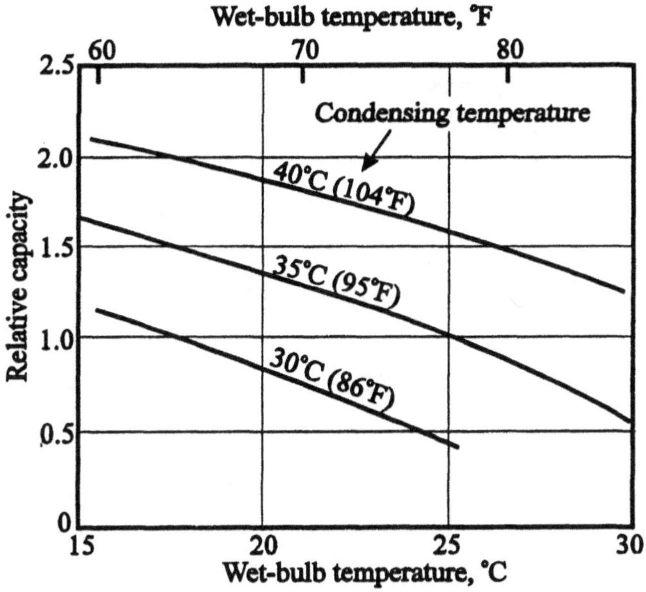

FIGURE 7.13
Relative heat-rejection capacity of an ammonia evaporative condenser as a function of the condensing and wet-bulb temperatures. The reference point is a 35°C (95°F) condensing temperature and a wet-bulb temperature of 25°C (77°F).

heat-rejection capacity with a temperature difference between condensing refrigerant and the ambient wet-bulb temperature, for example, 40°C to 25°C (104°F to 77°F), the capacity of the condenser will be less if the same temperature difference exists at a lower level, for example, 30°C to 15°C (86°F to 59°F).

The reason for this behavior lies in the evaporation process on which the evaporative condenser operates. The major heat-transfer mechanism in the evaporative condenser is due to the vaporization of water from the condenser tubes, and the rate of this vaporization is proportional to the difference of water-vapor pressure of the liquid water on the tube and the water vapor pressure in saturated air that surrounds the tube. Examination of a psychrometric chart (Fig. 6.18 or 6.20) shows that at the low level of temperature, the saturation curve flattens out so that a given difference in temperature translates to a lower difference in water-vapor pressure.

7.11 CATALOG SELECTION—TWO METHODS

Most catalogs of evaporative condensers show tables for selecting condensers by two different methods. One is the *condenser capacity* or *heat-rejection method* and the other is the *refrigeration capacity method*. The condenser capacity method

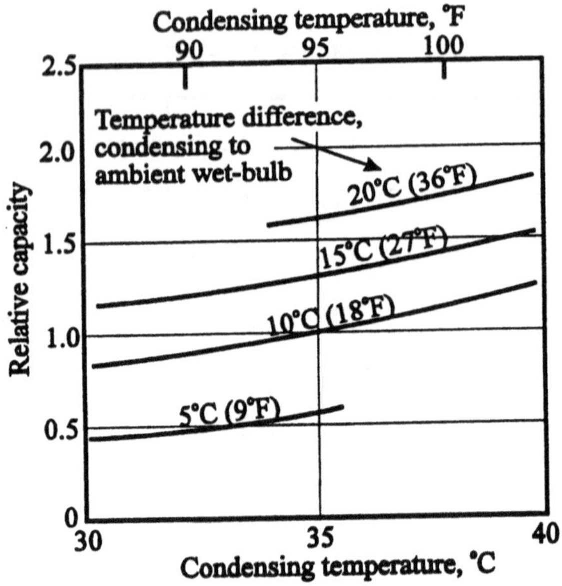

FIGURE 7.14
The level of temperature as well as the temperature difference between refrigerant and the air wet-bulb temperature influence the capacity.

is most straightforward because it deals with the actual heat-transfer rate at the condenser. The refrigeration capacity method is simply a tool to facilitate quick selections, because the user of the catalog will normally think in terms of the refrigeration capacity which then becomes the entry point of the catalog.

The condenser capacity method will be explored first. Two tables are associated with this method—one table presents the capacity factors as functions of condensing and wet-bulb temperatures and the other table is nominal heat rejection capacity of various models of condensers. Most manufacturers show the capacity factors in tabular form. The presentation in Figure 7.15 is graphical. Table 7.2 shows a excerpt of a manufacturer's table[11] of nominal capacity of a series of condensers.

The capacity factors shown in catalogs and in Fig. 7.15 may at first seem to present trends opposite of expectations. For example, the capacity factor would be expected to increase as the wet-bulb temperature drops and the condensing temperature increases. But the main purpose of the catalog is to facilitate selection of condensers, not to analyze their performance. The consistency of the method can be demonstrated by an example.

Example 7.3. Using Table 7.2 and Fig. 7.15, select a condenser to reject 586 kW (2,000,000 Btu/hr) while operating at a condensing temperature of 35°C (95°F) and a wet-bulb temperature of 25°C (77°F).

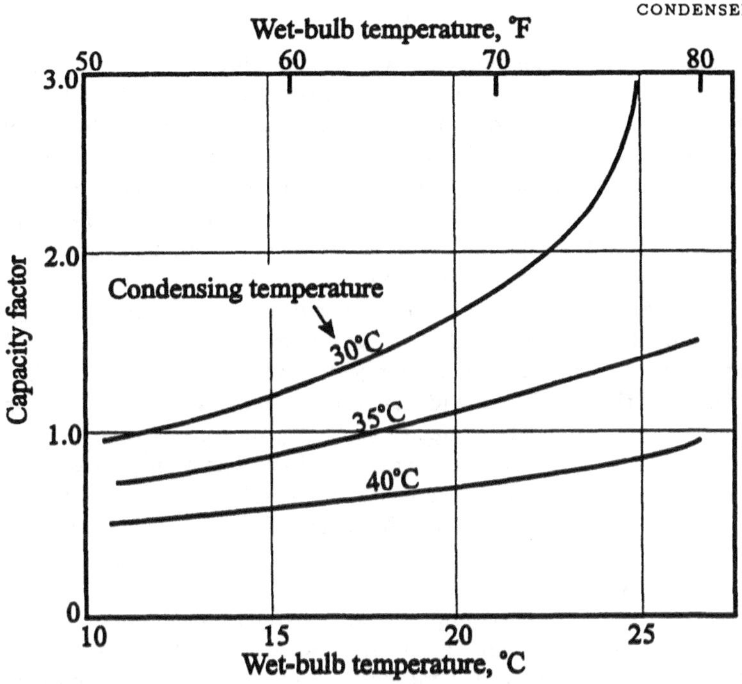

FIGURE 7.15
Capacity factors for selection of an evaporative condenser in conjunction with Table 7.2 using the condenser heat-rejection method.

TABLE 7.2
Nominal capacity of a line of evaporative condensers.

Model	Heat rejection rate		Model	Heat rejection rate	
	Thousands of Btu/hr	kW		Thousands of Btu/hr	kW
A	1544	452	F	2426	711
B	1764	517	G	2720	797
C	1911	561	H	3014	883
D	2058	603	I	3381	991
E	2205	646	J	3675	1077

Solution. Figure 7.15 shows a capacity factor of 1.4 at the prevailing condensing and wet-bulb temperatures and this factor is multiplied by the desired heat rejection rate:

$$586 \times 1.4 = 820 \text{ kW} \qquad \text{or} \qquad 2000 \times 1.4 = 2800 \text{ thousands of Btu/hr}$$

Referring to Table 7.2, Model G shows a capacity of 796 kW (2720 thousands of Btu/hr) which is low, but Model H with a capacity of 883 kW (3014 thousands of Btu/hr) would have adequate capacity.

The application of the capacity factors of Figure 7.15 in the example explains what might have appeared to be an incongruity. The capacity of a condenser is highest at high condensing temperatures and low wet-bulb temperatures, but Fig. 7.15 shows just the opposite trends. The purpose of the capacity factor is to aid in selecting the condenser and is multiplied by the specified heat-rejection rate. In so doing, when the capacity factor is low, a smaller condenser will be adequate, and the capacity factor is low when the condensing temperature is high and the wet-bulb temperature low.

If the refrigeration capacity method is chosen, a different capacity factor table and nominal rating table will be available from the manufacturer. Section 7.4 pointed out that the ratio of heat rejected at the condenser to the refrigeration capacity, the HRR, depends on the condensing temperature and the evaporating temperature. The capacity factor applicable to this method must therefore incorporate these two temperatures if the selection process is entered with the refrigerating capacity. A capacity factor table comparable to Fig. 7.15 offers the chance of incorporating the influence of the condensing temperature on the HRR, so a separate table correcting for the evaporating temperature will also be provided by the manufacturer.

Of these two selection techniques, the heat-rejection method is more powerful in accommodating system complexities, such as might occur in two-stage plants. The refrigeration capacity method is useful for quick estimates of the condenser size.

7.12 CAPACITY CONTROL

Capacity control of a condenser means reducing its capacity. This understanding of capacity control raises the question of why the condenser capacity should ever be reduced. When the condenser operates at full capacity, the condensing temperature will follow the wet-bulb temperature as it drops, and thus the compressor power will be reduced. In general the recommended strategy is to operate the condenser with full capacity, dropping the condensing temperature until limited by one or more of the following conditions:

- the condensing pressure is too low to adequately feed level-control valves and expansion valves
- the pressure of defrost gas is too low to achieve a satisfactory defrost,
- if the plant uses screw compressors with their oil cooled by direct injection of refrigerant, the pressure of the liquid must be high enough to force an adequate flow rate of liquid into the compressor
- savings in compressor power by further lowering of the condensing temperature are less than savings that would be possible in pump and fan motors of the condensers

If the pressure of liquid is too low, expansion valves and liquid-level controllers will not be able to pass enough refrigerant and capacity of evaporator coils will drop. In the case of a level controller to an intermediate-pressure subcooler/desuperheater, subcooling of the liquid will suffer and desuperheating of discharge gas from the low-stage compressor will degrade. It would be possible to use valves with larger ports, but this change could cause unstable feeding when the liquid pressure is high.

In order to defrost a coil the saturation temperature of the defrost gas must be well above 0°C (32°F). Tests[12] on an ammonia coil showed that successful defrosts could be achieved with a saturation temperature of the defrost gas of 15°C (59°F) resulting in saturation temperature inside the coil of approximately 10°C (50°F). Before a plant operator would set the minimum condensing temperature at 15°C (59°F) for unattended operation, defrosts of coils should be monitored for a period of time to be confident that no residual frost or ice remains on the coil following this defrost.

Manufacturers of compressors who apply direct injection for oil cooling recommend[13] minimum condensing temperatures of approximately 21°C (70°F) in R-22 and ammonia systems to provide adequate pressure for injection of the liquid.

The question of reducing air or water flow to save power required by the fan and pump motors will be addressed further in Section 7.15.

Still another situation which may influence the minimum condensing pressure occurs if the plant recovers heat from the high side of the system for uses elsewhere in the facility. A simple economic analysis may be able to establish whether it is cheaper to pay for the recovered heat from other sources and reap the saving on compressor power with a lower condensing pressure.

The two principal methods of reducing the condenser capacity are to reduce or cycle the flow of spray water or the air flow. Adjusting the flow rate of spray water will be be dismissed rather quickly in the next section, and the regulation of air flow evaluated in Section 7.14.

7.13 CAPACITY CONTROL— VARYING THE FLOW RATE OF SPRAY WATER

Reducing the flow rate of spray water by throttling the flow with a regulating valve or reducing the speed of the pump motor will lower the heat-transfer capacity of the condenser. Tests[8] suggest that the condenser capacity near its normal operating point varies as the flow rate to the 0.22 power:

$$\text{Condenser capacity} = (\text{Constant})(\text{spray- water flow rate})^{0.22} \qquad (7.7)$$

thus if the flow rate is reduced by 20% the capacity of the condenser would drop to 95% of its original value. At lower flow rates the drop in capacity is more precipitous until a complete interruption of the spray-water flow to dry operation drops the condenser capacity significantly.

Usually reducing the spray water flow rate is not recommended. If the rate is dropped much below the design value, areas of the tubes may become alternately dry and wet. The result is excessive scaling on that tube surface. Avoidance of scale is also one of the reasons for opposing cycling of the pump for capacity control. The second reason is that the frequent stopping and starting of the motor accelerates its wear.

7.14 CAPACITY CONTROL—VARYING THE AIR FLOW RATE

An equation[8] for the air-flow rate comparable to Equation 7.7 for the rate of spray-water flow is:

$$\text{Condenser capacity} = (\text{constant})(\text{air flow rate})^{0.48} \quad (7.8)$$

This equation matches closely some other data[14] that is shown in Fig. 7.16. If the air-flow rate is reduced by 50%, for example, the heat-rejection capacity of the condenser with a given combination of condensing and wet-bulb temperatures is 72% of the base capacity. One manufacturer[15], on the other hand, suggests that a reduction in air-flow rate will result in 58% of rated capacity. Some of the ways in which the air-flow rate can be regulated are:

- Variable-frequency drive of fan motor
- Two speed fan motors
- Pony motors
- Fan dampers
- Fan cycling on a single-fan unit
- Shutting down one fan in a multiple-fan condenser

Variable-frequency inverters driving the fan motor give the most precise regulation, but currently the first cost of the assembly makes this method the most expensive in overall first cost. Two-speed fan motors are available that operate with 1800/1200 rpm combination using a two-winding construction or 1800/900 rpm using a single winding. The 1800/1200 rpm combination requires an expensive motor but a low-cost starter, while the 1800/900 rpm combination offers a low-cost motor but an expensive starter. The pony motor arrangement mounts a different-speed motor on each end of the shaft and only one is powered while the other idles. Fan dampers are sometimes used, but in the hostile environment of the condenser the parts sometimes fail to move easily. Cycling the fan of a single-fan unit is a direct approach, but the condensing pressure oscillates and may cause control problems elsewhere in the system. One of the widely used methods of controlling the air flow is to cycle one or more fans in a multiple-fan unit. Such condensers must be equipped with baffles between the cells or much of the air delivered by one fan simply flows backward through an adjacent fan.

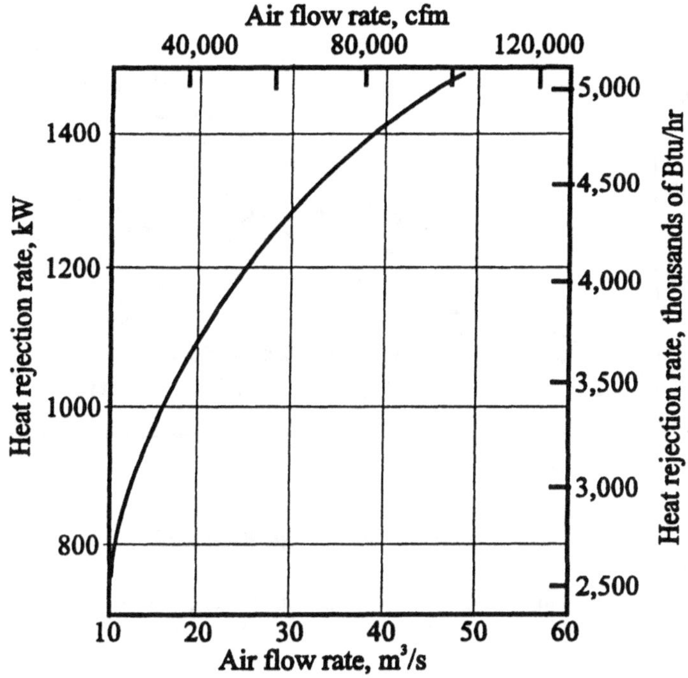

FIGURE 7.16
Effect of air-flow rate on the heat-rejection capacity of an evaporative condenser with given condensing and wet-bulb temperatures.

7.15 JUDICIOUS REDUCTIONS IN AIR-FLOW RATE

Most plants operate at the combination of maximum refrigerating capacity and design ambient temperature only a small percentage of the total time. In most cases, the refrigeration capacity is less and/or the ambient temperature is lower than design, offering the opportunity to reduce the condensing pressure. Up to a certain point, the plant operator should exploit the reduced condensing temperature to conserve compressor power. The catalog data for reciprocating and screw compressors in Chapters 4 and 5, respectively, indicate a power saving of the order of 3% per °C reduction in condensing temperature (1.7% per °F) in the range of 35°C (95°F) and 0°C (32°F) evaporating temperatures.

In operating regimes of perhaps 50% or more refrigerating capacity and wet-bulb temperatures above 15°C (59°F), the evaporative condenser would be operated with full fan and spray-water pump capacities. For lower refrigerating capacities and wet-bulb temperatures, the designer of the system may have specified controls that limit how low the condensing temperature (pressure) can drop. The setting of that minimum pressure is at the operator's discretion.

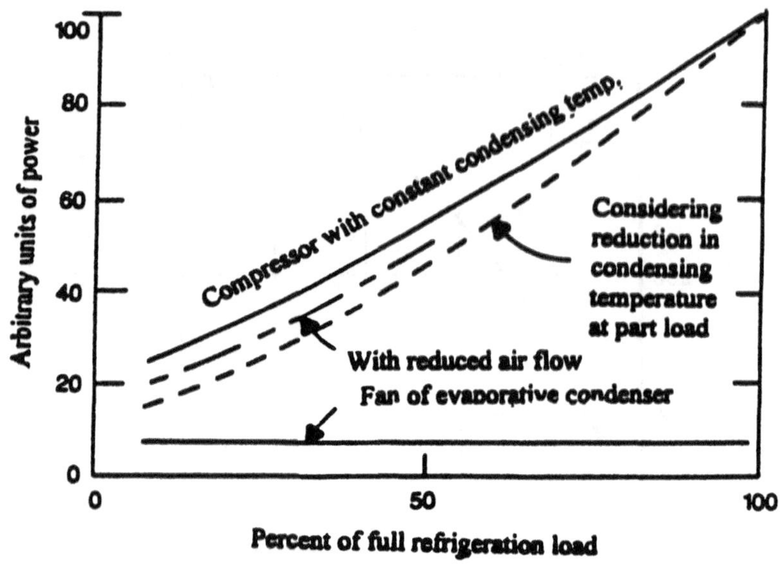

FIGURE 7.17
Relative power requirements of the compressor and the fan of an evaporative condenser. The evaporating temperature is in the range of 5°C (41°F) and the wet-bulb temperature is constant.

The conclusion has been reached that regulating the rate of air flow is the preferable method of reducing condenser capacity. An additional method in a multiple-condenser installation is to shut down completely one condenser and bring it back on line when necessary. With the exception of variable-speed drives, all of the methods of capacity control are a form of cycling wherein the condensing pressure drops low enough to reduce the capacity. When the condensing pressure passes through the control dead band, it rises to a point where full fan operation resumes. No good purpose is served by rapidly cycling the control, because it causes excessive wear on the affected motors and possibly erratic feeding of the evaporators serviced by expansion valves and pressure fluctuations downstream of level-control valves.

The next consideration is how to operate the fans to minimize combined compressor and fan power. The power required by the fans is small, relative to that required by the compressor, at least at full load. Figure 7.17 shows relative power requirements for the compressor and condenser fans as a function of the refrigeration load.

At full refrigeration capacity, the fan motors draw about 5%–8% of the power drawn by the compressor motor. As the refrigeration load decreases, the compressor power drops, but the fan power remains constant if the air-flow rate is not reduced.

Three curves are shown for the compressor power—the top one applies if the condensing temperature remains constant through the entire load range. The condensing temperature progressively drops as the load decreases because, as Fig. 7.13 shows, the condensing temperature falls as the load on the condenser drops. The intermediate compressor power curve in Fig. 7.17 represents the compressor power at reduced air flow rate, and suggests a trade-off between savings in fan power and compressor power.

The optimum conditions to shift from full fan operation to partial operation may be different for each plant. However, to conserve energy, many plant operators are overly influenced by the visibility of all the condenser fans operating and are not as conscious of the extra compressor power required.

7.16 SUBCOOLING THE REFRIGERANT IN THE CONDENSER

A reasonable desire is for the refrigerant liquid that flows out to the system to be subcooled. Conveying saturated liquid in pipes is always fraught with the drawback of flashing some liquid into vapor, which can restrict the mass flow when the refrigerant reaches an expansion valve or a level-control valve. Liquid can be subcooled in the tubes of the evaporative condenser by operating with liquid backed up into the tubes. This condition is not desirable, however, because heat-transfer area that should be available for condensation is reduced. The result will be a higher-than-necessary condensing temperature. The latter part of this chapter is devoted to proper drain piping, which has the objective of not backing liquid up into the condenser.

Even if subcooled liquid is produced in the condenser, this subcooled liquid flows to the receiver, as in Fig. 7.18, and here the liquid warms to a saturated state. Because both vapor and liquid are present in the receiver, the liquid is saturated. The heat-transfer process that takes place in the receiver is that of condensing some vapor to warm the subcooled liquid. In order to develop subcooled liquid a separate coil must be provided in the condenser that draws saturated liquid from the receiver and subcools it in the coil, as shown in Fig. 7.18.

7.17 POSITIONING THE CONDENSER

Not always is there complete freedom in where to place condensers, because the condensers should be close to the compressors that they serve, the walls of the machine room or other buildings may tend to obstruct air flow, and the condensers must be placed where their weight can be structurally supported. Within these limitations there are two major objectives to be kept in mind when siting the condensers. One is that the rate of air flow should not be restricted, or the reductions in capacity as discussed in Sec. 7.12 on capacity control will occur. The second precaution is to place the condensers so that there is a minimum of recirculation of discharge air from the same or other condensers entering a condenser. Recirculation results in a wet-bulb temperature of air entering the condenser that is higher than the ambient wet-bulb temperature.

Figure 7.19 shows several arrangements of condensers with Figs. 7.19a and

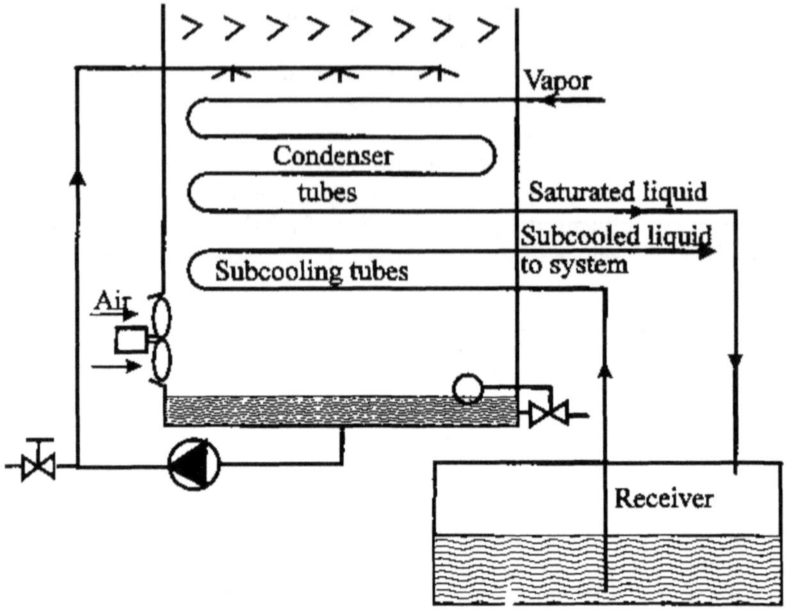

FIGURE 7.18
Liquid subcooler as auxiliary coil in the condenser case.

7.19b illustrating several problems while the placement in Fig. 7.19c offering a favorable placement. When the inlet of the condenser is close to a wall, as in Fig. 7.19a, the total air flow may be restricted. Furthermore some of the discharge air may be induced into the high-velocity air stream flowing down between the wall the condenser. In the arrangement of Fig. 7.19b, the condensers appear too close to one another so that the condenser on the right draws some discharge air into its inlet. Manufacturers of evaporative condensers often recommended minimum spacing distances between condensers. The positioning shown in Fig. 7.19c provides ready access of ambient air to both condensers.

7.18 WINTER OPERATION OF EVAPORATIVE CONDENSERS.

Many plants operate in geographic regions where ambient temperatures fall to near or below freezing. Two measures to prevent the spray water from freezing in such instances are:

(1) to locate the sump in a warm area, as in Fig. 7.20

(2) to drain the water from the condenser and operate the condenser dry.

The indoor sump in Fig. 7.20 must be able to accommodate all the water normally in suspension in the condenser during operation. The cost of pumping the spray water will be slightly higher than experienced when the sump is integral

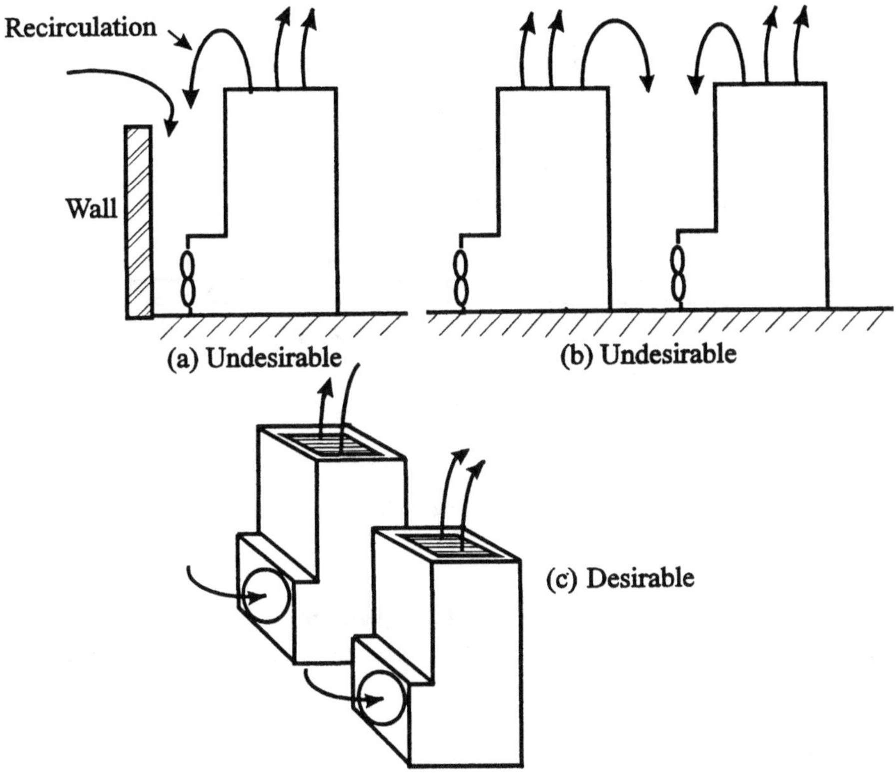

FIGURE 7.19
The condensers should be positioned so that the flow of entering air is not obstructed and as little as possible discharge air recirculates to the entrance.

to the condenser because of the additional head attributable to the difference in elevation between the condenser and the sump.

The indoor sump will prevent freezing of the main body of water in extremely cold weather, but the water droplets that drift out of the condenser may freeze close to the condenser causing icing conditions. An alternative to the indoor sump is to drain the condenser and operate dry. The heat-transfer capacity[16] of a condenser operating dry is strikingly less than when the condenser operates with the water sprays. Figure 7.21 shows relative capacities of a condenser operating dry compared to wet operation. It is significant that the dry condenser does not duplicate the wet capacity until the ambient dry-bulb temperature is about −30°C (−22°F). In many industrial refrigeration plants the load drops off as the ambient temperature drops, so the condenser can be shifted to dry operation when the ambient temperature drops below freezing. Such is not the case, however, for a plant with a predominantly process load which is only slightly affected by the ambient conditions. The sharp reduction

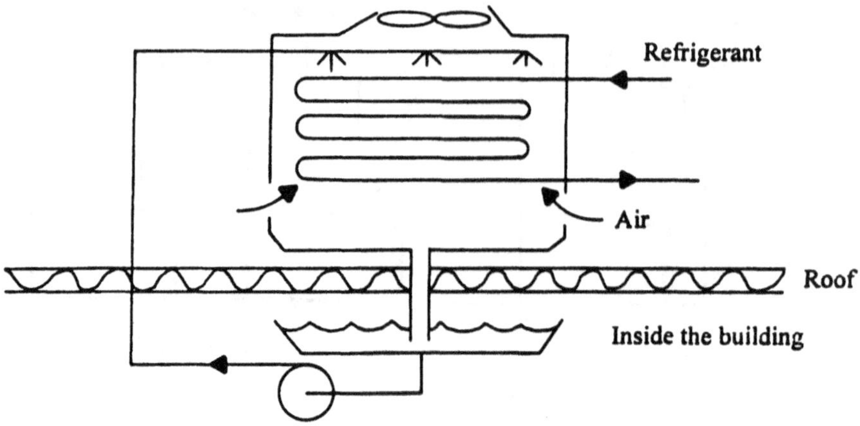

FIGURE 7.20
Locating the sump indoors to prevent freezing of the spray water during winter.

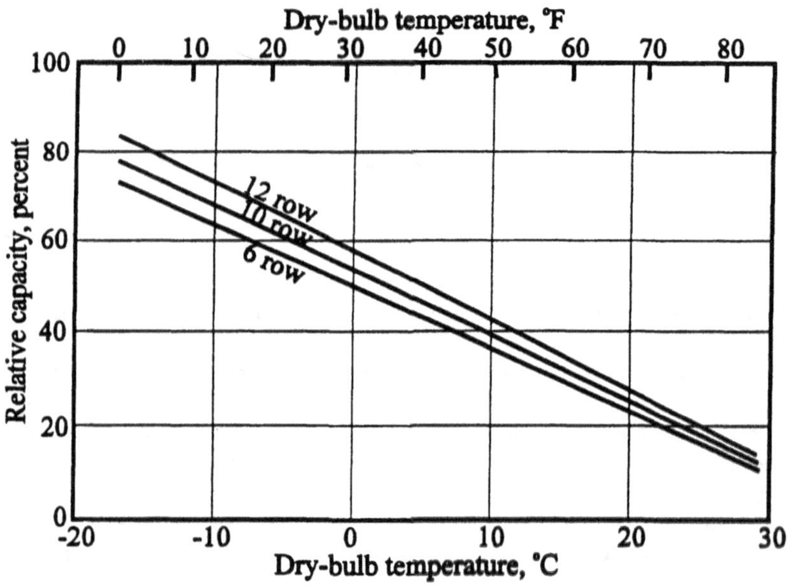

FIGURE 7.21
Heat-transfer capacity of a condenser operating dry compared to rated capacity operating wet with a condensing temperature of 35°C (95°F).

TABLE 7.3
Evaporating temperatures corresponding to standard atmospheric pressure of 101 kPa (14.7 psia).

Refrigerant	Evaporating temperature below which air could be drawn into the system through leaks
Ammonia	−33.5°C (−28.3°F)
R-22	−40.8°C (−41.5°F)
R-404a	−46°C (−50.8°F)
R-507	−46.7°C (−52°F)

in heat-transfer capacity under dry operation emphasizes that the evaporation process is the dominant one for transfer of heat to air in the evaporative condenser.

One incentive for a plant to shift to dry operation is that the costs of water treatment are usually a function of operating time, and this cost can be eliminated by dry operation.

7.19 PURGING THE CONDENSER OF AIR

Air and other noncondensible gases may enter a system through leaks in seals, gaskets, or uncapped valves. Air may also be present because of imperfect evacuation before the initial charging of the system or due to impurities in the refrigerant or oil. Another way that air gains access to the system is when an evaporator coil or a compressor is opened. Air could be drawn into the system through leaks in the low-pressure portion of the system when operating with refrigerant pressures below atmospheric, which will occur at evaporating temperatures shown in Table 7.3. Air drawn into the system on the low-pressure side is eventually pumped to the condenser where the liquid seal prevents it from traveling further.

The presence of noncondensables in condensers penalizes the system performance through the artificial elevation of the condensing pressure. As pictured in Fig. 7.22, the noncondensables add their partial pressure to that of the refrigerant vapor and thus increase the pressure against which the compressor must work. A further penalty is the reduction in the heat-transfer coefficient by requiring the refrigerant to diffuse through the noncondensables on its way to the tube surface where it condenses.

A test of the need for purging is to compare the actual pressure to the saturation pressure at the temperature of liquid at a location where liquid and vapor are in equilibrium, such as in the receiver of Fig. 7.22. If the actual pressure p is significantly higher than the saturation pressure corresponding to t, purging is warranted. Purging may be performed on rare occasions in small systems, but is often done frequently by automatic purgers on large systems. These automatic purgers proceed from one purge point to another to extract gas.

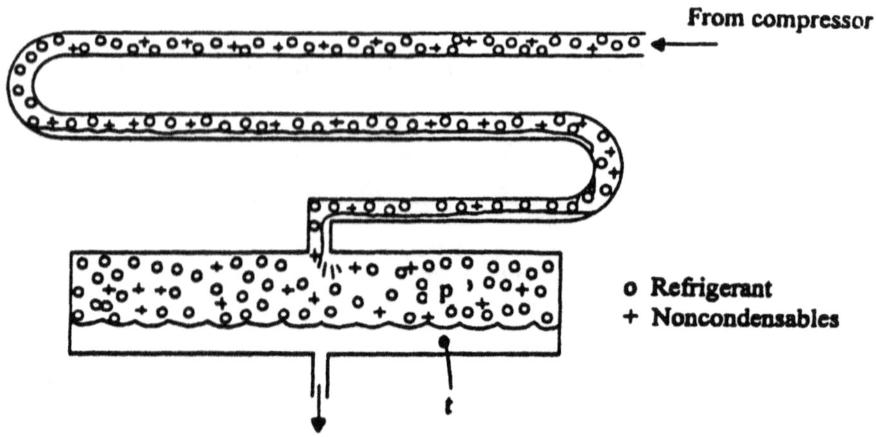

FIGURE 7.22
Noncondensables in a condenser.

There are preferable locations for purging, and basically these are (1) on the high-pressure side of the system, (2) where only vapor exists, and (3) where the vapor velocity is low. Air at a given pressure and temperature is more dense than ammonia, and not as dense as the halocarbon refrigerants, but no appreciable settling of one of the constituents can be anticipated. The air diffuses quite uniformly throughout slow-moving refrigerant.

The three principal concepts available for purging are

- direct venting of the air-refrigerant mixture
- compression of the mixture, condensing as much as possible of the refrigerant, and venting the vapor mixture that is now rich in noncondensables
- condensation of refrigerant using a small evaporator, followed by venting of the air-refrigerant mixture

Figure 7.23a shows the first method, a primitive, manual technique. Vapor is released from a high-pressure vessel, such as the receiver, and this vapor is mostly refrigerant but also contains a small amount of the noncondensables that are the target. In the case of ammonia, the discharge bubbles through a container of water to absorb the ammonia. As venting proceeds, more refrigerant liquid vaporizes, so the concentration of noncondensables decreases, but never drops to zero. This method wastes considerable refrigerant to expel a small amount of noncondensables.

The second method of purging, as shown in Figure 7.23b, consists of drawing a sample from the vessel with a small compressor that elevates the pressure and condenses some refrigerant on a water-cooled coil. The vapor vented from this after-condenser is higher in noncondensable content than at the original sam-

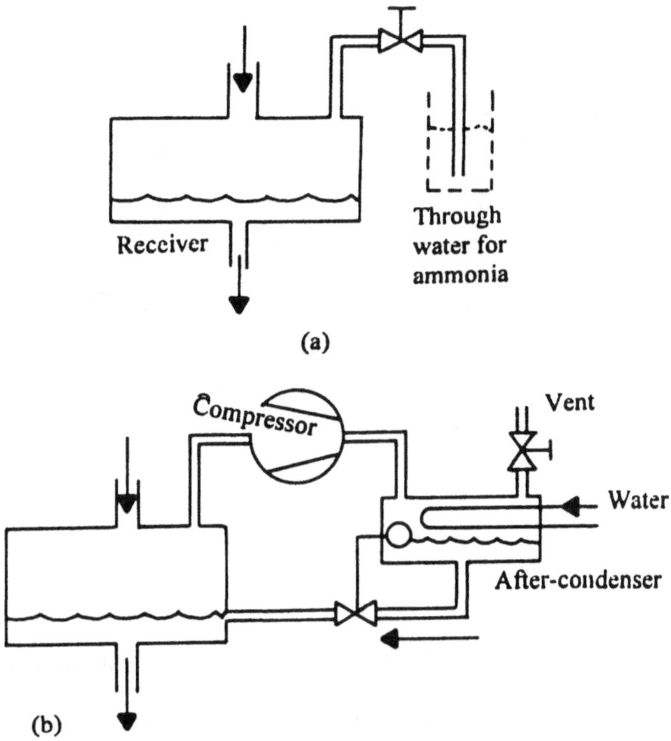

FIGURE 7.23
Two methods of purging noncondensables, (a) direct venting, and (b) compression of the refrigerant followed by condensation.

pling position. This purging concept is widely applied in centrifugal-compressor water-chiller systems using such low-pressure refrigerants as R-123, but sees limited application in industrial refrigeration.

The third concept in purging (Figure 7.24), which is widely used in industrial refrigeration, avoids the need of a separate compressor, and instead uses a low temperature developed in a small evaporator. The air- refrigerant mixture from the condenser or receiver bubbles through cold liquid and condenses most of the refrigerant. This concept is embodied in automatic purgers which move from one purge point to another allowing enough time at each for a satisfactory purge. Commercial models of refrigerated purgers employ refinements in handling the vented stream leaving the after- condenser. The proper procedure is to purge one point at a time, because if one solenoid control valve serves two or more purge points, the pressure at these positions will be equalized during purging. In the later sections of this chapter, the need to properly regulate pressure differentials will be emphasized.

A manual technique is sometimes used for a massive purge that would require a long time for automatic purgers to handle. The method applies to a

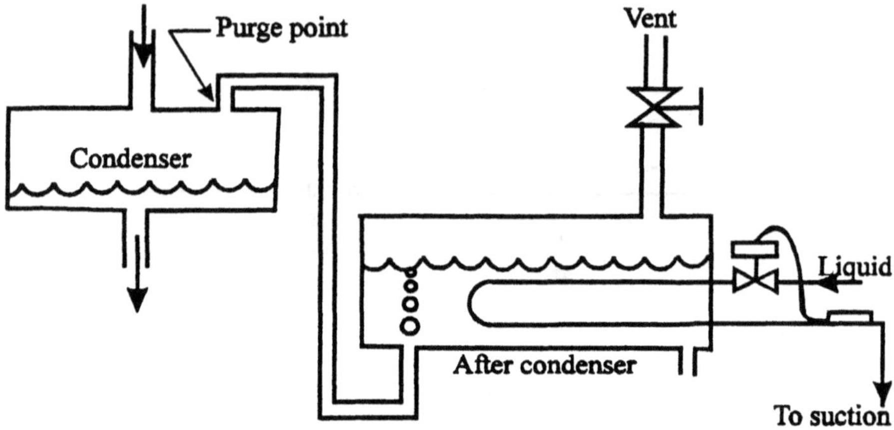

FIGURE 7.24
Purger for industrial refrigeration systems condenses refrigerant on a low-temperature evaporator.

multiple-coil condenser, such as shown in Fig. 7.25, that is equipped with individual valves from the discharge gas header and individual vents to atmosphere. If one coil is to be purged, the valve in the gas line is closed, but its fan and spray water continue to operate. Other parallel coils continue to operate normally, so the temperature in the coil being purged drops and ammonia vapor in the coil condenses. As the vapor condenses, the vapor volume decreases and liquid is drawn from the condensate line. After a period of time most of the ammonia vapor has condensed, leaving the small volume of mostly noncondensables which can be vented to the atmosphere or to a vessel of water.

7.20 INTENTIONAL SUBCOOLING IN A CONDENSER

An objective to be pursued in the next several sections is to effectively drain the condenser of liquid so that the maximum amount of surface area will be available for condensation. Before addressing that goal, however, a technique sometimes used in commercial refrigeration (e.g., supermarkets) that operate year-round will be shown. The condenser pressure must be prevented from dropping so low that the thermal expansion valves (typically used in that type of system) are unable to pass a sufficient flow rate of refrigerant. The condensing pressure must be artificially prevented from dropping during cold weather operation, and a procedure, such as the one shown in Fig. 7.26, backs liquid into the condenser to reduce its heat-transfer capacity and maintain the desired level of condensing pressure.

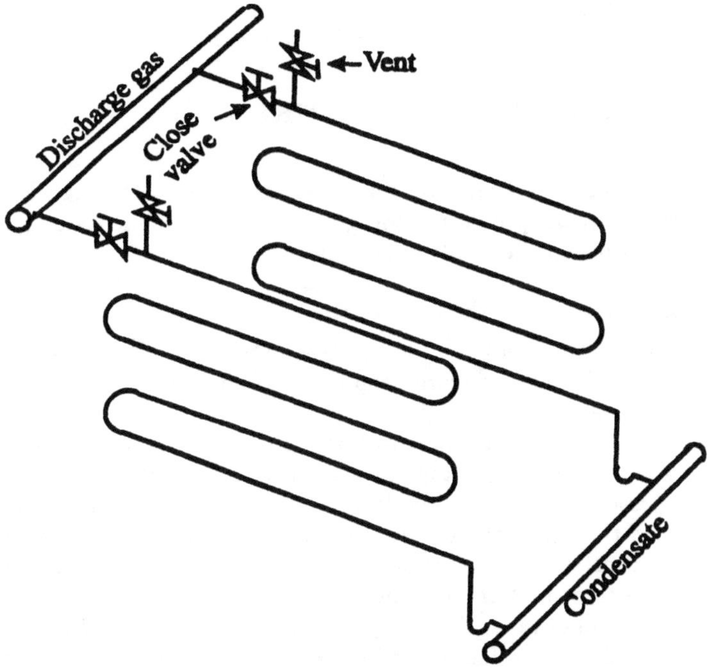

FIGURE 7.25
Manual purge of condenser tubes.

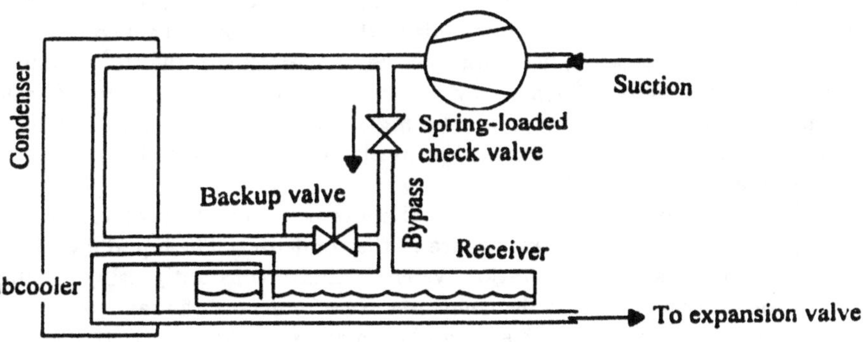

FIGURE 7.26
Condenser piping arrangement used on some commercial refrigeration systems to prevent drop of condensing pressure below a preset point.

The operating efficiencies offered by lower condensing pressures are thus not available, although the subcooling coil does reduce the enthalpy of liquid passing to the expansion valve and thereby increases the refrigerating effect.

7.21 PIPING OF HIGH-PRESSURE RECEIVERS

The condenser normally drains its liquid refrigerant into the high-pressure receiver, which is a vessel providing storage space for the condensate. The considerations that surround the condenser/receiver subsystem are:

- piping arrangements of the receiver
- connecting the vapor space in the receiver to vapor locations in the condenser
- the role of refrigerant pressure drop in the condenser

The two typical piping configurations of the receiver are shown in Figure 7.27. Figure 7.27a shows the top inlet or mixing-type receiver where all the condensate passes through the receiver. Usually the outlet of this receiver is as shown, with the tip of the outlet line not quite touching the bottom of the receiver. This construction avoids passing solid contaminants on to the system.

In the bottom inlet receiver of Figure 7.27b, most of the liquid passes directly to the evaporators. The only liquid flow in or out of the receiver is associated with the rise or fall in liquid level. The liquid level may shift because of transient differences in the condensation rate and the rate of liquid flow to the evaporators. Some designers prefer the bottom inlet receiver over the top inlet for two reasons: (1) the ability to use subcooled liquid when it is available from the condenser, and (2) inherent trapping of the liquid line which, as explained later, is a necessity for good drainage in multiple condenser installations. In the top inlet receiver of Figure 7.27a, the liquid and vapor are in equilibrium, so saturated liquid passes on to the low side of the system. Even if subcooled liquid enters the receiver, it will quickly assume the temperature of the stored liquid—a temperature heavily influenced by the machine room temperature. Because the machine room temperature will be higher in most cases than that of the liquid coming from the condenser, the available subcooling is lost in the top inlet receiver.

One of the considerations mentioned above is to provide a connection from the vapor in the receiver to vapor space in the condenser. The liquid level in the receiver of an industrial refrigeration system is almost constantly rising or falling because liquid flow rate from the condenser is not precisely the same as the rate demanded of the low side of the system. When an excess flow enters the receiver, there is a tendency to compress the vapor, which builds up the pressure in the receiver and temporarily restricts the flow of condensate from the condenser. If the flow rate demanded by the system exceeds temporarily the rate provided by the condenser, the pressure in the receiver drops and some of the liquid vaporizes. The requirement of a connection between the vapor in the receiver with vapor

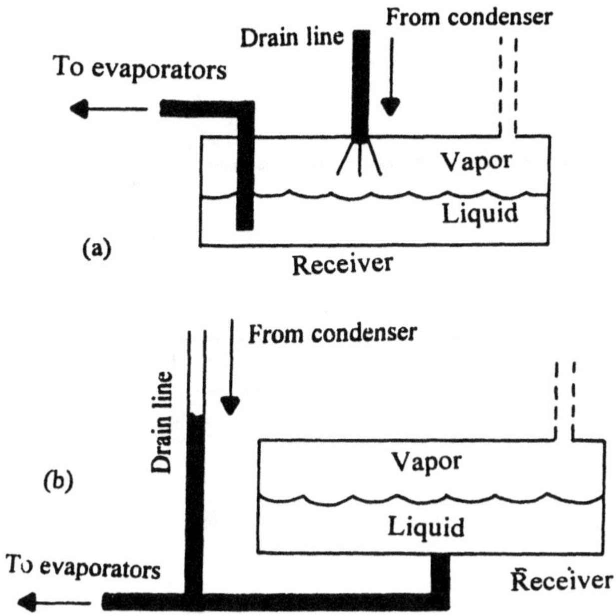

FIGURE 7.27
Piping arrangements for the receiver: (a) top inlet, mixing type, and (b) bottom inlet or surge receiver.

in the condenser will be a thread running through recommended procedures for drain piping that will be presented in the next several sections.

7.22 DRAINING CONDENSATE—SINGLE CONDENSER

The arrangement to insure proper draining of a single condenser depends on the type of receiver. If the receiver is top inlet, good drainage of the condensate results if the vapor in the receiver can flow freely counterflow to the condensate, as appears in Fig. 7.28. The provisions[15] for achieving open channel flow in the drain line are:

- avoid a horizontal drain line, if possible, sloping the line with a pitch of at least 1 in 50
- choose a large-size pipe based on liquid velocities no higher than about 0.5 m/s (100 fpm)
- do not place a valve in the sloping drain line with its stem upward. Instead use angle valves which introduce much less pressure drop than straight-through globe valves. If a straight-through valve is used, place it in the vertical portion

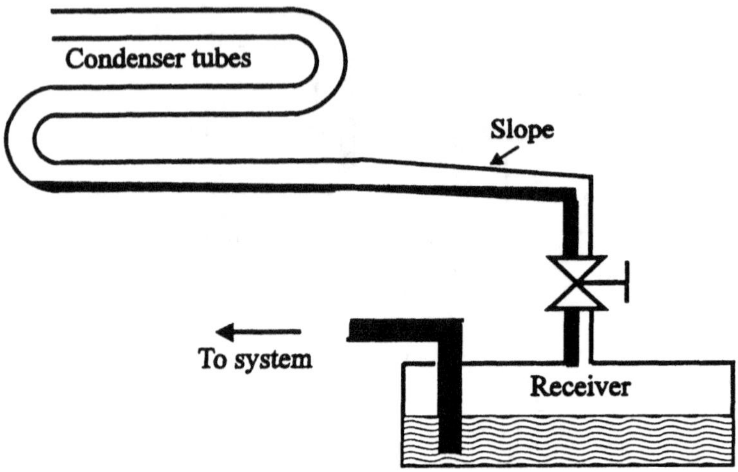

FIGURE 7.28
Proper draining of a single condenser to a top inlet receiver.

of the drain line

If the receiver is bottom inlet, as in Fig. 7.29, the liquid is inherently trapped and the vapor in the receiver has no access to the vapor space in the condenser. For this reason an equalizer line should be connected from the top of the receiver to the outlet of the condenser.

7.23 PRESSURE PROFILES IN A CONDENSER

Figure 7.30 shows changes in pressure that develop in a condenser when the equalizer line connects the top of the receiver to the inlet to the condenser. As the condensing refrigerant flows down through the condenser, a pressure drop occurs, $-\Delta p$. The equalizer line forces the pressure in the receiver to be the same as that of the condenser inlet, so some means must be found to recover the drop in pressure that occurs in the condenser. The $+\Delta p$ to cancel the $-\Delta p$ derives from the static head of a column of liquid refrigerant. In Figure 7.30 the liquid column is adequate to provide this gain in pressure. If the difference in elevation from the bottom of the condenser to the liquid level in the receiver is inadequate, the system achieves the necessary column of liquid by backing liquid into the condenser. Liquid that is forced back into the condenser is the root of many low-capacity problems with condensers. Incidentally, the arrangement shown in Figure 7.30 is a satisfactory alternate to the configuration of Figure 7.29, but Figure 7.29 is preferred because it does not require as much length of liquid column and the equalizer line is shorter.

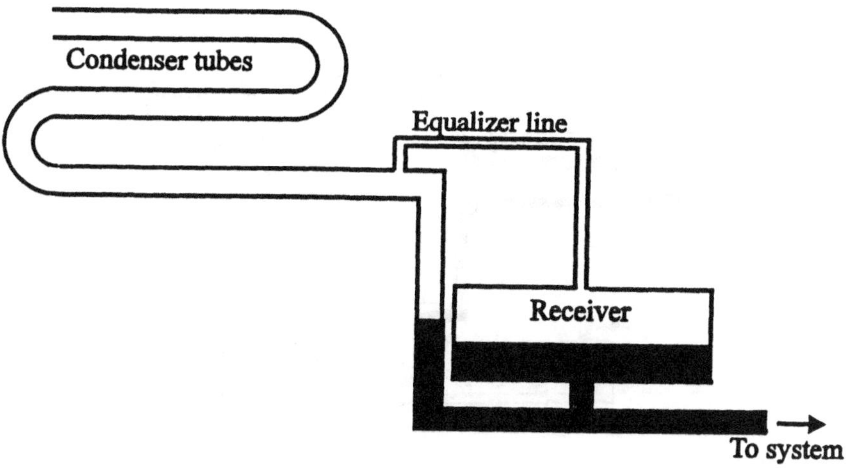

FIGURE 7.29
Proper draining of a single condenser to a bottom inlet receiver.

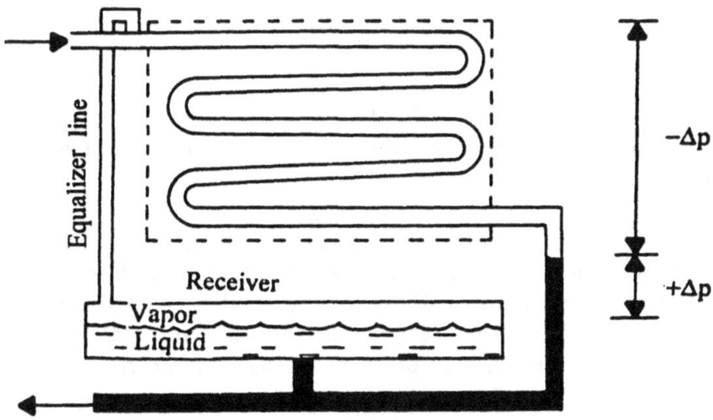

FIGURE 7.30
Pressure profiles when the receiver is equalized to the top of the condenser.

7.24 DRAINING CONDENSATE—PARALLEL CONDENSERS

A characteristic of industrial refrigeration plants is the use of parallel condensers and compressors. This fact is especially pertinent to ammonia systems, and not always true for halocarbon systems where the oil transfer between parallel units sometimes becomes a problem. Plants of even moderate size are usually designed for parallel condensers to offer flexibility in meeting a wide range of load

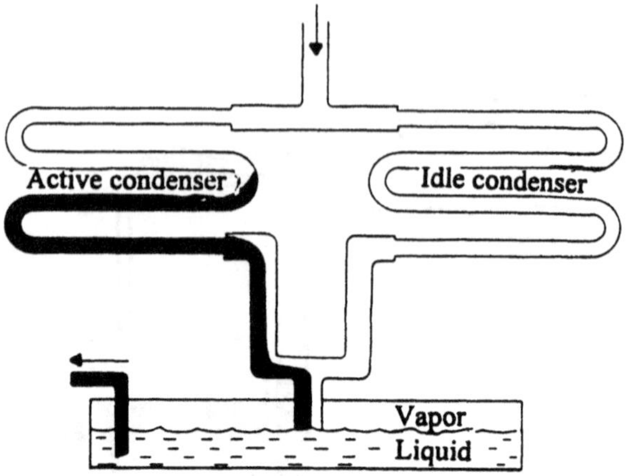

FIGURE 7.31
Backup of liquid into one of two parallel condensers.

variation. When condensers are piped in parallel, the following rules should be observed:

- trap the liquid drain lines
- provide a generous length of vertical drain line
- install an equalizer line between the receiver and the entrance of the condensers

The purpose of trapping the liquid drain lines is to aid in drainage of liquid from all condensers. The potential drainage problem with multiple parallel condensers is illustrated in Fig. 7.31 where the condenser on the right has, at least at this moment, a low flow rate of refrigerant. Some reasons for the low flow rate include: a different design of condenser than the other or the fans are completely or partially shut down.

The pressure drop through both condensers must be the same because there are two common points in the piping—at the inlet and the outlet of the condensers. The only way the active condenser on the left can operate with the low pressure drop of the condenser on the right is if liquid backs up into the condenser tubes. Assume that the drain lines are large, so the liquid head in the tubes on the left condenser supplements the available pressure difference.

To avoid the problem of liquid backup into one of the condensers, liquid lines from both condensers should be trapped, using arrangements such as those shown in Fig. 7.32. Once again, the pressure drop through the two condensers from their common inlet to their common outlet is identical, but because the liquid lines are trapped, the drain line is full, and the difference in liquid level in the vertical drain line compensates for the difference in pressure drop. Thus, the liquid head in the left condenser builds up in the drain line and not in the condenser tubes where it would adversely affect the condenser performance. The bottom inlet receiver in Figure 7.33 inherently provides liquid traps for each condenser.

Implied in the correct piping of Figures 7.32 and 7.33 is an adequate length of the vertical leg. An estimate of the needed length of liquid column can be derived from the knowledge that the maximum pressure drop in an operating ammonia condenser is usually about 3.4 kPa (1/2 psi), which is the maximum pressure difference to be compensated for between an idle and an operating condenser. A liquid column of 0.6 m (2 ft) would compensate for this pressure difference, but most designers attempt to place the condenser high enough to permit a 1.2-m (4-ft) column of ammonia. Liquid R-22 has twice the density of liquid ammonia which would suggest that only half the length of liquid column would be needed for R-22, but the pressure drop in an R-22 condenser is about four times that of an ammonia condenser. The reason for the higher pressure drop with R-22 is because of its low latent heat, requiring perhaps six times the flow of R-22 compared to ammonia for a given heat-rejection capacity of the condenser. The recommended length of liquid column in an R-22 installation is 2.4 m (8 ft).

An equalizer line is required since the vapor from the top of the receiver cannot vent back to the condensers through the drain line because of the liquid traps as it could in the single condenser in Figure 7.28. Also the pressure at the outlet of the condenser tubes may be different, so the equalizer line must be connected to the top of the condensers where the pressure is the same for both condensers.

7.25 DRAINING CONDENSATE—THERMOSYPHON OIL COOLING

When the oil for screw compressors is cooled by refrigerant circulating by means of a thermosyphon, some of the same principles of condenser draining apply, but there are some additional considerations. A vessel, called the thermosyphon receiver, is needed to separate the liquid from the liquid/vapor mixture that rises from the oil cooler. The vapor thus separated passes to the discharge gas header, as shown in Figure 7.33. The return of liquid/vapor from the oil cooler should flow to the thermosyphon receiver and not to the discharge header of the condenser. Were the liquid from the oil cooler to enter the condenser, it would overload it with liquid and reduce its heat-transfer capacity.

A significant fact is that there is appreciable flow of vapor through the vapor line from the thermosyphon receiver, so there is a pressure drop in this line. Thus the pressure in the thermosyphon receiver is higher than that of the condenser inlet and certainly higher than the condenser outlet. To overcome this pressure difference a liquid column must develop somewhere. Figure 7.33 shows bottom inlet to the thermosyphon receiver with individual liquid columns for each condenser. Entrance of the condensate to the top of the thermosyphon

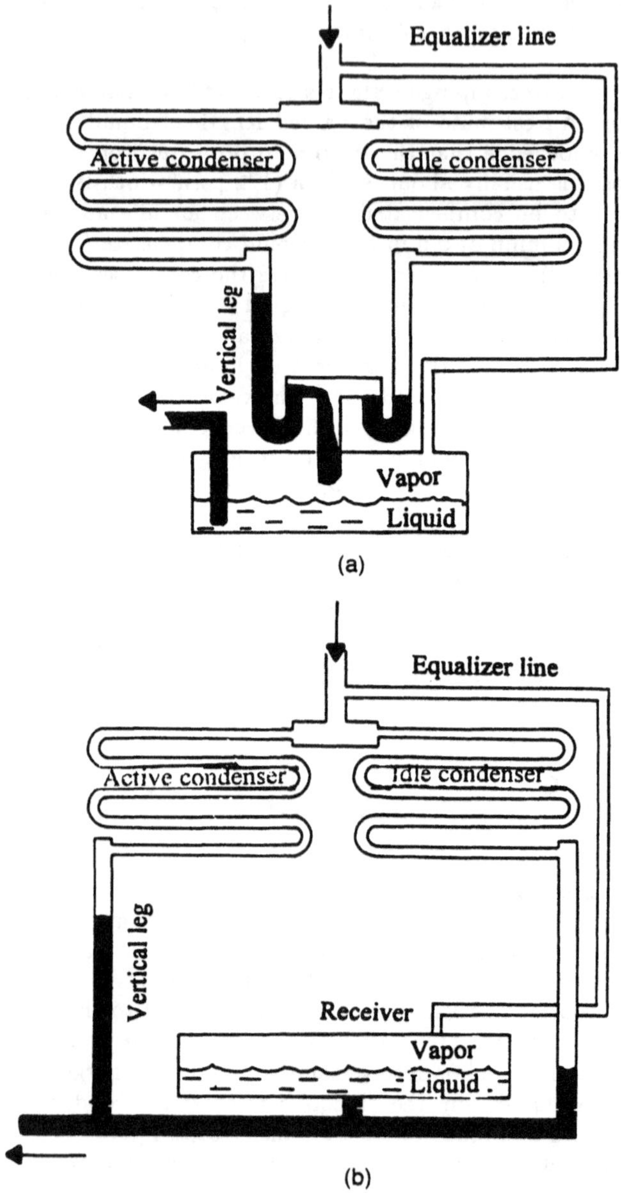

FIGURE 7.32
Trapping the liquid lines and using the liquid head in the vertical drain line to compensate for differences in pressure drop in the condensers in (a) a top inlet receiver and (b) a bottom inlet receiver.

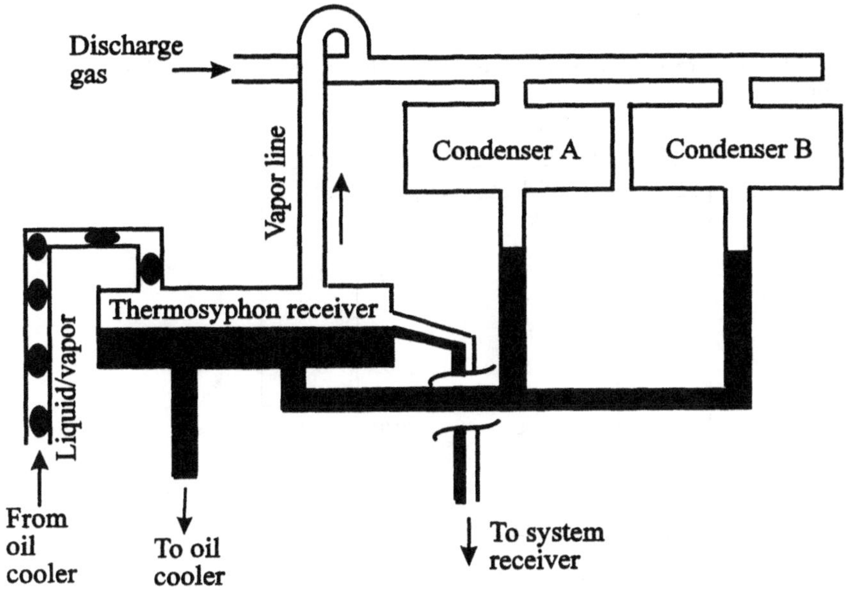

FIGURE 7.33
Trapping the liquid drain lines from multiple condensers that serve a thermosyphon receiver.

receiver would work satisfactorily if the drains were equipped with the P-traps of Figure 7.32a. Even if the installation consists of only one condenser, there is still the need to build up pressure from the outlet of the condenser to the thermosyphon receiver. A convenient means of providing a liquid seal to form a liquid trip is to immerse the drain line from the condenser below the liquid level maintained by the overflow to the system receiver.

7.26 SIZE OF EQUALIZER LINES AND THERMOSYPHON VAPOR LINES

Under normal circumstances equalizer lines between the receiver and the inlet or outlet of the condenser can be small, because it is not expected that they need carry high flow rates. They only need to convey a volume flow rate of vapor equal to the rate of change of liquid volume. There is one situation where the equalizer line should be generously sized and that is during a shutdown of a plant located in a cold ambient temperature but with the receiver located in a warm machine room. There will be a constant vaporization of the liquid in the receiver causing vapor to flow through the equalizer line to the discharge gas header where it will condense in the condenser. The pressure drop resulting from the vapor flow in the equalizer line must be compensated by a liquid column that will develop and back liquid into the condenser. For high pressure drops of, for example, 25 kPa (3.7 psi), an ammonia liquid column of 4.6 m (15 ft) may be needed. This elevation could be enough to push all of the liquid of the system

TABLE 7.4
Flow and refrigeration capacity of ammonia thermosyphon vapor lines and equalizer lines

Pipe size, inch	Thermosyphon vapor line		Equalizer lines	
	kg/s	lb/min	kW	tons
3/4			176	50
1			352	100
1-1/4			528	150
1-1/2	0.0529	7	792	225
2	0.0907	12	1056	300
2-1/2	0.166	22	1760	500
3	0.295	39	3520	1000
4	0.529	70	7040	2000
5	0.907	120		
6	1.66	220		
8	5.29	700		

into the condenser resulting in difficulty in restarting the plant.

As was pointed out in the previous section, the vapor line between the thermosyphon receiver and the discharge gas header (Figure 7.33) is more than a low-flow equalizer line, but carries a heavy flow of vapor and will be of larger size than an equalizer line. Table 5.3 in the chapter on screw compressors listed recommended pipe sizes for this line[17], which are reproduced here in Table 7.4 along with recommended equalizer line sizes[18].

7.27 SUPPLEMENTARY COOLING REQUIREMENTS PROVIDED BY EVAPORATIVE CONDENSERS.

In addition to the major task of condensing refrigerant, evaporative condensers are sometimes called on to cool some other fluid as well. The prime example of this supplementary cooling requirement, and a frequent one, is to cool the oil injected in screw compressors. To accomplish this assignment, pump water from the sump to the oil-cooling heat exchanger and return the warmed water to the sump. This additional cooling load reflects itself in the performance of the condenser, as shown in Figure 7.34, where the solid lines show the performance without, and the dashed lines with external cooling by the sump water.

When no supplementary cooling is demanded, the water temperature in the sump at A is the same as that sprayed over the tubes at the top of the condenser, point B. This equality of temperatures is inherent because, as Figure 7.1c shows, there is a direct connection from the sump to the sprays. When the sump water provides supplementary cooling, the spray water temperature at B′ will be higher than temperature A. The consequence of the higher spray-water temperature is a greater rise in air enthalpy through the condenser, and a higher

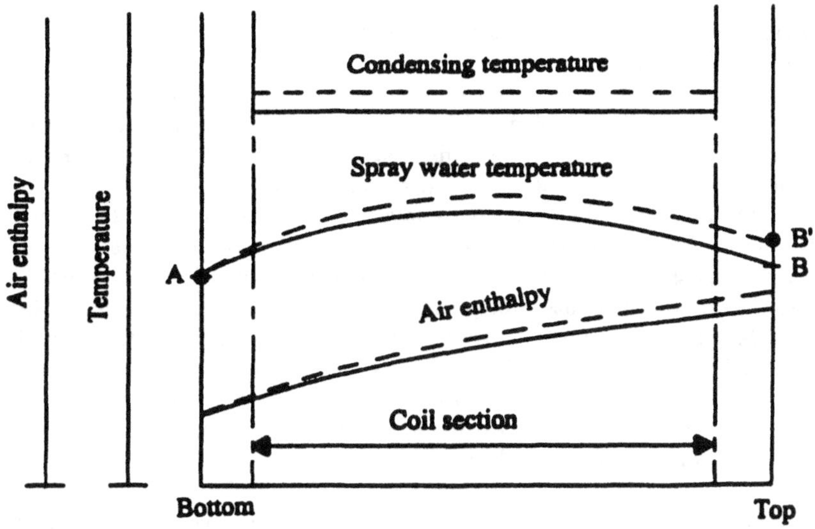

FIGURE 7.34
The solid lines show the temperatures of refrigerant and spray water and the enthalpy of air when the condenser provides no supplementary cooling, and the dashed lines show the performance when the sump water performs external cooling.

condensing temperature of the refrigerant.

If an additional load is imposed on the condenser by adding heat to the sump water, the condenser will accommodate the addition—a fact which is shown in Fig. 7.34 by the increased rise in air enthalpy through the condenser. The increased capacity does not come without a cost, however, and that cost is the increased condensing temperature of the refrigerant.

Manufacturers of evaporative condensers usually recommend a separate closed circuit evaporative cooler for the supplementary cooling duty to permit separate control of the refrigerant temperature and the fluid temperature, and to allow for dry operation during cold weather.

7.28 WATER TREATMENT IN EVAPORATIVE CONDENSERS

Evaporation of spray water is the dominant process of heat rejection in evapoartive condensers. The amount of heat transfer attributible to the difference in dry-bulb temperatures between the condenser tubes and the air is minor. Because the makeup water contains some minerals and other chemicals, and because the water vapor leaves the condenser with none of these impurities, the concentration of foreign materials in the spray water tends to increase. To keep the concentration of impurities under control, blowdown of some of the spray water should be provided whenever the condenser operates. Manufacturers of evaporative condensers typically recommend that the blowdown rate be approximately equal to the evaporation rate. The rate of makeup water is thus twice the blowdown rate.

Almost all sources of makeup water must be treated in some way to avoid one or more of the problems that beset the surfaces of the condenser that are in contact with water. These difficulties include scaling, fouling, and corrosion. Scaling is the deposit of a hard layer of minerals, usually calcium carbonate ($CaCO_3$), on the tube surfaces. A layer of scale 0.8 mm (1/32 in) thick on the tubes can be expectecd to reduce the capacity of the condenser by 30%. This layer of scale is usually due to high mineral concentration in the makeup water and if calcium carbonate precipitates it can form scale on the tubes. Generally a $CaCO_3$ concentration of less than about 170 ppm will be satisfactory. It should not be implied, however, that softened water with a mineral concentration of perhaps 30 ppm is desirable. Softened water may result in excessive corrosion.

Fouling usually refers to the accumulation of nonscale solids, such as dirt, silt, sand, algae, fungi, and bacteria.

Corrosion is a distressing problem in galvanized steel because it often takes the form of deteriorating the zinc coating, which exposes the steel to oxidation. Corrosion is an electrochemical process where an electrical potential develops between two different metals. When current flows as a result of this difference in voltage in the presence of an electrolyte, such as water with dissolved solids, one of the metals dissolves. A particularly sensitive situation is where the zinc coating on steel has deteriorated, because it is between these points that current will flow. Controlling the pH value of the spray water is particularly helpful in retarding corrosion. The pH value is an indicator of the alkalinity or the acidity of a solution, with a pH value of 7 being defined as neutral. Maintenance of a pH value of between 6 and 8 is usually recommended[19]. Corrosion inhibitors approved by the condenser manufacturer are also treatments to prevent corrosion.

A problem called *white rust* has appeared in the past decade and may be associated with the prohibition on the use of chromates for corrosion protection. White rust is the accumulation of a white, waxy, nonprotective zinc corrosion product on galvanized surfaces[20]. Typically it will appear suddenly and progress rapidly over the wetted, galvanized steel components of the condenser. If not corrected, white rust may lead to premature failure of the galvanized coating. One of the most effective methods of preventing white rust is to assure the *passivation* of the condenser during initial operation. Passivation is the natural formation of a protective, light crystalline film on the zinc surface. A recent trend in water treatment has been toward greater alkalinity (high pH), which some experts suspect promotes white rust. Indications are that soft water (less than 30 ppm total hardness) combined with high pH valves exacerbate the problem. Passivation is facilitated by initial use of untreated water and also by the use of phosphates. A water treatment expert should be consulted for the unique approach applicable to local conditions.

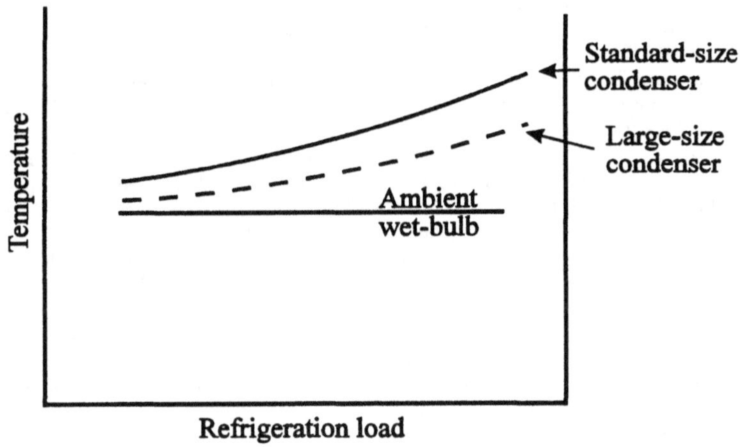

FIGURE 7.35
Condensing temperatures as affected by the refrigeration load and the condenser size.

7.29 THE CONDENSER AS A PART OF THE SYSTEM

The condenser operates according to its own rules, but it is also a part of the refrigeration system and therefore interacts with the other components. Except in the low-condensing temperature regions where certain operating limits may apply, it is always desirable to operate at as low a condensing temperature as possible. But, as Fig. 7.35 shows, an increase in the refrigeration load increases the condensing temperature in order to transfer a higher heat flow rate.

It is unfortunate that the condensing temperature increases and the compressor power per unit refrigerating capacity increases at just the time when the refrigeration rate is high. Figure 7.35 also shows that the condensing temperatures can be lowered at all ranges of refrigeration load if a large condenser is chosen. Certainly there would be additional first cost associated with the larger condenser, and perhaps additional fan power, but the compressor power drops throughout its life. A further advantage of the large condenser with its lower condensing temperature is that the peak refrigerating capacity of the compressor can be increased slightly.

REFERENCES

1. Nusselt, W. Z., "The Oberflaechenkondensation des Wasserdamples," *Ver. Deutsch. Ing.* 60: 541-549, July 1916.
2. White, R. E., "Condensation of Refrigerant Vapors: Apparatus and Film Coefficients for R-12," *Refrig. Eng.*, 55(5): 375, April 1948.
3. Goto, M, H. Hotta and S. Tezuka, "Film Condensation of Refrigerant Vapors on a Horizontal Tube," Proceedings of the 15th International Congress of Refrigeration,

Paper B1-20, Paris, France, 1979.

4. Murphy, R. W., and J. W. Michel, "Enhancement of Refrigerant Condensation," ASHRAE Transactions, 90(Part 1B): 72-79, 1984.
5. Tandon, T. N., H. K. Varma and C. P. Gupta, "An Experimental Study of Flow Patterns During Condensation Inside a Horizontal Tube ," *ASHRAE Transactions*, 89(Part 2A): 471-482, 1983.
6. Stoecker, W. F. and E. Kornota, "Condensing Coefficients when Using Refrigerant Mixtures," *ASHRAE Transactions*, 91(Part 2): 1351-1367, 1985.
7. Lee, S. H. and J. G. Knudsen, "Scaling Characteristics of Cooling Tower Water," *ASHRAE Transactions*, 85(Part I): 281-302, 1979.
8. Tezuka, S, T. Takada, and S. Kasai, "Performance of Evaporative Cooler," XIII International Congress of Refrigeration, Paper 2.86, International Institute of Refrigeration, 1971.
9. Finlay, I. C. and D. Harris, "Evaporative Cooling of Tube Tanks," *International Journal of Refrigeration*, 7(4): 214-224, July 1984.
10. Chapter 36, "Cooling Towers," *ASHRAE Handbook, HVAC Systems and Equipment*, American Society of Heating, Refrigerating, and Air-Conditioning Engineers, Atlanta, Georgia, 1996.
11. "Evaporative Condensers," Evapco, Inc. Taneytown, Maryland, 1984.
12. Stoecker, W. F., J. J. Lux, and R. J. Kooy, "Energy Considerations in Hot-Gas Defrosting of Industrial Refrigeration Coils," *ASHRAE Transactions*, 89(Part II): 549-573, 1983.
13. "What is the Lowest Possible Condensing Pressure at Which the Plant will Still Function Properly?" *NH_3 NEWS*, The International Institute of Ammonia Refrigeration, Washington, DC, July/August, 1995.
14. Brebner, R. and K. Free, "Design of Ammonia Condensers for Freezing Plants," Proceedings of Commission Meetings in Palmerston North (New Zealand), Internation Institute of Refrigeration, Paris, 1993.
15. "Evaporative Condenser Engineering Manual," Baltimore Aircoil Company, Inc., Baltimore, Maryland, 1984.
16. Benz, S. M., "Dry Operation of Evaporative Condenser," Proceedings of the 9th Annual Meeting of the International Institute of Ammonia Refrigeration, Washington, DC, 1987.
17. "Thermosyphon Oil Cooling," Frick/Frigid Coil/York International, 1991.
18. "ASHRAE Handbook, Refrigeration Volume," Chapter 3, American Society of Heating, Refrigerating and Air-Conditioning Engineers, Atlanta, Georgia, 1994.
19. "Maintenance Instricution for Evaporative Condensers, Closed Circuit Evaporative Coolers, and Cooling Towers," Evapco, Inc., Tanneytwon, Maryland, 1995.
20. "White Rust on Galvanized Steel Cooling Towers," Baltimore Aircoild Company, Baltimore, Maryland, 1992.
21. Cole, R. A., "Avoiding Refrigeration Condenser Problems," *Heating/Piping/Air Conditioning*, 58(7): 97-108 and 58(8): 123-131, July and August, 1986.

CHAPTER 8

LIQUID RECIRCULATION

8.1 LIQUID RECIRCULATION AND LIQUID OVERFEED

The terms *liquid recirculation* and *liquid overfeed* are sometimes used interchangeably and refer to the practice of delivering a greater rate of liquid refrigerant to the evaporator than evaporates. So at the exit of the coil a mixture of liquid and vapor flow out of the evaporator, as shown in Fig. 8.1. If there is desire to differentiate between liquid overfeed and liquid recirculation, it might be that liquid overfeed applies to the coil, while a liquid recirculation system incorporates the additional equipment to accommodate the overfeeding of the coil.

The overfeed operation contrasts to regulating with a controlled expansion valve the liquid flow rate to the evaporator so that it precisely matches the rate of evaporation.

The flooded evaporator, discussed in Chapter 6 and arbitrarily not included in the category of liquid recirculation, also operates in a liquid-overfeed manner, resulting in no superheat at the outlet of the evaporator. Flooded evaporators are supplied with high-pressure liquid like DX coils. Those configurations classified in this chapter as liquid recirculation have a central liquid-vapor separator capable of serving multiple evaporators. Furthermore, the liquid recirculation is forced by either mechanical pumps or by gas-pressure pumping.

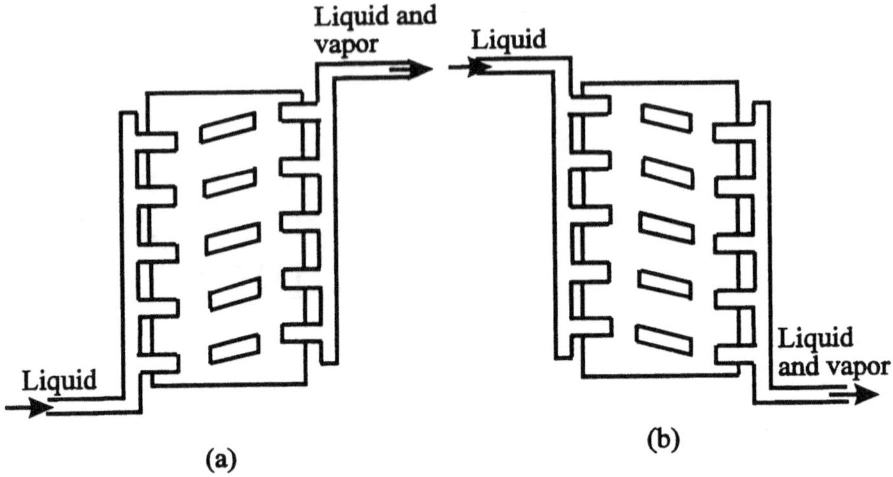

FIGURE 8.1
Evaporators operating with liquid overfeed in a liquid recirculation system: (a) bottom feed and (b) top feed of liquid.

8.2 MECHANICAL PUMPING AND GAS PUMPING

The essential arrangement of the mechanically pumped and the gas-pumped liquid recirculation systems are shown in Figs. 8.2 and 8.3, respectively. In either concept the mixture of liquid and vapor that floods out of the evaporators flows to a liquid-vapor separator. The vapor passes to the compressor, while the liquid is forced back to the evaporator, supplemented by the addition of more liquid feed.

8.3 ADVANTAGES AND DISADVANTAGES OF LIQUID RECIRCULATION

Some of the **advantages** of liquid recirculation systems over those operating with evaporators fed directly with high-pressure liquid are:

1. More effective use of evaporator heat-transfer surface. In nonliquid recirculation systems, some of the surface is used to superheat the leaving refrigerant with resultant low heat-transfer coefficients and reduced temperature differences between the refrigerant and the fluid being cooled. The heat transfer capability of the liquid recirculation system is often compared to the direct-expansion evaporator with liquid feed controlled by a superheat-control valve which is also called a thermostatic expansion valve and is described in more detail in Chapter 11 on refrigerant control valves. This expansion valve, as Fig. 8.4 shows, requires superheat in order to open, so the refrigerant near

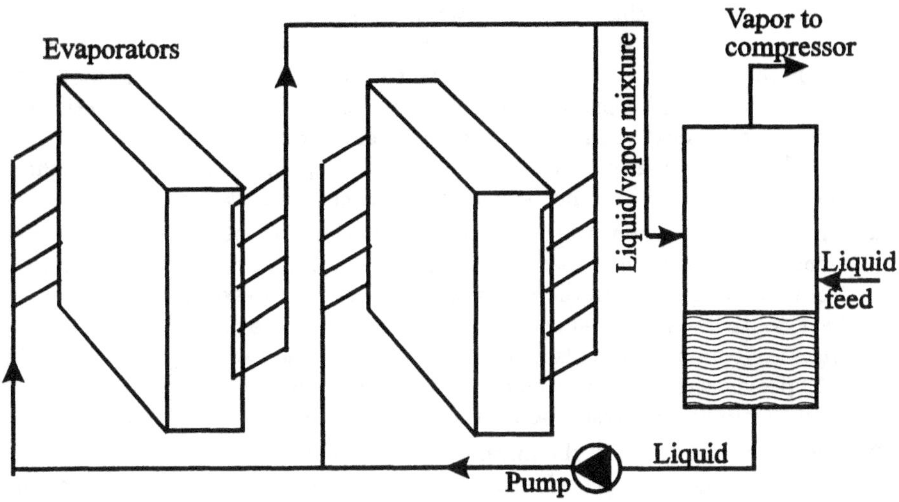

FIGURE 8.2
Conceptual arrangement of the mechanically pumped liquid recirculation system.

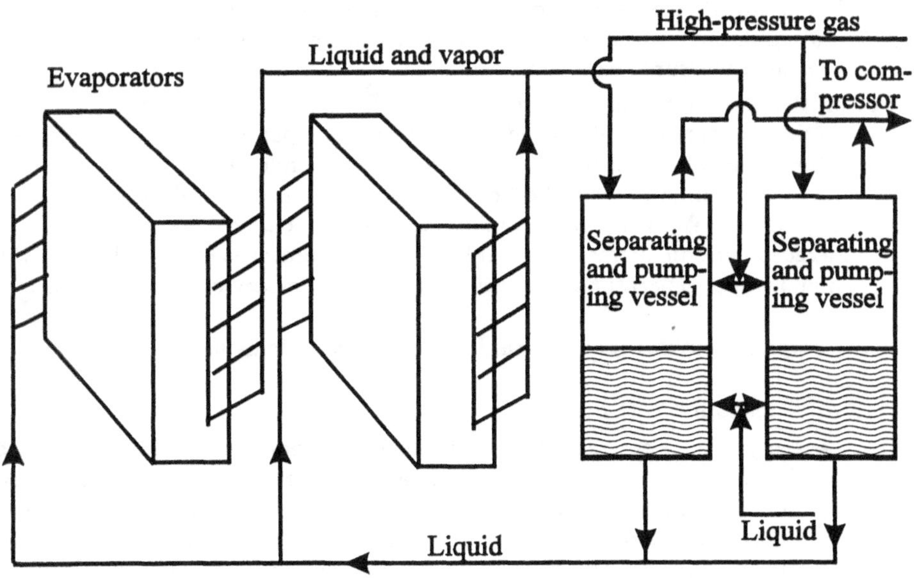

FIGURE 8.3
Conceptual arrangement of the gas-pumped liquid recirculation system. The piping is valved for alternate filling and pressurization of the vessels.

the exit of the evaporator is at a higher temperature than the saturation temperature prevailing in the upstream portion of the evaporator. In the superheating section of the evaporator the temperature difference between the fluid being cooled and the refrigerant is diminished, and furthermore, the heat-transfer coefficient between the evaporator surface and the superheated refrigerant is lower than the boiling coefficient elsewhere in the evaporator. In liquid recirculating systems the evaporating temperature prevails throughout the evaporator and the heat-transfer coefficient remains high.

2. The refrigerant entering the compressor is close to saturated vapor conditions. The separating vessel(s) in Figs. 8.2 and 8.3 protect the compressor from receiving liquid in the vapor, and the vapor possesses little or no superheat. With low superheat entering the compressor, the discharge temperature from the compressor is also moderated. Indeed, the two-stage system of Fig. 3.15 provides close to saturated vapor for the high-stage compressor, but the liquid recirculation system does the same for the low-stage compressor.
3. Flash gas is removed in the machine room rather than having it appear after the expansion valve at each evaporator. The pressure drop of the refrigerant in the evaporator tubes can thereby be reduced. The same advantage applies to the suction line from the evaporator (also refer to Disadvantage 1a.)
4. Valves controlling the liquid feed to the evaporator receive liquid at a uniform pressure, regardless of the condensing pressure. An expansion valve fed with liquid directly from the condenser experiences a high feed pressure in summer and a low one in winter. In a liquid recirculation system, the pump provides the same pressure year-round.
5. In the case of ammonia systems, oil which escapes the compressors carries to the low-pressure side of the system where it separates from the ammonia. This oil must be periodically drained and with individual feeding of evaporators with high-pressure liquid, oil must be drained periodically from each evaporator. With liquid recirculation, however, particularly if the coil undergoes periodic defrosts, the oil from the evaporator is brought back to the liquid/vapor separator so that oil draining is necessary at only one location.

Some **disadvantages** of liquid recirculation systems are:

1. Additional first cost due to:
   - Larger line sizes. The liquid line to the evaporators conveys the liquid destined for recirculation as well as that for evaporation. The return lines from the evaporators to the separating vessels carry liquid as well as vapor. Even though there is less flash gas in this suction line (Advantage 3), designers usually select a larger size than they would for conveying only the vapor of evaporation.
   - The liquid supply lines to the evaporators are cold, so they must be insulated.
   - Additional cost of mechanical or gas-pumping equipment.

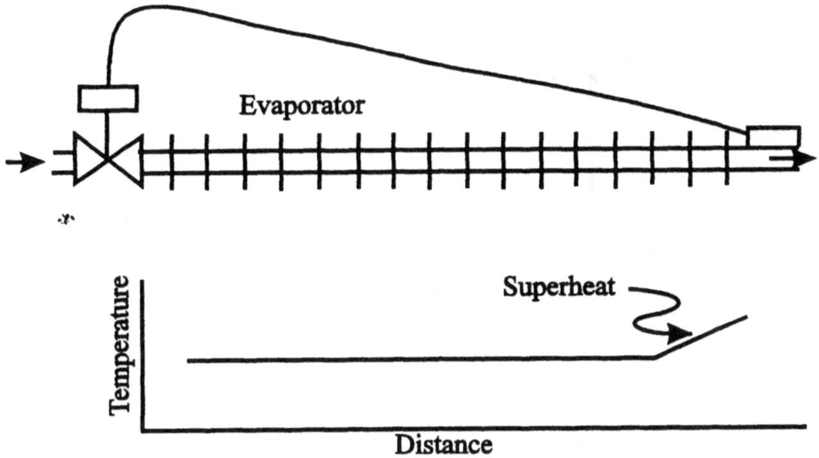

FIGURE 8.4
Reduced heat transfer near the outlet of an evaporator whose liquid feed is regulated by a superheat-controlled valve.

2. Greater refrigerant charge. The evaporator and suction line contain a greater fraction of liquid than is the case with direct expansion.
3. There is an operating cost associated with the pumping equipment.

The advantages of liquid recirculation generally dominate over the disadvantages in low-temperature, multiple-evaporator installations. When the plant serves low-temperature assignments, achieving good heat transfer in the evaporators is crucial, and the plant operates with high compression ratios where the quantitites of flash gas are appreciable and high suction superheat could be a problem. For moderate-temperature evaporators, such as those on docks, direct expansion should be considered. For a system with a small number of low-temperature evaporators, flooded coils may be the best choice. As the number of evaporators increases, a breakeven point is passed where the liquid recirculation system becomes the most economical choice. The critical number of evaporators is a function of the systems, but usually when the system includes more than three to five evaporators, liquid recirculation is chosen.

8.4 CIRCULATION RATIO

A fundamental benefit of liquid recirculation is the improvement of the heat-transfer coefficient on the refrigerant side of the evaporator. This improvement is attributable to improved wetting of the refrigerant-side surfaces and increasing the velocity of refrigerant. A measure of the extent of overfeed is called the recirculation number or circulation ratio, n

$$\text{Circulation ratio} = n = \frac{\text{refrigerant flow rate supplied to evaporator}}{\text{flow rate of refrigerant vaporized}} \tag{8.1}$$

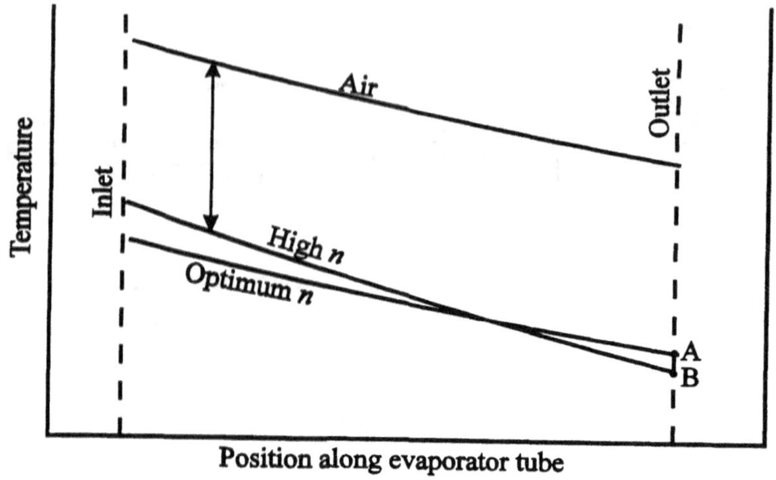

FIGURE 8.5
Temperature profiles of air and refrigerant for two values of n.

For there to be overfeed of liquid, n must exceed 1.0. The immediate question that arises is what is the optimal value of n? The value of n affects the performance of the evaporator in several ways. First, an increase in n increases the heat-transfer coefficient without limit. But the penalites of increased n are twofold: increased cost of pumping the liquid, and increased pressure drop through the evaporator. The increased refrigerant pressure drop results in a higher evaporating temperature for a given pressure leaving the coil.

The influence of n can be visualized as in Fig. 8.5. The basis of comparison of various values of n should be the combined power of the compressor and liquid pump to provide a given refrigeration duty. Refrigeration duty means the rate of heat transfer at the specified temperature of fluid or product being cooled. The air and refrigerant temperature distributions are profiled throughout the coil in Fig. 8.5. The distribution of the air and refrigerant temperatures are shown for two different values of n.

One value of n is the optimal one that was found by some means, and the other n is higher than optimum. The mean-temperature difference between the air and refrigerant is represented by the area between the curves of the air and refrigerant temperatures. The mean-temperature difference for the high n is less than for the optimum n, because the heat-transfer coefficient is higher with the high n. But the greater pressure gradient with the high n translates to a steeper slope of the evaporating temperature curve, and in this case, the outlet temperature and pressure from the evaporator will be lower with the high n. The result is that a lower compressor suction pressure must be provided which requires more compressor power for a given refrigerating capacity. Not shown in Fig. 8.5 is that the power for the liquid pump will be slightly higher with high n, and the pressure drop in the liquid/vapor return line from the evaporator to

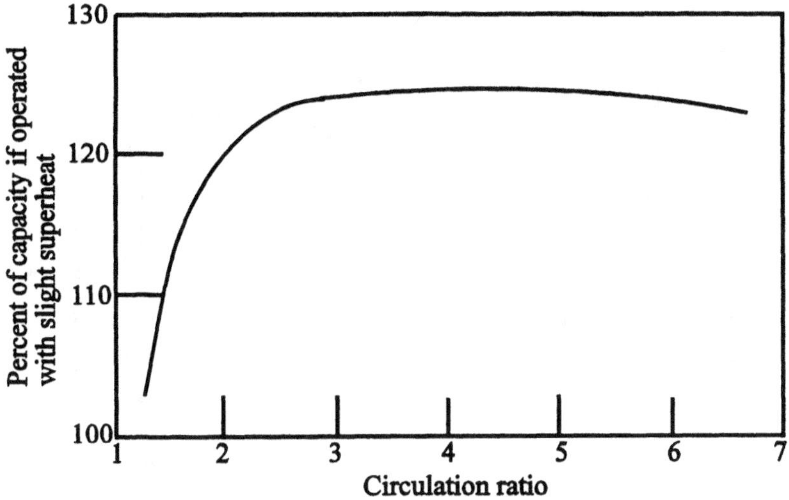

FIGURE 8.6
Effect of circulation ratio on the overall heat-transfer coefficient of an air-cooling coil[1].

the low- pressure receiver will also be higher. These two conditions are further penalties of the high value of n.

The above-mentioned general trends must be verified by laboratory measurements and field experience. Wile[1] measured the influence of the circulation ratio in laboratory tests of a steel, finned tube, ammonia evaporator cooling air, and the results are shown in Fig. 8.6. The outside diameter of the tubes of the coil was 16 mm (0.625 in). The overall heat-transfer coefficient is expressed as a ratio to that experienced with the coil operating under control of an expansion valve that maintains a small amount of superheat. Figure 8.6 indicates that a 25% improvement in heat-transfer capacity is possible with a circulation ratio of 3 or higher in comparison to a ratio of 1. Wile's tests showed little or no improvement when n increased beyond 4 or 5.

The coil in Wile's tests experienced a pressure drop of 10 kPa (1.5 psi) when $n = 7$ and when the evaporating temperature was −29°C (−20°F). Associated with this pressure drop is a 1.7°C (3°F) higher boiling temperature at the entrance of the evaporator than at the outlet. Wile observed that, while the typical practice is to specify the circulation ratio in selecting the coil, it is the refrigerant flow rate through the coil that more precisely expresses its optimum performance. The same coil could be selected for different capacities, depending upon the air-to-refrigerant temperature differences, resulting in widely different refrigerant flow rates for a given n.

Lorentzen[2] conducted some experiments similar to those of Wile, but with the additional parameter of heat flux. As the heat flux (the rate of heat transfer per unit area) increases, the U-value of the coil also increases, as illustrated in Fig. 8.7. This effect of the heat flux is generally observed in boiling heat transfer,

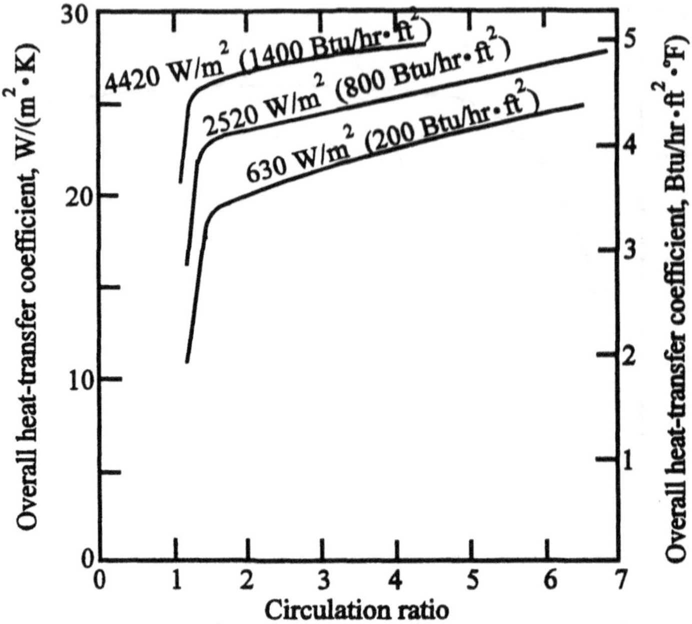

FIGURE 8.7
Influence of the circulation ratio and heat flux on the overall heat-transfer coefficient of an air cooler using a halocarbon refrigerant[2].

at least until the flux is so high that vapor blankets the surface. Lorentzen's studies showed a progressive increase in the U-value as n increased, in contrast to optimum U-value reached in Wile's experiments. Figure 8.7 shows an abrupt increase in U-value with the first liquid that overfeeds the coil. Indeed, continued increase in n improves the U-value, but the most important benefit occurs by converting the coil from direct expansion to a value of n that is even slightly greater than 1.0.

Richards[3] analyzed the work of several authors including the experimental data of Van Maale and Cosijn[4] that is shown in Fig. 8.8, and concluded that to achieve favorable heat-transfer coefficients the Froude number is a distinguishing characteristic. The Froude number is a dimensionless term reflecting the ratio of the inertia to the gravity forces and in its simplest form is:

$$\text{Froude number} = \text{Fr} = V^2/gD \tag{8.2}$$

where V = velocity, m/s (ft/s)
g = gravitational acceleration, m/s² (ft/s²)
D = diameter, m (ft)

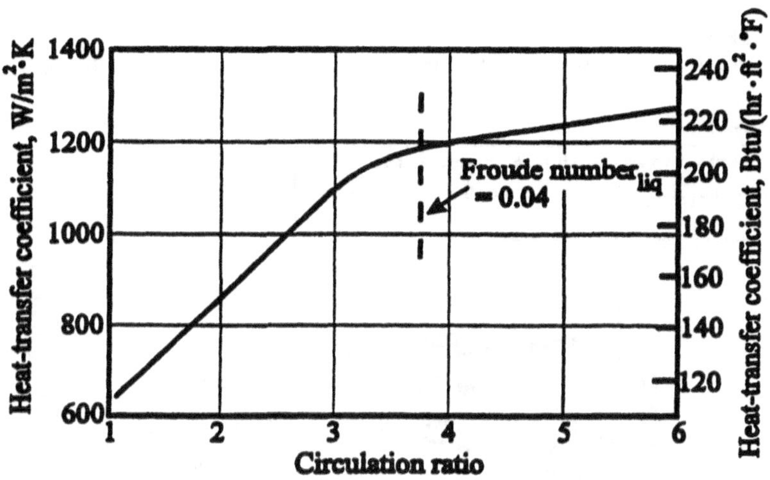

FIGURE 8.8
Heat-transfer coefficient of boiling ammonia in a liquid overfeed coil with tube diameters of 7/8 in. and operating at −20°C (−4°F) with a heat flux of 2825 W/m² (895 Btu/hr·ft²).

A Froude number for liquid, Fr_{liq}, is a modification of Fr in Eq. 8.2 and is

$$Fr_{liq} = \frac{\rho_L V_L^2}{(\rho_L - \rho_v) g D_i} \tag{8.3}$$

where ρ_L = density of liquid, kg/m³ (lb/ft³)
ρ_v = density of vapor, kg/m³ (lb/ft³)
V_L = liquid velocity, m/s (ft/s)
D_i = internal diameter of tube, m (ft)

The value of Fr_{liq} that appears to result in favorable heat-transfer coefficients is 0.04, which is associated with the achievement of annular flow in the tubes. This value of Fr_{liq} is noted on Fig. 8.8.

Some other recommendations of the circulation ratio are as follows. One manufacturer of coils[5] recommends: $n = 4$ for ammonia, and $n = 3$ for R-22. Geltz[6] recommends a higher n when the coil is circuited for top feed of refrigerant to achieve good wetting of the evaporating surfaces. That proposal is also apparent in the *ASHRAE Handbook*[7] whose recommendations are shown in Table 8.1. Another recommendation reflected in Table 8.1 is that n for the halocarbons can be less than for ammonia. The reason for choosing a lower value of n for halocarbons is also to avoid excessive pumping power. In comparison to ammonia, the latent heat of the halocarbon refrigerants is lower, but the liquid density is higher. These two reasons are partially compensating, but in general, for a given refrigeration capacity the power required by the liquid pump in an ammonia system is about one-third that in a halocarbon system.

While circulation ratios from 2 to 4 are standard, the possibility of choosing a circulation ratio much higher (between 20 and 40) also has been explored[8] on

TABLE 8.1
Recommended circulation ratios from the *ASHRAE Handbook of Refrigeration*.

Refrigerant	Circulation ratio
Ammonia, top feed with large-diameter tubes	6 to 7
Ammonia, normal feed with small-diameter tubes	2 to 4
R-22	3

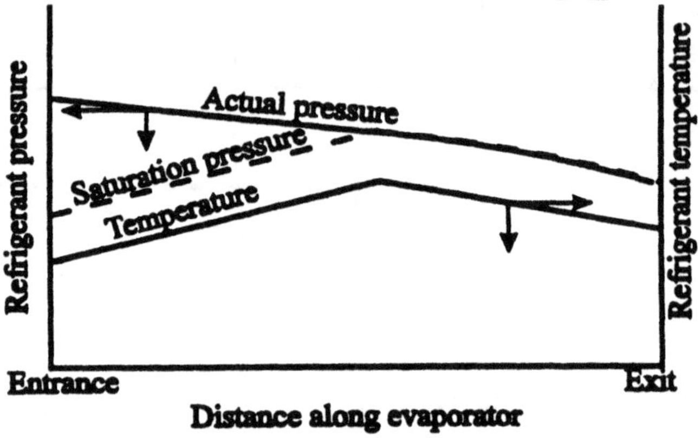

FIGURE 8.9
Temperature and pressure distribution in an evaporator with an extremely high circulation ratio.

an R-12 plate-type freezer. With conventional values of n, evaporation is likely to occur through the entire evaporator—from entrance to exit. The temperature of refrigerant thus drops in the direction of flow as the pressure drops. With very high values of n, the temperature profile is likely to be as shown in Fig. 8.9, where the liquid is under pressure at the evaporator entrance and absorbs the evaporator load through its increase in sensible heat, reflected in a rise in temperature. Somewhere in the evaporator, perhaps near the outlet, the rising saturation pressure meets the falling actual pressure and vaporization begins. The use of extremely high values of n may have merit, but only in special cases.

A conclusion from the foregoing discussion of optimum circulation ratios is that good evaporator performance will result with values of n between 3 and 4, and indeed these have been the traditional design values. In the past few years, however, attention has been directed toward low-charge systems. Systems with low inventories of ammonia are attractive for safety reasons. For halocarbon systems, new CFC replacements are expensive, so the cost of the refrigerant charge becomes a factor. One way to reduce the refrigerant charge in a liquid recirculating system is to reduce the value of n, which results in a greater fraction of vapor in the evaporator coils and the liquid/vapor return line. The density of vapor is less than liquid, so the mass will be less. Furthermore, the low-pressure receiver can be smaller and contain less charge.

The objective of high heat transfer is still valid, so application of the Froude number concept would seem to have merit. In order to achieve a minimum Froude number with a given flow rate of refrigerant, the velocity should be increased and/or the diameter decreased. Coils with smaller-diameter tubes, for example 5/8-in or 3/4-in rather than 7/8-in or 1- in, would appear to be a means of lowering the optimum n.

8.5 TOP FEED VERSUS BOTTOM FEED OF REFRIGERANT

The merits of the direction of refrigerant feed have been a continuing subject of debate, but the fact is that both directions of feed are in successful operation. The key requirement is proper design and application. Each type of feed has its inherent advantages.

Top-feed advantages

- smaller refrigerant charge permits a smaller low-pressure receiver
- natural draining of the coil prior or during defrost
- continuous transport of oil out of the coil

Bottom-feed advantages

- improved refrigerant-side heat-transfer coefficients for a given circulation ratio
- more uniform distribution of refrigerant through the various coil circuits.

When top feed is chosen, it may be advisable to design for a high circulation ratio, as Table 8.1 suggests.

Two of the many coil circuiting variations are shown in Fig. 8.10. Figure 8.10a shows horizontal headers with the tubes oriented in a vertical plane. This arrangement is often used for ice thermal storage units or in some cases for cooling air. Horizontal air flow should not be used because it would cause unequal loading of circuits. The circuiting in Fig. 8.10b has vertical headers with the planes of each tube circuit on an incline. With horizontal air flow, the lower circuits often draw a greater refrigerant flow rate than the upper circuits, so usually each circuit is equipped with an orifice (Sec. 6.9) having a smaller diameter in lower circuits than in the higher circuits.

The method of defrost also influences the direction of refrigerant feed. Hot-gas defrost combines well with bottom feed where the hot gas is supplied at the top, so the flow of refrigerant during defrost is opposite to that for refrigeration. Air, water, or electric defrost methods are more compatible with top feed of refrigerant because prior to defrost the liquid supply valve closes and much of the liquid refrigerant simply drains out of the coil prior to the external application of heat. Because the orifices are always placed at the inlets of the coil circuits

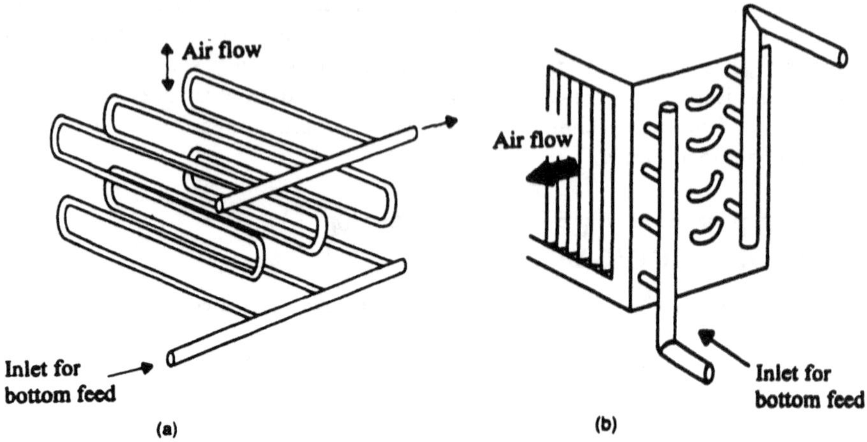

FIGURE 8.10
Colis with (a) horizontal headers and (b) with vertical headers.

they are at the top of top-feed coils. Thus, when a top feed coil is defrosted with hot gas, this hot gas first encounters the orifices which may restrict the flow of vapor. This is another reason for top-feed coils to be more adaptable to defrosting from external heat.

8.6 MECHANICAL PUMP SYSTEM

The liquid recirculation subsystem in which the liquid is motivated by a mechanical pump includes the following main components: pump, liquid-distribution piping, refrigerant controls at the coil, the coil with its defrost controls, the liquid/vapor return line, and the low-pressure receiver. Several types of pumps serve liquid recirculation systems, including positive displacement pumps(Section 8.7), open-type centrifugal pumps (Section 8.8), and hermetic centrifugal pumps (Section 8.10). The pump must be selected in conjunction with the design of the liquid- distribution system (Section 8.12) to provide the required flow rate of liquid refrigerant. Special attention must be dedicated to the inlet piping to the pump (Section 8.11) so that no vaporization of liquid occurs at the pump inlet (cavitation). A specific requirement to avoid cavitation is to provide adequate net-positive-suction head (Section 8.9).

8.7 POSITIVE-DISPLACEMENT PUMPS

Positive-displacement pumps are usually of the gear type where the meshing of gears conveys the liquid around the periphery of the gears, as in Figure 8.11a or in a combination of internal and external flow[9] as in Figure 8.11b. When the

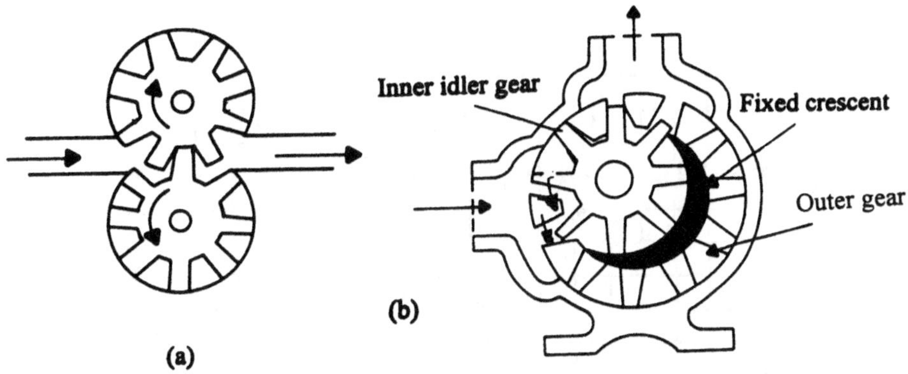

FIGURE 8.11
Two styles of positive-displacement gear pumps, (a) externally meshed; and (b) internally meshed gears.

pump speed is constant, the flow rate delivered by the pump is ideally constant, but because of leakage, which increases with an increase in pressure difference, the performance characteristics are typically as shown in Figure 8.12.

The positive-displacement pump delivers approximately a constant flow rate, so this type of pump is best suited to those liquid recirculation systems where the flow rate is essentially constant, and not, for example, to systems where thermostatically controlled solendoid valve open and close the liquid lines serving coils. If the liquid supply lines were restricted, the positive-displacement pump reacts by building up more pressure, so a pressure relief valve opens when the pressure rises above the pressure setting. The relief line discharges back to the low-pressure receiver. Because a shut-off valve in this line is necessary for servicing the pump, there is a possibility of accidentally trapping liquid between the shut-off valve and the relief valve. To avoid trapping cold liquid which could expand upon warm-up and possibly rupture the pipe or valve, some manufacturers provide a two-way relief valve which can relieve in the opposite direction at a very high pressure. Another approach to limiting the discharge pressure is to operate the pump with a variable-speed drive in which the speed is regulated by controlling to the desired discharge pressure[10].

The early liquid recirculation systems used positive-displacement pumps, but in recent years the centrifugal type has almost universally replaced them. This is not to suggest that the positive-displacement pump has no role in industrial refrigeration systems. It is ideally suited for liquid-transfer assignments, such as pumping liquid from a low-pressure vessel to one at high pressure. The positive-displacement pump can easily develop the high differential pressures that might be needed for this assignment, and normally this application does not require throttling to low flow rates which could overload the pump motor.

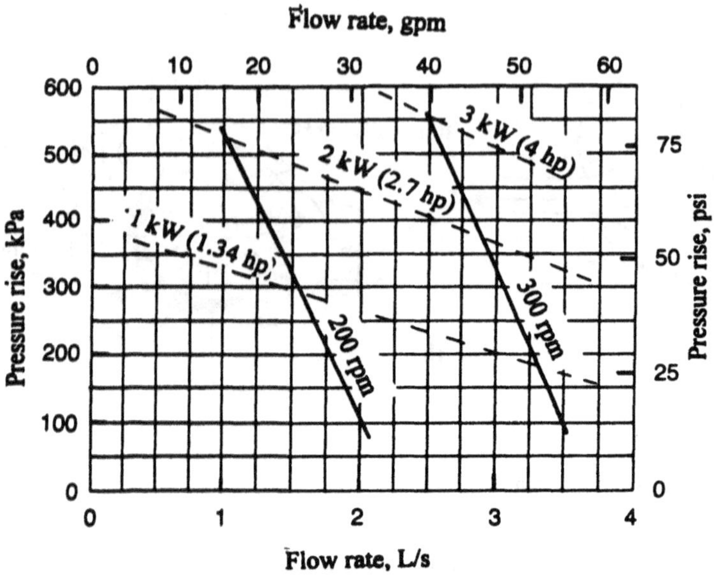

FIGURE 8.12
Pressure rise and shaft power versus flow rate of a positive-displacement ammonia liquid pump[7].

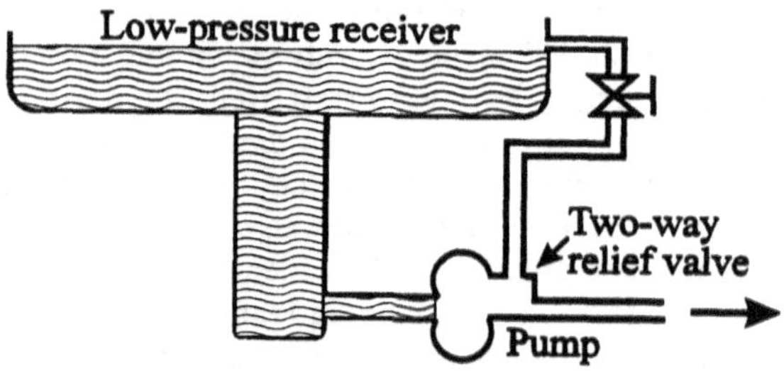

FIGURE 8.13
A built-in relief valve on a positive-displacement pump that permits relief in either direction.

8.8 CENTRIFUGAL PUMP—OPEN TYPE

The centrifugal pump functions by first setting the liquid in motion and then converting the velocity into pressure. This procedure is achieved with a pump as shown in Fig. 8.14 where a rotating impeller is contained within a progressively expanding housing that receives the liquid delivered by the impeller. The pressure differential that can be developed by the pump increases with an increase in rotative speed and with an increase in outside diameter of the impeller.

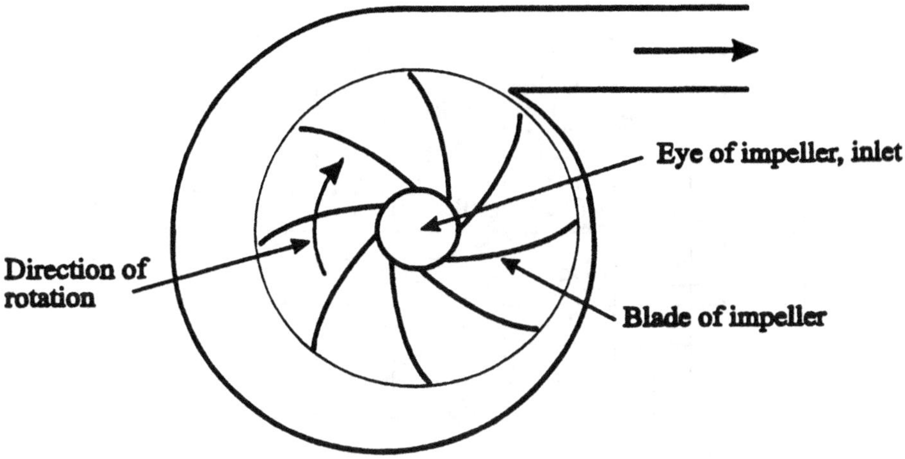

FIGURE 8.14
A centrifugal pump.

Typical pressure/flow characteristics of a centrifugal pump are shown in Fig. 8.15 for three pumps of different impeller diameters, all operating at the same rotative speed. The maximum pressure rise that the pump can deliver is at zero flow rate, as might occur when the discharge valve is closed. As valves in the system progressively open, the flow rate increases and eventually the pressure differential that the pump can develop drops. Also shown on Fig. 8.15 is the power requirement. A favorable characteristic of the centrifugal pump is the non-overloading characteristic. For example, the maximum power that the pump with the 230-mm (9-in) impeller diameter can draw is approximately 2.5 hp at a flow rate of 11 L/s (170 gpm). The power required by the centrifugal pump decreases as the flow rate is throttled off, in contrast to the characteristic of the positive-displacement pump.

The pressure-flow rate curves of Figure 8.15 apply to ammonia and do not apply to a fluid having a different density. Some manaufacturers present their performance curves in terms of *head* versus flow rate, because such curves can be used for any fluid. The head may be thought of as the elevation that the fluid would rise in a long vertical tube attached to the discharge pipe of the pump. The conversion of head to pressure is dependent upon the liquid density as expressed by the equations:

$$\text{Pressure rise, kPa} = \frac{(\text{head, m})(\text{density of liquid, kg/m}^3)(9.81\ \text{m/s}^2)}{1000} \tag{8.4}$$

$$\text{Pressure rise, psi} = \frac{(\text{head, ft})(\text{density of liquid, lb/ft}^3)}{144\ \text{in}^2/\text{ft}^2} \tag{8.5}$$

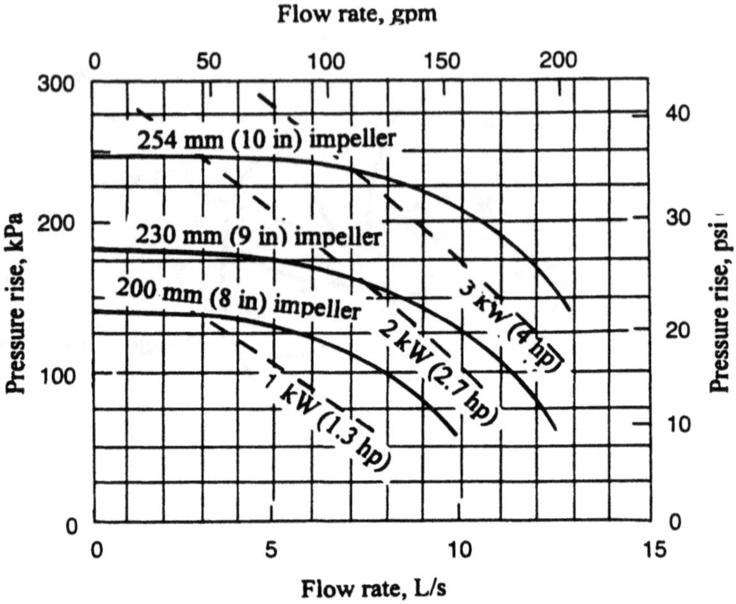

FIGURE 8.15
Pressure-rise/flow rate characteristics[11] of an ammonia centrifugal pump operating at 1750 rpm.

In the open-type pump the shaft driving the impeller passes through the casing of the pump. At this penetration, a seal retards the leakage of refrigerant out of the pump and inhibits air leakage into the pump if the pressure within the pump is below atmospheric. The seals are a potential source of leak of the volatile refrigerant, and pump manufacturers have continued to improve the quality of seals. To make the seal more effective, some pumps are equipped with a double seal, and the cavity between the two seals is filled with oil under pressure, as illustrated in Fig. 8.16. The high pressure is provided by the refrigerant, and the oil eventually leaks through the seals to the outside and/or into the system. When the oil level drops low, the reservoir is refilled by means of a manual oil pump.

An important precaution in the design, installation and operation of centrifugal pumps is to prevent *cavitation*. Cavitation is the conversion of liquid into bubbles of vapor because of an abrupt drop in pressure as the liquid flows into the pump housing or impeller. When these vapor bubbles move into a region of high pressure they collapse violently. Continued cavitation can result in erosion of the metal surfaces of the impeller, and reduction in flow rate. Because the centrifugal pump relies on the high density of liquid in the conversion of velocity to pressure, the presence of vapor can cause marked reductions in pumping rate. In the worst case the pump can become vapor bound and not pump any liquid. This problem is not so critical in positive-displacement pumps where the volume

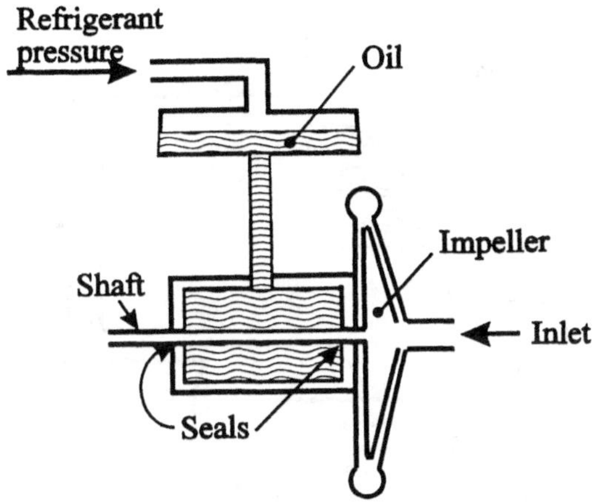

FIGURE 8.16
An oil seal of the shaft penetration in an open type centrifugal pumps.

flow rate will remain somewhat constant regardless of whether it is liquid or vapor. But even in the positive-displacement pump the mass flow rate will be degraded by cavitation.

Many of the installation practices surrounding the use of centrifugal pumps are directed toward avoiding cavitation. The next section, Section 8.9, concentrates on one of the most effective means of preventing cavitation— providing adequate *net-positive-suction head.* Section 8.11 deals with the recommended practices for piping the liquid from the low-pressure receiver to the pump.

At this point still another feature will be explained that is a necessary provision for virtually all centrifugal-pump installations. This requirement is to equip the system with continuous bypass of liquid, as Figure 8.17 shows. The purpose of this bypass is to always permit a flow of liquid through the pump, even if all the control valves out in the liquid system have closed off. The purpose of the bypass flow is to avoid trapping liquid in the pump which would be churned by the impeller and vaporized from the heat of friction. It would be possible to develop a control that opens the bypass when the system flow drops low, but this refinement is usually not warranted, and instead the hand control valve of Figure 8.17 is set slightly open.

Some pumps are equipped with what is called a volute vent which allows a line to be connected from the top of the pump housing to the low-pressure receiver. This vent may be opened when the pump starts to clear the pump of vapor and thus permit initiation of liquid pumping.

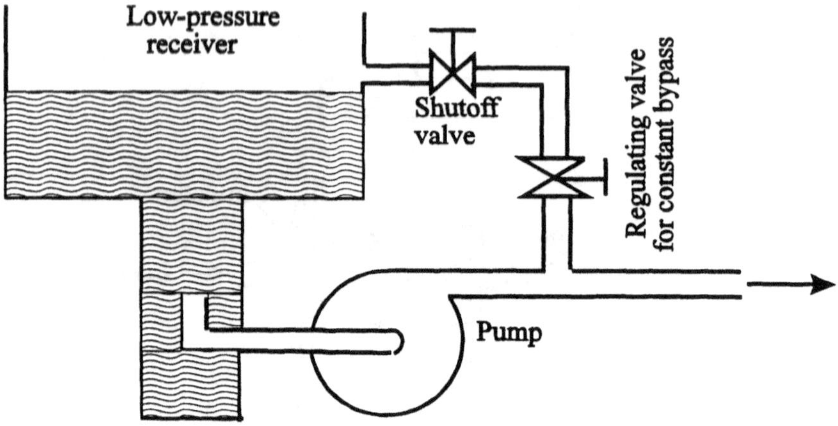

FIGURE 8.17
Continuous bypass of liquid from a centrifugal pump to avoid cavitation when the system flow drops low.

8.9 NET-POSITIVE-SUCTION HEAD (NPSH).

The single most effective method of preventing cavitation in the operation of a centrifugal pump when handling refrigerants close to their saturation condition is to provide an adequate amount of net-positive-suction head (NPSH). The NPSH is defined as the difference in elevation between the liquid level of the centerline of the pump and the liquid level in the vessel supplying the pump. The function of the column of liquid is to provide the static head that increases the pressure above saturation enough to inhibit vaporization of liquid in the passages of the pump.

The pump manufacturer will determine by calculation or by laboratory test the minimum NPSH. The individual factors[12] that would be included in a calculation are the following:

- pressure drop in the liquid leg, turn into the pump, and the open shutoff valve
- friction loss in the inlet connection of the pump
- acceleration of the liquid from entrance to just before the blade row
- influence of centrifugal force due to rotation of the impeller
- Coriolis force because of turn from axial to radial flow direction

The user of a pump can obtain from the manufacturer data such as shown in Figure 8.18, which provides insights in addition to the minimum values of NPSH at various operating points. The NPSH is essentially a function of flow

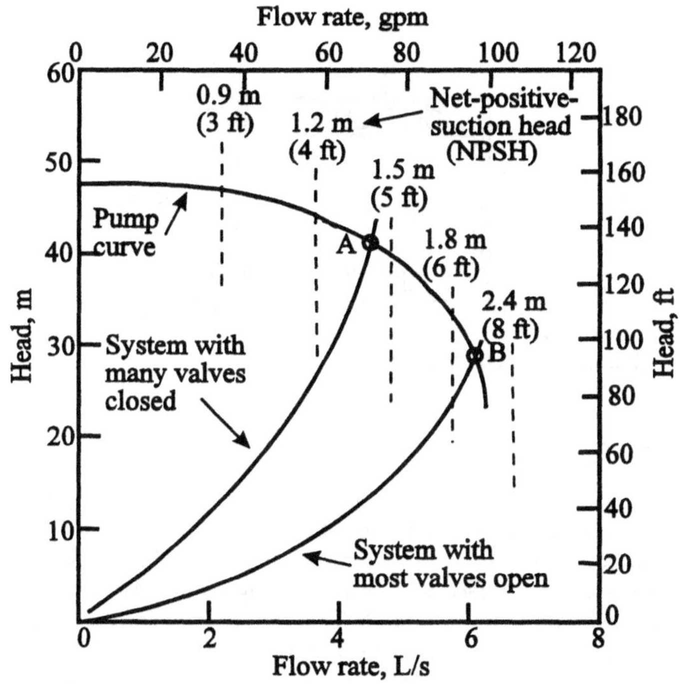

FIGURE 8.18
Performance of a centrifugal pump with an impeller diameter of 152 mm (6 in), operating at 3520 rpm.

rate and increases as the flow rate increases because all the individual terms mentioned above increase with flow rate. The first conclusion is that a pump does not have a unique NPSH, but the value depends where on its operating curve it is performing at the moment.

The pump whose performance is shown in Figure 8.18 can develop pressure differences that would typically be required in a liquid recirculation system, but the NPSH requirements are quite stringent, because some safety factor should be applied to the manufacturer's recommendations, perhaps doubling them since the pump manufacturer has no knowledge of the pressure drops in the piping leading up to the pump. Furthermore, achieving more than about 1.5 m (5 ft) of positive suction head because of the physical restrictions in the plant is usually difficult. The pump is small with an impeller diameter of 152 mm (6 in), which is appealing, but in order to develop the necessary pressure difference it must operate at 3520 rpm. This leads to the practice of normally choosing an operating speed of 1750 rpm, which necessitates a pump of larger impeller diameter. The NPSH values for the slower-speed pump would be 1/3 to 1/2 the values shown in Fig. 8.18.

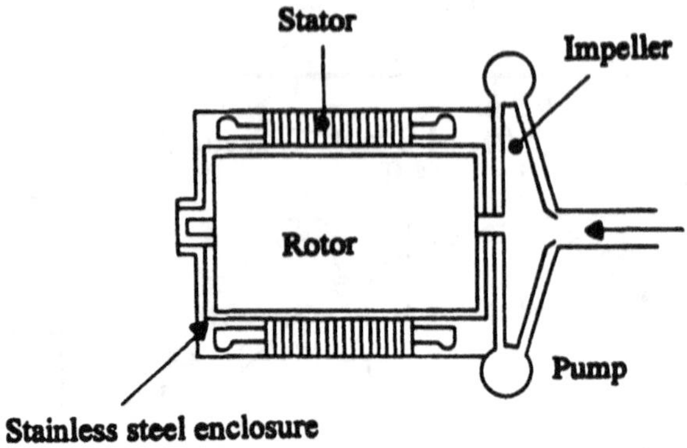

FIGURE 8.19
A cross-section of a semi-hermetic pump.

Also shown are the head-flow curves for two systems. Actually these curves may represent the same physical situation, but in one case many of the solenoid control valves are open and in the other most are closed. The balance points or operating points are the intersections of the piping curve with the pump curve and are represented by points A and B. The insight provided by the positions of A and B on the map is that when the piping network is drawing a high flow rate, operation moves to a condition requiring a high value of positive suction head. The pump may operate properly at the low-flow condition but cavitate at high-flow.

Usually designers try to select the maximum-flow operating point rather high on the performance curve, preferring Point A over Point B. Not only is the risk of cavitation reduced, but the pump will operate with greater efficiency higher on its curve. For the pump shown in Figure 8.18, Point A is close to its point of maximum efficiency (48%), while the efficiency at Point B is 35%. It is true that if maximum flow exists at Point A, which is at maximum efficiency, the efficiency drops off as the flow rate required by the piping network diminishes.

8.10 CENTRIFUGAL PUMP—HERMETIC

What are called *hermetic* refrigerant pumps are actually semihermetic, because many elements can be dismantled for servicing. What is distinctive about the pump is that there are no moving metal parts, such as the shaft, that must be sealed to prevent refrigerant leaks to the outside. In contrast to hermetically sealed compressors where refrigerant vapor is in direct contact with insulated electrical conductors, hermetic pumps prevent refrigerant liquid from contacting the conductors. As the diagram of Figure 8.19 shows, the rotor with its lamination is encapsulated in a stainless steel envelope. The stator which contains the windings is separated from the refrigerant pump by another pressure-tight

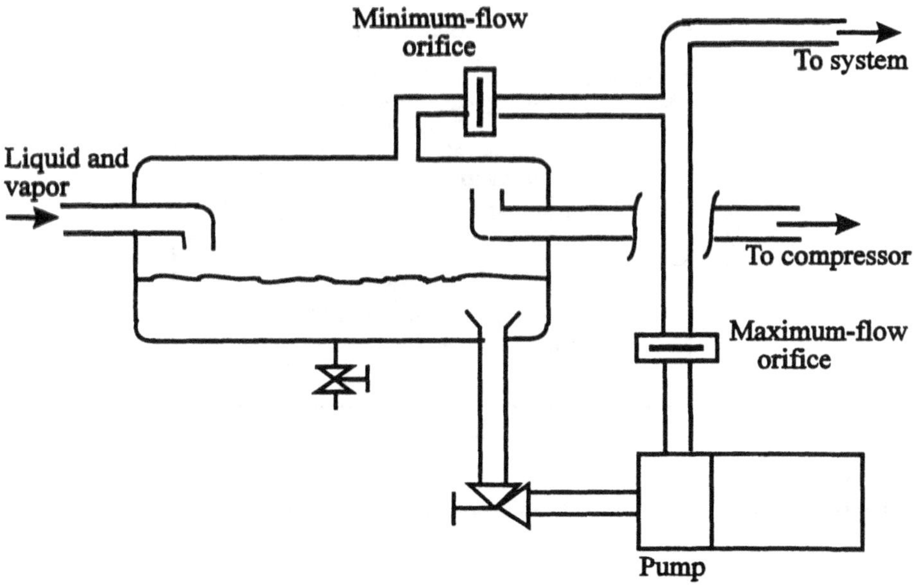

FIGURE 8.20
Maximum-flow and minimum-flow orifices in an hermetic pump installation.

housing and also enclosed within a steel housing. For brevity, the semi-hermetic pump will hereafter be called *hermetic*.

Hermetic pumps are usually higher in first cost than open-type pumps, even those with the oil seal. To compensate for the additional first cost, purchasers of hermetic pumps expect lower maintenance costs. Both hermetic and open-type pumps are widely used in liquid recirculation systems, especially those using ammonia as the refrigerant.

The basic diagram[13] of the installation of the hermetic pump assembly is shown in Figure 8.20. This arrangement shares most features of the layout of the open-type pump including the provisions for adequate NPSH and a bypass line back to the vessel to assure that there is always some liquid flow through the pump even if valves in the discharge pipe network are closed. Assuring this bypass flow is even more critical in the hermetic pump than it is in the open type. The reason is that a small flow rate of liquid passes outside the rotor envelope to cool the motor. Failure of this liquid flow could result in overheating of the motor and shutdown of the pump.

A hand valve could be used to regulate the bypass flow, but because of the critical nature of this stream and the possibility of improper adjustment, an orifice (the minimum-flow orifice) of the size specified by the pump supplier is preferred. This bypass flow is sometimes sent to the discharge line of the low-stage compressor before the intercooler/desuperheater to desuperheat the

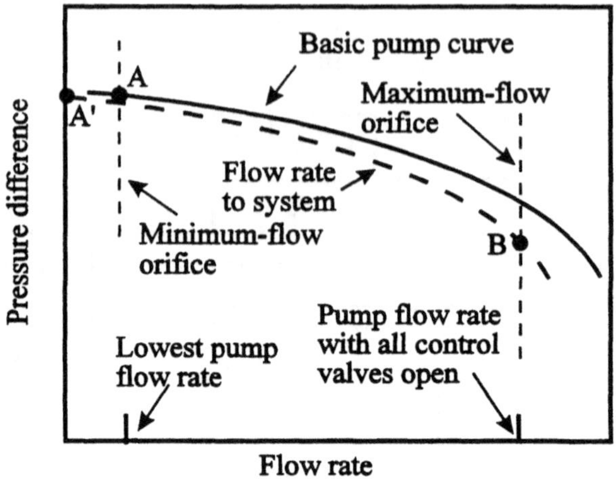

FIGURE 8.21
How the minimum-flow and maximum-flow orifices affect the performance of an hermetic pump.

discharge gas. Any liquid that fails to vaporize simply drains into the vessel which is designed for liquid/vapor separation. To enhance the desuperheating of the discharge gas, the bypass liquid can be atomized with a spray nozzle[14]. This spray nozzle, which behaves as an orifice, can be chosen to duplicate the pressure-drop characteristics of the specified minimum-flow orifice.

Here is how the two orifices affect the performance of the pump assembly. The size of the minimum-flow orifice must be equal or larger than the size specified by the manufacturer, while the maximum-flow orifice must be of equal or smaller size. As Figure 8.20 illustrates, the flow from the pump first reaches the minimum-flow orifice and if the pump flow is indicated by Point A (Figure 8.21), all of that flow rate bypasses to the low-pressure receiver. The flow rate delivered to the system is zero and indicated by Point A′ in Figure 8.21. Throughout the entire range of flow of the pump some liquid passes through the orifice, and the flow rate delivered to the system is shown by the dashed line. The fact that the pressure drops as the flow increases will result in reduced bypass flow, but that is no problem because the flow through the system is adequate to assure that the pump work does not vaporize the liquid.

The maximum-flow orifice limits the flow rate so that the NPSH provided by the installation is not exceeded. As Figure 8.18 showed, it is at high flow rates that large NPSHs are needed. Point B in Figure 8.21 is the maximum system flow which overcomes the resistance of the maximum-flow orifice and the piping network with all control valves open. This realization means that the pump manufacturer in specifying the diameter of the maximum-flow orifice should know the characteristics of the wide-open system, and combine that resistance with that of the orifice. This information is usually not available to the manufacturer, so the specification of the orifice normally results in a conservative choice. As control valves begin to close off, the performance rides up and to the left on the dashed performance curve and moves into the region of low values of required NPSH.

8.11 PIPING FEATURES SURROUNDING THE PUMP.

Some basic objectives in designing and installing the pump and its inlet piping are:

- provide generous NPSH
- install two pumps, one of which is standby
- provide a region for oil to separate from ammonia and be drained
- design with a minimum of fittings to avoid excessive pressure drop
- choose angle valves in preference to straight-through valves

Section 8.9 emphasized the need for adequate NPSH as the dominant method of guarding against cavitation. There are numerous successful piping configurations that provide positive suction head, and two are shown in Figures 8.22a and 8.22b. The difference in elevation between the liquid level in the low-pressure receiver and the pump should generously exceed the minimum NPSH. The arrangement in Figure 8.22a incorporates a large diameter liquid leg from which the liquid to the individual pumps is drawn. The configuration in Figure 8.22b shows separate lines running from the low- pressure receiver to each of the pumps.

The package should be equipped with two pumps, each capable of providing the required liquid flow rate by itself. In other words, one of these pumps is standby. If only one pump is available and that pump fails the entire recirculation system would be out of service. The discharge line of each of the pumps should be equipped with a check valve to prevent backflow through the idle pump.

The low-pressure receiver is the expected location for the accumulation of oil, and oil drain provision should be made. One of the advantages of the liquid recirculation system over individual flooded or direct-expansion coils is that the oil is brought back to one location—the low-pressure receiver. Figures 8.22a and 8.22b show different locations of the oil drain. In both cases the drain is from a region of stagnant motion where oil, being heavier than liquid ammonia, will separate.

Contributing to the relentless struggle to prevent cavitation is the effort to reduce the pressure drop in the inlet pipe and fittings as much as possible. High friction losses in these pipes only detract from the NPSH that has been provided. Methods for maintaining low friction include (1) keeping the horizontal distance short, (2) using angle valves in preference to straight-through valves wherever possible, and (3) avoiding high velocities because of pipes that are too small. Table 11.1 in the chapter on refrigerant control valves shows that the pressure

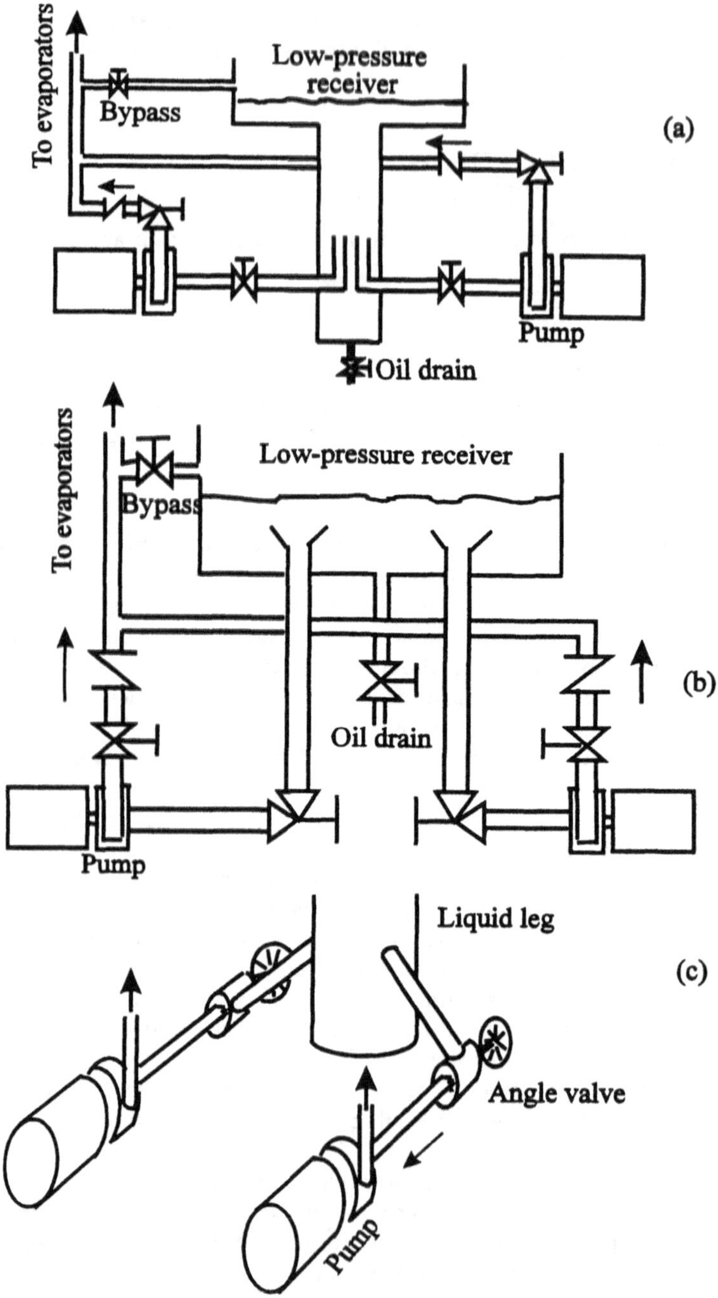

FIGURE 8.22
Three concepts for inlet liquid piping to the pumps in liquid recirculation systems.

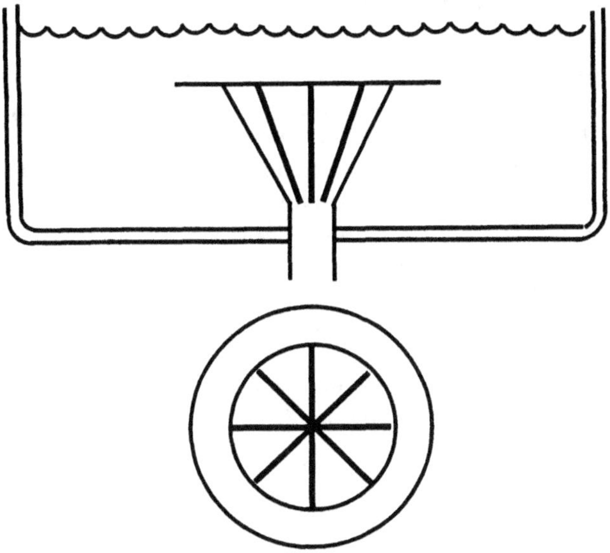

FIGURE 8.23
A vortex eliminator at the inlet of the suction pipe.

drop in an open angle valve is much less than in an open straight-through valve. The arrangement in Fig. 8.22a uses straight-through valves, so there must be adequate liquid head to overcome this pressure drop. The individual suction lines in Fig. 8.22b are adaptable to angle valves. In some recirculation packages, the orientation shown in Figure 8.22c is chosen to accommodate angle valves. A strong preference for angle valves in the discharge line does not prevail because the pressure of the liquid is high enough that vaporization is not a problem.

It may seem that the precaution against cavitation would be best served by choosing liquid suction lines as large as practical to achieve very low liquid velocities. This would normally be the case, except in instances where the pressure in the low-pressure receiver fluctuates. If the pressure falls rapidly, bubbles of vapor form almost immediately to absorb heat, which brings the liquid temperature back in conformity to the saturation temperature. If these bubbles form in the suction pipe, they have no opportunity of venting to the vapor space of the receiver. The objective becomes one, then, of passing the liquid out of the suction pipe in a reasonable amount of time. Lorentzen[2] recommended velocities between 0.8 and 1.0 m/s (160 to 180 fpm) as a compromise to combat cavitation due to pressure fluctuations and without encountering high pressure losses.

When drawing from the liquid leg as in Figures 8.22a and 8.22c, there is little likelihood of vapor bubbles entering the suction tube, but in the arrangement of Figure 8.22b a vortex may develop above the inlet which draws vapor into the pipe. To break up any such vortex, an eliminator[2] as shown in Figure 8.23 can be placed at the inlet.

8.12 THE LIQUID DISTRIBUTION SYSTEM AND PUMP SELECTION

Usually the path of the liquid mains and branches is dictated by the geometric configuration of the plant. After the layout is established, the designer must position the shutoff valves and decide what pressure relief provisions must be made in pipe sections between valves that could be shut off. Pressures could rise to unacceptable magnitudes if trapped liquid warms up. After the piping and valve layout has been determined, the major tasks are to:

(1) compute the rate of liquid flow
(2) select the pipe diameters
(3) compute the system pressure drop
(4) select the pump

Computing the liquid flow rate. The flow rate vaporized, $\dot{m}_{ev}$, is dictated by the total refrigeration rate, q_{total},

$$\dot{m}_{ev}, \text{kg/s (lb/min)} = \frac{q_{total} \text{ kW}}{h_{fg}, \text{ kJ/kg}} = \left(\frac{q_{total}, \text{ Btu/min}}{h_{fg}, \text{ Btu/lb}}\right) \tag{8.6}$$

The latent heat of vaporization, h_{fg}, is the difference between the enthalpy of saturated vapor, h_g, and the enthalpy of saturated liquid, h_f, at the temperature of the low-pressure receiver.

The flow rate of liquid delivered by the pump through the liquid distribution system, $\dot{m}_{pump}$, is

$$\dot{m}_{pump} = n\dot{m}_{ev} \tag{8.7}$$

where n is the circulation ratio as defined in Eq. 8.1.

Selecting the pipe diameters. Choosing the pipe diameters and selecting the pump becomes an iterative process, because the arbitrary choice of pipe diameters may result in a system pressure drop so high that a multistage pump is needed, or conversely the pressure drop is low enough that larger than necessary pipe is being used. Experienced design engineers can usually choose an appropriate size, perform a pressure-drop calculation, and if the pressure drop is unreasonable, can adjust the size. The target pressure drop for the system that results in a reasonably-sized pump is between 200 to 350 kPa (30 to 50 psi). The pressure drop in the pipe and fittings (elbows, tees, and open valves) is only one contributor to the total pressure drop, so perhaps less than 1/2 the total pressure difference available can be allocated to the pipe and fittings. The following equations can provide a rough guide for the liquid line size:

$$D, \text{ in.} = 2.25[(\dot{m}_{pump} \text{ kg/s})^2/(\Delta p, \text{ kPa})]^{1/5} \tag{8.8}$$

or

$$D, \text{ in.} = 0.22[(\dot{m}_{pump} \text{ lb/min})^2/(\Delta p, \text{ psi})]^{1/5} \tag{8.9}$$

Thus, if the liquid flow rate is to be 3 kg/s (400 lb/min) and 100 kPa (14.5 psi) is reserved for the pipe and fittings, the trial pipe size would be 1.40 or 1-1/2 inch.

Computing the system pressure drop. Knowledge of the pressure drop to be experienced in the liquid distribution system is required for selection of the pump. The components and geometry that contribute to the pressure drop include:

- straight pipe
- elbows, tees, and reducing fittings
- open shutoff valves
- balancing valves
- overcoming the head if the evaporators are at a higher elevation than the liquid level in the low-pressure receiver
- coil
- liquid/vapor return line
- back-pressure regulators on any evaporators

Chapter 9 on piping provides instructions on how to compute the pressure drop in straight pipe and fittings. Manufacturers' data are available to compute the pressure drop in open shutoff valves. The balancing valves, which many practitioners call expansion valves, should initially be partially closed in order to be able to open more if the liquid supply to the coil is inadequate. A reasonable estimate for the pressure drop through the balancing valves is between 35 and 70 kPa (5 to 10 psi).

Equations 8.4 and 8.5 may be used to compute the pressure difference associated with the rise in elevation of a line. In many closed liquid-pumping systems it is not necessary to compensate for the rise in elevation, because when the piping rises at one place it will drop somewhere else. The refrigeration system is different, however, when the liquid must be pumped up to an evaporator and this difference in pressure is not recovered in the liquid/vapor suction line. Some practicioners indulge in a philosopical debate as to whether the pump force provides all the impetus for the movement of refrigerant, or whether the compressor is also drawing vapor and thus assisting the circulation. It would appear, however, that the pump is responsible for the refrigerant from the liquid level in the low-pressure receiver and back to that point.

Sometimes one or more evaporators in a medium-temperature recirculation system are equipped with back-pressure regulators or evaporator-pressure regulators, as shown in Figure 8.24. This regulator prevents the pressure and temperature in the evaporator from falling so low that products would be damaged or that frost would form on the coil requiring defrost provisions. If the temperature of the low-pressure receiver and that of most of the evaporators is 2°C (35.6°F), but an evaporator-pressure regulator holds one evaporator at 5°C (41°F), the pressure in that evaporator will be, for ammonia, 53.3 kPa (7.7 psi)

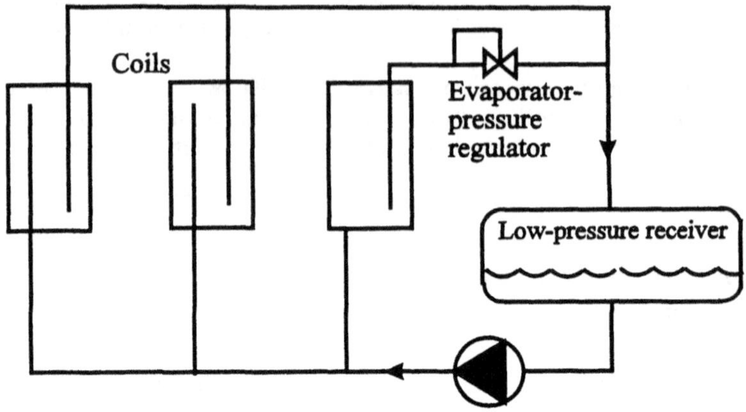

FIGURE 8.24
A evaporator equipped with a pressure regulator will require additional pump pressure to be supplied with refrigerant.

higher than the other evaporators. Unless the pump has been chosen to provide this excess pressure, the evaporator with the pressure regulator will not be supplied with refrigerant.

Selecting the pump. The two numbers needed to select the pump from a manufacturer's catalog are the flow rate and the pressure rise. Section 8.9 and Fig. 8.18 emphasized that the design point should be at the high-pressure part of the curve, rather than to the far right. For a given rotative speed, for example, 1800 rpm, the impeller diameter controls the pressure difference that the pump can develop, and the width of the impeller passages regulates the flow rate. Both of these influences are embodied in the pump performance curves.

8.13 PUMPING LIQUID REFRIGERANT WITH GAS PRESSURE—THE TWO-PUMPING VESSEL SYSTEM

As an alternative to mechanical pumps, liquid refrigerant may be circulated by means of gas pressure. There are several different concepts and methods of control in industrial use, but a common element is a pumping vessel held at low pressure during the portion of the cycle in which liquid drains from the low-pressure receiver, and is then subjected to high-pressure vapor to deliver the liquid. Both the J. E. Watkins and H. A. Phillips organizations were influential in developing gas pumping systems in the United States. Two gas-pumping concepts used in industrial practice will first be explained—the two-pumper drum system and the controlled-pressure receiver system. The controlled- pressure receiver is the most popular although the two-drum concept is also widely used. After the

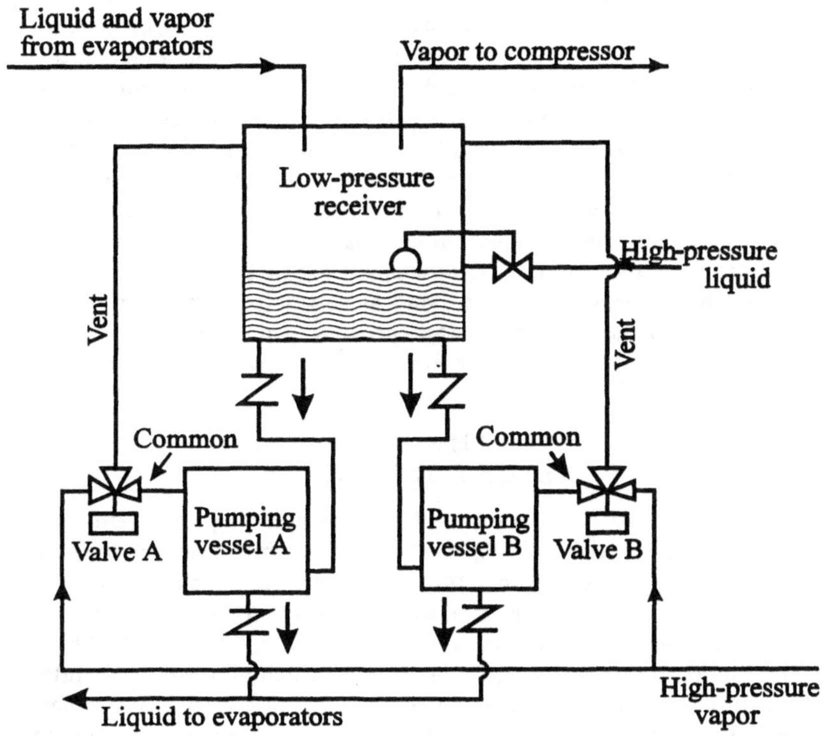

FIGURE 8.25
A gas-pumping system using two pumping vessels.

principles of these two systems are explained, an analysis of the pumping energy of gas pumping and mechanical pumping will be conducted. The advantages and disadvantages of gas pumping with respect to mechanical pumping will finally be explored.

In the two-pumper system of Figure 8.25, high-pressure vapor is imposed on one vessel while the other is filling. The functions of the two vessels then reverse. The switching operation is achieved through the use of two 3-way valves, Valve A and Valve B. The common ports of the valves are connected to their respective pumping vessel. While Valve A is positioned to connect the vent to the common port, Valve B allows high-pressure vapor to its pumping vessel. During this portion of the cycle, liquid from Vessel B is forced to the evaporators, and liquid drains from the low-pressure receiver to Vessel A.

To control the switching operation, it would be possible to sense the liquid levels in the pumping vessels, switching the positions of Valves A and B when the level drops to a low point in the vessel being emptied and/or rises to a high level in the vessel being filled. Instead, in gas pumping facilities such as this, a timer usually controls the switching operation. Time durations of two minutes for each of the modes of operation are typical. After deciding upon the pump/drain time, the pumping vessels can be sized to comfortably contain the maximum volume of liquid pumped from one vessel and drained into the other during the time increment. The maximum flow rate will generally occur when the evaporators are demanding the highest rate of refrigerant flow. The pressure of the pumping gas significantly affects the rate of liquid delivery, so for low-temperature evaporators in a two-stage system, vapor at the intermediate pressure should be chosen for the pumping gas. The intermediate pressure is likely to hold quite constant. The pumping gas for a recirculation system operating at the intermediate pressure uses gas at the condensing pressure, and since this pressure may vary widely, one approach is to maintain a constant pressure using a downstream pressure regulator in the high-pressure vapor line.

8.14 GAS PUMPING—THE CONTROLLED-PRESSURE RECEIVER

What is probably the most popular of the gas-pumping arrangements uses a controlled-pressure receiver (CPR), one version of which is shown schematically in Fig. 8.26. A key concept is that the receiver no longer operates at essentially the condensing pressure, but is maintained at a lower pressure, the controlled pressure, with a saturation temperature of approximately 15°C (59°F). Liquid from the condenser flows through a float valve which allows only liquid to pass and also drops the pressure from the condensing pressure to that of the CPR. The desired pressure in the CPR is maintained by the pressure-regulating valve which vents to the low-pressure receiver. Since essentially saturated liquid enters the float valve, some of the liquid flashes to vapor, and this vapor passes through the pressure regulator to the low-pressure receiver.

Liquid from the CPR flows to the evaporators in a quantity that overfeeds all the coils, and the liquid/vapor from these evaporators returns to the low-pressure receiver. The vapor is drawn out by the compressor, and the liquid drains to the pumping vessel. During the draining process the three-way valve connects the pumping vessel to the low-pressure receiver, allowing gas to vent from the pumping vessel. Check valve Ckv A prevents higher-pressure gas from the CPR from flowing into the pumping vessel while draining of liquid is taking place.

After the pumping vessel has accumulated some liquid, the three-way valve shifts its status to allow high-pressure discharge gas to enter the pumping vessel, forcing liquid through Ckv A into the CPR. Check valve Ckv B prevents access of the pressure of the pumping vessel into the low-pressure receiver. When the transitions between pumping and draining are controlled by a timer, less liquid accumulates in the pumping vessel during draining when the refrigeration capacity and refrigerant flow rate through the evaporators are low. If the transitions are controlled by the liquid level in the pumping vessel, the cycle time shortens at high rates of liquid transfer.

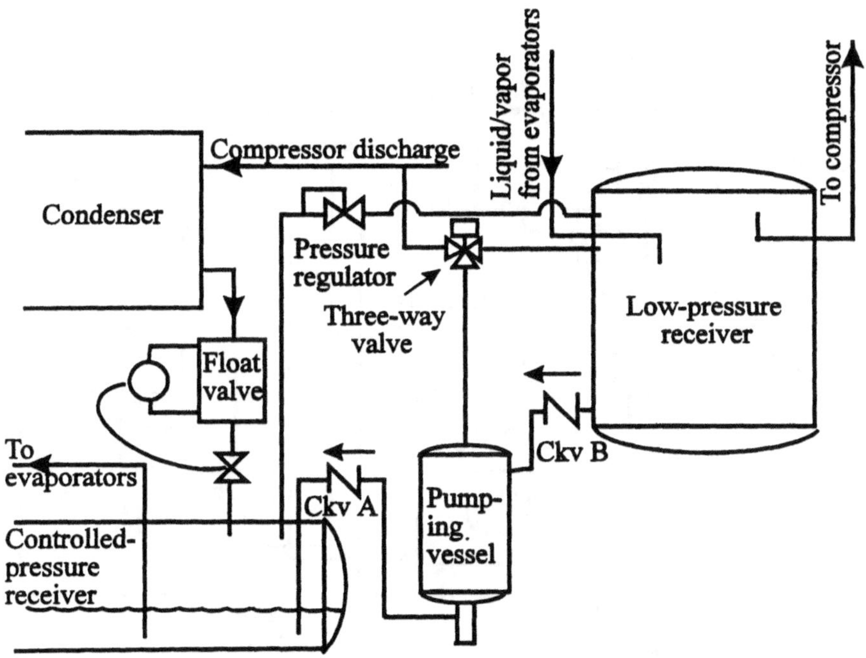

FIGURE 8.26
Liquid recirculation by gas pumping using a controlled-pressure receiver.

The CPR recirculation system shown in Figure 8.26 serves a single-stage plant. The CPR concept is adaptable to two-stage systems as well, and a typical flow diagram is shown in Figure 8.27. The main vessels are the CPR, the flash-tank/intercooler, the low-pressure receiver, and two pumping vessels. The CPR feeds the intermediate-temperature evaporators and the intercooler supplies the low-temperature evaporators. The system incorporates three pressure regulators with Regulator A feeding high-pressure vapor to the CPR if its pressure drops, and Regulator C venting vapor to the intercooler if the CPR pressure rises too high. Regulator B provides high enough pressure to the high-pressure pumping vessel to force liquid into the CPR. The liquid level is not controlled in the intercooler, so the option is not available to bubble discharge vapor from the low-stage compressor through liquid for desuperheating.

8.15 ENERGY ANALYSIS OF GAS PUMPING

The comparison of mechanical pumping and gas pumping that will be presented in Section 8.16 will contend that greater mechanical energy must be expended in a gas-pumping system than in a system with mechanical pumps. While the overall merits of gas pumping and mechanical pumping are debated endlessly,

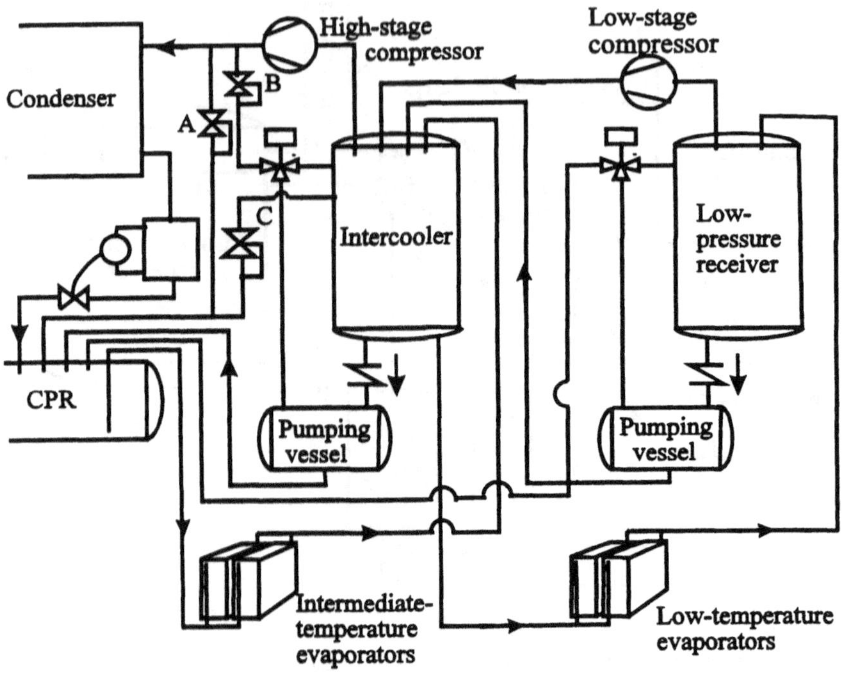

FIGURE 8.27
A two-stage system operating with controlled pressure receivers.

TABLE 8.2
Organization of a two-stage CPR recirculation system.

Function	For high stage	For low stage
Typical temperatures with low-pressure receiver at −33°C (−28°F)	CPR at 15°C (59°F)	Flash-tank/intercooler operating at, for example, -10°C (14°F)
Control of pressure	CPR, source from compressor, vent to intercooler	Intercooler, by compressor capacities
Source of gas for pumping vessel	High-stage compressor	High-stage CPR
Destination of liquid from pumping vessel	High-stage CPR	Flash-tank/intercooler

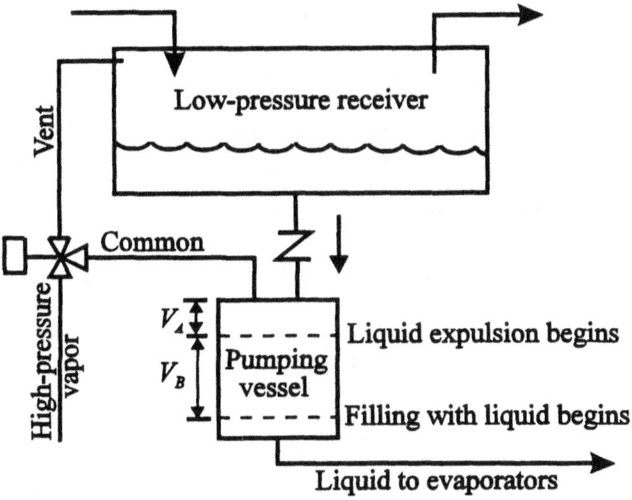

FIGURE 8.28
Energy analysis of pumping a quantity of liquid to a higher pressure using gas pumping.

there seems to be general agreement that gas pumping requires more power than mechanical pumping. A brief analysis follows to demonstrate not only the comparative power, but illustrates several of the factors that influence the power when gas pumping is employed.

The analysis will focus on the delivery of a given quantity of liquid at low pressure to a higher pressure in an arrangement as shown in Figure 8.28. The assignment is to elevate the pressure of a liquid volume V_B from that of the low-pressure receiver to that needed to deliver the liquid to the evaporators. In Figure 8.28, liquid drains from the low-pressure receiver to the pumping vessel until only a vapor volume of V_A remains in the pumping vessel. At this point the three-way valve switches, allowing high-pressure vapor to begin forcing liquid out of the pumping vessel. When the liquid level has been reduced to the lower dashed line, the three-way valve reverses and the vapor at high pressure vents into the upper vessel. Liquid can then once again drain from the low-pressure receiver to the pumping vessel. None of the energy of the high-pressure vapor is recovered, and the pumping energy for the cycle is that required to elevate the low-pressure vapor back to that of the high-pressure pumping gas. The analysis that follows will determine the energy required to pump a liquid volume V_B. The mass of vapor used to pump liquid volume V_B is:

The mass of high-pressure vapor used for pumping is:

Mass of high-pressure vapor occupying $V_A + V_B = (V_A + V_B)\rho_{hp}$ less the mass in $V_A + V_B$ following venting, $(V_A + V_B)\rho_{lp}$. Thus the mass of high-pressure vapor required is:

$$\text{Mass of pumping vapor} = (V_A + V_B)(\rho_{hp} - \rho_{lp})$$

where ρ_{hp} = density of high-pressure vapor, kg/m³ (lb/ft³)
ρ_{lp} = density of low-pressure vapor, kg/m³ (lb/ft³)

The energy required to pump liquid volume V_B is, thus

$$\text{Energy, kW (Btu/lb)} = (V_A + V_B)(\rho_{hp} - \rho_{lp})\Delta h$$

where Δh = work of compression, kJ/kg (Btu/lb)

To bring the comparative costs into better focus, a specific case will be explored.

Example 8.1. Compute the energy required to pump a volume V_B in an ammonia system whose evaporators and low-pressure receiver operate at −26°C (−15°F), which corresponds to a pressure of 145.1 kPa (21 psia) if the liquid must be elevated in pressure at least 200 kPa (29 psi). Compare the energy requirements when pumping by the following methods:

(a) mechanical pumping just meeting the pressure rise of 200 kPa (29 psi).

(b) gas pumping using pumping gas from the high-pressure receiver at a saturation temperature of 28°C (82.4°F), corresponding to a pressure of 1100 kPa (159.5. psia).

(c) gas pumping using pumping gas from an intercooler at a saturation temperature of −2°C (28.4°F) corresponding to a pressure of 399 kPa (57.9 psia).

Solution. Assume that the motors driving the pump and compressor both have efficiencies of 90%, that the efficiency of the compressor is 75% and that of the liquid pump is 45%.

(a). *Mechanical pump.* The specified pressure rise of 200 kPa (29 psi) is typical for a liquid recirculation system. The energy required by the mechanical pump to elevate the pressure of a volume V_B by 200 kPa (29 psi) is:

$$\text{energy} = (200 \text{ kPa})(V_B \text{ m}^3) = 200V_B \text{ kJ with } V_B \text{ in m}^3$$

or

$$= (29 \text{ lb/in}^2)(144 \text{ in}^2/\text{ft}^2)(V_B \text{ ft}^3)$$

$$= 4176(V_B \text{ ft}^3) \text{ ft-lb} = 5.37V_B \text{ Btu with } V_B \text{ in ft}^3$$

Applying the motor and pump efficiencies,

energy = $(200)(V_B \text{ m}^3)/[(0.90)(0.45)] = 494(V_B \text{ m}^3) \text{ kJ} = 13.3(V_B \text{ ft}^3)$ Btu.

This calculation of energy does not include an allowance for liquid bypass, as was discussed in Sec. 8.8.

(b). *Pumping with gas from the high-pressure receiver.*

Vapor densities:

high pressure, 1/0.1171 = 8.54 kg/m³ (0.533 lb/ft³)

low pressure, 1/0.804 = 1.24 kg/m³ (0.776 lb/ft³

The mass of vapor to be compressed is:

$$(8.54 - 1.24)(V_A + V_B) = 7.3(V_A + V_B \text{ m}^3) \text{ lb}$$

$$(0.533 - 0.0776)(V_A + V_B) = 0.4554(V_A + V_B \text{ ft}^3) \text{ lb}$$

The isentropic work of compression between 145 and 1100 kPa (21 and 159.5 psia) is:

$$1735 - 1429 \text{ kJ/kg} = 306 \text{ kJ/kg (131.5 Btu/lb)}$$

which becomes, after inclusion of motor and compressor efficiency

$$(306 \text{ kJ/kg})/[(0.90)(0.75)] = 453 \text{ kJ/kg (195 Btu/lb)}$$

so the work to pump V_B is:

$$\begin{aligned} 7.3(V_A + V_B \text{ m}^3)(453 \text{ kJ/kg}) &= 3307(V_A + V_B \text{ m}^3) \text{ kJ} \\ &= 88.7(V_A + V_B \text{ ft}^3) \text{ Btu} \end{aligned}$$

c. *Pumping with gas from the intercooler.*

Use the same procedure as in part b except that the density of the high-pressure pumping gas is 3.223 kg/m^3 (0.201 lb/ft^3) and the ideal work of compression between pressure of 145 and 400 kPa (21 and 58 psia) is 135 kJ/kg (58.0 Btu/lb). Including the efficiencies, the work to pump V_B is:

$$396(V_A + V_B \text{ m}^3) \text{ kJ or } 10.7(V_A + V_B \text{ ft}^3) \text{ Btu}$$

Some conclusions can be drawn from the three results in Example 8.1:

- The pumping power of the mechanical pump is proportional to V_B while that of gas pumping is proportional to $(V_B + V_A)$. Thus the volume in the pumping vessel above the maximum liquid elevation must be filled with high-pressure vapor and then emptied with each cycle. Volume V_A should thus be kept to a minimum that still provides a vapor cushion in the pumping vessel.
- Another observation is the sensitivity of gas pumping to the pressure of the pumping gas. The pumping work is reduced by a factor of 8 by using intermediate-pressure pumping gas compared to that at the condensing pressure. The intermediate-pressure vapor in this case could generate a pressure difference of $400 - 145 = 255$ kPa (37 psi) on the liquid. A pressure difference of 200 kPa (29 psi) was specified for the mechanical pump, so if intermediate-pressure vapor is available and its pressure is high enough, the efficiency of the pumping process could be essentially the same as mechanical pumping, except for the burden of V_A.

  When the only practical source of pumping gas is at a high pressure, some improvement in efficiency can be achieved by passing this high-pressure pumping gas through a pressure-regulating valve that reduces the pressure in the pumping vessel. The benefit derives from the fact that the pressure in the pumping vessel is lower after expelling the liquid from the pumping vessel. Thus, the mass of refrigerant venting to the low-pressure receiver and compressed back to high-pressure is reduced.
- If the readily available source of pumping gas is just slightly higher than needed for the pumped liquid, the pumping power is the same order of magnitude as with mechanical pumping. The additional power required to mechanically pump the bypass liquid and the parasitic effect of V_A somewhat cancel one another. On the other hand, if high-pressure vapor is the only

source of pumping gas, the energy required by gas pumping will far exceed that of mechanical pumping.

There is, however, another loss associated with gas pumping that must be charged, and this loss is occasioned by the warm vapor heating the cold liquid and the walls of the pumping vessel. The magnitude of this loss is difficult to estimate, but some experimental results[15] suggest an order of magnitude. This research showed a temperature stratification in the liquid in the pumping vessel with the liquid close to the liquid-vapor surface being warmed to essentially the same temperature as the saturation temperature of the pumping gas. The temperature of the liquid in the lower section of V_B was essentially unaffected. An order of magnitude of 10% of the liquid was elevated in temperature, which has the net effect of decreasing the refrigerating effect of each unit mass delivered to the evaporators. Comparative refrigerating effects would be:

no liquid heating: 1429 − 81.5	=	1347.5 kJ/kg (579 Btu/lb)
heating of 10% to −2°C: 1429 − 92	=	1337 kJ/kg (575 Btu/lb)
heating of 10% to 28°C: 1429 − 107	=	1322 kJ/kg (568 Btu/lb)

The reduction in refrigerating effect is only several percent, but it applies to all the liquid pumped, not only that which evaporates, so this loss is approximately 10%.

Usually the amount of time that high-pressure gas is applied to the liquid will be greater in the concept of Figure 8.25 where gas is continuously imposed on one or the other pumping vessels. In the controlled-pressure receiver systems of Figures 8.26 and 8.27 the pumping process is only intermittent, so this system is less subject to the gas heating loss.

All the foregoing may be summarized by estimating that the cost of gas pumping may be 50% to 100% higher than mechanical pumping, if low-pressure pumping vapor is available and yet is high enough in pressure to provide the needed liquid pressure to the evaporators. The designer and plant manager must weigh this additional pumping cost against potential advantages of the gas-pumping system.

8.16 RELATIVE MERITS OF GAS AND MECHANICAL PUMPING

While the first cost of the two systems are usually considered about equal, the gas- and mechanical-pumping systems have advantages and disadvantages.

Advantages of the gas-pumping system

- lower repair costs
- lesser potential for refrigerant leak
- no rigid requirements for NPSH

Advantages of the mechanical pump system

- usually lower pumping energy requirements
- no parasitic losses due to leaking three-way and check valves
- greater flexibility in changing pressures and flow rates of liquid delivered
- less space requirements

Even though logical points can be cited as advantages of one system over the other, the personal preference of the designer or owner continues to be an important factor in the choice. This reason may appear trivial, but when a designer or operator has the experience to operate one or the other system successfully, there is a persuasive reason for choosing that system on the next project.

8.17 WHEN TO CHOOSE LIQUID RECIRCULATION

The advantages and disadvantages of liquid recirculation were listed in Section 8.3. Liquid recirculation competes against flooded-coil evaporators and direct expansion. The advantages of liquid recirculation become dominant in low-temperature installations and where there are multiple evaporators, particularly if they are located some distance from the machine room. In a two-stage system, the low-temperature evaporators might be served by a liquid recirculation system, and the intemediate-temperature evaporators may operate in direct expansion. It is also possible to install liquid recirculation on the intermediate-temperature evaporators with the flash-tank/intercooler serving as the low-pressure receiver.

In addition to low-temperature applications where liquid recirculation now predominates, the concept may have advantages in higher-temperature systems used for air conditioning[16]. Improved heat-transfer coefficients in the evaporator and the positive feeding of multiple evaporators are advantages of liquid recirculation in these cases too. Furthermore, such air-conditioning systems are likely to use a halocarbon refrigerant, and a recirculation system will convey oil on the low-pressure side to the low-pressure receiver where a central oil return can be provided, rather than running the risk of oil accumulation in some evaporators.

Liquid recirculation is more advantageous when serving tube-type evaporators in contrast to evaporators where boiling takes place outside tubes in the shell of the evaporator. Properly applied flooded evaporator coils operate with the same effectiveness of coils served by liquid recirculation. Since oil return provisions must be made for each coil in systems with flooded evaporators, when the number of coils exceeds three to five, it is usually more effective to choose liquid recirculation which has only one point of oil removal—the low-pressure receiver.

Some manufacturers fabricate a liquid recirculation package which includes the low-pressure receiver, pump(s), liquid-level control, and oil-drain or return provisions. The piping interfaces to the other segments of the system are:

- high-pressure liquid supply

- low-temperature liquid supply to the evaporators
- liquid/vapor return from the evaporators
- vapor connection to the compressors.

REFERENCES

1. Wile, D. D. "Evaporator Performance with Liquid Refrigerant Recirculation," Proceedings of Meeting of Annex 1962-1, p 281, Washington D.C., 1962.
2. Lorentzen, G "How to Design Piping for Refrigerant Recirculation," *Heating/Piping/Air Conditioning*, 37(6): 139-152, June 1965.
3. Richards, W. V., "Circuiting and Circulating Numbers for Ammonia Evaporators," Seminar Presentation at Annual Meeting of the American Society of Heating, Refrigerating, and Air-Conditioning Engineers, San Diego, June 1995.
4. Van Maale, J. and E. A. Cosijn, "Cooler Output as a Function of the Recirculation Number of the Refrigerant," Proceedings of the 12th Congress of Refrigeration, International Institute of Refrigeration, Paris, 1967.
5. Niederer, D. H., "Liquid Recirculation, Top or Bottom Feed, What Rate of Feed/" *Air Conditioning and Refrigeration Business,*, December 1963 and July 1964.
6. Geltz, R. W. "Pump Overfeed Evaporator Refrigeration Systems," *Air Conditioning, Heating and Refrigeration News*, p. 42, March 13, 1967.
7. "Liquid Overfeed Systems," *ASHRAE Handbook, Refrigeration Systems and Applications*, American Society of Heating, Refrigerating and Air-Conditioning Engineers, Atlanta, Georgia. 1994.
8. "Lorentzen, G. "Design of Refrigerant Recirculating Systems with a Special View to Their Use for Contact Freezers on Board Fishing Vessels,: Proceedings of Meeting of Commissions B1 and D3, Tokyo, International Institute of Refrigeration, Paris, 1974.
9. Viking Pump, Inc. a Unit of IDEX Corporation, Cedar Falls, Iowa, 1997.
10. Whitmire, K. "Installing and Maintaining Gear Pumps," *Chemical Engrg.*, 103(3): 96-98, March 1996.
11. Cornell Refrigerant Pump, Cornell Pump Company, Portland Oregon, 1995.
12. Kramer, R and Neumaier R., "Centrifugal Pumps and Rotary Positive- Displacement Pumps of Hermetic Design," Hermetic-Pumpen GmbH, Gundelfingen, Germany, 1985.
13. Hansen, C. C.,"Design and Application of Semi-hermetic Refrigerant Pumps," *Heating/Piping/Air Conditioning*, 58(11): 155-161, November 1986.
14. Cole, R. A., "Energy Efficient Gas Cooling for Two-Stage Ammonia Refrigerating Plants," Proceedings of Commision Meeting of the International Institute of Refrigeration, Purdue University, July 1992.
15. Lorentzen, G and O. D. Baglo, "An Investigation of a Gas Pump Recirculation System," Proceedings of the 10th International Congress of Refrigeration (Copenhagen), International Institute of Refrigeration, Paris, 1959.
16. Bongio, V. J., "Industrial Studies in Energy Retrofit," *Heating/Piping/Air Conditioning*, 50(2): 51-56, February 1978.

CHAPTER 9

REFRIGERANT PIPING

9.1 SIZE AND CONFIGURATION OF PIPE

One of the tasks of a designer is to select the size of the various pipe runs in the systems and to specify the pipe materials, the placement, slope, and support of this pipe. The usual concern is to be sure to make the pipe large enough, but there are several situations where a minimum refrigerant velocity must be maintained in the pipe which specifies a maximum pipe size.

This chapter explains how to compute the pressure drop, both by the use of equations from fluid mechanics as well as by special-purpose charts. Most often the pipe carries either liquid or vapor in horizontal or vertical runs. Some exceptions are when a mixture of liquid and vapor flow in the pipe. Sometimes the vapor is responsible for lifting liquid in a vertical riser, and in other cases the pipe must be inclined so that open-channel flow results.

9.2 CATEGORIES OF PIPING

Somewhat different requirements apply to the various section of pipe, depending on the location and purpose served in the system. Table 9.1 lists the various categories of piping in an industrial system, the state of the refrigerant, the importance of pressure drop, and whether pitching or trapping is required. The pipes listed in Table 9.1 carry either liquid or vapor, except for several lines in liquid recirculation systems that convey a mixture of liquid and vapor. The

TABLE 9.1
Categories of piping.

Role	State of refrigerant	Pressure drop permitted	Geometric requirements
Compressor discharge	Vapor	Moderate	
From high-pressure receiver	Liquid	Moderate	Limited increase in elevation
In liquid recirculation systems:			
From pump to evaporators	Liquid	Moderate	
Return to low-pressure receiver	Liquid/vapor	Low	Pitch downward
Vertical risers	Liquid/vapor	Low	
Suction to compressor	Vapor	Low, except for oil return	Trap for oil return in direct-expansion systems
Hot-gas lines for defrost	Vapor	Moderate	Trap for liquid removal

almost universal criterion for selecting the pipe size is the pressure drop, and this calculation is used in one form or another in practically all pipe selections.

9.3 PRESSURE DROP OF FLUIDS FLOWING IN CIRCULAR PIPES

Designers of refrigeration systems make most of their pipe-size selections using special-purpose pressure-drop charts. A separate chart is available for each different refrigerant and there are usually only a limited number of refrigerants with which to deal. For vapors, the pressure drop depends on the pressure level, as indicated by the saturation temperature. There may be situations, however, that are not covered by standard charts, such as for new refrigerants or for secondary coolants. For these occasions, it is valuable to fall back on such basic pressure-drop equations as Darcy-Weisbach. The equation also indicates qualitatively how the fluid properties and pipe geometry influence the pressure drop. The pressure drop equation takes a slightly different form in SI and I-P units,

$$\text{In SI units}: \ \Delta p = f\frac{L}{D}\frac{V^2}{2}\rho \tag{9.1}$$

$$\text{In I–P units}: \ \Delta p = f\frac{L}{D}\frac{V^2}{2g}\rho \tag{9.2}$$

where			SI	I-P
Δp	=	pressure drop	Pa	lb_{force} per ft^2
f	=	friction factor	dimensionless	dimensionless
L	=	length of pipe	m	ft
D	=	diameter of pipe	m	ft
V	=	velocity	m/s	fps
ρ	=	density	kg/m^3	lb/ft^3
				$g = 32.2\ ft/s^2$

The velocity V is the volume rate of flow divided by the cross-sectional area of the pipe,

$$V \text{ m/s (fps)} = \frac{Q \text{ m}^3/\text{s}}{\pi(D, \text{ m})^2/4} \quad \text{or} \quad \frac{Q \text{ ft}^3/\text{s}}{\pi(D, \text{ ft})^2/4} \tag{9.3}$$

The friction factor f is a function of the Reynolds number and the roughness of the pipe, and can be read from a Moody chart[1] such as Fig. 9.1. The Reynolds number, Re, is a dimensionless group,

$$Re = \frac{VD\rho}{\mu}, \quad \frac{(\text{m/s})(\text{m})(\text{kg/m}^3)}{\text{Pa}\cdot\text{s}} \quad \text{or} \quad \frac{(\text{ft/s})(\text{ft})(\text{lb/ft}^3)}{(\text{lb/ft}\cdot\text{hr} \div 3600 \text{ s/hr})} \tag{9.4}$$

where μ = viscosity, Pa·s (lb/ft·hr)

The units cancel in both of the representations of Eq. 9.4. In the SI version the pascal = newton/m^2, and the force in newtons = (mass)(acceleration) or kg×m/s^2. The magnitude of the Reynolds number is important in its own right because it indicates the pattern of flow: laminar if $Re < 2100$ and turbulent if $Re > 3000$. Different equations for heat-transfer coefficients apply to turbulent flow compared to laminar flow, and the same is true for pressure-drop equations. In refrigeration practice, laminar flow is rare. About the only time it occurs is in the flow of viscous oil. The individual terms in Eq. 9.4 are available as follows:

- D will be known
- V from Eq. 9.3
- ρ either directly from the tables of refrigerant properties in the appendix, or as the reciprocal of the specific volume
- μ from Table 9.2.

The Reynolds number is one of the terms with which to enter Fig. 9.1. The other quantity needed to determine the friction factor from Fig. 9.1 is the relative roughness, ϵ/D. The roughness ϵ of the surfaces of two different materials used for refrigerant pipe and tubes are:

Material	Roughness, ϵ
copper	0.0000015 m (0.000005 ft)
steel	0.000046 m (0.00015 ft)

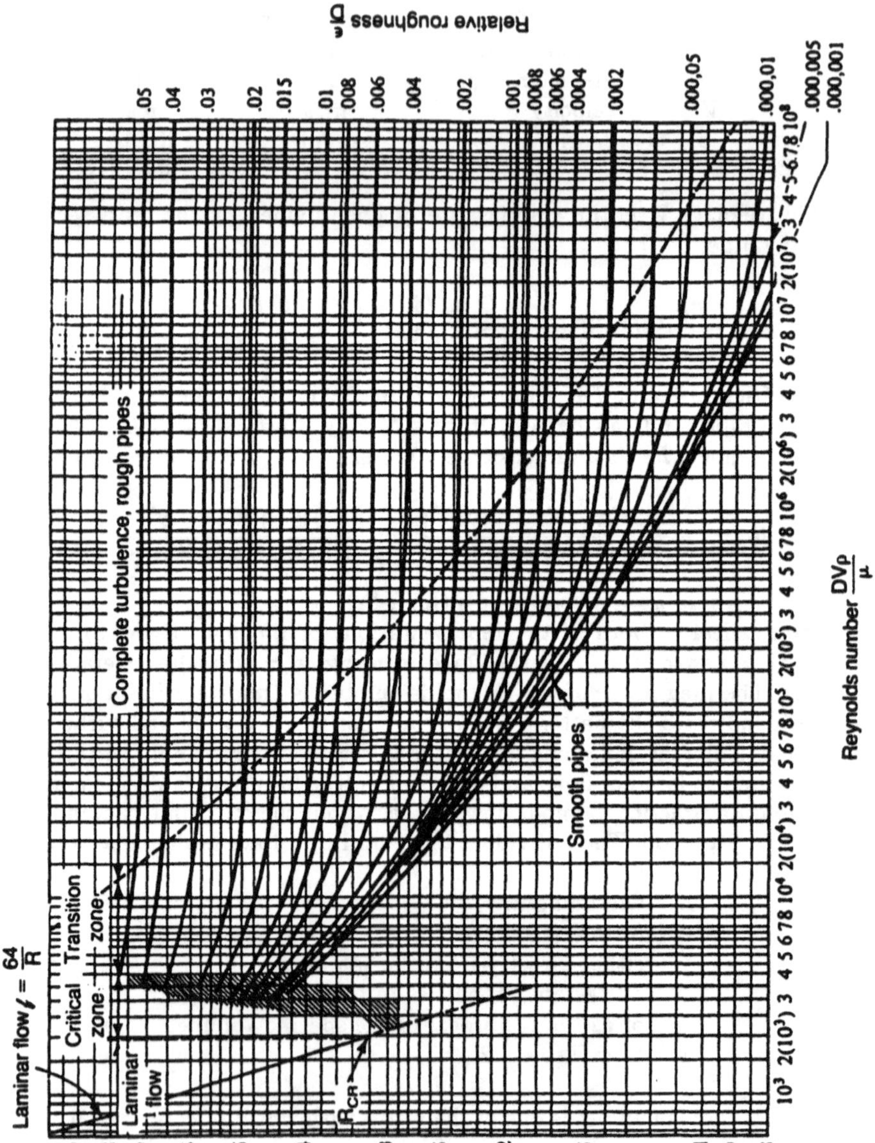

FIGURE 9.1
Moody chart for determining the fraction factor f.

TABLE 9.2
Viscosities of several refrigerants in their saturated liquid and saturated vapor states[2,3].

Refrigerant	SI, Pa·s			I-P, lb/(ft·hr)		
	Temp., °C	Liquid	Vapor	Temp., °F	Liquid	Vapor
R-22	−40	0.000330	0.0000101	−40	0.798	0.0245
	−20	0.000275	0.0000110	0	0.654	0.0269
	0	0.000237	0.0000120	40	0.533	0.0295
	20	0.000206	0.0000130	80	0.480	0.0325
	40	0.000182	0.0000144	120	0.426	0.0362
	60	0.000162	0.0000160			
R-134a	−20	0.000373	0.0000101	0	0.876	0.0245
	0	0.000287	0.0000109	30	0.705	0.0263
	20	0.000226	0.0000119	60	0.576	0.0283
	40	0.000178	0.0000131	90	0.473	0.0305
	60	0.000139	0.0000147	120	0.387	0.0332
R-404A	−40	0.000320		−30	0.653	0.0252
	−20	0.000240	0.0000107	0	0.556	0.0264
	0	0.000180	0.0000118	30	0.440	0.0285
	20	0.000137	0.0000132	60	0.351	0.0310
	40	0.000105	0.0000152	90	0.266	0.0358
R-507	−40	0.000273	0.0000093	−40	0.658	0.0225
	−20	0.000220	0.0000101	0	0.520	0.0247
	0	0.000178	0.0000110	40	0.411	0.0271
	20	0.000142	0.0000121	80	0.317	0.0305
	40	0.000108	0.0000139	120	0.227	0.0365
Ammonia	−20	0.000236	0.0000097	−20	0.629	0.0227
	0	0.000190	0.0000104	0	0.558	0.0237
	20	0.000152	0.0000112	40	0.437	0.0256
	40	0.000122	0.0000120	80	0.341	0.0277
	60	0.000098	0.0000129	120	0.268	0.0299

The calculation of the pressure drop in both SI and I-P units will be illustrated by Examples 9.1 and 9.2, respectively, for identical situations.

Example 9.1. What is the pressure drop per m of tube length when 1.51 kg/s of R-22 vapor at −6.7°C flows through copper tubing that is 74.8 mm ID?

Solution. At −6.7°C the specific volume is 0.05855 m^3/kg ($\rho = 17.08$ kg/m^3) from the R-22 Table A-13 in the appendix, so the volume flow rate is:

$$1.51\ (0.05855) = 0.0884\ m^3/s.$$

With the cross-sectional area of the tube of 0.00439 m^2, the velocity V is 0.0884 ÷ 0.00439 = 20.14 m/s. From Table 9.1 at −6.7°C, $\mu = 0.0000117$ Pa·s. The Reynolds number is:

$$Re = \frac{(20.14 \text{ m/s})(0.0748 \text{ m})(17.08 \text{ kg/m}^3)}{0.0000117 \text{ Pa·s}} = 2.2 \times 10^6$$

The roughness ϵ of copper tubing is 0.0000015 m, so

$$\frac{\epsilon}{D} = \frac{0.0000015}{0.0748} = 0.00002$$

From the Moody chart, Fig. 9.1, at the above values of Re and ϵ/D, the friction factor f is found to be 0.0108. Finally substitute applicable terms into the pressure drop equation, Eq. 9.1,

$$\Delta p = (0.0108)\left(\frac{1 \text{ m}}{0.0748 \text{ m}}\right)\frac{(20.14 \text{ m/s})^2}{2}(17.08 \text{ kg/m}^3)$$

$$\Delta p = 500 \text{ Pa/m}.$$

Example 9.2. What is the pressure drop per 100 ft of pipe length when 200 lb/min of R-22 vapor at 20°F flows through type L copper tubing that is 3-1/8 inch OD (2.945 inch ID)?

Solution. At 20°F the specific volume from Table A-14 is:

$$0.9369 \text{ ft}^3/\text{lb} \ (\rho = 1.067 \text{ lb/ft}^3)$$

so the volume flow rate is:

$(200 \text{ lb/min})(0.9369 \text{ ft}^3/\text{lb}) = 187.4 \text{ ft}^3/\text{min}$ or $3.123 \text{ ft}^3/\text{s}$.

With the cross-sectional area of the tube of 0.0473 ft², the velocity:

$V = 3.123 \div 0.0473 = 66.0$ fps.

From Table 9.1 at 20°F, $\mu = 0.282$ lb/(ft·hr) = 0.00000783 lb/(ft·s).

The Reynolds number is:

$$Re = \frac{(66 \text{ ft/s})(0.245 \text{ ft})(1.067 \text{ lb/ft}^3)}{0.00000783 \text{ lb/(ft·s)}} = 2.2 \times 10^6$$

From the Moody chart, Fig. 9.1, at the above values of Re and ϵ/D, the friction factor is found to be 0.0108. Finally, the appropriate terms are substituted into the pressure-drop equation, Eq. 9.2:

$$\Delta p = (0.0108)\left(\frac{100 \text{ ft}}{0.245 \text{ ft}}\right)\frac{(66.0 \text{ fps})^2}{2(32.2 \text{ ft/s}^2)}(1.067 \text{ lb/ft}^3)$$

$$\Delta p = 317.6 \text{ lb/ft}^2 \text{ or } \frac{317.6}{144} = 2.2 \text{ psi}$$

The pressure drop of 500 Pa per m computed in Example 9.1 is identical to 2.2 psi per 100 ft in Example 9.2.

Equations 9.1 and 9.2 show influences of length, density, and diameter on the velocity and pressure drop. The pressure drop is proportional to the length—an obvious expectation. The pressure drop is also proportional to the density of the fluid, which means that high-pressure discharge gas experiences a greater pressure drop than does suction vapor in the same size pipe with a given velocity. This statement must be qualified: it does not imply that discharge lines would be larger than the suction lines. Just the opposite is true, because the high density of the discharge gas also results in a lowered velocity for a given mass rate of flow.

The pipe size exerts a dramatic influence on the velocity and pressure drop. Equation 9.2 can be modifed to show that

$$\frac{V_2}{V_1} = \left(\frac{D_1}{D_2}\right)^2$$

so if the diameter if reduced by 50%, the velocity increases to four times its original value for a given volume flow rate. When the velocity effect is inserted in the pressure drop equation and combined with the appearance of D in the fL/D group, the ratio of the pressure drops varies inversely as the diameter to the 5th power,

$$\frac{\Delta p_2}{\Delta p_1} = \frac{D_1}{D_2}\left(\frac{D_1}{D_2}\right)^4 = \left(\frac{D_1}{D_2}\right)^5$$

Reducing the pipe size by 50% causes a 32-fold increase in the pressure drop, showing that the pressure drop is extremely sensitive to the pipe size.

9.4 PRESSURE-DROP DETERMINATIONS USING CHARTS

Designers are called on to make repeated pressure-drop calculations, but for a limited number of refrigerants and for specified pipe sizes. The situation is ready-made for the construction of charts to duplicate the calculations using the equation. Typical of these charts is a set developed by D. D. Wile for ASHRAE[4,5] for R-12, R-22, R-502 and ammonia. Three charts are presented for each refrigerant: liquid, vapor at suction conditions, and vapor at compressor discharge conditions. A sample page for low-pressure R-22 vapor is shown in Fig. 9.2.

Example 9.3. Using Fig. 9.2, determine the pressure drop in a 75-ft length of nominal 3-1/8-inch type L copper tubing through which 200 lb/min of R-22 vapor at 20°F is flowing.

Solution. In the right-hand block of lines in Fig. 9.2 locate the point for 200 lb/min with 3-1/8-inch tubing. Next move left horizontally to the left-hand block of lines until reaching the 20°F evaporating temperature line. The pressure drop indicated is 2.3 psi per 100 ft length.

The pressure drop in 75 ft of length is:

$$\Delta p \text{ for } 75 \text{ ft} = 2.3\frac{75}{100} = 1.7 \text{ psi}$$

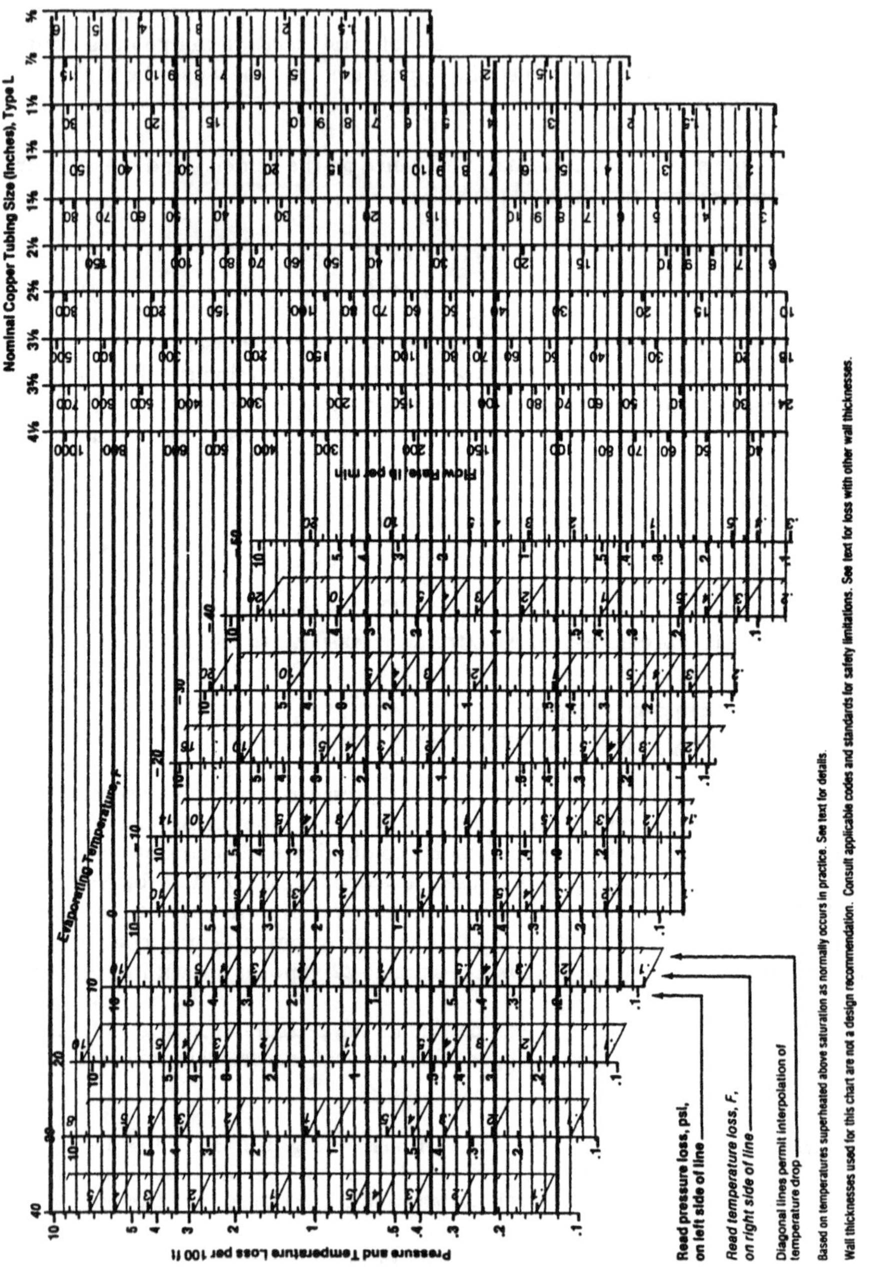

FIGURE 9.2
Pressure drop in suction lines with R-22 flowing in copper tubing[4].

The numbers on the right side of the vertical 20°F evaporator line show a drop in saturation temperature of 2.1°F which corresponds to the pressure drop of 2.3 psi.

The conditions in Example 9.3 are identical to those in Example 9.2 where a pressure drop of 2.2 psi was calculated, so the two approaches agree approximately, but the chart in Fig. 9.2 is based on 20°F superheat which increases the specific volume over that for saturated vapor that was assumed in Example 9.2. Reference 4 states that for 20°F of superheat, the pressure drop will be 5.2% higher than for saturated vapor, so reducing 2.3 psi by 5.2% results in good agreement between Examples 9.2 and 9.3. It is no surprise that the chart in Fig. 9.2 and the pressure-drop equation Eq. 9.2 agree well, because the charts in Reference 4 were based on this standard equation.

Equations 9.1 and 9.2 apply to incompressible flow, so certainly are valid for liquid, but are sufficiently accurate for gases as well, so long as the change in pressure are very small with respect to the pressure level. At some of the high-velocity regions of the charts, the compressibility of gas should have been taken into consideration[6]. This situation is not serious, because in the normal design of refrigeration systems, practical choices do not normally drift into the extremely high-velocity regions.

9.5 OPTIMUM PIPE SIZE

Computation of the pressure drop of the refrigerant flowing in a pipe is only one step in the decision process of selecting the pipe size. Ultimately the decision of the size of a vapor pipe is economic, trading off the additional cost of a larger pipe against the saving in compressor energy over the equipment lifetime. For a given situation, the cost trends are as shown in Fig. 9.3 where all costs reflect present worth.

It may seem at first that for a given flow rate and refrigerant condition the optimum diameter of a long pipe will be greater than that of a short one. Richards[7] showed, however, that by setting to zero the derivative of the total cost, the length cancels. A summary form of the equation representing the costs shown in Fig. 9.3 is:

$$\text{Total cost, } \mathrm{C} = \mathrm{C}_1 DL + \mathrm{C}_2 \Delta p \tag{9.5}$$

where C_1 incorporates the material and labor cost of the pipe installation and assumes that these costs are proportional to the pipe diameter. The $\mathrm{C}_2 \Delta p$ term is the present worth of the lifetime costs to overcome Δp. The constant C_2 thus incorporates the number of hours of operation per year, the life of the facility, and the cost of money. When the equation for Δp, Eq. 9.1 or 9.2, is substituted into Eq. 9.5 for the existing flow rate, the new equation is:

$$\mathrm{C} = \mathrm{C}_1 DL + \frac{\mathrm{C}_3 L}{D^5} \tag{9.6}$$

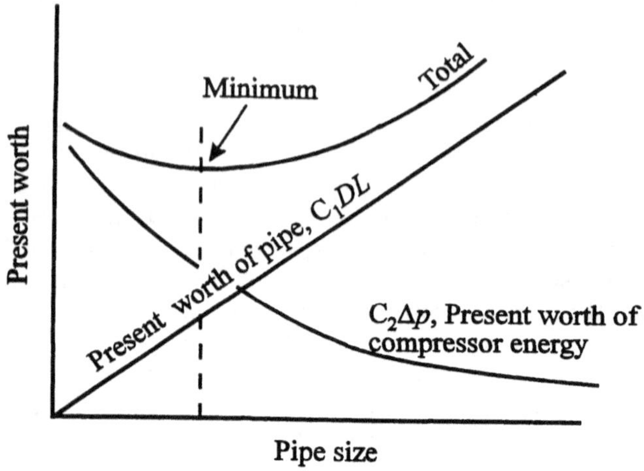

FIGURE 9.3
Optimum vapor pipe size at the minimum of the sum of the first cost of the pipe and present worth of the lifetime compressor energy costs.

To find the optimum diameter, differentiate Eq. 9.6 and set equal to zero:

$$\frac{dC}{dD} = C_1 L - \frac{C_3 L}{D^6} = 0$$

$$\text{Optimum } D = \left(\frac{5C_3}{C_1}\right)^{1/6} \tag{9.7}$$

The length L cancels, which shows that the optimum diameter is independent of length.

In principle, the optimization calculation, subject to such constraints as minimum diameter to achieve a certain velocity or maximum diameter to meet space limitations, could be performed on each design. Such an effort is not practical, and the best that can be hoped for is a periodic check of the optimum to accommodate shifts in cost of materials and energy.

9.6 PRESSURE DROP IN FITTINGS

In addition to the pressure drop that occurs in straight pipe, fittings such as elbows and tees also cause pressure drop. The physical layout of the piping system determines where elbows and branch tees are located. In addition there are noticeable drops in pressure through valves, even when they are wide open. Judgment must be used on where to install valves and what type of valve to choose. Even valves that are rarely shut off may be invaluable in isolating a certain component or even another valve in rare instances. On the other hand, extra valves placed in vapor lines may represent a persistent demand for extra compressor power whenever the system operates.

TABLE 9.3
Pressure-drop coefficients for some frequently used valves and fittings—values of PDC in Eq. 9.8.

Pipe dia. mm (in)	90° Elbow	Tee, branch	Globe valve	Gate valve	Angle valve	Check valve
25 (1)	1.5	1.8	9	0.24	4.6	3.0
50 (2)	1.0	1.4	7	0.17	2.1	2.3
67 (2-1/2)	0.85	1.3	6.5	0.16	1.6	2.2
100 (4)	0.7	1.1	5.7	0.12	1.0	2.0

The magnitude of the pressure drop in fittings or open valves is proportional to the density of the fluid flowing and to the square of the velocity. The pressure-drop coefficients, PDC, in SI and I-P are applied as indicated in Eq. 9.8.

$$\Delta p = \text{PDC}\left(\frac{V^2\rho}{2}\right) \quad \text{or} \quad \Delta p = \text{PDC}\left(\frac{V^2\rho}{2g_c 144}\right) \tag{9.8}$$

where Δp = pressure drop, Pa or Δp = pressure drop, psi
V = velocity, m/s; V = velocity, fps
ρ = density, kg/m^3; ρ = density, lb/ft^3
g_c = 32.2 ft-lb/lb$_f$-s^2

When used in the expressions of the form in Eq. 9.8, the same values of PDC apply to both SI and I-P units. Values[8] of PDC are presented in Table 9.3.

An important observation that can be made from Table 9.3 is the relative pressure drops of globe and angle valves. An angle valve causes anywhere from 1/2 to 1/6 the pressure drop of a globe valve (depending upon the valve and pipe size) and should be considered if the physical arrangement permits.

Many designers find it convenient to express the pressure drop of fittings and open valves as equivalent length of straight pipe. This equivalent length, L_{eq}, can then simply be added to the actual length of straight pipe to determine the total pressure drop. The form of Eq. 9.8 and the data in Table 9.3 provide the tools to compute L_{eq}, since the PDC term for fittings is comparable to the fL/D group in Eqs. 9.1 and 9.2. A typical value of the friction factor f for refrigeration applications is approximately 0.02. Thus,

$$L_{eq}, \text{m} = \frac{\text{PDC}(D, \text{m})}{0.02} \quad \text{and} \quad L_{eq}, \text{ft} = \frac{\text{PDC}(D, \text{ft})}{0.02} \tag{9.9}$$

Applying Eq. 9.9 to the data in Table 9.3 yields the equivalent lengths of Table 9.4.

The equivalent lengths increase slightly as the pipe diameter increases, which is due to change in ratio of pressure drop of straight pipe and fittings. Several important observations can be made from Table 9.4. One is that the pressure drop in the fittings can be a significant contributor to the total pressure drop. Another is the relative pressure drops of globe and angle valves. A globe

TABLE 9.4
Equivalent length of straight pipe in m (ft) resulting in the same pressure drop as the fitting.

Pipe dia. mm (in)	90° Elbow	Tee, branch	Globe valve	Gate valve	Angle valve	Check valve
25 (1)	1.9 (6.3)	2.3 (7.4)	11.2 (37)	0.3 (1)	5.8 (19)	3.8 (12)
50 (2)	2.5 (8.2)	3.5 (11.5)	17.5 (57)	0.4 (1.4)	5.2 (17)	5.8 (19)
67 (2-1/2)	2.8 (9.3)	4.3 (14)	22 (71)	0.5 (1.8)	5.4 (18)	7.4 (24)
100 (4)	3.5 (11)	5.5 (18)	29 (94)	0.6 (2)	5 (16)	10 (32)

valve causes anywhere from 2 to 6 times the pressure drop of an angle valve (depending upon the valve and pipe size). When the physical arrangement permits, an angle valve should be chosen over a globe valve.

9.7 DROP IN SATURATION TEMPERATURE AS SELECTION CRITERION

It is expected that the size of the pipe will be chosen such that the pressure drop is appropriate, but what is appropriate is often best indicated by the drop in saturation temperature in °C (°F). For example, a 20-kPa (3-psi) pressure drop may be of little concern in the discharge line, but could be significant in a low-pressure suction line. Also in moving from one refrigerant to another the drop in saturation temperature remains a reliable indicator of the penalty. The reason that the drop in saturation temperature is such a good indicator can trace back to the expression for the coefficient of performance of a Carnot cycle (Chapter 2) which is a function only of saturation temperatures, regardless of the pressures.

9.8 SELECTING THE PIPE SIZE

The ability to determine the pressure drop in a refrigerant line may be crucial, but there remains the decision of how much pressure drop (or drop in saturation temperature) to specify. While the optimization process discussed in Sec. 9.5 would be ideal, designers usually resort to some conventions that at least give reasonable pipe sizes. The various pipe sections are addressed individually:

- **Suction to compressor.** The total drop in saturation temperature is usually chosen to be 0.5 to 2°C (0.9 to 3.6°F). The exceptions are vertical risers both for halocarbon direct expansion and for ammonia liquid overfeed coils. For halocarbon direct-expansion systems the velocity of the refrigerant vapor must be high enough to convey oil back to the compressor. For ammonia liquid-overfeed coils the vapor velocity in the riser must be high enough to blow the liquid out so that it doesn't fill the riser.
- **Discharge from compressor to condenser.** The total drop in saturation temperature is usually chosen from 1.0 to 3.0°C (1.8 to 5.4°F). A given drop

in saturation temperature in the discharge pipe is slightly less penalizing in compressor power than the same drop in temperature on the suction side.

- **High-pressure liquid.** A pressure drop in this section may exact no penalty whatsoever on system performance, because the pressure drop that does not occur in the pipe will take place in the expansion device or level-control valve. The expansion device provides the final reduction of pressure to the intermediate pressure (in two-stage compression) or to the low pressure (in single-stage compression). The concern about pressure drop in this line arises more in assuring that the pressure does not drop to the saturation pressure corresponding to the existing refrigerant temperature. Were the pressure to drop to that point, the liquid would flash into vapor, aggravate the pressure gradient, and possibly restrict the flow through the expansion device. Refrigerant velocities chosen for liquid lines range from 1 to 2.5 m/s (3 to 8 ft/s).
- **Liquid/vapor return from evaporators to low-pressure receiver.** The line from the evaporators back to the low-pressure receiver in liquid recirculation systems carries a mixture of liquid and vapor. Calculations of pressure drops in the flow of liquid/vapor mixtures, while possible, are complex. To avoid cumbersome calculatons, yet still make allowances for the presence of liquid, some designers choose the line size, first by determining the appropriate size if the pipe were carrying only vapor, then step up to the next pipe size to allow for the combined flow of liquid.
- **Hot-gas defrost lines.** To make an intelligent choice of pipe size, the required flow rate of hot gas as a function of the evaporator size should be known. A rough estimate of the hot-gas flow rate is that it is twice the refrigerant flow rate used during refrigeration service. With this assumption, the recommended sizes of ammonia hot-gas branch lines proposed by Hansen[9] use as a basis a velocity of 15 m/s (3000 fpm) with 21°C (70°F) hot gas. This velocity would be appropriate for hot-gas branch lines serving a single evaporator of a cluster of evaporators defrosted at the same time. The hot-gas mains may be sized for carrying half the total of all connected evaporators on the assumption that no more than half the evaporators would be defrosted at one time.

  Recent efforts to operate plants with as low a condensing temperature as possible impacts the desired size of hot-gas lines. The ultimate criterion is the saturation temperature at which the defrost gas can condense in the evaporator being defrosted, so the drop in saturation temperature in the hot-gas line reappears as the most appropriate basis for selecting the pipe size. As the condensing temperature of the plant drops, the defrost gas becomes less dense, and when the condensing temperature of the plant drops from 35°C (95°F) to 15°C (59°F), for example, the drop in saturation temperature for several common refrigerants doubles[10].

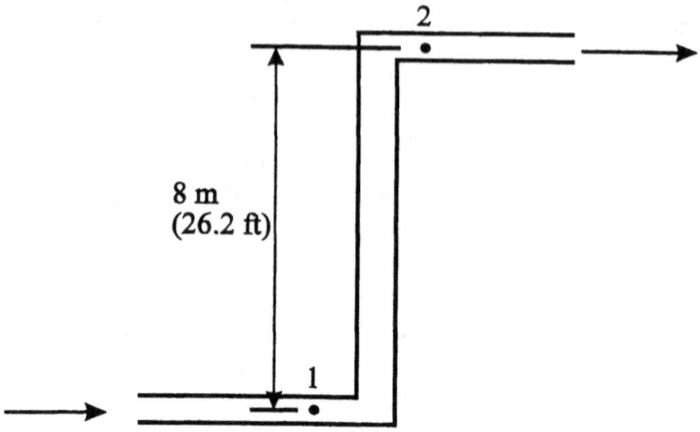

FIGURE 9.4
Liquid line riser in Example 9.4.

9.9 RISE IN ELEVATION OF LIQUID FLOWING IN A PIPE

Equations 9.1 and 9.2 as well as charts such as the one in Fig. 9.2 predict pressure drops due to friction in horizontal pipes. When a pipe carrying liquid slopes upward in the direction of flow, an additional pressure drop results. When sloping downward, the liquid head contributes a pressure increase in the direction of flow.

The liquid line from the condenser or receiver to an expansion valve or level-control valve is one location where an increase in elevation of the pipe can cause trouble.

Example 9.4. The liquid line, shown in Fig. 9.4, in an R-507 system rises 8 m (26.2 ft). At point 1 before the rise, the temperature is 30°C (86°F) and the pressure is 1525 kPa (221 psia). What is the temperature and the pressure at point 2?

Solution. Liquid R-507 at point 1 is subcooled as indicated by the fact that the pressure is higher than the saturation pressure at 30°C (86°F) which is 1465 kPa (212.5 psia). At 30°C (86°F) the density of liquid is 1022 kg/m^3 (63.8 lb/ft^3). In rising 8 m (26.2 ft) the pressure drop due to the loss in liquid head is:

$$\frac{(8\text{ m})(1022\text{ kg/m}^3)(9.81\text{ m/s}^2)}{1000\text{ Pa/kPa}} = 80.2\text{ kPa}$$

or

$$\frac{(26.2\text{ ft})(63.8\text{ lb/ft}^3)}{144\text{ in}^2/\text{ft}^2}$$

Neglecting the friction in the riser, the pressure at 2 is

$$1525\text{ kPa} - 80.2\text{ kPa} = 1445\text{ kPa (209.5 psia)}$$

TABLE 9.5
Reduction in saturation temperature associated with a rise in a liquid line at liquid temperatures of 30°C (86°F).

Refrigerant	Drop in saturation temperature °C/m	°F/ft
R-134a	0.53	0.29
R-22	0.37	0.20
R-507	.262	0.14
R-717, Ammonia	0.17	0.092

The pressure at point 2 has dropped slightly below the saturation pressure at 30°C (86°F) which is 1465.4 kPa (212.5 psia), so some liquid will flash into vapor and the temperature will drop to the saturation temperature at this pressure. Vapor in the liquid entering an expansion valve may severely restrict the mass flow rate that the valve can pass.

A critical parameter in evaluating effects of rises in liquid lines is the drop in saturation temperature. Table 9.5 shows this parameter for several refrigerants and indicates that, of the four refrigerants shown, R-134a is the most sensitive and ammonia the least sensitive to rises in the liquid line. The physical arrangement of the plant usually dictates the need for a rise in the liquid line, and the only responses open to the designer or contractor if a flashing problem is anticipated is to install a heat exchanger to subcool the liquid or to elevate the pressure of the liquid with a pump.

9.10 DOWNWARD SLOPING PIPE

When a liquid line slopes downward in the direction of flow, the gain in pressure due to the drop in elevation works counter to the drop in pressure due to friction. A borderline condition exists when the increase in pressure due to the drop in elevation just equals the drop in pressure due to friction.

Example 9.5. What is the downward pitch of an ammonia liquid line carrying 35°C (95°F) liquid at a velocity of 0.51 m/s (100 fpm) in a 50-mm (2-in.) pipe so that the gain in pressure attributable to the liquid head just cancels the pressure drop due to friction?

Solution. The pitch of the pipe, as in Fig. 9.5a, is to be computed so that the gain due to the liquid head cancels the friction pressure drop. Equation 9.1 (9.2), which expresses the pressure drop in Pa (psf), can be revised to compute the loss of head due to friction using either of the relationships:

$$\Delta p, \text{ Pa} = (\text{head loss, m})(\rho, \text{ kg/m}^3)(\text{g, m/s}^2)$$

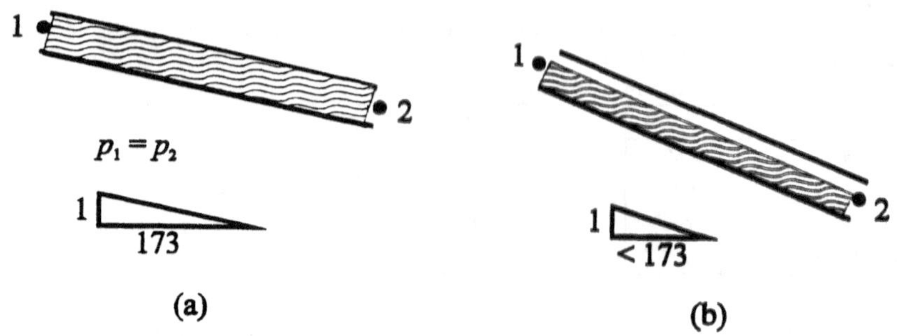

FIGURE 9.5
Downward pitch of the liquid ammonia line in Example 9.5: (a) where the pressure drop due to friction is precisely canceled and (b) where the slope is greater and the line runs partially filled with liquid.

$$\Delta p, \text{ psf} = (\text{head loss, ft})(\rho, \text{lb/ft}^3)$$

so the head loss in either system of units is:

$$\text{Head loss, m/m (ft/ft)} = f\frac{L}{D}\frac{V^2}{2g} \tag{9.10}$$

The Reynolds number Re is:

$$Re = \frac{(0.51 \text{ m/s})(0.05 \text{ m})(\rho = 587 \text{ kg/m}^2)}{\mu = 0.00013 \text{ Pa·s}} = 115,100$$

$$\frac{\epsilon}{D} = \frac{0.000046}{0.05} = 0.00092$$

From Fig. 9.1, $f = 0.022$, so

$$\text{Head loss} = 0.022\left(\frac{1 \text{ m}}{0.05 \text{ m}}\right)\left[\frac{(0.51 \text{ m/s})^2}{2(9.81 \text{ m/s}^2)}\right] = 0,0058 \text{ m/m(ft/ft)}$$

The pitch of the pipe resulting in the gain of pressure due to a drop in elevation canceling the friction is:

$$\frac{1}{0.0058} \text{ or 1 in 173.}$$

If the line in Example 9.5 is pitched more steeply than 1 in 173 with the same mass flow rate, the line will run partially filled with liquid. The situation of Fig. 9.5b is the one sought in the draining of a single condenser, Fig. 7.28, where for an ammonia liquid velocity of 0.51 m/s (100 fpm), the same as in Example 9.5, a pitch of 1 in 50 is recommended. Compared to the pitch of 1 in 173 calculated in Example 9.5, enough vapor space will be provided to permit counterflow of the vapor to the liquid. A slope of less than 1 in 173 results in the pipe flowing full.

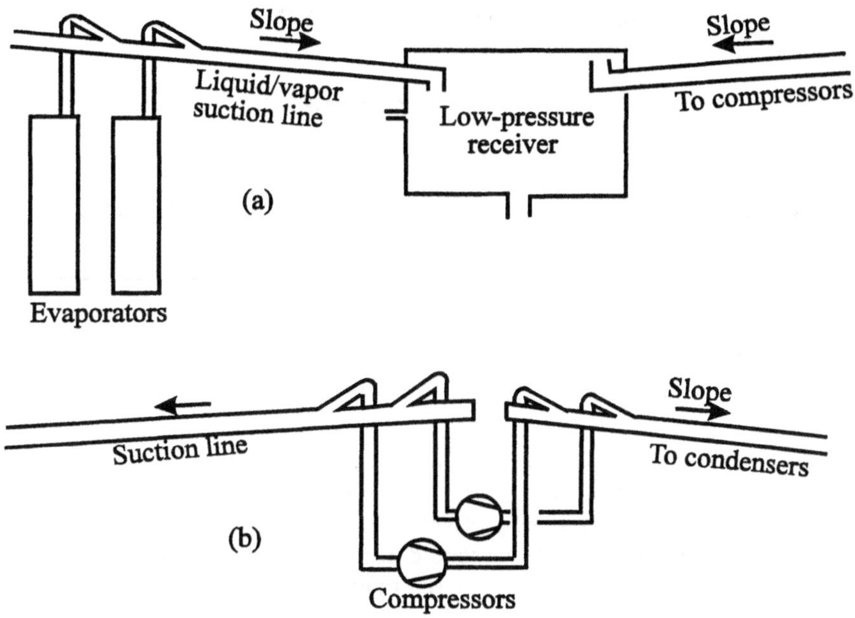

FIGURE 9.6
Slope of vapor lines and branch connections from evaporators and compressors.

9.11 AVOIDING DRAINAGE TO COMPRESSORS AND TO EVAPORATORS

If the physical arrangement will permit it, certain vapor lines should be sloped in order to prevent liquid from draining back to the evaporator outlets and to the suction of the compressors. Figure 9.6 shows several vapor lines and the recommended directions of their slopes. Figure 9.6a shows a section of a liquid recirculation system where the liquid/vapor line is inclined toward the low-pressure receiver. This direction of slope aids in the return of liquid during normal operation and is also important should there be a power outage. In that case the compressors cease operating, interrupting the flow of vapor, but the liquid in the line continues to drain to the low-pressure receiver. The suction line of the compressor is sloped away from the compressor, but the discharge line, as in Fig. 9.6b, slopes away from the compressor toward the condenser.

The recommended magnitude of slope is often specified as 1:240, and while it may be easy to show slope on the design drawings, it is sometimes difficult in a long line to physically achieve the desired slope. Usually there is no problem for lines running above the roof, but sometimes within a building headroom may be sacrificed.

Figure 9.6 also shows the recommended orientation of the branches into and out of the vapor lines. The connections are made to the top of the vapor

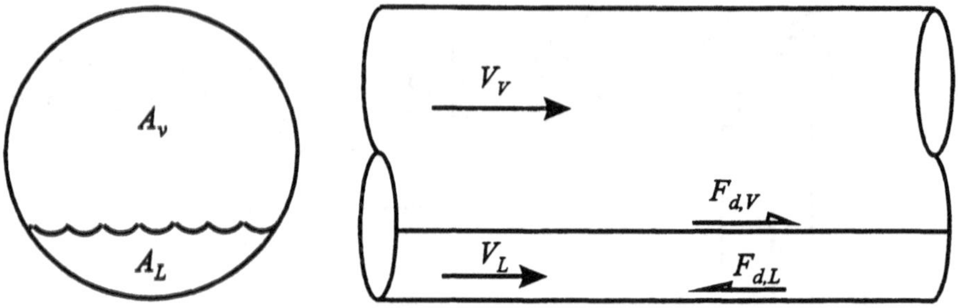

FIGURE 9.7
Liquid flowing along the bottom of a liquid/vapor return line.

line, so that any liquid that inherently exists in a liquid recirculation system or could be formed by condensation does not back up into the evaporators or accumulate at a compressor, particularly an idle one.

While branch connections should normally be to the top of the main, in the case of liquid lines, they can enter the main at the side or even at the bottom.

9.12 LIQUID CONTENT IN HORIZONTAL LIQUID/VAPOR RETURN LINES

Computing the liquid content in the liquid/vapor return line of a recirculation system is a frequent assignment. One need arises when the inventory of refrigerant in the system is to be estimated. A more frequent requirement, as is explained in Chap. 10 on vessels, is to determine how much volume must be provided in the low-pressure receiver to receive the liquid in the liquid/vapor return line should the power be interrupted. The estimate of the mass of liquid is a crude estimate at best, but two models will be shown that hopefully can bracket the actual amount.

Several relationships are basic to quantifying the calculation.

$$\dot{m}_L + \dot{m}_V = n\dot{m}_V \quad \text{so} \quad \dot{m}_L = (n-1)\dot{m}_V \tag{9.11}$$

$$\dot{Q} = VA \quad \text{so} \quad \dot{Q}_V = V_V A_V \quad \text{and} \quad \dot{Q}_L = V_L A_L \tag{9.12}$$

$$\dot{m} = \dot{Q}\rho \quad \text{so} \quad \dot{m}_V = \dot{Q}_V \rho_V \quad \text{and} \quad \dot{m}_L = \dot{Q}_L \rho_L \tag{9.13}$$

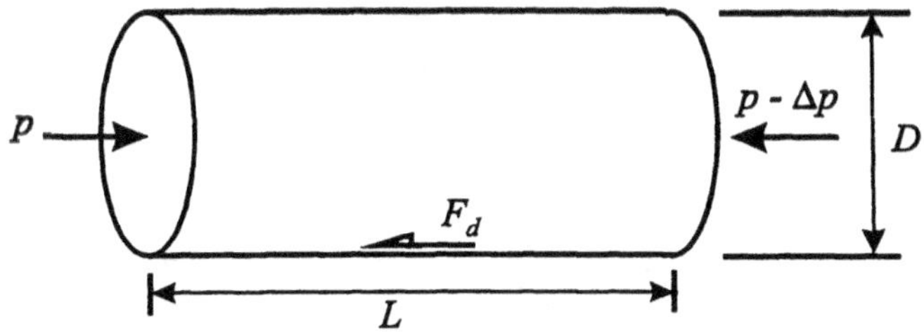

FIGURE 9.8
Solving for F_d by a force balance.

where $\dot{m}$ = mass rate of flow, kg/s (lb/s)
n = circulation ratio as defined in Eq. 8.1
$\dot{Q}$ = volume rate of flow m^3/s (ft^3/s)
V = velocity, m/s (ft/s)
A = area, m^2 (ft^2)

Subscript L refers to liquid and subscript V refers to vapor.

Equations 9.11 through 9.13 can be combined to form a working equation for the ratio of cross-sectional areas of the liquid to that of vapor,

$$\frac{A_L}{A_V} = \frac{(n-1)V_V \rho_V}{V_L \rho_L} \tag{9.14}$$

The final term of interest is usually the fraction of the pipe cross-section occupied by liquid, FL

$$FL = \frac{A_L}{A_L + A_V} = \frac{1}{1 + A_V/A_L} \tag{9.15}$$

Two models of the liquid/vapor flow pattern in the pipe are proposed to describe the limits of FL.

- Model 1: the liquid and vapor move with identical velocities, so $V_L = V_V$.
- Model 2: the liquid moves with a velocity such that the drag force of the vapor on top of the liquid balances the drag of the pipe on the underneath side of the liquid.

Model 1 is simple, but does not acknowledge that *slip* occurs wherein the velocity of the vapor is higher than that of the liquid. Model 2 attempts to evaluate the V_L relative to V_V through the introduction of the drag force, F_d. F_d has units of newton/m^2 (lb_{force}/ft^2), and in the section of pipe shown in Fig. 9.8, the total force of drag on the fluid exercised by the pipe is:

$$\text{Drag force in length } L = F_d \text{ (area of cylinder)} = F_d \pi D L$$

This is the same magnitude of force acting on the cross-sectional area of the fluid, $\Delta p \pi D^2/4$, so the forces can be equated and the expression for Δp from Eqs. 9.1 introduced,

$$F_d \pi D L = f \frac{L}{D} \frac{V^2 \rho}{2} \frac{\pi D^2}{4}$$

Canceling,

$$F_d = \frac{V^2}{8} \rho \quad \text{so} \quad F_{d,V} = \frac{V_V^2}{8} \rho_V \quad \text{and} \quad F_{d,L} = \frac{V_L^2}{8} \rho_L \tag{9.16}$$

If for the time being the area of the vapor/liquid interface is assumed to equal that of the curved boundary of the liquid with the pipe, the balance of drag forces acting on the liquid is

$$f \frac{(V_V - V_L)^2}{8} \rho_V = f \frac{V_L^2}{8} \rho_L \tag{9.17}$$

The relative velocity of the vapor acting on the liquid in Eq. 9.17 is $V_V - V_L$. If the friction factors for the liquid flow and vapor flow are assumed to be equal, Eq. 9.17 can be reduced to:

$$\left(\frac{V_V}{V_L} - 1 \right)^2 = \frac{\rho_L}{\rho_V}$$

which reduces to the velocity ratio using Model 2,

$$\frac{V_V}{V_L} = 1 + \sqrt{\frac{\rho_L}{\rho_V}} \tag{9.18}$$

Example 9.6. In an ammonia liquid recirculation system operating at −30°C (−22°F) with $n = 3$, compute the fraction of the cross-sectional area of the liquid/vapor return line occupied by liquid (a) using Model 1, and (b) using Model 2.

Solution. At −30°C (−22°F) the liquid density ρ_L is 677.5 kg/m^3 (42.3 lb/ft^3) and the vapor density ρ_V is 1.035 kg/m^3 (0.0646 lb/ft^2), so $\rho_L/\rho_V = 655$.

(a) *Model 1.* $V_L = V_V$, so from Eq. 9.14 with $n = 3$, $A_L/A_V = 2/655 = 0.00305$.

Then the fraction of the cross section occupied by liquid, from Eq. 9.15, is

$$FL = \frac{1}{1 + 1/0.00305} = 0.003 \text{ or } 0.3\%$$

(b) *Model 2.* Using Eq. 9.18

$$\frac{V_V}{V_L} = 1 + \sqrt{655} = 25.6$$

which when substituted into Eq. 9.14 gives

$$A_L/A_V = (3 - 1)(25.6)(0.00153) = 0.078$$

and from Eq. 9.15,

$$FL = \frac{1}{1 + 1/0.078} = 0.072 \text{ or } 7.2\%$$

Model 2 in Example 9.6 shows a considerably lower value of V_L than the no-slip assumption in Model 1 resulting in V_L being 1/25th of V_V. Model 2 further indicates that 7.2% of the cross section is occupied by liquid in contrast to 0.3% indicated by Model 1. Model 2 probably overestimates the fraction of area occupied by liquid because of several conditions usually prevailing in the liquid/vapor suction line. One is that the line is often pitched, as recommended in Sec. 9.11, so the vapor flows faster than Model 2 indicates and the volume occupied by the liquid is less. Another factor is the assumption of the flow pattern of Fig. 9.7 where the liquid flows peacefully at the bottom of the tube. When flowing at a high velocity, the vapor may pick up some of the liquid and blow it through the tube as a mist. The liquid volume in this situation will also be less than Model 2 indicates. A purpose served by the two models is that they provide the outer limits of the liquid content.

9.13 SUCTION RISERS IN DIRECT-EXPANSION HALOCARBON SYSTEMS

This section concentrates on the selection of the pipe size of suction risers in halocarbon systems that are designed for direct expansion (not liquid recirculation). When the physical arrangement requires vertical risers at the evaporator outlet, the dominant factor influencing the selection of the riser size is oil return. In halocarbon systems, the oil generally remains in solution in the refrigerant in the evaporator. This is in contrast to ammonia systems where the oil and ammonia separate. As the halocarbon refrigerant progressively vaporizes in flowing through the tubes of the evaporator, the oil concentration increases until the refrigerant leaves as small droplets of liquid with a high oil concentration. The objective is to return this oil through the suction line to the compressor, which can be accomplished by maintaining sufficiently high refrigerant vapor velocities in the riser.

The required vapor velocity to transport oil in a vertical riser is a function of the refrigerant, its density (thus the evaporating temperature and amount of superheat), the viscosity of oil, and the pipe diameter. As the evaporating temperature drops and with it the density of the vapor, a higher velocity must be provided. Higher vapor velocities are also needed as the diameter increases.

Figure 9.9 shows recommended design vapor velocities computed from the equation adapted from Reference 11:

$$\text{Velocity, m/s} = 0.723 \left[\text{g}(D,\ \text{m}) \left(\frac{\rho_L}{\rho_V} - 1 \right) \right]^{1/2} \qquad (9.19)$$

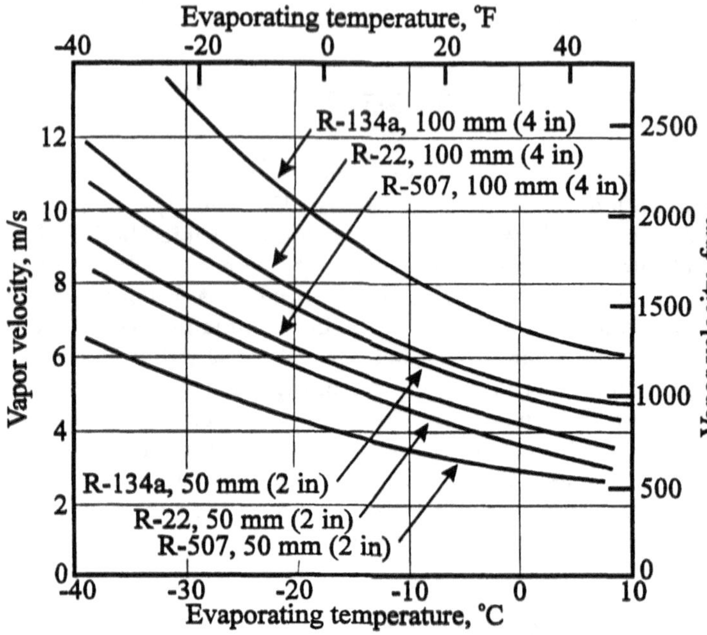

FIGURE 9.9
Recommended vapor velocities for oil return in Type L copper tubing suction risers.

$$\text{Velocity, ft/s} = 0.723\left[g(D,\ \text{ft})\left(\frac{\rho_L}{\rho_V} - 1\right)\right]^{1/2} \tag{9.20}$$

where g = gravitational acceleration = 9.807 m/s² (32.2 ft/s²)
ρ_L = density of oil-refrigerant solution, kg/m³ (lb/ft³)
ρ_V = density of refrigerant vapor, kg/m³ (lb/ft³)

The original research that developed the data shown in Fig. 9.9 was performed on CFCs and HCFCs with mineral oil. Equations 9.19 and 9.20 are assumed to also apply to new HFC refrigerants using synthetic oils, and these curves are also shown in Fig. 9.9.

Suction risers that are sized for full-capacity operation may be too large and unable to return oil at part load. The riser could be sized for the minimum flow rate, which of course, imposes a pressure drop penalty at higher velocities. Another approach is to use a double-suction riser, as illustrated in Fig. 9.10. When the refrigerant flow rate drops, oil will not be lifted into the suction line and collects in the oil trap. The oil blocks refrigerant flow in the right riser, diverting refrigerant to the left riser, which is sized to lift oil, even at the low flow rate. When the high flow rate resumes, the oil in the trap is carried up to the suction line. This quantity of oil may be excessive for the compressor to handle instantaneously, so an oil trap in the suction line is recommended.

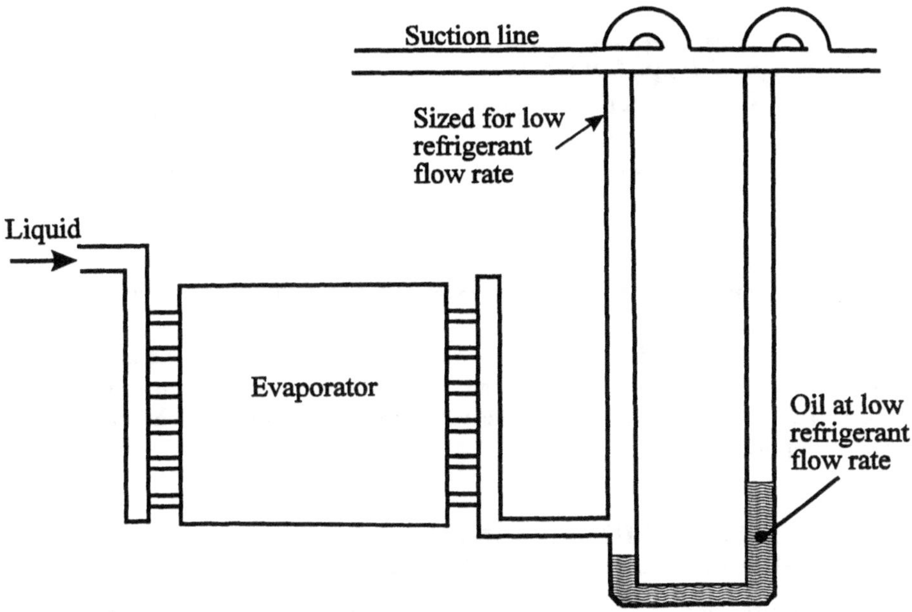

FIGURE 9.10
A double suction riser for oil return which shifts all refrigerant to the left riser during low-flow conditions.

9.14 SIZING THE LIQUID/VAPOR RETURN LINE

Mixtures of liquid and vapor flow through the tubes of condensers and evaporators and in the return line from the evaporator to the low-pressure receiver in liquid recirculation systems. Sizing the pipe and specifying the pressure drop in evaporators and condensers is under the jurisdiction of the component manufacturer, but the system designer is responsible for the liquid/vapor return lines. In horizontal or near-horizontal liquid/vapor return lines, as in Figs. 9.11a and 9.11b, the pressure drop is influenced by the degree of slope of the pipe. A high velocity of vapor will carry the liquid along in a mist. At lower vapor velocities, and/or if the line has sufficient pitch in the direction of flow, the liquid and vapor may separate, with the liquid flowing along the bottom of the tube.

The designer must select the size of and compute the pressure drop in the liquid/vapor suction lines shown in Fig. 9.11. The classic method for computing the pressure drop of two-phase, liquid/vapor flow in horizontal lines is the Lockhart and Martinelli correlation[12]. When using this method, the pressure gradient of each phase is computed as though it flowed separately in the pipe. Then, a graph provided by Lockhart and Martinelli combines those two gradients to predict the two-phase gradient. Most designers of refrigeration systems do not expend the effort to make this calculation. Designers often select the pipe

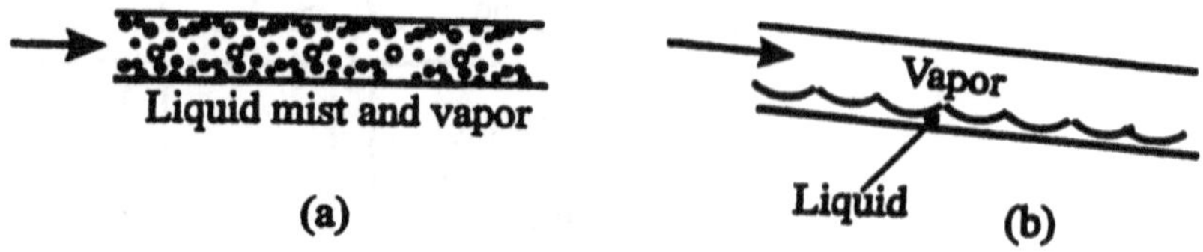

FIGURE 9.11
Liquid/vapor return line that is (a) horizontal and (b) sloped slightly in the direction of flow.

by first determining the size that gives the desired pressure gradient were the vapor alone flowing in the pipe. Then the next larger pipe size is selected with the extra cross-sectional area of the pipe assumed to provide the area for the flow of liquid.

9.15 LIFTING LIQUID FROM THE EVAPORATOR IN VERTICAL RISERS

A special situation is encountered frequently in the return line from a liquid-overfeed evaporator, as shown in Figs. 9.12a and 9.12b. An example of the situation is a spiral freezer chamber where the liquid and vapor leaving the evaporators must be elevated to the liquid/vapor return line. Typically several evaporators are stacked vertically such that their combined height might be 3 m (10 ft) and then the riser extends another 3 m (10 ft) or more to reach the return line. Liquid accumulates in the riser and this column of liquid exerts a static head on the evaporators, raising the pressure and thus the evaporating temperature. Field measurements indicate that evaporators operating in the −30 to −40°C (−22 to −40°F) temperature may lose 15 to 20% of their capacity because of the column of liquid.

It is appropriate at this point to indicate that there is an effective means of avoiding the static head problem by the installation shown in Fig. 9.13. The liquid flowing out of the evaporators separates from the vapor in a vessel. When the liquid level rises to the setpoint in the vessel, the pump turns on and delivers the liquid into the liquid/vapor return line. Now the relatively dry vapor flows up the riser without the impediment of the accumulated liquid. The process of the separate transfer of liquid is not without its own problems. Some plant managers prefer not to have ammonia vessels in their processing area for safety reasons. Also, as was stressed in the discussion of pumps in liquid recirculation systems, Chapter 8, some suction head is required to avoid pump cavitation. The evaporators may be close to floor level, so the vessel and pump must be placed in a pit. Servicing a pump in a pit is not a welcome task for a mechanic. There is an alternative, however, and this is to use gas pumping and thus avoid the, net-positive-suction-head requirement. This choice is often made, but that arrangement has the disadvantage of blocking the flow from the evaporator dur-

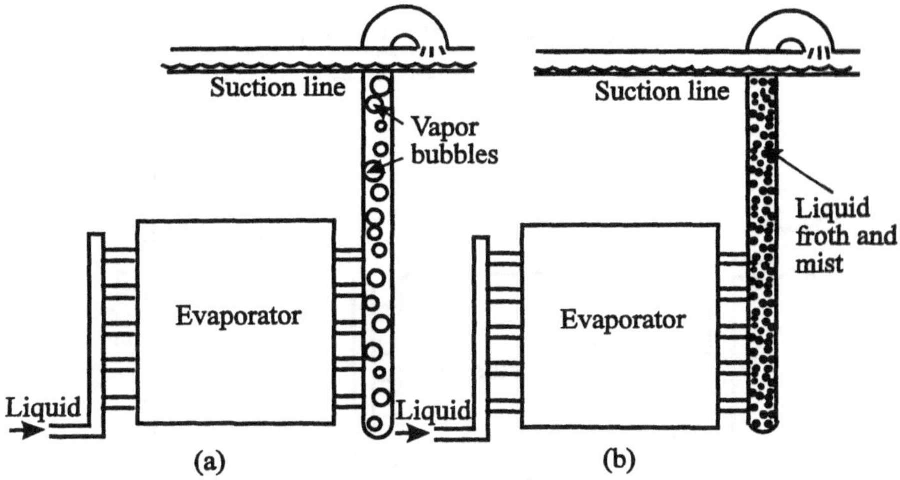

FIGURE 9.12
(a) Low vapor velocity in a liquid/vapor riser resulting in a high liquid fraction and (b) vapor velocity high enough to blow most of the liquid out of the riser.

ing the pumping phase, which causes some system transients. In summary, if the physical situation of the plant permits and the plant manager approves, the liquid head problem can be addressed by the process of liquid removal and separate pumping.

In many situations it becomes necessary for the vapor to carry the liquid to a higher elevation in the riser, and the selection of the size of the riser significantly influences the pressure drop. A point to stress is that lowering the velocity does not necessarily lower the pressure drop. As Fig. 9.12a illustrates, a low velocity may permit liquid to accumulate in the riser and in the extreme case the bubbles of vapor rise due to their own bouyancy. With a riser of length h m (ft), 80% of which is occupied by liquid, for example, the pressure drop would be

$$\text{Pressure drop, Pa} = (0,80)(h,\ \text{m})(\rho_L,\ \text{kg/m}^3)(9.81\ \text{m/s}^2)$$

$$\text{Pressure drop, psi} = (0.80)\frac{(h,\ \text{ft})(\rho_L,\ \text{lb/ft}^3)}{144}$$

Thus, a riser that is 6 m (20 ft) tall could contribute a pressure drop 32 kPa (4.6 psi).

On the other end of the scale, as Fig. 9.14 shows, with a very small riser the pressure drop due to friction is high. Richards[13] translated experimental results into recommendations for riser sizes that will provide the minimum pressure drop. Table 9.6 shows these recommendations for various circulation ratios with ammonia evaporating at −40°C (−40°F). If the size of the riser is selected for full or design flow, then at part load with lower velocities, the vapor will fail to adequately blow out liquid from the riser. For this reason the assembly recommended for conveying oil shown in Fig. 9.10 is often used.

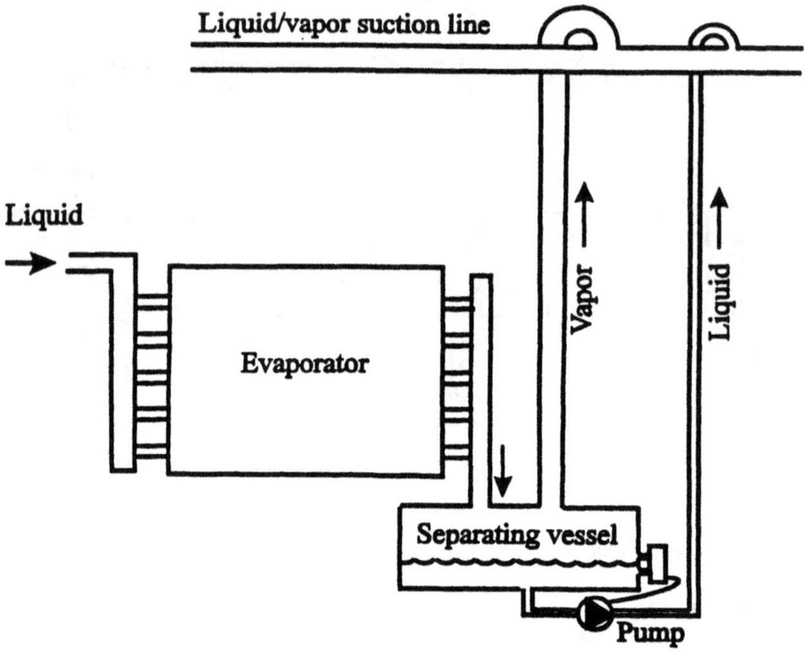

FIGURE 9.13
Lifting liquid from the outlet of the evaporator by first separation, then pumping.

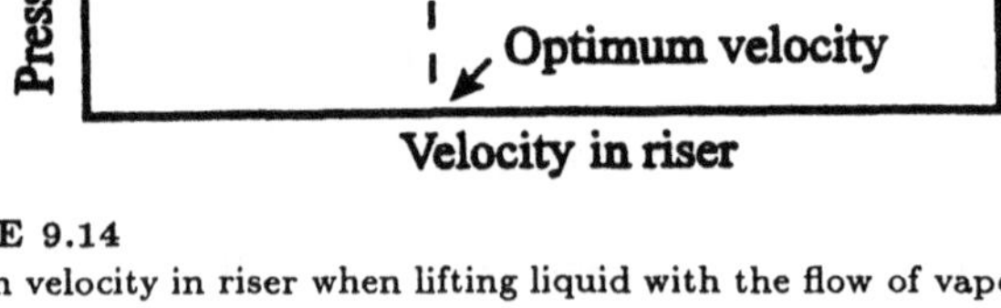

FIGURE 9.14
Optimum velocity in riser when lifting liquid with the flow of vapor.

The next assignment after the pipe size has been determined for the design refrigeration capacity is to compute the pressure drop in the riser. For the case of −40°C (−40°F) ammonia when lifting liquid in a 3-m (10-ft) riser with a recirculation ratio of 5, Richards estimates the drop in saturation temperature to be of the order of 1.6°C (2.9°F) when operating at the vapor velocity for minimum pressure drop.

TABLE 9.6
Evaporator capacities resulting in minimum pressure drops for vertical risers with various pipe sizes and circulation ratios. The refrigerant is ammonia at −40°C (−40°F).

Circulation ratio	Pipe diameter of riser, mm (inch)						
	37 (1.5)	50 (2.0)	63 (2.5)	75 (3.0)	100 (4.0)	125 (5.0)	150 (6.0)
2.5	18.6 kW (5.3 tons)	34.8 (9.9)	53.4 (15.2)	90.0 (25.6)	184.2 (52.4)	324 (92.2)	513 146)
3.15	17.6 kW (5.0 tons)	32.7 (9.3)	50.3 (14.3)	87.9 (25.0)	173 (49.3)	305 (86.7)	482 (137)
4.0	16.9 kW (4.8 tons)	31.3 (8.9)	48.5 (13.8)	84.4 (24.0)	167 (47.4)	293 (83.4)	464 (132)
5.0	16.2 kW (4.6 tons)	30.2 (8.6)	46.8 (13.3)	81.2 (23.1)	160 (45.6)	282 (80.1)	447 (127)

9.16 PIPE SAFETY

Probably more accidents connected with refrigeration systems are associated with the piping system than with any other subsystem. To avoid the chance of damage, particularly damage that would result in a release, care should be exercised in several categories: (1) choice of pipe and fittings, (2) support and anchoring of pipe, (3) location of pipe runs and valves where they are least likely to suffer damage, and (4) cleanliness of the pipe.

- **Choice of pipe and fittings.** Regardless of pressure, the pipe should be Schedule 80 for 1-1/2-in size and smaller. Use Schedule 40 pipe for 2- to 6-in lines and Schedule 20 for 8- to 12-in lines. It may seem illogical to specify the thick Schedule 80 pipe for small diameter, but the reason is not to better contain internal pressure, it is to provide more rigidity should a heavy weight fall on or against the pipe. Joints in pipe of size 1-1/2-in and larger should be welded, not threaded. In fact many contractors weld all joints in 3/4-in pipe and larger.
- **Support and anchoring of pipe.** Pipe hangers should be of the trapeze type and all hanger materials should be galvanized or painted. Hangers should be attached to the building structure or primary supports. Pipe hangers should be placed not more than 3 m (10 ft) apart and should be located not more than 0.7 m (2 ft) from each change of direction. Supports of piping on roofs should be sturdy, painted, or galvanized. The support should rest on a concrete paver or wood sleeper with the roof beneath protected by a membrane material.
- **Location of pipe runs and valves.** When possible, long lines between the machine room and refrigerated space should be run on the roof. When pipes must be inside storage or process areas, they should be in protected locations. Valves should be located in and out of each component of the

same size as the connecting pipe. Hand valves should close against the flow, and the stems should be oriented horizontally. Screwed connections to valves should not be used for sizes larger than 1 in. Horizontal piping shall have only ececctric reducers with the straight sides on the bottom, except at a pump inlet where the straight side should be on top.

- **Cleanliness of pipe.** For installation, the pipe must be clean, new and free of rust, scale, sand, and dirt. Pipe should be stored inside, out of the weather and the ends should be capped. Some contractors install a liberal number of filter/driers which are replaced periodically.

9.17 CHOICE OF MATERIALS FOR REFRIGERANT PIPING

Copper is the material almost universally used for halocarbon piping, with no low-temperature limitations in industrial refrigeration practice. For ammonia systems, since copper is not acceptable, all pipe is steel in one form or another. While aluminum is often used for the tubes of air-cooling coils, it is hardly ever used for liquid and vapor piping in a plant. Cast iron pipe should not be used and wrought iron pipe should not be used for liquid lines.

The most commonly used steel for ammonia systems is carbon steel of type A53 or A106. Type A53 F is butt-welded and should not be used, on the other hand A106 is seamless, so is acceptable. The ANSI/ASME Refrigeration Piping Standard[14] permits the use of A53 or A106 down to temperatures of −29°C (−20°F). One of the steels permitted by the ANSI/ASME standard below this temperature is type A333. Compared to A53 or A106, A333 pipe is approximately three times as expensive, the the cost factor is even higher for valves and fittings. A53 and A106 steels may be used below −29°C (−20°F) provided they pass impact tests at the temperature at which they will be operated. Obviously, impact testing adds to the cost. Another option offered by the standard is to design for the circumferential or longitudinal tensile stress to be 40% of the allowable stress given for the material. Refrigeration interests in recent years have been attempting to get A53 and A106 steels approved for below −29°C (−20°F) operation on the basis that with ammonia the pressure of the refrigerant will be in the range of atmospheric, so that pressure stress on the pipe will be negligible.

At temperatures lower than −46°C (−5°F) many designers uses stainless steel for ammonia and hydrocarbons.

9.18 PIPE INSULATION

The cost of the insulation of low-temperature pipe is usually of the same order of magnitude as that of the pipe itself. Also, since the diameter of the insulated pipe may be twice or three times that of the pipe itself, the designer must provide sufficient space for the complete assembly.

Vapor barriers are essential for low-temperature pipes. In contrast to the water vapor path through the insulation of a refrigerated building which enters the insulation from the warm outdoors and eventually is extracted by the coils in the refrigerated space, there is no outlet for the water vapor that enters pipe insulation. For this reason, the water vapor retardation should be as effective as possible for prolonging the life of the insulation. Some specifications[15] for insulation on low-temperature pipe include.

- Cover the insulation with vapor-retarding jackets with a perm rating of 0.02 or less.
- The supports must not be in contact with the piping.
- The thickness of insulation must be sufficient to maintain the insulation surface temperature above the dewpoint. Isocyanurate or Styrofoam are recommended.
- Since the cladding, insulation, and the pipe all have different coefficients of thermal expansion, double-layer application is suggested for extremely low temperature insulation with staggered joint construction.
- Pack contraction joints with flexible insulation material.

9.19 PIPING IN PERSPECTIVE

One of the objectives of this chapter has been to present methods of installation that result in safe and effective operation of the refrigeration plant. The major thrust, however, has been on procedures to determine the recommended size of pipe, which usually means the minimum recommended size. As a general rule, it is an operating advantage for pipes to be larger than the minimum recommended. The exception to the general rule is that vertical lines from the evaporator should not be oversized; otherwise, oil may not return satisfactorily in a halocarbon system, and liquid refrigerant may accumulate in an ammonia evaporator and raise the evaporating temperature. Certainly there is a price to pay for installing large-size pipe—the increased cost of materials and installation. The payback for these extra costs is a low pressure drop in the pipes that yields modest improvements in efficiency for the life of the plant and increased refrigeration capacity.

All those involved in the planning of a system may be convinced that the capacity specified for the plant is final and will suffice for eternity. Five years later, however, another refrigerated space or process may need to be added, and large-sized pipes can accommodate a certain amount of expansion before approaching their minimum design sizes.

REFERENCES

1. Moody, L. B., "Friction Factors for Pipe Flow," *ASME Transactions*, 66: p 671, 1944.
2. *ASHRAE Handbook, Fundamentals*, American Society of Heating, Refrigerating and Air-Conditioning Engineers, Atlanta, Georgia, 1993.
3. "Transport Properties of SUVA^R Refrigerants," DuPont Chemicals, 1993.
4. Wile, D. D., "Refrigerant Line Sizing," American Society of Heating, Refrigerating, and Air-Conditioning Engineers, Altanta, Georgia, 1977.
5. Wile, D. D., "Refrigerant Line Sizing," *ASHRAE Transactions*, 88(Part I): 117-126, 1982.
6. Timm, M. L., "An Improved Method for Calculating Refrigerant Line Pressure Drops," *ASHRAE Transactions*, 97(Part I): 194-203, 1991.
7. Richards, W. V., "Refrigerant Vapor Line Sizing Not Dependent on Length," Proceedings of Commission B2, 16th International Congress of Refrigeration, International Institute of Refrigeration, pp 240-244, Paris, 1983.
8. "Engineering Data Book," The Hydraulic Institute, Cleveland, Ohio, 1979.
9. "Ammonia Valve Capacities and Sizing," Hansen Technologies, Inc., Burr Ridge, Illinois, 1984.
10. Stoecker, W. F., "Selecting the Size of Pipes Carrying Hot Gas to Defrost Evaporators," *International Journal of Refrigeration*, 7(4): 225-228, July 1984.
11. Jacobs, M. J., F. C. Scheideman, S. M. Kazem, and N. A. Macken, "Oil Transport by Refrigerant Vapor," *ASHRAE Transactions*, 86(Part 2): 318-328, 1976.
12. Lockhart, R. W. and R. C. Martinelli, "Proposed Correlation of Data for Isothermal, Two-Component Flow in Pipes," *Chemical Engineering Progress*, 45: 39-48, 1949.
13. Richards, W. V., "Improved Freezer Operation Using Secondary Circulating Methods," Proceedings of the 13th Annual Meeting of the International Institute of Ammonia Refrigeration, Washington, DC, 1991.
14. ANSI/ASME *Code for Pressure Piping, B31.5-1992*, American Society of Mechanical Engineers, New York, 1992.
15. "National Commercial and Industrial Insulation Standards," Midwest Insulation Contractors Association, Omaha, Nebraska, 1993.

CHAPTER 10

VESSELS

10.1 VESSELS IN INDUSTRIAL REFRIGERATION SYSTEMS

Vessels in industrial refrigeration systems serve either or both of the following functions: (1) storage of liquid, and/or (2) separation of liquid from vapor. The major categories of vessels are:

- high-pressure receivers
- flash tank (or subcooler)/desuperheater
- low-pressure receiver for liquid recirculation
- surge drum on a flooded coil
- suction-line trap or accumulator
- thermosyphon receiver

Vessels are much more common in industrial refrigeration systems than in air-conditioning and there are several reasons for this difference:

Parallel refrigerant circuits. Air-conditioning systems usually are built with a single refrigerant circuit, while industrial systems incorporate parallel compressors, condensers, and evaporators. With a system of multiple components, liquid is likely to move from one condenser or evaporator to another. Also the liquid content in these components varies with time, so a vessel should be available to provide a reservoir for these changes in liquid content.

Liquid recirculation and flooded coils. Liquid along with vapor leaves both liquid overfeed and flooded coils, and this liquid must be separated for return

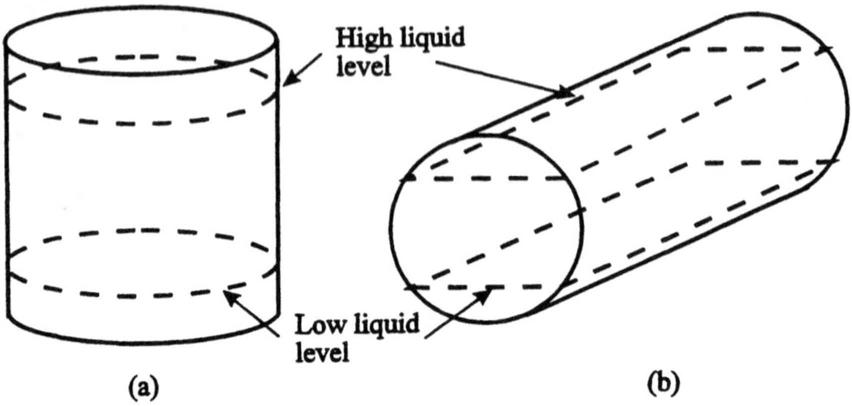

FIGURE 10.1
Useful storage volumes in (a) a vertical vessel and (b) in a horizontal vessel.

to the evaporator. The vapor that passes on to the compressor must be free of liquid, so a vessel performs a process of separating the liquid from the vapor.

Defrosts. Industrial systems often refrigerate air at low temperatures which brings with it the need to periodically defrost evaporator coils. During hot-gas defrosts, refrigerant liquid shifts locations.

Frequent expansions. When air-conditioning systems require additional capacity, it is usually provided by the installation of an additional single-circuit package. The expansion of industrial systems, on the other hand, is usually accomplished by the installation of additional evaporators, compressors, and condensers. Additional refrigerant inventory is required for such enlargements, and generously sized storage vessels facilitate such expansions.

10.2 LEVELS IN LIQUID RESERVOIRS

A fundamental principle in selecting the size of liquid vessels is to choose them large enough that during operation they never become completely full of liquid nor completely empty. As Fig. 10.1 shows, there must always be some vapor space above the highest liquid level to be experienced. A vessel completely filled with liquid may inadvertendly be valved off, and should the temperature of the liquid increase, which would increase the liquid volume, pressures could develop so enormous that the vessel could rupture. Also, the liquid should not be permitted to completely drain from the vessel, because this would result in carrying vapor along with the liquid to the next component. Vapor bubbles adversely affect the performance of control valves and liquid pumps, for example.

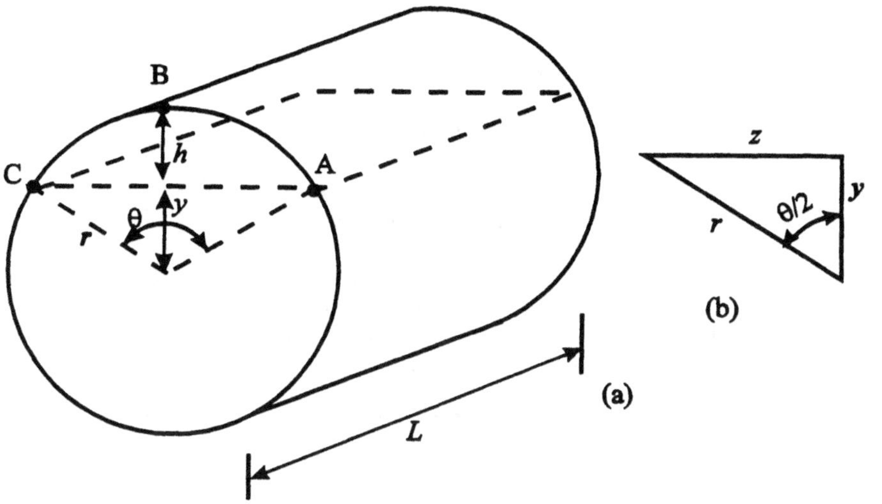

FIGURE 10.2
Volume of liquid in a partially filled horizontal vessel.

10.3 VOLUME OF LIQUID IN A PARTIALLY FILLED HORIZONTAL VESSEL

Several occasions arise where the volume of liquid in a partially-filled vessel must be calculated. One need will be encountered in this chapter in the sizing of low-pressure receivers. Another is when the inventory of refrigerant must be computed, perhaps for regulatory purposes. The volume in a vertical vessel can be computed easily by multiplying the cross-sectional area by the height of the the liquid. For a horizontal vessel it is more complex. In the vessel of Fig. 10.2 which has a radius r, the liquid level is at the plane indicated by AC. A formula for the volume above the liquid level in Fig. 10.2 is:

$$\text{Vapor volume} = \frac{r^2}{2}(\theta - \sin\theta)L \tag{10.1}$$

where θ is in radians $=$ (θ in degrees) $\div$ 57.3

Usually the term sought is the volume of liquid or perhaps the fraction of the vessel volume that is occupied by liquid, Fr_{vol}. It would be convenient to have an expression for Fr_{vol} as a function of the fraction of liquid height, Fr_{ht}. Fr_{vol} is the total volume less the vapor volume divided by the total volume,

$$Fr_{vol} = \frac{\pi r^2 - \frac{r^2}{2}(\theta - \sin\theta)}{\pi r^2} = 1 - \frac{1}{2\pi}(\theta - \sin\theta) \tag{10.2}$$

From the triangle in Fig. 10.2b,

$$\frac{y}{r} = \cos\frac{\theta}{2} \quad \text{so} \quad \theta = 2\cos^{-1}\frac{y}{r} \tag{10.3}$$

and since

$$Fr_{ht} = \frac{y+r}{2r} = \frac{y}{2r} + \frac{1}{2}$$

$$\frac{y}{r} = 2Fr_{ht} - 1 \qquad (10.4)$$

Equation 10.4 for y/r may be substituted into Eq. 10.3 to find θ, which can then be substituted into Eq. 10.2 to calculate Fr_{vol}.

Example 10.1. For a horizontal cylindrical vessel 1.2 m (3.94 ft) in diameter and 3.5 m (11.5 ft) long that is 2/3 full of liquid,
(a) what fraction of the vessel does the liquid occupy, and
(b) what is the surface area of liquid?

Solution. (a) Fraction of vessel occupied by liquid. The vessel is 2/3 full of liquid, so from Eq. 10.4

$$y/r = 2(2/3) - 1 = 1/3.$$

From Eq. 10.3, θ can be computed:

$$\theta = 2\cos^{-1}(1/3) = 2.46 \text{ rad}$$

and finally, from Eq. 10.2

$$Fr_{vol} = 1 - \frac{1}{2\pi}(2.46 - \sin 2.46) = 0.709$$

The volume of liquid in the vessel is

$$\text{Volume of liquid} = (\pi r^2 L)Fr_{vol} = \pi(0.6^2)(3.5)(0.709) = 2.81 \text{ m}^3 = 99.1 \text{ ft}^3$$

(b) The liquid surface area is important in designing vessels that separate liquid and vapor,

$$z = \sqrt{r^2 - y^2} = \sqrt{0.6^2 - 0.2^2} = 0.566 \text{ m } (1.86 \text{ ft})$$

$$\text{Area} = 2zL = 2(0.566 \text{ m})\ (3.5 \text{ m}) = 3.96 \text{ m}^2 = 42.6 \text{ ft}^2$$

To bypass the calculation chore, Table 10.1 lists values of Fr_{vol} as a function of Fr_{ht}.

10.4 LIQUID/VAPOR SEPARATION—VERTICAL VESSEL

Along with providing storage for liquid, a chief function of most vessels used in industrial refrigeration is to separate liquid from vapor to assure that the vapor reaching the compressor is free of significant amounts of liquid. The vessel should be sized for whichever of the two functions, storage or separation, controls. The process of separation that occurs in vertical vessels is different than that in horizontal vessels, so design of the two orientations will be addressed separately.

TABLE 10.1
Liquid volume fractions (Fr_{vol}) in horizontal vessels as a function of the liquid height fraction, Fr_{ht}.

Height F_{ht}	Volume F_{vol}	Height F_{ht}	Volume F_{vol}	Height F_{ht}	Volume F_{vol}	Height F_{ht}	Volume F_{vol}
0.01	0.0017	0.26	0.207	0.51	0.513	0.76	0.815
0.02	0.0048	0.27	0.218	0.52	0.525	0.77	0.826
0.03	0.0087	0.28	0.229	0.53	0.538	0.78	0.837
0.04	0.0134	0.29	0.241	0.54	0.551	0.79	0.847
0.05	0.0187	0.30	0.252	0.55	0.564	0.80	0.858
0.06	0.0245	0.31	0.264	0.56	0.576	0.81	0.868
0.07	0.0308	0.32	0.276	0.57	0.588	0.82	0.878
0.08	0.0375	0.33	0.288	0.58	0.601	0.83	0.887
0.09	0.0446	0.34	0.300	0.59	0.614	0.84	0.897
0.10	0.0520	0.35	0.312	0.60	0.626	0.85	0.906
0.11	0.0599	0.36	0.324	0.61	0.639	0.86	0.915
0.12	0.0680	0.37	0.336	0.62	0.651	0.87	0.924
0.13	0.0764	0.38	0.349	0.63	0.664	0.88	0.932
0.14	0.0851	0.39	0.361	0.64	0.676	0.89	0.940
0.15	0.0941	0.40	0.374	0.65	0.688	0.90	0.948
0.16	0.103	0.41	0.386	0.66	0.700	0.91	0.955
0.17	0.113	0.42	0.399	0.67	0.712	0.92	0.963
0.18	0.122	0.43	0.411	0.68	0.724	0.93	0.969
0.19	0.132	0.44	0.424	0.69	0.736	0.94	0.976
0.20	0.142	0.45	0.436	0.70	0.748	0.95	0.981
0.21	0.153	0.46	0.449	0.71	0.759	0.96	0.987
0.22	0.163	0.47	0.462	0.72	0.771	0.97	0.991
0.23	0.174	0.48	0.474	0.73	0.782	0.98	0.995
0.24	0.184	0.49	0.487	0.74	0.793	0.99	0.998
0.25	0.196	0.50	0.500	0.75	0.805	1.00	1.000

While it is usual to refer to liquid separation as though it is a process of complete removal of liquid, such is not the case. In the vessel there is a spectrum of droplet sizes, and the separation techniques are successsful in removing only the largest droplets. The small droplets are carried out, but some are evaporated in the suction line and others vanish immediately on entering the suction of the compressor. The droplet size is a defining characteristic in the separation principles that will be explained, which immediately raises the question of what is the largest droplet size that should be permitted to escape. This question cannot be answered by analytical means alone and must be decided on the basis of field experience. Thus, if values used for the separation criteria in the design of the vessel result in liquid carryover problems, the criteria must be tightened.

Gravity is the fundamental force used for separating liquid from vapor and retaining the liquid in the vessel. When a drop of liquid falls freely in a motionless

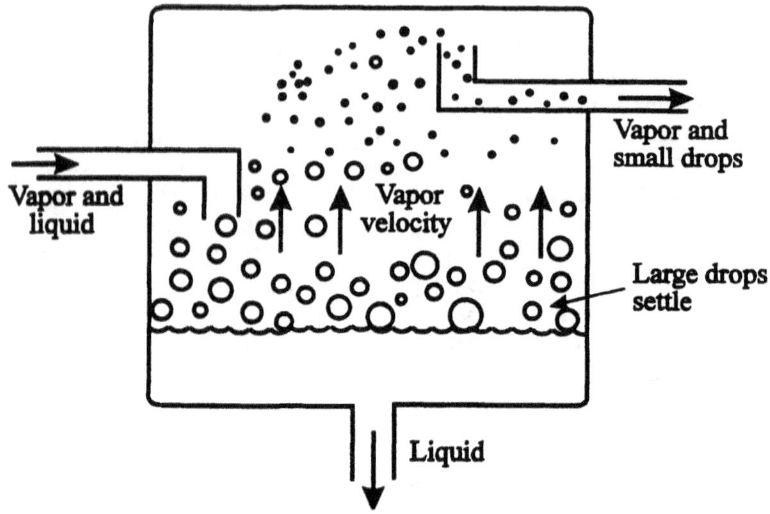

FIGURE 10.3
Liquid separation in a vertical vessel.

vapor, its maximum velocity occurs when the force of gravity just equals the drag force, so no net force is available for acceleration. The upward velocity in the vertical separator of Fig. 10.3 must be low enough that all but the small-diameter drops settle. It is reasonable to expect that separation will not be perfect, but if the only drops carried out are the small ones, the total mass of liquid carried out will be small and the small drops can be vaporized more easily than the large ones.

When a drop of liquid falls freely in a motionless vapor, its maximum velocity is called the terminal velocity. In the separating vessel if the vapor moves upward at the terminal velocity, a drop of the critical diameter remains suspended, the small drops are carried out, and the large ones settle. When a falling drop or particle reaches its terminal velocity, the gravitational force equals the drag force. The gravitational force is:

$$\text{in SI} \quad F_{grav} = (\text{volume})(\rho_f - \rho_v)\text{g} = \frac{\pi d^3}{6}(\rho_f - \rho_v)\text{g} \tag{10.5}$$

$$\text{and in I–P} \quad F_{grav} = \frac{\pi d^3}{6}(\rho_f - \rho_v) \tag{10.6}$$

where F_{grav} = gravitational force, newtons (lb)
ρ_f = density of liquid, kg/m^3 (lb/ft^3)
ρ_v = density of vapor, kg/m^3 (lb/ft^3)
d = diameter of drop, m (ft)
g = gravitational acceleration = 9.81 m/s^2 (32.2 ft/s^2)

TABLE 10.2
Maximum separation velocities in m/s (fpm) for ammonia and R-22 under steady-flow conditions[1].

Refrigerant	Vertical separation distance mm (in)	Saturation temperature °C (°F) −57 (−70)	−40 (−40)	−23 (−10)	−7 (20)	10 (50)
R-22	250 (10)	0.33 (65)	0.21 (41)	0.14 (27)	0.10 (19)	0.07 (13)
	610 (24)	1.36 (267)	0.88 (173)	0.61 (120)	0.44 (86)	0.31 (62)
	910 (36)	1.57 (310)	1.04 (205)	0.72 (141)	0.52 (102)	0.39 (77)
NH_3	250 (10)	0.80 (158)	0.48 (95)	0.31 (61)	0.21 (42)	0.15 (29)
	610 (24)	3.30 (649)	1.99 (392)	1.29 (253)	0.879 (172)	0.64 (125)
	910 (36)	3.54 (697)	2.17 (428)	1.43 (281)	0.99 (195)	0.71 (139)

The drag force is:

$$\text{in SI} \quad F_{drag} = C_D(\text{frontal area of drop})\left(\rho_v \frac{V^2}{2}\right) = C_D \frac{\pi d^2}{4} \rho_v \frac{V^2}{2} \tag{10.7}$$

$$\text{in I–P} \quad F_{drag} = C_D \frac{\pi d^2}{4} \rho_v \frac{V^2}{2g} \tag{10.8}$$

where F_{drag} = drag force, newtons (lb)
C_D = drag coefficient, dimensionless
V = relative velocity of vapor and liquid drop, m/s (ft/s)
g = gravitational acceleration = 9.81 m/s^2 (32.2 ft/s^2)

The expression for the terminal velocity V_t is found by equating the gravitational force of Eq. 10.5 or Eq. 10.6 to the drag force of Eq. 10.7 or Eq. 10.8,

$$V_t = \sqrt{\frac{4gd(\rho_f - \rho_v)}{3\rho_v C_D}} \tag{10.9}$$

The drag coefficient C_D is a function of the Reynolds number Re

$$Re = \frac{V_t d \rho_v}{\mu_v}$$

where μ_v is the viscosity of the vapor. Several specific values of C_D are 30, 4, 1.1, and 0.5 at Reynolds numbers of 1, 10, 100 and 1000, respectively.

For a number of years the *ASHRAE Handbook*[1] has recommended separation velocities, and these velocities are shown for ammonia and R-22 in Table 10.2. The table shows another dimension, which is the separation distance between the liquid level and the vapor outlet. As expected, if there is only a short separating distance, the permitted velocity is lower.

Reverse engineering can be applied to the separating velocities of Table 10.2 in order to determine critical drop diameters that correspond to the recommendations. Figure 10.4 shows the result of that process with separating distances

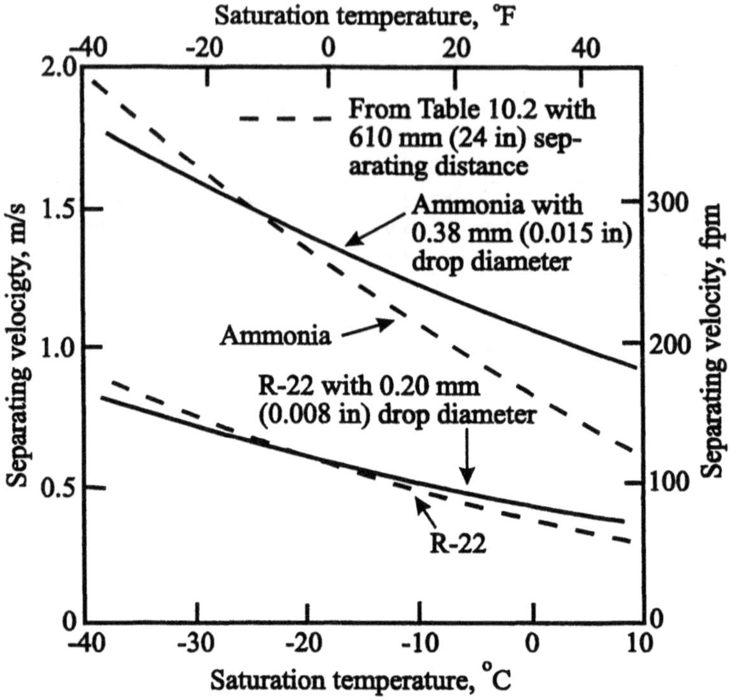

FIGURE 10.4
Comparison of recommended separating velocities by two methods.

of 610 mm (24 in) chosen in Table 10.2. The slopes of the separating velocity curves are not identical using the two methods, but an ammonia drop diameter of 0.38 mm (0.015 in) and an R-22 drop diameter of 0.20 mm (0.008 in) yield magnitudes in the same order of magnitude. A consistent trend is that the maximum velocity permitted decreases as the saturation temperatures increase. The reason is that as the saturation temperature increases, so does the density of vapor which, in accordance with Eq. 10.9, results in a reduction of the terminal velocity, V_t.

Examination of Fig. 10.4 raises the question of why the drop diameter selected for ammonia is so much larger than that chosen for R-22. The real criterion is whether liquid carryover problems result when the drops of diameters smaller than a certain value escape the vessel. The ability of a drop to vaporize in the suction line and in the entrance sections of the compressor is a deciding factor, so is ammonia more readily vaporized than R-22? The liquid density of ammonia is half that of R-22, so half the mass of liquid ammonia exists in a drop of a given size. The thermal conductivity of liquid ammonia is five times that of liquid R-22, so heat can flow more easily in ammonia. But on the other hand, the latent heat of ammonia is five times that of R-22, and all of this additional heat must pass through the convection coefficient at the surface of the drop, and this coefficient should be about the same for ammonia and R-22.

The comparison of Fig. 10.4 raises a question whether the recommended separating velocities from ammonia are too high. There are occasional reports from the field of liquid carryover of ammonia when the values from Table 10.2 are used, and most designers use velocities approximately 0.5 to 0.8 m/s (100 to 150 fpm). Providing the desired separating velocity is one influence on the vessel size and the other, which will be addressed in Sec. 10.8, is to provide adequate volume for storage of liquid. In most cases the size of low-pressure receiver that accommodates the liquid storage will also result in low vertical velocities.

A question might be raised about the terminal velocity method in that it does not acknowledge separating distances as does Table 10.2. Certainly a minimum separating distance must be provided, and 450 to 600 mm (18 to 24 in) is recommended. The feature of the separating distances emerges because the original technical paper[2] on which the values of Table 10.2 are based introduced principles applicable to the design of fractionating columns[3]. One factor included is the surface tension of the liquid, which influences the ability of vapor to tear itself away from liquid. This process occurs in the shell-side of an evaporator and probably too when vapor is bubbled through liquid in a desuperheater. It is not known how much the surface tension influences the separation of liquid and vapor returning from overfeed coils.

Lorentzen[4] has influenced European practice, basically using the terminal velocity technique and recommending separating velocities between 0.5 m/s and 1 m/s (100 to 200 fpm). The higher of the recommended values corresponds to an ammonia droplet diameter of 0.25 mm (0.010 in) at a temperature of −40°C (−40°F). At 0°C (32°F) with a drop diameter of 0.25 mm (0.010 in), the recommended velocity is 0.3 m/s (60 fpm). Wiencke[5] supports the recommendations of Grassmann and Reinhart[6] that drop diameters of 0.2 mm (0.008 in) for both ammonia and the halocarbons should not be exceeded. Figure 10.5 shows separating velocities for several drop sizes of ammonia and for R-22, R-134a, and R-507 with drop diameters of 0.20 mm (0.008 in). Before using high separating velocities for ammonia, the designer might want to be assured that unique features of the facility will mitigate against liquid carryover problems.

10.5 LIQUID/VAPOR SEPARATION—HORIZONTAL VESSEL

Both horizontal and vertical vessels are widely used, and it is often the physical characteristics of the machine room that has a major influence on the choice of orientation. Many designers prefer a vertical vessel for a mechanically pumped recirculation system, because the net-positive-suction head is more easily achieved. The vertical vessel usually requires less floor space than the horizontal, but does require greater vertical space in the machine room. For recirculation packages that incorporate the vessel and the pumps on a skid, the vessels are usually oriented horizontally to meet head room limitations on shipping.

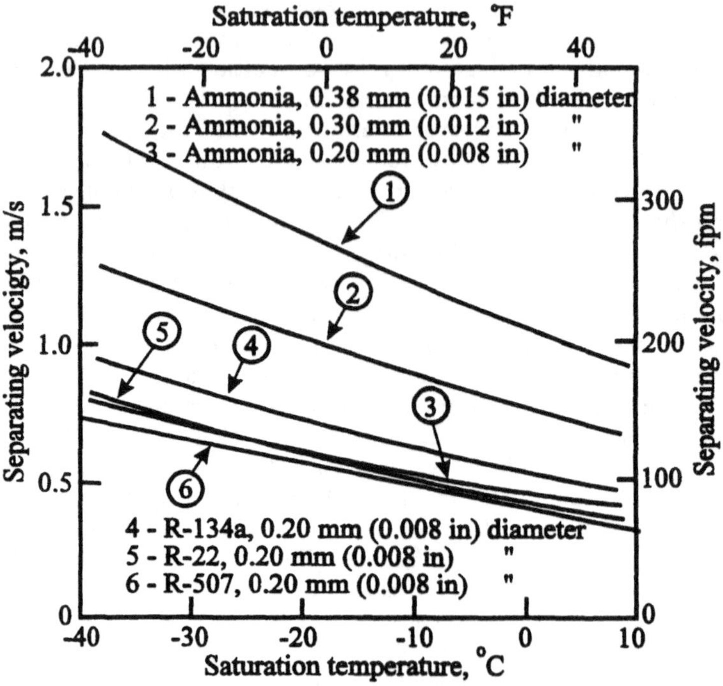

FIGURE 10.5
Recommended maximum liquid/vapor separating velocities for various refrigerants and the drop diameters on which the recommendations are based.

The mechanism for separation of liquid from vapor in horizontal vessels is somewhat different than the process in a vertical vessel, but some of the same principles apply. In the horizontal vessel of Fig. 10.6, the horizontally flowing vapor carries liquid drops while at the same time the drops have a vertical component of velocity because of the gravitational force. If the drops are initially assumed to have no horizontal or vertical velocity when they enter the vessel, the drops accelerate horizontally due to drag force of the vapor on the drops while they accelerate vertically due to gravity. If a drop descends to the liquid surface before being carried out, it will be captured.

A key term is thus the time T for the drop to fall a distance y and settle to the liquid level. This time must not exceed the time to traverse the separating length L in Fig. 10.7. Thus arises the concept of T as the *minimum residence time*. Furthermore,

$$T = \frac{L}{V} = \frac{LA}{\text{volume flow rate}} \tag{10.10}$$

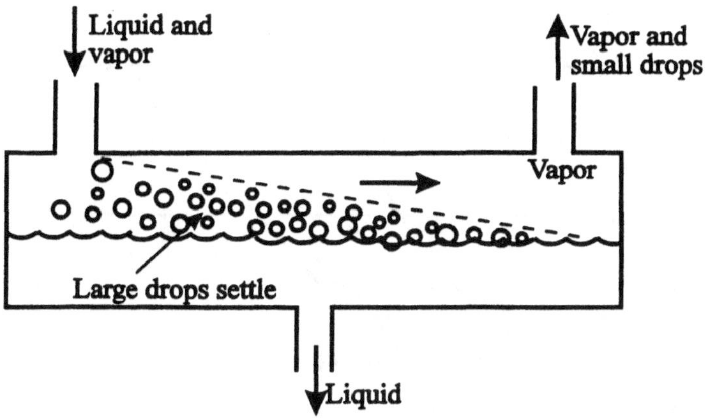

FIGURE 10.6
Separation of liquid from vapor in a horizontal vessel.

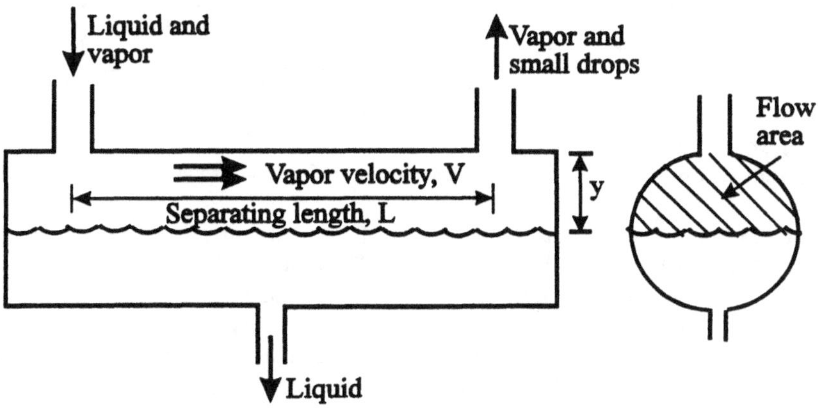

FIGURE 10.7
Horizontal velocity and separating length which combine to yield the concept of residence time for a horizontal vessel.

where T = minimum residence time, s
L = separating length, m (ft)
V = horizontal vapor velocity, m/s (ft/s)
A = flow area, m^2 (ft^2)

The liquid level carried in the horizontal vessel has a significant impact on the ability to separate liquid from vapor, because as the level rises the flow area decreases, which reduces the actual residence time. For this reason, many designers try to avoid operation with liquid levels much above the midheight of the vessel. The vessel should be designed, then, so that drops of larger than the critical diameter fall a distance y, in Fig. 10.7, before the vapor with its velocity V travels the distance L. The time required for a drop to settle due to the force

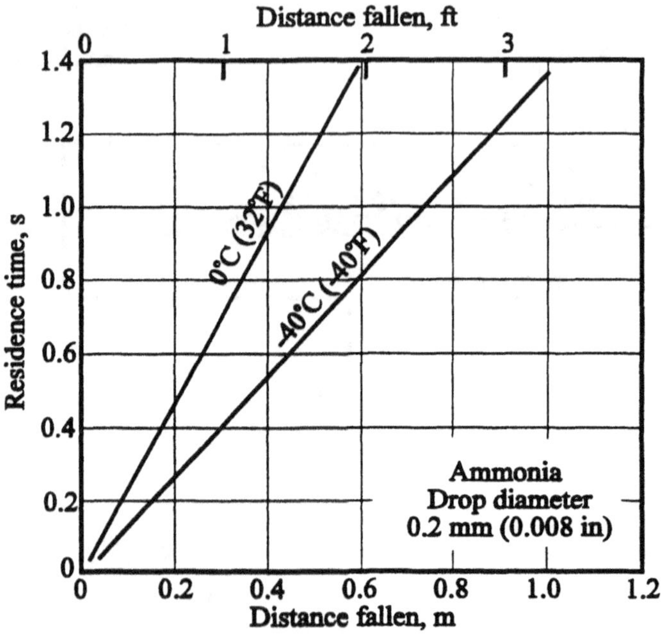

FIGURE 10.8
Residence times with ammonia for various values of y in horizonal vessels.

of gravity for ammonia, R-22 and R-507 at saturation temperatures of −40°C (−40°F) and 0°C (32°F) are shown in Figs. 10.8, 10.9, and 10.10, respectively.

The calculations assume that the drop starts its descent with zero vertical velocity, but when the liquid/vapor enters vertically from the top, as in Fig. 10.7, there will be an initial velocity which makes the recommendations in Figs. 10.8 to 10.10 conservative. The lines in the figures are essentially straight, because the terminal velocity is reached very quickly—usually within 0.1 s. The values are of the same order of magnitude quoted by Richards,[7] who suggests for ammonia 0.7 s at −1°C (30°F) and 0.5 s at −18°C (0°F). The critical drop diameter for the three refrigerants has been chosen as 0.2 mm (0.008 in) and had a drop diameter of 0.3 mm (0.012 in) been selected, the required residence times would have been approximately 30% to 40% shorter.

Once the residence time is determined, the area A in Eq. 10.10 can be computed for the prevailing volume flow rate. Table 10.1 facilitates the area calculation.

10.6 HIGH-PRESSURE RECEIVER

The high-pressure receivers in small plants that operate on a seasonal schedule are typically sized to contain all the refrigerant existing in the system. During pump-down operation, the king valve (Fig. 10.11) in the liquid line from the

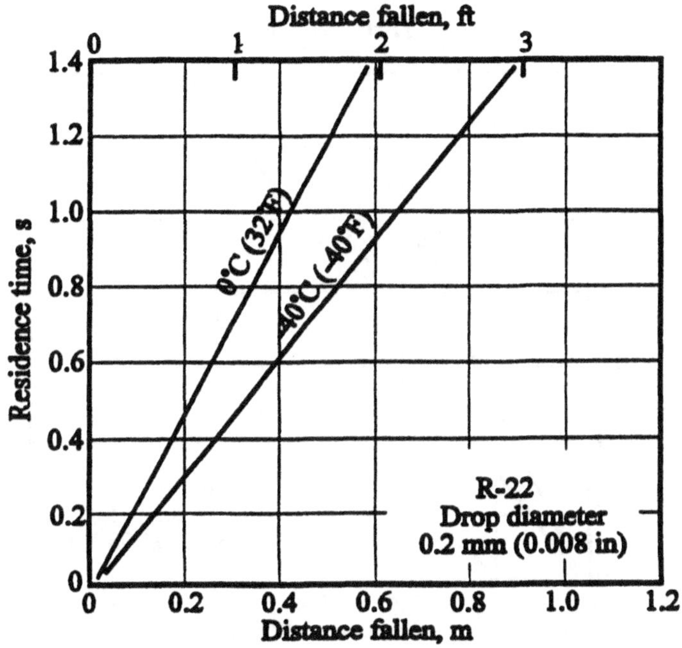

FIGURE 10.9
Residence times with R-22 for various values of y in horizonal vessels.

receiver is closed and the refrigerant delivered by the compressor(s) condenses and drains into the receiver. When all liquid has been withdrawn from the low side, the valves between the condenser(s) and receiver are closed to confine all liquid refrigerant in the receiver. In these plants, the receiver must accommodate all the liquid of the system in the storage volume between the high and low liquid levels, as shown in Fig. 10.11. Some vapor volume is always required above the highest liquid level, and a reserve of liquid should always prevail in the receiver, even when the remainder of the system is fully supplied. The piping of the receiver illustrated in Fig. 10.11 is top inlet, and the other concept in piping, as was shown in Fig. 7.27, is the bottom inlet.

Many large plants operate all year, and receivers for these plants are never expected to contain all the liquid in the system. Two of several bases used by designers to select the size of receivers are:

- storage volume to pump down the largest refrigerated room or unit served by the plant
- store full refrigerant flow for a specified duration of time, 30 minutes, for example.

The rationale behind the first basis is that any of the refrigerated spaces, including the largest, may need to be taken out of service. The second basis

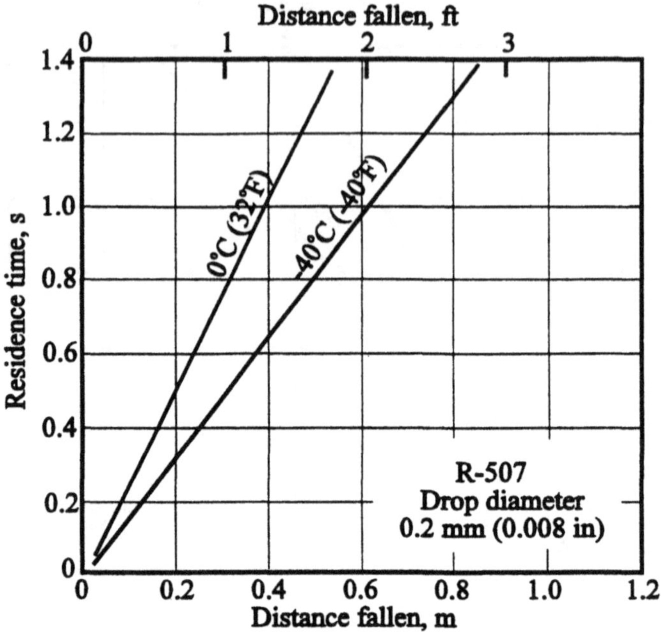

FIGURE 10.10
Residence times with R-507 for various values of y in horizonal vessels.

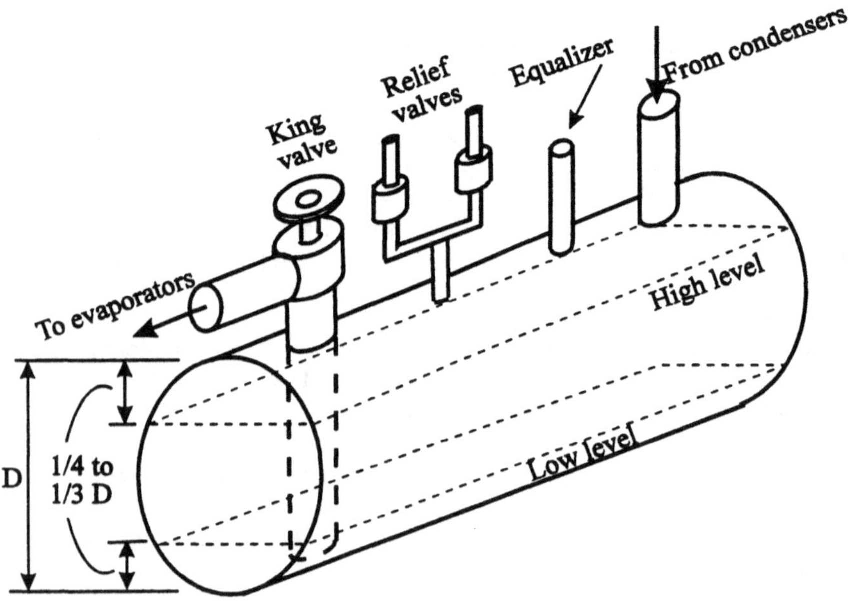

FIGURE 10.11
Liquid storage volume between the high and low levels in a high-pressure receiver.

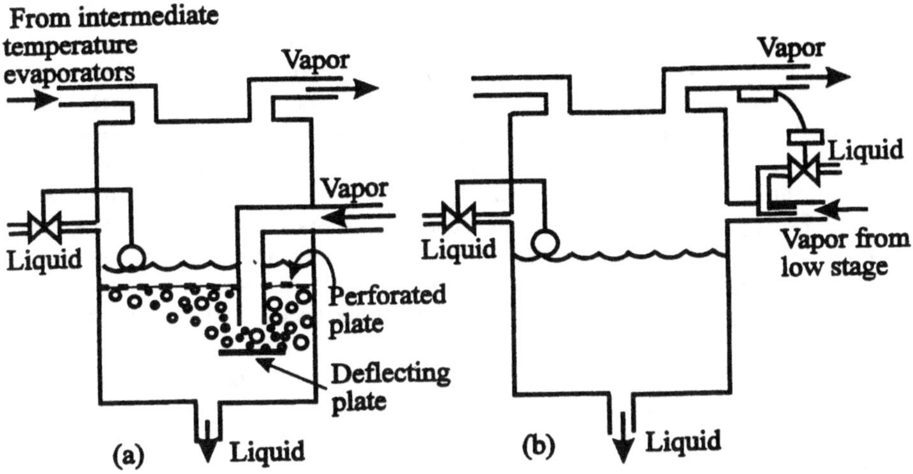

FIGURE 10.12
A flash-tank/desuperheater with (a) desuperheating by bubbling vapor through liquid and (b) desuperheating by spraying liquid into vapor.

is predicated on being able to interrupt the liquid supply to the plant for a short period of time, meanwhile continuing compressor operation. Standard shell diameters in North America are 0.51 m (20 in); 0.56 m (22 in); 0.61 m (24 in); 0.76 m (30 in); ... 1.22 m (0.48 in); 1.37 m (54 in), etc. The cost of vessels larger in diameter than 1.52 m (60 in) increases abruptly.

10.7 FLASH-TANK/DESUPERHEATERS

The flash-tank/desuperheater is the vessel in a two-stage system operating at the intermediate pressure that provides removal of flash gas in direct cooling of liquid (Fig. 3.2) or subcooling liquid with a heat exchanger (Fig. 3.5). The other purpose of the vessel is to desuperheat discharge vapor from the low-stage compressor by bubbling it through the liquid in the vessel (Fig. 3.9). Section 10.1 indicated that vessels in industrial refrigeration systems have either or both the function of liquid storage and/or separation of liquid from vapor. Usually the flash-tank/desuperheater must be sized only for separation of liquid and vapor. The exception to that limited role is when other streams in addition to the liquid supply and discharge vapor from the low-stage compressor flow into the vessel. If there are intermediate-temperature evaporators discharging into the vessel, for example, surge volume should be provided to allow for the rush of liquid coming from the evaporators during defrost.

Desuperheating of vapor, which is a unique function of this vessel, is not required in other vessels of the system. One method of desuperheating is to bubble the vapor through the liquid, a process that is effective, but requires two simultaneous functions that are somewhat conflicting. One process is agitation for good desuperheating and the other is an orderly flow pattern for good liquid/vapor separation. Some vessel designers choose a configuration like that shown in Fig. 10.12a which injects the vapor below a perforated plate to spread out the vapor bubbles, thus providing more intimate contact of vapor and liquid and avoiding a geyser which would make the liquid and vapor separation more difficult. Still another feature used by some designers is the installation of a plate against which the vapor is directed. The purpose of this plate is to prevent the vapor bubbles from blowing to the liquid outlet and being drawn out of the vessel. This precaution would be especially appropriate if the vessel is part of a liquid recirculation systems where vapor bubbles in the liquid could vapor- bind the pump.

The other concept of desuperheating, as shown in Fig. 10.12b, is to spray liquid into the incoming vapor. The diagram shows a control valve regulating the liquid flow, but, as was discussed in Chap. 9, the bypass from a centrifugal pump may serve as the liquid source.

10.8 LOW-PRESSURE RECEIVER

This vessel is one that performs both the role of liquid/vapor separation and liquid storage. The dashed lines in Fig. 10.13 show five distinct liquid levels of interest. The level controller, whether it is a capacitance level sensor as shown in Fig. 10.13, or a float switch, regulates the solenoid valve in the liquid line. If the level in the vessel drops below the control point, the solenoid valve in the liquid supply line opens. When the level in the vessel reaches the control point, the solenoid valve closes. Volumes above and below this controlled level accommodate *surge volume* and provide *ballast volume*. The surge volume serves the purpose of accommodating liquid that might be forced out of evaporators during defrost. Another source of liquid surge in some plants is from the liquid/vapor line when pitched downward to the low-pressure receiver. Should the electric power in the plant be interrupted, the liquid continues to drain from the liquid/vapor line.

The controlled level is not the lowest operating level, because some liquid supply should be available between the controlled level and level that actuates the low-level alarm. The reason for needing this ballast volume is that during startup or resumption of operation of one or more evaporators, the pump may withdraw refrigerant from the vessel at a greater rate than is supplied at that moment by the combination of the controlled liquid supply and from the return from the liquid/vapor line. If the vessel empties to the low-level alarm status, operation continues but an operator or the security system is informed of the fact. On a further drop in level the low-level cutout is reached, which is usually set to stop pump operation.

There are also two designated control levels above the controlled level: the high-level alarm and the high-level cutout. The alarm simply notifies the operator or the security system to summon an on-site investigation. Should the liquid level reach the point of the high-level cutout, the compressors are shut down

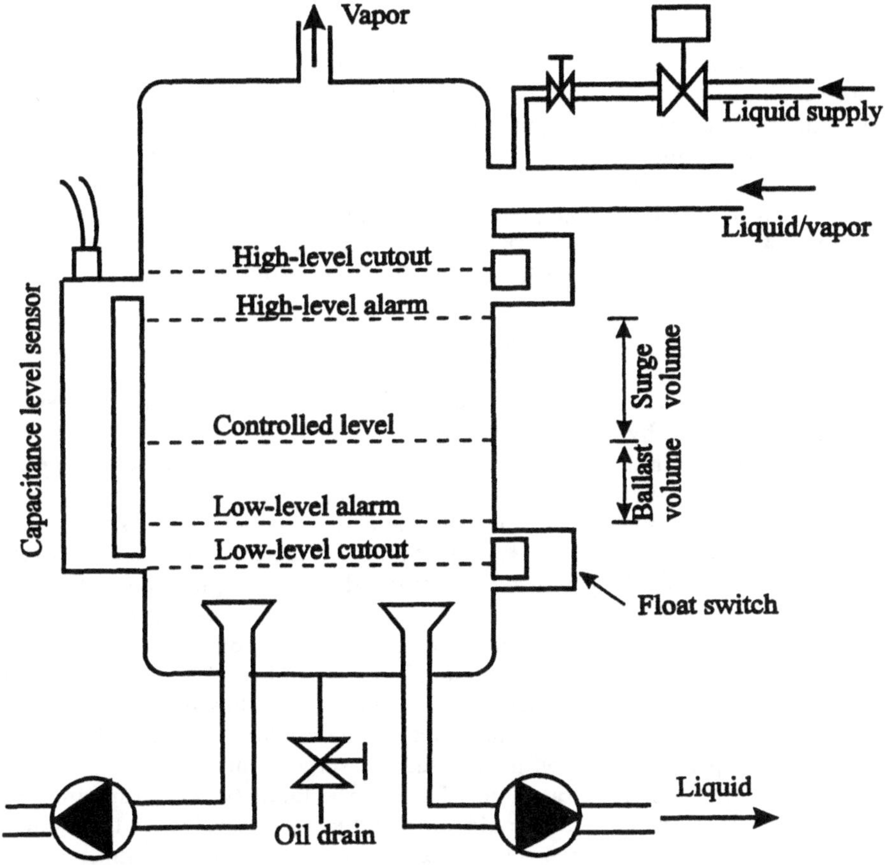

FIGURE 10.13
Liquid levels maintained in a low-pressure receiver.

for their protection. Figure 10.13 shows for controllers a popular combination of the capacitance level controller and float switches. Several electrical current values can be picked off the 4-20 mA output of the capacitance level sensor that correspond to the controlled level, the high-level alarm, and the low-level alarm. To provide an independent backup, the ultimate safety provisions of the high- and low-level cutouts are actuated by separate float switches.

Computing the ballast volume. The ballast volume is provided to permit the pumps to draw liquid from the low-pressure receiver for a short interval to bring the liquid content of evaporators up to the steady-state amount following a shutdown. Typically, a five-minute time period is assumed adequate for this purpose, so the ballast volume is the design pump flow rate in volume flow per minute multiplied by 5.

Computing the surge volume. Two major contributors to momentary excess flow into the low-pressure receiver are flooding of abnormal rates of liquid out of the evaporator due to defrost or to sudden increases of refrigeration load, and the liquid in the liquid/vapor return line that drains back to the low-pressure receiver in the event of pump or power failures. During a defrost it is assumed that the entering defrost gas pushes all liquid in the coil out to the return line. The fraction of liquid in the coil during operation depends on whether the coil is top or bottom fed. For a top-fed coil the percentage of liquid is often assumed to be about 30%. For bottom-fed coils the percentage is sometimes chosen as high as 80%. Another approach to the estimate[5] that incorporates the circulation ratio n but does not distinguish between top and bottom feed is

$$\text{Liquid volume percent} = 1 - \frac{1}{1.2n^{0.2}} \tag{10.11}$$

Equation 10.11 gives with a circulation ratio of 3, for example, the percentage of volume occupied by liquid to be 33%.

The premise that the liquid in a coil is pushed out during a hot-gas defrost and sent to the low-pressure receiver is now being questioned by some engineers. In times past, it was certainly true that the defrost gas was sent to the coil immediately after interrupting the supply of liquid to the coil, and the liquid would flow rapidly through the pressure-regulating valve or through a float drainer. As was emphasized in the treatment of hot-gas defrost in Chapter 6, the coil should first be emptied of liquid by closing the liquid supply solenoid but continuing refrigeration until all or most of the liquid in the coil has been drawn off by the compressor. The liquid that was contained in the coil thus never leaves as liquid. Only experience will determine whether the modern recommended defrost sequence will permit reduction in the required surge volume.

To estimate the fraction of liquid FL in the liquid/vapor return line, Sec. 9.12 presented relations that should bracket the correct value. If the liquid is assumed to move at the same velocity as the vapor, thus with no slip,

$$FL = \frac{(n-1)\rho_v}{(n-1)\rho_v + \rho_L} \tag{10.12}$$

On the other hand, the other extreme described in Sec. 9.12 is if the liquid flows along the bottom of a horizontal pipe, dragged by the faster-moving vapor. The expression is:

$$FL = \frac{(n-1)R}{(n-1)R+1} \tag{10.13}$$

$$\text{where } R = \frac{\rho_v}{\rho_L} + \sqrt{\frac{\rho_v}{\rho_L}}$$

Equation 10.12 underestimates the fraction of liquid occupying the pipe, because slip will always occur. Equation 10.13 overestimates the value, if the vapor velocity is high enough to generate mist flow rather than stratified flow, and also if the pipe is sloped in the direction of flow.

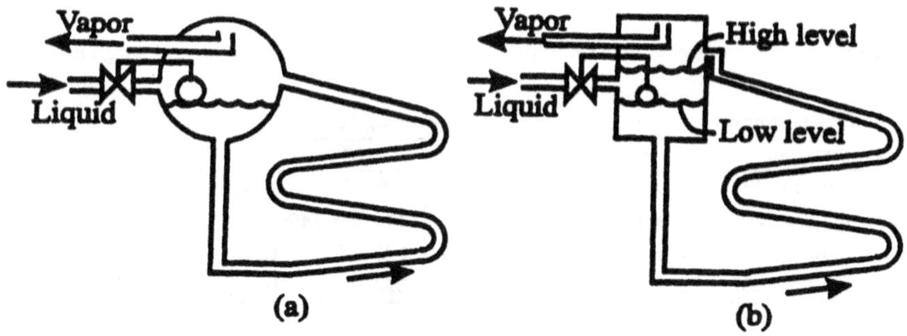

FIGURE 10.14
Surge drums for flooded coils, (a) oriented horizontally, and (b) oriented vertically.

10.9 SURGE DRUM FOR A FLOODED EVAPORATOR

Flooded evaporators, first described in Sec. 6.7, are equipped with a surge drum at a level slightly above the evaporator coils. Either a horizontal or vertical surge drum is possible, as shown in Fig. 10.14, with the choice usually dependent upon the space restrictions. The surge drum serves the purpose of liquid storage and liquid/vapor separation. A need for surge volume arises when the evaporator is subjected to a sudden heavy heat load, in which case the rate of boiling abruptly increases and the vapor thus developed pushes liquid out of the evaporator. In the case of air coils during hot-gas defrost the surge drum provides storage during the transient conditions.

The principles for disengaging vapor from liquid are the same ones applicable to low-pressure receivers, as discussed in Secs. 10.4 and 10.5. Usually the manufacturer of the coil provides the surge drum as part of the package and thereby makes the decision on the dimensions of the surge drum. As a suggestion of the order of magnitude of the volume of the surge drum, one recommendation[8] for finned evaporator coils is that horizontal surge drums have a free volume as great as the internal volume of the evaporator coils. The free volume is defined as the vapor space from the controlled level to the level at which liquid would carry out of the vessel.

One manufacturer[9] offers guidelines for dimensions of horizontal surge drums, such as shown in Fig. 10.14a, as expressed in Table 10.3. For vertical surge drums, as shown in Fig. 10.14b, the permitted refrigeration capacity of the coil is about twice that shown in Table 10.3 for a given vessel size and evaporating temperature.

TABLE 10.3
Appropriate refrigerating capacities of flooded coils corresponding to several surge drum sizes.

Dimension of surge drum	Evaporating temperature, °C (°F) −40°C (−40°F)	−20°C (−4°F)	0°C (32°F)
0.35 x 1.07 m (14" x 42")	42 kW (12 tons)	67 kW (19 tons)	105 kW (30 tons)
0.51 x 1.22 m (20" x 48")	88 kW (25 tons)	140 kW (40 tons)	176 kW (50 tons)
0.61 x 1.52 m (24" x 60")	105 kW (30 tons)	183 kW (52 tons)	246 kW (70 tons)
0.76 x 1.52 m (30" x 60")	158 kW (45 tons)	323 kW (92 tons)	404 kW (115 tons)

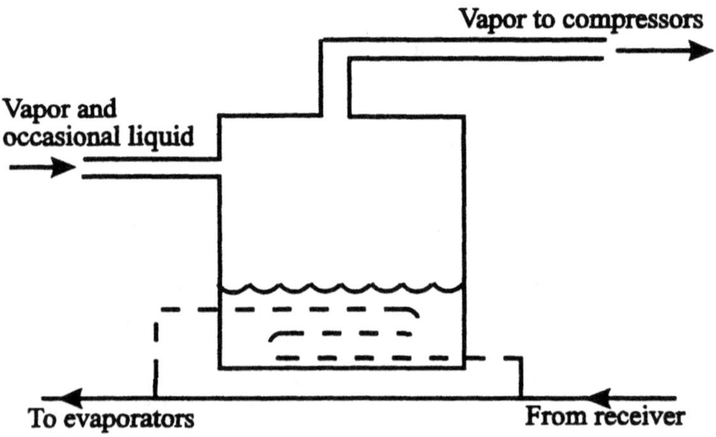

FIGURE 10.15
A suction-line accumulator with the option of passing warm liquid through the accumulator to vaporize any trapped liquid from the suction line.

10.10 SUCTION-LINE ACCUMULATOR

The vessel shown in Fig. 10.15 should normally not be necessary, and in newly designed and constructed systems it usually is not found. In a two-stage liquid recirculation system the low-pressure receiver should be designed to prevent liquid from reaching the low-stage compressors, and the flash-tank/intercooler should prevent liquid from reaching the high-stage compressors.

The operators of many plants that employ a suction-line accumulator, however, are comforted by having one. In plants that through the years have been expanded, modified, and restructured there is sometimes liquid carryover and the suction-line accumulator is the final protector of the compressor from slugging with liquid. Some of the reasons that liquid carryover occurs are because of improper design of some other separating vessel, or occasional liquid carryover from a leaking expansion valve on a direct-expansion coil or the imposition of a sudden heat load on a flooded coil. The installation of a suction-line accumulator in a plant that is carefully designed could permit smaller surge drums on all the flooded coils and the occasional carryover of liquid could be accommodated by the suction-line accumulator.

Figure 10.15 shows the option of directing warm liquid from the high-pressure receiver to a coil in the bottom of the accumulator. When liquid does collect in the accumulator the vaporization by heat from the machine room is a slow process, and the supply of heat from the warm high-pressure liquid accelerates vaporization and recovers a small amount of refrigeration as well. The warm liquid coil in the accumulator will have the undesirable effect of superheating the suction vapor when no liquid is present in the accumulator. Some plants equip the accumulator with a transfer system employing mechanical or gas pumping, which is initiated by a rise in the liquid level in the accumulator. These arrangements deliver the liquid to the high-pressure receiver.

10.11 THERMOSYPHON RECEIVER

A small but important vessel is the thermosyphon receiver which is an integral part of that type of oil cooling concept for screw compressors. The procedure for selecting the size of the thermosyphon receiver is explained in Sec. 5.14 and the key requirement is that it provides a reserve for five minutes of flow to the oil cooler if the supply of liquid from the condenser is interrupted. Separation of liquid and vapor indeed also takes place in the thermosyphon receiver, but the requirements are not stringent. Some liquid mist could pass through the vent line to the condenser inlet without an adverse effect, but it should be emphasized that the return of liquid and vapor from the oil cooler should not be delivered to the condenser inlet, because such a large fraction of liquid would degrade the condenser performance.

Some designers and contractors combine the thermosyphon receiver and the system receiver as shown in Fig. 10.16. Rather than transferring the liquid from a separate thermosyphon receiver to a separate system condenser, the liquid simply spills over from the oil cooler reserve into the system receiver. When multiple condensers feed the receivers, the individual condensate drain lines are trapped as was described in Chapter 7.

10.12 OIL POTS

A small vessel frequently installed beneath the low-pressure receiver and other larger vessels is an oil pot whose purpose is to accumulate oil to facilitate periodic draining. Figure 10.17 shows oil pots with their typical connections, including the line from the bottom of the low-pressure receiver, the drain line, and the equalizer line, which is connected to the vapor section in the upper portion of the low-pressure receiver. Without the equalizer line it may be difficult to achieve drainage into the oil pot because of the vapor pressure that could build up. In addition, since the oil pot is usually classified as a vessel, it must be protected by relief valves.

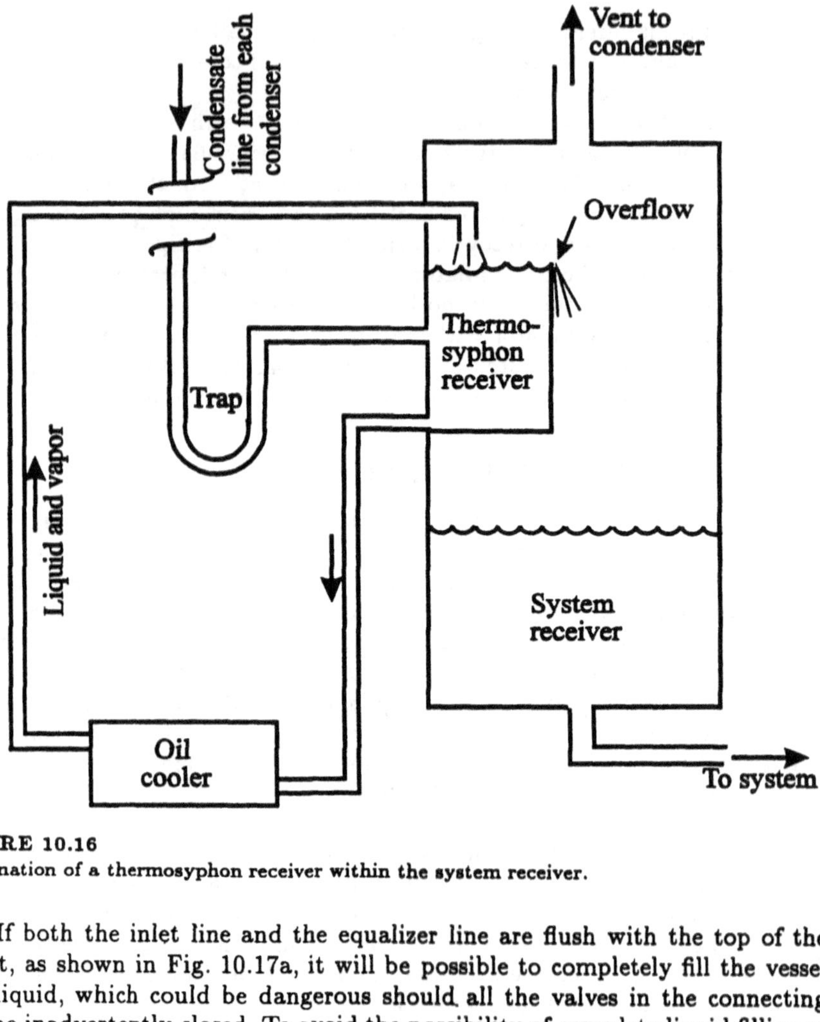

FIGURE 10.16
Combination of a thermosyphon receiver within the system receiver.

If both the inlet line and the equalizer line are flush with the top of the oil pot, as shown in Fig. 10.17a, it will be possible to completely fill the vessel with liquid, which could be dangerous should all the valves in the connecting lines be inadvertently closed. To avoid the possibility of complete liquid filling of the vessel, the equalizer line can be extended into the oil pot a short distance. When the level of oil (and refrigerant) rises in the oil pot to the bottom of the equalizer line, liquid is forced up the equalizer line but the vapor space at the top of the vessel is preserved.

10.13 SEPARATION ENHANCERS

Certain refinements of the basic inlets and outlets to the vessel are usually used to facilitate the separation of liquid and vapor. Any of the entrance lines that are carrying liquid are usually turned downward. The vapor line leaving for the compressor usually has an upward turn, as shown in Fig. 10.18a, which requires

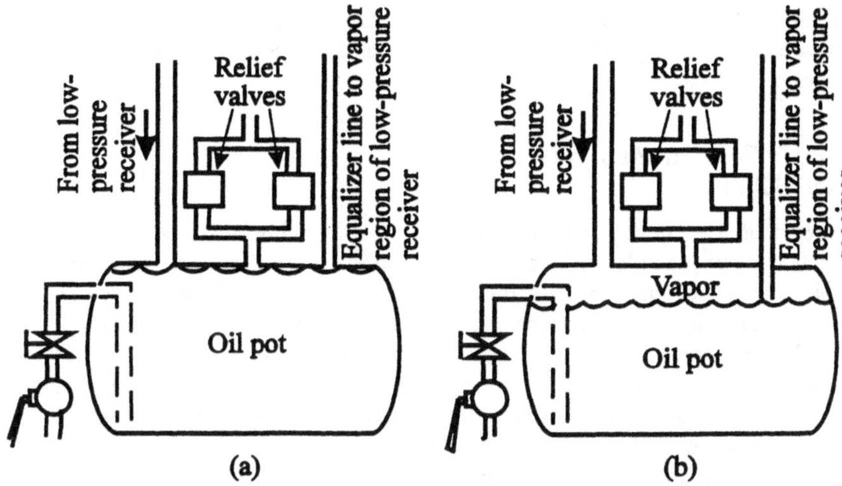

FIGURE 10.17
Oil pots with (a) the equalizer line flush with the top of the oil pot, and (b) extended into the oil pot to provide a vapor space at the top.

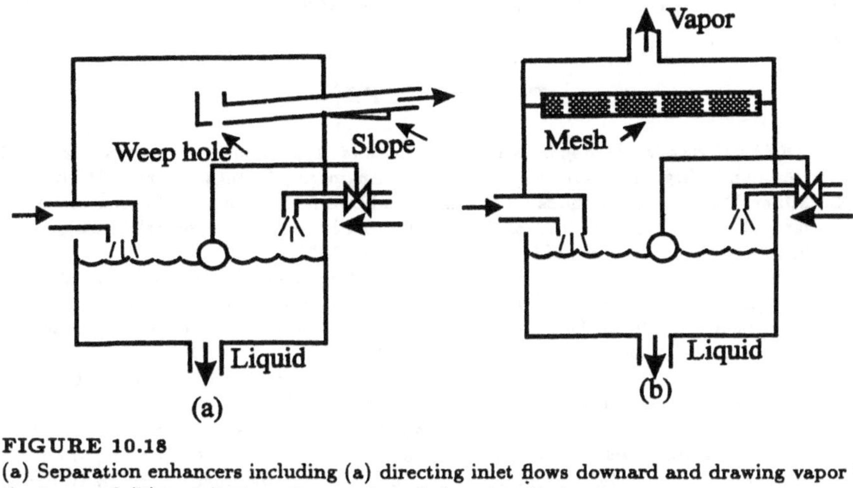

FIGURE 10.18
(a) Separation enhancers including (a) directing inlet flows downard and drawing vapor from the top, and (b) installation of a metal mesh for mist elimination.

any liquid drops to travel upward and make a 180° turn to inhibit capture by the suction line. This suction line should be sloped back to the vessel, so a liquid trap develops which can be prevented by drilling a weep hole at the elbow of the pipe.

Another approach to improved liquid separation is through coalescing the small drops into larger ones by passing the liquid/vapor mixture through eliminator baffles or a mesh, as is shown in Fig. 10.18b. These devices are commonly used for vessels in chemical plants[10,11] and surely permit the choice of a smaller-size vessel. They have the disadvantage of introducing pressure drop which in the view of many refrigeration system designers is a penalty that can be avoided by a larger-size vessel with no obstructions.

10.14 VESSEL SIZING IN PERSPECTIVE

The thrust of this chapter has been to compute the minimum sizes of vessels. In this task the simple rule is, "Never scrimp on the size of the vessel." Certainly there is a price to pay for the installation of a generously sized vessel—the increased cost of materials and installation. Furthermore, additional space will be occupied and larger vessels may entail a larger refrigerant charge. The payback for this extra cost is improved separation of liquid and vapor and the ability to accommodate surges in the shifts of liquid between components, both of which provide additional protection for the compressor.

All those involved in the planning of a system may be convinced that the capacity specified for the plant is final and will suffice for eternity. Five years later, however, another refrigerated space or process may need to be added, and if the vessels were generously sized at the time of the original installation, they may be able to accommodate a moderate-sized expansion without the need for replacement.

REFERENCES

1. *ASHRAE Handbook, Refrigeration Systems and Applications*, American Society of Heating, Refrigerating and Air-conditioning Engineers, Atlanta, Georgia, 1994.
2. Miller, D., "Recent Methods for Sizing Liquid Overfeed Piping and Suction Accumulator Receivers," Proceedings of the XIII International Congress of Refrigeration, vol. 2, International Institute of Refrigeration, Paris, 1971.
3. Brown, G. G., *Unit Operations*, Chapter 24, "Vapor-Liquid Transfer Operations—Design and Control of Fractionating Columns," John Wiley & Sons, 1951.
4. Lorentzen, G., "On the Dimensioning of Liquid Separators for Refrigeration Systems," *Kaeltetechnik-Klimatisierung*, 18(3): 89-97, 1966.
5. Wiencke, B., "Richtlinien fuer die Dimensionierung von Schwerkraftfluessigkeitsabscheidern in Kaelteanlagen,: *Die Kaelte und Klimatechnik*, 9: 496-508, 1993.
6. Grassmann, P. and A. Reinhart, "Zur Ermittling der Sinkgeschwindigkeiten von Tropfen und der Steiggeschwindigkeiten von Blasen," *Chem. Ing. Techn.*, pp 348-349, 1961.
7. Richards, W. V., "Old Habits in Ammonia Vessel Specification," *Air Conditioning, Heating and Refrigerating News*, p 28, May 13, 1985.
8. Stoecker, W. F., "How to Design and Operate Flooded Evaporators for Cooling Air and Liquids," *Heating/Piping/Air Conditioning*, 32(12): 144-168, December 1960.
9. "Pressure Vessel Sizing,: Automatic Refrigerant Control, Section 1, Niagara Blower Company, Buffalo, New York, 1980.
10. Gerunda, A. "How to Size Liquid-Vapor Separators," *Chemical Engineering*, 88(9): 81-84, May 4, 1981.
11. Wu, F. H., "Drum Separator Design—A New Approach," *Chemical Engineering*, 91(7): 74-80, April 2, 1984.

CHAPTER 11

VALVES AND REFRIGERANT CONTROLS

11.1 TYPES OF VALVES

All devices considered in this chapter, even those called controls, are valves in that they are placed in refrigerant lines and can restrict or even completley block the flow of refrigerant. Two-position shutoff valves are expected to operate either in a completely open or a completely closed position. Other valves modulate the flow rate of refrigerant in response to some variables, such as temperature, pressure, or liquid level.

The specific types of valves explored in this chapter are:

- manual shutoff valves
- manual expansion valves
- check valves
- solenoid valves
- level controls
- pressure-regulating valves
- superheat controlling expansion valves

A discussion of safety relief valves appears in Chapter 13, Safety.

11.2 MANUAL SHUTOFF VALVES

A basic valve type distributed liberally throughout an industrial refrigeration system is a manual shutoff valve. In the completely open position, this valve should allow a free flow of refrigerant and when closed completely block the flow. The usual function of the shutoff valve is to isolate a component or a section of the system. Some major categories of manual shutoff valves are globe, angle, inline, and ball valves, as shown schematically in Fig. 11.1. Three desirable characteristics of manual shutoff valves are:

- that they permit no passage of refrigerant when closed
- that they cause only a low-pressure drop of refrigerant flowing through them when they are open
- that they do not leak to atmosphere

Several other types of valves are gate and butterfly valves which meet the low-pressure-drop requirement, but in general do not seal as well as other valves when closed. Consequently they are not widely used in industrial refrigeration service. All valves shown in Fig. 11.1 have accessible handles, but in recent years a strong preference has developed for capped valves when the valve does not need to be opened and closed often.

Shutoff valves are oriented so that they close against the flow which usually means that they close against the high pressure. With this orientation the upstream pressure assists in the opening of the valve. If the valve must open against high pressure, cases have been reported where the pressure holds the disc with such force that the stem may pull away from the disc in an attempted opening. When the valve closes against the flow the highest pressure is kept off the stem and bonnet when the valve is shut off. Globe valves should be mounted with the stem horizontal so that any vapor in the line cannot form a pocket at the valve inlet which would periodically release, causing noise and unsteady flow.

Ball valves have become very popular in the past few years, primarily because of the low pressure drop that they cause in their completely open position. A further advantage of ball valves in certain situations is that they are *quarter-turn* valves so that a quarter turn of the handle permits quick opening or closing of the valve. An undesirable characteristic of the basic ball valve is that of trapping liquid within the ball when the valve is shut off. A ball valve in a cold liquid line traps cold liquid inside the ball when the valve is closed, and this liquid is likely to warm up when the flow is interrupted. The trapped liquid expands which could blow out the valve seat or even rupture the valve body.

Two methods[1] used most commonly to relieve pressure of trapped liquid in the ball and prevent damage are upstream-venting and self-relieving seats. In upstream venting a small hole is drilled through one side of the ball, connecting the upstream line with the cavity when the valve is in its closed position. This configuration bypasses the upstream seat, and provides a continuous vent path for cavity pressure. In the self-relieving seat design, the seats act as internal relief

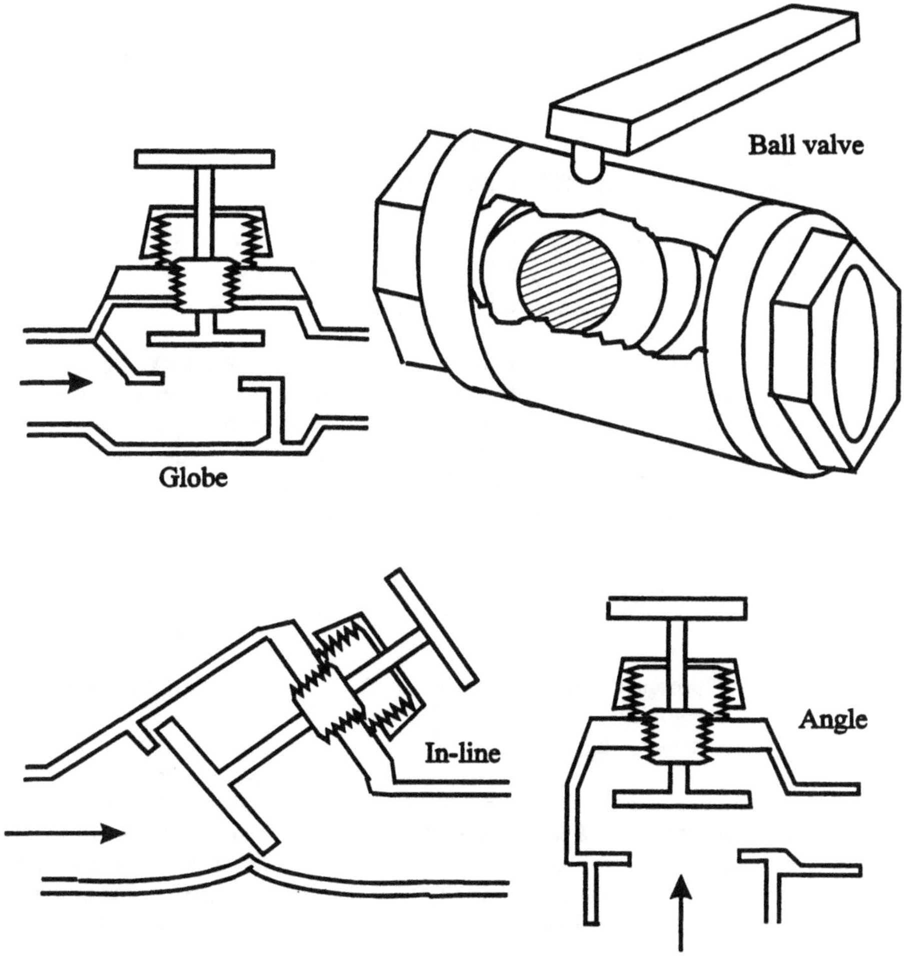

FIGURE 11.1
Some types of manual shutoff valves.

valves to open a vent path from the valve body cavity to the line. Self-relieving valve seats serve as normal valve seats unless the pressure within the ball rises to an extreme level, in which case they permit leakage of a few drops of liquid.

Judgment should be used in whether and what type of valves should be incorporated in the lines. Even valves that are rarely shut off may be invaluable in isolating a certain component or even another valve on rare occasions. On the other hand, extra valves, particularly those placed in vapor lines, may represent a persistent demand for extra compressor power when ever the system operates. One estimate[2] calculated that a fully open valve in the liquid/vapor line between the evaporator and the low-pressure receiver causing 7.5-kPa (1.1-psi) pressure

TABLE 11.1
Pressure drop coefficients for frequently used valves and fittings[3]—values of c in Eq. 11.1.

Pipe dia. mm (in)	90° Elbow	Tee branch	Globe valve	Gate valve	Angle valve	Check valve
25 (1)	1.5	1.8	9	0.24	4.6	3.0
50 (2)	1.0	1.4	7	0.17	2.1	2.3
67 (2.5)	0.85	1.3	6.5	0.16	1.6	2.2
100 (4)	0.7	1.1	5.7	0.12	1.0	2.0

drop could add $9,400 to the annual operating cost of a 2100 kW (600 ton) system. A fully open ball valve would add only $43 to the annual operating cost.

Some pressure drop is expected when refrigerant flows through open valves, and this pressure drop adds to the pressure drop occurring in straight sections of pipe. Methods for calculating the pressure drop in straight pipes were presented in Chapter 9, and Table 11.1 provides data for computing the pressure drop in fittings and valves. Table 11.1 gives the values of the c-terms in the equation

$$\Delta p = c\left(\frac{V^2\rho}{2}\right) \quad \text{or} \quad \Delta p = c\left(\frac{V^2\rho}{2g_c\,144}\right) \qquad (11.1)$$

where Δp = pressure drop, Pa or Δp = pressure drop, psi
V = velocity, m/s; V = velocity, fps
ρ = density, kg/m^3; ρ = density lb/ft^3
g_c = 32.2 ft/s^2

An important observation that can be made from Table 11.1 is the relative pressure drops of globe and angle valves. An angle valve causes anywhere from 1/2 to 1/8 the pressure drop of a globe valve (depending upon the valve and pipe size) and should be considered if the physical arrangement permits. The pressure-drop coefficients for ball valves are likely to be approximately the same as gate valves.

11.3 MANUAL EXPANSION OR BALANCING VALVES

Manual regulating valves are designed to adjust the flow rate through their entire stem travel. Shutoff valves, on the other hand, are not intended for use as regulating valves since they provide most of their regulation in the first turn of the valve from its closed position. Two frequent applications of manual expansion or balancing valves are at the evaporator coils of liquid-recirculation systems and in conjunction with on-off liquid level control valves, as illustrated schematically in Fig. 11.2. In the liquid recirculation system of Fig. 11.2a the function of the valves is to throttle the flow rate to coils whose unthrottled coil-and-piping circuit has a lower pressure drop than others. The liquid supply pressure ahead of

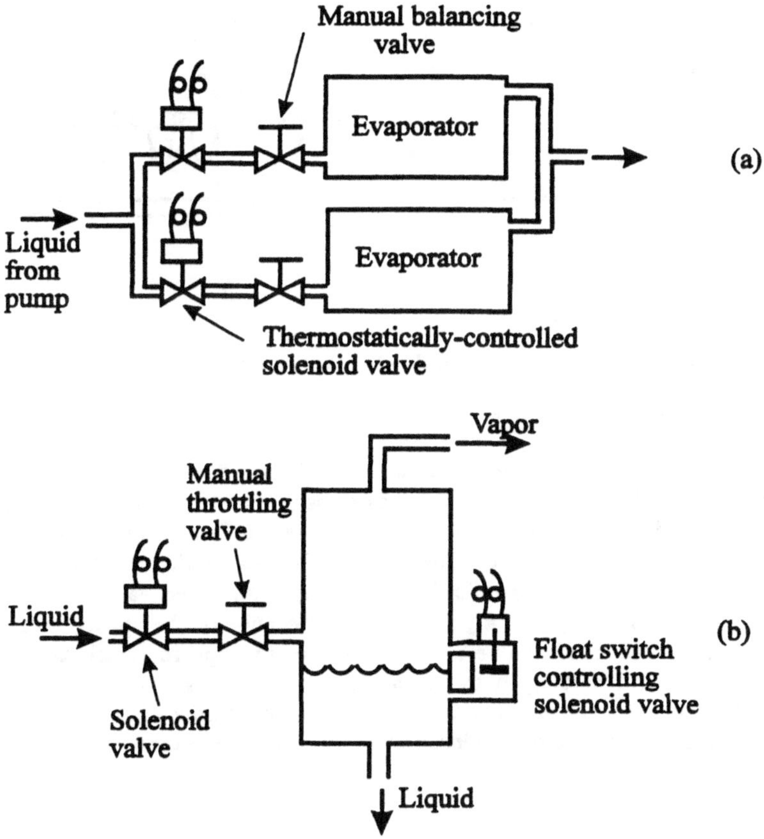

FIGURE 11.2
Manual control valves in (a) liquid recirculation systems, and (b) in conjunction with an on-off liquid level control valve.

the valves is increased, which diverts liquid to those coils that otherwise might be inadequately fed. In liquid recirculation systems, the drop in pressure through the valve is small relative to that occurring when an expansion valve separates condensing and evaporating pressures.

Valves regulating liquid flow into vessels where the level is controlled are often electrically operated on-off valves. Such solenoid valves are combined with manual control valves, as illustrated in Fig. 11.2b, to prevent wild pressure fluctuations in the vessel as the solenoid valve opens and closes. Pressure drop occurs in both the manual valve and the solenoid valve, even when it is open, but approximately 2/3 of the pressure drop should be taken in the manual valve.

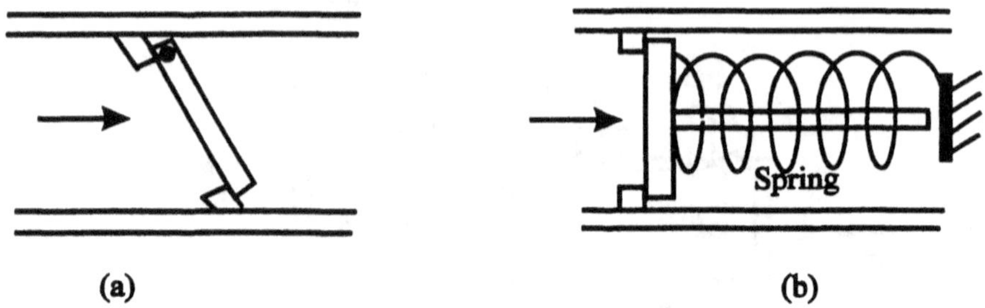

FIGURE 11.3
Check valves of the (a) gravity type, and (b) of the spring-loaded type.

11.4 CHECK VALVES

Check valves allow the flow of refrigerant in only one direction by automatically closing when fluid attempts to flow in the opposite direction. The moving element may be assisted by gravity so that the valve closes unless a slight pressure drop in the permitted flow direction opens the valve, as in Fig. 11.3a. This swing type of check valve is not as reliable as the spring-actuated valve of Fig. 11.3b, but this check valve imposes a pressure drop because of the spring pressure. A more sophisticated check valve is the gas-powered type in which the valve is normally open and held open by spring force. When the downstream pressure rises above the entering pressure to the valve, vapor from a high-pressure source closes the valve, which remains closed as long as the downsteam pressure exceeds the upstream pressure.

11.5 SOLENOID VALVES–DIRECT-ACTING

Solenoid valves are electrically operated shutoff valves. Probably the most common is the normally closed (NC) valve, but normally open (NO) valves are also available[4]. With both types, system pressure works to keep the valve closed when that position is desired. Solenoid valves thus can hold against high upstream pressures, but will not restrain much pressure in the reverse direction. Two classifications of solenoid valves are:

1. direct-acting
2. pilot-operated (to be discussed in the next section)

In the direct-acting solenoid valve, as shown in Fig. 11.4, the magnetic force developed by the electric coil draws the stem and connected plunger off the valve port when the coil is energized. Some solenoids are designed to allow the stem to start its motion before engaging the plunger which is seated against the system

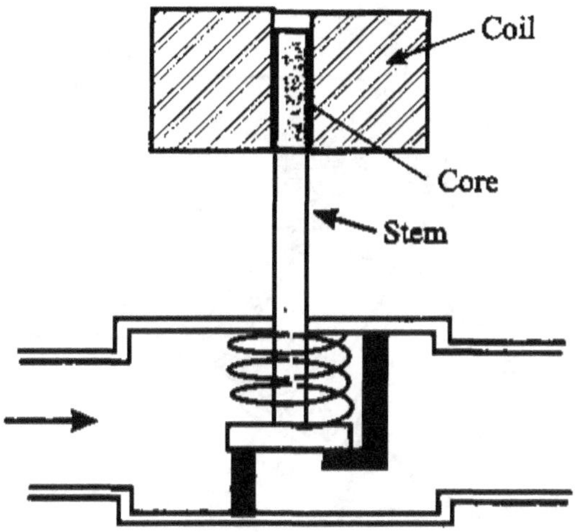

FIGURE 11.4
A direct-acting solenoid valve.

pressure. The momentum of the stem thereby helps open the valve. When the coil is de-energized, the plunger either drops into the closed position by gravity, and/or a light spring assists the closing.

A solenoid valve must be selected to be able to open against the maximum operating pressure differential (MOPD), a characteristic of the valve listed in the manufacturer's catalog. Because direct-acting valves must generate sufficient force in the coil to open the valve against the system pressure, they are limited to line sizes of perhaps 6 to 25 mm (1/4 to 1 in).

11.6 SOLENOID VALVES—PILOT-OPERATED AND GAS-PRESSURE-OPERATED

To make the solenoid valve practical for large pipe diameters, concepts must be applied other than direct force on the valve stem from the magnetic coil. Two approaches are pilot-operated and gas-pressure-operated valves. In the pilot-operated valve, a small solenoid opens to apply high upstream pressure on a power piston. The power piston has a greater area than that of the valve plug, which is also subjected to the upstream pressure, so the force of the power piston opens the valve. One design of a pilot-operated solenoid[5] is shown in Fig. 11.5 where the opening of the pilot solenoid permits upstream pressure at M to pass through the passage N to the power piston.

The pilot-operated solenoid valve requires a minimum pressure drop across it, even when completely open. There are situations where the pressure drop through the open valve is to be kept as low as possible, such as in low-pressure

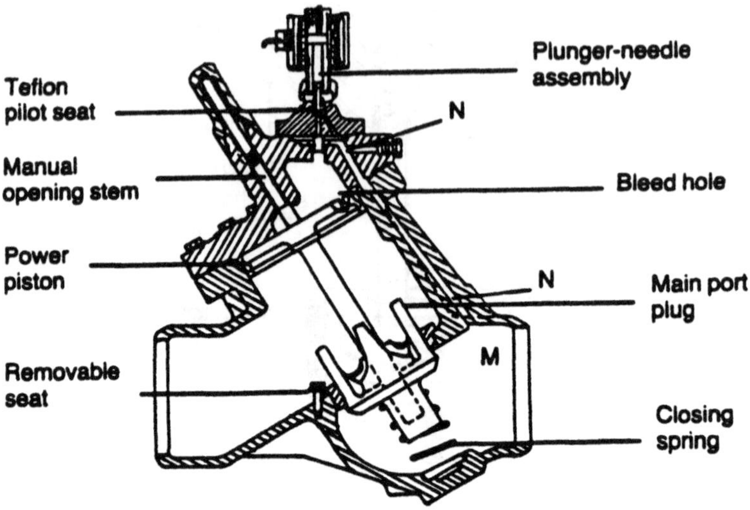

FIGURE 11.5
A pilot-operated solenoid valve. (*Courtesy, Refrigerating Specialties Div. of Parker-Hannifin Corp.*)

vapor lines, or in the liquid leg or the return line of a flooded evaporator. In other situations, viscous oil may be coating the moving parts of the valve, which requires more positive action to open and close the valve. The gas-pressure-operated valve[5], such as ahown in Fig. 11.6, may be considered in such cases. The construction is similar to the pilot-operated solenoid, except that pressure from the condenser or another high-pressure source is applied to the power piston. Usually a pressure of the high-pressure gas of 69 kPa (10 psi) higher then the inlet pressure to the valve is adequate to open the main valve. When the valve is to be opened, the pressure pilot solenoid opens to allow the high-pressure gas access to the power piston. To close the valve, the pressure pilot solenoid valve closes and the bleed pilot solenoid valve opens. With this valve status, the pressure above the power piston vents to the low-pressure side, permitting the spring to close the main valve.

11.7 GAS-POWERED SUCTION STOP VALVES

In low-temperature installations the shutoff valve between the evaporator and the suction line should offer as little flow resistance as possible during refrigeration. A normally-open valve meets this requirement, but in addition the valve must be capable of closing positively, for example, during a hot-gas defrost. A valve designed to meet these requirements, and first shown in Fig. 6.49 among the defrost piping arrangements, is the gas- powered suction valve, as shown in Fig.

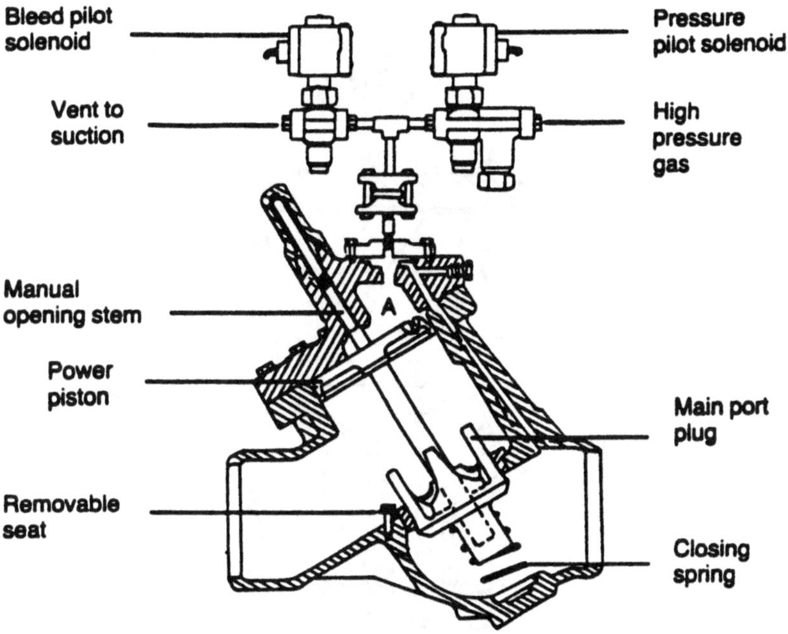

FIGURE 11.6
A gas-pressure-operated solenoid valve. (*Courtesy, Refrigerating Specialties Div. of Parker-Hannifin Corp.*)

11.7. If the external gas pressure is removed, the pressure above the piston vents through the bleed hole and the spring opens the valve. When the valve is to be closed, high-pressure gas from an external source is applied to the piston, which closes the valve, working against the spring compression.

11.8 PRESSURE-REGULATING VALVES: DIRECT-ACTING, PILOT-OPERATED, AND EXTERNALLY COMPENSATED

There are frequent situations where refrigerant flow controls must regulate pressure; several are illustrated in Fig. 11.8. Controlling the pressure in the evaporator, as in Fig. 11.8a, may be required in two different situations. It may be necessary to prevent the evaporating temperature from dropping too low in order to protect the product being refrigerated. Another application of a control that regulates the pressure in the evaporator is used during hot-gas defrost, as explained in Sec. 6.22. A different technique to prevent evaporating pressure from dropping too low, shown in Fig. 11.8b, is called hot-gas bypass. The valve controls its downstream pressure and opens if the evaporator pressure drops below

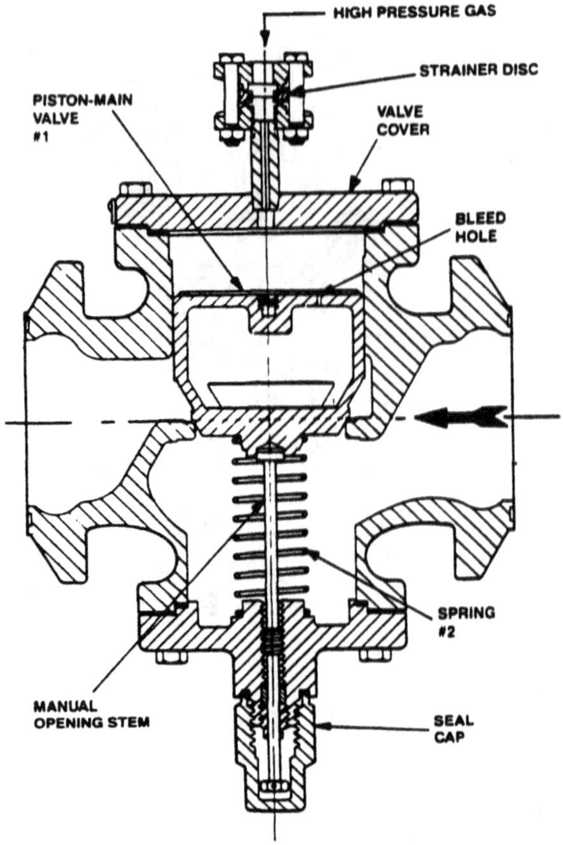

FIGURE 11.7
A gas-powered suction stop valve. (*Courtesy, Refrigerating Specialties Div. of Parker-Hannifin Corp.*)

the outlet pressure setpoint. Another pressure-regulating valve that controls its downstream pressure is shown in Fig. 11.8c. This valve might be used to prevent the suction pressure to the compressor from rising so high that the compressor motor overloads.

A cutaway view of a small valve that regulates the upstream pressure is shown in Fig. 11.9a. The valve remains closed until the upstream pressure rises to the setpoint whereupon the diaphragm lifts off the port. For the valve to open wider and pass a greater flow rate, the upstream pressure rises somewhat higher than the setpoint as the diaphragm works against the spring force. The setpoint is fixed by adjusting the screw that regulates the spring precompression.

Pilot-operated valves are available in larger sizes, as in Fig. 11.9b, where in an operation similar to pilot-operated solenoids[5] the upstream pressure controls a flow of pilot fluid to the power piston.

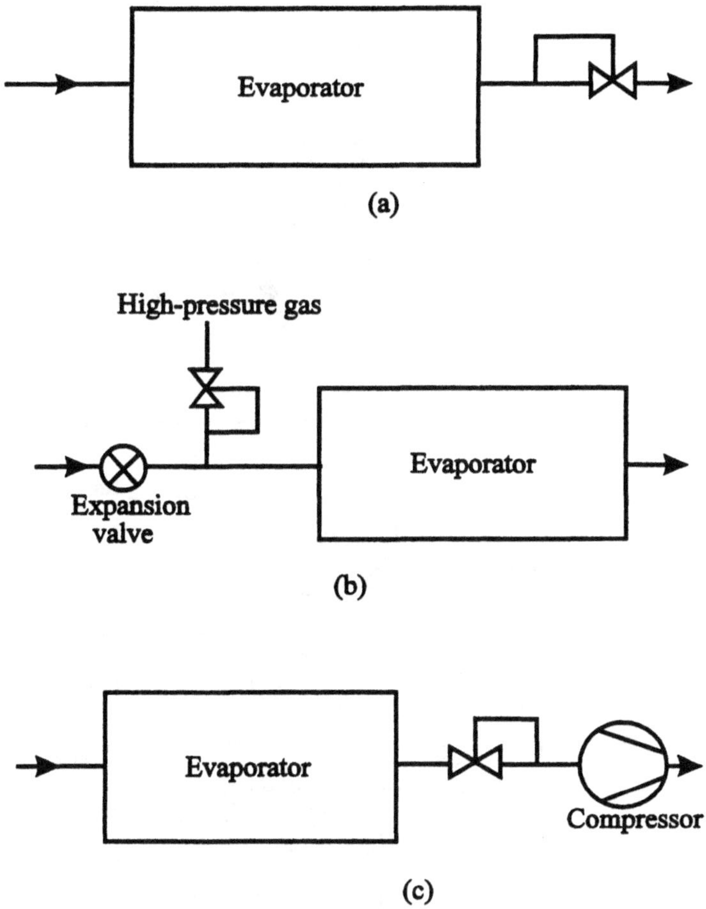

FIGURE 11.8
(a) Inlet pressure regulator preventing the evaporator pressure from dropping below the setting, (b) outlet pressure regulator preventing the evaporator pressure from dropping below the setting, and (c) an outlet pressure regulator that prevents the compressor suction pressure from rising above the setting.

While both the valves shown in Fig. 11.9 are manually adjustable, in some installations it would be convenient or necessary to provide automatic adjustment of the settings. Additional capabilities are available[6] that allow adjustment of the setting through a pneumatic signal, a signal based on temperature, or an electric signal. In addition, pressure-regulating valves are available that will control one of two different pressures, depending on whether a small pilot solenoid is open or closed.

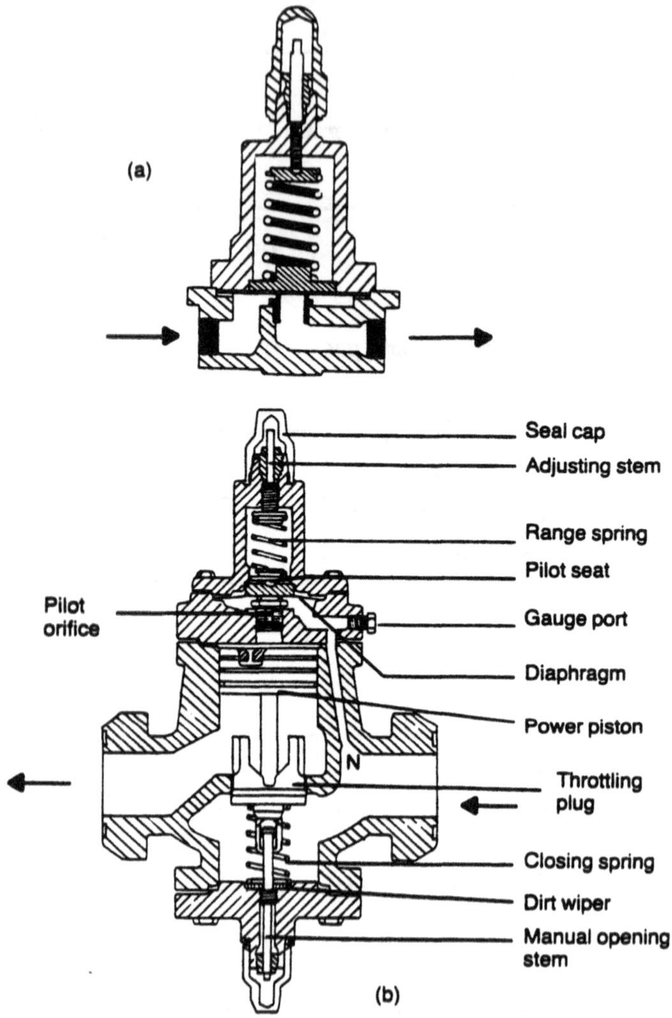

FIGURE 11.9
Controlling the upstream pressure with (a) a direct-acting valve and (b) a pilot-operated valve.

11.9 LIQUID LEVEL CONTROLS—HIGH- AND LOW-SIDE FLOAT VALVES

Two major categories of liquid-level controllers used in industrial refrigeration systems are (1) high-side float valves, and (2) low-side float valves. These classes of valves, as shown schematically in Fig. 11.10, differ in the placement of the valve. The flow regulation in a high-side float (Fig. 11.10a) is downstream of the

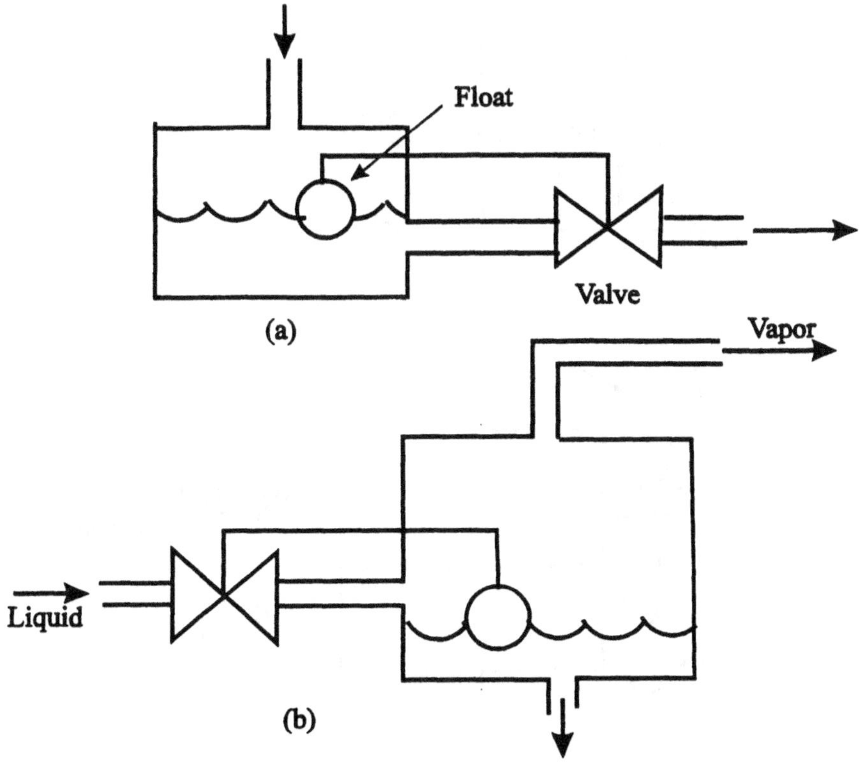

FIGURE 11.10
(a) A high-side liquid level controller and (b) a low-side level controller.

controlled liquid level and the flow regulation in the low-side float (Fig. 11.10b) is upstream of the controlled liquid level.

The low-side level controllers are much more common in industrial refrigeration systems than are high-side controllers, but there are roles for the high-side float valve, which are primarily to allow liquid to pass but prevent the flow of vapor. Three applications for the high-side float valve are (1) to drain condensed liquid from a coil during hot-gas defrost as is shown in Fig. 6.56, (2) to drain condensed liquid from a hot-gas main, and (3) to be placed at the outlet of a condenser in a critically charged system. The function of the valve in the latter situation is to allow condensed refrigerant to leave the condenser, but prohibit the passage of vapor. These critically-charged systems sometimes have no high-pressure receiver, but there must be a vessel somewhere on the low side to accommodate changes in liquid volume in the evaporators.

The low-side liquid level controller is the type commonly encountered in industrial refrigeration systems, because it is the class used for control of the liquid level in low-pressure receivers and flash-tank/desuperheaters.

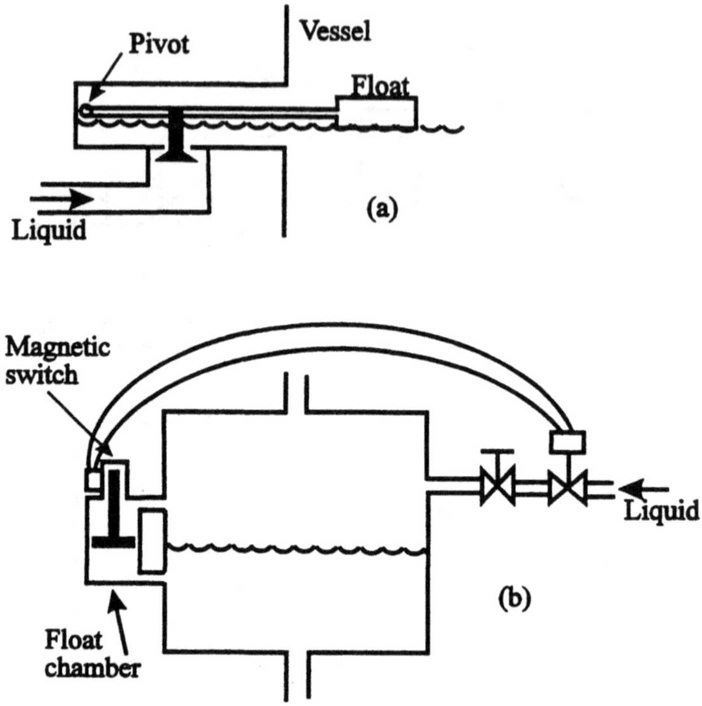

FIGURE 11.11
(a) A mechanical float valve, and (b) a float-switch level controller.

11.10 LIQUID LEVEL CONTROLS—MECHANICAL AND FLOAT SWITCH

Numerous types of level controllers are used in industrial refrigeration practice, two important types of which are shown in Fig. 11.11. The mechanical type in Fig. 11.11a is a simple one applicable for smaller systems. As the liquid level rises, the valve stem closes the port to restrict the flow of liquid. This type of valve provides a modulated rate of flow. A standard type that has been used for many years is the float switch of Fig. 11.11b where the rise in liquid level elevates the float in the float chamber which discontinues the flow of electric current to the solenoid valve to interrupt the supply of liquid to the vessel. The float chamber serves a useful purpose in damping some of the splashing likely to exist in the liquid level in the vessel. The manual expansion valve, as was shown in Fig. 11.2, is always combined with the solenoid valve to reduce the flow rate when the solenoid is open to avoid short pulses of high flow rates which would disturb the pressure in the vessel.

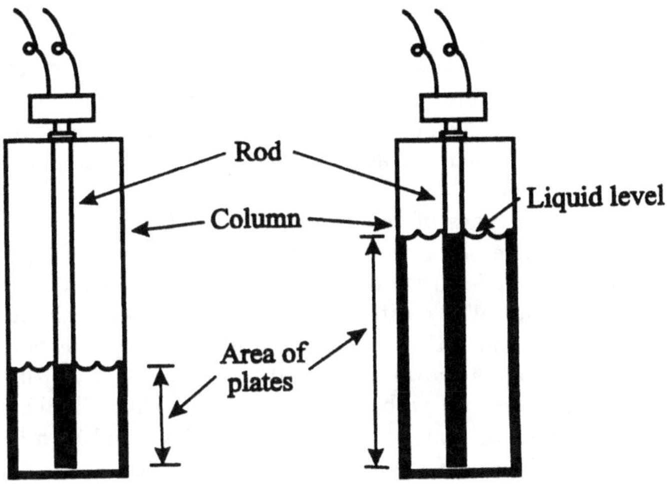

FIGURE 11.12
The liquid level in the column will change the capacitance between the rod and the column.

11.11 LIQUID LEVEL CONTROLS—CAPACITANCE LEVEL SENSOR

Electrical capacitance between two plates is the electrical charge divided by the voltage difference needed to maintain this charge. A capacitance level sensor measures the capacitance between the center rod and the column in Fig. 11.12. When the liquid level is different in the two columns, the capacitance in them will also be different.

The total capacitance[7] for a given clearance between the rod and the column is expressed by the equation

$$\text{Total capacitance} \simeq D_v A_v + D_l A_l \tag{11.2}$$

where D_v = dielectric constant of vapor
A_v = area of plates surrounded by vapor
D_l = dielectric constant of liquid
A_l = area of plates surrounded by liquid

Furthermore, the approximate dielectric constants of vapors and several liquids are:

Material	Dielectric constant
Most vapors	1.0
Ammonia liquid	15.5
R-22 liquid	6.1
Oil	2.0

Vapor has a dielectric constant appreciably less than liquids, so the application of Eq. 11.2 shows that as the liquid level in the probe chamber rises, so does the capacitance between the rod and the column. This capacitance is sensed and usually converted into an electric current varying between 4 and 20 mA. A comparison of the dielectric constant of oil and ammonia, for example, suggests that if the probe chamber has accumulated oil, that the sensed level will be in error.

There are two types of capacitance level indicators, the single-point and the continuous. Single-point probes indicate the presence of liquid at one level, and thus are comparable to the float switch. This application of the capacitance principle fails to capitalize on its full capabilities, because circuitry is available to open or close a contact at multiple values of the probe output. A typical set of level indicators is shown in Fig. 10.13 applied to a low-pressure receiver where the capacitance level probe not only manages the normal control level, but signals two extreme positions to activate the low-level and high-level alarm. The high- and low-level cutouts are usually part of a separate system, perhaps float switches, so that the plant is protected in the event of a failure of the capacitance level controller.

An advantage of the capacitance level controller is that a setting of a level can easily be adjusted by changing the pickoff point in the 4 to 20 mA range that triggers the action. Changing the level setting with a float switch is not so simple, because the connections of the float chamber on the vessel must be relocated. Capacitance probes have no moving parts in the sensor to wear out, so they are not subject to sticking and failure of electrical contacts in the probe.

11.12 PROPER PLACEMENT OF THE FLOAT CHAMBER

It would be possible to install a float or a capacitance probe directly in the vessel, but it is highly preferable to attach a float or capacitance chamber external to the vessel. One reason is that the level in the vessel may be subject to considerable agitation, whereas the level in the chamber will be much more tranquil. The float chamber requires two connections to the vessel, one above and one below the desired liquid level. The connections to the vessel should be made to the side of the chamber, rather than from the bottom. As Fig. 11.13a shows, in an ammonia system any oil in the vessel could collect in the line between the vessel and the chamber. Since oil has a different density than liquid ammonia, the float would settle at a level different than that in the vessel. As was mentioned in the previous section, the capacitance level sensor would be distorted by oil for an additional reason, which is that the dielectric constant of oil differs from that of liquid ammonia or the halocarbons.

If the liquid in the vessel is boiling or has compressor vapor bubbling through it to desuperheat the vapor, the density of the mixture in the vessel will be less than in the float chamber. The result, as illustrated in Fig. 11.13b, is that the liquid level in the vessel will be higher than in the float chamber.

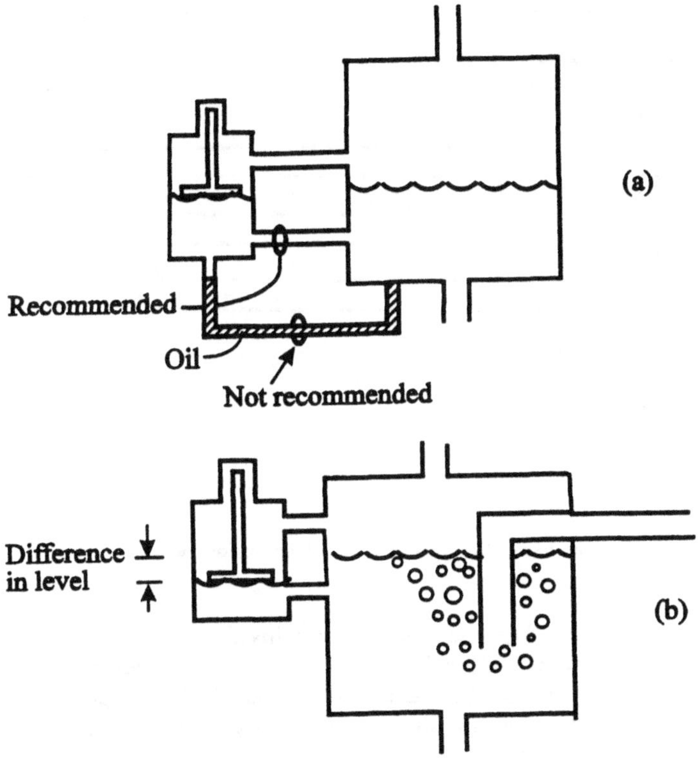

FIGURE 11.13
Float chambers: (a) with one leg connected to the bottom of the vessel and collecting oil, and (b) difference in levels in the float chamber and vessel when liquid in the vessel contains vapor.

Equalizer pipe lines should be a minimum of 30 mm (1-1/4 in) pipe for low- pressure vessels[7]. Chambers for low-pressure vessels should be a minimum of 75 mm (3 in) in diameter.

11.13 SUPERHEAT-CONTROLLED EXPANSION VALVES (THERMO OR TX VALVES)

Probably the most widely used flow control device in general refrigeration practice, including air conditioning, is the superheat-controlled expansion valve. The valve is popularly called a thermostatic expansion valve, a name that implies controlling a fixed evaporating temperature. The valve does no such thing, as will be explained later, so we shall call the valve a *superheat-controlled valve*, although another acceptable name is *thermovalve*. The valve is also called a *TXV* by some refrigeration professionals. Even though the superheat-controlled valve is used

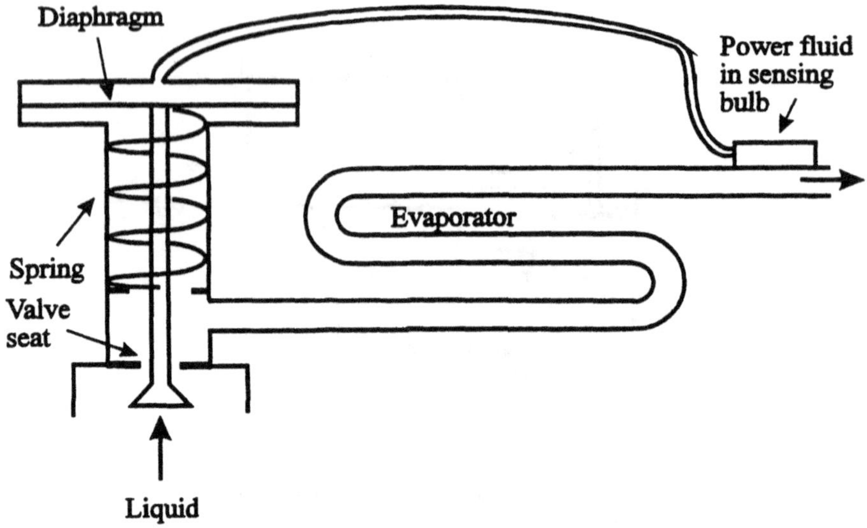

FIGURE 11.14
A schematic diagram of the basic superheat-controlled expansion valve.

so widely, one objective of this section is to distinguish where the valve can be applied effectively in industrial refrigeration and where it is not recommended.

A schematic diagram of the basic superheat-controlled expansion valve is shown in Fig. 11.14. The essential objective of the expansion valve is to regulate a flow rate of refrigerant to the evaporator that matches the rate boiled off. The valve accomplishes this objective by controlling the amount of refrigerant superheat leaving the evaporator.

The valve stem is positioned by the pressure difference on opposite sides of the diaphragm. The pressure under the diaphragm is provided by the refrigerant at the entrance to the evaporator, and the pressure on the top side of the diaphragm by what is called the power fluid. In the basic expansion valve, the power fluid is the same refrigerant used in the system. It is in vapor form, except for a small amount of liquid in the sensing bulb.

A slight force exerted by the spring on the valve stem keeps the valve closed until the pressure above the diaphragm overcomes the combined forces of the spring and the evaporator pressure. For the pressure above the diaphragm to be higher than the evaporator pressure below the diaphragm, the power fluid temperature must be higher than the saturation temperature in the evaporator. The suction gas must, therefore, be superheated to bring the power fluid up to the valve-opening pressure.

The superheat-controlled expansion valve is a type of proportional controller in that the response of the valve stem is proportional to the error between the sensed and set values. The stem position thus bears a relationship to the superheat at the evaporator outlet as shown in Fig. 11.15. Conceivably the valve

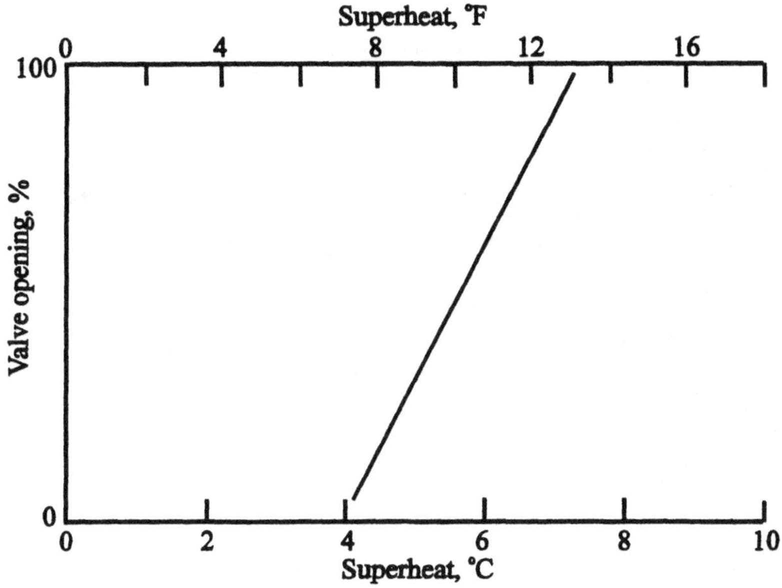

FIGURE 11.15
Range of superheat controlled by the expansion valve.

could be adjusted to operate with no superheat when just closed and thus hold about 3°C (5.4°F) superheat when completely open. But such a setting provides no protection against a surge of liquid moving through and out of the evaporator more rapidly than the valve could respond. As a precaution, then, the application of a precompression on the spring assures that even at its nearly closed position, the valve provides some superheat leaving the evaporator, as is the case in Fig. 11.15.

If the amount of superheat at the evaporator outlet is to be truly regulated by the valve, the stem position should be controlled by the pressure difference of the power fluid and the pressure at the outlet of the evaporator, rather than at the inlet as is true of the basic valve of Fig. 11.14. Because of the pressure drop between the inlet and outlet of the evaporator of perhaps 15 to 40 kPa (2 to 6 psi) which varies as a function of evaporator load, the performance of the basic expansion valve would be distorted. To correct this problem, a standard expansion valve is available that incorporates an external equalizer, as shown in Fig. 11.16. This valve is provided with an additional connection which permits the pressure at the outlet of the evaporator to be applied to one side of the diaphragm.

We stated earlier that the name *thermostatic* erroneously suggests that this type of expansion valve controls the evaporating temperature. Instead, the evaporating temperature and pressure of a refrigeration system served by a compressor of a constant displacement rate will ride up and down, as shown in Fig. 11.17.

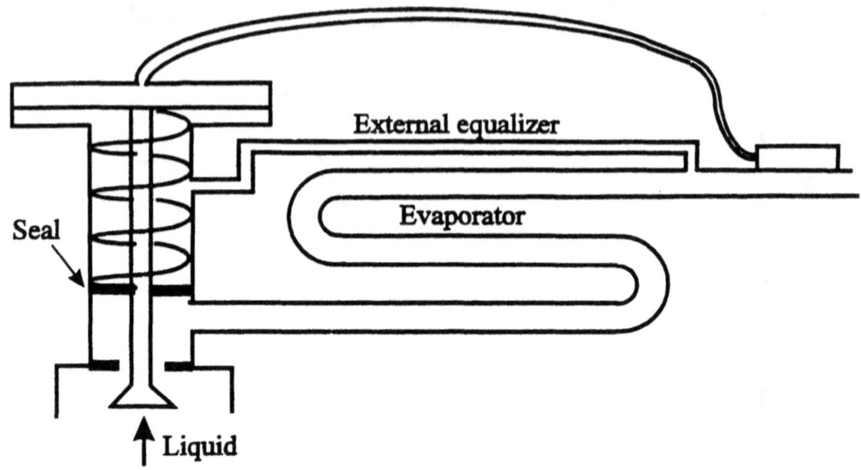

FIGURE 11.16
Superheat-controlled expansion valve with external equalizer.

Two balance points are shown, one at a heavy refrigeration load and the other at a low load. The valve admits a flow rate of refrigerant equal to that evaporated, and at low loads the compressor/valve combination settles on a lower evaporating pressure and temperature than at heavy refrigeration loads.

11.14 SUPERHEAT-CONTROLLED VALVES FOR INDUSTRIAL REFRIGERATION

The superheat-controlled expansion valve needs no defense in its use in comfort air conditioning applications and in commercial refrigeration, but it is appropriate now to examine its applicability to industrial refrigeration systems. Many respected industrial refrigeration professionals have in the past made flat assertions that the superheat-controlled valve should not be used in low-temperature, ammonia refrigeration systems. The next few paragraphs will present some reasons[8] why those two conditions present challenges for this type of expansion valve. In addition, certain applications will be proposed where this type of expansion valve could be properly applied.

One drawback of superheat-controlled valves for low-temperature applications is the high superheat that occurs at low temperatures with the consequent reduction in the effectiveness in the use of the evaporator surface. This characteristic is particularly applicable to systems where the evaporating temperature is frequently pulled down from a warm state. Fig. 11.18a, for example, shows a valve serving an R-22 system with R-22 as the power fluid. The valve is set at 5°C (9°F) superheat at a high temperature of 5°C (41°F) which provides a 100-kPa

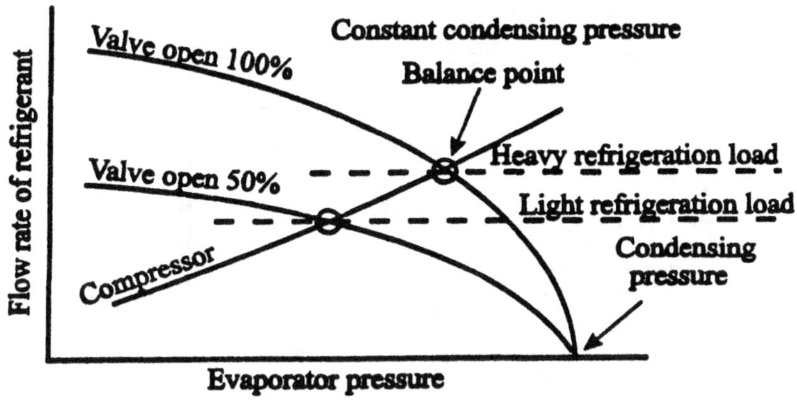

FIGURE 11.17
Shift of evaporating pressure and temperature as the refrigeration load changes on a system with a superheat-controlled expansion valve.

(14.6-psi) pressure difference across the diaphragm. To open the valve the same amount at a low temperature of −30°C (−22°F), a superheat of 12°C (22°F) is required. This large amount of superheat will penalize the performance of the evaporator. The situation can be avoided by selecting a type of valve that uses what is called a cross charge for the power fluid, as illustrated in Fig. 11.18b. This cross charge is a specially blended mixture whose pressure-temperature curve is precisely displaced from the curve of the system refrigerant. In this way a constant relationship of Δp to superheat over a wide range of temperatures prevails.

One of the critical drawbacks of the superheat-controlled expansion valve for low-temperature systems is that superheat in the evaporator becomes a progressively greater deterrent to the maintenance of capacity and coefficient of performance as the evaporating temperature drops. In fact, the improved evaporator heat-transfer characteristic, as shown in Fig. 8.6 and 8.7, was one of the reasons for applying liquid recirculation. The penalty on the power and capacity of each degree of temperature becomes more prominent in low-temperature systems.

The superheat-controlled valve encounters another difficulty when operating at low temperatures, and that is that there may be no temperature available that is high enough to provide the superheat needed to open the valve. As Fig. 11.15 shows, a superheat of perhaps 7°C (12.6°F) may be required to fully open the valve. An air coil in a frozen-food storage facility may be designed for a 5.5°C (9.9°F) temperature difference between the inlet air and the refrigerant. The entire coil is surrounded by air at the entering temperature, so there is no source of air at a high enough temperature to provide the required superheat. The drawbacks of the superheat-controlled valve in industrial refrigeration systems that

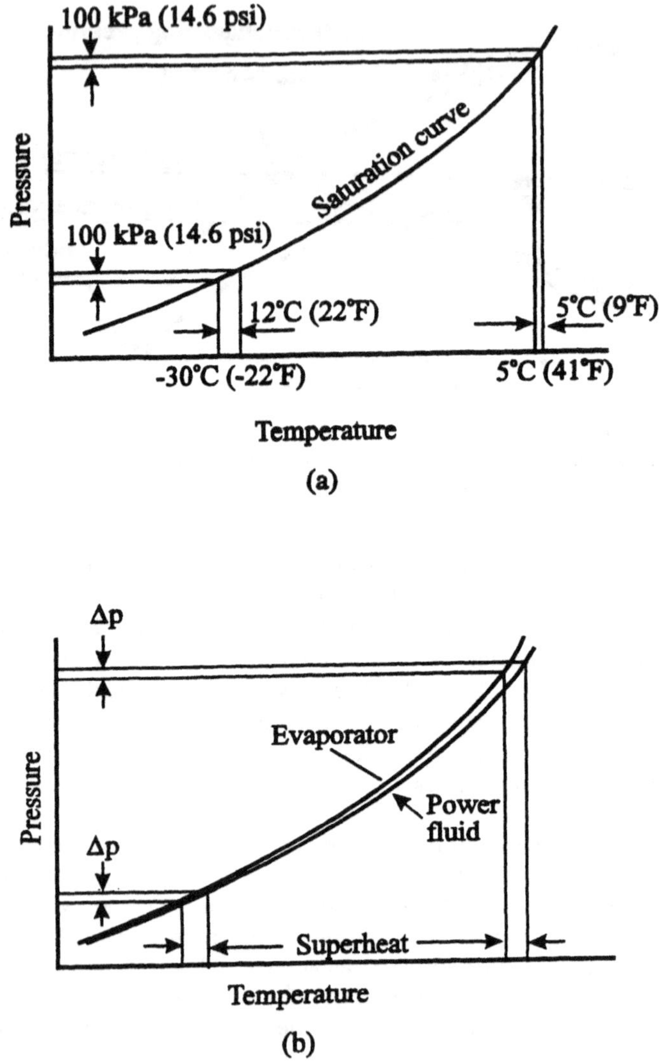

FIGURE 11.18
(a) If the power fluid is the same as the refrigerant in the system, high superheat results at low temperatures, which is corrected by (b) using a cross charge as the power fluid.

operate at low temperature are formidable, but at medium temperature levels—in the neighborhood of 0°C (32°F)—the temperature level alone is not a major issue.

Consider next the question of the compatability of ammonia with direct-expansion systems in general and with superheat-controlled expansion valves in particular. Refrigeration practitioners occasionally report attempts to install a superheat-controlled valve on a coil designed for liquid recirculation. Invariably the coil is not able to provide as high a heat transfer rate as with liquid recirculation, and the reason is that the liquid ammonia flows along the bottom of the tubes without wetting the entire tube wall. Section 6.27 when discussing direct-expansion ammonia air-cooling coils stressed the need of small-diameter tubes with long circuit lengths in order to achieve sufficient agitation of the boiling refrigerant to adequately wet the surfaces of the tubes. Interest is high in successful application of direct expansion to ammonia coils, because of the lower refrigerant charge and the reduction in first cost by eliminating liquid pumping equipment.

Ammonia possesses five or six times the latent heat of halocarbon refrigerants, and this property is almost always a valuable asset. With control valves, however, the high latent heat may be a drawback, because the flow rate of ammonia for a given refrigerating capacity will be low. Ammonia expansion valves are built with much smaller ports than those for halocarbons refrigerants, and at their nearly closed position a minute change in stem position results in large percentage changes in flow rate. Expansion valves in ammonia systems are thus more prone to instability than in halocarbon systems. Another feature of ammonia superheat-controlled expansion valves is that the outlet line of the evaporator to which the bulb is strapped is likely to be steel. The thermal conductivity of steel is about one-eighth that of copper, so the response of the ammonia valve is likely to be more sluggish.

An approach[9] to circumventing the control instability that occurs with a nearly closed control valve is to resort to pulse-width modulation of a valve that is either completely open or completely closed. A signal, such as the superheat of the refrigerant leaving the evaporator, is fed to a microprocessor which regulates the pulse width. Figure 11.19 shows three different percentages of flow rates that are accomplished by varying the fraction of time that the valve is open during the uniform pulse width. A typical cycle time is 6 s. This valve is essentially a rapidly opening and closing solenoid valve which will survive for millions of cycles.

11.15 VALVE AND CONTROL SELECTION

Valves and controls may constitute only a small fraction of the first cost of a plant, but they are the focus of considerable attention during the operating life of the plant. Careful consideration should be given to the location of shutoff valves, because valves at convenient locations facilitate isolation of a component

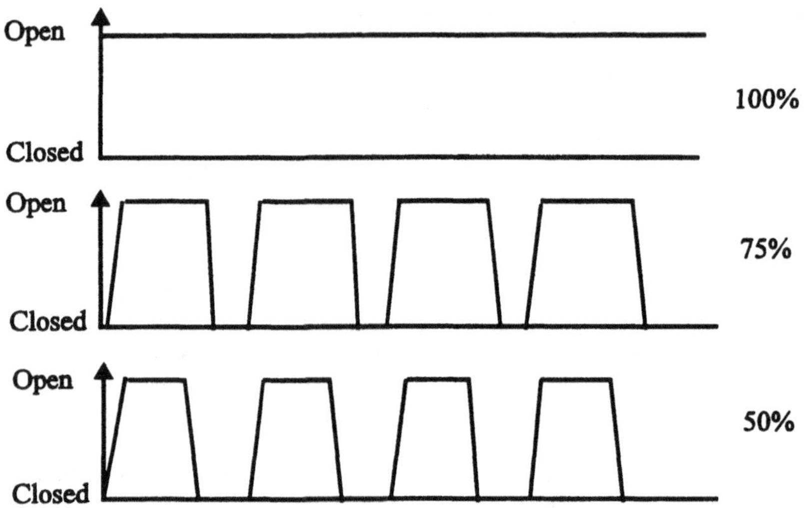

FIGURE 11.19
A pulse-width-modulated valve.

for servicing. The pressure drop of an unneeded valve, however, adds to operating cost of the system. Proper installation of solenoid shutoff, pressure-regulating, and level-control valves should be followed up with regular maintenance.

REFERENCES

1. Ridey, M., "Relieving Overpressure," *Chemical Engineering*, 98(12): 98-101, December 1991.
2. Stamm, R. H. "Troubleshooting Industrial Refrigeration Systems," *Heating/Piping/Air Conditioning*, 76(3): 93-95, March 1984.
3. *Engineering Data Book,*, The Hydraulic Insitute, Cleveland, OH, 1979.
4. Solenoid Valves, Bulletin 30-10. Sporlan Valve Company, Washington, Missouri, 1993.
5. Solenoid Valves and Pressure Regulators, Bulletins 30-05 and 31-90, Refrigerating Specialties Division of Parker Hannifin Corp., Broadview, IL., 1986 and 1976.
6. Pressure Regulators Compensating Controls, Catalog 90, Hubbell Corporation, Mundelein, Illinois.
7. "Techni-Briefs," 1(1), Hansen Technologies Corporation, Burr Ridge, Illinois, February 1997.
8. Johnson, R. J., "Thermostatic Expansion Valve Liquid Feed with Ammonia Refrigerant," Presentation at American Meat Institute Energy Cost Control Workshop, October 1981.
9. Jessen, L. M., "Use of Pulse Width Modulated Valves in Industrial Refrigeration," Proceedings of the 19th Annual Meeting of the International Institute of Ammonia Refrigeration, Washington, D.C., pp 263-280, 1997.

CHAPTER 12

REFRIGERANTS

12.1 CFCS AND THE UPHEAVAL IN THE REFRIGERATION INDUSTRY

The most radical revolution that the refrigeration and air-conditioning industry has experienced is probably the one that began a few years ago with the phase-out of chlorofluorocarbons (CFCs). After having about 50 years to develop and learn to properly apply a certain assortment of refrigerants, the industry needed quickly to replace some widely used refrigerants. Manufacturers of domestic refrigerators and auto air conditioners, for example, have been forced to find a substitute for R-12, which has been used effectively in those products for many years. The industrial refrigeration industry had settled primarily on the use of two refrigerants—ammonia and R-22. A third refrigerant, R-502, which is a dominant refrigerant in supermarket refrigeration and other commercial applications as well as occasional industrial applications, is scheduled for retirement.

The premise of this chapter is that ammonia will continue to be a dominant refrigerant in the industrial refrigeration field. Also, the halocarbon refrigerant that is most prominent in the industrial refrigeration field, R-22, will be vitally important for the next several decades until it is discontinued. R-502 will be phased out because it contains a CFC. A crystal ball is needed to predict which halocarbon or halocarbons will eventually replace R-22 and R-502. A few industrial refrigeration systems used R-12, but most of these systems have been retired by now. If the system served moderate-temperature evaporators, R-134a could serve the application. In fact, the most successful replacement so far has been of replacing R-12 with R-134a, sometimes in the same equipment, provided that the mineral oil is completely removed and replaced with synthetic oil and elastomer seals and hoses replaced. The vapor density of R-134a is relatively low, so the compressor displacement required for a given refrigerating capacity is high, which requires a larger and more expensive compressor.

It is not possible at this time to predict with certainly which halocarbon refrigerants will become the favored substitutes for R-22 and R-502. There certainly is no shortage of candidates, and the commercial refrigeration field in supermarkets is wrestling with the large number of potential alternatives. In almost all cases these refrigerants are not single refrigerants, such as R-134a, but a mixture—either an azeotropic or a zeotropic mixture. The variety of replacements for R-22 and R-502 is so extensive that it is difficult to envision that all the candidates can survive. A technician servicing supermarkets will not carry ten different cans of refrigerant on the truck and will not want to repair and recharge equipment with a large variety of refrigerants. It is expected that the competing manufacturers will practice a certain amount of cross-licensing to reduce the number of available refrigerants to a manageable group. The list given in Section 12.5 presents a number of the halocarbon refrigerant mixtures that are currently available. In preparing this chapter one refrigerant, R-507, typical of the class of azeotropic mixtures is chosen for coverage and R-404a as an example of a zeotropic, but near-azeotropic mixture. Both of these refrigerants could actually emerge as halocarbons that are companions to ammonia in industrial refrigeration.

Another important consequence of the banning of CFCs is that R-11, which has been a blowing agent in foamed insulation products, must be replaced. This conversion is important in the construction of refrigerated buildings and will be addressed in Chapter 19.

In the comparisons of refrigerants presented in this chapter, some of the CFCs will be included as a point of reference, but properties of R-12 and R-502 will not be included in the appendix, because their properties have appeared in many other references for years. The properties of the following refrigerants will be included in the appendix: R-22, R-23, R-134a, R-404a, R-507, and ammonia.

12.2 THE OZONE LAYER AND GLOBAL WARMING

The molecules of refrigerants in the halocarbon refrigerant family are made up of some or all of the following elements: carbon, hydrogen, chlorine, and fluorine. Examples of the molecular structures of several classes of halocarbons are shown in Fig. 12.1. Because some of the halocarbons are considered environmentally safe and some others damaging, the industry has adopted a practice of distinguishing among them by the designations of CFC, HCFC, and HFC.

Figure 12.1a shows the structure of a chlorofluorocarbon (CFC) such as R-12, which is a nonhydrogenated (no hydrogen) halocarbon. The chemical is extremely stable, which is a desirable feature for a refrigerant, but when released to the atmosphere it ultimately diffuses to the upper atmosphere. In the upper

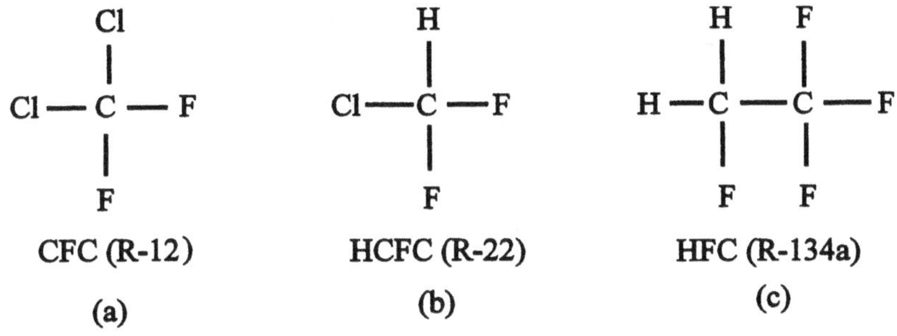

FIGURE 12.1
Molecular structure of several groups of halocarbon refrigerants having differing environmental effects, (a) a CFC, (b) an HCFC, and (c) an HFC.

atmosphere it breaks down, and the chlorine combines with ozone that exists there, depleting the ozone concentration. While ozone has harmful effects at the earth's surface, ozone in the upper atmosphere has the beneficial role of blocking some of the sun's ultraviolet rays. Too high an intensity of ultraviolet rays could result in a greater incidence of skin cancer among the earth's occupants.

The second group of halocarbons is hydrogenated, because it contains a hydrogen atom, as Fig. 12.1b shows. This group is called the hydrochlorofluorocarbons (HCFCs), of which R-22 is an example. Because of the hydrogen atom, the chemical is not quite as stable as a CFC, so when released to the atmosphere, most of it breaks down before reaching the ozone layer. An HCFC, therefore, is likely to be much less damaging to the ozone layer.

Yet another group of halocarbons, called the hydrofluorocarbons (HFCs), is illustrated in Fig. 12.1c. This chemical not only breaks down before reaching the stratosphere, it contains no ozone-depleting chlorine. The particular refrigerant shown in Fig. 12.1c is R-134a, which has similar saturation properties to R-12 and is the front runner for a replacement of R-12.

Molina and Rowland[1] (1974) first hypothesized that the presence of CFCs in the upper atmosphere contributed to depletion of stratospheric ozone (O_3). Further tests and analyses confirmed the perils of continued discharge of CFCs into the atmosphere. A further detriment of the CFCs in the atmosphere is the global warming potential due to the greenhouse effect. Perhaps the discharge of CO and CO_2 represents a more signifcant risk than does the discharge of the CFCs, but the threat must be attacked on all fronts. Table 12.1 shows the ozone depletion potential (ODP) and the global-warming potential (GWP) of some CFC, HCFC, and HFC refrigerants. Ammonia, which is not listed, neither depletes ozone nor causes global warming.

TABLE 12.1
Ozone-depletion potential, ODP, and global-warming potential, GWP (relative to CFC-11) of several refrigerants.

Refrigerant	Formula	ODP	GWP
CFC-11	$CFCl_3$	1.00	1.00
CFC-12	CF_2Cl_2	1.0	3.20
CFC-115	$CClF_2CF_3$	0.6	10.6
HCFC-22	$CHClF_2$	0.055	0.3
HFC-32	CH_2F_2	0	0.12
HCFC-123	$CHCl_2CF_3$	0.02	0.02
HFC-125	CHF_2CF_3	0	0.63
HFC-134a	CH_2FCF_3	0	0.31
HFC-143a	CH_3CF_3	0	0.76

Sources: *Report No. 25*, World Meteorological Organization, Global Ozone Research and Monitoring Project, 1991. Lund et al. 1988, UNEP 1988, Makhijani et al. 1988, du Pont 1988.

12.3 NUMERICAL DESIGNATIONS OF REFRIGERANTS

Standard 34-92 of the American Society of Heating, Refrigerating and Air-conditioning Engineers[2] has categorized and numbered all refrigerants, including air and water. The industrial refrigerants, including those in current as well as future possible use, fall into one of five chemical groups:

- Halocarbons
- Azeotropes
- Zeotropes
- Organic compounds
- Inorganic compounds

Examples of acceptable usage of number designations are R 22, R-22, Refrigerant 22, or (Trade name) 22. The convenience of a numbering system is especially apparent for the halocarbons that have complex chemical names. Some popular industrial refrigerants are listed in Table 12.2.

The halocarbon family consists of halogenated hydrocarbons that contain one or more of the halides—chlorine, bromine, fluorine, or iodine. Hydrogen may or may not be present. The numbering system applicable to the halocarbons and some hydrocarbons is as follows:

- The first digit on the right is the number of fluorine atoms
- The second digit from the right is one more than the number of hydrogen atoms
- The third digit from the right is one less than the number of carbon atoms

TABLE 12.2
Numerical designation of some refrigerants.

Family	Numerical designation	Chemical name	Chemical formula
	12	Dichlorodifluoromethane	CCl_2F_2
Halocarbons	13	Chlorotrifluoromethane	$CClF_3$
	22	Chlorodifluoromethane	$CHClF_2$
	23	Trifluoromethane	CHF_3
	32	Difluoromethane	CH_2F_2
	125	Pentafluoroethane	CHF_2CF_3
	134a	Tetrafluoroethane	CH_2FCF_3
Azeotrope	R-507		R-125/R-143a
Hydrocarbons	170	Ethane	C_2H_6
	290	Propane	C_3H_8
	600	Butane	C_4H_{10}
Inorganic compounds	717	Ammonia	NH_3
	744	Carbon dioxide	CO_2

When the third digit from the right is zero, it is omitted from the number. The 600 series has been assigned to miscellaneous organic compounds and the 700 series to inorganic compounds. Azeotropes that have been commercialized are assigned identifying numbers in the 500 series in order of their commercial appearance.

12.4 REFRIGERANT BLENDS

A summary statement of the position of industrial refrigeration with respect to refrigerants is that ammonia is likely to continue to be dominant in the field, but there is a need for refrigerants of low toxicity to supplement ammonia and to take the place immediately of R-502 and in the future R-22. While it is a suitable replacement for R-12, R-134a will be limited in its applications in industrial refrigeration, so the need remains for a halocarbon appropriate for low-temperature refrigeration. The industry has been forced to turn to mixtures or blends.

Refrigerant mixtures or blends are often characterized as azeotropic mixtures or nonazeotropic (or correctly, zeotropic) mixtures. Azeotropic blends are precise mixtures of substances that have properties differing from either of the two constituents. The diagram of the temperature-pressure-concentration relationship of an ideal zeotropic mixture might appear as shown in Figure 12.2. In a zeotropic mixture, the concentration of the two substances in the vapor is different from that in the liquid at a given pressure and temperature. There are certain applications where the properties of a zeotrope are advantageous, such as in the autocascade system for ultra low temperatures as explored in Chapter 21. For

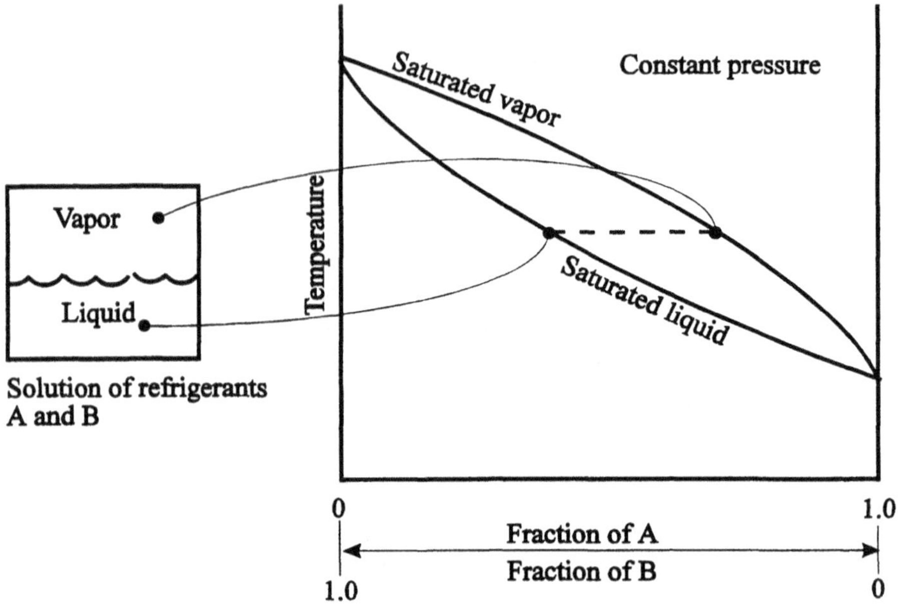

FIGURE 12.2
An ideal zeotropic mixture of substances A and B.

conventional industrial refrigeration systems, however, the zeotrope of Fig. 12.2 has the drawback that if a leakage from a system occurs that is in vapor form, for example, the composition of the lost vapor will be different than the original charge. Replacing the lost refrigerant thus requires some analysis of the concentration of the refrigerant mixture remaining in the system. Another charcteristic of the zeotrope when boiling or condensing is that its temperature changes, in contrast to a single refrigerant whose temperature remains constant during a constant-pressure evaporation or condensation. This characteristic could be an advantage on some systems which are designed for this behavior, but it is not the standard expectation in an industrial refrigeration system.

12.5 AZEOTROPIC MIXTURES

An azeotropic mixture, in contrast to the zeotropic mixture, has a temperature-pressure-concentration diagram where the saturated vapor and saturated liquid lines coincide at a range of concentrations, as shown in Fig. 12.3. At the point or region where the saturated vapor and liquid lines merge, the mixture of the two substances behaves with the properties of a single substance, having properties different from either of its constituents.

Even combinations that are azeotropic at certain concentrations at one pressure may not be perfectly azeotropic at another pressure, as Figure 12.4

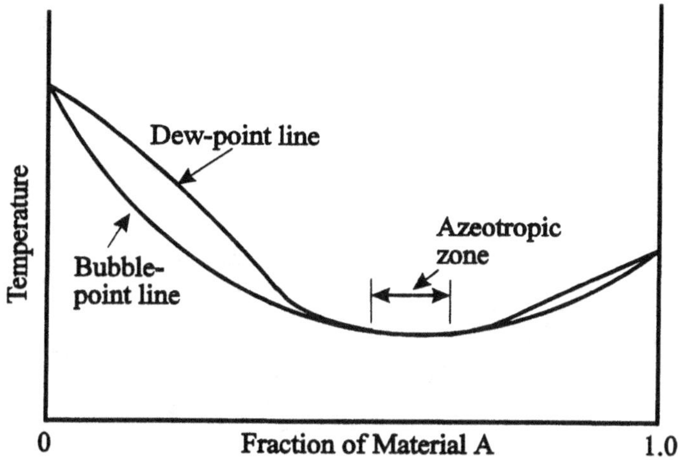

FIGURE 12.3
An azeotropic mixture of Materials A and B in the range of 50-to-60% of A.

shows. The low-pressure temperature-concentration at low pressure, which is that of Figure 12.3 may change as the pressure increases. Usually the azeotropic region shifts toward the high concentration of the low- temperature boiler (material A in this case). Even the azeotrope R-502, which has been used successfully for many years, was subject to fractional distillation at certain operating pressures.

Most of the new substitute halocarbons that have become available in the past few years are blends, most of which are near-azeotropic, but a few are labeled azeotropic, such as R-507 which will be used as an example in this book. As an example of a near-azotropic refrigerant, R-404a will be chosen. In contrast to single refrigerants and azeotropes, the saturation tables (see Appendix) show a slightly different pressure for the vapor compared to the liquid at a given temperature.

Table 12.3 lists the major blends intended as substitutes for R-502 and/or R-22.

12.6 SATURATION PRESSURES AND TEMPERATURES

The magnitudes of the evaporating and condensing pressures at the expected operating temperatures strongly affect the choice of refrigerant. Figure 12.5 shows the saturation pressures as functions of temperature for most of the refrigerants that are listed in Table 12.2. It is desirable that the operating pressures be low enough to use pipe and vessels of standard wall thicknesses. On the other hand, pressures below atmospheric have the disadvantage of drawing in air along with water vapor should there be any leaks. If a leak in the system does exist, it is usually preferable for refrigerant to leak out rather than for air to leak in. The

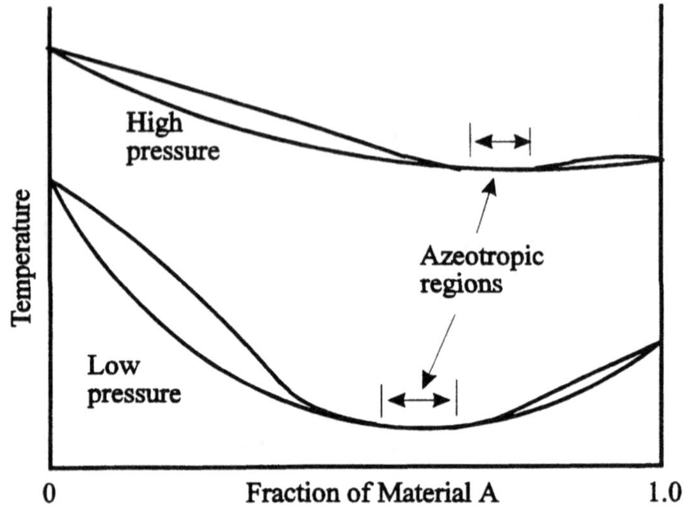

FIGURE 12.4
A shift in the azeotropoic region as the pressure changes.

TABLE 12.3
Zeotropic and azeotropic blends and their composition intended as substituted for R-502 and R-22.

Number	Type	Constituents	Composition
R-402a	Zeotrope	R-125/290/22	60/2/38
R-402b	"	"	38/2/60
R-403b	"	R-290/22/218	5/56/39
R-404a	"	R-125/143a/134a	44/52/4
R-407a	"	R-32/125/134a	20/40/40
R-407b	"	"	10/70/20
R-407c	"	"	23/25/52
R-408a	"	R-125/143a/22	7/46/47
R-410a	"	R-32/125	50/50
R-410b	"	"	45/55
R-507	Azeotrope	R-125/143a	50/50

graph in Fig. 12.5 explains why R-23, ethane, and carbon dioxide are usually reserved for low-temperature systems, and why R-134a might be chosen in the future for systems in which the condensing temperature is high, such as in an industrial heat pump.

Carbon dioxide (R-744) possesses some thermodynamic characteristics that make it advantageous in certain applications. The pressure-enthalpy diagram of carbon dioxide is shown in Fig. 12.6 and demonstrates that solid carbon dioxide, or dry ice, sublimes from a solid to a vapor at −78.4°C (−109.1°F) when the pressure is atmospheric. At pressures above 518.1 kPa (75.1 psia), carbon dioxide

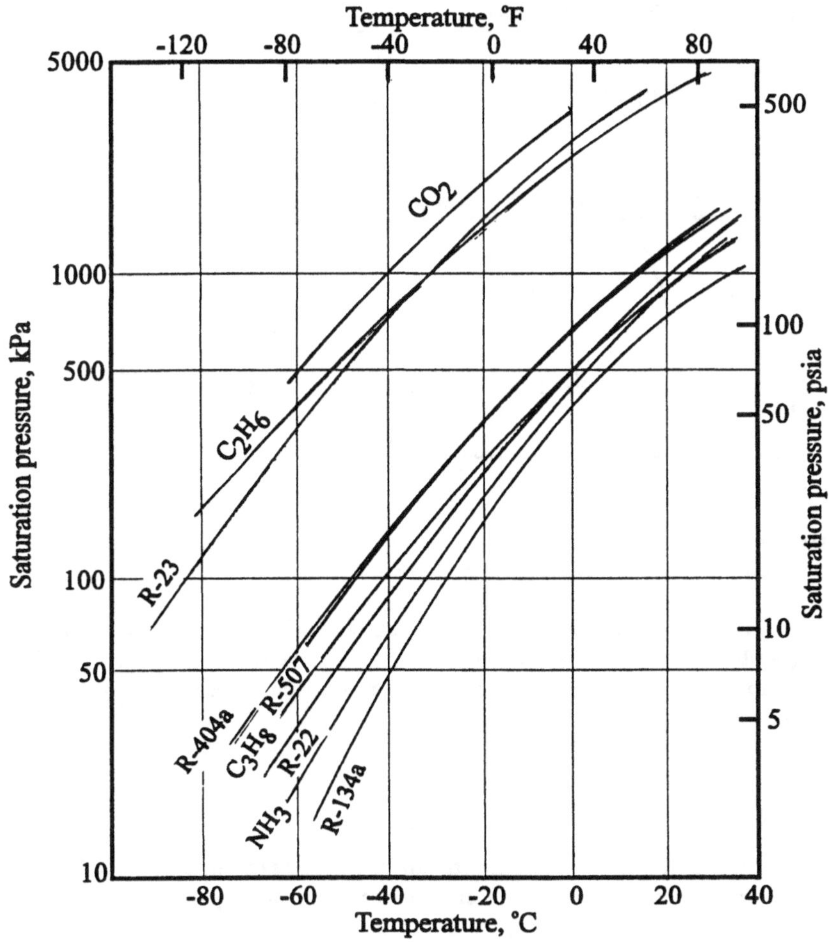

FIGURE 12.5
Saturation pressure as a function of temperature for some industrial refrigerants.

does not solidify and can only exist in liquid and/or vapor phases.

12.7 PHYSICAL PROPERTIES

Several important physical properties[3] of the refrigerants studied in this chapter are listed in Table 12.4. Some of these properties reinforce the distinction between low-temperature and moderate-temperature refrigerants that is apparent from the saturation pressures shown in Figure 12.5. The freezing temperatures of all

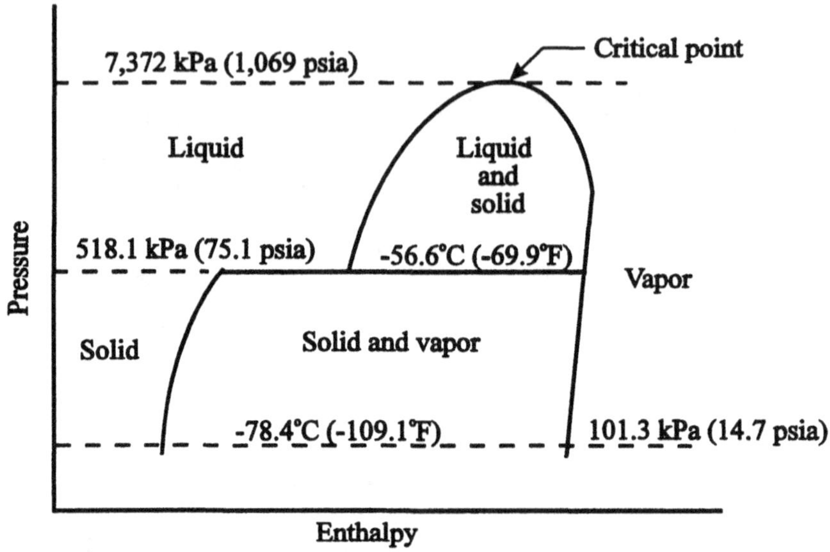

FIGURE 12.6
Skeleton pressure-enthalpy diagram of carbon dioxide.

the refrigerants shown are adequately low for operating temperatures down to −100°C (−148°F), with the exception of ammonia and carbon dioxide. Table 12.4 does not show freezing temperatures for the azeotropes, because they have no precise freezing point.

12.8 LOW-TEMPERATURE HALOCARBON REFRIGERANTS

The halocarbon in the high-pressure group of refrigerants in Fig.12.5, R-23, is a candidate for the low-temperature circuit of the cascade system which will be examined in greater detail in Chapter 21. R-23 may replace R-503, which was a popular low-temperature refrigerant but also an azeotropic mixture, composed of 59.9% R-13 and 40.1% R-23. Since R-13 is a CFC, R-503 has been phased out. R-23 now remains as a single refrigerant adaptable to low-temperature circuits. Water is fairly soluble in R-23, which is an advantage because R-23 is not extremely susceptible to separation and freezing during low-temperature operation. The solubility of oil in R-23 may not be as good as it was in R-503, but this problem is currently being addressed. In one retrofit where R-503 was replaced by R-23, the time for drawing down the temperature was increased. Furthermore, the R-23 discharge temperature from the compressor was high—132°C (270°F)—necessitating the injection of a controlled amount of liquid refrigerant at the suction of the compressor.

Lately, an azeotropic blend of R-23 and R-116 (hexofluoroethane) has been explored. Preliminary tests suggest the use of polyol ester (POE) oils to achieve

TABLE 12.4
Several physical properties of industrial refrigerants.

Refrigerant	Molecular mass	Boiling point at std. atm. °C (°F)	Freezing point °C (°F)	Critical temperature °C (°F)	Critical pressure kPa (psia)
R-22	86.48	−40.76 (−41.37)	−160 (−256.0)	96.0 (204.8)	4,974 (721.1)
R-23	70.02	−82.1 (−115.8)	−155 (−247)	25.6 (78.1)	4,833 (701)
R-125	120.02	−48.5 (−55.3)	−103 (−153)	66.3 (151.3)	3,518 (510.3)
R-134a	102.03	−26.16 (−15.1)	−96.6 (−141.9)	101.1 (214.0)	4,067 (589.9)
R-503	87.5	−88.7 (−127.7)		19.5 (67.1)	4,182 (606.4)
R-507	98.9	−46.7 (−52.1)		70.9 (160)	3,793.6 (550)
R-170	30.07	−88.8 (−127.8)	−183 (−297.4)	32.2 (90.0)	4,891 (709.2)
R-290	44.10	−42.07 (−43.73)	−187.7 (−305.9)	96.8 (206.2)	4,254 (616.8)
R-600	58.13	−0.5 (−31.1)	−138.5 (−217.3)	152.0 (305.6)	3,794 (550.1)
R-717	17.03	−33.3 (−27.94)	−77.7 (−107.9)	133.0 (271.4)	11,417 (1,655)
R-744	44.01	−78.4 (−109.1)	−56.6 (−69.9)	31.1 (88.0)	7,372 (1,069)

lubricant/refrigerant miscibility. This refrigerant offers lower compressor discharge temperatures than R-23.

12.9 CYCLE PERFORMANCE

Various thermodynamic properties of a refrigerant combine when operating within a cycle to influence the performance. Tables 12.5 and 12.6 show several important quantities that give added insight into the choice of refrigerant. To provide a common basis of comparison, the standard vapor-compression cycle with an evaporating temperature of −15°C (5°F) and a condensing temperature of 30°C (86°F) has been chosen for the refrigerants in Table 12.5. For the low-temperature refrigerants in Table 12.6, −50°C (−58°F) and −15°C (5°F) are the evaporating and condensing temperatures, respectively.

The refrigerants listed in Table 12.5 often serve large systems, so their

TABLE 12.5
Performance of several refrigerants in a standard vapor-compression cycle operating at −15°C (5°F) evaporating and 30°C (86°F) condensing temperatures.

Variable	HFC-134a	HCFC-22	R-507	R-717
Latent heat at −15°C, (5°F), kJ/kg (Btu/lb)	207.79 (89.33)	215.93 (92.8)	175.3 (75.4)	1,312.6 (564)
Vapor specific volume at −15°C (5°F), L/kg (ft^3/lb)	119.91 (1.921)	77.34 (1.238)	51.0 (0.816)	508.0 (8.14)
Mass flow rate per unit refrigerating capacity, kg/s per kW (lb/min per ton)	0.00681 (3.17)	0.00615 (2.87)	0.00885 (4.12)	0.000907 (0.422)
Volume flow rate leaving evaporator per unit refrigerating capacity L/s per kW (ft^3/min per ton)	0.8165 (6.08)	0.476 (3.54)	0.452 (3.37)	0.461 (3.43)
Coefficient of performance	4.60	4.66	4.31	4.77

energy characteristics are important. A high latent heat at the evaporating temperature reflects itself in the mass rate of refrigerant flow per unit refrigeration capacity. The latent heats of the halocarbons shown in Table 12.5 are of the same order of magnitude, but the latent heat of ammonia is much higher. Another quantity that affects the size of components is the volume flow rate leaving the evaporator per unit refrigerating capacity. Ammonia, HCFC-22, and R-507 show comparable magnitudes of the volume flow rate per unit refrigerating capacity in comparison to the higher value of HFC-134a. The consequence of this comparison is that the compressor required in an HFC-134a unit must have greater volume-pumping capacity than compressors serving systems with the other refrigerants.

Systems with different refrigerants perform with somewhat unique COPs, so it is fair to say that some refrigerants are more efficient than others. A comparison of the COPs for the refrigerants shown in Table 12.5 indicates that all these refrigerants are about equally efficient, with the possible exception of R-507, which has a COP of about the same order of magnitude as R-502, for which it is a potential replacement.

The choice of low-temperature refrigerants is now quite limited, as was discussed in Section 12.8. Table 12.6 shows a comparison of three candidate refrigerants, HFC-23, ethane (R-170), and carbon dioxide (R-744). The use of HFC-23 as a single refrigerant is just now proving itself, but the results so far have been promising. Carbon dioxide has certain desirable properties, but cannot be used with evaporating temperatures lower than −56.6°C (−69.9°F), which is its triple point.

TABLE 12.6
Performance of several low-temperature industrial refrigerants in a standard vapor-compression cycle with a −50°C (−58°F) evaporating temperature and a −15°C (5°F) condensing temperature.

Characteristic	HFC-23	Ethane	CO_2
Latent heat at −50°C (−58°F), kJ/kg (Btu/lb)	209.6 (90.1)	431.8 (185.2)	336.5 (144.4)
Vapor specific volume at −50°C (−58°F), L/kg (ft^3/lb)	48.49 (0.776)	99.4 (1.592)	55.7 (0.892)
Volume flow rate leaving evaporator per unit refrigerating capacity L/s per kW (ft^3/min per ton)	0.274 (2.04)	0.301 (2.24)	0.209 (1.56)
Coefficient of performance	4.96	4.80	4.66

TABLE 12.7
Matrix of flammability and toxicity for classifying refrigerants in ASHRAE Standard 34-92.

	Lower toxicity	Higher toxicity
Highly flammable	Group A3 (propane)	Group B3 (vinyl chloride)
Moderately flammable	Group A2 (HFC-142a, HFC-152a)	Group B2 (ammonia)
No flame propagation	Group A1 (HFC-134a)	Group B1 (HCFC-123)

12.10 SAFETY OF REFRIGERANTS

Some factors on which the safety of refrigerants are judged include toxicity, carcinogenicity, mutagenicity, and flammability. Recommendations as to where various refrigerants should and should not be used and specification of toxic levels and flammability limits are available from several sources. The data from the different sources do not always agree, although general conclusions can be drawn about, for example, their relative toxicity and flammability.

One safety classification of refrigerants that has been in a state of flux as the writers of the standard attempt to keep pace with the development of new refrigerants is ANSI/ASHRAE Standard 34-92, *Number Designation and Safety Classification of Refrigerants*[2]. The safety classifications established in Standard 34-92 are carried over into ANSI/ASHRAE Standard 15-94, the premier standard for industrial refrigeration in North America. Standard 34-92 establishes a matrix formed by levels of flammability and toxicity, as shown in Table 12.7. The table also lists an example or two of refrigerants classified in each category.

Underwriters Laboratory[4,5] has classified some of the industrial refrigerants with respect to their hazardous effects on life. Several of the groups are:

TABLE 12.8
Permissible exposure during eight hours (weighted average) specified by OSHA[6].

Refrigerant	Maximum allowable, by volume, ppm
Ammonia	35 (short-term exposure limit)
Carbon dioxide	10,000
CFC-12	1,000
Propane	1,000

TABLE 12.9
Threshold limit values of several industrial refrigerants[7].

Refrigerant	TLV, ppm	
	Time-weighted value	Short-term exposure limit
Ammonia	25	35
Butane	800	–
Carbon dioxide	5,000	15,000
HCFC-22	1,000	1,250

- **Group 2**. Gases or vapors which in concentrations of about 1/2 to 1% (5,000 to 10,000 ppm) for durations of exposure of about one-half hour are lethal or produce serious injury[4]: ammonia.
- **Group 5**. Gases or vapors which in concentrations somewhere between 2% and 20% by volume for durations of exposure of about 2 hours are lethal or produce serious injury[5]: HFC-22 and carbon dioxide.
- **Group 6**. Gases or vapors which in concentrations of 20% by volume for durations of exposure of about 2 hours do not appear to produce injury. It is hoped that HFC-134a will fit into this category.

The U.S. Occupational Safety and Health Administration (OSHA)[6] lists maximum permissible exposures, as shown in Table 12.8.

The American Conference of Governmental Industrial Hygienists defines two Threshold Limit Values (TLVs). One is the TLV-TWA, which is the time-weighted average concentration for a normal 8-hour day, 40-hour week, to which nearly all workers could be exposed without adverse results. The TLV Short-Term Exposure Limit (TLV-STEL) is the maximum exposure for a 15-minute period repeated no more than four times a day. Adopted TLVs are shown in Table 12.9.

Threshold limit values for the new HFC refrigerants are not yet established, but tentatively HFC-134a will have TLVs that are similar to HCFC-22.

TABLE 12.10
Physiological response to ammonia vapor[8].

Exposure	Concentration, ppm
Least detectable odor	53
Least amount for prolonged exposure	100
Maximum for 1/2 to 1 hour exposure	300-500
Least amount causing throat irritation	408
Least amount causing eye irritation	698
Least amount causing coughing	1,720
Dangerous for even 1/2 hour exposure	2,500-4,500
Rapidly fatal for short exposure	5,000-10,000

TABLE 12.11
Effect of short-term exposure to ammonia[9].

Concentration, ppm	Effects
150-200	Affected eyes to a limited extent after about one-minute exposure, but vision not seriously impaired. Breathing not affected.
440	Affected eyes more quickly, but not sufficiently to seriously impair vision.
600	Eyes streaming within about 30 seconds. Still breatheable.
700	Tears to eyes in a few seconds. Still breatheable.
1,000	Eyes streamed instantly and vision impaired, but not lost. Breathing intolerable to most participants. Skin irritation after several minutes exposure.
1,500-greater	Instant reaction was to get out of the atmosphere.

An older set of exposure standards[8] for ammonia that is reproduced in Table 12.10 was not quite so stringent as the two standards mentioned above.

Table 12.11 presents results of tests of 1- to 3-minute duration conducted on seven volunteer subjects are presented in Table 12.11.

Some refrigerants are flammable. The ranges of percentages by volume in air constituting flammable limits of several industrial refrigerants are:[4]

- Butane 1.6% to 6.5%
- Propane 2.3% to 7.3%
- Ammonia 16.0% to 25%

Carbon dioxide, HFC-134a, and HCFC-22 are not considered to be flammable refrigerants, although the combustion of a 50% mixture of air and HCFC-22 at pressures higher than 1380 kPa (200 psi) can be triggered by high temperatures.[10]

Many halocarbon refrigerants will decompose when exposed to high temperatures from flames or electric resistance heaters.[11] The products of the decomposition, such as hydrogen chloride and hydrogen fluoride, may be irritating and even toxic.

In summary, all refrigerants should be handled carefully. Even the halocarbon refrigerants, which are the safest, can be injurious and even fatal in high concentrations. The hydrocarbons are quite flammable and should be used only in plants already outfitted to avoid flames and electrical sparks. Ammonia is the most toxic of the industrial refrigerants and has flammability limits somewhere between those of the hydrocarbons and the halocarbons. The flammability limits of 16% to 25% by volume in air correspond to 160,000 to 250,000 ppm, which is 500 to 1000 times the concentration considered toxic. No human could function in an environment where ammonia could burn. Some refrigeration professionals suspect that many accidents reported as ammonia explosions were actually ruptures of pipes or vessels. Leaks of natural gas may also cause an explosion that results in discharges of ammonia.

12.11 COMPATIBILITY WITH THE MATERIALS REFRIGERANTS CONTACT

Most metals may be used in refrigeration systems with the following exceptions. Magnesium, zinc, and aluminum alloys containing more than 2% magnesium are not recommended for use in halocarbon systems, particularly if even traces of water are present.[3] Ammonia should not be used in any system containing copper, brass, or other alloys containing copper.

In certain cases, elastomers such as rubber, Buna N, and Buna S will be weakened by certain halocarbons. Elastomers are often used as seals, and the manufacturer of the refrigerant should be consulted for recommendations of what materials to use.

12.12 WHICH REFRIGERANTS WILL SERVE INDUSTRIAL REFRIGERATION IN THE FUTURE?

This chapter is being written at a time when the phase out of CFCs and the potential phaseout of some HCFCs has resulted in an upheaval in the choice of refrigerants. A capsule summary of the status in the past is that for low-temperature industrial refrigeration systems, HCFC-22 and ammonia have been the strongest competitors. Refrigerant-502 had found a role for some low-temperature applications, particularly when operating at high pressure ratios where its low adiabatic discharge temperature was an advantage. Statistically, CFC-12 has not been widely used in industrial refrigeration. Even though CFC-12 at low temperatures has desirable mutual solubility with oil, its requirement of high volume rates of flow at the compressor place it at an economic disadvantage.

CFC-12 is no longer in the picture and its replacement is HFC-134a, which might find applications in systems requiring a refrigerant of low toxicity with evaporating temperatures above −5°C (23°F). There is a possibility that R-507, some of the new HFC blends, and even HFC-125 will be important in the future. For the next several decades, HCFC-22 will continue to compete vigorously with ammonia as the refrigerant of choice, until HCFC-22 is phased out. The manufacturers of halocarbon refrigerants are not idle, and new refrigerants are already available. Unfortunately, all the refrigerants seem to have one or more drawbacks, and their applications must be directed to minimizing their disadvantages.

12.13 AMMONIA VS. HCFC-22

The decision to be made in the next several decades by designers and owners of industrial refrigeration plants in the choice of refrigerant is primarily between ammonia and HCFC-22. The decision may be a quick one if ammonia is not permitted at the location or its use is inadvisable. Codes in certain cities or counties may restrict the type of facility in which ammonia can be used. In general, its use is permissible in locations physically separated from spaces to which the general public has access. Even if no codes legislate against it, placing a large ammonia system close to a school, hospital, or similarly occupied building would not be prudent. Progress is being made in the development of packaged ammonia liquid chillers with low refrigerant charges that open opportunities for applications not previously considered, such as water chillers for air conditioning.

The first point of comparison will be cost. The prices of refrigerants fluctuate and are a function of the quantity bought, but a rough comparison on a mass basis of ammonia, HCFC-22, and HFC-134a is shown in Figure 1.9. The costs are relative to that of HCFC-22 and show ammonia with the lowest cost and HFC-134a the highest. The cost comparison becomes a factor in plants containing tens of thousands of kilograms or pounds. The cost comparison is even more striking when it is realized that to fill a plant, a certain amount of volume (not mass) of liquid must be supplied. Because the density of liquid ammonia is about half that of the halocarbons, only half the mass needs to be purchased.

Table 12.5 showed comparisons of mass and volume flow rates for HFC-134a, HCFC-22, R-507, and ammonia. The volume flow rates that must be handled by the compressor in systems of the latter three refrigerants are comparable, so there is no advantage for one of the refrigerants. The mass flow rate for a given refrigerating capacity of ammonia is 1/7 that of HCFC-22, and this comparison exerts itself both in required pipe sizes and in requirements for liquid recirculating systems. Only 1/7 of the liquid need be pumped for a given refrigerating capacity, which means that the liquid pump can be much smaller and pumping power will be much smaller in an ammonia system.

Table 12.5 also showed that the volume flow rate in the suction lines is

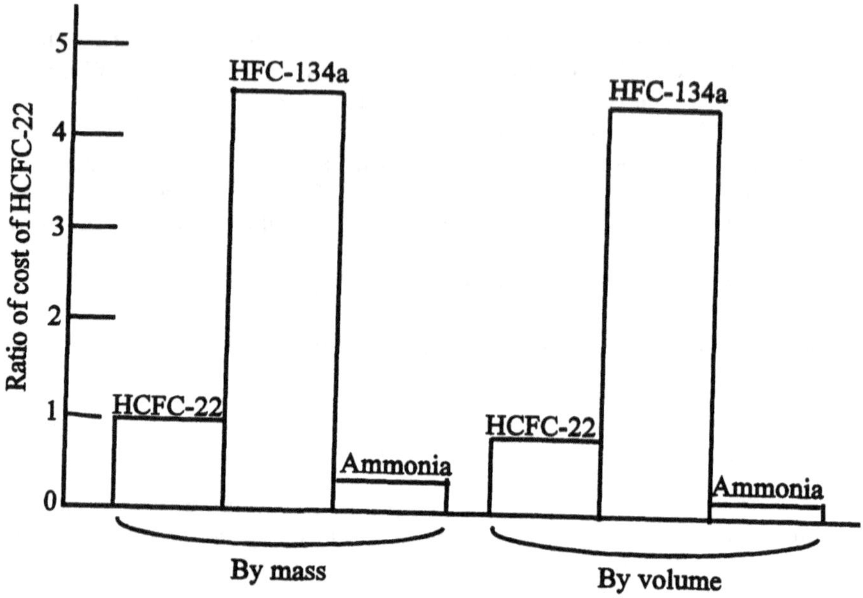

FIGURE 12.7
Comparative costs of several industrial refrigerants.

comparable for HCFC-22, R-507, and ammonia, which may suggest that the pipe sizes for low-pressure vapor should be of the same order of magnitude. Another factor is concealed, however. The principal criterion in selecting vapor line sizes is the drop in saturation temperature. Table 12.12 compares the refrigeration capacities for several different pipe sizes and drops in saturation temperature for HCFC-22 and ammonia. This advantage for ammonia shows itself particularly in industrial facilities spread over a large area where the suction lines are long. Either the pipe size for ammonia could be smaller than that for HCFC-22 or for a given pipe size the pressure drop penalty will be less.

Ammonia is more tolerant of water that might inadvertently get into the system. In most halocarbon systems, water remains separate from the refrigerant and could freeze, especially immediately after an expansion valve. With ammonia, however, water stays in solution and contaminations less than 100 ppm, for example, cause no penalty to system operation.

Ammonia enjoys higher heat-transfer coefficients than HCFC-22, primarily because most of the thermodynamic and transport properties that affect heat transfer are favorable to ammonia. The ratios of these properties for ammonia relative to HCFC-22 are the following:

- Specific heat of liquid and vapor, 4 to 1
- Latent heat of vaporization, 6 to 1
- Liquid thermal conductivity, 5.5 to 1

TABLE 12.12
Refrigeration capacities served by pipes of various diameters using HCFC-22 and ammonia. The quantities apply to suction vapor lines 30.5 m (100 ft) long operating at −17.8°C (0°F) saturation temperature.

Refrigerant	Refrigerating capacity, kW (tons)			
	Drop in saturation temperature of 0.56°C (1°F)		Drop in saturation temperature of 1.1°C (2°F)	
	4 in. dia.	6 in. dia.	4 in. dia.	6 in. dia.
HCFC-22	199 (56.7)	587 (167)	281 (80)	833 (237)
Ammonia	545 (155)	1,583 (450)	731 (208)	2,110 (600)

TABLE 12.13
Range of heat-transfer coefficients for ammonia and HCFC-22 used by several designers, expressed in W/m²·K (Btu/hr·ft²·F).

	Ammonia	HCFC-22
Condensation outside tubes	7,500-11,000 (1,300-2,000)	1,700-2,800 (300-500)
Condensation inside tubes	4,200-8,500 (750-1,500)	1,400-2,000 (250-350)
Boiling outside tubes	2,300-4,500 (400-800)	1,400-2,000 (250-350)
Boiling inside tubes (liquid recirculation)	3,100-5,000 (550-850)	1,500-2,800 (250-500)

- Viscosities, 0.8 to 1
- Liquid density, 0.5 to 1

A survey[12] conducted among designers of heat exchangers who work with both ammonia and HCFC-22, as shown in Table 12.13, indicates that heat-transfer coefficients used in industrial practice are two to three times higher for ammonia.

It is fair to point out by way of additional perspective that the copper tubes used for halocarbon refrigerants lend themselves conveniently to heat-transfer enhancement, so that the values that apply to bare tubes shown in Table 12.13 for HCFC-22 can be improved. On the other hand, research results are showing that heat-transfer enhancement is possible with the steel or aluminum tubes used in ammonia heat exchangers as well. So far, the possibilities have not generally been commercialized.

A capsule conclusion of the foregoing discussion is that ammonia has many advantages as a refrigerant. Its major disadvantage is the low concentration considered toxic. The safety provisions to be discussed in Chapter 13 are particularly crucial for maintaining and expanding the realms of ammonia applications.

REFERENCES

1. Molina, M. J. and F. S. Rowland, "Stratospheric sink for chlorofluoromethanes: Chlorine-atom catalyzed destruction of ozone," *Nature*, 249: 810-812, 1974.
2. ASHRAE Standard 34-92, *Number Designation and Safety Classification of Refrigerants*, American Society of Heating, Refrigerating and Air-Conditioning Engineers, Atlanta, Georgia, 1992.
3. American Society of Heating, Refrigerating and Air-Conditioning Engineers, Atlanta, Georgia, from the 1993 ASHRAE *Handbook—Fundamentals*, Chapter 16.
4. Nuckolls, A. H., "The Comparative Life, Fire, and Explosion Hazards of Common Refrigerants," *Miscellaneous Hazard No. 2375*, Underwriters Laboratory, Northbrook, Illinois, 1933.
5. Nuckolls, A. H., "The Comparative Life, Fire, and Explosion Hazards of Difluoromonochloromethane ('Freon-22')," *Miscellaneous Hazard No. 3134*, Underwriters Laboratory, Northbrook, Illinois, 1940.
6. Standard 1910.1000, *Air Contaminants Subpart Z-Toxic and Hazardous Substances*, Occupational Safety and Health Administration, March 1990.
7. "Threshold Limit Values and Biological Exposure Indices for 1990-1991," American Conference of Governmental Industrial Hygienists, Cincinnati, Ohio, 1991.
8. Henderson and Haggard, "Noxious Gases," Chemical Catalog Co., 1927.
9. Wallace, D. P, "Atmospheric Emissions and Control," *Ammonia Plant Safety*, 21: 51-56, 1979.
10. Sand, J. R. and D. L. Andrjeski, "Combustibility of Chlorodi- fluoromethane," *ASHRAE Journal*, 24(5): 38-40, 1982.
11. "Safety of Freon^R Refrigerants," Bulletin S-38, E. I. du Pont de Nemours & Company, Wilmington, Delaware, 1985.
12. Stoecker, W. F., "Growing Opportunities for Ammonia Refrigeration," *Heating, Piping, and Air Conditioning*, 61(9): 93-108, September 1989.

CHAPTER 13

SAFETY

13.1 SAFETY IS PARAMOUNT

Of all the facets in the design and operation of refrigeration facilities, none may be more critical than safety. Providing the required refrigeration capacity at low energy costs may affect the profit of a business, but energy efficiency only becomes a factor after safe operation is ensured. The primary motivation for a safe plant is to protect people, including employees and the neighboring public. In addition, accidents often result in loss or damage to stored products as well as requiring repairs—all of which may be costly.

The emphasis on safety impacts the refrigeration plant in three major categories:

- Design
- Construction and installation
- Operation

A fundamental responsibility of the engineer in the *design* of the facility is to observe pertinent safety codes. Most code requirements are written to assist in achieving a safe plant. The designer is also compelled to conform to codes for reasons of professional liability. A fundamental objective of the designer is to specify a system that is efficient and meets the capacity and temperature requirements of the client. Then, beyond obeying the stipulations of the codes, the designer should choose components and arrange the equipment for convenient access for service and maintenance. One provision that enhances maintenance is the effective monitoring of the equipment, so proper instrumentation, as will be discussed in Chapter 14, should be planned in the design stage.

Following design, another phase in the life of a refrigeration plant is its *construction and installation*, either for the modification of an existing plant or in building a new one. Design documents never specify all necessary details, so the installer must make many decisions during construction. The installer, therefore, has a key role in providing good workmanship and making judgments that prevent failures once the plant is in operation. A critical time for accidents in the process of installing or repairing a facility is after the components have been charged with refrigerant but before operation becomes routine.

The third important sector where safety must be stressed is in the *operation* of the plant. The operating personnel should understand the plant thoroughly—they should be very familiar with how the system works and why it works the way it does. With such an understanding, when a nonroutine problem arises they can analyze the situation to decide the best course of action. The person most vulnerable to an accident is the operator or service technician. A significant number of injuries are suffered by the mechanic making an adjustment or trying to correct a problem. Some accidents occur even though the worker takes proper precautions and uses proper tools and procedures. Sadly, however, many accidents are caused by improper practices. These accidents often could have been prevented if the technician had been properly trained—a responsibility of the supervisor.

Because the conformity to safety codes and standards assists in arriving at a safe plant, this chapter steps through what is probably the premier U.S. industrial refrigeration standard, ANSI/ASHRAE Standard 15-94, *Safety Code for Mechanical Refrigeration*[1]. When the plant is designed and constructed according to applicable codes, the assignment is only half done. Numerous opportunities remain in the arrangement and placement of components as well as in the construction details to enhance the safety of the plant. A sampling of these features will be described in this chapter.

13.2 SAFETY CONSIDERATIONS FOR AMMONIA PLANTS COMPARED TO THOSE USING HALOCARBONS

In much of this chapter, there is no special mention of which refrigerant is being assumed. The fact is, however, that safety requirements are heightened for ammonia. Certainly, all the specifications in ANSI/ASHRAE Standard 15-94 dealing with the testing and protection of pressure vessels are equally applicable to all refrigerants. But the appreciably lower concentration considered toxic for ammonia compared to that for most of the halocarbons, as pointed out in Chapter 12, reinforces the need for special care and attention to safety details in ammonia systems.

Unique precautions, however, also apply to halocarbon systems. In a large halocarbon plant, a significant quantity of refrigerant could escape without the operators being aware of it, because halocarbons are odorless. A further characteristic of halocarbons is that they are heavier than air and may collect in low levels of the building, such as a basement. High concentrations can asphyxiate a worker, so this potentially dangerous situation should be prevented by suitable detectors and alarms. With ammonia, the existence of a leak is evident because of the odor, and the challenge is to control the magnitude and correct the leak to avoid injury to people and damage to products.

13.3 OVERVIEW OF ANSI/ASHRAE STANDARD 15-94

In addition to routine editorial changes and rearrangements, the 1994 revision of the standard contains some significant changes from past versions, particularly Standard 15-78 which was in force for many years. One of the major revisions is in the occupancy standards as influenced by the refrigerant classifications adopted in the companion standard ANSI/ASHRAE Standard 34-92, which was discussed in Chapter 12. Along with the accommodation of the new refrigerant classifications, the standard specifies the types of building occupancies where the refrigerants of the various classifications can be applied. Furthermore, new ventilation requirements for machine rooms are specified, and these new specifications are reviewed later in Section 13.8.

ANSI/ASHRAE Standard 15-94 consists of 14 sections and three appendices. The first three sections cover preliminary explanations. Sections 4 through 7 combine the refrigerant classifications with indications of where and how each of these classifications can be applied. Sections 8 through 10 deal especially with vessels and their protection from high pressures. Section 11 is titled Installation Requirements and includes material on machine room ventilation that is especially important for ammonia. Appendices A and C address the new types of refrigerants—the blends of nonazeotropic mixtures. Appendix B is pertinent to industrial refrigeration systems, because it deals with emergency releases of refrigerants.

13.4 ADDITIONAL CODES

ANSI/ASHRAE Standard 15-94 refers to other codes, and thus makes those references a part of the original code. Referenced codes include the ANSI/ASME *Boiler and Pressure Vessel Code*[2] of the American Society of Mechanical Engineers, which is applicable to unfired pressure vessels and the ANSI/ASME B31.5 *Refrigeration Piping Code*[3] for piping. Pressure vessels exceeding 152 mm (6 in) I.D. with design pressures greater than 103.4 kPa gauge (15 psig) must comply with the requirements of the *Boiler and Pressure Vessel Code* in design, fabrication, inspection, and testing during construction. In a similar manner, refrigerant piping for design pressures 103.4 kPa gauge (15 psig) or higher must comply with the ANSI/ASME B31.5 Code. Another standard referenced in ANSI/ASHRAE Standard 15-94 that is particularly applicable to flammable refrigerants, such as those in Groups A2, A3, B2, and B3, as described in Chapter 12, is the *National Electrical Code*.[4]

A standard[5] developed by the International Institute of Ammonia Refrigeration, Standard ANSI/IIAR-2-1992, addresses only ammonia mechanical refrigeration systems, so it can be ammonia-specific. Some of the sections, such as those dealing with pressure relief, piping, and vessels, parallel ANSI/ASHRAE Standard 15-94, but a number of other specifications are unique. ANSI/IIAR-2-92 requires 99.95% purity for ammonia. The standard also requires nameplates on the major components that state the manufacturer, year of manufacture, manufacturer's identification or model number, and design pressure. The purposes of these nameplates are to identify that the equipment has been safety-tested and to aid in detecting improper application. The standard specifies two levels of design pressure, high- and low-side. An alternate approach to specifying machine room ventilation is contained in the standard, and this subject is explored in more detail in Section 13.8 of this chapter.

Several other codes and standards that may apply in certain jurisdictions are ANSI/UL 207-1986, *Safety for Refrigerant-Containing Components and Accessories, Nonelectrical*[6] of the Underwriters Laboratory and the BOCA *Basic National Mechanical Code*[7]. ANSI/UL 207 parallels ANSI/ASHRAE Standard 15-94, omitting some of the Standard 15-94 specifications but including others. Some governmental units have adopted such codes as the *Basic National Mechanical Code* of BOCA (Building Officials and Code Administrators) and the *Uniform Mechanical Code*[8] of the International Conference of Building Officials. The BOCA Code cites ANSI/ASHRAE Standard 15-94 for permitted quantities of refrigerant in the system. For the discharge of ammonia from a system, the BOCA Code parallels ANSI/ASHRAE 15-94 and IIAR-2-92 in stating that the preferred discharge is to atmosphere, but also describes an optional discharge to a tank of water, as will be discussed further in Section 13.12.

Engineers in the United Kingdom are guided by two different codes[9,10], one for ammonia and the other for halocarbons. These codes contain some of the same specifications that are parts of IIAR-2-92 and ANSI/ASHRAE Standard 15-94.

13.5 VESSELS

The suggestion that any refrigerant-containing component is a vessel is too broad a definition, because ANSI/ASHRAE Standard 15-94 excludes tube-type condensers and evaporators from the category of vessels. However, Section 9.11.1, which does apply to condensers, requires that the component shall have an ultimate strength sufficient to withstand three times the design pressure for which it is rated. Furthermore, auxiliaries such as gauges and bodies of valves are not considered vessels. The usual interpretation of Standard 15-94 is that a vessel is a component containing liquid refrigerant (except for the tube-type components described above). Thus, an oil separator is not considered to be a vessel. Standard 15-94 groups some of the components into different categories, for example:

TABLE 13.1
Design pressure for several refrigerants.

Description	Saturation temperature	Design pressure, kPa gauge (psig)		
		R-22	R-502	R-717
Low sides	26.5°C (80°F)	993 (144)	1,117 (162)	958 (139)
High side, water- or evaporative-cooled	40.5°C (105°F)	1,445 (211)	1,600 (232)	1,482 (215)
High side, air-cooled	51.5°C (125°F)	1,917 (278)	2,082 (302)	2,020 (293)

- Components smaller than 152 mm (6 in) diameter with a design pressure less than 103.4 kPa gauge (15 psig) are exempt from the requirements listed below.
- Pressure-containing components smaller than 152 mm (6 in) diameter with a design pressure greater than 103.4 kPa gauge (15 psig) must be listed with an approved testing laboratory or shall meet the requirements of the ANSI/ASME *Boiler and Pressure Vessel Code.* For a component to be *listed,* it must have been tested and identified as acceptable by an approved, nationally recognized testing laboratory.
- Vessels larger than 152 mm (6 in) diameter with a design pressure greater than 103.4 kPa gauge (15 psig) must conform to Section VIII of the ANSI/ASME *Boiler and Pressure Vessel Code.*

ANSI/ASHRAE Standard 15-94 provides specifications for pressure-relief devices on vessels and what their settings should be. A pressure vessel must be equipped with a pressure-relief device if there is any possibility that the vessel could be valved off from another part of the system where a pressure-relief device exists.

A key characteristic of a vessel is its design pressure. The design pressure depends on the refrigerant and whether the vessel is on the high- or low-pressure side of the system. As Table 13.1 shows, the design pressure (as specified in ANSI/ASHRAE Standard 15-94) is the saturation pressure at various maximum temperatures of refrigerant liquid.

The pressures listed in Table 13.1 are the minimum, and designers often request a rounded-off pressure slightly higher than those listed. The fabricator, in turn, must work with standard material thicknesses which, in combination with the vessel diameter, determine the pressure capability of the vessel. Therefore, the vessel from the fabricator may be rated at a pressure slightly higher than requested, or the fabricator may contact the designer to ask whether a slightly lower design pressure than requested could be permitted. The vessel is hydrostatically tested by the fabricator to 1.5 times the design pressure, which provides a simultaneous test for leaks. The test pressure does not mean that the safety factor on the vessel is only 1.5, because the calculated shell thickness is such that the safety factor is usually about 4 or 5.

The welding operations on the vessel are tightly prescribed in order to *qualify* the vessel, which refers to both the welder and the procedures. For the procedures to be qualified, specific materials must be used for the weld, and the welder must be qualified by passing tests on whatever type of weld is in question (pipe, couplings, shell, heads, etc.).

The buyer of the vessel expects a certification, often called the stamp, which is a stainless steel nameplate with the symbol *U* (for unfired) within the ASME emblem, as shown in Fig. 13.1a.

The owner and operator of a plant usually want the stamp to be visible, even if the vessel is ultimately insulated. The fabricator can be requested to mount an extension bracket, as in Fig. 13.1b, so that the nameplate is not concealed by the insulation.

Table 13-1 shows that a lower design pressure would be permitted for a low-side vessel. Many designers, nevertheless, specify a high-side design pressure even for low-pressure receivers and constant-pressure receivers, so that if ever there is a need to store liquid during shutdown, the vessels will be pressure-rated for this purpose.

13.6 PIPING AND VALVES

The ASME *Code for Pressure Piping*[3] is the applicable code to follow in designing and installing piping. In addition to some specific requirements, the code lists some general precautions, such as the following:

- "Consideration must be given to expansion of liquid refrigerant trapped in or between closed valves and a means provided to prevent overpressure."
- "Impact forces (including hydraulic shock and liquid slugging) caused by either external or internal conditions shall be considered in the design of piping components."
- "Piping shall be arranged and supported with considerations to vibration."
- "Piping systems shall be designed, arranged and supported so as to withstand reaction forces due to let down or discharge of fluids."

These issues and several additional ones related to piping and valves will be addressed through an exploration of the following individual topics: (1) expansion of trapped liquid, (2) impact forces and shock caused by sudden deceleration of liquid, (3) liquid propelled by high-velocity vapor resulting from sudden reduction in pressure, (4) valving the oil drain line in an ammonia system, and (5) additional recommendations for safe piping practices. The first three of the above subjects are also explored in Reference 11.

Expansion of trapped liquid. A frequent cause of pipe rupture is due to trapping liquid between two valves. If this liquid is cold when trapped, when it warms it can develop enormous pressures that may be high enough to rupture the

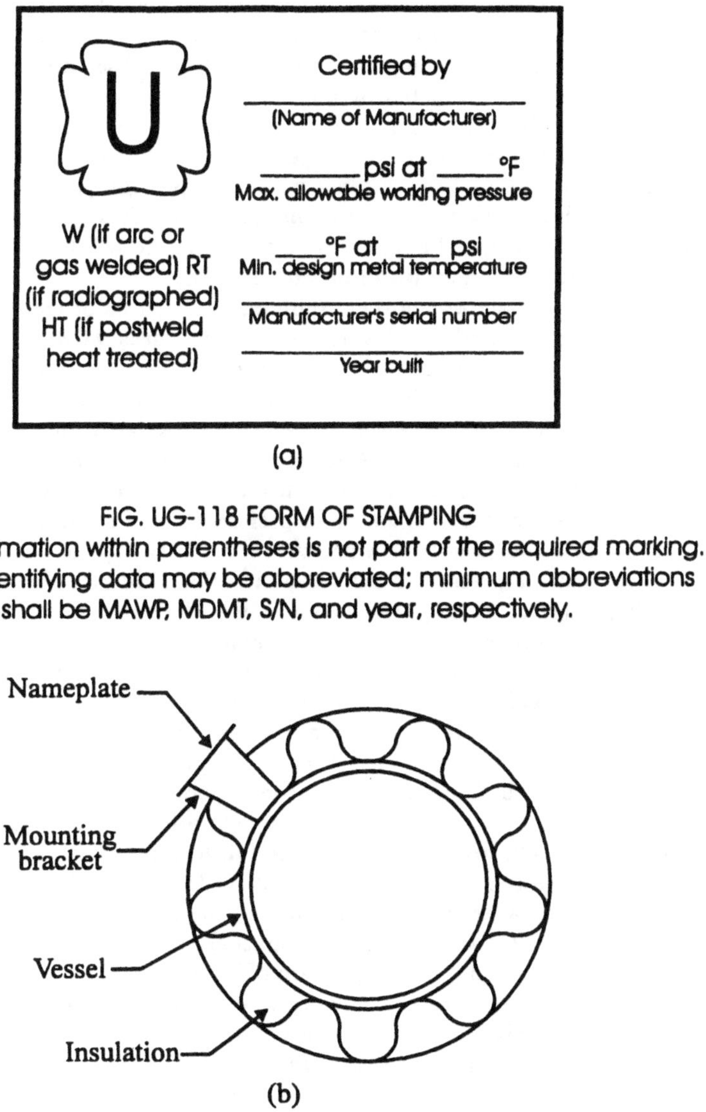

FIGURE 13.1
(a) Replica of the certification stamp for a shell-and-tube heat exchanger, and (b) extension bracket for an insulated vessel.

pipe or valves. A section of pipe with two shutoff valves at either end, as shown in Fig. 13.2a, is an obvious example of the case where a pressure-relief valve must be installed. The discharge from this relief valve could be vented directly to atmosphere, but it is adequate and preferable to discharge to a section of the system that is itself protected by a relief valve:

A more subtle situation where liquid can be trapped is between a check valve and a shutoff or solenoid valve, as in Fig. 13.2b. This arrangement is common in liquid-recirculation systems with multiple pumps, where each pump is equipped with a check valve on the discharge side. If solenoid valves regulating liquid flow to the coils all happen to be closed, liquid will be trapped and during the no-flow situation may warm up and expand.

Figure 13.2c shows another combination capable of trapping liquid. Here an outlet pressure regulator downstream from a check valve can act as a shutoff valve. If the outlet pressure is high, the regulator will remain closed and the check valve will not permit liquid to pass back through it. Some commercial regulators are designed, however, to relieve against abnormally high pressures.

As Figs 13.2a through 13.2c show, the section of pipe in which liquid refrigerant might be trapped is equipped with a pressure-relief valve. Some designers specify an inlet pressure regulator in preference to a standard relief valve. The reason for this choice is that a pressure-relief valve opens based on a pressure difference. Thus, a pressure-relief valve rated for 1724 kPa gauge (250 psig) normally discharges to atmospheric pressure of 0 kPa gauge (0 psig) and thus opens with a pressure difference of 1724 kPa (250 psig). If the discharge of a relief valve shown in Fig. 13.2 vents to a pressure vessel operating at 1,034 kPa gauge (150 psig), it will actually require a pressure of 1724 + 1034 = 2758 kPa (400 psi) to open the valve. Inlet pressure regulators, on the other hand, open only against a spring pressure and are independent of the pressure at the valve outlet.

The ball valve of Fig. 13.2d is becoming popular because of its low pressure drop when completely open. If a ball valve in a liquid line is closed, it will trap liquid within itself. Most ball valves are fabricated in such a way that the seal is capable of venting to the upstream side of the valve when it is closed. To be certain, some installers drill a small hole in the ball to vent upstream.

Sudden deceleration of liquid. Impact forces can be generated by quickly stopping the motion of a moving stream of liquid in a pipe. An initial estimate of the magnitude of the pressure spike can be made by visualizing a plug of liquid in motion, as in Fig. 13.3. When the valve closes, a pressure must build up on the liquid cross section at the valve (and thus throughout the liquid) such that the force equals the rate of change of momentum of the moving rod of liquid.

$$pA = \frac{\text{(mass)(original velocity} - 0)}{\text{time needed to bring liquid to rest}}$$

where:

p = pressure, Pa (psf)
A = area, m^2 (ft^2)

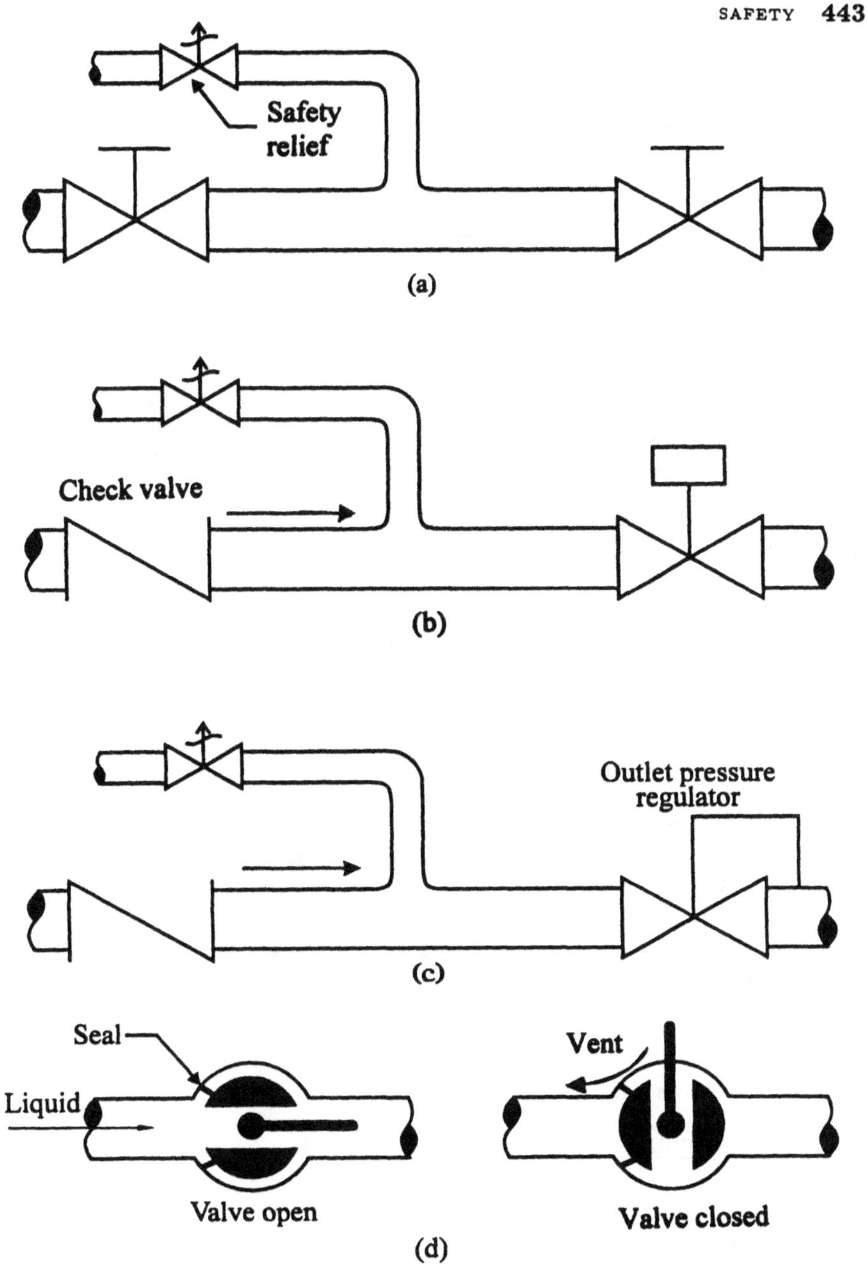

FIGURE 13.2
Trapping liquid (a) between two shutoff valves, (b) between a check valve and a manual or solenoid valve, (c) between a check valve and an outlet pressure regulator, and (d) within a ball valve.

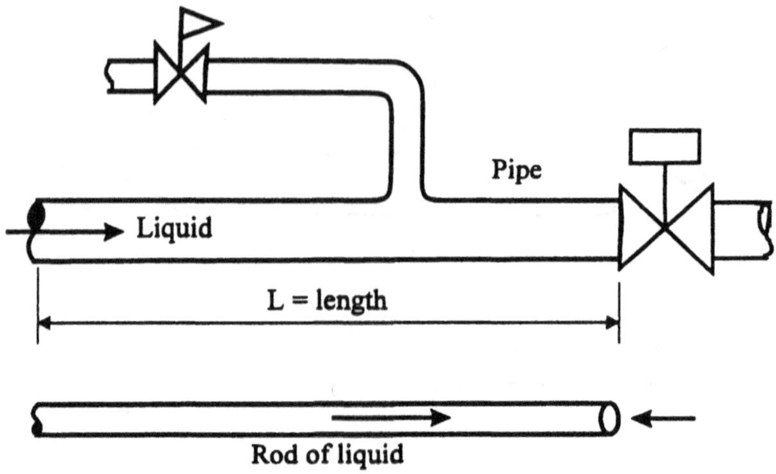

FIGURE 13.3
Sudden closing of a valve in a liquid line develops high pressure in the liquid.

The mass is the volume of the liquid rod multiplied by the density of the liquid, so:

$$pA = \frac{(LA\rho)V}{closing\ \ time,\ \mathrm{s}} \tag{13.1}$$

where:

L = length of rod, m (ft)
ρ = density of liquid, kg/m^3 (lb/ft^3)
V = velocity, m/s (ft/s)

Example 13.1. Refrigerant-22 with a density of 1286 kg/m^3 (80.3 lb/ft^3) moving at 2 m/s (6.6 ft/s) in a 20 m (66 ft) length of pipe is brought to rest by the closing in 0.02 s of a solenoid valve. How much overpressure is developed?

Solution. Applying Eq. 13.1:

$$p = \frac{(20\ \mathrm{m})(1,286\ \mathrm{kg/m^3})(2\ \mathrm{m/s})}{0.02\ \mathrm{s}} = 2,572,000\ \mathrm{Pa} = 2572\ \mathrm{kPa}\ (373\ \mathrm{psi})$$

The high-pressure pulse that first develops at the valve transmits back upstream through the liquid refrigerant at a sonic velocity[12]. For ammonia liquid at −34°C-(−29°F), the sonic velocity is 1715 m/s (5625 ft/s). The associated noise is attributable to the metal pipes becoming sound generators. The pressure pulse calculated in Example 13.1 is not sufficient to rupture the pipe. However, in plants where the liquid velocity is high or a solenoid valve closes quickly, pipes may shake each time a valve closes, which is an indication to the operator that an abnormal situation prevails. Also, the pressure-relief valve might open momentarily at each pressure pulse and eventually start leaking. Designing for low velocities in the liquid lines is one method of reducing this impact force.

Liquid propelled by high-velocity vapor. Some disastrous incidents appear to have as their root cause the sudden expansion of high-pressure vapor into a low-pressure space. If liquid refrigerant is in the path of the high-velocity vapor, this liquid can be driven at high velocity against the end of the pipe, possibly causing a rupture. Two typical situations where there could be a rapid release of pressure are both associated with the hot-gas defrost process:

1. Opening of the valve admitting high-pressure defrost gas into the coil that is to be defrosted
2. Opening of the suction solenoid valve between the coil and the suction line when defrost is complete

In both of these situations, vapor rushes into the low-pressure region and high vapor velocities exist during the pressure equalization. In the first situation, a coil may be damaged at the initiation of defrost. During refrigeration service the coil, particularly a liquid-overfeed coil, contains considerable liquid, as illustrated in Fig. 13.4a. Furthermore, additional liquid may enter the coil because of condensed refrigerant in the hot-gas line. When the defrost process starts, high-pressure gas rushes into the coil that is at a low pressure and drives some of the liquid with high velocity against U-bends or headers of the coil, as shown in Fig. 13.4b. The force may be enough to rupture a tube in the coil or the drain pan.

Another possible cause of hammer-like blows and the knocking sound within the coil at the start of defrost may be attributable to the sudden condensation of bubbles of vapor in the subcooled liquid[12]. When the liquid is subcooled by more than about 35°C (63°F), the rapid condensation of vapor causes the liquid to close on itself with such velocity that the impact generates high localized pressures.

Some steps that may be taken to avoid liquid hammer at the initiation of defrost are the following:

- Limit the pressure of the entering defrost gas by means of a pressure-regulating valve, as shown in Fig. 13.4b. In summer operation when the condensing pressure and thus the pressure of the defrost gas is high, the entering pressure permitted is reduced. Dropping the defrost gas pressure from, say, 1275 kPa gauge (185 psig) to 950 kPa gauge (140 psig), does not cause a noticeable lengthening of the defrost time.
- Employ a pumpout process. In this operation, the first step on initiation of defrost is to close the solenoid in the liquid supply line to the coil. The fan continues to operate giving the liquid in the coil time to vaporize. Only after this pumpout operation (perhaps 5 to 10 minutes) does the fan stop, the valve in the suction line closes, and the defrost gas valve opens. The

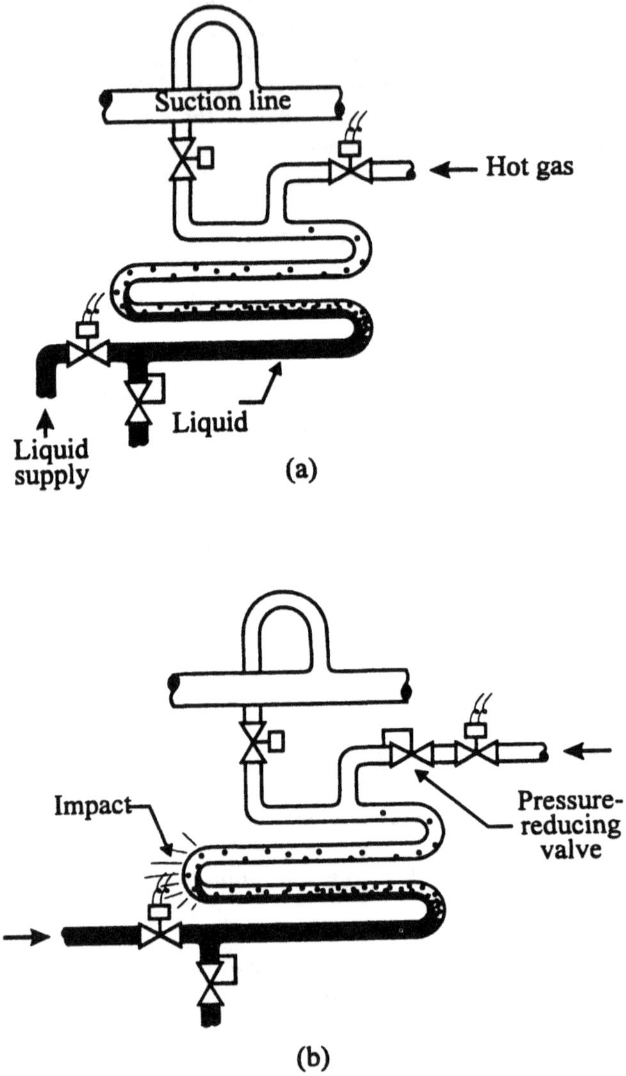

FIGURE 13.4
(a) Liquid in the coil immediately before defrost, and (b) high-pressure defrost gas driving liquid at high velocity against a U-bend or header of the coil.

pumpout operation rids the coil of most of its liquid so the entering defrost gas finds little or no liquid to drive against the ends of tubes.

- Keep the defrost gas main as free of condensed refrigerant as possible. The defrost gas line often passes through low-temperature spaces which condense the refrigerant; but some means of preventing liquid accumulation are to: install a main solenoid valve in the machine room that opens only when one of the solenoids controlling flow to a coil opens, and/or install a small high-side float valve near the coils and pitch the defrost gas pipe to this float valve. The float valve drains liquid to a wet suction line.
- Choose slow-opening solenoid valves for the defrost gas valve.

Another critical time when there is danger of liquid hammer is upon termination of defrost. Near the end of the hot-gas defrost the pressure in the coil builds up to that of the pressure-regulating valve relieving the refrigerant condensate. If the return to refrigeration operation consists of no more than the sudden opening of the solenoid valve between the coil and suction line and reopening of the liquid supply, the high-pressure vapor in the coil rushes into the suction line. In a liquid-recirculation system, there will surely be some liquid in the liquid/vapor suction line. The high-velocity gas can drive some of this liquid to the end of the suction line, as shown in Fig. 13.5, or to the first elbow where an enormous impact is generated.

To prevent this situation[11], a bleed valve can be installed to bypass the pressure-regulating valve on the refrigerant condensate line. When defrost is terminated, close the solenoid valve in the defrost gas line, and open the solenoid in the bleed line. This allows the pressure in the coil to drop slowly, and when the large valve between the coil and suction line opens there is little pressure to vent from the coil.

Here is a word of caution. The bleed valve is small and may become blocked with foreign matter. If this happens, the valve might open as required, but the pressure equalization does not occur. A periodic check is advisable to be sure that the coil pressure subsides at the desired rate.

A pattern that emerges from inspections of the incidents of rupture occurring in a coil or in the suction line at the initiation or termination, respectively, of defrost is that (1) the metal in the region of rupture is at a very low temperature, and (2) the rupture usually occurs at a weld. The metal at these low temperatures may be exhibiting brittleness, and high-quality welds are mandatory, especially in these sections of the system.

Valves in an oil drain line. Accidents sometimes occur when draining oil in an ammonia system. What can happen is that the operator cracks the shutoff valve, shown in Fig. 13.6, but nothing happens because the oil is cold and stiff. The operator then slowly continues to open the shutoff valve, which may require about eight or ten turns to open completely. At some point during the opening process, a tunnel breaks through the oil, allowing ammonia liquid to rush out the drain. The operator may at first try to close the shutoff valve, but the required time is so long that the operator is forced to try to escape. Drain lines should be

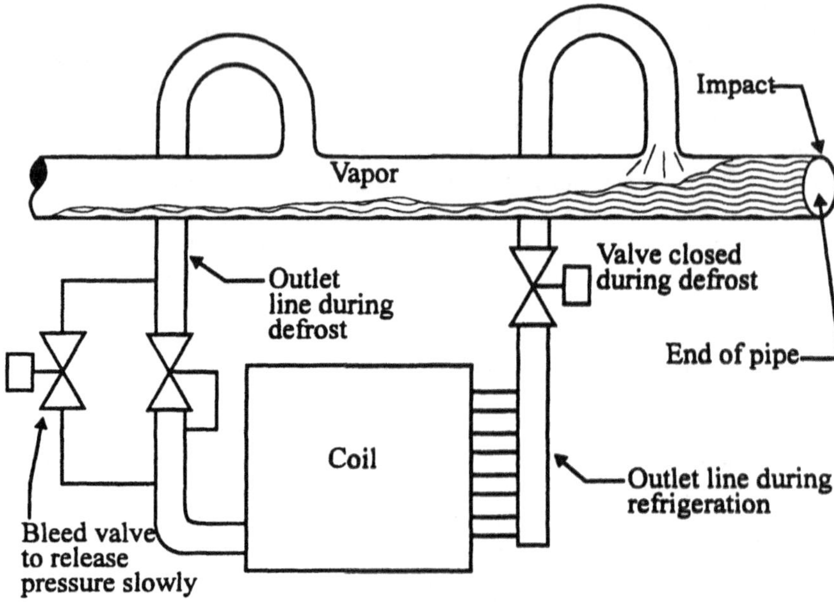

FIGURE 13.5
Rush of high-pressure vapor into a low-pressure suction pipe upon termination of defrost drives liquid at high velocity against the end of the line.

equipped with a quarter-turn ball or butterfly valve that could quickly interrupt the rush of ammonia and oil. Instead of the quarter-turn valve, some plants are equipped at this position with a spring-closing valve that automatically closes if the manual opening force is withdrawn.

Additional recommendations for safe piping practices.

Several technical specifications[13] that result in safer ammonia piping are the following:

- Use A106B or A53B pipe for low-temperature service. Some engineers also specify seamless pipe for liquid and hot-gas lines.
- Use 3000 lb socket weld fitting (A-181) for up to 37 mm (1-1/2 in) schedule 80 pipe.
- Use long radius weld ells and A-105 butt weld fitting on 50-mm (2-in) and larger pipe.
- Use no threaded joints on liquid lines larger than 25 mm (1 in) or on gas or vapor lines larger than 37 mm (1-1/2 in). If at all possible, seal weld the joint and paint after assembly to prevent external corrosion and leaks.
- Use a minimum of bolted flange joints. Use seal caps on the stems of all manual valves that are not frequently operated.

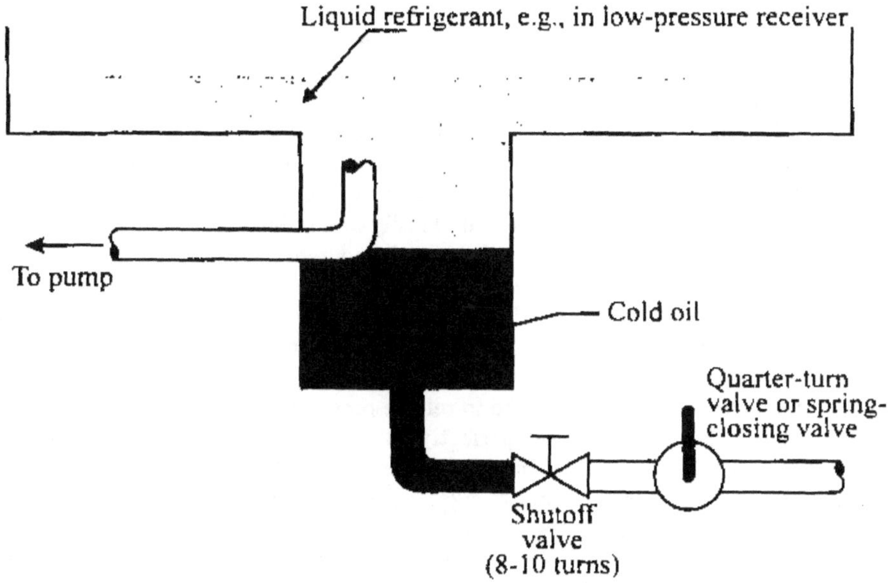

FIGURE 13.6
Quick-closing valve on an oil drain line.

13.7 RELIEF DEVICES

Questions that arise and decisions that must be made with respect to relief devices include the following: Where must a relief device be installed? Which of the several available devices should be chosen? How many should be installed at a given position? To where should the device be relieved? What pressure setting should be chosen? What is the required discharge capacity of the device? If outlet vent piping is called for, what size should it be?

It is convenient first to categorize several types of devices, such as those shown in Fig. 13.7. There is a distinction between a pressure-limiting and a pressure-relief device. An example of a pressure-limiting device is the high-pressure cutout, Fig. 13.7a, which should be installed at the outlet of a reciprocating or screw compressor and before the discharge stop valve. If the pressure rises too high, the device shuts off the compressor but does not vent refrigerant.

A fusible plug, as shown in Fig. 13.7b, is a fitting whose port is sealed by an insert composed of an alloy with a low melting temperature. When installed on a vessel in contact with liquid refrigerant, the fusible material melts when the liquid temperature (which is the saturation temperature corresponding to the pressure) rises too high. Fusible plugs may be appropriate for small refrigeration appliances, but are not used in industrial systems, because the entire charge is vented when the fusible plug melts. Furthermore, if no liquid is in the vessel for some reason, the fusible plug cannot sense a high pressure.

The most widely used pressure-relief devices are spring-loaded valves for vapor, as shown in Fig. 13.7c, and the relief valves for liquid, as shown in Fig. 13.7d. The latter are usually small and no special attention is devoted to specifying their capacity, because relief of even a minute amount of liquid is sufficient to reduce the pressure.

Relief valves for vapor must be installed on all vessels, which includes shell-and-tube condensers and evaporators. The connection on the vessel should be above the liquid level, because the discharge of liquid would be especially dangerous and furthermore would have little effect on reducing the pressure. No stop valves may be installed between the vessel and relief valve, with the exception of a three-way valve serving a dual relief valve, Figure 13.8. The three-way valve should be positioned so that only one of the relief valves is subject to the vessel pressure, as is shown in Figure 13.8, and not in the mid-position, which would allow both valves to relieve at the same time. The reason for this practice is that often if a valve opens once or twice to relieve pressure it does not reseat perfectly. If only one valve fails to reseat perfectly, it can be removed and replaced while the other provides the safety function. The sizes of both valves are selected so that either can handle the total refrigerant flow by itself.

From a liquid relief valve in a section of pipe where liquid could be trapped by valves, the discharge is directed to a section of the system where another relief valve is located. Vapor relief valves often vent to the atmosphere, although it is also possible to relieve from a high-pressure vessel to a low-pressure vessel that vents to the atmosphere. One section of the plant eventually must be vented to the atmosphere, and the valve controlling this relief must be of a size that it can accommodate the total flow rate of the vessels served. The flow capacity of the valve is influenced by the size and length of the discharge line from the valve. The minimum diameter of the discharge pipe is that of the outlet fitting of the valve. If the discharge line is unusually long, it is advisable to choose a pipe diameter one size larger than the outlet fitting.

Each pressure vessel with a volume larger than 0.085 m^3 (3 ft^3) must be provided with a pressure-relief valve, and vessels larger than 0.283 m^3 (10 ft^3) must be equipped with two parallel valves, as in Fig. 13.8. Some designers and plant operators always use double and never single relief valves. With a single valve there is no way to remove it for servicing. Each of these valves must have a venting capacity that prevents the pressure from rising more than 10% above the setting of the relief valve.

The size of relief valves is related to the discharge capacity of air, which is calculated[1] by the formula:

$$C = fDL \tag{13.2}$$

where:

C = minimum discharge capacity of relief device, lb/min of air
D = outside diameter of the vessel, ft
L = length of the vessel, ft
f = factor depending on the refrigerant

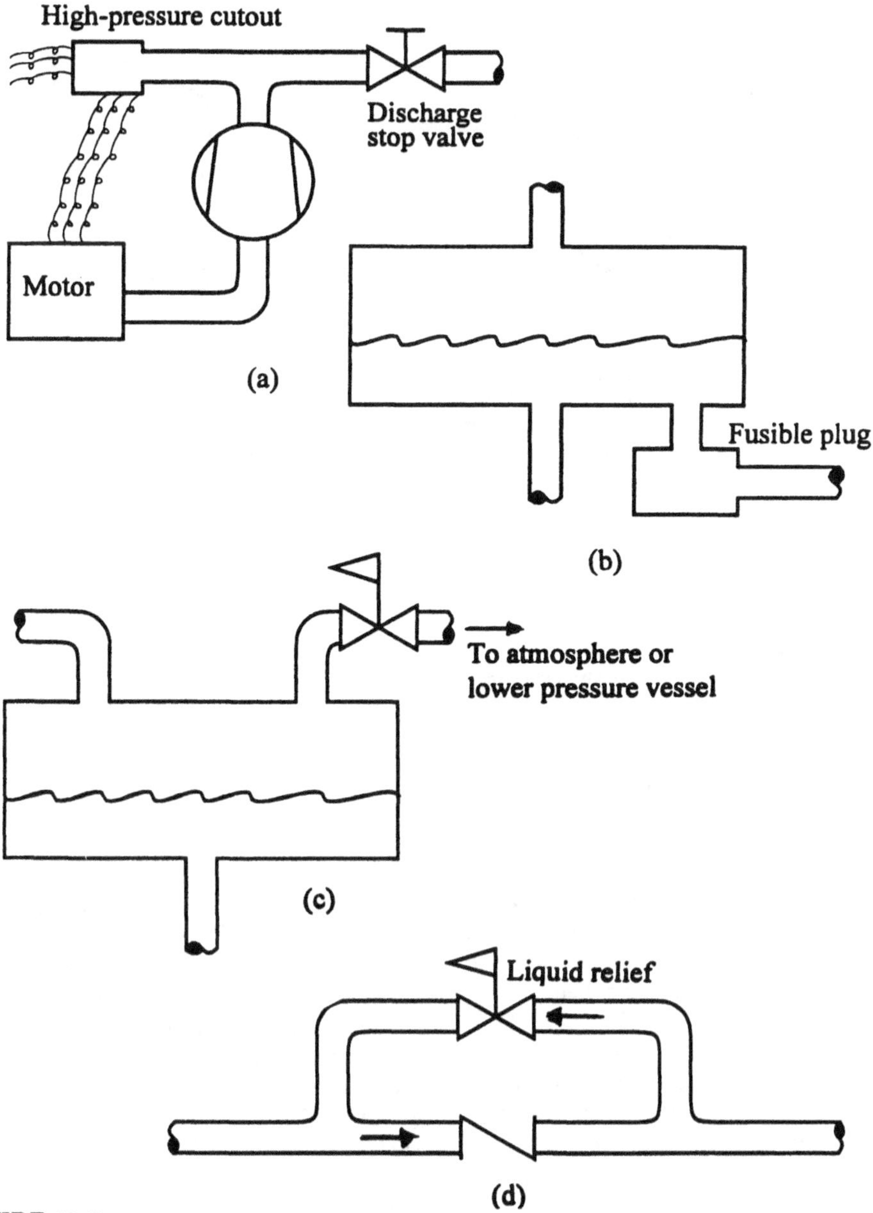

FIGURE 13.7
Some pressure-type safety devices including (a) pressure-limiting cutout, (b) fusible plug, which is not normally used in industrial refrigeration systems; (c) pressure-relief valve for vapor, and (d) pressure-relief valve for liquid.

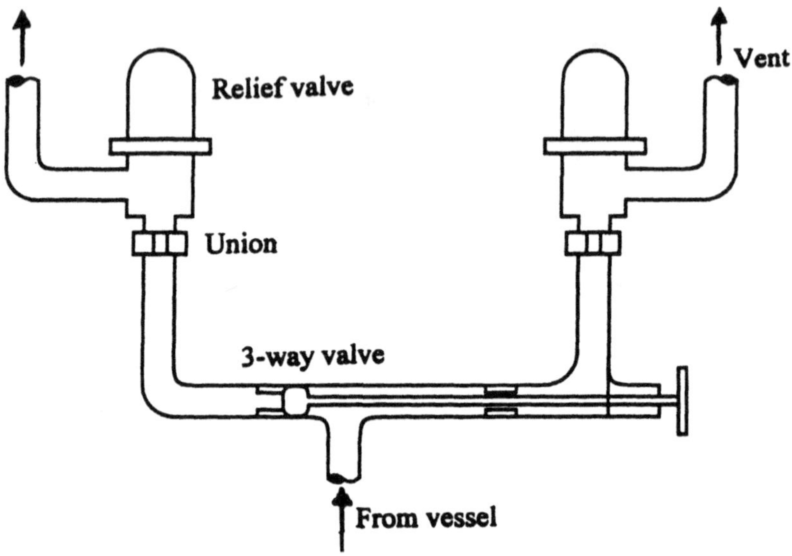

FIGURE 13.8
Three-way valve serving two relief valves.

The values of f for several refrigerants are:

Ammonia, f = 0.5
R-12, f = 1.6
R-22, f = 1.6
R-502, f = 2.5

In SI units, the terms in Eq. 13.2 are: C in kg/s, D in m, L in m, for which the f-factors are 0.041 for ammonia, 0.13 for R-12 and R-22, and 0.20 for R-502.

It may have been expected that the discharge capacity of the valve in Eq. 13.2 would have been related to the volume of the vessel, thus equal to (constant)$(\pi D^2/4)L$, where D is the diameter and L the length of the vessel. The rationale[14], however, is that the relief valve shall be able to control the pressure in the event of a fire where the heat radiated to the vessel is the projected area of the vessel, and thus the product of D and L. The relief valve must be able to release refrigerant at a rate such that the rate of vaporization of liquid in the vessel provides sufficient cooling effect to limit the saturation pressure. The heat flux used as the basis of the specification is 10 kW/m^2 (3,171 Btu/hr·ft^2). The f-factor is unique for each refrigerant and accounts for the differences in latent heats of the refrigerants. Ammonia has a smaller value of f than the other refrigerants listed because of its high latent heat, which reduces the flow rate needed for the cooling effect.

The value of C in Eq. 13.2 is the flow rate of air, and there is no reference to refrigerant. An examination of catalogs of manufacturers of relief valves also shows that the ratings are specified in flow rate of air for various rated pressures. This procedure simplifies the task of selecting a relief valve but also means that the flow rate of refrigerant that the relief valve is able to pass may typically not even be calculated. It is possible, however, to translate the data for air to that of refrigerant flow if, for example, it is important to know how much ammonia refrigerant has been released during a discharge of a certain time duration. Appendix 11 of Section VII of the ASME *Boiler and Pressure Vessel Code*[2] provides an equation that permits the conversion:

$$\frac{\dot{m}_{amm}}{\dot{m}_{air}} = \frac{K_{amm}\sqrt{M_{amm}/T_{amm}}}{K_{air}\sqrt{M_{air}/T_{air}}} \tag{13.3}$$

where:

K	=	constant that is related to the ratio of specific heats
K_{amm}	=	350
K_{air}	=	356
M	=	molecular mass (28.97 for air and 17.03 for ammonia)
T	=	absolute temperature
$\dot{m}$	=	flow rate

T_{air} is arbitrarily chosen as 289 K (520·R), and T_{amm} is the saturation temperature (in absolute) at the rated pressure. Equation 13.3 resolves to the following conversions for two frequently used pressure ratings:

For a valve rated at 1,825 kPa absolute (250 psig):

$$\dot{m}_{amm} = 0.717\dot{m}_{air}$$

For a valve rated at 1,135 kPa absolute (150 psig):

$$\dot{m}_{amm} = 0.733\dot{m}_{air}$$

13.8 VENTILATION OF MACHINERY ROOMS

Ventilation requirements for machinery rooms are specified by various codes and standards such as those of BOCA[7], Uniform Mechanical Code[8], IIAR[5], and ASHRAE[1], and in general all the specifications differ. A new member of the family of documents making ventilating recommendations is Bulletin 111, *Ammonia Machinery Room Ventilation*[15] of the International Institute of Ammonia Refrigeration. While carrying no weight as a code or standard, Bulletin 111 seems to be backed with strong logic. Furthermore, the ventilation rates specified by Bulletin 111 are often the most generous and safest. The recommended strategy, then, is to select the ventilation rate based on Bulletin 111, but also calculate the rate(s) specified by whatever code(s) is applicable at the plant location to be sure that those requirements are met. This section on the ventilation of machinery rooms first summarizes the procedure outline in Bulletin 111. Later on, the rationale on which Bulletin 111 and the other codes and standards are based will be compared. Finally, several related requirements of the ventilation equipment will be explained.

Bulletin 111 defines two ventilation rates, one a continuous rate and the other the emergency rate:

- *Continuous rate*, 5.08 to 10.16 L/s per m^2 of floor area (1 to 2 cfm/ft^2)
- *Emergency rate*, 50.8 L/s per m^2 of floor area (10 cfm/ft^2) with a minimum of 9,440 L/s (20,000 cfm)

The dominant reason for ventilating machine rooms is to prevent excessive concentrations of refrigerant, regardless of the refrigerant. A further function of the ventilating system is to minimize discomfort to the operators by removing heat from motors and other electrical equipment during hot weather. The flow rate of ventilation air for this purpose is recommended in ANSI/IIAR-2-92 and Bulletin 111 such that ambient ventilation air brought through the machinery room does not rise in temperature more than 10°C (18°F). If, for example, compressor motors of 900 kW could be in operation in addition to 20 kW of additional energy usage in the machine room, the total kW of heat would be 20 kW + (0.1)(900 kW), or 110 kW. Only 10% of the compressor motor power is estimated to be released into the machinery room, based on about 6% due to motor inefficiency and about 4% heat loss to the ambient from the compressor and its auxiliary equipment. To limit the temperature rise of the ventilating air to 10°C, the required flow rate Q would be:

$$Q = \frac{\text{rate of heat release}}{(\text{air density})(\text{specific heat})(\text{temperature rise})}$$

$$Q = \frac{110\,\text{kW}}{(1.2\,\text{kg/m}^3)(1.0\,\text{kJ/kg}\cdot\text{K})(10\,\text{K})}$$

$$Q = 9,170\,\text{L/s}\,(19,420\,\text{cfm})$$

The air flow rate requirement for cooling the machine room is nearly as large as the 9,440 L/s (20,000 cfm) minimum for controlling the refrigeration concentration.

The shift from the flow rate chosen for continuous to the emergency flow rate should be activated by a high-ammonia concentration alarm, and there shall also be a manual switch located outdoors. The power supply for the ventilating fans should be separate from that supplying the major equipment in the machinery room which might need to be shut down in the event of an emergency. It is quite acceptable to install two ventilating fans with a two-speed motor on one of them. For continuous ventilation, one fan operating at low speed is sufficient.

As was previously mentioned, there are different ventilation requirements of the various codes and standards. The Uniform Mechanical Code specifies twelve air changes per hour, and the BOCA Mechanical Code requires six air changes per hour. ANSI/ASHRAE 15-94 bases its recommended ventilation rates on the mass of refrigerant, G, in the largest system, any part of which is located in the machinery room. The applicable equation is:

$$Q\text{, cfm} = 100\sqrt{G\text{, lb}} \tag{13.4}$$

or

$$Q\text{, L/s} = 70\sqrt{G\text{, kg}} \tag{13.5}$$

The recommendation of Bulletin 111 of a minimum flow of 9,440 L/s (20,000 cfm) for emergency ventilation may be preferable to basing the ventilation rate on air changes or mass of refrigerant in the system. The emergency ventilation rate in Bulletin 111 is based on maintaining an ammonia concentration below 4% (a value to be explained in the next paragraph) if a small pipe, for example, a high-pressure liquid pipe 25 mm (1 in) in diameter, ruptures. Incidentally, using Eq. 13.4 or 13.5 does not yield 9,440 L/s (20,000 cfm) until the refrigerant charge is 18,140 kg (40,000 lb). Bulletin 111 almost always recommends higher and more conservative ventilation rates.

The emergency ventilation rate is to be activated by an ammonia sensor that also sounds an alarm. The next question is what setting to choose of the concentration sensor to trigger the emergency rate. The concentration of 40,000 ppm or 4% was mentioned previously, and the significance of this value is that the National Fire Protection Association usually adopts one-fourth of the lower flammability concentration as the alarm setting. The lower flammability limit of ammonia is 16% and one-fourth of this concentration is 4%. So certainly, the emergency level should be lower than 40,000 ppm for fire protection, but that concentration would be fatal to workers. For this reason, the emergency concentration level should be set at perhaps 100 or 150 ppm. Too high a setting could result in added risk for the personnel, and too low a setting would result in nuisance alarms. If a ventilating system that maintains the ammonia concentration less than 4% is *not* installed, the machinery room is no longer considered *nonhazardous* by the National Electrical Code, and more expensive electrical switches and other gear are required.

Here are a few final recommendations for the ventilation system. Because ammonia vapor released in the machinery room tends to rise to the ceiling, the exhaust should be at the high sidewall or the ceiling, as illustrated in Fig. 13.9. The inlet should be on the opposite side of the room at a low level. Any fan that is off for the continuous flow rate but is brought into service for the emergency rate should be equipped with louvers that are closed when the fan is off and open when the fan operates. The fail-safe status of the louvers should be the open position. The required area for passage of air at the inlet and exhaust is recommended to be such that the air velocity is of the order of 3.0 m/s (600

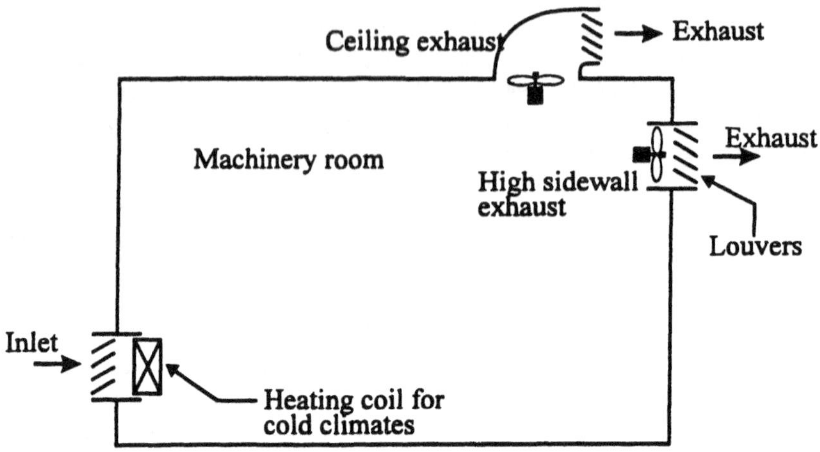

FIGURE 13.9
Placement of inlet and exhaust openings in the machine room.

fpm). For a ventilation air flow rate of 9,440 L/s (20,000 cfm), therefore, the openings in the wall or ceiling should have an area of 3.06 m^2 (33 ft^2).

13.9 FIRE PROTECTION IN REFRIGERATED WAREHOUSES

Fires in refrigerated structures are rare, but they do occur and can be costly. The major loss associated with a fire may be the product, whose value may be 5 to 10 times that of the building. It may at first seem that there would be negligible flammable material in a low-temperature warehouse, but for a certain 20,000 m^3 (700,000 ft^3) building, the estimated content of wood in the pallets was 100,000 kg (220,000 lb), 150,000 kg (330,000 lb) of cardboard cartons, and 7,000 kg (15,400 lb) of plastic wrapping material[16]. Many types of insulation, including many cellular plastic foams, will also burn.

The fundamental approach of the designer and operator should be to prevent a fire in the first place. Toole[16] and Duiven and Twilt[17] cite the causes of most fires in their surveys as attributable to:

- Welding
- Heating cable at doors and defrost drain lines
- Faults in other electrical equipment, such as transformers, battery chargers, or solenoids
- Hot bitumen from roof work
- Poor housekeeping (rubbish fire started by a cigarette butt)

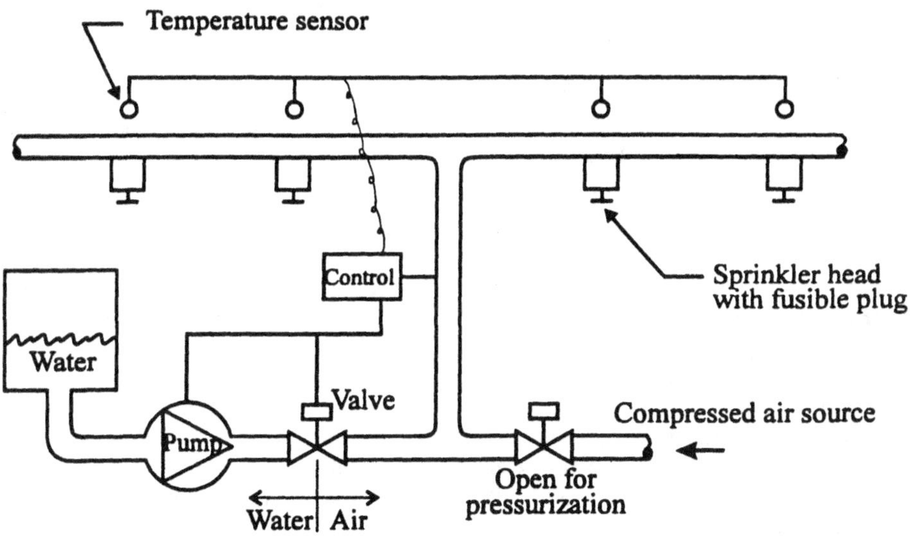

FIGURE 13.10
A dry sprinkler system for fire protection in a low-temperature space.

To the above list could be added fires starting in adjacent areas, such as the machine rooms or shops. Some fires may, unfortunately, be attributable to arson or sabatoge by a disgruntled employee or former employee. Many insurance underwriters require water sprinkler systems for refrigerated storage areas, even in those spaces maintained at subfreezing temperatures. For such applications a dry-type sprinkler system, as illustrated in Fig. 13.10, may be used. Some insurance companies require the dry system at temperatures lower than about 7°C (45°F). In this type of system, the distribution pipes are filled with compressed air. If heat from a fire melts the fusible element in a sprinkler head, the air pressure drops. This drop in air pressure is interlocked with a temperature sensor at each of the sprinkler heads, and if both indicate abnormal temperatures, the control starts the pump and opens the valve in the water line. Earlier systems were actuated by the melting of the fusible plug alone, and an errant opening of the plug allowed water to fill the distribution system. The water would soon freeze in the pipes, and the entire system had to be dismantled.

All refrigerated buildings are equipped with insulation, and the combustibility characteristics of various insulations should affect the choice of material and/or how the insulating material is applied. Polyurethane and polystyrene can burn and emit smoke and noxious fumes. These materials now can be purchased with self-extinguishing additives, although in one experience[18] the self-extinguishing sulfate was vaporized by the heat of the fire and the remaining polystyrene became a burning, molten mass. There are factors in addition to the flammability characteristics that influence the choice of insulating material, but the following are broad characterizations of some frequently used insulations:

- Polyurethane—combustible
- Polystyrene—combustible, but is treated with a flame retardant that is effective against small flames
- Polyisocyanurate—combustible
- Cellular glass—resistant to combustion

More precise descriptions of the fire characteristics are available from the Material Safety Data Sheet for each of the insulation materials. The fact that a material may burn under certain conditions does not necessarily preclude its use. It may simply require that the material be enclosed or otherwise protected by a noncombustible material. Polystyrene may be used in floors, for example, because the insulation material is protected by the concrete. Cellular glass has excellent fire resistance, but tends to be heavier and more expensive than some of the other insulations. Insulated built-up panels now predominate in construction and have either metal or plastic skins. The metal is fire-resistant but is harder to clean than the plastic, which might melt in the event of a fire.

13.10 LEAK DETECTION AND AMMONIA SENSORS

Both for reasons of safety and to prevent interruption of refrigeration service, it is important to sense leaks of refrigerant. Since halocarbons are odorless, sensors are important in installations using such refrigerants to ensure that a leakage of a major portion of the refrigerant charge does not go undetected. Also, leakage of large quantities of a halocarbon represents a danger to humans who are not aware of being engulfed in high concentrations. When an ammonia leak occurs where people are present, the odor of ammonia provides its own warning. But for unoccupied spaces, such as refrigerated storage areas and unattended machine rooms, automatic sensors are crucial to providing an early warning of a potential problem.

The principles of operation[19] of the most popular ammonia sensors are: (1) detector tubes, (2) solid state, (3) electrochemical, and (4) infrared (or ultraviolet). The concentrations of ammonia and the applicable temperature ranges of these types of sensors are enumerated in Table 13.2.

The most widely used concept for sensing ammonia is the solid-state type because of its relatively low cost and ease of installation and maintenance. The solid-state sensor works well for initiating alarms in the range of ammonia concentration of several hundred ppm, but it is normally not very precise in the range of 20 to 50 ppm, which might be of interest for personnel exposure regulations. One experience report[20] suggests that solid-state sensors employed in low-temperature spaces should be allowed several weeks to age before taking seriously the indications from the sensor.

TABLE 13.2
Some characteristics of several types of ammonia sensors.

Sensor type	Concentration range, ppm	Operating temperature, °C (°F)	Purpose	Calibration frequency
Detector tube	5 to 1,000	20 (70)	Personnel, detect leaks	Long term
Electro-chemical	0.7 to 1,000	−5 to 45 (23 to 113)	Personnel, leak alarm	Frequent
Solid state	300 to 5,000	−40 to 90 (−40 to 190)	Leak alarm	Moderate frequency
Infrared or UV	10 to 10,000	−10 to 50 (14 to 122)	Personnel	Long term

13.11 PERSONAL PROTECTION AGAINST AMMONIA

It is economically crucial to protect products from damage by exposure to ammonia, but an even more critical requirement is that any escape of ammonia be prevented from causing personal injury. Any worker in ammonia refrigeration plants should be familiar with the following two documents:

- Material Safety Data Sheet (MSDS) for ammonia
- Poster on first aid for ammonia exposure, such as one available from the International Institute of Ammonia Refrigeration

The MSDS provides data on some of the physical properties of ammonia, but those items that deal primarily with personal safety are as follows:

- If at all possible, avoid inhalation, skin contact, and especially contact with eyes.
- The general rule is to deluge affected areas with water.
- Remove clothing, except if it is so saturated with liquid ammonia that it is frozen to the skin.
- Contact lenses must not be worn while working with ammonia.
- Use face shield when making repairs.
- Protect hands with rubber gloves.

There are two types of masks—canister and self-contained breathing apparatus (SCBA). The canister masks are the most convenient and manageable of the two, but the SCBA is the most effective.

If a spill has occurred and no person is in danger in the room, start ventilation of the room before entering it, even if protected by a mask.

13.12 DISCHARGE OF AMMONIA—DIRECTLY TO ATMOSPHERE

In the routine operation of the plant there should be no discharge to atmosphere. In well-maintained plants, it is not possible for a visitor to detect from the smell that ammonia is the refrigerant. Slight smells of ammonia sometimes occur because of leakage at valve packings, pump seals, venting from freshly drained oil, or from an air purger. A strong trend has developed for the use of capped valves, particularly those needing only occasional adjustment, in preference to the wheel type in order to avoid leakage at the packing. The quality of pump seals has improved over the past decade with the introduction of oil-pressurized seals and the use of hermetic pumps to dramatically reduce leakage at pump seals (see Chapter 8).

A significant quantity of ammonia may be released when a safety relief valve opens, but the purpose of this relief of pressure is to prevent a serious rupture of a pipe or vessel. Nevertheless, some special measures for treating releases from safety valves may be warranted, and providing these measures may be manageable, because the precise location where the release will occur is known. Three approaches to handling discharges from such point sources as relief valves are as follows:

- Direct discharge to atmosphere
- Absorption in water and subsequent disposal of the ammonia-water solution
- Flaring

The direct discharge of ammonia to atmosphere may at first appear to be a cop-out, but many refrigeration professionals on occasion discharge ammonia to atmosphere with a clear conscience, provided the concentrations that come into contact with plants, animals, and humans are maintained low. This position is justified by several factors[21] such as that the release of ammonia into the atmosphere does not contribute to ozone depletion or global warming, and that ammonia is a naturally occurring compound in air and is a necessary ingredient for many biological processes.

One realization that became apparent in a study of concentrations in the vicinity of ammonia vapor releases[22] was that, to reduce the ground-level concentrations, two measures should be attempted: discharge the ammonia at as high an elevation as possible and discharge the ammonia upward at a high velocity. Figure 13.11 shows ground-level concentrations for a release of 0.15 kg/s (20 lb/min) with a wind velocity of 4.5 m/s (10 mph). Curves B and C may be compared to the base case represented by Curve A, which represents concentrations when the stack height is 6 m (20 ft) above ground level and the diameter of the stack is 50 mm (2 in). The conditions shown for Curve A result in a maximum ground-level concentration of 115 ppm at a downwind distance of 40 m (130 ft). If the diameter of the stack pipe were reduced to 25 mm (1

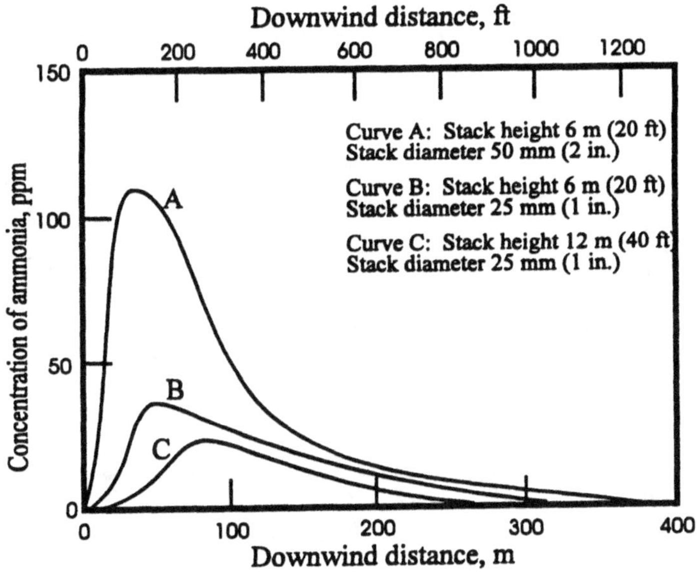

FIGURE 13.11
Ground-level concentrations of ammonia in the vicinity of a vapor release of 0.15 kg/s (20 lb/min) with a wind velocity of 4.5 m/s (10 mph).

in), as Curve B shows, the exit velocity is roughly four times that of the base case. The peak concentration drops to 36 ppm at a downwind distance of 60 m (200 ft). The second approach, shown in Curve C, demonstrates that elevating the stack height to 12 m (40 ft) reduces the maximum concentration to 21 ppm at a downwind distance of 75 m (250 ft).

While the skillful treatment of direct discharges of ammonia into the atmosphere may give satisfactory results in the view of the plant operator, the discharge of a significant amount of ammonia usually must be reported for information purposes to environmental authorities. In the United States, a release of 45 kg (100 lb) or more in a 24-hour period must be reported. Furthermore, regulatory agencies sometimes do not permit even the careful direct discharge to atmosphere. In this situation, absorption of ammonia in water or flaring of ammonia might be required.

13.13 DISCHARGE OF AMMONIA—ABSORPTION IN WATER

Another concept for handling discharges of ammonia from a refrigeration system is to absorb the ammonia in water. ANSI/ASHRAE Standard 15-94 has traditionally listed the discharge of ammonia into a tank of water as an option. The recommended proportions of water to ammonia are 1 m^3 of water for each 120 kg of ammonia (1 gallon of water per lb of ammonia). Field tests[23] of the absorption

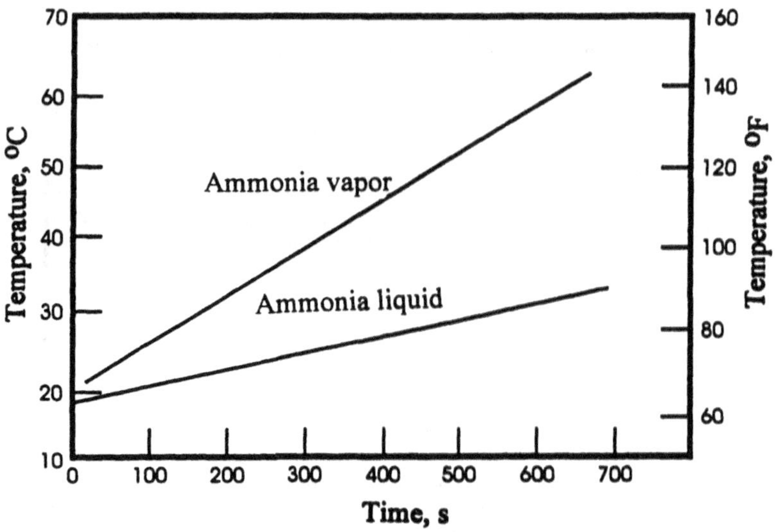

FIGURE 13.12
Absorption of ammonia in water, ending with the final composition of 120 kg of ammonia per m^3 (1 lb of ammonia per gallon of water).

of ammonia in vapor form in a vessel of water showed the effectiveness of this process, as illustrated in Fig. 13.12. During a time interval of approximately 11 minutes, 90 percent of the released ammonia was absorbed to bring the ammonia concentration in the water to the end condition of 120 kg of ammonia in 1 m^3 of water (1 lb of ammonia per gallon of water). Figure 13.12 shows the temperatures of the solutions during absorption of both vapor and liquid. In absorbing ammonia vapor, the temperature of the solution rises from the starting temperature of about 20°C (68°F) to 60°C (140°F) in the 11-minute absorption period. Water in the vessel will absorb liquid ammonia equally well, as Fig.13.12 also shows, and the temperature rise is only about 15°C (27°F). The greater rise in temperature when vapor is absorbed is explained by the need to condense the vapor in the absorption process.

Sometimes regulating agencies prefer a facility absorb the entire charge of the plant in a vessel of water. This requirement is normally not practical for a large plant with a charge of, for example, 5,000 kg (11,000 lb) or greater. Even for water vessels capable of absorbing modest-sized discharges, the water in these vessels must be kept from freezing in cold climates.

If a water tank for absorption of the ammonia is used, a question that arises is what to do with the aqua-ammonia solution after ammonia has been absorbed in it. In some localities, firms are available to haul away the solution for reclamation. If the solution is relatively free of oil, it may be used as agricultural

TABLE 13.3
Concentrations of ammonia in water corresponding to various pH values.

Percentage of ammonia in water, mass basis	pH
10	12.0
*1.7	11.6
1.0	11.5
0.1	11.0
0.01	10.5
0.001	10.0
Concentration with normality = 1	

fertilizer. If direct discharge to atmosphere is permitted, the ammonia can be vented off to the atmosphere naturally, sometimes even heating the solution slightly to speed the release. In connection with the absorption studies reported in Fig. 13.12, some desorption tests showed that when a vessel was vented through a tube, it required about 50 days to bring the ammonia concentration down to one-half of its starting concentration.

One of the methods of disposing of the aqua-ammonia solution is to dilute and discharge the mixture into either the sanitary or storm sewer. This action should be taken only after consultation with the appropriate authorities who are likely to specify a maximum permitted pH value of the diluted mixture. It is appropriate, then, to explain pH values and also to emphasize what some magnitudes of pH mean with respect to the mass concentration of ammonia in the water. The pH value, which means *potential of hydrogen*, is defined as:

$$pH = \log(1/H^+)$$

where:

H^+ = number of gram-ions of hydrogen per liter

Pure water contains 0.00000001 gram-ions of H^+ per liter, so pure water possesses a pH value of log(10,000,000) or 7. Absorption of ammonia in water decreases the number of hydrogen ions, so the pH value of an ammonia/water solution exceeds 7, and high pH values indicate a high concentration of ammonia. An important point is that only a small amount of ammonia can rapidly raise the pH value from 7. Table 13.3 shows the percentage concentrations by mass resulting in various pH values. The table shows that even one part of ammonia in 100,000 parts of water elevates the pH to 10. The consequence of this distribution of pH values is that if an ammonia/water mixture can be discharged only if the pH value is below 11, for example, then enormous quantities of dilution water will be required.

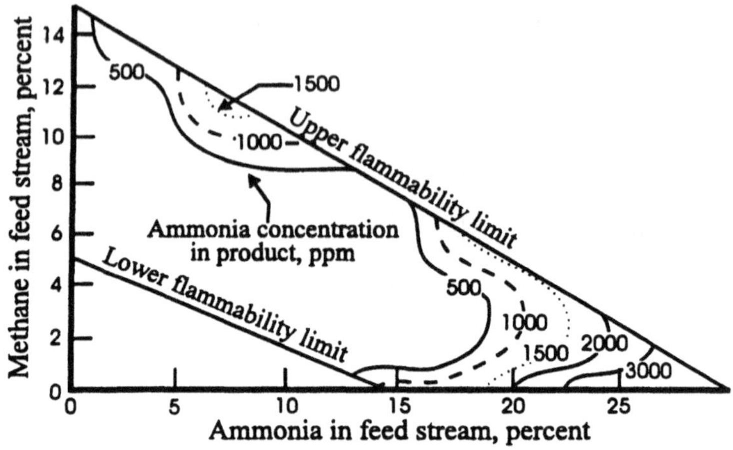

FIGURE 13.13
Concentration of ammonia in combustion products when flaring an ammonia/air mixture with the aid of methane.

13.14 DISCHARGE OF AMMONIA—FLARING

The third method of disposing of ammonia discharges is by flaring or burning the ammonia. This method is particularly applicable when the release to be treated will be located at an identifiable point, such as the discharge vent of a relief valve. It is recognized that mixtures of ammonia and air with ammonia concentrations between about 15 percent and 27 percent of ammonia will burn. This combustion can be sustained, however, only through the use of a flame holder, such as an iron screen. This iron screen is maintained at a high temperature by the burning ammonia. Practical flares must provide a support flame from the burning of a hydrocarbon such as methane. Tests[24] on ammonia flares show that the methane concentrations of between 4 and 10 percent in the ammonia/air mixture are effective in facilitating the combustion. Figure 13.13 shows the effectiveness of combustion as a function of the percentages of methane and ammonia in the feed stream. Bracketed within the two straight lines is the flammability region of the mixture, and the contours within this region show the concentration of ammonia in ppm in the combustion products.

The map shows that with an initial concentration of less than 15 percent ammonia in air, most of the ammonia can be oxidized by providing a methane concentration of between 4 to 10 percent. With initial ammonia concentrations higher than 15 percent, there will still be unburned ammonia in the combustion products, but significantly less than the original 150,000 to 250,000 ppm.

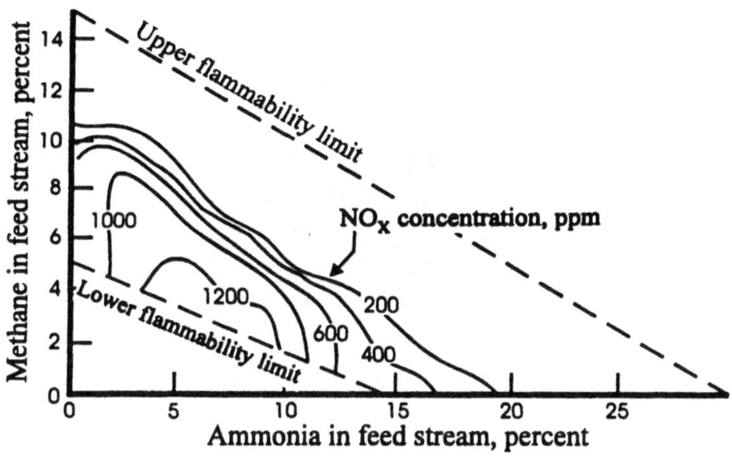

FIGURE 13.14
NO_x concentrations in the combustion products of an ammonia flare.

The complete and ideal chemical reaction in the combustion of ammonia is:

$$4NH_3 + 3O_2 \rightarrow 2N_2 + 6H_2O$$

When the flame temperatures are especially high, however, some of the NH_3 dissociates and along with a small quantity of nitrogen in the air combines to yield NO and NO_2, which are the NO_x compounds in Fig. 13.14. Figure 13.14 shows contour lines of the NO_x concentration indicating that the concentrations can be reduced by tuning the flare to operate toward the upper flammability limit. The region of lower NO_x concentrations is that portion of the map associated with reduced flame temperatures in contrast to the high flame temperatures near the lower flammability limit.

One of the conclusions from the curves is that the flaring of pure ammonia is not possible. The ammonia must be diluted with combustion air, preferably with an incoming mixture of less than 15 percent ammonia.

13.15 LARGE RELEASES OF AMMONIA

There are few situations where discharges of ammonia are planned. An exception might be where a plant is being decommissioned or reduced in size. Even when ammonia must be removed from a section of the plant being repaired or retrofitted, every effort should be made to recover the ammonia. Discharges from relief valves are not normal operation, although the specific function of the relief

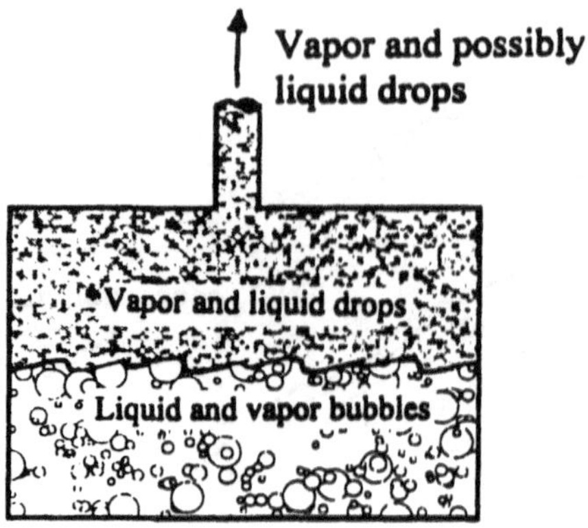

FIGURE 13.15
Release of ammonia vapor from a vessel containing liquid.

valve is to discharge ammonia to prevent a much more serious failure of a vessel. The designation of large releases that are discussed in this section primarily refers to those resulting from a rupture of a pipe, fitting, or vessel. The occurrence of such ruptures is almost always the result of an external impact such as a fork-lift truck striking a pipe or its support. The three major forms of release are: (a) as a vapor, (b) as an aerosol, and (c) as a liquid spill on the floor of the facility or earth outdoors. Several forms of these releases may occur simultaneously in a given incident.

One form of release, illustrated in Fig. 13.15, is when ammonia vapor escapes from a vessel containing liquid. In order to supply the vapor, liquid in the vessel must evaporate, and in doing so drops the temperature of the remaining liquid. As the temperature of the liquid drops, so does the pressure in the vessel, so the rate of discharge progressively decreases. For the idealization of no heat transfer to the vessel during discharge, calculations show that 20 percent of the liquid in a vessel initially at a temperature of 27°C (80°F) will vaporize to reduce the temperature to −33°C (−28°F), at which point the pressure inside the vessel is atmospheric, so flow stops. Actually, there will always be heat transfer to the vessel, so more than 20 percent of the liquid vaporizes and a small flow rate will continue to boil off.

A dramatic demonstration of an effective response to a vapor release from a liquid tank has been shown by the school of Hydro Care in Landskrona, Sweden, which is dedicated to the safe handling of ammonia. The technique used was to throw a canvas cover over the entire vessel. The cover provides thermal insulation such that the temperature of the liquid drops rapidly and the flow rate of ammonia vapor diminishes perceptively. Workers, using masks, then remove the canopy to repair the leak. If a long time is necessary, the vessel may have to be recovered.

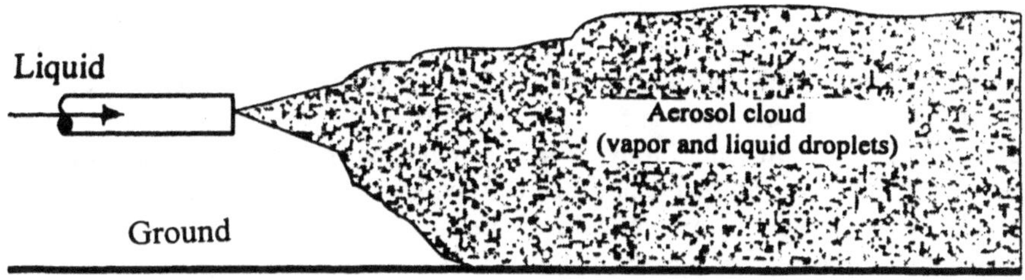

FIGURE 13.16
A release of ammonia as an aerosol.

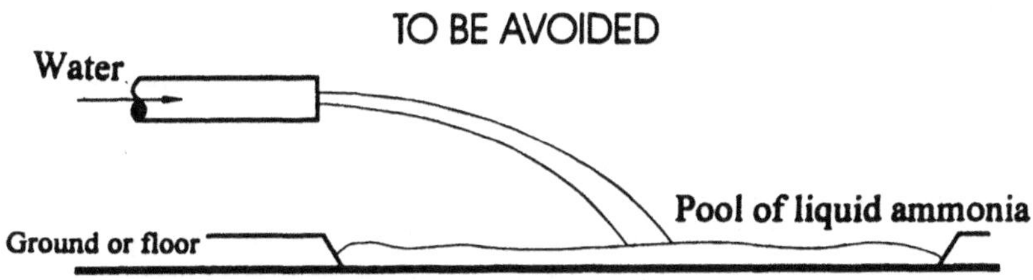

FIGURE 13.17
Water should normally not be applied to a pool of liquid ammonia resulting from a spill.

Probably the most treacherous release of ammonia is in aerosol form, shown schematically in Fig. 13.16. Tiny droplets of liquid are dispersed thoughout the vapor, and the vaporization of this liquid develops a cold, dense combination which usually clings to the ground rather than rising quickly and away from people and vegetation.

Another form of release is one that might occur if a liquid line is ruptured and liquid ammonia spills onto the floor of a refrigerated space or machine room or onto the earth if the release is outdoors. The recommendation is that water normally *not* be applied to the liquid, Fig. 13.17. Immediately following the spill, the rate of vaporization will be high, but as the remaining liquid and the floor or ground on which it has spread becomes cold, the rate of vaporization diminishes rapidly. Ventilation of the enclosed space may be adequate to disperse the ammonia vapor until all has been evaporated. Experiments are going on currently with foams that could cover the liquid to retard the rate of vaporization even more.

The final topic associated with large releases is the concept of positive-pressure ventilation. A machine room should be equipped with permanently-installed fans of sufficient ventilating capacity as specified in Section 13.8. If a release occurs in a refrigerated space, however, installed ventilating capability will normally not be available. Large-capacity portable fans should be present on site for such emergencies. It may at first seem that one placement of the fan at an opening of the space is as good as another. The attempt to use the fan to exhaust air and ammonia fumes from the space, as shown in Fig. 13.18a, is likely not to be effective, however, because the bypass air from the exterior to the fan intake reduces the withdrawal rate of the fan from the space. Positive-pressure ventilation, however, as shown in Fig. 13.18b, takes advantage of the jet issuing from the outlet of the fan to seal the opening.

13.16 BEYOND THE CODES

The adherence to codes and standards is a giant step toward designing and constructing a safe refrigeration system. There are also numerous provisions that contribute to a safe environment that are not covered by codes. A fundamental requirement is that the equipment be arranged in an orderly manner and sufficient space be provided to allow access to the equipment for frequent visual examinations and for servicing. The trend to locate coils that serve refrigerated storage spaces in penthouses is commendable, because it then permits piping to be placed above the roof line and out of reach of fork-lift trucks. Every effort should be made to avoid placing pumps in pits. The reason that pumps are sometimes installed in pits is to provide adequate net-positive-suction head to prevent cavitation of liquid in the pump. This location of pumps should be avoided, if possible, because the location places a technician working on the pump at risk should a release of ammonia occur during servicing.

Careful thought should be given to the location of valves. Valves should ideally be placed where they can be reached conveniently by the operator, but should not be located where they could be damaged by a passing fork-lift truck. The choice of penthouse enclosures for evaporator coils usually permits valve groups for shutoff and hot-gas defrost to be placed outside the penthouse. This location provides access for servicing and usually less hostile temperatures than inside the space. Many designers are striving toward placing only welded connections inside the refrigerated space.

The choice and location of king valves has received much attention in the past few years. The king valve is located in the liquid line from the receiver and can be closed to interrupt the supply of liquid to the system in case of an emergency. In large systems, there may be several high-pressure receivers and also several liquid supply lines in which a shutoff valve would be advisable. A basic provision is to either locate the control or the valve itself outside the machine room. A hand valve in the machine room may not always be accessible in an emergency. If the high-pressure receiver is located outdoors, it is logical for the king valve to be outdoors as well. If the king valve is located in the machine room, it should be a normally-closed solenoid valve with the control outdoors in a locked box to which only authorized personnel have access.

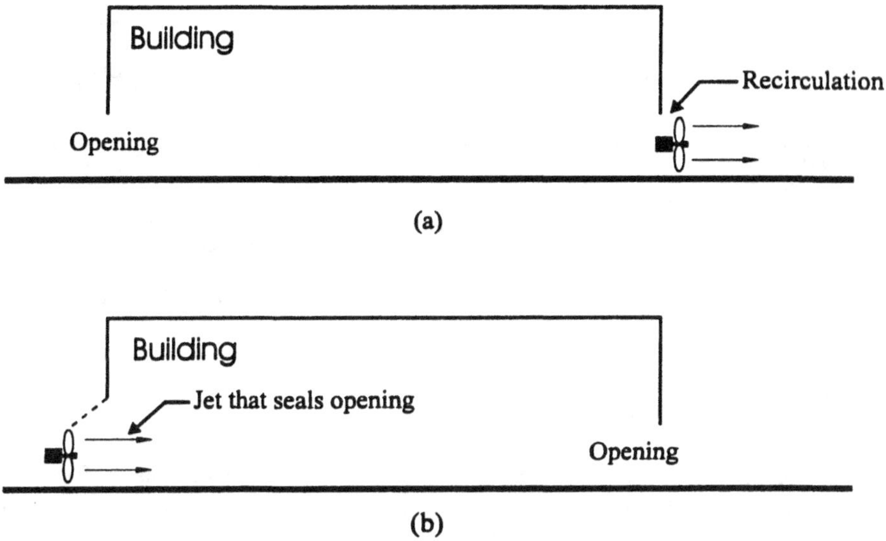

FIGURE 13.18
(a) Exhausting from a contaminated room where the effectiveness is diminished because of recirculation, and (b) the preferred method of positive-pressure ventilation that uses the outlet jet to seal the opening.

Both liquid and vapor are present in many pipes, and such pipes should be inclined or pitched such that liquid will drain to locations where it can be handled properly. Pipes carrying both liquid and vapor include liquid/vapor return lines in liquid recirculation systems, but also in some others as well where condensation of vapor to liquid may occur unintentionally. Several lines where pipes should be provided with slope and the direction of pitch are the following:

- Liquid/vapor suction line—toward low-pressure receiver
- Vapor line connecting the low-pressure receiver to the low-stage compressor—toward low-pressure receiver
- Line carrying vapor between the flash-tank/intercooler and high-stage compressor-toward intercooler
- Horizontal portion of discharge line between compressor and condenser—toward condenser
- Defrost gas line—back toward machine room

Pipes should be clearly labeled to indicate in the first place that they carry ammonia, the function in the system, whether the ammonia is liquid or vapor, the pressure level, and the direction of flow. Pipe labels are valuable in helping new operators or outside technicians understand the configuration of the system. They also help an operator avoid opening or closing the wrong valve during routine operation, as well as helping avoid incorrect responses in the

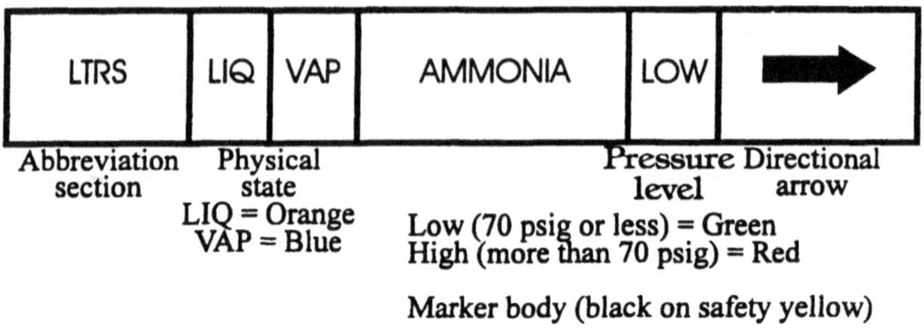

FIGURE 13.19
Labels for ammonia pipes.

event of an emergency. The International Institute of Ammonia Refrigeration has standardized[25] the form, color, and information to be contained on pipe labels, as illustrated in Fig. 13.19.

Still another crucial administrative action is to maintain up-to-date as-built plans and drawings that are readily accessible to operators and repair personnel, but stored where they remain in a clean, readable condition. For refrigeration plants with an ammonia charge of 4,500 kg (10,000 lb) or greater, the OSHA Process Safety Management regulation[26] requires that a process flow diagram as well as a piping and instrumentation diagram be readily available in the plant.

13.17 A PLANT SAFETY PLAN

It is the duty of management to ensure that an emergency plan is in place for the facility and that employees know their responsibilities should an emergency arise. Many plants on a regular basis conduct dry runs of simulated emergencies to help acquaint staff with their role(s). Training should be an ongoing process for both the safe and efficient operation of the plant. These training sessions can be conducted by a qualified staff member within the organization or an outside specialist can be brought in. In addition, there are now short courses available in design, management, and operation of industrial refrigeration plants that are sponsored by technical and trade organizations. Some of the benefits of these courses are that the instructional staff usually includes top people in the industry. Furthermore, the interchange of knowledge among the participating firms serves a valuable purpose.

REFERENCES

1. ANSI/ASHRAE Standard 15-94, *Safety Code for Mechanical Refrigeration*, American Society of Heating, Refrigerating and Air-Conditioning Engineers, Atlanta, Georgia, 1994.
2. ANSI/ASME *Boiler and Pressure Vessel Code*, American Society of Mechanical Engineers, New York, New York, 1992.
3. ANSI/ASME *Code for Pressure Piping, B31, B31.5-1992*, American Society of Mechanical Engineers, New York, New York, 1992.
4. ANSI/NFPA 70-90, *National Electrical Code*, National Fire Protection Association, Quincy, Massachusetts, 1990.
5. ANSI/IIAR-2-1992, *Equipment, Design, and Installation of Ammonia Mechanical Refrigeration Systems*, International Institute of Ammonia Refrigeration, Washington, DC, 1992.
6. ANSI/UL 207-1986, *Safety for Refrigerant-Containing Components and Accessories, Nonelectrical*, Underwriters Laboratories, Inc., Northbrook, Illinois, 1986.
7. BOCA *Basic National Mechanical Code*, Building Officials and Code Administrators, Country Club Hills, Illinois, 1990.
8. *Uniform Mechanical Code*, International Conference of Building Officials, Whittier, California, 1991.
9. *Mechanical Integrity of Vapour Compression Refrigerating Systems for Plant and Equipment Supplied and Used in the United Kingdom, Part I—Design and Construction of Systems Using Ammonia as the Refrigerant*, The Institute of Refrigeration, 76 Mill Lane, Carshalton, Surrey, SM5 2JR, United Kingdom, 1989.
10. *Safety Code for Refrigeration Systems Utilizing Chlorofluorocarbons, Part I—Design and Construction*, The Institute of Refrigeration, 76 Mill Lane, Carshalton, Surrey, SM5 2JR, United Kingdom, 1984.
11. Bulletin No. 116, *Avoiding Component Failure in Industrial Refrigeration Systems Caused by Abnormal Pressure or Shock*, International Institute of Ammonia Refrigeration, Washington, DC, 1992.
12. Loyko, L. L., "Hydraulic Shock in Ammonia Systems," Proceedings of Annual Meeting of the International Institute of Ammonia Refrigeration, Austin, Texas, March 1989.
13. Richards, W. V., Personal communication, 1990.
14. "Safety Relief Devices for Refrigerant Pressure Vessels," Henry Valve Company, Melrose Park, Illinois, 1975.
15. Bulletin No. 111, *Ammonia Machinery Room Ventilation*, International Institute of Ammonia Refrigeration, Washington, DC, 1989.
16. Toole, C. W., "Fire Prevention and Safety in Refrigerated Warehouses," Proceedings of Commission D1, International Institute of Refrigeration, pp 21-26, Paris, France, 1982.
17. Duiven, J. E. and L. Twilt, "Policies for Fire Safety of Cold Stores in the Netherlands," Proceedings of Commission D1, International Institute of Refrigeration, pp 39-45, Paris, France, 1982.
18. Webber, J. F., "Special Design Considerations Regarding Fire Protection for High-Rise Cold Stores," ASHRAE Seminar, San Francisco, California, January 1986.
19. Perez-Blanco, H. and W. F. Stoecker, "Ammonia Sensors for Refrigeration Application," Proceedings of Annual Meeting of the International Institute of Ammonia Refrigeration, St. Louis, Missouri, March 1994.
20. Lindborg, A., "Ammonia Safety in Refrigerated Warehouses," Proceedings of Meeting of Commission D1, International Institute of Refrigeration, Paris, France, March 1982.
21. "Chapter 3, Ammonia and the Environment," Ammonia Data Book, International Institute of Ammonia Refrigeration, Washington, DC, 1993.
22. Stoecker, W. F., "Expanded Applications for Ammonia—Coping with Releases to Atmosphere," *International Journal of Refrigeration*, 13:86-94, March 1990.

23. Fenton, D. L., A. F. Noeth, and R. L. Gorton, "Absorption of Ammonia into Water," *ASHRAE Transactions*, 97:(Part I) 204-213, 1991.
24. Kelley, R., A. S. Khan, K. S. Chapman, and D. L. Fenton, "Operating Characteristics of a Flare/Oxidizer for the Disposal of Ammonia from an Industrial Refrigeration Facility," technical paper submitted to ASHRAE, 1995.
25. Bulletin No. 114, *Identification of Ammonia Refrigeration Piping and System Components*, International Institute of Ammonia Refrigeration, Washington, DC, 1991.
26. "Process Safety Management of Highly Hazardous Chemicals, 29 CFR 1910.119," Occupational Health and Safety Administration, Washington, DC, 1992.

CHAPTER 14

ELECTRICAL CONTROL AND INSTRUMENTATION

14.1 INSTRUMENTATION AND CONTROL

This chapter concentrates on the electrical equipment associated with the refrigeration system. It focuses on controlling the electrical supply to the system and on electrical and/or electronic sensors and actuators. Because the full range of electrical topics is so vast, a selection has been made, as illustrated in Fig. 14.1, such that the subjects treated in this chapter are those most intimately connected with the refrigeration system. Those topics not covered or only mentioned in passing include, for example, electrical circuits (single- and three-phase, wye and delta), conductor sizing, circuit protection, power factors and their corrections, and transformers. Another important body of knowledge that is not covered is that of motors—the types of industrial motors and their characteristics.

The electrical principles that will be explained are those related to controlling the power to electric motors and other equipment and the electrical types of instrumentation and actuators. The first major topic presented will be ladder diagrams, which are the widely accepted means of showing control logic and are the plans followed by electricians. Means of executing ladder diagrams include electromechanical relays, programmable controllers, and computer controllers. Instruments providing visual indications (pressure gauges, thermometers, etc.) are standard and should continue to be installed. The shift in recent years to

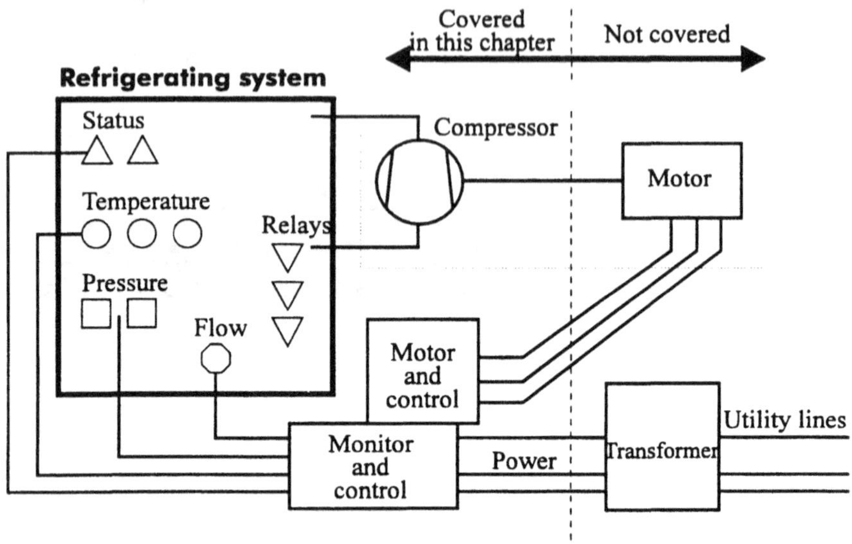

FIGURE 14.1
Full range of electrical considerations and those addressed in this chapter.

electronic and/or computer control, however, stimulates a greater need for electric and electronic sensors, transducers, and actuators. These devices will be explained in the latter sections of this chapter.

14.2 LADDER DIAGRAM SYMBOLS

Ladder diagrams serve two purposes—they represent the plan for hardwiring a panel of electromechanical devices and they also represent the logic of the control plan. Logic means the conditions that must be met before a certain action is taken. The ultimate action of most ladder diagrams is to provide electric power or to interrupt power to motors and other electrical devices. The symbols[1] that are probably the most used in ladder diagrams are shown in Fig. 14.2 for manual switches, Fig. 14.3 for switches controlled by physical variables and other conditions, Fig. 14.4 for timing switches, Fig. 14.5 for symbols referring to the controls for the power portion of the electric system, and Fig. 14.6 for miscellaneous symbols.

A toggle switch, as shown in Fig. 14.2, retains its position (open or closed) until manually changed. Push buttons are designed for momentary contact or interruption. A dashed line indicates a mechanical linkage between two push buttons, which in Fig. 14.2 shows one push button opening and the other closing a contact when the button is pressed.

In Fig. 14.3, the switch changes from its normal position when the sensed variable increases above its setting. For example, the normally closed (NC) temperature switch set for 40°C (104°F) is closed when the sensed temperature is

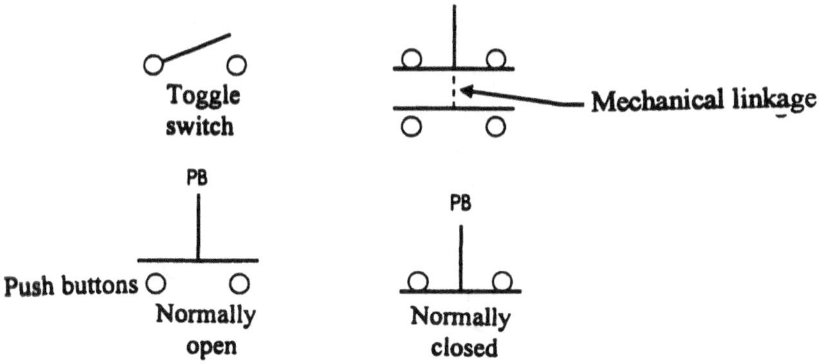

FIGURE 14.2
Some symbols for manual switches used in ladder diagrams.

below 40°C (104°F) and open when the temperature is above the setting. If a low-temperature cutout is to open a switch when the temperature drops below 0°C (32°F), for example, a normally open (NO) switch would be chosen and set for the temperature. During satisfactory operation above 0°C (32°F), the switch is in its non-normal state (closed).

The symbols for another class of components used in ladder diagrams are shown in Fig. 14.4 and apply to timing switches. Most timing switches are single-throw, but double-throw switches are also available, as shown in Figs. 14.4e and 14.4f. One class of timing switches is indicated by the upward-pointing arrow ↑ and another by a downward-pointing arrow ↓, representing a delay *on* and a delay *off*, respectively. The energizing of the coil initiates the time delay of a delay-on switch, while the denergizing of the coil initiates the time delay of a delay-off timing switch.

Figures 14.4a and 14.4c show NC switches, and Figs. 14.4b and 14.4d show NO switches. The NC, timed-open, delay-on switch of Fig. 14.4a begins the timing upon energizing of the coil and opens the contacts following the specified delay. When the coil is deenergized, the contacts immediately return to their NC status. Should the coil be deenergized during the delay period, the switch remains closed and the timer is reset to zero. The NO, timed-closed, delay-on timing switch of Fig. 14.4b begins the timing operation upon energizing of the coil and closes the switch following the delay. When the coil is deenergized, the contacts immediately return to their NO status. Should the coil be deenergized during the delay period, the contacts remain open and the timer is reset to zero.

The down arrow ↓ timer switches in Fig. 14.4c and 14.4d are in the status shown (normally closed or normally open, respectively) when the coil has been deenergized for some time. When the coil is energized, the switch changes instantly to its nonnormal status, at which condition it remains so long as the coil is energized. When the coil is deenergized, timing begins, and following the specified delay, the contacts revert to their normal position. If the coil should be

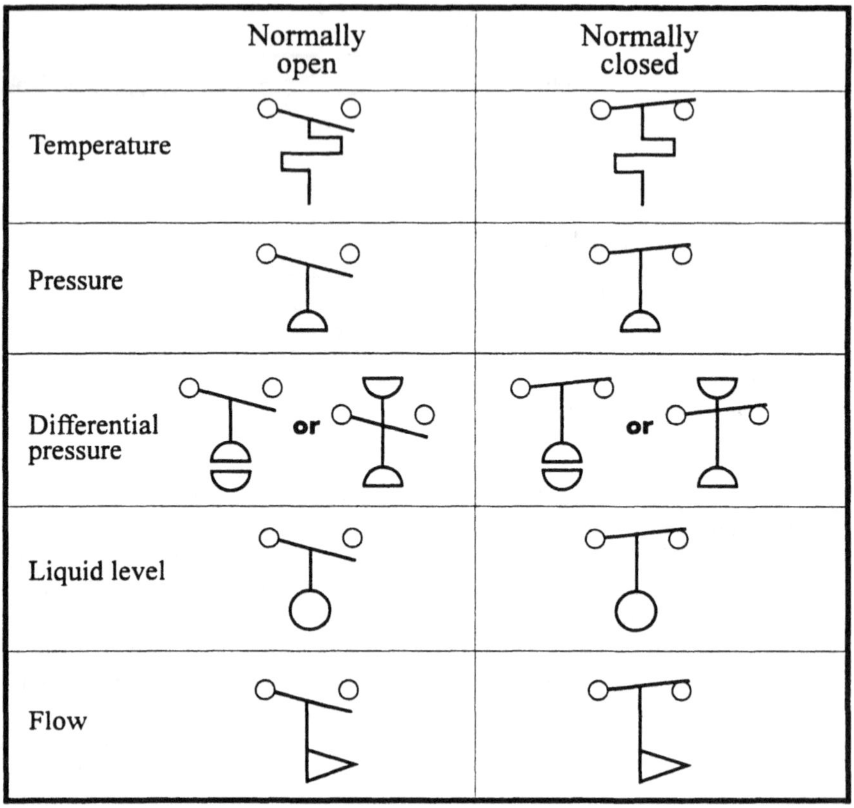

FIGURE 14.3
Switches activated by physical variables or conditions.

energized during the delay period, the contacts return to their nonnormal status and the timer resets to zero.

The double-throw timing switch of Fig. 14.4e is a combination of the switches in Figs. 14.4a and 14.4b. The status of the blade shown occurs when the coil has been deenergized for a period of time. When the coil is energized, the timer begins, and following the specified delay, the blade changes from the NC contact to the NO contact, where it remains so long as the coil is energized. Deenergization of the coil returns the blade to the NC position instantly. Some commercial timing switches of this type are supplied with power continuously, and what is referred to as energizing and deenergizing of the coil is achieved by closing and opening, respectively, external contacts.

The double-throw, delay-off timing switch of Fig. 14.4f combines the functions of the switches in Figs. 14.4c and 14.4d in the following manner. The position of the switch shown is what occurs when the coil has been deenergized

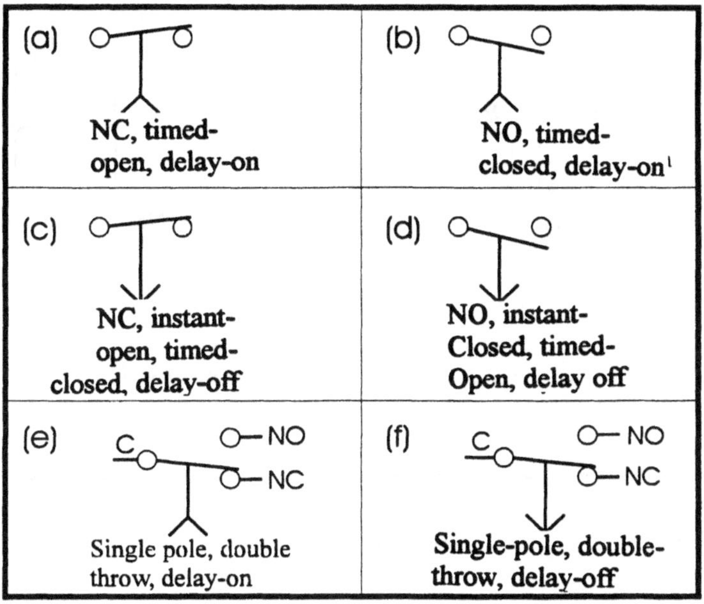

FIGURE 14.4
Timing switches.

for a period of time. Energization of the coil changes the blade instantly from the NC post to the NO post. Deenergization of the coil initiates the timer. Following the specified delay, the blade changes to contact the NC post.

Several other symbols sometimes associated with ladder diagrams that refer to the power portion of the electrical network are shown in Fig. 14.5. To show that two crossing electrical lines are not physically connected, it is only necessary to show the cross without a heavy dot. This practice is standard in the electronic industry since the diagrams sometimes become massive. For drawings applicable to refrigeration systems often used in industrial workplaces, the semicircular bypass avoids the risk of misreading a soiled drawing.

The symbols for coils of relays, solenoids, contacts, and several other components used in ladder diagrams are shown in Fig. 14.6.

14.3 LADDER DIAGRAMS

The ladder diagram gets its name from the appearance of the network, which consists of the horizontal *rungs* extending from one vertical *rail* to the other. The left rail is designated $L1$ and the right rail $L2$. Usually, $L1$ is hot and is at a voltage of 115 V ac, for example, and $L2$ is usually neutral. In some cases, however, $L2$ is 115 V ac of opposite polarity to $L1$, so that 230 V prevails across the ladder. When a voltage is applied across a control relay (CR), any contact associated

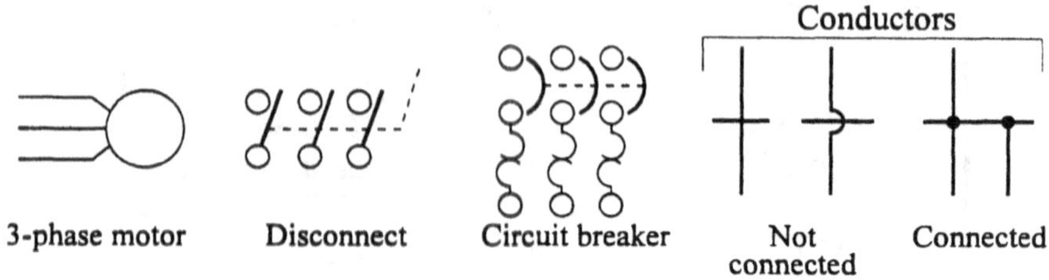

FIGURE 14.5
Several symbols applicable to the power portion of the electrical network.

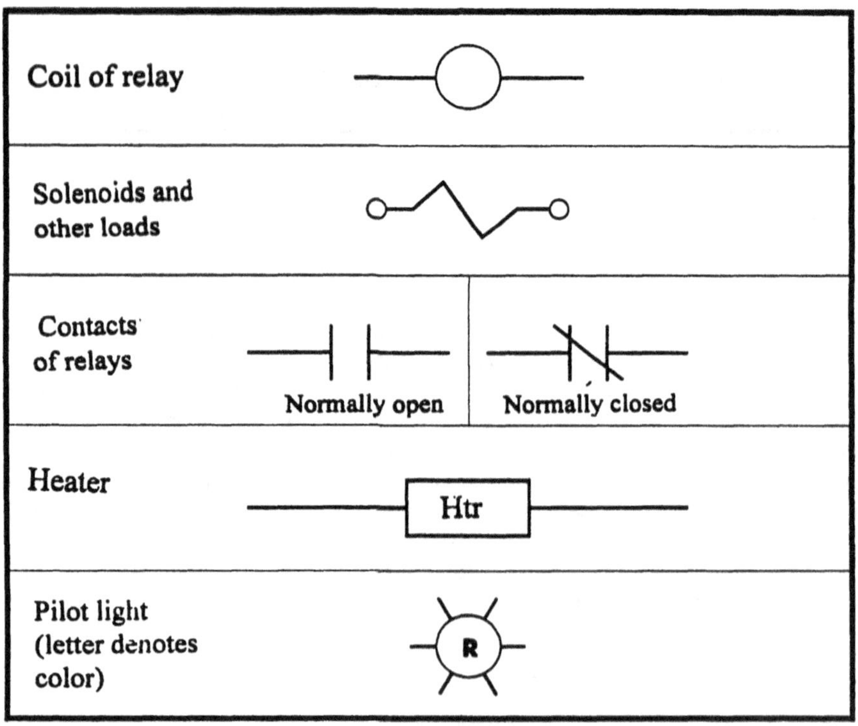

FIGURE 14.6
Some miscellaneous symbols.

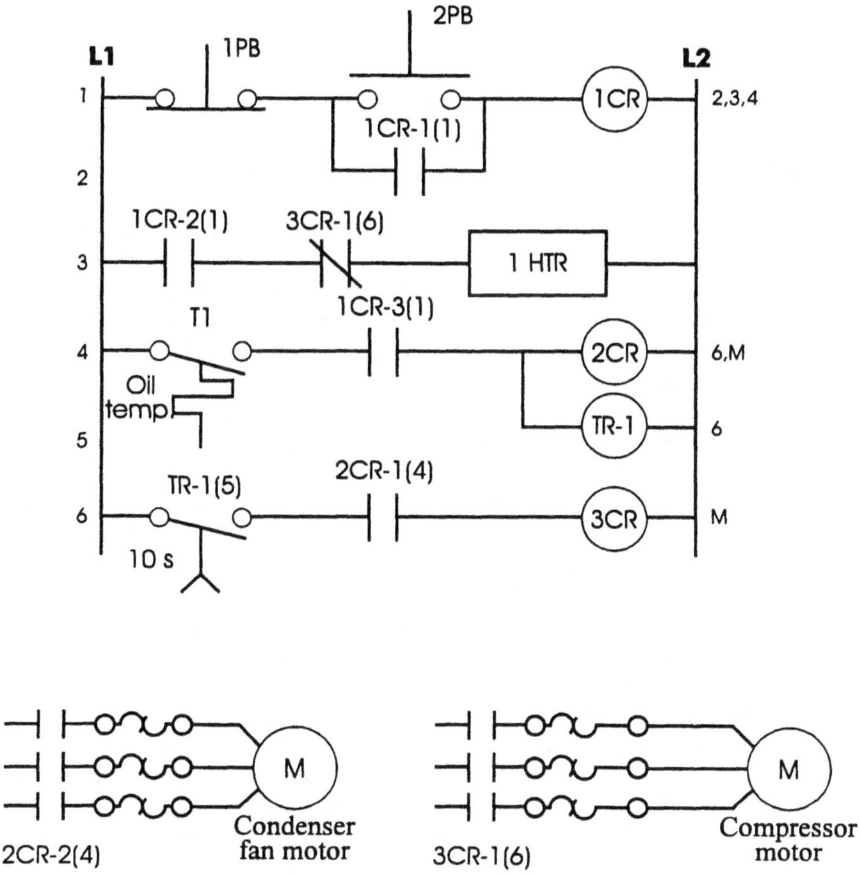

FIGURE 14.7
Ladder diagram for starting and stopping a condenser fan motor and a compressor motor in conjunction with its crankcase heater.

with this relay is switched to the opposite of its normal position. A simple ladder diagram depicting the operation of the compressor and condenser fan motors and the heater for the compressor crankcase is shown in Fig. 14.7. The compressor is started by pressing a push button, designated $2PB$, and stopped by pressing a different button, $1PB$. When the start button is pressed, the crankcase heater comes on if needed until the temperature of the oil rises to a temperature of 30°C (86°F). When the desired oil temperature has been reached, the condenser fan starts and runs for 10 s before the compressor motor starts. Once the compressor is operating, there is no further need to warm the oil, and the heater shuts off.

The plan expressed by the ladder diagram in Fig. 14.7 is as follows. To start the sequence, press push button $2PB$, which feeds power to the left side of $1CR$.

Control relay $1CR$ changes the status of all contacts associated with it. One of these contacts is $1CR$-1, which is wired in parallel to $2PB$. Closing this contact permits $1CR$ to remain energized even after $2PB$ is released. Another contact associated with $1CR$ is $1CR$-2, which closes. Because $3CR$-1 is normally closed, power is fed to the crankcase heater, $1HTR$. The final contact associated with $1CR$ is $1CR$-3, which closes. If the oil temperature is not yet up to the desired value of 30°C (86°F), temperature switch $T1$ is open and no power reaches $2CR$ and TR-1. When the oil temperature rises to the point that $T1$ closes, $2CR$ and TR-1 are energized. The energizing of $2CR$ closes the contacts in the 3-phase lines to the condenser fan motor and also closes the contacts in line 6. Energizing TR-1 starts the timing switch, which after a 10 s delay energizes $3CR$. Energizing $3CR$ closes the contacts in the 3-phase lines to the compressor motor, thus starting that motor. Another contact served by $3CR$, namely $3CR$-1, turns off the crankcase heater.

To stop the compressor, press $1PB$, which interrupts power to $1CR$. Contact $1CR$-1 reverts to its NO position and deenergizes $1CR$ even after $1PB$ is released. Contact $1CR$-2 returns to its normally-open position, but the heater had been off anyway, so the status of the heater remains unchanged. Contact $1CR$-3 opens so that $2CR$ is deenergized, which opens the contacts in the 3-phase lines to the condenser motor and also opens $2CR$-1, which deenergizes $3CR$ to stop the compressor motor.

Also illustrated in Fig. 14.7 is a recommended numbering pattern. Each of the rungs is numbered in sequence down the left rail, and each of the control relays, contacts, and other switches are given unique numbers. The designation $2CR$-1(4), for example, indicates a contact whose status is changed by control relay $2CR$. The number 1 indicates that this contact is the first one associated with $2CR$, and the number 4 in parentheses designates the rung number in which $2CR$ is located. The numbers down the right rail indicate the rungs in which contacts are located that are associated with any CR of that rung. In rung 1, for example, $1CR$ activates contacts in rungs 2, 3, and 4.

Example 14.1. At the moment that $3CR$ is energized, starting the compressor motor and shutting off the crankcase heater, some remaining liquid refrigerant in the crankcase oil evaporates to cool the oil and open the temperature switch $T1$. What is the response of the control logic thereafter?

Solution. The opening of $T1$ deenergizes $2CR$ and TR-1. The condenser fan stops, and the contacts of $2CR$-1 change to their NO status, which deenergizes $3CR$, stopping the compressor motor and allowing the contacts of $3CR$-1 to return to their NC status. Coil $1CR$ has remained energized, so the heater resumes operation and another startup is attempted.

14.4 LADDER DIAGRAM FOR A SCREW COMPRESSOR

To illustrate a more complex control logic, Fig. 14.8 shows a ladder diagram for the implementation of many of the functions needed to start, stop, and reset a screw compressor package. Most of the standard features are incorporated, but to provide a gradual progression to full realism, the control of the slide valve and the capability of automatic remote control have been omitted. A complete example is presented in Reference 2, which in turn is taken from a manufacturer's manual.[3] Those functions that are incorporated are the following:

1. Push-button start.
2. Cutouts to stop the compressor and oil pump in the event of:
   - oil pressure failure lasting longer than 6 s.
   - low oil temperature.
   - high temperature of discharge refrigerant.
   - low suction pressure.
   - high discharge pressure.
   - overloads of motors serving compressor and oil pump.
3. Push-button stop and reset. The function of the reset is to prevent the compressor from automatically starting up again after a cutout has stopped operation.
4. Operation of the oil pump for 30 s prior to starting the compressor in order to allow the slide valve to move to its unloaded position (the slide valve operation is not shown in Fig. 14.8).
5. Anti-recycle timing, which prevents a restart of the compressor motor within 20 min of the previous start.
6. Keeping the oil warm during compressor shutdown and turning the heater off during compressor operation.

The following is a brief, line by line description of each switch, contact, and relay in Fig. 14.8:

1. $1TR$-1 is normally open (NO) but closed if $1TR$ is energized, and it opens 6 s after $1TR$ is deenergized. $1TS$ is NO and closes on satisfactory high oil temperature. $2TS$ is normally closed (NC) but opens on high refrigerant discharge temperature. $1PS$ is NO but closes on adequately high suction pressure. The motor overload cutouts are NC but open on high motor current. $2PS$ is NC but opens on high discharge pressure. $1PB$ is a NC push button for stopping. $2PB$ is a NO push button for starting. $1CR$ is a relay coil.
2. The pilot lights are in parallel with each cutout and are off when the cutout is closed, because most current passes through the switch. If the switch is open, the light is illuminated. $1CR$-1 continues to energize $1CR$ after the start button is released.

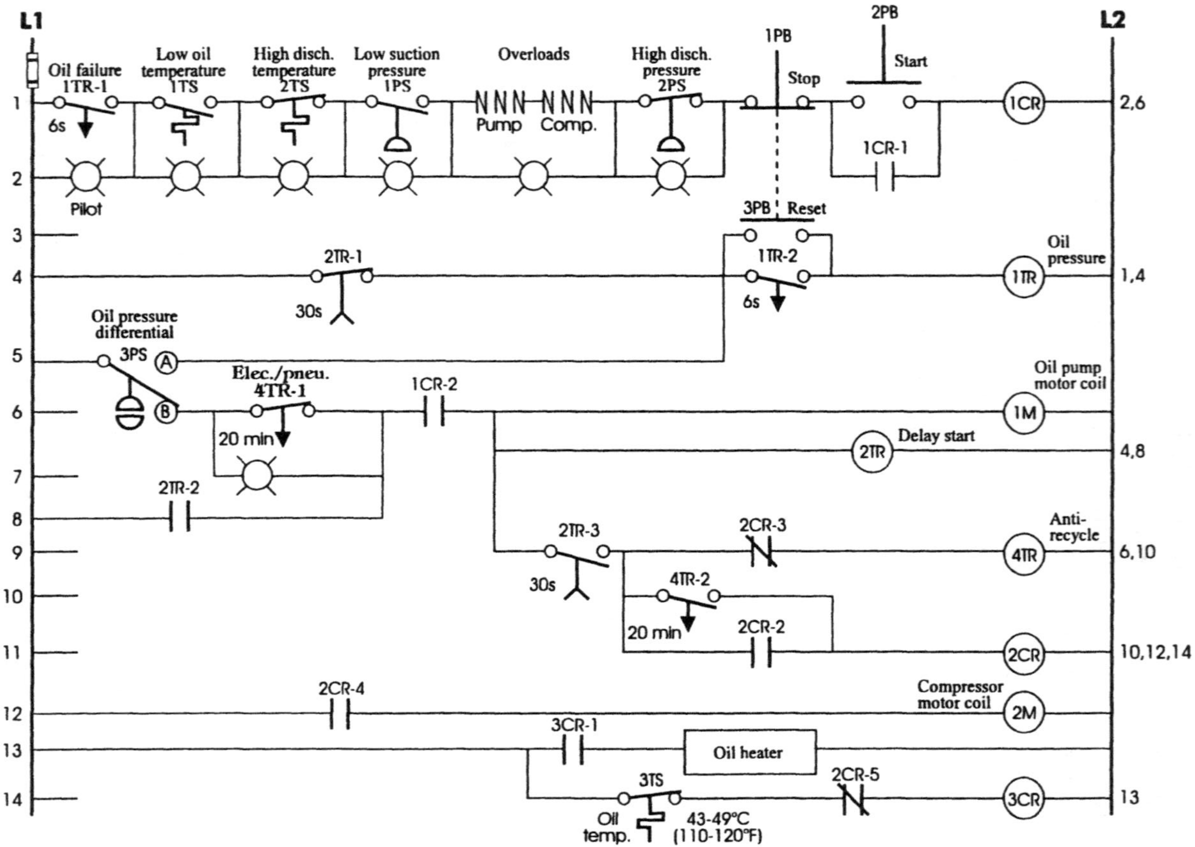

FIGURE 14.8
Ladder diagram for the control of a screw compressor.

3. 3*PB* is a NO push button reset mechanically connected to the STOP button.
4. 2*TR*-1 is NC but opens 30 s after energizing 2*TR*, providing power to 1*TR* on startup. 1*TR* is a relay coil that remains energized if the oil pressure differential is above 140 kPa (20 psi).
5. If the oil pressure differential is greater than 140 kPa (20 psi), then the contact of 3*PS* is with *A*; but if the differential pressure is very low, the blade is in contact with *B*.
6. 4*TR*-1 is a special timing switch, combining both an electrical and pneumatic capability. It is NC and opens when 4*TR* is energized. The switch remains open for 20-min before reclosing. Because of the pneumatic capability, the 20 min timing does not require electric power. 1*M* is the oil pump motor relay coil.
7. A pilot light, which is in parallel with 4*TR*-1, lights when 4*TR*-1 is open. 2*TR* is a relay coil initiating the two 30 s time delays of 2*TR*-1 and 2*TR*-3, as well as closing contact 2*TR*-2.
8. 2*TR*-2 contact is closed by energizing 2*TR* and continues to provide power to 2*TR* as long as 1*CR*-2 is closed.
9. 2*TR*-3 is NO but closes 30 s following the energization of 2*TR* to provide power to the compressor relay coil, 2*CR*. 2*CR*-3 is NC and opens by relay coil 2*CR*. 4*TR* is an anti-recycle relay coil that is energized only momentarily on startup of the compressor.
10. 4*TR*-2 is NO and closes at the momentary energizing of 4*TR* and remains closed for 20 min.
11. When 4*TR*-2 is closed, 2*CR* relay coil is energized, which closes 2*CR*-2 to maintain power to 2*CR* even after 4*TR*-2 reopens.
12. 2*M* is the compressor motor relay.
13. 3*CR*-1 is a NO contact serving the oil heater.
14. 3*TS* is a thermostatic switch that closes when the oil temperature drops lower then 43°C (110°F) and opens when the oil temperature rises above 49°C (120°F).

Example 14.2. Is the heater on or off when the oil temperature is (a) 40°C (104°F), (b) 50°C (122°F), and (c) 45°C (113°F)?

Solution. If the compressor is running, 2*CR* is energized and the NC contact in line 14 is open, deenergizing 3*CR*. Therefore, the heater will be off, regardless of the oil temperature. If the compressor is not operating (a) 3*TS* is closed, 3*CR* is energized, and the heater is on; (b) 3*TS* is open, 3*CR* is deenergized, and the heater is off; (c) could be either on or off depending upon whether the oil is cooling off following the heater switching off or warming up following the heater switching on.

Example 14.3. What are the states of relay coils, switches, and contacts when the compressor and oil pump have been operating for more than 20 min?

Solution. Relay coils energized: $1CR$, $1TR$, $1M$, $2TR$, $2CR$, and $2M$
Relay coils deenergized: $4TR$ and $3CR$
Switches closed: all cutouts in line 1, $1TR$-2, $3PS$ to A, $4TR$-1, and $2TR$-3
Switches open: $2TR$-1 and $4TR$-2
Contacts closed: $1CR$-1, $1CR$-2, $2TR$-2, $2CR$-2, and $2CR$-4
Contacts open: $2CR$-3 and $3CR$-1.

Example 14.4. What are the states of relay coils, switches, and contacts when the compressor and oil pump have been operating for less than 20 minutes?

Solution. Same as solution to Ex. 14.3, except: $4TR$-1 is open and $4TR$-2 is closed.

Example 14.5. What are the states of relay coils, switches, and contacts after the compressor and oil pump have been operating for more than 20 minutes and then the STOP/RESET button is pressed?

Solution. $1PB$ interrupts power to $1CR$. The contacts of $1CR$-2 open to interrupt power to the oil pump and compressor relays to stop those motors. In addition, $2TR$ is deenergized and $4TR$ remains deenergized. Switch $4TR$-1 remains in its NC position, because the compressor had been operating more than 20 min. Contact $2CR$-5 reverts to its NC status, permitting thermostat $3TS$ to maintain the temperature of the oil between the prescribed limits. When the oil pump stops, $3PS$ changes from the A to the B contact. A section that remains energized is line 4, because switch $2TR$-1 is in its NC position and $1TR$-2 remains closed when $3PB$ is pressed. Thus, $1TR$ remains energized, which keeps switch $1TR$-1 closed. All the cutouts and overloads in line 1 remain closed because $3PB$ made contact when the STOP/RESET button was pressed.

Example 14.6. After the compressor and oil pump have been stopped, as in Ex. 14.5, the START button is pressed. Describe the sequence.

Solution. Relay $1TR$ has been left energized, so cutout $1TR$-1 is closed as are all the other cutouts in line 1. Pressing $2PB$ energizes $1CR$, which closes $1CR$-1 to maintain $1CR$ energized even after $1PB$ is released. There is no oil pressure yet, so $3PS$ is in contact with B, feeding power to line 6. Because $4TR$-1 is closed and $1CR$-2 is now closed, the oil pump motor relay $1M$ and the start delay relay $2TR$ are energized. Energizing $2TR$ closes $2TR$-2. The oil pump starts and in several seconds the minimum differential oil pressure is established and $3PS$ switches to contact A. Power continues to reach $1CR$-2, however, because $2TR$-2 is now closed. The compressor motor does not start immediately, because $2TR$-3 does not close until 30 s have elapsed after the oil pump starts. The purpose of this delay is to permit the slide valve to drive to the minimum-load position. After the 30 s delay, $2TR$-3 closes and $2TR$-1 opens and remains open during operation of the compressor and pump, but power to the right side of line 4 comes through the A contact of $3PS$. When $2TR$-3 closes, the anti-recycle timer $4TR$ is energized, opening $4TR$-1 and closing $4TR$-2 for 20 minutes thereafter. The closure of $4TR$-2 energizes $2CR$, closing $2CR$-2 and $2CR$-4, starting the compressor motor. The anti-recycle relay $4TR$ is energized only momentarily, because as soon as $2CR$ is energized $2CR$-3 opens.

The system should now be in full operation and continue to run until the STOP button is pressed or a safety switch interrupts the operation.

Example 14.7. A trace of liquid refrigerant in the oil flashes to vapor when passing through the pump, causing a momentary (approximately 2 s) reduction in oil pressure differential. Describe the sequence of events that follows.

Solution.

The reduction in oil pressure differential allows the blade of $3PS$ to drop from position A so that $1TR$ is deenergized. The timers of $1TR$-1 and $1TR$-2 are activated, but the timers are set for 6 s. Before switches $1TR$-1 and $1TR$-2 open, the blade of $3PS$ returns to A, which zeros the timers of $1TR$-1 and $1TR$-2 which remain closed. This feature has thus functioned as planned and avoided a nuisance shutdown.

Example 14.8. The compressor has been operating for more than 20 min when the high discharge pressure cutout switch opens because of an unintentional, momentary shutdown of the condenser.

(a) Will $1CR$, $2CR$, $1TR$, $2TR$, and $2M$ be energized or deenergized, and will $1TR$-1 be open or closed?

(b) If the START button, $2PB$, is pressed, what will be the response?

(c) If the STOP/RESET button, $1PB$, is pressed, what action occurs?

(d) If after pressing STOP/RESET the START button is then pressed, what will be the response?

Solution.

(a) All the coils will be deenergized, and $1TR$-1 will open after 6 s.

(b) Because $1TR$-1 is open, no power reaches $1CR$, so no coils are energized.

(c) Closing the RESET switch energized $1TR$, closing $1TR$-2, which maintains power to $1TR$. $1TR$-1 closes. The blade of $3PS$ is on B, but because $1CR$-2 is open, no power reaches the relay coils that would start the motors.

(d) The normal startup sequence will take place, as in Ex. 14.6.

14.5 IMPLEMENTING THE LADDER DIAGRAM

The traditional equipment used to execute the logic expressed in ladder diagrams is a panel of electromechanical relays. When power is applied to the coil, the contacts associated with that relay shift to their nonnormal status. Delay timers that change the status of the contacts following the delay for which the timer is set are also available.

Currently, programmable controllers compete vigorously with relay panels. The logic of the ladder diagram is incorporated in the software of the microprocessor-based programmable controller. The relative costs of the electromechanical relay panel and the programmable controller are shown in Fig. 14.9.

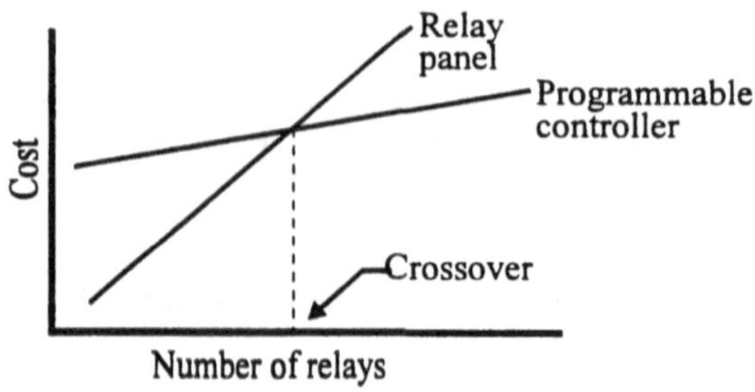

FIGURE 14.9
Relative costs of programmable controllers and electromechanical relay panels.

The cost of the relay panel increases roughly in proportion to the number of relays, while a given model of programmable controller can accommodate any number of relays up to its maximum. The cost of each additional relay on a programmable controller is thus negligible. The crossover number of relays is now quite low—perhaps from three to five.

Because industrial electricians are familiar with ladder diagrams, many programmable controllers are equipped with either a permanent or portable terminal and a screen that shows the image of the ladder diagram as it exists or as it is developed.

14.6 INSTRUMENTATION FOR MONITORING A REFRIGERATION PLANT

This section supports the contention that adequate and even extensive instrumentation of a refrigeration facility can enhance efficiency, reliability, and safety. Future sections will focus on electronic and computer monitoring and control, but this section allows for low technology exemplified by visual indicators such as pressure gauges and thermometers. The refrigeration plant of several decades ago was typically equipped with an instrument panel with some large pressure gauges, pilot lights, cutouts, alarms, and sometimes even switchboxes. Through the years, the instrument panel has virtually disappeared, and pressure gauges, for example, are often only small gauges mounted on the compressor package.

The purchase and installation of pressure, temperature, flow, and status sensors entails a cost that must be justified through the information provided by the measurement. Ideally, the value of knowing the magnitude of an operating variable has a monetary value, and this value must justify the cost of the instrument. While it is possible to roughly estimate the cost of buying and installing an instrument, it is unfortunately almost impossible to assess the value of the reading. To try to provide at least a qualitative value of knowing the magnitude of various operating variables, the author has conducted surveys of more than a hundred refrigeration professionals during industrial refrigeration workshops.

The results are shown in Fig. 14.10, and in addition, a tabulation of comments of why the variable is important is shown below. The type of plant assumed is a two-stage liquid recirculation system.

1. Refrigerant pressure leaving compressor. This pressure is one of the basic variables indicating performance conditions of the plant. Along with the pressure entering the condenser, a blockage in the oil separator can be detected. The condensing pressure relative to the wet-bulb temperature of the ambient air indicates how well the condenser is performing and along with item 8 indicates the presence of noncondensibles.

2. Refrigerant pressure entering the high-stage compressor. This measurement indicates the intermediate pressure, which is important when there are intermediate-temperature evaporators.

3. Refrigerant pressure entering low-stage compressor. This is another fundamental indicator of system performance. In combination with item 4, excessive pressure drop in the suction lines can be detected.

4. Refrigerant pressure at coil outlets. This pressure can be translated into the evaporating temperature of the refrigerant.

5. Pressure at outlet of recirculating pump. No pressure rise across the pump suggests pump failure, and fluctuations of pressure indicate pump cavitation.

6. Pressure difference across the oil pump. This pressure difference must be provided to ensure good lubrication of bearings. An unusually high pressure difference may indicate a blocked filter, and fluctuations across the pump serving a reciprocating compressor could detect liquid refrigerant in the crankcase.

7. Temperature of refrigerant at compressor discharge. On a reciprocating compressor, an unusually high discharge temperature may disclose that a discharge valve is broken. On a screw compressor, a high discharge temperature may hint that a bearing is failing or that the oil-cooling system is malfunctioning.

8. Liquid temperature from the condenser. In conjunction with the condenser pressure (item 1), the combination can disclose the presence of noncondensibles. On the other hand, excessive subcooling may be indicated, which shows that liquid is backing up into the condenser.

9. Temperature of suction gas entering the low-stage compressor. This measurement helps show that the vapor is dry, which is desirable, and it also helps show that the vapor does not possess excessive superheat, which is undesirable.

10. Temperature of suction gas to the high-stage compressor. This measurement is not necessary if the presence of liquid in the intercooler can be ensured.

11. Temperature of air in refrigerated spaces. This is the primary purpose of many systems, so satisfactory temperatures should be verified.

12. Total refrigerant flow rate of the plant. This measurement is not nearly as useful as knowing the flow rate of individual compressors.

13. Refrigerant flow rates of individual compressors. Deterioration of performance can be detected. If the flow rates of all individual compressors are known, the total of subgroups and/or total plant flow rate can be computed.

14. Total compressor power. This measurement aids in watching demand charges for electricity.

Measurement or instrument	Importance or usefulness					
	Little			Moderate		Very
PRESSURE						
1. Refrigerant pressure leaving compressor						xxx
2. Refrigerant pressure entering high stage					xxx	
3. Refrigerant pressure entering low stage				xxx		
4. Refrigerant pressure at coil outlets			xxx			
5. Pressure at outlet of recirculating pump					xxx	
6. Pressure difference across oil pump					xxx	
TEMPERATURE						
7. Refrigerant at compressor discharge						xxx
8. Liquid from condensers or receiver					xxx	
9. Suction gas to low stage						xxx
10. Suction gas to high stage			xxx			
11. Air temperature in refrigerated spaces						xxx
FLOW RATE						
12. Total plant liquid flow rate	xxx					
13. Individual compressor flow rate				xxx		
ELECTRICAL POWER						
14. Total compressor power		xxx				
15. Individual compressor power					xxx	
16. Condenser fan power	xxx					
STATUS						
17. Condenser fans				xxx		
18. Evaporator coil fans					xxx	
19. Hot-gas defrost						xxx
20. Refrigerant leak detection						xxx
LIQUID LEVEL						
21. In high-pressure receiver						xxx
22. In low-pressure receiver	xxx					
23. Central panel for instruments						xxx

FIGURE 14.10
Relative importance placed on instrumented variables through a survey of over 100 refrigeration professionals.

15. Power drawn by individual compressors. Shows the extent of the loading for each compressor. If the refrigerant flow rate is known, the efficiency of each compressor can be determined.

16. Condenser fan power. Important information when energy consumption is being watched carefully.

17. Status of condenser fans. The ability to determine the status of condenser fans can save repeated trips to the roof. Whether the fan is actually delivering air is the crucial determination, because sometimes a contact may be closed that should indicate fan operation, but because of some problem the fan motor is not operating.

18. Status of fans on evaporator coils. If the plant has a large number of coils and fans, one or more could be out of service without the operator knowing. If a fan fails, liquid refrigerant may accumulate in the coil, and when this coil defrosts, excessive liquid may be returned to the low-pressure receiver.

19. Status of hot-gas defrost. This measurement aids in staggering the defrosts of the various coils. The information also helps check whether the length of time for defrost is excessive.

20. Refrigerant leak detection. This is important information to obtain immediately with ammonia in order to prevent damage to the product. In the case of odorless halocarbons, the knowledge may help avoid the loss of the entire charge, which could shut down the plant.

21. Liquid level in the high-pressure receiver. Loss of refrigerant that is unknown to the operator can be detected. One means of this indication is a gauge glass, which is protected from damage and equipped with valves at the top and bottom that automatically close should the glass be broken. For electric or electronic indication, a capacitive-type liquid level indicator is available. To allow for the possibility of an electronic failure, some operators like to combine the capacitive level indicator with a float chamber and magnetic switch to provide an alarm.

22. Liquid level in the low-pressure receiver. In general, only the extremely low or the extremely high level is to be avoided, and low-pressure receivers are equipped with alarms to detect these problems.

23. Central panel for instruments. Such a panel, as stated in the introduction to this section, has to a large extent passed out of fashion. The respondents in the survey strongly supported a central panel or station, however, for the following reasons: aids in troubleshooting because simultaneous readings and indications are available, sequences of operation can be followed more easily, aids in better use of operator's time, attracts more attention than remote indicators, and it is possible to see the big picture from a central instrumentation facility. Even with a central instrument panel, inspections by the operator of remote equipment are still crucial to listen and observe for nonnormal operation. Also, the mounting of some instruments directly on the components is valuable.

The general pattern of the respondents to this survey was to rate most of the items in Fig. 14.10 as either moderately or very important. One element of the cost/benefit equation missing in Fig. 14.10, however, is that the participants were not forced to pay for the instrument in a competitive environment. The system designer should be aware of the benefits of each item of data, so that he or she can make an intelligent decision whether or not to install a particular item of instrumentation. Even if the complete installation of an instrument cannot be justified, a strong case can be made to at least install pressure taps with their small capped valves so that a gauge could be installed later, either permanently or temporarily during troubleshooting. Another provision with a relatively miniscule cost is the installation of thermometer wells at crucial locations, which could pay big dividends during system diagnosis.

14.7 THE RISE OF COMPUTER MONITORING AND CONTROL

Electronics and microprocessors have invaded almost every crevice of technical life. It is no surprise, then, that microcomputer systems have penetrated the control functions of industrial refrigeration plants.[4,5,6,7,8] The power of a computer control system can provide, at an affordable cost, a monitoring and control capability that dwarfs the potential of mechanical, pneumatic, and even noncomputer electronic control.

Several questions of strategy face any author in planning an explanation of computer monitoring and control systems. A fundamental question is whether the refrigeration professional needs to know anything at all about the technical elements of computer control. After all, one hears freqently, "You don't need to know how a television set works in order to turn it on and enjoy it." This challenge forces an early distinction among types of computer monitoring and control systems. One type is exemplified by the now widely used computer controllers of compressors, especially screw compressor packages. Assemblers of these packages report that customers choose the microprocessor controller option over the electromechanical relay panel about 90% of the time. The microprocessor control system for a compressor package is an example where spending time and effort learning the internal details of the controller may not be warranted. This control system is factory assembled by a manufacturer who knows from experience the needs and capabilities of this controller and has considerable experience in the operation of this dedicated computer controller.

Another class of computer monitoring and control systems is one that concerns itself with broad plant operations. Each application of such a control system is unique, and each plant manager may have different emphases and needs. Indeed, it is possible to hire computer control specialists who can install and set into operation a system that will function and provide a certain amount of benefit. We contend, however, that as much knowledge as possible possessed by the user is valuable, stopping short of the internal electronics within integrated circuits and microprocessors. The subjects of electronics and computer control are not part of the traditional body of knowledge studied and acquired by refrigeration professionals, so this study will be new territory for some readers.

The case can be made that the person designing and operating the computer monitoring and control system should be one who understands the plant being controlled. Here are a few examples why. An instrumentation specialist may be able to select a temperature sensor that performs well electronically, but the refrigeration plant engineer might be aware of potential corrosion of the sensor material and also that there is temperature stratification in the fluid, which requires the sensor to be placed at a precise position. The computer control specialist may specify the gear that would properly start and stop a motor, but the refrigeration plant manager is aware that frequent starts of the motor must be avoided. A further reason for the refrigeration plant person having an intimate knowledge of the monitoring and control system is that with experience comes the desire to change the list of variables being monitored and modify the control protocols. When these revisions and routine troubleshooting can be handled by plant personnel, with consultation as needed with the computer control specialist, the work is more likely to be performed properly and expeditiously.

14.8 LEVELS OF UNDERSTANDING AND PERFORMANCE

Before concentrating on the technical features of the computer controller, it may be productive to approach the question of what a refrigeration professional should learn about computer monitoring and control from still another viewpoint. This approach establishes three levels of conceptual understanding that parallel a possible subdivision of the hardware/software in the field installation. These three levels are:

- Components
- Integration of hardware components and software into a working assembly
- Combining the operation of a properly functioning monitoring and control system with the plant operation, including effective human interface

The three levels represent a sequence of increasing conceptual difficulty. The component level addresses the individual elements, such as transducers, transmitters of information, converters between signals in the analog and digital form, actuators that start and stop and regulate the performance of equipment, and the computer itself. The principles of operation and the knowledge of the proper application of the devices are reflected in the component level. Included in this level are how to input and process data, communicate through the screen or printer, and send out signals. The requirements of the component level are fulfilled when all of the elements perform their required functions. The transducers must sense their respective variables with appropriate accuracy, the hardware of the computer must function properly, and the actuators must respond with precision to their respective control signals. The proper performance of the components is a necessary, but not sufficient, requirement for an effective monitoring and control system.

The second level of understanding and performance includes the proper interfacing of all the components that are external to the computer. Some of the capabilities incumbent on the integration level include conversion, transmission, and conditioning of signals. For example, a temperature transducer may provide an electrical resistance that is a function of the temperature being sensed. This resistance may need to be converted to a voltage which is transmitted from the remote location to the computer interface. Here, when commanded by the computer, the voltage is converted to a binary number for further processing. The processing might be a translation to the Celsius (Fahrenheit) scale for printing or display, or for deciding on the magnitude of an associated control signal. In the latter case, the digital signal must be processed, converted to an appropriate form for the actuator, and transmitted to the actuator.

Because the computer can perform only one operation at a time, even though these operations are performed at lightning speed, a sequence must be established in which all the tasks are performed. Some duties must be executed on frequent intervals, while others require only occasional updating. The computer software sets up a priority for the tasks.

The existence of the third and highest level of understanding and performance arises from the necessity of realizing that it is possible for a computer monitoring and control system to operate with precise functional fidelity and yet exert little or no impact on plant operations. The monitoring and control system could operate in a closed world and still accomplish some good; but it does not achieve its full potential, because it does not monitor and control the most significant functions or facilitate communication with the manager and operators of the plant.

Some of the elements of this level of understanding and performance are the following:

- The plan for the system
- The use of the information provided by the system to improve plant facilities and procedures
- The application of information provided by the system to improve plant operations

The planning function incorporates the exercise examined in Section 14.6 of deciding what points to monitor. There is a cost associated with each additional variable that is to be monitored, primarily in the transducer, communication line, and converter. The additional cost of the computer hardware and software of each additional variable, however, is usually moderate.

Another set of decisions concerns what to do with the information acquired. Should it be used as the basis to send out a control signal, simply stored in memory for possible future access, displayed on the screen, printed out, or checked to see if it is within specified limits and possibly activate a message or initiate an alarm? The functions to be controlled must also be determined, including decisions about whether the computer itself should execute the action or whether an operator should be informed and required to take the action, even through the computer.

The computer monitoring and control system should serve the plant manager responsible for the facilities and procedures by providing the information needed to operate a safe, efficient, and reliable plant and even indicate improvements that can be made in these areas. A manager might be interested in trends, such as energy requirements, indications of needed maintenance, and hints of potential safety problems. The manager should, for example, be able to extract defrost records to know whether the frequency and duration of hot-gas defrosts are appropriate. In the planning of the computer system, ease of access by the manager to information is mandatory, otherwise time constraints will discourage the manager from analyzing the plant operations with the necessary thoroughness.

The operator should be able to determine from the monitoring and control system that the plant is performing its duties, as least to the extent to which it is capable. Are the space temperatures being maintained, are refrigerating components developing their expected capacity, are there any safety problems imminent? Many plant operators do not have prior experience with electronic and computer equipment and may be intimidated by the new system. Asking the operators to participate in some of the planning meetings for the computer control system and also providing special training helps incorporate the operators as supportive members of the team. Perhaps an operator can become so enthusiastic that he or she wants to spend all of his or her time at the computer terminal. The indoctrination of the operator has then gone too far, because it is still vitally important for the operator to devote time inspecting the plant with his or her eyes, ears, and nose.

14.9 HOW A MONITORING AND CONTROL SYSTEM FUNCTIONS

The configuration of the computer control system of one refrigeration plant probably differs from those of most other plants, but a generic monitoring and control system is likely to incorporate the elements shown in Fig. 14.11. Information from the status sensors or transducers must first be converted into digital form to be recognized by the computer. The on/off status sensors are already in digital form, but data from the transducers must first be converted by an analog-to-digital converter. Based on the input information and the control program stored in its memory, the computer sends out instructions to the actuators or switches. In addition to these input and output ports, the computer can communicate with plant operators through the keyboard, screen, and printer. Further communication options may be through serial communication, perhaps through modems, to other computers. These options will be explored later in this chapter.

Not all of the components in the input and output streams may be visible. Input/output boards are available to plug into some of the most popular desktop computers, and these I/O boards often incorporate multiplexing and conversion. Only the lines between the board and its transducers and actuators are seen.

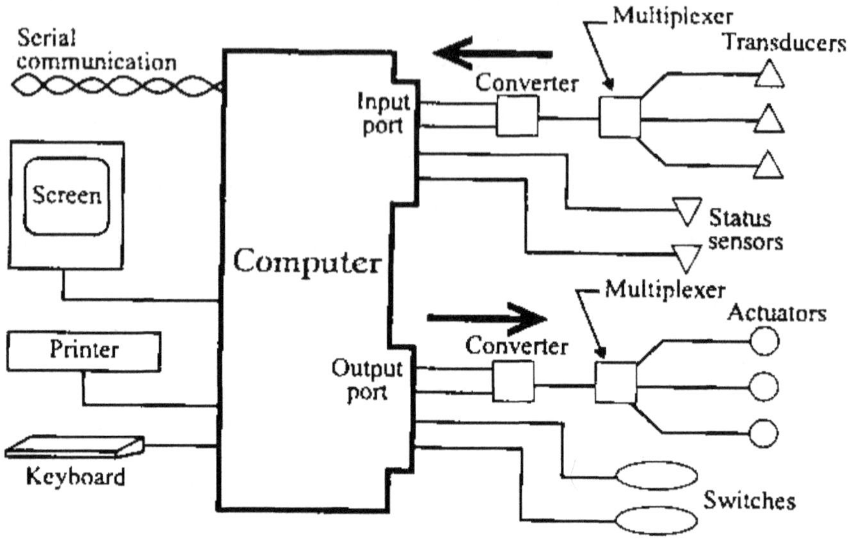

FIGURE 14.11
Overview of a computer monitoring and control system.

The converters and multiplexers must be present, however, and the way they function will also be described below.

14.10 TEMPERATURE TRANSDUCERS

A transducer is a device designed to translate a magnitude in one form into that of another. A temperature transducer, for example, translates a temperature into an electrical voltage or current. Probably the most widely monitored and controlled variable is temperature. Four of the most important types of temperature transducers are: (1) thermocouples, (2) metal or alloy as a resistance-temperature device, (3) thermistor, and (4) integrated circuit sensor.

The thermocouple circuit, shown in Fig. 14.12, has been used for decades as a laboratory tool and is still used to some extent in computer control systems. When a difference in temperature exists between the sensing junction and the reference junction, a voltage develops between the two copper leads. The magnitude of voltage output is approximately 0.043 mV/°C (0.024 mV/°F).

Resistance-temperature devices (RTDs) utilize materials whose resistance changes with temperature. Figure 14.13 shows some ratios of resistances of two metals, tungsten and platinum, as well as that of the third type of temperature transducer, the thermistor. The thermistor is an RTD using a solid-state material rather than a metal or alloy.

While the thermocouple provides an output in the desired form (electrical voltage), another step is required when using the RTDs, because the resistance

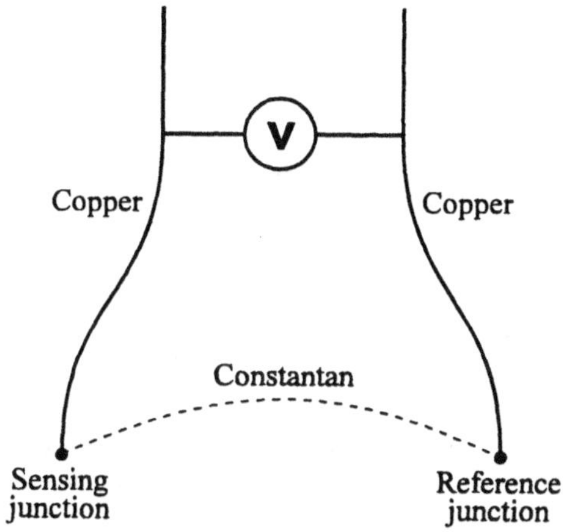

FIGURE 14.12
A thermocouple circuit.

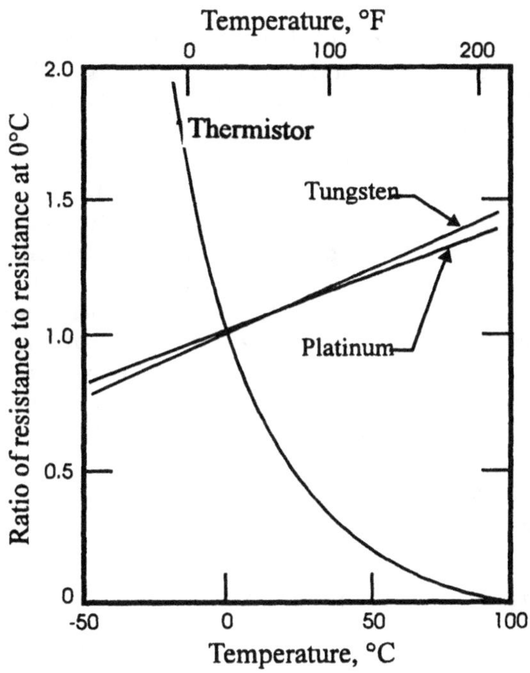

FIGURE 14.13
Ratios of resistances to the resistance at 0°C (32°F).

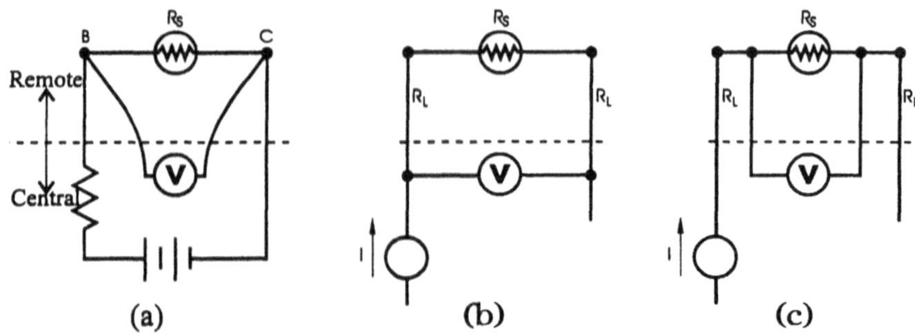

FIGURE 14.14
Converting the resistance of a temperature sensor to a voltage using (a) a series circuit; (b) two-wire, constant current circuit; and (c) four-wire, constant current circuit.

must be converted into a corresponding voltage or current. It would be possible to place the sensor in series with a resistance and supply the combination with a constant voltage, as shown in Fig. 14.14a. The resistance between points B and C indicates the temperature, and this resistance is found by dividing the voltage drop by the current.

The more common procedure is to supply the sensor with a known constant current and measure the voltage drop across the sensor, as in Figs. 14.14b and 14.14c. Figure 14.14b shows a two-wire circuit, and Fig. 14.14c shows the superior four-wire circuit, which eliminates possible distortion of the reading caused by varying lengths and resistances of the current-carrying wires.

One of the newest members of the temperature transducer family is the integrated circuit transducer. One version of this type of transducer is supplied with 5 V across its two leads. The magnitude of current in microamperes that the device passes equals the absolute temperature in Kelvin, as shown in Fig. 14.15a. A simple application of this transducer is shown in Fig. 14.15b where, by measuring the voltage drop across the known resistance, the current and thus the temperature can be determined. Even though the applied voltage to the transducer is specified as 5 V, the device is tolerant to variations in that voltage. If, for example, the applied voltage were 10 V instead of 5 V, only an error of 1 microampere and thus 1°C results.

A comparison of the advantages and disadvantages of these four classes of transducers is shown in Table 14.1.

14.11 PRESSURE TRANSDUCERS

There are hundreds of manufacturers of pressure transducers which use many different operating principles. Two major categories are: (1) mechanical, and (2) material with an electrical property sensitive to pressure. The mechanical type pressure transducer uses a bellows, membrane, or some other device that moves in response to a change in pressure. The change in position is then transduced

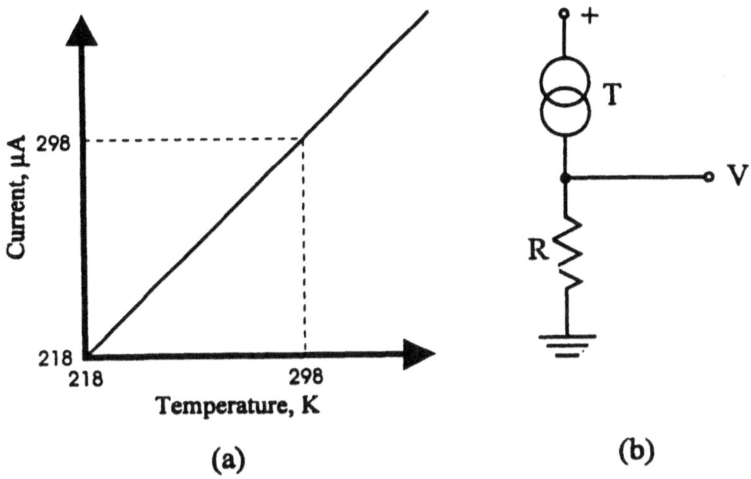

FIGURE 14.15
(a) Current-temperature characteristics of integrated circuit transducers, and (b) conversion of current into a voltage reading.

to a voltage or electric current. In the past several years, transducers using pressure-sensitive materials have become prominent. In these transducers, the pressure-sensitive element is usually a semiconductor whose electric resistivity or capacitance changes as the pressure imposed on the material changes.

Figure 14.16 shows a schematic diagram of a mechanical-type pressure transducer, which is a commonly used product in the industry. The motion of a bellows, in response to a change in pressure, moves the wiper of a potentiometer to change the voltage output, which then becomes the transducer signal. Also shown in Fig. 14.16 is a pressure tap and capped valve, which permit the mounting of a pressure gauge to calibrate the pressure transducer. Pressure transducers can fail, and it is worse for the transducer to give an incorrect reading than if it gives no reading at all. In industrial refrigeration systems where ammonia is widely used as the refrigerant, the purchaser of a transducer for the refrigerant pressure must be sure the ammonia comes into contact only with materials compatible with ammonia, e.g., steel, stainless steel, and aluminum.

14.12 STATUS INDICATORS

The schematic diagram of a generic computer control system shown in Fig. 14.11 indicated status signals as one of the classes of inputs to the computer. Some examples of status that are useful and sometimes crucial are: (1) whether a motor (of a fan, pump, or compressor) is on or off, (2) whether a solenoid valve is open or closed, and (3) whether a particular coil is in the process of being defrosted. One form of status is whether or not there is a voltage, and another status form is whether a contact is closed or not.

TABLE 14.1
Advantages and disadvantages of some types of temperature transducers.[9]

Transducer	Advantages	Disadvantages
Thermo-couple	Self-powered, simple, rugged, inexpensive, wide temperature range	Nonlinear, low voltage reference required, amplification needed
RTD	Stable, accurate, more linear than thermo-couple	Expensive, slow, current source required, small resis-tance change, low absolute resistance, self heating
Thermistor	High output, fast, two-wire ohm measurement	Nonlinear, limited temp-erature range, fragile, current source required, self heating
Integrated circuit	Very linear, highest, output, inexpensive	Not applicable to high temperatures, power supply required, slow, self heating

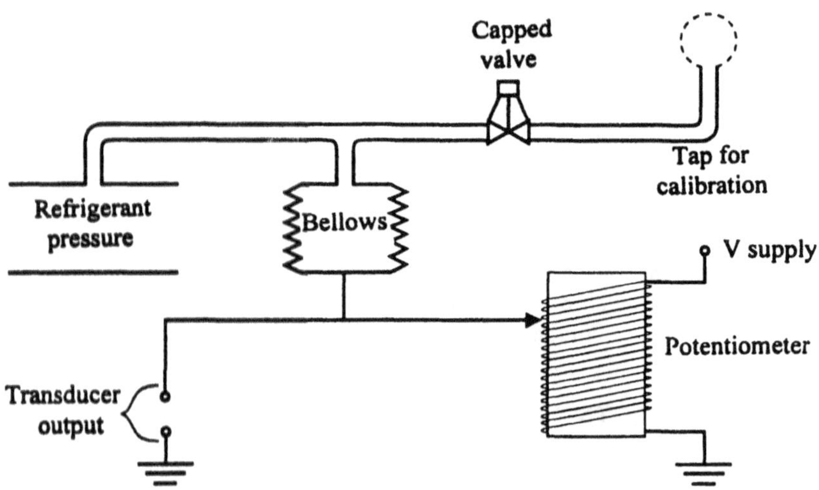

FIGURE 14.16
A pressure transducer with a mechanical pickup.

The most manageable form of status from the standpoint of the computer is receiving either a 0-V or 5-V signal. The computer can conveniently supply 5 V to an external status contact and then receive either 0 or 5 V in return, depending upon whether the contact is open or closed. The on-or-off status of power often controls a relay that opens and closes the contact sensed by the computer.

Plant operators point out that verification of power at a motor is not the same as ensuring that the motor is operating. When possible, the operators prefer that the existence of air flow in the case of a fan or liquid flow in the case of a pump be the condition that is sensed.

One provision that sometimes needs to be made is to *debounce* the switch in cases where a contact closure is sensed. Most mechanical switches do not close cleanly, and instead exhibit an instantaneous contact and then chatter on and off the contact, Fig. 14.17a. The computer is quick enough to interpret this one closure as a succession of closures.

A solution is the debounced switch circuit that employs two NAND gates, as shown in Fig. 14.17b. The characteristic of a NAND gate is that its output is low (0 V) if both the inputs are high (5 V). For any other combination of inputs, the output is high. In using this circuit, the contact actuates a double-pole switch between X (which gives a low output at Q) and Y (which gives a high output at Q). Furthermore, on the first contact of the switch to Y, the output Q goes high and remains high, even if the switch chatters between Y and an open position. The output Q goes low on the first contact of the switch with X and remains there even during nonconstant contact.

14.13 FORMS OF COMMUNICATION BETWEEN THE COMPUTER AND SENSORS OR ACTUATORS

The sensors and actuators may be some distance from the computer, so some form of electrical communication must transmit varying magnitudes from the sensors to the computer and from the computer to the actuators. Typical forms of communication are the following:

- From the sensor to the converter and from the converter to the actuator, 0-10 V or 4-20 mA
- Between the converter and computer, multiconductor digital signal

Using the temperature converter of Fig. 14.14c as an example in which the sensor is supplied with a constant current, the level of temperature is reflected in the magnitude of voltage. If the 4-20 mA current convention has been chosen, the voltage must be converted to a proportional current in this range. The 4-20 mA convention, incidentally, is very common in industrial applications.

The multiplexer in Fig. 14.11 is controlled by several lines from the computer and selects the particular sensor value that is to be read. For the analog

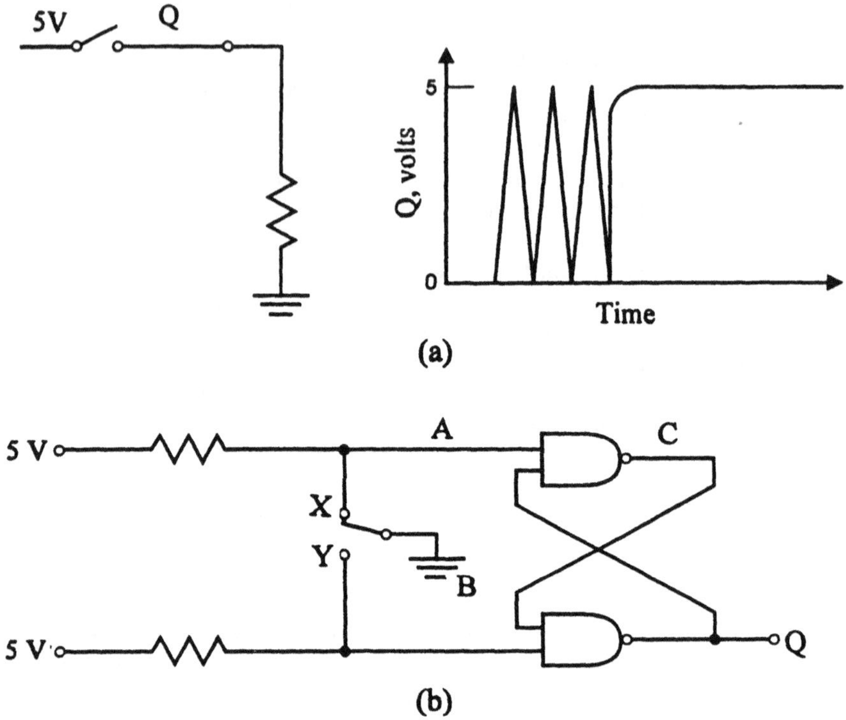

FIGURE 14.17
(a) Bouncing of a mechanical switch used for status detection, and (b) a debouncing circuit.

values directed from the computer to the actuators, a multiplexer (or more precisely a demultiplexer) under control of the computer sends the desired analog voltage or current to the selected actuator.

Between the converter and computer, digital signals of either 0 or 5 V on a multiconductor line form a binary representation of the analog value being passed. Normally, the length of this cable is kept short and might even be enclosed in the computer interface.

14.14 DIGITAL-TO-ANALOG AND ANALOG-TO-DIGITAL CONVERSION

Two forms in which numerical values can be expressed are *analog* and *digital*. Examples of analog representations are those of temperatures, pressures, and flow rates in which there is an infinite variation possible within the range experienced. When these physical variables are transduced to a voltage (for example, a range of 0 to 10 V) or to a direct current (4 to 20 mA, for example), the voltage or current is also an analog variable because of the possibility of infinite gradations within the range.

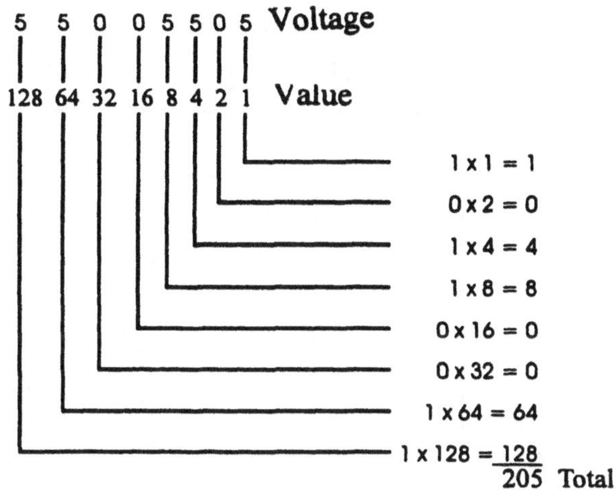

FIGURE 14.18
An eight-bit binary number.

A computer, however, does not inherently work with analog variables but instead uses groups of digital signals where the voltages on the conductors of the cluster are either 0 V or 5 V. Each of these electric lines constitutes a *bit*, where a 0-V level is understood as a zero and a 5-V level is a 1. An example of an 8-bit number is shown in Fig. 14.18, wherein the bits progress in magnitudes of values, namely 1, 2, 4, ..., 128—each value doubling the previous one.

Computers process these binary numbers using combinations of simple arithmetic operations to build up to the computation of complicated functions. Since even the smallest personal computer performs these operations at a high level of sophistication, no explanation of the internal workings of the computer will be attempted. The refrigeration professional may be confronted, however, with some decisions in specifying signal-processing equipment external to the computer, so this section continues with a discussion of conversions between analog and digital numbers—a process that takes place outside the central processor of the computer.

The first of these devices is the digital-to-analog converter (DAC), as shown in Fig. 14.19a. As the binary number coming to the DAC changes, the output voltage changes almost instantaneously.

The analog-to-digital converter (ADC), shown in Fig. 14.19b, is more complex than the DAC. In fact, the DAC is one component of the ADC. The strategy used by one type of ADC is to find the binary number that, when converted by the DAC to a voltage, matches the input voltage. The ADC needs a *start of conversion* pulse from the computer whereupon the controller turns on one bit at a time starting at the most significant bit. If a bit causes the output voltage from the DAC to exceed the input voltage, that bit is turned off. Otherwise, the bit is left on. The controller proceeds to the next lower bit and repeats the test.

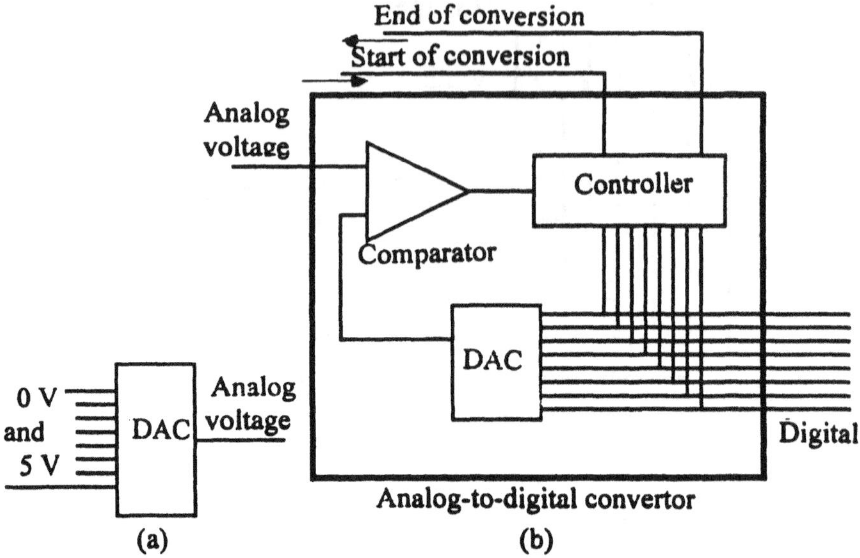

FIGURE 14.19
(a) Digital-to-analog converter, and (b) analog-to-digital converter.

When all bits have been examined, the *end of conversion* signal is given, which the computer could use as the signal to read the digital output of the ADC.

The greater the number of bits available in the digital signal, the greater is the resolution. For example, suppose that a temperature sensor has a range of 0 to 50°C (32 to 122°F) and the interval over which the temperature is to be controlled is from 2 to 3°C (35.6 to 37.4°F). If an 8-bit digital signal that possesses 256 different discrete numbers corresponds to an output of the sensor of 0 to 10 V, the control range incorporates five numbers, so the resolution is 0.2°C (0.35°F). If a more expensive 10-bit converter is used and can be handled by the computer, 20 numbers are available in the control range with a resolution of 0.05°C (0.09°F).

14.15 PARALLEL COMMUNICATION—NUMBERS AND CHARACTERS

Figure 14.18 suggests that numerical values can be transmitted by the zeros and ones (0 or 5 V) on a bundle of wires. The status of all the wires changes simultaneously when a new numerical value is imposed on the lines. This mode is called *parallel* communication and is the high-speed procedure used within the computer. Letters and other characters, as well as numbers, can be transmitted by means of binary representations. A convention used for translating letters and other characters is the American Standard Code for Information Interchange, or

ASCII for short. The ASCII code is a group of 7-bit binary numbers representing all the letters of the alphabet, numbers, punctuation, and certain carriage control instructions. In order to transmit the letter C in ASCII, for example, the binary number 1000011 is communicated.

14.16 SERIAL DATA TRANSMISSION, RS-232 AND RS-422

A companion to the parallel mode of data transmission is the serial mode, wherein one bit at a time is sent along a single conductor. The process of transmitting an 8-bit character from a computer and converting it into serial data for transmission is shown schematically in Fig. 14.20a. The register is first fully loaded from the computer, and when transmission begins, the end bit sets the voltage of the line to its corresponding value of 0 or 5 V. In a timing sequence controlled by the clock, all the remaining bits are shifted by one and the next voltage loaded on the line. The result of the transmission represented on the time scale is shown in Fig. 14.20b.

Parallel communication requires eight or more conductors, while a rudimentary system transmitting serial data requires only one line plus a ground line. Both parallel and serial modes of communication have their roles. Within the microcomputer and to closely coupled peripherals, parallel data transmission has the advantage of speed. For transmission over longer distances, the bundles of a large number of conductors would be expensive. Furthermore, serial data communication is mandatory for transferring data on or off discs and for sending data over telephone lines using modems. The disadvantage of serial data transmission is that it is slow relative to parallel communication.

It would be possible to apply serial communication using the convention of 0 and 5 V for short distances, but the signals would be susceptible to distortion caused by electrical noise, differences in ground potential, and damage from short circuits. Instead, the 0- and 5-V signals are converted into one of the widely used conventions, such as RS-232 or RS-422.

The convention for serial data transmission that has been most popular for many years is RS-232-C, as established by the Electronic Industrial Association.[10] The standard specifies the necessary electrical, mechanical, and functional characteristics of the data exchange; however, these characteristics will not be discussed here. Figure 14.21 shows a basic feature of RS-232-C in translating the zeros and ones (0 and 5 V) to greater differences in voltage. Typically, the high voltage of RS-232-C is of the order of 12 to 15 V and the low voltage is −12 to −15 V. The convention converts a zero (0 V) to a positive voltage and a one (5 V) to a negative voltage. Equipment operating on RS-232-C must be capable of identifying a voltage between 3 and 15 V as a zero and a voltage between −3 and −15 V as a one.

The RS-422-A convention[11], and still another new protocol, RS-480, are supplanting the RS-232-C protocol. Instead of the single line of the RS-232-C convention in combination with a global ground, the RS-422-A uses two lines

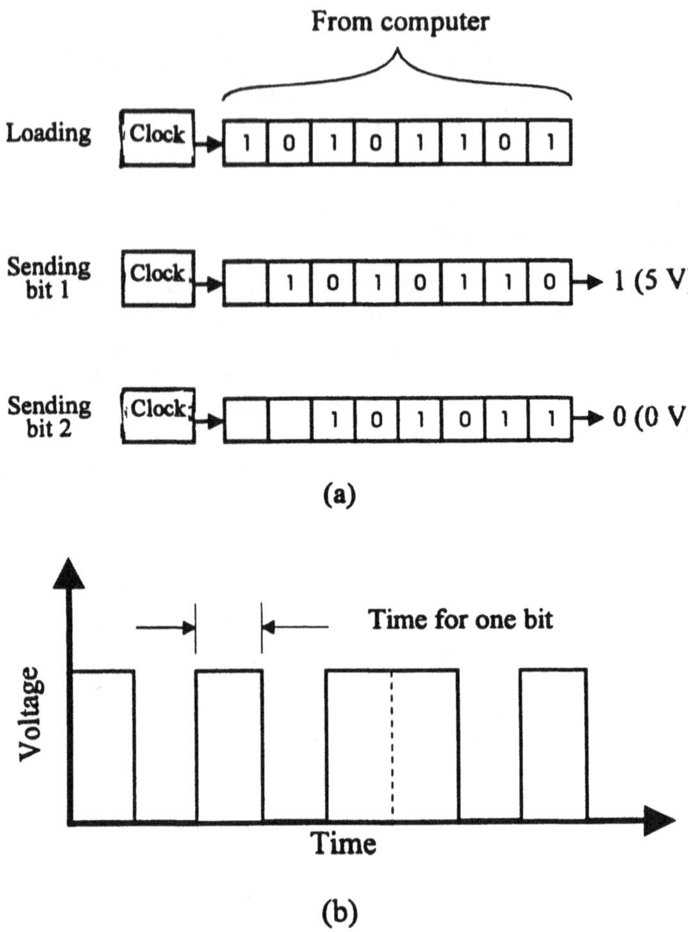

FIGURE 14.20
Serial data transmission with (a) the loading and shifting, and (b) the voltage history shown on a time scale.

with the relative voltage indicating a zero or one. If the voltage of line A is greater than line B a zero is indicated, and if the relative voltage is reversed a one is indicated. RS-422-A is thus not susceptible to differences in the reference voltage. Further comparisons are as follows:

Protocol	Maximum transmission distance	Maximum transmission speed	Susceptibility to errors
RS-232	15 m (50 ft)	20 kbits/s	Highly
RS-422	1,200 m (4,000 ft)	10,000 kbits/s	Partially

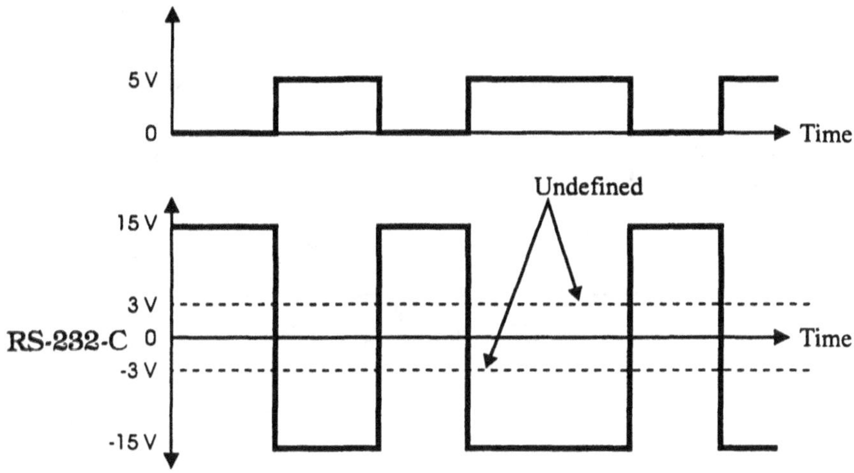

FIGURE 14.21
Conversion of zeros and ones to the voltages of RS-232-C.

14.17 MODEMS

Transmitting by RS-232-C or even RS-422-A over long distances poses some problems. After a certain distance, degradation of the signal occurs because of noise. Furthermore, connections by hard wires become prohibitive if the lines must run through other people's property, across the city, or from one city to another. Surmounting the transmission problem leads to using conductors that already exist, namely telephone lines.

The telephone system is structured to accommodate conversation between people, and telephone companies speak of "voice-grade" channels that transmit signals satisfactorily in the 300- to 3000-Hz frequency range. The sound of the voice to which phone line transmission is adaptable is composed of mixtures of various frequencies of sine waves. The phone line would have difficulty transmitting tones coming directly from the voltages of RS-232-C or RS-422-A, because it is unable to make the sharp changes in tone in passing from one bit to another. A device is called for, then, that converts RS-232-C or RS-422-A into sine waves for telephone line transmission and then reconverts to the RS protocol. This device is called a modulator-demodulator or modem, shown schematically in Fig. 14.22. The modulator converts a zero bit into sine waves of a specified frequency that prevail for the length of time needed to transfer a single bit. A one bit is converted into sine waves of a different frequency. This compilation of bursts of sine waves carries over the phone lines, and at the other end the demodulator converts the signal back to the RS protocol. A modem package incorporates both a modulator and a demodulator, so that the same unit can convert between RS protocol and sine waves in either direction.

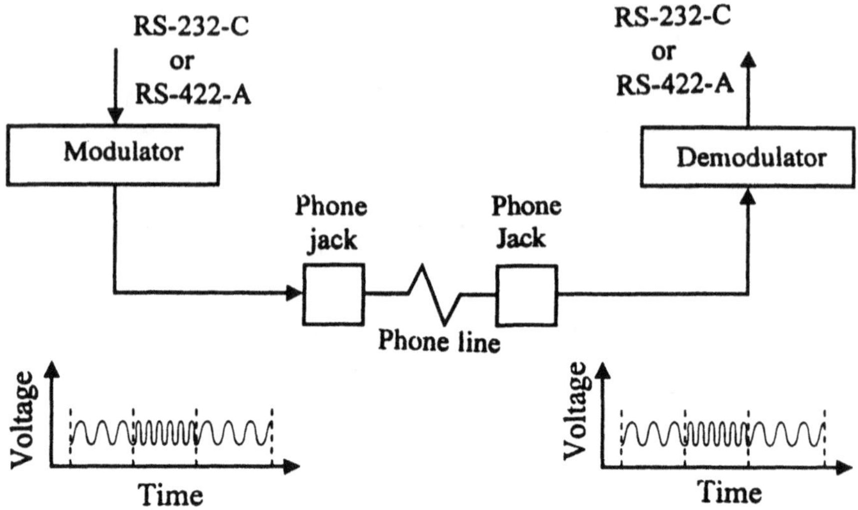

FIGURE 14.22
A modem that converts between RS-232-C or RS-422-A and sine waves which are transmittable over phone lines.

14.18 THE COMPUTER SYSTEM AS A COMMUNICATOR

The previous six sections have dealt primarily with communication issues—conversion of physical variables to electric signals by transducers and sending information in binary form and in the serial mode. These subjects are details of a broad-scope function of computer systems, and that role is that of a *communicator.* One of the valuable purposes of computer systems is to transfer information from one place to another. A computer system would be a valuable asset in a refrigeration plant even if its only function were to provide a means of communication of information from widely dispersed locations to a central position.

Computer systems in refrigeration plants and in those of the process industry are sometimes referred to as monitoring and control systems. The communication function corresponds primarily to monitoring. We now progress to the control operation where actuators open, close, and modulate electrical switches and valves.

14.19 ACTUATORS

The most likely actions taken by the control system are to open and close electric switches, valves, and dampers and to transmit alarms. The signal that prompts any of these actions is normally electric, and this control signal is normally low in voltage and/or current. One of the first requirements is to magnify these weak signals to a level sufficient to move elements of a switch or valve.

TABLE 14.2
Overview of actuators.

Actuator type	Two-position or modulating	Input	Controlled variable	Device
Switch	Two-position	d.c. or a.c.	d.c. or a.c.	Electromagnetic relay
"	"	d.c.	d.c	Transistor switch
"	"	d.c.	a.c.	Optically isolated switch
"	"	d.c. or a.c.	a.c	Solid state relay
"	Modulating	Resistance	Power	Triac
Valve	Two-position	Usually a.c.	Open/close	Solenoid valve
"	Modulating	a.c.	Position	Modulating solenoid
"	"	Air pressure	Position	Pneumatic
"	"	a.c.	Position	Reversible electric motor
"	"	Pulses	Position	Stepper motor

One way to classify actuators is as two-position or modulating. Electric switches are commonly two-position, namely on or off, and most valves are two-position as well, either open or closed. But in many instances, electric actuators are needed to modulate a voltage or current, and many valves and dampers must be able to adjust themselves in a partially open position. To present an overview of actuators, Table 14.2 proposes that switches and valves (including dampers) are the major categories of actuators important in industrial refrigeration systems. Available means of activating switches and valves are then categorized depending on whether they are two-position or modulating.

Another form of valve actuator that appears frequently in industrial refrigeration systems is the pilot-operated valve, which was explained in Chapter 11 in the valves and controls chapter. The pilot-operated valve applies a small solenoid valve to control refrigerant from a high-pressure source to assist in changing the position of a large valve.

Pneumatic valves and dampers are very common in medium- and large-size air-conditioning systems for commercial buildings and have the advantage of simplicity and ease of service, and they can provide considerable power. When pneumatic actuators are combined with the electric and electronic modes of the computer controller, a device called an electric-to-pneumatic transducer converts a modulating voltage to a modulating air pressure. In the past, pneumatic actuators dominated over electric motorized drives, which were not as long-lasting due to their incessant starting, stopping, and changing of direction of rotation. Motorized valves have undergone continuous improvement, however. Stepper motors, which rotate several degrees with each electric pulse, have wide application in electrical and mechanical equipment and have found a niche in controlling precisely the position of valves.

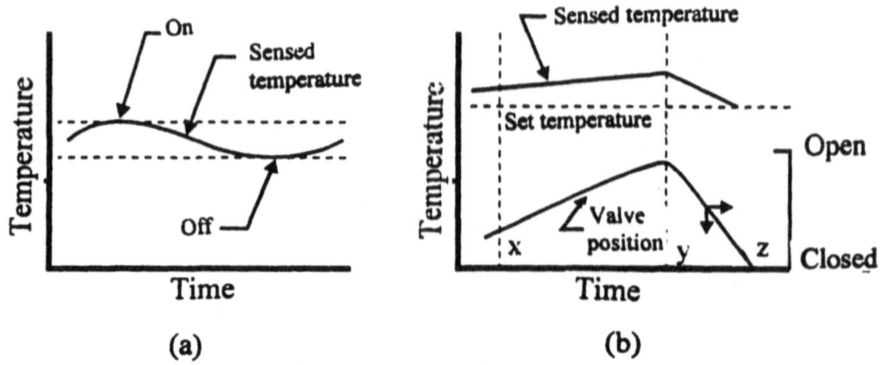

FIGURE 14.23
Two modes of control, (a) two-position or on-off, and (b) proportional.

14.20 PROPORTIONAL-INTEGRAL CONTROL MODE

The majority of control loops in industrial refrigeration plants are two-position, or on-off. In most storage spaces for frozen food, for example, when the temperature in the space rises to the setpoint, a solenoid valve opens allowing liquid refrigerant into the coil. The additional refrigeration capacity thus provided will drop the temperature. Most two-position controls either by design or inherently possess a dead band, such that there is a finite difference in temperature between the initiation and termination of refrigeration, as shown in Fig. 14.23a. In certain applications, this variation of temperature is not objectionable.

Another mode is *proportional control* where, for example, to open a valve further, a progressively greater error between the sensed variable and the control point is required. The standard pneumatic controllers used in air conditioning are almost all of the proportional mode. Figure 14.23b shows that at time x there is a deviation between the sensed (controlled) temperature and the setpoint, which is necessary to open the valve controlling the rate of refrigeration. If the refrigeration load increases between times x and y, the offset of the deviation of the sensed temperature must increase further in order to open the valve fully. As the load drops off as time z is approached, the sensed temperature can draw close to the setpoint. This is the typical operation of a proportional controller.

To hold the sensed temperature closer to the setpoint, the integral mode may be introduced to provide combined proportional-integral (P-I) control. The integral mode, as illustrated in Fig. 14.24, compares the sensed temperature with the desired value and calculates the area on the graph, thus integrating the error with respect to time. This area or integral readjusts the pseudo-setpoint to match the sensed temperature with the desired value, which has occurred by point x in Fig. 14.24. Between points x and y, the value of the pseudo-setpoint is such that it holds the sensed temperature precisely at the desired value. At point y the load begins to drop off, as does the sensed temperature, so the integral mode begins

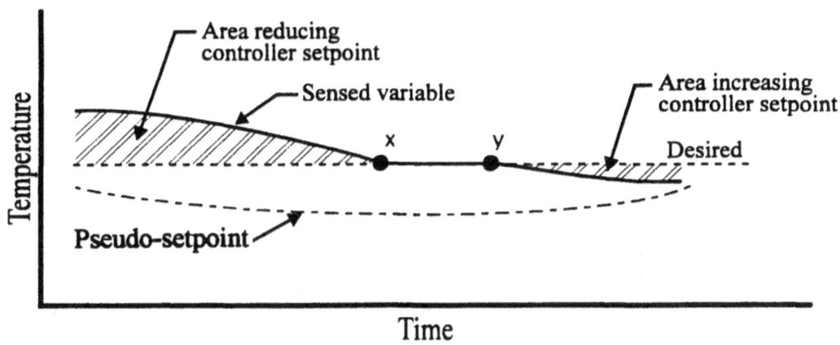

FIGURE 14.24
Introducing the integral mode to hold the controlled variable closer to the desired value.

an integration that results in values negative to that occurring before point x. This integration below the line of the desired value moves the pseudo-setpoint higher to forestall a droop in temperature below the desired value.

Proportional-integral control is not new and has been readily available in the chemical and process industries for many years, even in combination with still another mode, the derivative, to form P-I-D control. The derivative mode operates like a skilled human controller who can make a larger change in the setpoint if the sensed variable is changing rapidly. The derivative mode has not normally proved to be necessary in industrial refrigeration. In the past, the integral or reset mode has been accomplished by gears and other mechanical elements in the controller resulting in a costly controller. With computers or programmable logic controllers, the integral mode can be incorporated inexpensively with a little software and is thus readily available.

The final question is whether this refinement is even needed, because the industry has survived for decades without P-I control. The temperatures in spaces storing frozen food vary up and down with the refrigeration load, held within a several-degree band by the thermostat clicking on and off. But requirements for more-precise control are beginning to appear, such as in the storage of so-called superchilled food products in the neighborhood of −1°C (30°F). Dropping the temperature below this target by a degree of two will freeze the product, and allowing the temperature to rise above the target reduces the shelf life of the product. The precision of the P-I control mode is ideal for such an application.

14.21 PROGRAMMABLE CONTROLLERS

The basic programmable controller is a device that executes electronically the logic displayed on a ladder diagram. While a modern programmable controller incorporates many additional capabilities, the elementary programmable controller

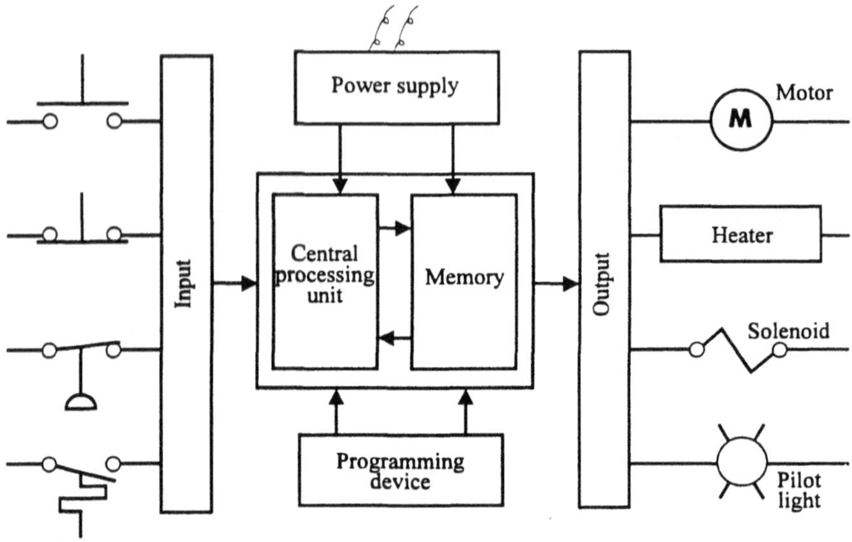

FIGURE 14.25
Structure of a basic programmable controller.

replaces electromechanical components with a microprocessor and digital electronic devices, as shown symbolically in Fig. 14.25. The origin of programmable controllers can be traced[12] to the automotive industry of the late 1960s when the electromechanical relay panels controlling manufacturing operations were becoming massive and inflexible. Calling on the new microprocessor technology of that era, auto manufacturers, working with computer and electronic specialists, developed the first programmable controllers. From the automotive industry, the application of programmable controllers spread to other industries in manufacturing and process fields.

Progressive developments from the early basic programmable controllers have led to units equipped with the following features:

- Peripherals: Keyboards and monitors that display the status of variables and I/O devices, as well as the ladder diagram that is being entered.
- Setpoints and control sequences that are changeable from such input devices as hand-held keyboards and programs downloaded from a central computer or a master programmable controller.
- Self-contained analog control loops that receive an analog input variable and deliver an analog output for an actuator. Controls applying the P-I-D mode, as described in Section 14.20, are usually available. The magnitude of the variable is monitored to prompt other actions, if desired.
- Hierarchical structure permitting a master programmable controller to communicate with and rule over slave controllers.

• Serial communication applying such protocols as RS-232-C or RS-422-A.

Several ways in which the capability of a programmable controller can be characterized include the number of input/output (I/O) connections, scanning speed, computational speed, and the memory size. Programmable controllers are commercially available from the small, low-cost units with several dozen input/output points to large ones with several thousand I/O points. The scanning speed dictates the time to complete one cycle, and during one cycle the status of all the input points are scanned and passed to the program stored in memory. Based on the combination of inputs and the instructions in the program, the outputs are reset as needed. The cycle time varies from about one second to a fraction of a second, depending on the number of I/Os and the computational speed.

The early programmable controller, as typified by Fig. 14.25, implemented a ladder diagram, receiving on/off inputs and providing on/off outputs. Personal computers began appearing about the same time, and developments through the years have brought the capabilities and characteristics of the two machines closer. From the beginning, the programmable controller was designed for an industrial environment in which electrical noise, vibration, and extremes of temperature and humidity prevail. The early personal computers could not survive long in such an environment and have progressed toward factory hardening. The personal computer is inherently a computer and the programmable controller a controller. The personal computer implements floating-point mathematics and high-level languages, such as BASIC, FORTRAN and C-language, and is effective in data collection and storage. Data-acquisition hardware/software is now readily available for personal computers, and serial communication is standard for programmable controllers, so the distinction between the two devices is becoming blurred.

In contrast to a panel of electromagnetic components, the programmable controller permits a change in the logic with no change in hardware, only an update of the program. Another strength of the programmable controller is that numerous control sequences can be accommodated internally without another relay. Referring to the ladder diagram of Fig. 14.7, for example, the $1CR$ coil and contacts that work off it must be implemented by a physical coil and contacts in a relay panel. In a programmable controller, on the other hand, the $1CR$ coil and contacts are accommodated by logic internal to the programmable controllers, and only the true inputs and outputs require external connections.

14.22 REFRIGERATION APPLICATIONS VS. BUILDING AUTOMATION SYSTEMS

An intriguing question arises frequently about the similarities and contrasts between the computer control applications in industrial refrigeration and those addressed by building automation systems (BASs). A different group of well-known companies develop, market, and service BASs than those who provide the equipment and install computer control systems in industrial refrigeration plants. Since the comfort air-conditioning field from which BASs developed is related to industrial refrigeration, it could be productive to ask why computer control has taken different paths in the two fields.

Some differences and possible reasons for the differences are:

- BASs grew out of predominantly analog controls which were pneumatic in the U.S. and electric in Europe. The traditional needs in industrial refrigeration have been discrete on/off inputs and outputs as represented by ladder diagrams. When air conditioning plants converted to direct digital control (DDC), the computer upgraded the proportional control mode to PI control. In industrial refrigeration, the panels of electromechanical relays gave way to programmable controllers.
- While machine rooms for medium- and large-size buildings are not generally luxurious, the environments are usually not as hostile as in the machine rooms of industrial refrigeration plants. The operators of industrial refrigeration plants therefore sought suppliers of control equipment of an industrial character such as that used in manufacturing plants and in the chemical industries. Manufacturers of BASs were able to adapt personal computers to their needs.
- With the increasing popularity of the screw compressor and its natural adaptation of a sophisticated ladder diagram, as shown in Fig. 14.8, to a microprocessor, the screw compressor provided a natural interface to programmable controllers. Another common control package served well by a ladder diagram is defrost control. These two local control packages, as well as some others that might exist in the plant, drive the type of master controller toward a discrete on/off type rather than one adapted to analog controls.
- The modern BAS fulfills other functions in addition to regulating the temperature control system. In many buildings, the BAS provides security monitoring and in certain cases is integrated with the fire control system. These requirements are rare in industrial refrigeration plants.
- The pressure for low first cost is probably more intense for the BAS, so standardization to reduce the cost is possible and common. In industrial refrigeration, the variety of requirements is extensive because each plant seems to be unique. From the standpoint of designing, manufacturing, and marketing the control system, there seems to be such disparity that it is difficult for any supplier to operate comfortably in both worlds.

There remain some differences in the needs of industrial refrigeration plants and building air conditioning systems, but many features are common to both classes of controllers. Furthermore, the developments in the BAS field should be closely watched to evaluate which might have effective application to industrial refrigeration.

14.23 POWER OUTAGES

Power outages occur on occasion, and most plants have a sequence of steps to restore operation. In a generic two-stage, liquid-recirculation plant, the fan motors on the evaporators will usually resume operation as soon as power returns. The condenser fans are often regulated by the discharge pressure, so if that pressure is high enough, the fans and circulating water pumps in the condenser restart. If the condensing pressure has dropped lower than the setting, the fans do not begin operating until one or more compressors are in operation and the discharge pressure increases.

The compressors and liquid pumps will also shut down during the outage, and solenoid valves controlling the flow of liquid to vessels should be of the normally-closed type so the flow of liquid is interrupted in the event of a power outage. The operation of the pumps and valves is interlocked with that of the compressors, so that some compressor capacity is available before liquid is supplied to vessels and before pumps deliver liquid to the evaporators. Compressors are then started one at a time in an unloaded condition and brought up to the capacity needed to control the evaporating and intermediate pressures. Simultaneously, the liquid pumps begin operating and the liquid line solenoid valves open as needed.

The startup of compressors is perhaps the only intervention needed to restore operation of a standard system, and this function can be performed manually in plants where operators are on full-time duty. Since the practice of full-time operators is declining, the computer controller can be programmed to bring the compressors back into service following a power outage.

14.24 PIPING AND INSTRUMENTATION DIAGRAMS

A piping and instrumentation diagram, usually abbreviated P&ID, is precisely what the name implies. It is a diagram showing all the piping, control and shutoff valves, and instrumentation. Display of the instrumentation incorporates the indication of the type of instrument (temperature, pressure, liquid level, etc.), its location, and its function. Each item of instrumentation is given a designation on the diagram. The diagram thus shows, for example, the indicator or transducer and a line to a master controller or directly to an actuator. P&IDs have been widely used for many years in the chemical industry, and now in the United States under the rules of Process Safety Management of the Occupational Safety and Health Adminstration they must be readily available and understood by operators of large refrigeration plants.

P&IDs show components and piping of the refrigeration system and the location of instrumentation. Symbols suggested for use in the diagrams are provided by such groups as ASHRAE[13] and the Instrument Society of America[14]. Samples of these symbols extracted from References 13 and 14 are shown in Figure 14.26. The incorporation of some of these symbols into a diagram for a pump liquid recirculation subsystem is shown in Figure 14.27.

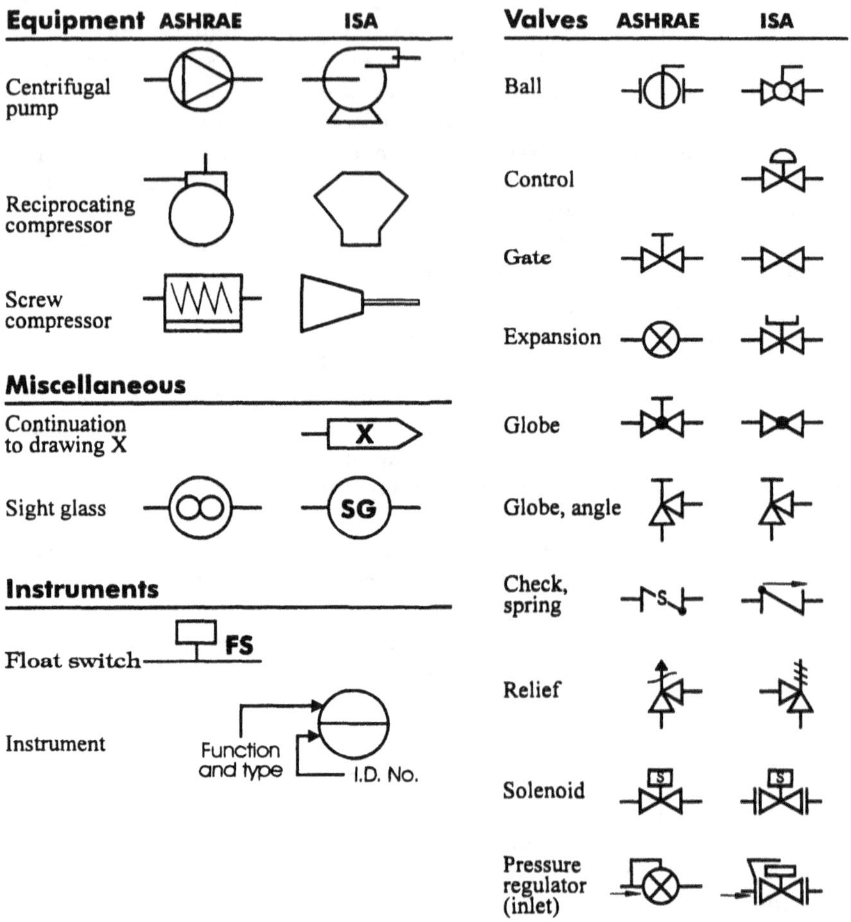

FIGURE 14.26
Sample symbols used in piping and instrumentation diagrams.

A schedule of designations is:

XXA	An alarm	XXS	Switch
HLX	High level	LLX	Low level
LPX	Low pressure	DPX	Differential pressure
PI	Pressure indicator		

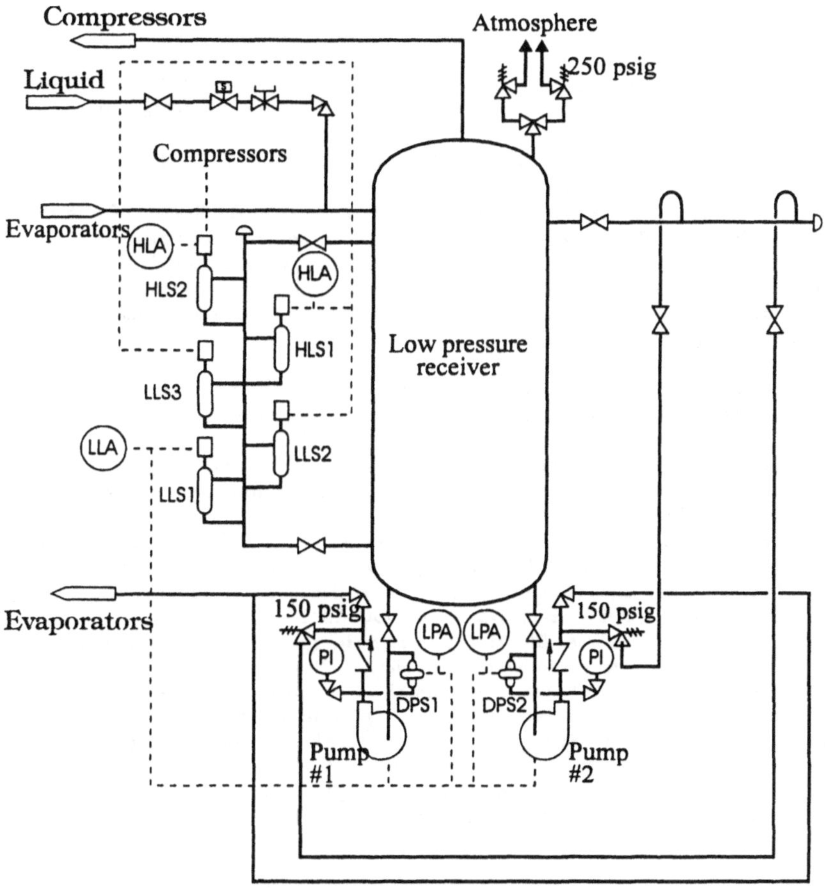

FIGURE 14.27
A P&ID for a pump recirculating subsystem.

In recent years, when refrigeration plant operators were asked to become familiar with the P&ID for their plant, many discovered that such a diagram did not even exist. Some plants that had P&IDs found them out-of-date. The focus on P&IDs is part of the worthwhile emphasis that instrumentation and controls are now receiving.

REFERENCES

1. *Electrical Standard EMP-1 / EGP-1-1967*, Joint Industrial Council, McLean, Virginia, 1967.
2. Woods, D., P. Kollasch, and F. F. Snyder, "Basic Electricity II, Ladder Diagrams," Refrigerating Engineers and Technicians Association, Chicago, Illinois, 1986.
3. FES, Inc., York, Pennsylvania.
4. Baillie, J. D. and P. L. Langford, "Microprocessor Control of a System," *ASHRAE Journal*, 27(9): 22-27, 1985.
5. Alyohin, N. B. and V. L. Persianov, "Micro-computer in the System of Automatization Research of Marine Refrigeration Plants," Proceedings of Commission D, 17th International Congress of Refrigeration, pp 340-344, Vienna, Austria, 1987.
6. Bjorklund, S. and J. Oleszko, "A Computerized Supervision System for Refrigerated Cargo Ships," Proceedings of Commission D, 17th International Congress of Refrigeration, pp 331-339,
Vienna, Austria, 1987.
7. Pearson, S. F., "Computer Control of a Large Refrigerated Distribution Centre with Remote Monitoring of C.O.S.P," Proceedings of Meeting of Commission B2, International Institute of Refrigeration, pp 179-184, Bristol, England, 1989.
8. Lunn, D., "Overview of Interfacing for Monitoring and Controlling," Proceedings of 9th Annual Meeting of the International Institute of Ammonia Refrigeration, San Diego, California, March 1987.
9. "Practical Temperature Measurements," Hewlett-Packard Application Note 290.
10. EIA Standard RS-232-C, *Interface Between Data Terminal Equipment and Data Communication Equipment Employing Serial Binary Interchange*, Catalog 3, Electronic Industrial Association, Washington DC, 1971.
11. EIA Standard RS-422-A, *Electrical Characteristics of Balanced Voltage Digital Interface Circuits*, Electronic Industrial Association, Washington, DC, 1978.
12. Jones, C. T. and L. A. Bryan, *Programmable Controllers, Concepts and Applications*, International Programmable Controls, Inc., Atlanta, Georgia, 1983.
13. American Society of Heating, Refrigerating and Air-Conditioning Engineers, Atlanta, Georgia, from the 1993 *Handbook—Fundamentals*.
14. "Instrumentation Symbols and Identification," Instrument Society of America, Research Triangle Park, North Carolina, 1984.

CHAPTER 15

LUBRICATION AND OIL HANDLING

15.1 OIL IN THE REFRIGERATION SYSTEM

Special compressors are available that need no lubrication, and refrigeration systems using this type of compressor can be free of oil. The cost to buy and operate an oil-free compressor is much greater than that of the lubricated type, so practically all industrial refrigeration plants use lubricated compressors. The need for oil is only to provide lubrication between moving parts and to seal certain openings in the compressor. No situation can be imagined where oil is beneficial elsewhere in the refrigeration system.

This chapter emphasizes the knowledge that is useful and even necessary in many cases for the designer, manager, or operator of a refrigeration system. This knowledge includes the types of oils encountered in industrial refrigeration, the terminology describing some physical properties of oils and the significance of these properties, how to prevent oil from passing into the system beyond the compressor, and how to remove from the system the oil that does escape the compressor package.

The manufacturer of the lubricating oil is an obvious resource for specific information on the oil sold by that firm. The relative quantity of oil sold for lubricating refrigeration systems is small in comparison to other uses of oil, so there is only a limited number of suppliers selling oil for refrigeration purposes.

The manufacturer of the compressor is one of the best sources of advice on which oil to choose, because the compressor builder accumulates the experience of many users. A plant manager who elects to use an oil contrary to the recommendation of the compressor manufacturer runs risks.

The basic function of refrigeration oil is to provide lubrication between sliding metal surfaces, but oil often serves other purposes as well. Oil often serves as a seal to prevent refrigerant from leaking between high- and low-pressure regions of the compressor, performs a cooling function that is significant in some types of compressors, removes debris from contacting surfaces, and often reduces the noise generated within the compressor.

The ASHRAE *Handbook—Refrigeration*[1] contains a quite thorough treatment of the subject of lubricating oil for refrigeration purposes, and this source should be consulted. The objective of this current chapter is to complement the ASHRAE *Handbook* by providing an introduction to the subject and concentrating on topics unique to industrial refrigeration. If this chapter can help equip the designer, manager, or operator of a refrigeration plant to discuss intelligently with the oil supplier and compressor manufacturer the proper choice of oil, as well as how to handle the oil properly once in the system, the chapter will have served its purpose.

15.2 SOME TYPES OF REFRIGERATION OIL

The major categories of oils[2] used for lubricating refrigeration compressors are: (1) mineral oils, and (2) synthetic oils. Mineral oils are extracted from crude oil and, for lubrication purposes, are usually either of a paraffinic or naphthenic base. Some years ago, mineral oils were virtually the only type available, but in recent years synthetic oils extracted from such feedstocks as natural gas have become popular. Several important bases for synthetic oils are: dialkylated benzene, polyalphaolefin, and polyalkylene glycol.

Some of the applications for the various oils are as follows:

- Mineral oil, paraffin base—not recommended for applications at low temperature
- Mineral oil, naphthenic base—the standard oil used in ammonia plants
- Synthetic, dialkylated benzene base—good miscibility with halocarbons at low temperature; sometimes used in ammonia plants where oxidation problems might occur
- Synthetic, polyalphaolefin base—used when discharge temperatures are high and also in ammonia systems operating at low temperatures, because the oil does not solidify easily at low temperatures
- Synthetic, polyalkylene glycol base—normally recommended only for systems using a halocarbon as a refrigerant or in special cases with ammonia (Section 15.13)

If a mineral oil is available with all the desired properties, it is usually chosen, because it costs approximately one-third that of synthetic oil. On the other hand, synthetic oils are tailored to achieve certain properties and may provide more satisfactory service.

15.3 SOME TERMS USED TO CHARACTERIZE OILS

Some important properties of oil that influence their behavior in refrigeration systems include viscosity, miscibility and solubility with the refrigerant, pour point, flash and fire point, vapor pressure, and floc point. The two dominant properties are the viscosity and solubility, so the next two sections are devoted to each of them. The meaning and importance of the other properties are as follows:

POUR POINT. The pour point is the lowest temperature at which oil will pour or flow when tested according to the standard method prescribed in ASTM D-97. In order to provide lubrication, the pour point of the oil must be below the operating temperature of the oil. The pour point is not as important for the oil in halocarbon systems as it is for oil in an ammonia system. The reason is that, in a halocarbon system, some refrigerant is likely to be dissolved in the oil, so the oil/refrigerant will flow at a lower temperature than that of the oil alone. On the other hand, oil and ammonia do not mix, so the pour point of the oil is critical. Mineral oils, used so much in ammonia systems, have pour points of about −40°C (−40°F) for naphthenic oils and about −18°C (0°F) for paraffinic oils. The lower pour point of most naphthenic oils is one reason for choosing this type over paraffinic oil for ammonia systems.

FLASH AND FIRE POINT. The flash and fire point derived from ASTM D-92 is not a critical property for usual refrigeration service. The flash and fire points of most lubricating oils are generally higher than 175°C (350°F), which is a temperature not normally encountered in refrigeration systems. The flash and fire point along with the vapor pressure is more an indication of the volatility of the oil. When high compressor discharge temperatures occur in a system using an oil with a low flash and fire point, the oil may carbonize and coat the valves of a reciprocating compressor.

VAPOR PRESSURE. In the modern, high-efficiency, coalescing type of oil separators (Section 15.9), virtually the only oil that escapes the separator is that which leaves in the form of vapor. For this reason, an oil with a low vapor pressure is preferred. Precise data on the vapor pressure of a particular oil might be difficult to obtain. An indication of the order of magnitude of the vapor pressure of two oils[1], one a mineral and the other a synthetic, is shown in Fig. 15.1.

FLOC POINT. The floc point is the highest temperature at which wax-like materials or other solid substances precipitate when a mixture of 10 percent oil and 90 percent R–12 is cooled under specific conditions. The test[3] is limited to R–11 and R–12, in which oil readily dissolves. Attempts to apply the test to

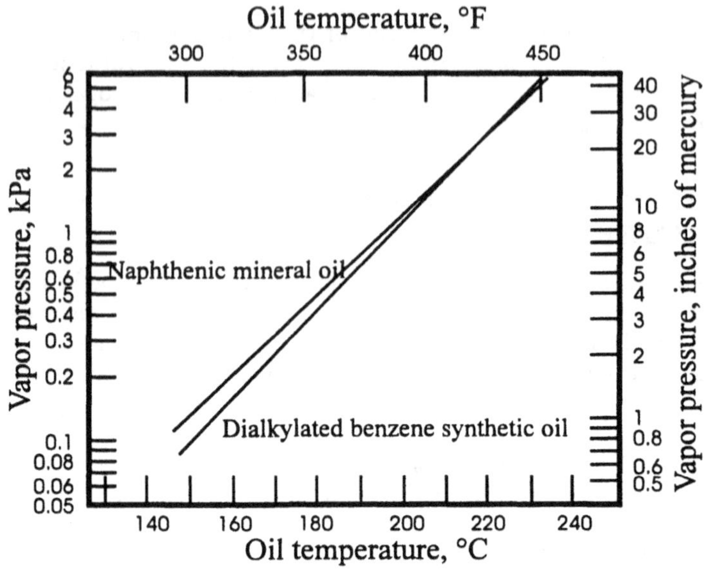

FIGURE 15.1
Relationship of the vapor pressure to temperature for two different oils having similar viscosities.

R–22 have been unsuccessful and presumably would be with ammonia as well. So for these refrigerants that are so important in industrial refrigeration, the standard floc test is not applicable, even though the temperature at which wax begins to precipitate would influence the choice of oil.

15.4 VISCOSITY

The viscosity of an oil is one of the key properties influencing its lubricating characteristics, but a pragmatic question is whether the user of an oil needs to know anything about viscosity so long as the choice of oil is that recommended by the compressor manufacturer. Two important functions of oil that are influenced by the viscosity are: sealing between metal surfaces to prevent leakage of refrigerant and lubrication between rubbing and rolling surfaces. In general, the desirable viscosity is the lowest value that provides an adequate seal to prevent refrigerant leakage. A low viscosity oil results in lower compressor power requirements, but if the viscosity is too low, inadequate sealing results, which manifests itself as a loss in volumetric efficiency. At very low viscosities the lubricating property is impaired, but usually the loss of sealing effectiveness is the controlling factor. Another important characteristic of the viscosity of oil that may be valuable to the user is how temperature influences viscosity. This topic will be addressed in this section.

Unfortunately, there are many units of viscosity swirling about the industry, so the next several paragraphs explain the meaning of the frequently-encountered

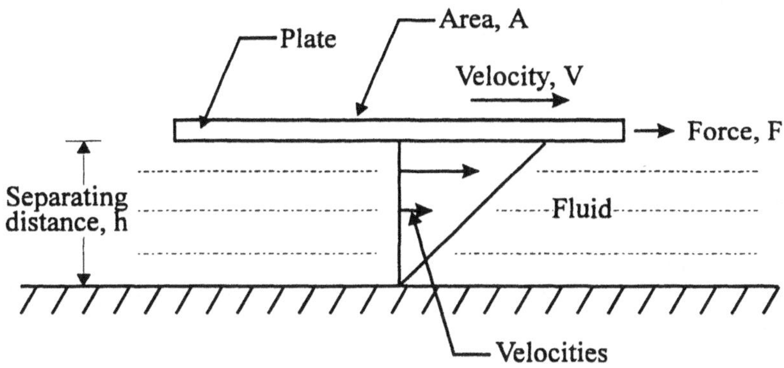

FIGURE 15.2
Terms to define viscosity in Eq. 15.1.

terms that characterize viscosity and how they are related.

The viscosity of a fluid is defined as follows:

$$\text{Viscosity} = \frac{\text{shearing stress}}{\text{rate of shearing strain}}$$

The viscosity can be defined further and quantified in terms of the variables illustrated in Fig. 15.2 as:

$$\mu = \frac{F/A}{V/h} \tag{15.1}$$

where:

μ = viscosity, Pa·s (lb_m/ft·s)
F = force, N (lb_f)
A = area in contact with fluid, m^2 (ft^2)
V = velocity, m/s (ft/s)
h = separating distance, m (ft)

The viscosity defined by Eq. 15.1 is called the ***absolute viscosity*** and is used in heat-transfer and fluid-flow correlations. The refrigerant viscosity tables frequently used to compute Reynolds numbers are absolute viscosities in units of Pa·s (lb_m/ft·s). Another set of units for absolute viscosity is *poise*, which results if F is in dynes when $A = 1$ cm^2, $V = 1$ cm/s,.and $h = 1$ cm. Thus, 1 poise = 1 dyne·s/cm^2. The more commonly used term is *centipoise*, which is 1/100 poise. To convert viscosity in:

- centipoise to lb_m/ft·s, multiply by 0.000672.
- lb_m/ft·s to Pa·s, multiply by 1.4882.
- Pa·s to centipoise, multiply by 1000.

Another version of viscosity that is frequently used is the ***kinematic viscosity***, which is the absolute viscosity divided by the density:

$$\nu = \mu/\rho \tag{15.2}$$

where:

ν = kinematic viscosity, m^2/s (ft^2/s)
ρ = density of lubricant, kg/m^3 (lb_m/ft^3)

Still another set of units for kinematic viscosity is the ***stoke***, which is a poise divided by density in g/cm^3, which becomes cm^2/s. The centistoke, which is 1/100 of a stoke, is more commonly used than the stoke. To convert kinematic viscosity in:

- centistokes to ft^2/s, multiply by 0.000010764.
- ft^2/s to m^2/s, multiply by 0.0929.
- m^2/s to centistokes, multiply by 1,000,000.

While it may seem that we already have more types of viscosity and sets of units than are needed, there remains still another. The viscosities that have been explained so far are the standard ones used for refrigerants, water, air, and many other fluids. For lubricating oils, however, a viscosity most widely used takes its name from the measuring apparatus. This viscosity is called the ***Saybolt Seconds Universal*** or SSU and is related to the time in seconds required for a standard volume of oil to pass through a tube of specified dimensions with the oil at a given temperature. The SSU is most closely related, but not proportional, to the kinematic viscosity, and a conversion that is reasonably accurate is:

$$\nu, \text{ centistokes} = 0.22(SSU) - 180/(SSU) \tag{15.3}$$

The foregoing presentation on units and conversions of viscosities will hopefully be useful reference data, but of more immediate importance to a user is to know what viscosity characteristics are desirable for oils that lubricate industrial refrigeration compressors. The oil with the lowest viscosity that gives the necessary sealing properties with the refrigerant used for the entire range of temperatures and pressures should be chosen[1]. When the viscosity is too low to provide adequate sealing, the volume pumping capacity of the compressor drops. A viscosity that is too high causes high power requirements, so the choice of a desirable viscosity is a compromise. This choice is normally made by the manufacturer of the compressor who has test facilities available. Some typical viscosities[1] of refrigerant oils used in the industry are as follows:

- Ammonia screw compressors, 280–300 SSU
- Ammonia reciprocating compressors, 150–300 SSU
- Refrigerant-22 reciprocating compressors, 150–300 SSU

Another viscosity-related characteristic is the *viscosity index*, which is an empirical number indicating the change in viscosity as a function of temperature. The viscosity of oil with a high viscosity index changes less with a change in temperature than an oil with a low viscosity index. In general, an oil with a high viscosity index is preferable, because its behavior is less subject to variations in the operating temperature of the refrigerant.

15.5 SOLUBILITY OF REFRIGERANTS IN OILS

All refrigerant gases are soluble at least to a limited extent in mineral oils. Ammonia is not significantly soluble, while some halocarbons are highly soluble in oils. In all cases, the solubility of the refrigerant increases with pressure of the gas and decreases with temperature of the oil. Some of the implications of solubility are as follows:

- Since refrigerants have much lower viscosities than oils, a large amount of refrigerant in the oil decreases its viscosity.
- Oil that is carried out of the compressor and not removed by the oil separator ultimately reaches the low-pressure side of the system. Different procedures must be employed to remove and/or return the oil to the compressor in ammonia systems compared to halocarbon systems because of the differing solubilities.
- Although a small amount of oil in evaporators using certain halocarbon refrigerants may increase the heat-transfer coefficient slightly, the presence of oil is generally detrimental to good performance. In ammonia systems, the oil may coat the heat-transfer surfaces and reduce the rate of heat transfer. In halocarbon systems, the vapor-pressure of an oil-refrigerant solution is lower than that of the pure refrigerant, resulting in a higher evaporating temperature for a given suction pressure.
- When the refrigerant is highly soluble in the oil and when the plant is not in operation, the refrigerant may enter into solution with the oil in the crankcase (reciprocating compressor) or oil separator and storage sump (screw compressor). A heater should keep the oil warm during idle periods or at least before operation, so that the refrigerant does not boil rapidly when the pressure drops upon startup, which could cause foaming of the oil.

Briefly stated, because of the differences in oil solubilities of ammonia and the halocarbons, oil is handled differently depending on the refrigerant. In an ammonia system, the oil that passes on to the low side separates from the liquid refrigerant, and because the oil has a greater density than liquid ammonia it may be drained or transferred in a closed circuit back to the compressor, as will be described in Section 15.10. In halocarbon systems, the oil is likely to be in solution in the evaporator and/or low-pressure receiver, so the basic process is to remove a sample continuously or intermittently, boil off the refrigerant, and return the remaining oil to the compressor. Facilities that accomplish this oil return are examined further in Section 15.11. A statement heard frequently is that ammonia vapor is not soluble in oil. For practical purposes this statement is correct, but it may be useful in certain situations to know that even ammonia is slightly soluble in oil. In the pressure range of 100-1,000 kPa (14.7–147 psia) and oil temperatures from 0 to 150°C (32 to 302°F), the solubility[4] of ammonia in oil is between 0.25 to 0.75% by weight. The solubility of ammonia in oil is so low that neither the boiling point of the ammonia nor the viscosity of the oil is affected by the solubility.

The terms *solubility* and *miscibility* are often used interchangeably, but miscibility is particularly related to ***mutual solubility***. In some regions of the refrigerating system, the concern is how much refrigerant vapor is dissolved in liquid oil, while in other sections liquid refrigerant predominates and small proportions of oil are dissolved in the refrigerant. These two cases represent mutual solubility. Completely miscible refrigerants and oils are mutually soluble in all proportions and at any temperature. R-11 and R-12 with mineral oil exhibit complete miscibility. Partially miscible refrigerant-oil systems are mutually soluble above a certain temperature and separate below that temperature. An immiscible system is exemplified by ammonia and oil.

15.6 SOLUBILITY OF R-22 IN MINERAL OILS AND ITS IMPLICATIONS

Since R-22 is an important industrial refrigerant, its behavior with oil deserves special attention, because the oil in an R-22 system is handled differently than oil in an ammonia system. Oil and R-22 are partially miscible, but that miscibility is dependent upon temperature, oil concentration, and type of oil. A general pattern of oil/R-22 miscibility is shown in Fig. 15.3. The curves apply to different oils, and the variation shows that it is not appropriate to expect one solubility curve for all oil/R-22 solutions. At temperatures higher than a given point on the line, the two substances are miscible. Below the point there is a separation, and because the oil is less dense than the solution of halocarbon refrigerant and oil, it rises to the top of the liquid mass, as shown in Fig. 15.4. The view in Fig. 15.4 is an idealization, because in an evaporator, for example, there is considerable turbulence and mixing.

One implication of the presence of oil in solution with R-22 is that it affects the heat-transfer coefficient, as will be discussed in the next section. Removal of oil from R-22 evaporators will be investigated in Section 15.11.

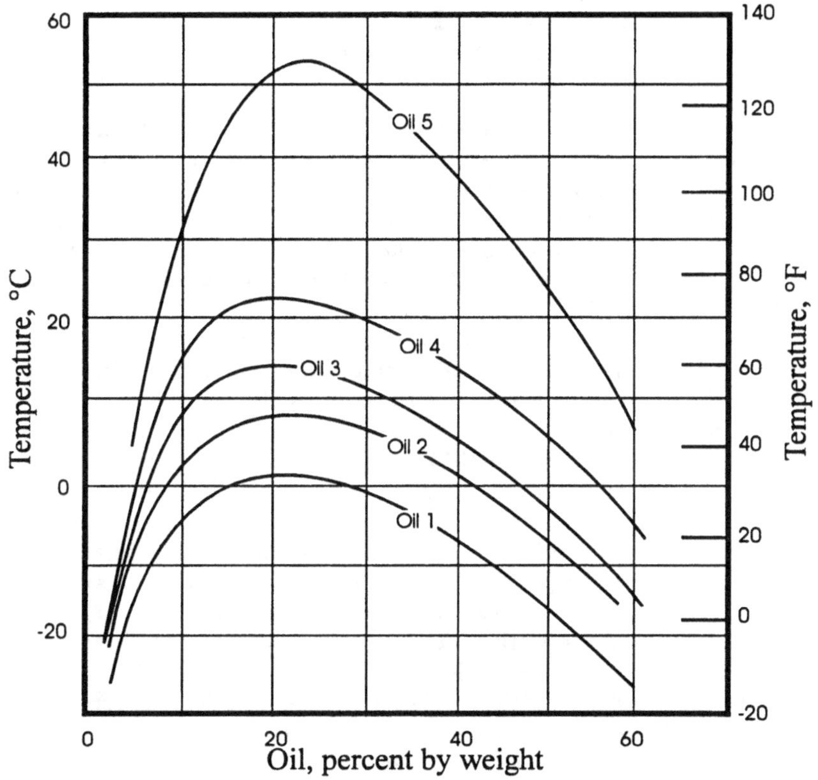

FIGURE 15.3
Miscibility of several different types of naphthenic oils with R-22.[1]

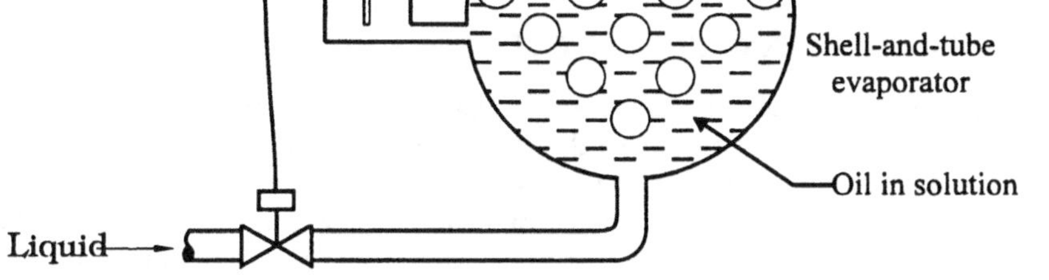

FIGURE 15.4
Stratification of oil and oil/R-22 solution in an evaporator.

15.7 EFFECT OF OIL ON HEAT TRANSFER IN AMMONIA AND R-22 EVAPORATORS

Understanding how oil in evaporators influences heat-transfer coefficients is important when deciding which measures to take to prevent oil from reaching the evaporator. While the major interest is on the effect of oil on the performance of evaporators, several comments on the effect on condensation are: (1) with halocarbon condensers, oil always decreases the heat-transfer coefficient[5], and (2) in the case of ammonia, the presence of oil might actually increase the condensing heat-transfer coefficient[6].

The remainder of this section concentrates on both R-22 and ammonia evaporators. The behavior differs depending on which is the refrigerant and whether the boiling occurs within the tubes or pool boiling outside tubes. In industrial refrigeration practice, boiling of refrigerant within tubes occurs primarily in air-cooling coils. When R-22 is the refrigerant in liquid recirculation systems, most of the oil is carried out of the coil along with liquid refrigerant. In the low-pressure receiver, however, oil will accumulate and must be constantly or periodically removed from there. Much more information is available on heat transfer during pool boiling of R-12 and oil in comparison to that available for R-22 and oil. But similar behavior is expected for R-22 and R-12 up to a point. Chaddock summarized the work of Tschernbylski[7] and Stephan[8], as shown in Fig. 15.5. The usual expectation is that the first several percent of oil actually increases the heat-transfer coefficient before the continued degradation. The high concentrations of dissolved oil shown for R-12 may not be possible for R-22, because of the solubility limitations shown in Fig. 15.3.

In the case of ammonia boiling in the shell of an evaporator, oil normally has little influence on the heat-transfer coefficient, because it separates from the ammonia and collects in an oil accumulator that must be provided. Shah[9] showed qualitatively that oil can reduce ammonia heat-transfer coefficients when boiling within tubes, as seen in Fig. 15.6. This reduction is normally not a problem in air-cooling coils that defrost frequently to warm any accumulated oil and reject it from the evaporator. Accumulated oil in some ammonia tube-type evaporators used for process cooling can be a problem if the evaporator must operate for a long time without a warmup to drain the oil.

15.8 CONTAMINANTS IN OIL

Contaminants are undesirable substances that develop in the system and are carried along with the lubricating oil. All reliable suppliers of oil endeavor to avoid introducing contaminants with new oil added to the system. The term *contaminants* should be distinguished from *additives*, which are introduced for such purposes as anti-foaming and anti-wear. Most contamination of the oil takes the ultimate form of metals, varnishes, sludge, dirt, and sand[10]. Some contaminants originate from improperly cleaned pipe and components dating from the installation or repair of the system.

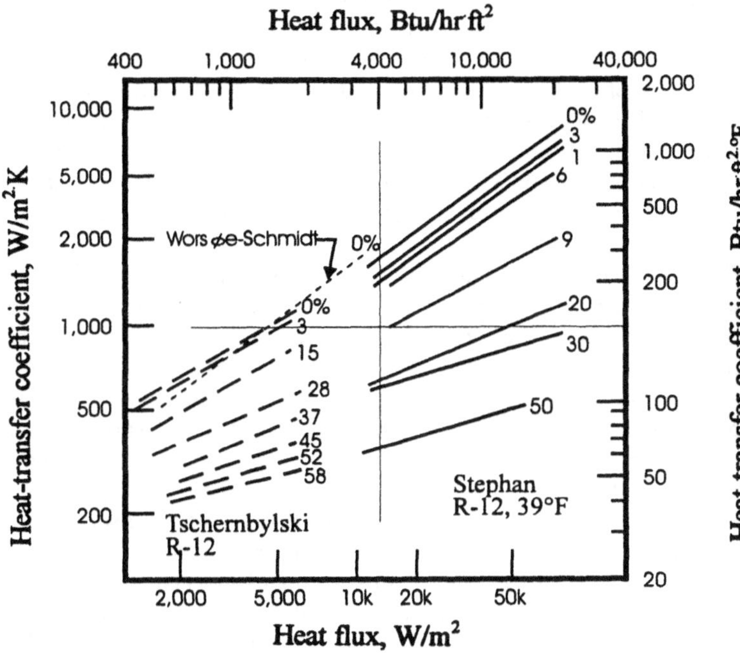

FIGURE 15.5
Effect of oil concentration on the heat-transfer coefficients of R-12.

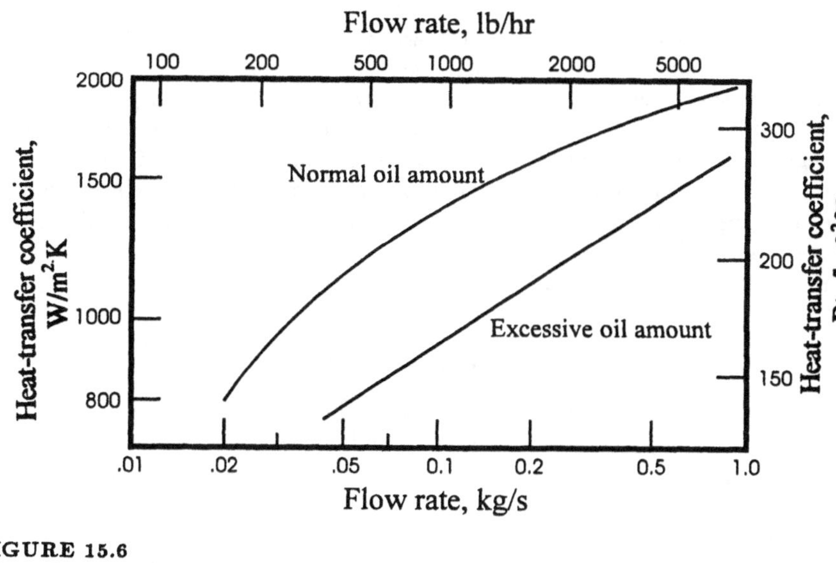

FIGURE 15.6
Effect of oil when ammonia boils inside tubes.

Of particular interest at the moment are the contaminants attributable to changes in the oil. The major and almost exclusive source of solid contaminants is oxidation of the oil, which in turn is caused by the presence of air, water, and/or fine metal particles. Oxygen must be present for oil to oxidize, and the usual source of oxygen is air that has entered the system. Oxidation increases the viscosity of the oil and produces organic acids, which increase the total acid number (TAN) of the oil. The TAN should normally be lower than 1.0. Water contamination of ammonia forms ammonium hydroxide, which is corrosive to steel.

The operators of well-maintained plants perform analyses of the oil in their systems. In this procedure, a sample of oil is periodically extracted and subjected to a spectographic and other analyses. The used-oil analysis can indicate whether the iron content is high, which could indicate corrosion. Another tell-tale sign that can alert an operator to a potential problem is an unusually high content of copper, lead, or tin, which could indicate bearing wear. Another routine measurement in the oil analysis is the viscosity. The various undesirable processes that affect the oil frequently result in increased viscosity. A high viscosity could result in increased power requirements of the compressor and serve as a signal to change the oil. This act is also an answer to an oft-asked question, "Should the oil in the system ever be changed?" In systems using reciprocating compressors without coalescing separators, there is usually enough oil carryover to the low side that periodic draining of the old oil and adding of the new oil makes an oil change unnecessary. In systems equipped with coalescing separators, which means all systems using screw compressors and many new systems using reciprocating compressors, the oil may need to be changed periodically, and the used-oil analysis is the best indication of that need.

Another practical question that operators frequently ask is, "Should I filter and reuse oil drained from an ammonia system?" The answer given by most oil and compressor specialists is, "No." As oil works in the system, the additives are gradually depleted[11]. If left in the machine too long, it is possible that the oil begins to leave deposits on internal parts. The viscosity of the oil could progressively increase due to the depletion of the additives as well as from the slow but progressive vaporization of aromatics. While filtration would remove solids from the oil intended for reuse, chemical contaminants would remain and the reclaimed oil would not contain the additives of new oil.

15.9 COALESCING OIL SEPARATORS

It is only in the compressor that oil is absolutely necessary. Its purpose is lubrication of the metal parts of the compressor and, in the case of screw compressors, to seal the spaces between the rotors. Oil is not needed in the rest of the system and, in fact, is undesirable in such components as the evaporator because it usually reduces the heat-transfer coefficient. The exception to the previous statement is that some operators have reported that following the installation of a high-efficiency oil separator, some control valves in the low-temperature section

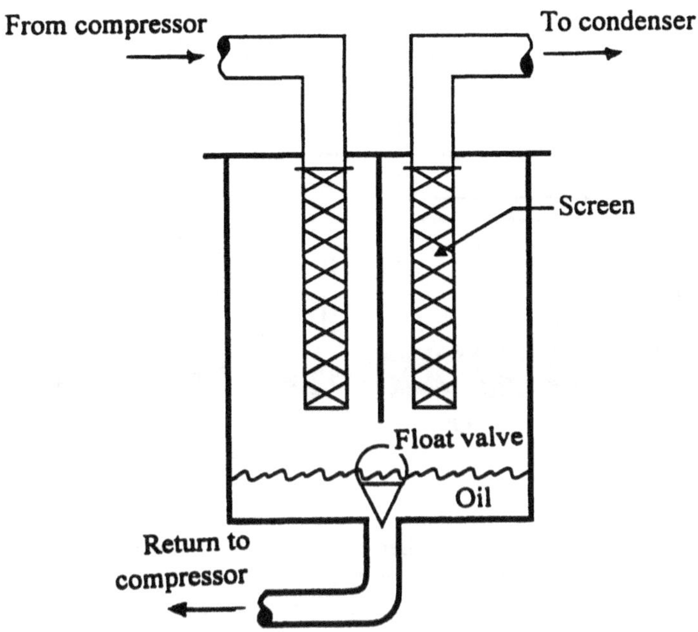

FIGURE 15.7
A simple discharge-line oil separator.

of the system began to stick. This occasional situation is not sufficient to deviate from the objective of avoiding the escape of oil into the condenser/evaporator portion of the refrigeration system.

A traditional discharge-line oil separator, as shown in Fig. 15.7, forces all the discharge vapor from the compressor to pass through several screens, which remove much of the oil from the refrigerant stream. The oil that is removed returns to the compressor under the regulation of a float valve. Oil separators of this type are commonly used with reciprocating compressors, and a few reports indicate that the oil concentration in the refrigerant leaving the separator is between 30 and 80 ppm. These values were the result of measurements of the amount of oil drained from the system in a measured number of hours of operation at known compressor capacity.

When screw compressors appeared on the scene with their oil injection for sealing the passages between the rotors, the amount of oil that the discharge-line separator would pass on to the rest of the system was objectionable. To improve the efficiency of oil removal for screw compressors, coalescing separators were chosen. The design of one configuration of coalescing separator is shown in Fig. 15.8 and consists of two major sections. The first section encountered by the oil-laden refrigerant allows the large drops of oil to fall to the sump. This separating section removes a larger quantity of oil than the coalescing section that follows, but would by itself be no improvement over the simple separator of

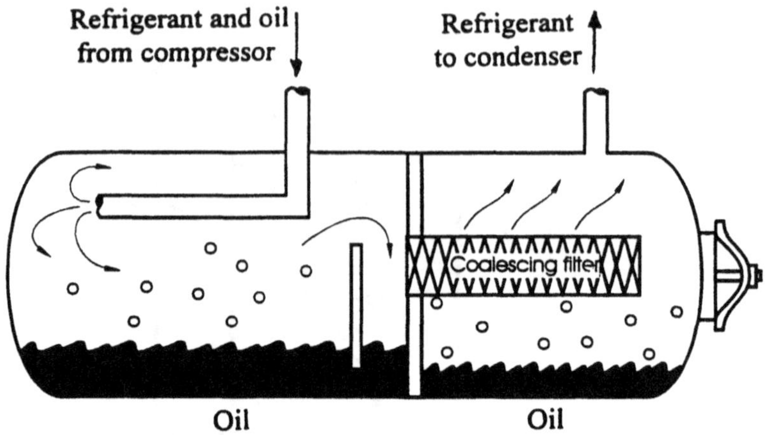

FIGURE 15.8
A coalescing oil separator.

Fig. 15.7. The refrigerant vapor then passes to the coalescing section and flows through the mesh of fine metal. Here the tiny oil droplets impinge on the mesh and consolidate with other small droplets until the size is sufficient for them to fall to the bottom of the coalescing sump.

Good coalescing separators, when operating at their recommended gas flow rate, remove virtually all oil in liquid form, and the only oil passing to the remainder of the system is vapor. The expected oil concentration in the discharge gas leaving a coalescing separator can therefore be predicted from the typical vapor pressures of mineral oil shown in Fig. 15.1. The concentration[12] will be a mild function of the refrigerant pressure, as shown in Fig. 15.9. When the success of coalescing separators applied to screw compressors became evident, the industry developed and is now applying the coalescing separator with good results to reciprocating compressors as well.

15.10 REMOVING OIL FROM THE LOW SIDE OF THE AMMONIA SYSTEM

The preceding section states that even with high-efficiency separators, some oil will carry over to the evaporators and to the low-side vessels. In the case of ammonia, this oil will separate from the refrigerant, and this tendency to separate is exploited in the arrangements designed to drain the oil from the ammonia system. One of the arrangements that is applicable to removing oil from a low-pressure receiver is shown in Fig. 15.10. The pump inlets are slightly above the bottom of the receiver, so the refrigerant below those inlets will have little motion, which permits the oil to settle and migrate down the oil leg. From the oil leg, the oil moves to the oil pot where it is contained until drained. The drain line is equipped with a spring-closing valve, which snaps shut in the event the

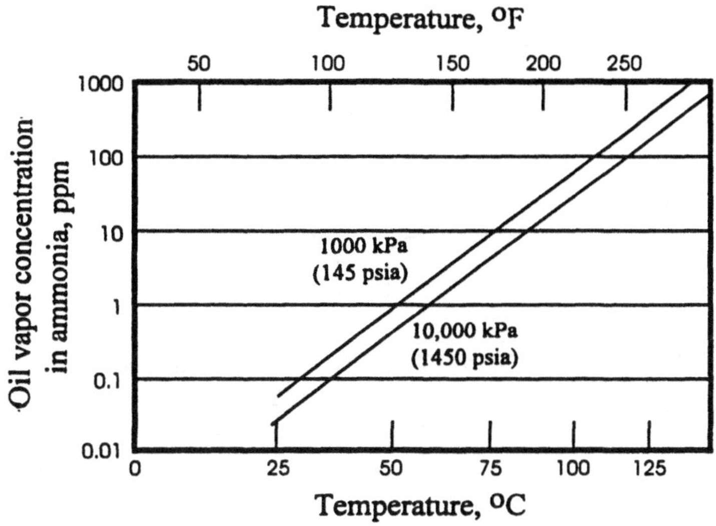

FIGURE 15.9
Oil concentration attributable to the oil vapor content of refrigerant leaving an ammonia coalescing separator.

operator performing the draining process is driven from the scene by the escape of a large bubble of ammonia. The oil pot is capable of being completely valved off, so it must be protected by relief valves.

The oil pot is equipped with an equalizer line that is connected to the top of the vessel or to an inlet vapor line. The purpose of this line is to vent vapor during the flow of oil into the pot. Without this equalizer line, the oil may not drain properly. Another situation may arise when the valves in the equalizer line and the line from the oil leg are open. A thermosyphon effect may develop in which liquid ammonia continuously flows into the relatively warm pot, vaporizes, and vents to the high elevation.

15.11 REMOVING THE OIL FROM THE LOW SIDE OF A HALOCARBON SYSTEM

The approach to returning oil from the low side of halocarbon systems differs from that used for ammonia, because the oil and halocarbon are at least partially mutally soluble. Furthermore, liquid halocarbons are more dense than oil, so any oil that does separate in a flooded shell-and-tube evaporator or a low-pressure receiver tends to rise in the vessel rather than settle. The two basic steps in returning oil from the low side of a halocarbon system are: (1) continuously or periodically extract a small quantity of liquid refrigerant, and (2) return the oil in this sample either by lifting it into the suction line or by boiling off the refrigerant in a heated still and pumping the oil that remains back to the compressors.

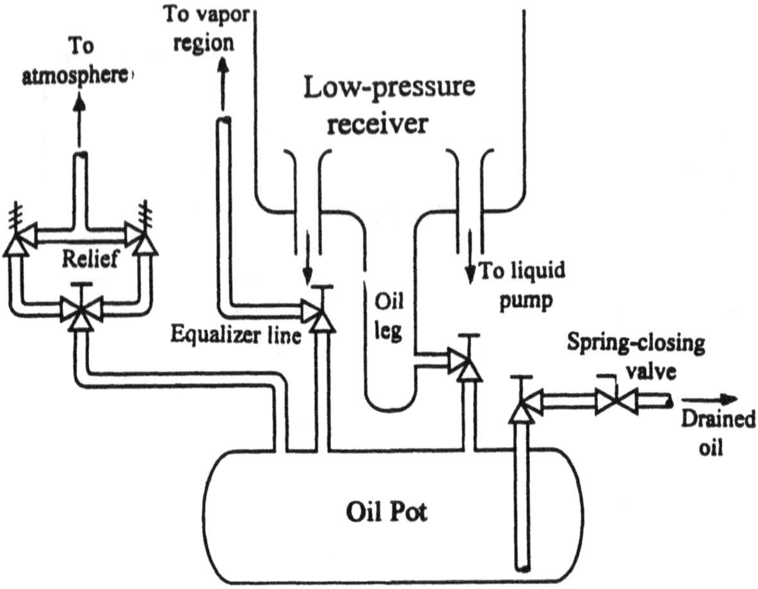

FIGURE 15.10
Provisions for draining oil from an ammonia vessel.

There are numerous variations of oil return arrangements for halocarbon systems, but two popular methods are shown in Figs. 15.11 and 15.12, both of which are applicable to pumped liquid recirculation systems. In Fig. 15.11, a sample of liquid refrigerant and its dissolved oil is extracted from the pumped discharge where its pressure is higher than the suction pressure. This sample is directed to a trap in the suction line where the vapor velocity lifts the mixture to the elevated suction line. Because there is liquid refrigerant in this mixture that should not be permitted to return to the compressor, a liquid/suction heat exchanger vaporizes the liquid refrigerant. The diameter of the suction line must be chosen to provide a high enough vapor velocity to carry the oil back to the compressor, just as in direct expansion systems.

Another approach to oil return, shown in Fig. 15.12, uses an electrically-heated still to vaporize the refrigerant liquid. A small positive-displacement pump periodically delivers the oil that remains back to the compressor(s).

The sampling of oil and refrigerant mixtures in the systems of Figs. 15.11 and 15.12 is a positive process, because the pressure in the discharge line of the pump is higher than in the suction line. The same concepts will function, however, in non-pumped flooded shell-and-tube heat exchangers and flooded evaporators. It is necessary in those arrangements that the trap (Fig. 15.11) or the still (Fig. 15.12) be at an elevation sufficiently lower than the sampling point to provide adequate static head to achieve flow of the mixture.

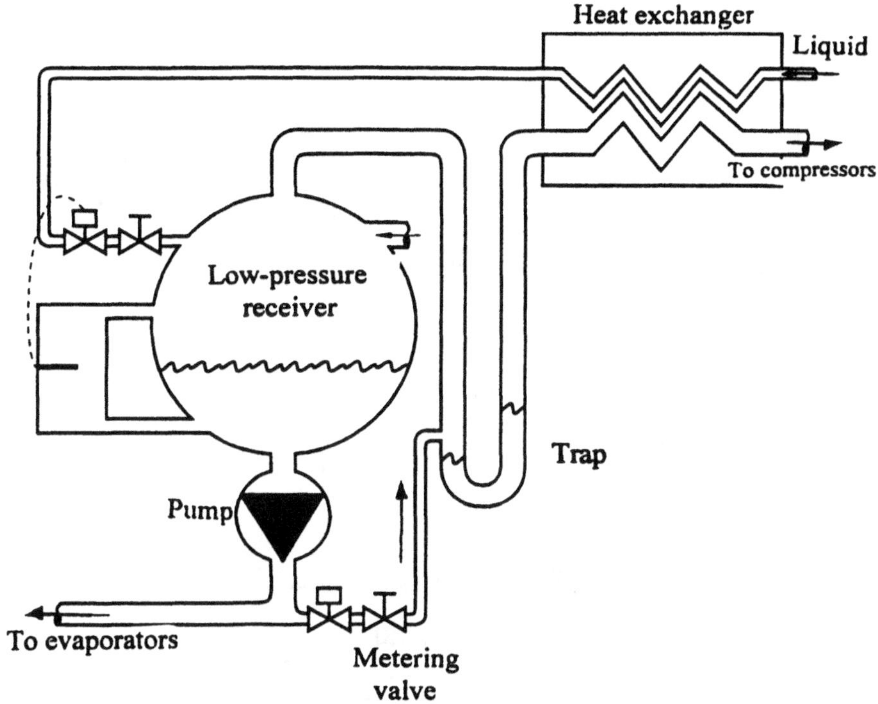

FIGURE 15.11
Oil return in a halocarbon pumped recirculation system by trapping the mixture of liquid refrigerant and oil.

15.12 OIL THAT IS SOLUBLE IN AMMONIA

Ammonia and mineral oil are not mutually soluble. The tendency of oil to separate from liquid ammonia is usually listed as an advantage of ammonia, and the property is used as a means of removing oil from the low side of an ammonia system. Whether this characteristic of ammonia and oil is truly an advantage may be debated, but users of ammonia have adapted to the property and achieved successful oil separation.

A departure from the acceptance of the separation of oil and liquid ammonia is to choose an oil that is miscible with ammonia[13]. The POE oils should not be considered, because they react chemically with ammonia. Polyalkylene glycols (PAGs) are possible choices, however. A range of properties are possible depending on the ratio of ethylene oxide and propylene oxide in the oil.

Figure 15.13 shows the miscibility of the 1:1 ethylene oxide:propylene oxide composition of PAG oil with ammonia. Ammonia and oil are miscible at temperatures above the line in Fig. 15.13, which shows that there is miscibility down to

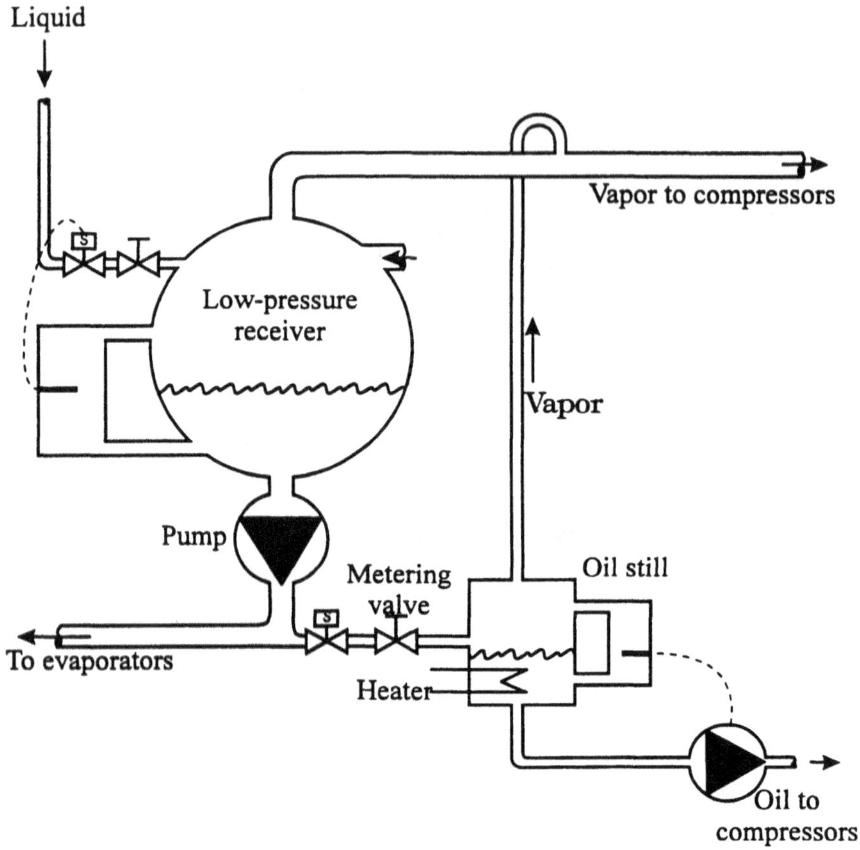

FIGURE 15.12
Oil return in a halocarbon pumped recirculation system using a heated still.

reasonably low temperatures. It is relevant to ask why there would be any interest in oils soluble in ammonia. The answer is that such an oil would remove one of the barriers to wider application of ammonia to direct expansion systems. Oil could be returned to the compressor from medium-temperature evaporator coils in the same manner as is done in halocarbon systems. The oil would be carried along with the liquid ammonia through most of the length of tubes and finally conveyed with high vapor velocity in the suction line back to the compressor.

15.13 OIL FOR CFC AND HCFC ALTERNATIVES

Mineral oils, usually including some additives, have performed successfully with the halocarbon refrigerants of the past, such as in R-12, R-22, and R-502 systems.

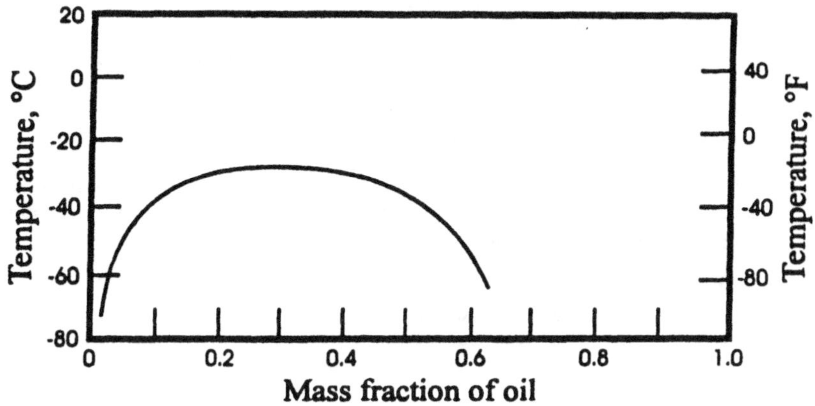

FIGURE 15.13
Miscibility of a PAG oil with ammonia.

But R-12 and R-502 will have to be replaced because of their CFC content. Like other HCFCs, R-22 will need to be retired in several decades or less. Because mineral oil has low solubility in the replacement refrigerants, new oils had to be found. The principal candidates for replacement oils are members of the following families: polyalkylene glycols (PAGs), polyalphaolefins (PAOs), and polyol esters (POEs). It is the latter group, the POEs, that seem to be finding greatest industrial acceptance as the oil for the new HFC refrigerants[14,15].

The POE oils, while front runners for replacements, do have their drawbacks. The oils are expensive, hygroscopic, and have strong solvent tendencies. They also are non-foaming, and the current design of many compressors takes advantage of the foaming of oil for effective lubrication.

Even though CFC-12 never enjoyed extensive use in industrial refrigeration, it should be mentioned that its principal substitute, HFC-134a, functions well with POEs. No system changes are needed for this conversion, but the mineral oil serving the CFC-12 system must be completely removed and replaced with the POE oil.

Refrigerant-502 has been used to a certain extent in industrial refrigeration, and optional halocarbon refrigerants for applications that would have been served in the past by R-502 include R-404A and R-507. These new refrigerants are compatible with POE oils.

A more difficult decision regarding oil is if a new HCFC-22 system is installed with the expectation that in a decade the refrigerant will have to be replaced. The temptation to plan for the future by charging the R-22 system with POE oil is not recommended, because this combination does not provide the expected oil foaming. Also, the color-type of moisture indicators do not function reliably. The replacement of R-22 remains one of the difficult decisions, and large plants are now seriously considering ammonia, while the small plants are considering such refrigerants as MP39 and HP80/81 with POE oil.

REFERENCES

1. American Society of Heating, Refrigerating and Air-Conditioning Engineers, Atlanta, Georgia, from the 1994 ASHRAE *Handbook—Refrigeration*, Chapter 7.
2. Villadsen, V. and F. V. Boldvig, "Oil in Refrigeration Plants," Proceedings of the Annual Meeting of the International Institute of Ammonia Refrigeration, San Francisco, California, 1984.
3. ANSI/ASHRAE Standard 86–1993, *Methods of Testing the Floc Point of Refrigeration Grade Oils*, American Society of Heating, Refrigerating and Air-Conditioning Engineers, Atlanta, Georgia, 1983.
4. Friedmann, J. R., *Kholodilnaia Technika*, 20, 1948 and H. Steinle, *Kaltemaschinenoele*, p 81, Springer-Verlag, Berlin, Germany, 1950.
5. Tichy, J. A., N. A. Machen, and W. M. B. Duval, "An Experimental Investigation of Heat Transfer in Forced Convection Condensation of Oil-Refrigerant Mixtures," *ASHRAE Transactions*, 91(Part 1A): 297-308, 1985.
6. Abdul-Manov, K. A. and N. I. Mirmov, "Experimental Study of Heat Transfer from Oil-Contaminated Ammonia Vapor Condensing on Horizontal Tubes," *Heat Transfer–Soviet Research*, 3(6): 176-180, 1971.
7. Tschernbylski, I. and G. Ratiani, *Kholodilnaia Technika*, 32(3): 48-51, 1955.
8. Stephan, K., "Influences of Oil on Heat Transfer of Boiling Freon 12 and Freon 22," Proceedings of XI International Congress of Refrigeration, Munich, Germany, 1963.
9. Shah, M., "Visual Observations in an Ammonia Evaporator," *ASHRAE Transactions*, 81(Part I): 295-306, 1975.
10. Landry, J. F., "The Effects of Contaminants on Compressor Oils in Ammonia Systems," Proceedings of the 9th Annual Meeting of the International Institute of Ammonia Refrigeration, San Diego, California, 1987.
11. Rossignol, R., "Hydrotreated Oils for Ammonia Refrigeration," *Air Cond., Htg., Refrig. News*, pp 3-4, December 14, 1992.
12. Reinhart, A., "Ammonia in Liquid Chillers and Heat Pumps," Proceedings of Commission Meeting, International Institute of Refrigeration, Hannover, Germany, 1993.
13. Burke, M. and H. Kruse, "Solubility and Viscosity of New Oil/ Ammonia Systems," Proceedings of Commission Meeting, International Institute of Refrigeration, Ghent, Belgium, 1993.
14. Short, G. D. and R. Cavestri, "High-Viscosity Ester Lubricants for Alternative Refrigerants," *ASHRAE Transactions*, 98(Part I): 789-795, 1992.
15. "CFC Update IV," *Phoenix Pflyer*, Phoenix Refrigeration Systems, Inc., Conyers, Georgia, March 1994.

CHAPTER 16

CONSERVATION, RECLAIM, AND STORAGE OF ENERGY

This chapter is divided into three parts: (1) conservation of energy in refrigeration plants, (2) transfers of thermal energy from one location in the plant to another, and (3) shifting the time when refrigeration is produced and when it is used through the application of thermal storage. The time will probably never return when the energy costs of the refrigeration plant can be ignored. Twenty years ago in a facility such as a refrigerated warehouse, the electricity and fuel costs would have constituted between 5 and 8% of the total annual expenses. Today the percentage would likely be between 15 and 25%.

It is a team effort to develop and operate an energy-effective plant. A skillful operator may be frustrated in his or her attempts to operate the plant efficiently because of the inflexibility of the design or construction. Once having designed and built a plant capable of efficient operation, the features that can be exercised to conserve energy must be communicated to the operator.

TABLE 16.1
Some decisions in the design and construction of the plant that affect energy requirements.

Increased capital investment	Energy saving
Thicker insulation on the structure	Less heat gain to refrigerated space
Thicker insulation on pipe and vessels	Less heat gain to refrigerant
Larger pipe diameters, especially vapor lines	Reduced drop in saturation temperature
Increased heat-transfer areas of evaporators and condensers	Less compressor power (higher evap. and lower cond. temps.)
Higher circulation ratio in liquid overfeed coils	Improved heat-transfer coefficients in evaporators
Fewer coils on a hot-gas defrost cluster	More flexibility in starting and terminating defrosts
Quick opening/closing doors on refrigerated structures	Reduced infiltration
High-efficiency compressor motors	Reduced electrical power for a given amount of shaft power
Two-speed motors on fans serving air coils	Reduced power and internal heat generation at part load

PART I. ENERGY CONSERVATION

16.1 DESIGNING THE PLANT TO CONSERVE ENERGY

A variety of decisions affecting energy must be made in the design and construction of the plant, and most of them require capital investments in order to conserve energy during operation. Table 16.1 lists some of the decisions made in plant design—several of which are of the yes/no type while others are questions of extent (thickness of insulation, diameter of pipes, etc.).

The decisions listed in Table 16.1 all impact the energy requirements of the plant and, as Fig. 16.1 shows, all exhibit common trends. The common pattern is that the energy penalty, shown by the ordinate, decreases as the capital expenditure, shown by the abscissa, increases. The *rate* of energy reduction decreases, however, as the capital investment increases.

One of the classic decisions in the design of refrigerated spaces is choosing the thickness of insulation in the walls, roof, and ceiling. This subject will be addressed agaonin Chapter 19 in connection with the refrigerated structure, but the expected trend in heat gain is shown in Fig. 16.1a. The slope of the curve at

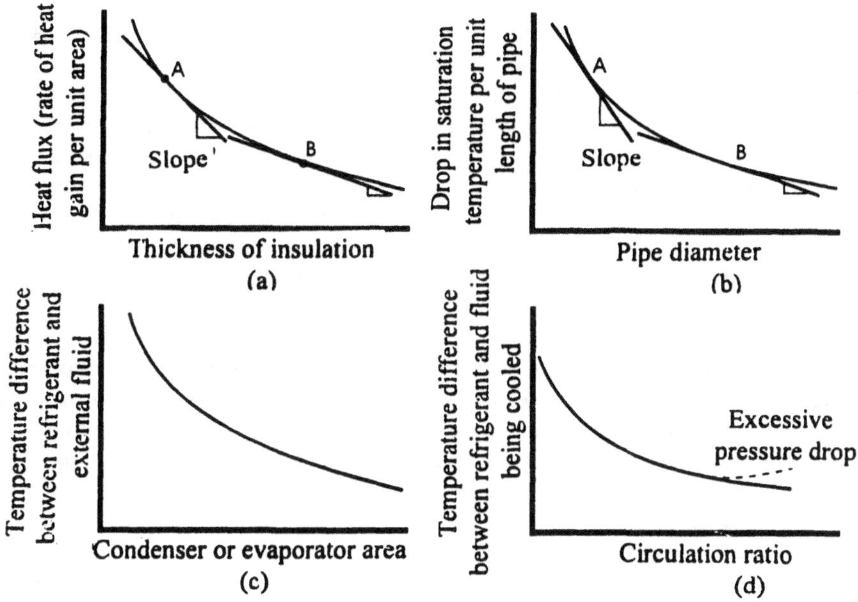

FIGURE 16.1
Impact of some design decisions on energy requirements of the refrigeration plant: (a) insulation thickness, (b) diameter of vapor lines, (c) heat-transfer area of the condenser or evaporator, and (d) circulation ratio in a liquid overfeed coil.

any point indicates the rate of decrease of heat flux per unit increase in insulation thickness. At point A, the absolute magnitude of the slope is greater than at B, so as the thickness increases, the benefit of a further increase in the insulation thickness diminishes.

Pipes and vessels that contain low-temperature brines and refrigerants are almost always insulated, but high-temperature pipes in a refrigeration system, such as discharge lines from the compressor, are rarely insulated. Insulation on low-temperature pipes and vessels, in addition to inhibiting heat gain, prevents condensation or frosting and the resultant dripping of water.

Choosing large-diameter pipes for refrigerant vapor lines decreases the drop in saturation temperature, as Fig. 16.1b shows, which ultimately permits the compressor to operate with a reduced difference in pressure between suction and discharge. As the pipe size becomes larger, each unit increase in pipe diameter produces less benefit.

When evaporators and condensers are chosen with large heat-transfer areas, the temperature difference between refrigerant and the external fluid is reduced, which results in lower compressor power for a given rate of heat transfer, Fig. 16.1c. As is true of all the decisions shown in Fig. 16.1, the return progressively diminishes as the heat-transfer area increases.

The fourth example of a decision that is one of degree is the choice of circulation ratios in liquid recirculation plants. Figure 16.1d shows a progressive decrease in temperature difference between the liquid or air being cooled and the refrigerant, thus permitting a higher suction pressure. In contrast to the other three illustrations in Fig.16.1, the increase in circulation ratio may entail only small increases in capital investments associated with successive step-ups in sizes of the motor and pump. Increasing the rate of refrigerant pumped demands more power at the pump, which at high flow rates can overshadow the power saved at the compressor. Furthermore, at high flow rates the pressure drop of refrigerant in the coil elevates the mean boiling temperature for a given leaving pressure of refrigerant.

Another feature of the design and construction of a plant cited in Table 16.1 that can either permit or preclude an efficient mode of operation is the clustering of several evaporators onto one set of defrost controls. In an arrangement in which one set of hot-gas defrost controls serves more than one evaporator, the first cost and possible maintenance costs can be reduced. The ganging of coils on one set of defrost controls does, however, reduce the flexibility, because the defrost of all the coils in the cluster must start and stop at the same time. The frequency and duration of defrost are thus dictated by the coil experiencing the worst frosting condition. The losses attributable to hot-gas defrost can be reduced by individual defrost controls, but only if each individual coil operates under some type of demand defrost (coil defrosted only when needed and defrost terminated when the coil is free of frost and the melted frost fully drained).

One of the sources of loads imposed on refrigerated spaces is air infiltration through doors when they are open. Since doors must be open at certain times in order to move products in and out of the room, an approach to reducing this load is to install doors that open and close quickly. Doors of this type are likely to be more expensive than slower-acting doors, so once again the designer and/or manager is faced with a capital investment decision[1] that can conserve energy.

The designer may choose between a standard and a high-efficiency motor. Typically, the high-efficiency motor costs 20% more than the standard motor and achieves only a 2% increase in efficiency. Such an investment may not seem attractive until it is realized that the annual electricity cost for a high-use motor may be ten times the first cost of the motor. In such an instance, the additional investment offers a one-year payback.

Two-speed motors are often chosen for the fans on evaporator coils. Two-speed fan motors may be advantageous at low refrigeration loads, because they directly reduce the power required by the fan, as well as the internal heat gain in the refrigerated space.

16.2 OTHER DESIGN DECISIONS AFFECTING ENERGY REQUIREMENTS

Several additional decisions that frequently confront the designer are: (1) the choice of small packaged refrigeration units or one large central unit, (2) the mix of types and sizes of compressors, (3) single-stage or two-stage compression, and (4) the extent of instrumentation.

Packaged refrigeration units typically use a halocarbon refrigerant and have air-cooled condensers. They are mounted through an opening in the roof or wall with the evaporator inside the refrigerated space and the condenser and compressor outside. The usual expectation is that small packaged units can provide the needed refrigeration capacity with a lower first cost, but with higher operating and maintenance costs. Packaged units predominate in refrigerated spaces of less than about 1,000 m^2 (10,000 ft^2) area and in central plants with areas greater than 2,000 m^2 (20,000 ft^2).

Some of the considerations affecting the type and number of compressors are: (1) the installed cost of the compressors and motors, (2) maintenance and service costs, and (3) power requirements. No universal pattern applies to all installations, but showing some relative costs may suggest approaches to the decision process. First, consider the typical trends of costs of reciprocating and screw compressors, including their motors, as shown in Fig.16.2. Costs like those shown in Fig. 16.2 vary somewhat, so even these relative costs are different from one situation to another, but at lower capacities the first cost of the reciprocating compressor is less than that of the screw compressor. The crossover point is shown at approximately 800 kW (250 tons) refrigerating capacity, which corresponds roughly to a volumetric pumping capacity of about 0.3 m^3/s (600 cfm).

The choice of compressors involves selecting the number and size to provide the total refrigeration capacity. In general, one compressor capable of delivering full capacity is the choice giving the lowest first cost. This choice, however, results in inflexibility, because should that one compressor need service or fail, there is no reserve. Also, at part-load operation it is beneficial to be able to shut down one compressor and continue to operate the other closer to its full load, particularly in the case of screw compressors. Thus, it is usually advisable to select two or more smaller compressors for a given assignment, even though the total first cost will be greater than if only one were chosen. The part-load efficiency of reciprocating compressors equipped with cylinder unloaders is normally higher than that of screw compressors operating at part load. Combining screw and reciprocating compressors is often favorable, with the screw compressor handling the base load and the reciprocating compressor serving to trim the capacity.

Another design decision affecting the energy requirements of the refrigeration facility is the choice of single-stage or two-stage compression. Figure 16.3 shows that the reduction in compressor power achieved by using two-stage compression is a function of the evaporating temperature. Energy considerations are crucial in making the decision whether to choose single- or two-stage compres-

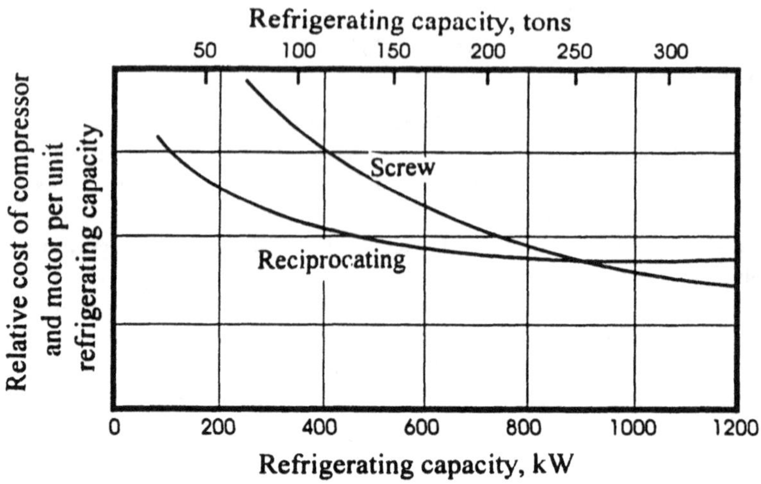

FIGURE 16.2
Relative first costs of compressor and motor as a function of size for compressors operating with −5°C (23°F) evaporating and 35°C (95°F) condensing temperatures.

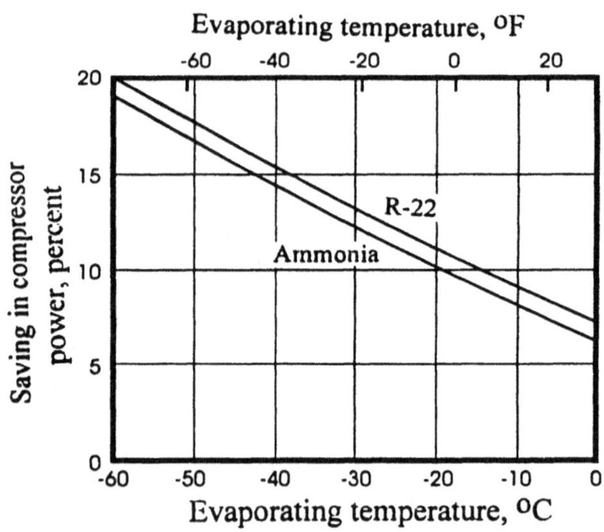

FIGURE 16.3
Compressor power savings through the use of two-stage compression with flash-gas removal and intercooling compared to single-stage compression. The condensing temperature is 35°C (95°F).

sion, but in addition, other influences are the limitations of the compressors with respect to pressure ratios and whether evaporators operating at the intermediate-temperature level are part of the system. Since reciprocating compressors are usually limited to pressure ratios of approximately 8 or 9, this limit is reached at evaporating temperatures in the range of −25°C (−13°F) if the condensing temperature is approximately 35°C (95°F).

Another situation where two-stage systems are attractive is when the plant incorporates intermediate-temperature level evaporators, which serve produce storage rooms or loading docks of frozen food warehouses. The evaporating temperature of these intermediate-level loads fixes the intermediate-temperature level.

Chapter 14 enumerated some of the electrical and electronic measuring devices that can be applied to refrigerating systems. Whatever the type of instrument, there is a cost associated with its installation and maintenance. Without knowledge of pressures and temperatures throughout the system, the operator is in the dark as to how the plant is performing. In order to diagnose operating problems and to maintain energy-efficient operation, certain important variables must be measured. Virtually all plants are equipped with pressure gauges at the compressor suction and discharge. Additional positions throughout the system where the indication of pressure and temperature should also be considered are:

Pressures

- Liquid leaving the condenser or high-pressure receiver
- Outlet of evaporators
- At all pressure-regulating valves
- Inlet and outlet of liquid recirculation pumps

Temperatures

- Compressor suction and discharge
- Liquid refrigerant leaving the condenser or receiver
- Refrigerant entering and/or leaving evaporators
- Liquid leaving the low-pressure receiver in a liquid recirculation system

The entering and leaving temperatures of the fluid that is chilled in an evaporator should also be measured. Although air temperatures can readily be measured with thermometers held manually at the desired position, it is convenient to have thermocouples or resistance-temperature devices installed that can be read from a central location.

A crucial quantity needed for energy analysis is the refrigerant flow rate, but its measurement is usually difficult. Orifices in vapor lines are generally reliable but introduce pressure drops, which may waste more energy than can be saved by knowing the flow rate. Measuring the liquid flow rate by any means that introduces a pressure drop may also cause flashing of liquid into vapor and erroneous readings. Some of the newer concepts of measuring flow rate or velocity, such as sensing the rotative speed of a turbine, measuring the rate at which vortices are shed from a specially designed obstruction, or applying ultrasonic principles, are opening up convenient means of flow measurement.

16.3 OPERATION AND MAINTENANCE OF THE SYSTEM TO CONSERVE ENERGY

The previous two sections presented some of the challenges facing the designer when planning a facility that is to operate efficiently. The operating personnel inherit a plant and attempt to provide the refrigeration duty at minimum cost. Lowest-cost operation is generally the same as minimum-energy operation, assuming that measures to conserve energy do not shorten the life of the equipment or entail greater maintenance costs.

Several dominant guidelines for conserving energy in operation are as follows:

1. Maintain suction pressures at the compressors as high as possible.
2. Maintain discharge pressures of the compressor as low as possible.
3. Operate the plant efficiently within the desirable suction and discharge pressures.

While rules 1 and 2 are desirable general guidelines, care must be exercised that they are not achieved at the expense of more power used by the auxiliaries (fans and pumps) than saved at the compressors. Even the ideal Carnot cycle pointed the direction for efficient operation through the expression of the coefficient of performance (COP):

$$COP = \frac{T_{evap}}{T_{cond} - T_{evap}} \tag{16.1}$$

The temperatures in Eq. 16.1 are on the absolute Kelvin (Rankine) scale, which is t°C + 273.15 (t°F + 459.7). A given increase in T_{evap} is slightly more effective in increasing the COP than the same reduction in T_{cond}.

An even more realistic picture of how changes in evaporating and condensing temperatures influence operating efficiency can be derived from compressor performance data, as shown in Fig. 16.4. The graph presents the improvement in the COP per degree increase in evaporating temperature and degree decrease in condensing temperature and applies to a system using an 8-cylinder, R-22, reciprocating compressor. Some of the trends predicted by Eq. 16.1 are apparent in Fig. 16.4, for example, the percentage increase for a one-degree change is more prominent when operating with small differences in condensing and evaporating

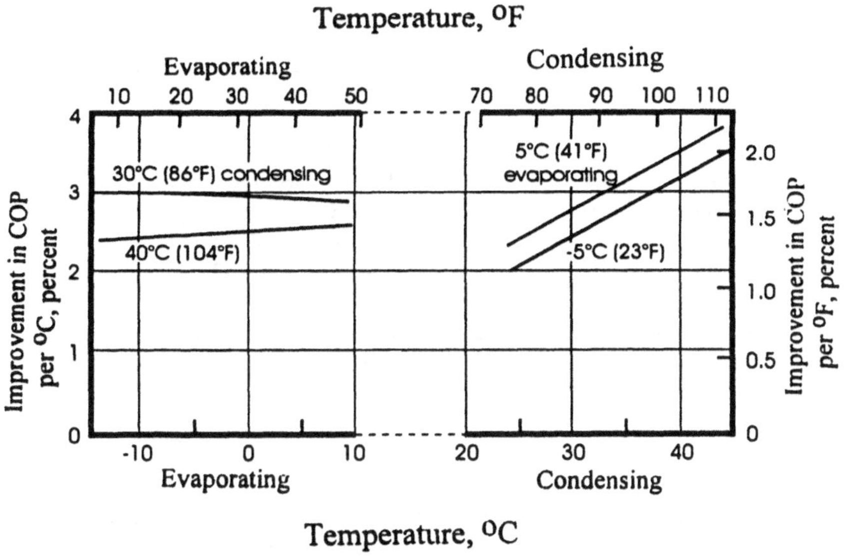

FIGURE 16.4
Percent savings in power for a given refrigerating capacity through a one-degree increase in evaporating temperature or decrease in condensing temperature. The system uses an 8-cylinder, R-22, reciprocating compressor.

temperatures. There are some discrepancies between Eq. 16.1 and Fig. 16.4, however, in that Eq. 16.1 predicts a progressive decrease in the percent saving as the condensing temperature increases at a given evaporating temperature, which is in contrast to the increases shown in Fig. 16.4. The reason for the difference is that the influence of compression efficiency appears in Fig. 16.4 in addition to the cycle efficiency. At a constant evaporating temperature, a reduction in condensing temperature also increases the compression efficiency.

A conclusion to carry away from this discussion is that the reduction in compressor power for a given refrigerating capacity, by increasing the evaporating temperature or decreasing the condensing temperature, is of the order of 2.5 to 3.0% per °C (1.5 to 2.0% per °F).

Several approaches to elevating the evaporating temperatures and reducing condensing temperatures are shown in Table 16.2. To operate the plant efficiently, the total power, not just the compressor power, must be minimized. The auxiliaries, such as the fans serving the condensers and air-cooling coils, require an amount of power that is not negligible. Perhaps the power required by the auxiliaries constitutes 10-15% of the total full load power. However, when those auxiliaries continue to draw a constant power while the compressor unloads, they may require 30% or more of the total power. The challenge is one of optimization, and shutting off or reducing the operating speed of some of these auxiliaries at part load may save more power than the resulting increase in compressor power. The application of two-speed motors on the fans of evaporators and condensers

TABLE 16.2
Actions in the design and operation of refrigeration plants to increase the evaporating temperature and decrease the condensing temperature.

	In design	In operation
Increase evaporating temperature	larger area of evaporators; larger suction pipes	oil removal from refrigerant; optimum coil defrost; raising setting of suction pressure at part load
Decrease condensing temperature	larger condensers; larger compressor discharge pipes	allowing condenser pressure to drop at low ambient temperatures; purging air and non-condensibles; cleaning air side of evaporative condenser tubes

is discussed more thoroughly in Chap. 7. In both cases, a tradeoff exists between the power required by the fans and that of the compressors.

The final point addressed under the subject of energy-efficient operation of the plant is the strategy of choosing the mix and loading of compressors to deliver the required refrigerating capacity with the minimum total compressor power. The compression efficiency of screw compressors drops off at part load, so it is advantageous to operate this type of compressor near its full load. The part-load efficiency of reciprocating compressors with cylinder unloading is quite high, so many plants combine the operation of screw and reciprocating compressors, allowing the screw type to handle the base load and the reciprocating compressor to provide the trimming of capacity.

A comparison of total power requirements of a plant having two screw compressors, each with 700 kW (200 tons) refrigerating capacity, and one 350 kW (100 tons) reciprocating compressor operating in various combinations is shown in Fig. 16.5. In general, the screw compressor(s) should be operated as near as possible to full capacity or shut off completely.

PART II. HEAT RECLAIM

16.4 HEAT RECLAIM

If at one position in the plant heat is being rejected and at another location heat is required, the possibility of transferring this heat is attractive. High-temperature vapor from the compressor enters the condenser to be desuperheated and condensed. Instead of rejecting heat to the atmosphere through the condenser, why not use the heat to supply a need? There are many situations in which heat recovery is economical, and many others in which it is not. The decision of how and even whether to use heat reclaim can be a tricky one that is based on relative

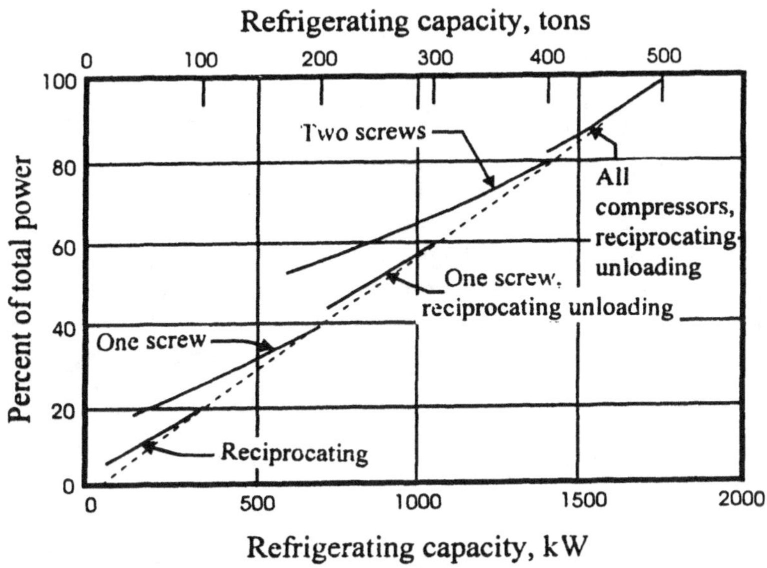

FIGURE 16.5
Total power requirements of the mix of two 700 kW (200 ton) screw compressors and one 350 kW (100 ton) reciprocating compressor covering the full load range.

fuel costs, how much of a capital investment is required, how much operating cost will be saved, and whether the timing of the availability and requirement for the heat is coincident. Another dimension complicating the implementation of heat reclaim is the temperature level—the temperature level of the rejected heat and the temperature level of the heating need.

16.5 SOURCES AND POTENTIAL NEEDS OF HEAT

Before attempting to integrate sources of heat with needs for heat, consider each category separately. Figure 16.6 shows a skeleton two-stage system and the potential sources of heat. The temperature levels of these various sources of heat are dependent on the design of the plant, its refrigerating load, the ambient conditions as they affect the condensing temperature, and the type of compressor. Some representative temperatures from the sources shown in Fig. 16.6 are presented in Table 16.3.

The sources of heat and their temperature levels are dependent upon the type of compressor. For example, heat from cooling of the cylinder heads applies only to reciprocating compressors and heat from cooling the injected oil pertains only to screw compressors. The temperatures of oil and refrigerant leaving screw compressors, shown in Table 16.3, are representative of oil cooled by a

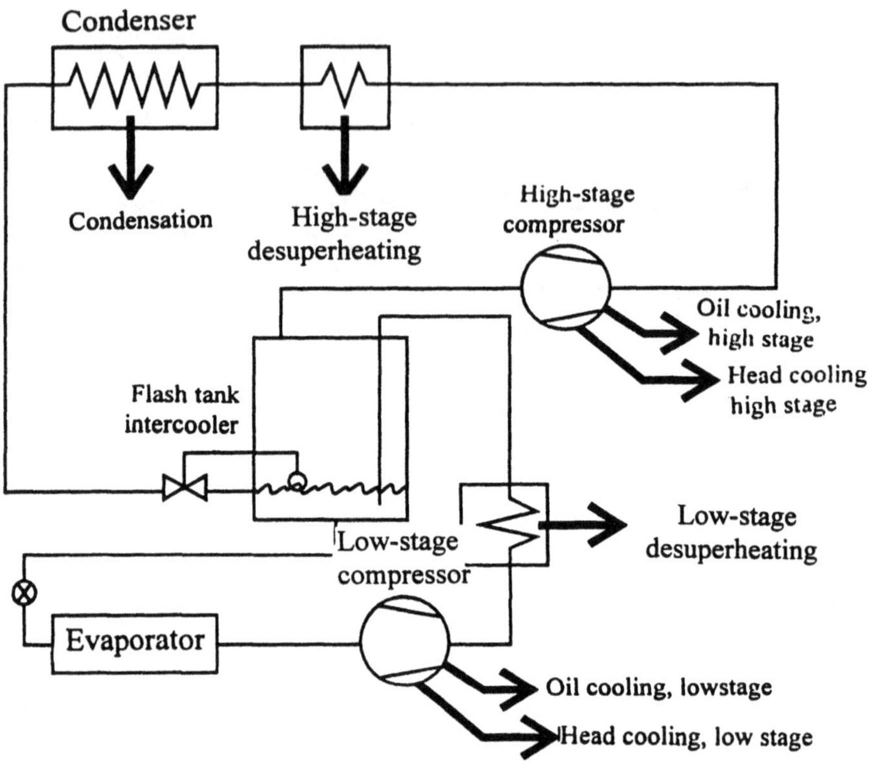

FIGURE 16.6
Heat rejection from a two-stage refrigeration system.

thermosyphon or a liquid-cooled heat exchanger. When direct injection of refrigerant is used to cool oil, the oil and refrigerant leave the compressor at lower temperatures than shown in Table 16.3—refrigerant leaving at 45°C (110°F), for example.

TABLE 16.3
Typical temperatures of the heat sources in a two-stage refrigeration plant.

Source	Type of compressor	Source temperature
LOW STAGE		
Water from head cooling	reciprocating	30-35°C (85-95°F)
Oil to be cooled	screw	55-65°C (140-150°F)
Desuperheating refrigerant	reciprocating	80-90°C (175-195°F)
Desuperheating refrigerant	screw	50-60°C (120-130°F)
HIGH STAGE		
Water from head cooling	reciprocating	30-35°C (85-95°F)
Oil to be cooled	screw	60-70°C (150-170°F)
Desuperheating refrigerant	reciprocating	95-115°C (200-240°F)
Desuperheating refrigerant	screw	65-75°C (150-170°F)
Condensing refrigerant	reciprocating	16-38°C (60-100°F)
Condensing refrigerant	screw	16-38°C (60-100°F)

Next, consider the uses of heat that might be scavenged from the refrigeration plant. The needs for heat are dependent on the nature of the plant and are different for refrigerated warehouses, food distribution centers, and processing plants (dairies, meat and poultry processors, and chemical plants). Table 16.4 lists some potential heating needs and the temperature ranges in which these needs might exist.

16.6 MATCHING THE HEAT SOURCE WITH THE NEED

Numerous factors influence the possible matchup of heat sources, as listed in Table 16.3, with one or more of the needs in Table 16.4. These factors include: (1) the temperature level, (2) the magnitude of energy rate, and (3) the timing of availability and needs (daily or seasonal). The unique combination of these factors for each plant encourages an individual analysis of each situation.

There are some general conclusions that apply most of the time. One such conclusion, which pertains to the heat sources from Table 5-3, is that the magnitude of heat flow available and/or its temperature usually eliminate from consideration the water from cooling of the heads of reciprocating compressors and the heat from oil cooling of screw compressors. The remaining sources are thus the refrigerant itself—either through desuperheating or condensation. Any heat supplied by means of desuperheating vapor from the low-stage compressor has a double benefit, since it displaces a heating cost but also reduces the refrigeration load on the high stage with the resultant reduction in compressor power.[2]

TABLE 16.4
Some needs of heat and their temperature levels.

Potential use	Minimum refrigerant temperature
Underfloor heating of subfreezing spaces	10-15°C (50-60°F)
Defrost water when water defrost is used	22-27°C (72-80°F)
Space heating in winter	
— docks	27-32°C (80-90°F)
— dry warehouses	32-38°C (90-100°F)
— offices and shops	38-43°C (100-110°F)
Reheat coils for humidity control	32-38°C (90-100°F)
Preheat tap water for washing equipment and for process purposes	43-49°C (110-120°F)
Preheating boiler feed water	43-49°C (110-120°F)

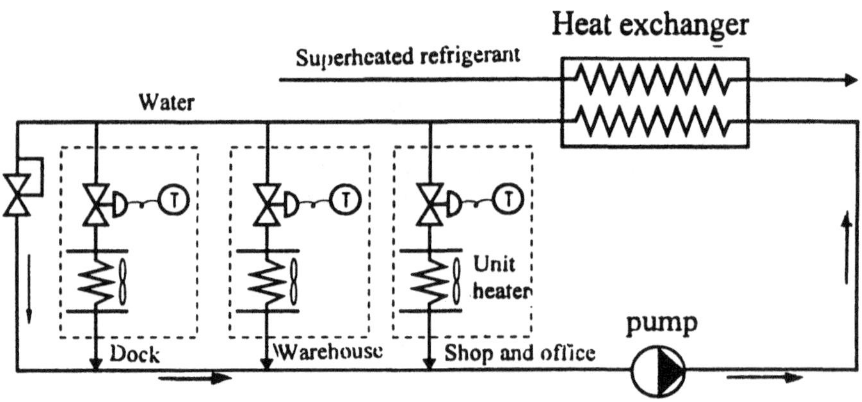

FIGURE 16.7
Transferring refrigerant heat to air through an intermediate fluid, such as water.

Examination of the needs for heat in Table 16.4 shows the lowest temperature requirement to be that of underfloor heating in frozen food storage facilities. The magnitude of this requirement is of the order of 6 W per m^2 of floor area (2 Btu/hr per ft^2), so it is usually a small magnitude but at a temperature level that is easily achieved.

The next higher levels of temperatures for space heating listed in Table 16.4 have approximate target air temperatures of 16°C (60°F) for loading docks handling nonrefrigerated products, 18°C (65°F) for nonrefrigerated warehouses, and

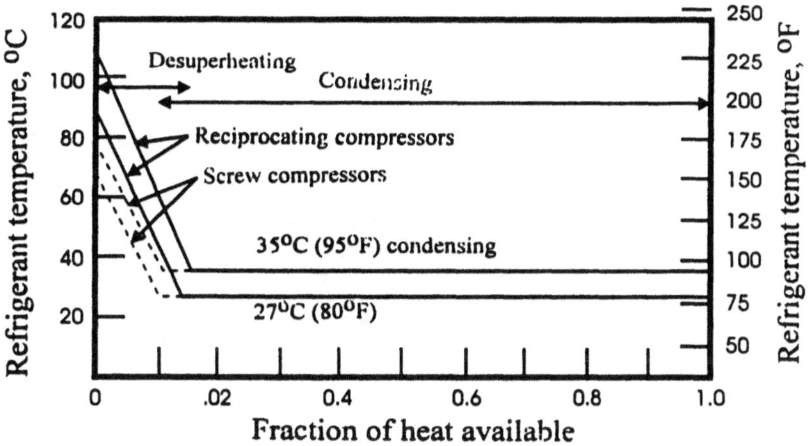

FIGURE 16.8
Typical temperatures and fractions of heat available from ammonia. The total rate rejected is 1.2 to 1.3 times the refrigerating capacity.

24°C (75°F) for shops and offices. To respond to the typical squeeze of a low-temperature difference between the refrigerant providing the heat and the air to be heated, generous areas should be chosen for the heat exchangers. A further consideration is that the refrigerant lines are likely to be remote from the spaces to be heated, so an intermediate heat-transfer fluid, as shown in Fig.16.7, is often used. This intermediate fluid allows flexibility in transferring the heat to remote locations but introduces another temperature difference between the refrigerant heat supply and the space air. A backup source of heat may be advisable, particularly for the offices and shops, in the event the rate of heat flow available from the refrigerant is inadequate.

The next factor influencing the attractiveness of heat reclaim is the magnitude of heat flow—how much is needed and how much is available. Figure 16.8 shows the fraction available from the desuperheating and condensing portions of heat rejection. The highest temperatures occur in the superheated refrigerant, but most of the heat is available from the condensation process. Figure 16.8 indicates that between 10 and 15% of the total heat rejection is available from the higher-temperature superheated refrigerant. The other quantity controlling the magnitude of available heat is the refrigeration rate, because the total rate of heat rejection is typically 1.2 to 1.3 times the existing refrigeration load.

The third important factor affecting the attractiveness of heat reclaim is timing, both daily and seasonal. If the operation of the refrigeration plant and the needs for heating coincide well, as is true for the meat processing facility[3] whose daily profiles of source and need are shown in Fig. 16.9, then the situation may be favorable for heat reclaim. Further exploration would be required of the temperature levels and an analysis of investment costs and operating savings to

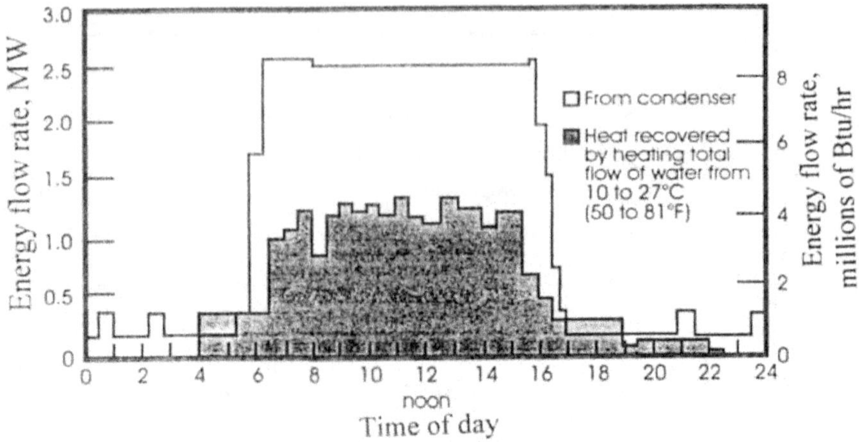

FIGURE 16.9
Daily profiles of heat rejected from the refrigeration system and heat needed for processing in a meat plant.

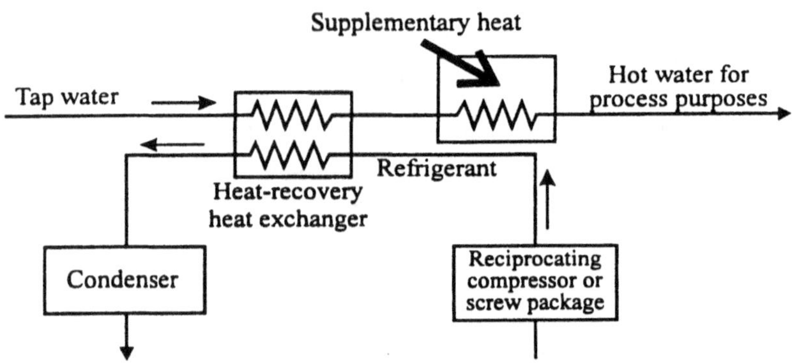

FIGURE 16.10
Preheating process water with superheated refrigerant.

determine the economic viability of the project.

In most situations the seasonal profiles are unfavorable. More energy for space heating is required during cold weather, for example, when the refrigeration load may be light. Meat and poultry processing plants and dairies often are only moderately sensitive to outdoor temperatures, however, so for such facilities the seasonal imbalance is not severe.

16.7 PREHEATING WATER WITH HEAT FROM BOTH DESUPERHEATING AND CONDENSATION

A potentially attractive opportunity for heat recovery is that of preheating water used for boiler makeup, washing, and other process purposes, as shown in Fig. 16.10. The operation is self-regulating in that the incoming water extracts whatever heat it can from the refrigerant, and the balance of heat is provided by a boiler or whatever other type of water heater is available. The outlet water temperature from a given heat-reclaim heat exchanger is a function of a number of variables. It depends on the inlet temperature, the condensing temperature, and the flow rate of refrigerant as well as the inlet temperature and flow rate of water.

Figure 16.11 shows temperature profiles through the heat exchanger and the influence of some variables on these profiles. Superheated ammonia enters the heat exchanger at a temperature of 90°C (194°F). The inlet temperature of the water is 10°C (50°F), and the right terminus of the profile is the outlet water temperature. The breaks in the curves mark the location where desuperheating of the refrigerant has been completed and condensation is beginning. The highest outlet water temperature of 22°C (71.6°F) occurs with the 35°C (95°F) condensing temperature and the high refrigerant flow rate of 0.8 kg/s (106 lb/min). This ammonia flow rate roughly corresponds to a plant with a refrigerating capacity of 900 kW (250 tons). The water temperature is brought from its entering temperature of 10°C (50°F) to a temperature of 17°C (62.6°F) by condensation of refrigerant, and the remaining elevation of the water temperature to 22°C (71.6°F) is accomplished by the superheated ammonia in 16.4% of the heat exchanger area.

Figure 16.11 also shows the water-temperature profile for a low ammonia flow rate, namely 0.4 kg/s (53 lb/min). The profiles have similar shapes in the condensing portion of the heat exchanger for all flow rates, so long as condensing continues. With a low flow rate the superheated refrigerant is cooled to the condensing temperature in a small fraction of the heat exchanger (8% in this case).

The heat-reclaim heat exchanger has a lower potential for heat recovery at reduced condensing temperatures, as Fig. 16.11 shows. Despite the fact that the temperature of the entering superheated refrigerant is the same for all cases shown in the figure, the capability of preheating the water is reduced when the condensing temperature drops.

Figure 16.11 shows that the rate of heat transferred from the refrigerant to the water is influenced by a number of factors, even though the water flow rate and the temperature of entering refrigerant are constant. The combined effect of all these variables makes it difficult for a manufacturer of these heat exchangers to present catalog information in a simple form. For this reason, some manufacturers have programmed the performance of their products on a

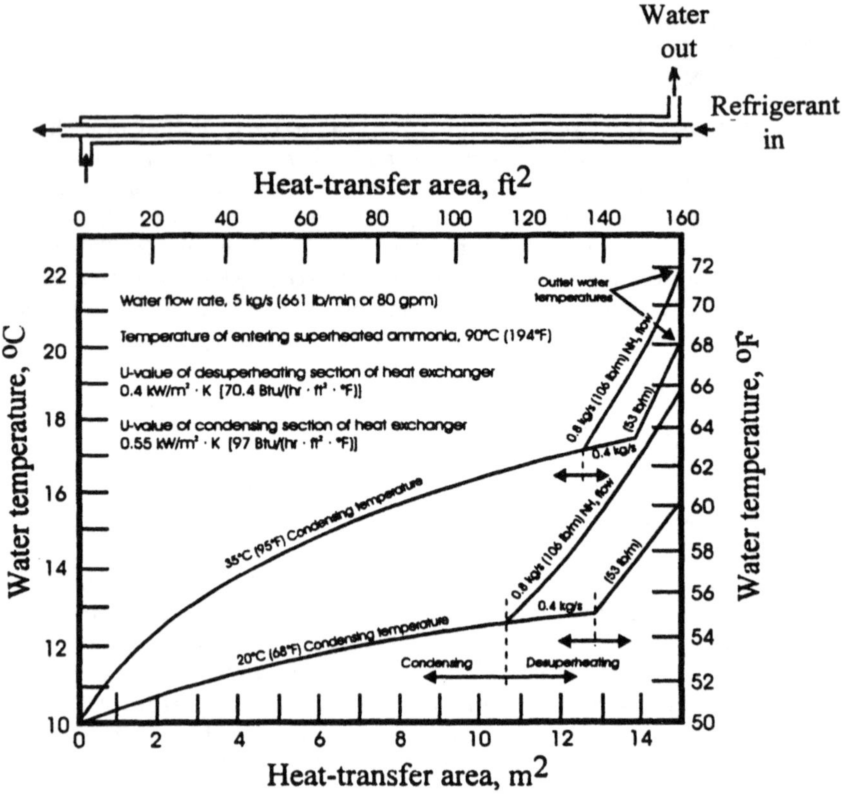

FIGURE 16.11
Water-temperature profiles for a heat-reclaim heat exchanger with combinations of two different condensing temperatures and flow rates of ammonia.

computer so that the potential customer can request capacity information for various combinations of conditions.

16.8 THE INFLUENCE OF THE HEAT EXCHANGER AND ITS DESIGN ON THE PERFORMANCE OF THE SYSTEM

Several different configurations of heat-reclaim heat exchangers are used in practice, depending upon the requirements. A conventional shell-and-tube heat exchanger, as shown in Fig. 16.12a, is often used in situations where the temperature of the fluid being heated does not need to be brought up to a high value, for example, glycol for underfloor heating. When the temperature of the outlet fluid must be increased as much as possible, it is desirable to place the superheated

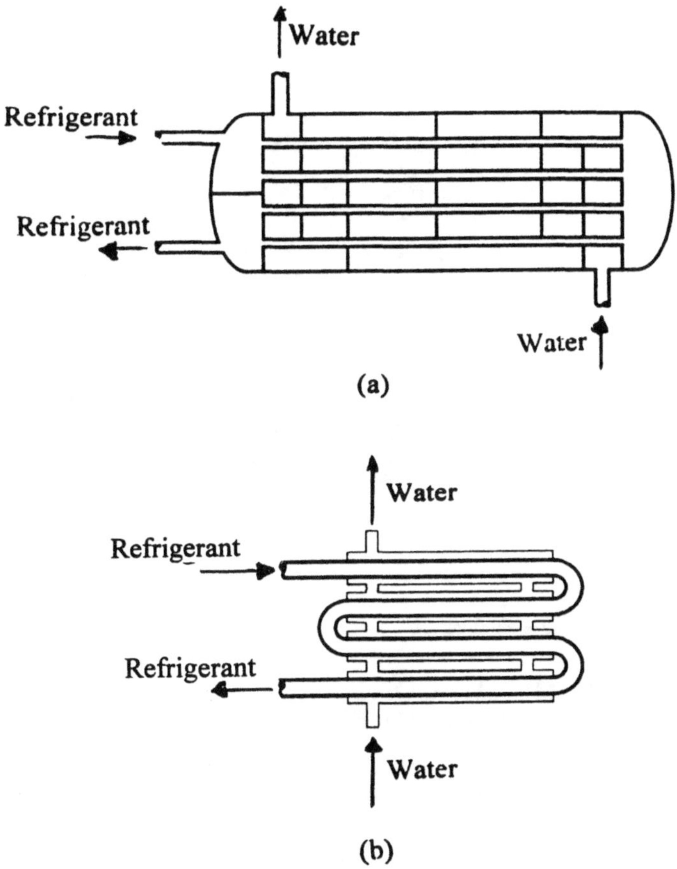

FIGURE 16.12
(a) Shell-and-tube, and (b) tube-in-tube heat exchangers.

refrigerant opposite the leaving fluid. One of the heat-exchanger designs that accomplishes this is the counterflow tube-in-tube heat exchanger, Fig. 16.12b.

Some codes prohibit heating potable water in heat exchangers such as those in Fig. 16.12, because if a leak should develop in the metal between the refrigerant and water, the water would be contaminated. To meet this problem, special *double-wall vented* heat exchangers are used, as illustrated in Fig. 16.13. The objective of the double-wall vented heat exchanger is to permit a leak of either refrigerant or water to pass into the intermediate space. Good thermal contact must be provided between the inner surface of the middle tube and the outer surface of the inner tube, a condition satisfied by constructing the vented space as a helical passageway, as in Fig. 16.13.

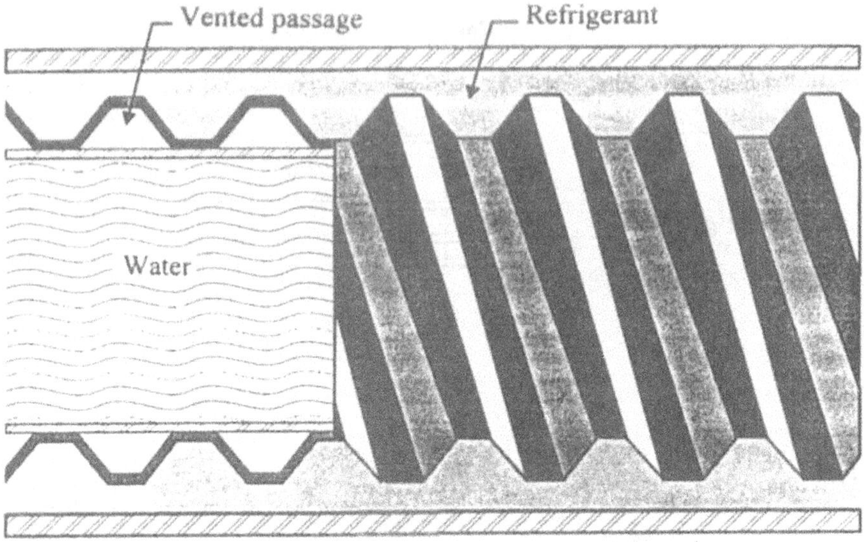

FIGURE 16.13
A double-wall vented heated exchanger.

16.9 ELEVATING THE CONDENSING TEMPERATURE TO RECOVER HEAT

When the temperatures of the superheated and condensing refrigerant are lower than needed for the heating requirement, the proposal is sometimes made to artificially elevate the condensing temperature. Two of the possible arrangements of equipment for elevating the condensing temperature of the refrigerant are shown in Fig. 16.14. In Fig. 16.14a, the refrigerant from the compressor passes first through the heat-reclaim heat exchanger before flowing to the condenser. A temperature control of the outlet process water regulates the capacity of the condenser, perhaps by cycling the fan(s) on the condenser. Another concept, shown in Fig. 16.14b, is to provide an auxiliary compressor that further compresses a portion of the discharge gas to serve the heat-reclaim heat exchanger. The concept in Fig. 16.14a entails lower first cost but is more costly to operate than the concept in Fig. 16.14b. In Fig. 16.14a, even the refrigerant not used in the heat-reclaim heat exchanger must be pumped to the high condensing pressure.

For the proposal of elevating the condensing temperature to be viable, the cost of electricity to drive the compressor must be less than the cost of fuel used for direct heating. As Fig. 16.15 shows, the cost of recovered heat when using an auxiliary compressor, as in Fig. 16.14b, is a function of the saturation temperature of the compressed refrigerant as well as the power cost. On the other hand, when the heat is supplied from a boiler, the target temperature of the heat supplied has little influence on the cost, at least in this range of temperatures.

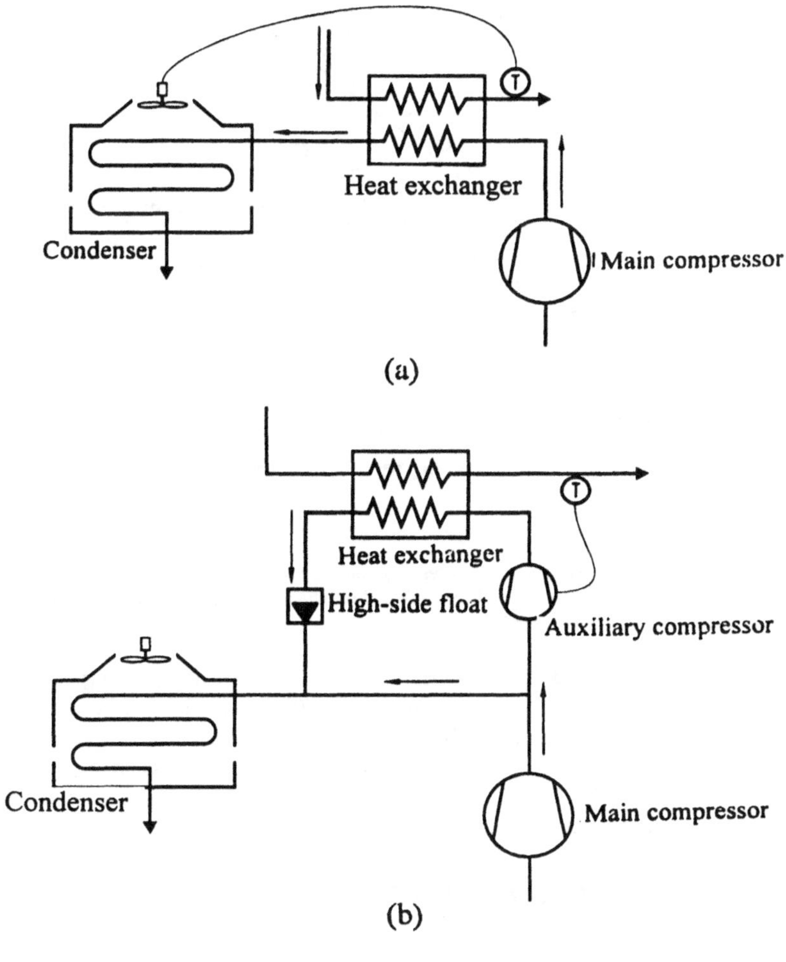

FIGURE 16.14
Elevating the condensing temperature to meet heating requirements with (a) the heat-reclaim heat exchanger in series with the condenser, and (b) an auxiliary compressor pumping only the vapor flow rate needed for heat recovery.

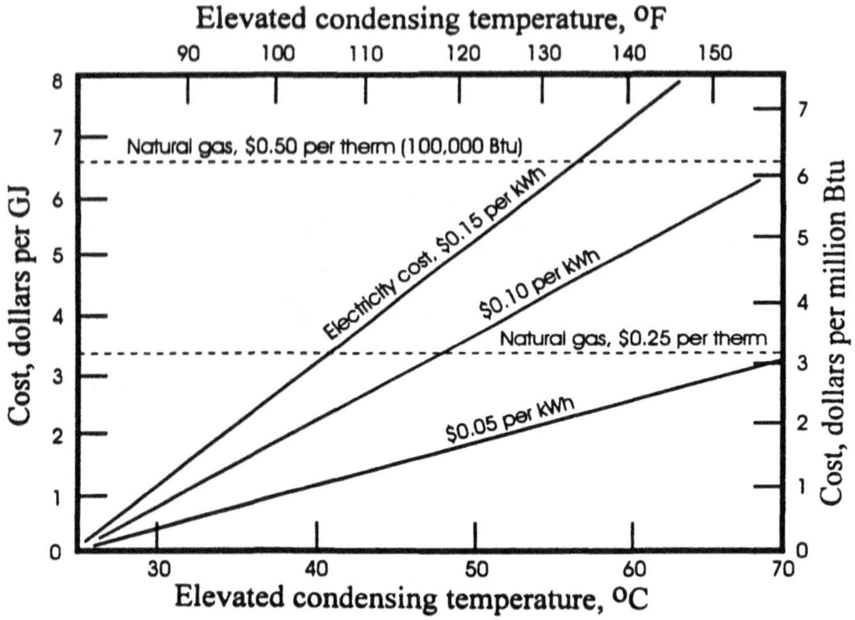

FIGURE 16.15
Cost of reclaiming heat by elevating the condensing temperature of ammonia from 25°C (77°F) in order to achieve a useful temperature level. (Based on the following efficiencies: compression, 75%; motor, 90%; and combustion, 80%).

The cost of reclaimed heat in Fig. 16.15 is based on pumping ammonia vapor from a condensing temperature of 25°C (77°F). If the condensing temperature is lower, the cost will be higher than that shown. If the condensing temperature of the entire flow rate of refrigerant is elevated, as in Fig. 16.14a, the cost shown in Fig. 16.15 should be multiplied by $1/F$, where F = fraction of total heat rejection that is reclaimed.

16.10 EVALUATING A POTENTIAL HEAT-RECLAIM PROJECT

Part II of this chapter on heat reclaim pointed out some of the considerations affecting the decision of whether or not to apply heat reclaim. Even though a superficial observation may suggest that *free heat* is available, the timing of the availability of this heat and that of the need must match, and the temperature levels must be compatible.

Virtually each situation is unique and must be analyzed individually. Some steps in the analysis include: (1) constructing plots of the daily profiles of magnitudes of individual sources and needs for heat, (2) for those sources and needs where the timing coincides, determining whether the temperature of the source(s)

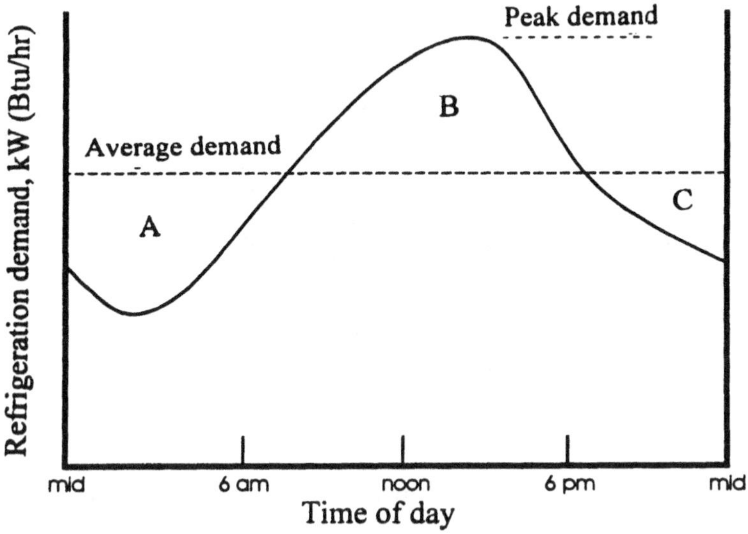

FIGURE 16.16
Typical refrigeration load variations during a day.

sufficiently exceeds the temperature level required by the need(s), and (3) for the surviving candidates analyzing the performance and the costs (investment cost, refrigerant pressure-drop penalty, etc.) and comparing the cost to the heating energy savings.

PART III. THERMAL STORAGE

16.11 LEVELING THE REFRIGERATION LOAD

It is a rare plant operation where the demand for refrigeration is constant throughout the entire daily cycle. Instead, as shown in Fig. 16.16, there is a variation in load with peaks and valleys. If there is no refrigeration storage, the plant must have the refrigerating capacity to meet the peak demand, and then it operates at reduced capacity the remainder of the day. To even out these peaks and valleys, thermal storage may be applied in such a way that the refrigeration plant operates at constant capacity throughout the entire day. In this way, the excess refrigeration capacity represented by A and C in Fig. 16.16 is stored and shifted to the time of high demand to help meet the requirements of period B.

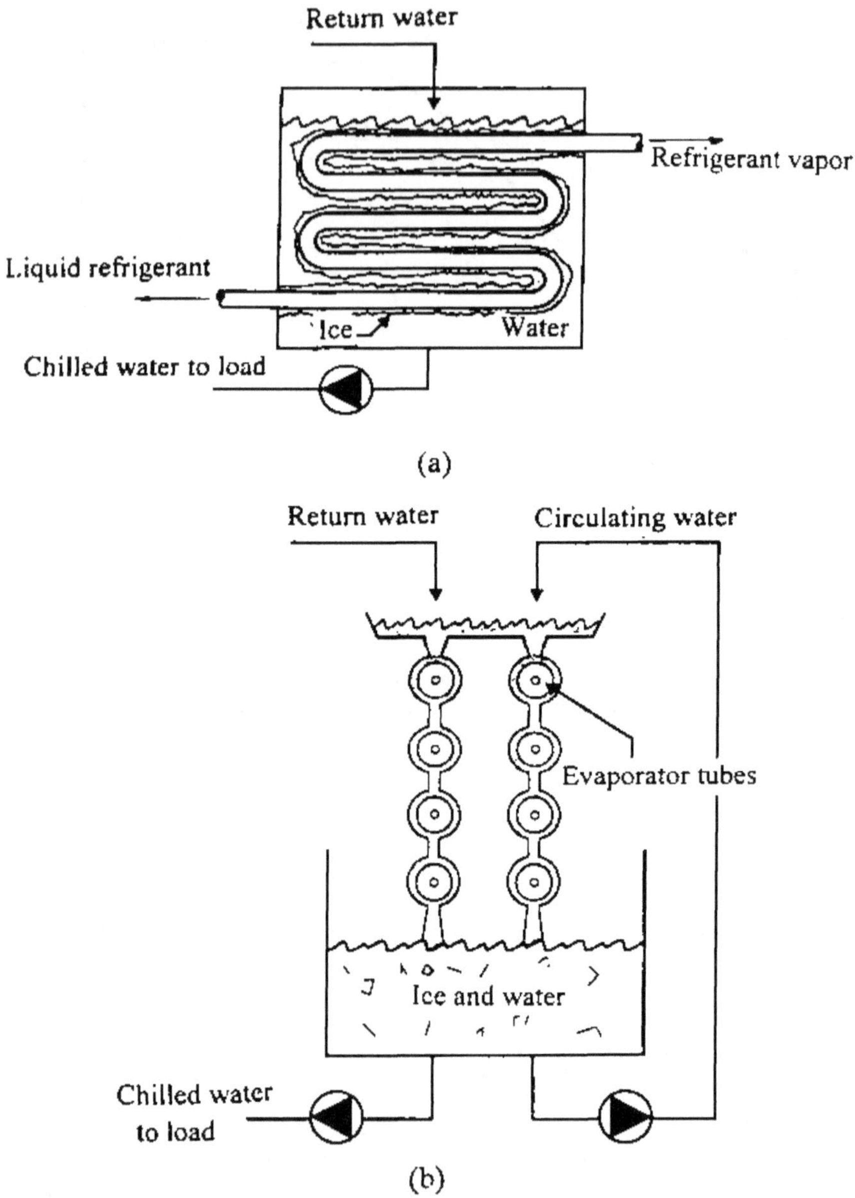

FIGURE 16.17
Two thermal storage concepts using ice: (a) the ice builder, and (b) the ice harvester.

Because electrical demand charges are a dominant component of the electricity expense, another strategy of operation is to run the refrigeration plant at high capacity during periods of low demand charge and reduce the refrigeration rate during periods of high demand charge.

Several concepts of thermal storage are shown schematically in Figs. 16.17 and 16.18. The two prominent thermal storage devices that use ice are shown in Fig. 16.17—the ice builder and ice harvester.

The ice builder of Fig. 16.17a consists of a tank of water in which tubes carrying evaporating refrigerant or cold brine are immersed. During the freezing phase, the tubes are cold enough that liquid water in the tank freezes on the outside surface of the tubes. This freezing usually takes place when the flow rate of water and/or its return temperature are low. The ice typically builds up to thicknesses of 50 or 60 mm (2 to 2-1/2 in) using evaporating temperatures that may range between −10 and −3°C (14 to 27°F). During periods of high refrigerating load, the return water temperature and/or its flow rate are so high that ice melts from the tubes and in so doing provides some or all of the refrigerating capacity, depending on how much capacity is supplemented by the refrigeration plant. The temperature of supply water leaving the ice builder during the thawing phase is typically from 1 to 3°C (34 to 37°F), depending on the heat load on the chilled water.

The ice harvester[4] concept, shown in Fig. 16.17b, achieves thermal storage by operating what is essentially an ice maker. A layer of ice 3 to 10 mm (1/8 to 3/8 in) thick freezes on cold plates in which evaporator tubes are imbedded. To harvest the ice, refrigeration in the plates is interrupted and hot gas is supplied to the coils to melt a thin layer of ice at the metal surface, permitting the ice to slide to the bin below. The return water flows down the cold tubes and during heavy heat load melts the ice in the bin to recover the stored refrigeration. When the flow rate and/or the temperature of the return water are high, no freezing of ice occurs and the water is chilled by the evaporator tubes. A recirculating stream of water is available when ice is to be accumulated and there is little or no flow of return water.

In a refrigerated warehouse storing frozen food, shown schematically in Fig. 16.18a, the reduction in temperature of a room full of products by even one degree represents a sizeable energy quantity. Thermal storage can be accomplished by dropping the temperature of the space and products during periods of low electrical demand charge and allowing the temperature to climb to the original level during periods of high demand charge. The products themselves thus become the medium for thermal storage. Daily changes of a degree or two in the temperature of a product wrapped in vapor-impermeable film seem to cause no noticeable deterioration in quality[5]. The practice is not recommended for fish and ice cream, and the level of temperature during the fluctuations must remain below the level at which the quality of food degenerates. Varying storage temperatures seems to be more applicable for short-term storage of frozen food such as occurs in grocery distribution centers than it is in long-term storage that might occur in many refrigerated warehouses.

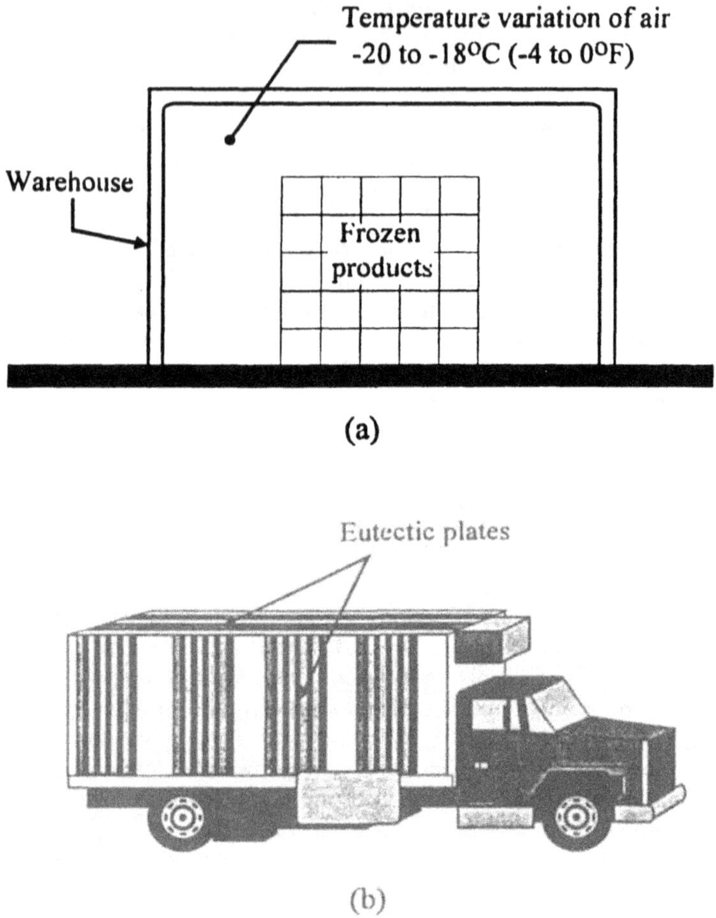

FIGURE 16.18
Thermal storage through (a) the daily variation of temperature of stored products, and (b) eutectic plates.

To provide thermal storage at temperatures other than the freezing point of water, eutectics may be applied. A traditional use of eutectics has been in truck refrigeration[6], as shown in Fig. 16.18b, where the roof and/or walls of the inside of the truck are lined with thin packs containing eutectic material. This material is chosen to freeze and melt at a temperature of perhaps −18°C (0°F). After deliveries have been made and the truck has returned to the depot, the eutectic material is frozen by supplying the evaporator tubes in the plates with low-temperature refrigerant. During delivery of the load of products, the melting of the eutectic material provides the refrigeration.

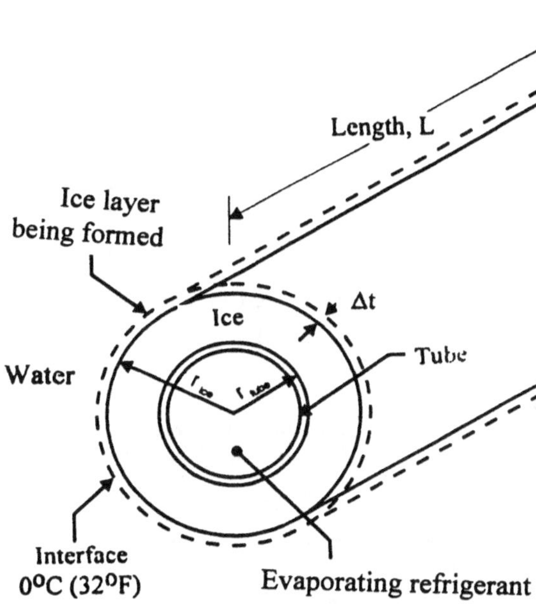

FIGURE 16.19
The heat removed to freeze the thin layer of ice flows by conduction through the ice to the tube and then to the evaporating refrigerant.

16.12 THE ICE BUILDER IN ITS FREEZING PHASE.

We return now to the popular ice builder shown in Fig. 16.17a. Achieving the required magnitude of thermal storage in the specified time to meet the anticipated refrigeration load constitutes a delicate balance of choices. The magnitude of thermal storage is proportional to the mass of ice frozen and is therefore determined by the tube diameter, ice thickness, and length of tubes in the ice builder. To freeze this quantity of ice in the specified time requires certain combinations of evaporating temperatures and refrigerating capacity. Then, during the meltoff phase, the design of the ice builder, the flow rate of water, and the temperature of the return water all influence the rate of melting. A further complication in predicting the rate of melting is that it is possible and sometimes efficient to continue operation of the refrigeration plant during part or all of the meltoff phase.

The complications in predicting the operation of an ice builder make it advisable to consult the manufacturer, especially since manufacturers now have accumulated design and field data. One of the predictions that can be made analytically that compares quite well with field performance is the rate of ice

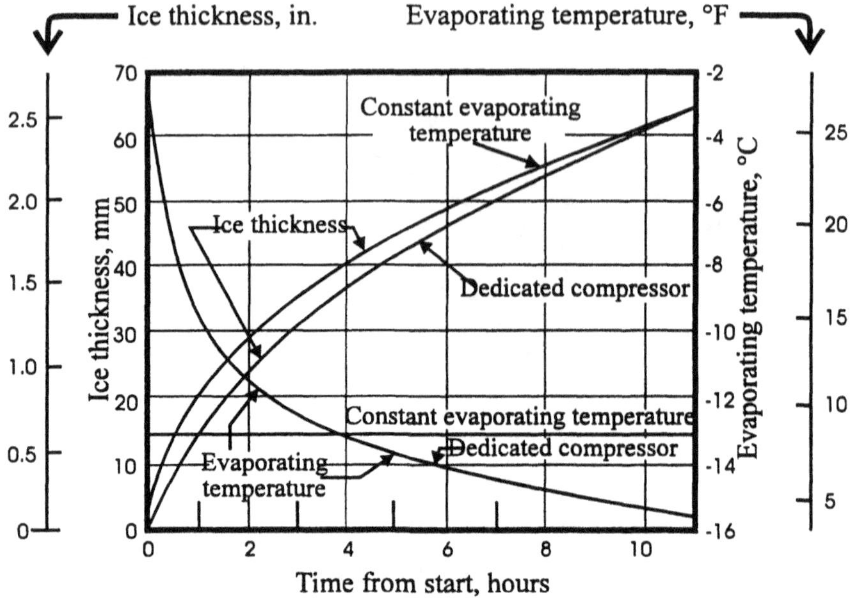

FIGURE 16.20
Rates of ice buildup and profiles of evaporating temperatures for an ice builder in one case served by a dedicated compressor and in the other by a constant evaporating temperature of −13.2°C (8.2°F).

buildup[7]. This calculation is a numerical integration performed on a computer and would not be a computation every designer or purchaser could justify making. Our purpose will be to show the rates of ice buildup, both using a dedicated compressor and a constant evaporating temperature. Knowledge of these trends can aid in selecting an appropriate ice builder.

The mathematical model for predicting the rate of ice buildup is illustrated in Fig. 16.19 and is basically a procedure of calculating the increase in ice thickness, Δr, during a short time increment. The temperature of the interface between water and ice is assumed to be 0°C (32°F). The rate of heat removal from the ice must match the rate conducted through the ice and also match the rate of heat removal by the evaporating refrigerant. After calculating one time increment, the radius of the ice cylinder r_{ice} has increased, and the new value of r_{ice} applies to the calculation in the next time increment.

The model can be used to compare the performance of an ice builder served by a dedicated compressor with an ice builder connected to a large refrigeration system in which the evaporating temperature remains essentially constant. Figure 16.20 depicts the growth of the ice thickness and the profile of the evaporating temperature during the 11-hour buildup phase. The final thickness of ice is 63.5

mm (2.5 in). The compressor is a 4-cylinder unit and the condensing temperature is 30°C (86°F). The tubes in the ice builder have an outside diameter of 31.26 mm (1.23 in) and are a total of 900 m (2,950 ft) long. To build up 63.5 mm (2.5 in) in 11 hours would require a constant evaporating temperature of −13.2°C (8.2°F). Several observations can be made from Fig. 16.20. The rate of ice buildup is rapid at first and then drops off as the thickness increases. The major thermal resistance in the heat-transfer process is in the ice.

One of the tradeoffs in designing the thermal storage system is whether to operate for more hours with an ice builder with shorter tube length or to buy a larger ice builder and obtain the desired total mass of ice with a smaller final ice thickness. The dedicated compressor required several percent less compessor energy during the buildup phase, namely 390.2 kWh vs. 395.9 kWh for the constant temperature operation. The reason is that the dedicated compressor freezes a large portion of the ice in the early stages at a high evaporating temperature. The freezing rate decreases as the ice becomes thick, because of the high thermal resistance of the ice.

While the dedicated compressor is slightly more efficient than the operation at constant evaporating temperature, the improvement is normally not enough to justify providing a dedicated compressor when refrigeration capacity is available from a central plant at an appropriate evaporating temperature. The designer would start with the available evaporating temperature and work with the manufacturer of the ice builder to achieve the desired magnitude of thermal storage in a length of time that fits the load profiles of the plant.

16.13 SAVINGS IN ENERGY COSTS TRANSLATABLE TO PROFITS

An industrial refrigeration plant requires energy to operate, but it is the magnitude of that cost that deserves the attention of the designer, manager, and operator of the facility. When the plant is operating on a low profit margin, as is the usual case, a given percentage saving in energy may amplify to several times that percentage improvement in profits. Two requirements for an effective energy program in the plant are: (1) knowledge of the methods of conserving energy, and (2) diligence in implementing these measures. In the case of electrical energy, the demand charge may constitute one-fourth to one-third of the total bill. For this reason, creative ways of shifting the time of energy consumption will also pay dividends.

REFERENCES

1. Balbach, G. C., "Life-Cycle Cost of a Cold Storage Door," Proceedings of the Annual Meeting of the International Institute of Ammonia Refrigeration, March 1982.
2. Cole, R. A., "Heat Recovery from Industrial Refrigeration Systems," *Heating/Piping/Air Conditioning*, 56(11): 39-44 and 91-98, November 1983.

3. Villadsen V. and F. V. Boldvig, "Heat Recovery and Heat Pump Applications in Ammonia Plants," Proceedings of the Annual Meeting of the International Institute of Ammonia Refrigeration, March 1982.
4. Knebel, D. E., "Economics of Harvesting Thermal Storage Systems: A Case Study of a Merchandise Distribution Center," *ASHRAE Transactions*, 94(Part 2): 1894–1904, 1988.
5. Moleeratanond, W., A. Kramer, B. H. Ashby, W. A. Bailey, and A. H. Bennett, "Effect of Temperature Fluctuation on Energy Consumption and Quality Changes of Palletized Foods in Frozen Storage," *International Journal of Refrigeration*, 2(6): 221-227, November 1979.
6. Lauterbach, W. E., "Plate/Blower-Coil Systems for Low Temperature Trucks," *ASHRAE Journal*, 15(3): 55-60, March 1973.
7. Richards, W. V., "Performance Rating of Static Ice Builder with Dedicated Compressor," Proceedings of Commission B, 17th International Congress of Refrigeration, International Institute of Refrigeration, Vienna, Austria, 1987.

CHAPTER 17

THE REFRIGERATION AND FREEZING OF FOOD

17.1 WHAT THE REFRIGERATION ENGINEER NEEDS TO KNOW

The refrigeration and freezing of food are probably the largest applications of industrial refrigeration. The field of food processing starts at the point of harvest and extends to its consumption, embracing such operations as transportation, processing, sanitation, and packaging, as well as refrigeration. A wide range of specialists contribute to the effective processing of foods, and while additional knowledge is always helpful, the refrigeration technologist may not have depth in such fields as the microbiology of foods, which is the strength of food technologists and scientists.

Maintaining certain conditions is a fundamental responsibility of the refrigeration engineer, and the conditions over which he or she may have some control are: (1) temperature of the air or other fluid surrounding the product, (2) the relative humidity of air, particularly for the refrigeration of unfrozen foods, (3) the air velocity, and perhaps (4) the chemical environment, such as a CO_2 atmosphere. The refrigeration engineer should know what conditions are optimum for food products and should be skilled in maintaining these conditions. Another important capability is how to compute the refrigeration load contributed by the product—an activity that leads naturally to the load calculations studied in the next chapter.

Knowledge of desirable storage conditions (especially temperature and relative humidity of air in the storage space) is fundamental information, and it is also quite extensive. It is more practical to point out where this information can be readily obtained than to try to reproduce the entire body of information, which would extend over a hundred pages.

A widely used source of basic data is the ASHRAE *Handbook—Refrigeration Systems and Applications*[1], hereafter referred to as the ASHRAE *Refrigeration Handbook.* This reference is readily available to most practitioners in the field. The specifications of storage conditions provided in this chapter are examples and excerpts from the complete tables. An indication of the extent of data available in the ASHRAE *Refrigeration Handbook* is that there are separate chapters devoted to commercial freezing methods, microbiology of foods, methods of precooling fruits and vegetables, meat products, poultry products, fishery products, dairy products, deciduous tree and vine fruits, citrus fruits, vegetables, fruit juice concentrates, precooled and prepared foods, bakery products, candies, meats, beverages, and egg and egg products.

This chapter first addresses the required storage conditions of unfrozen food products, especially the temperature and relative humidity, and points to sources of such information. This section is followed by a presentation of recommendations on how to predict cooling rates. The chapter then moves to the freezing process and the equipment used for commercial freezing. Next are guidelines for predicting freezing rates of products, and finally freezing with cryogenic materials is described.

17.2 STORAGE TEMPERATURES RECOMMENDED FOR UNFROZEN FRUITS AND VEGETABLES

The optimum storage temperature of the product being stored is fundamental information to the refrigeration engineer. A recommended source is the ASHRAE *Refrigeration Handbook*, which provides information on a variety of fruits and vegetables, a sample of which is shown in Table 17.1. The selection of products shown in Table 17.1 indicates that each product has a unique ideal storage temperature. Also shown in such tables are the approximate storage life.

In addition to the key data on storage temperatures, humidities, and storage life, as shown in Table 17.1, the ASHRAE *Refrigeration Handbook* also presents one or more paragraphs of individual recommendations for each of a number of fruits and vegetables, such as the samples shown below. The original source for most of this information is a U.S. Department of Agriculture publication[2].

Green peas, 0°C (32°F) and 95-to-98% rh.

Green peas lose part of their sugar rapidly if they are not refrigerated

TABLE 17.1
Recommended storage conditions for some fruits and vegetables[1].

Commodity	Storage temperature, °C (°F)	Relative humidity, percent	Approximate storage life
	Fruits		
Apples	0 to 3 (32 to 37)	90–95	3–8 months
Honeydew melons	5 to 10 (41 to 50)	90–95	2–3 weeks
Peaches	−0.5 to 0 (31 to 32)	90–95	2–4 weeks
Pears	−1.6 to −0.5 (29 to 31)	90–95	2–7 months
Strawberries	−0.5 to 0 (31 to 32)	90–95	5–7 days
	Vegetables		
Beans, green	4 to 7 (40 to 45)	90–95	7–10 days
Broccoli	0 (32)	95–100	10–14 days
Lettuce, head	0 to 1 (32 to 34)	95–100	2–3 weeks
Peas, green	0 (32)	95–98	1–3 weeks
Potatoes, early	3 to 4 (37 to 39)	90–95	4–5 months
Seed, vegetable	0 to 10 (32 to 50)	50–65	10–12 months
Tomatoes, ripe	7 to 10 (45 to 50)	90–95	4–7 days

promptly after harvest. They should keep in salable condition one or two weeks at 0°C (32°F). Top icing is beneficial in maintaining freshness. Peas keep better unshelled than shelled.

Apricots, 0 to 1°C (32 to 34°F) and 90-to-95% rh.

Apricots are not stored for a prolonged time but may be held for two or three weeks if they are picked at a maturity firm enough that they will not bruise. Unfortunately, this maturity does not yield a good dessert-quality fruit. Care must be used in sizing and packing the fruit going into storage, as small surface bruises can become infected with disease-producing organisms.

A general observation from Table 17.1 is that the optimum temperature and humidity vary from one commodity to another, so the refrigeration plant serving facilities that store a variety of products throughout the year must be flexible. Also, the optimum storage temperatures for several of the fruits listed in the table are below 0°C (32°F), which requires rather precise control, because if the temperature of the product drops several degrees below the target temperature, the product could freeze.

Bananas are an important product that require environments different from most other fruits. Bananas must be green when shipped to market so that they will not soften or sustain injury during handling[2]. Basically, green bananas are held at 13°C (55°F), while temperatures below 12°C (54°F) are to be avoided. When the bananas are to be ripened, the temperature is brought up to about

18°C (64°F), held for several days, and then brought back to 13°C (55°F) for any continuation of storage. During ripening the relative humidity should be between 90 and 95%.

17.3 RELATIVE HUMIDITY

For the storage of unfrozen fruits and vegetables, the relative humidity is the second most important condition the refrigeration engineer must maintain. Rapid degradation of the product can occur if the relative humidity is appreciably different from the recommended value. Table 17.1 presents desired relative humidities for the several example products. If the relative humidity is too low, most fruits, vegetables, and cut flowers will suffer wilting or shriveling, as well as a loss of weight. Also, the vitamin C content of green vegetables decreases more readily when the product is stored at conditions that result in wilting. At a given relative humidity, the dehydration process is more rapid at higher storage temperatures compared to lower ones. The reason is that the driving force for moisture removal is the difference in water vapor pressure between the liquid water in the product and that of the air. When, for example, the product and surrounding air are both at a temperature of 2°C (35.6°F) and the relative humidity of the air at 90%, the driving force is [705.5 − 0.9(705.5)] = 70.6 Pa (0.010 psi). These vapor pressures are found from pressure-temperature tables of water. If the product and air are at a temperature of 8°C (46.4°F), the driving force is [1,072 − 0.9(1,072)] = 107.2 Pa (0.0155 psi). This calculation assumes that the water in the product is pure; actually it is in solution and therefore exerts slightly lower water vapor pressures.

Desirable relative humidities for most fruits and vegetables are in the 90–95% range. The higher bracket of this humidity range is preferable for some fruits, such as apples, pears, and berries. Some vegetables that are preserved better with high humidity, even approaching 100%, are leafy vegetables, such as lettuce, cabbage, and celery, as well as sweet corn, radishes, melons, and many others. Some vegetables require humidities lower than the 85–95% range, such as dried onions, beans, squash, and pumpkins.

Another application where high humidities are not permitted is in meat-cutting rooms. The problem to be avoided is water vapor from the moist air in the room condensing on cool structural members and then dripping on the product[3].

Based on their research, fruit and vegetable specialists recommend high relative humidities for certain products. There is a strong suspicion that many storage facilities are operating at lower humidities than optimum, because the system is not properly designed and operated. Table 17.2 summarizes recommendations for the selection and operation of evaporators when either high or low relative humidities are desired.

TABLE 17.2
Strategies for achieving high or low relative humidities.

Objective	Coil selection and operation	Extreme measure
High humidity	Large coil area and small difference in temperature between air and refrigerant	Moisten floors or operate humidifier
Low humidity	Small coil area and large difference in temperature between air and refrigerant	Add sensible heat

17.4 COOLING RATES WITH FORCED AIR COOLING

The period of time that unfrozen fruits and vegetables remain in storage is usually short compared to that for frozen foods. The short storage period is associated with frequent transfers of product in and out of the refrigerated space. Ideally, the temperature of the incoming product will be the same as the recommended storage temperature. Rarely is this the case, so the initial treatment of the incoming product is to drop its temperature. Most storage rooms designed for holding product have neither the refrigeration capacity nor the intensity of air movement for rapid removal of field heat[2]. Instead, rapid precooling, as will be explained in this and the next several sections, is the first step in good temperature management. Delays at high temperature between harvest and precooling cause deterioration of the product.

An examination of the cooling process identifies variables over which the refrigeration system has some control and those over which it has no control. The critical location in the product is its center, because if the temperature there can be brought down to the recommended level, the remainder of the product will be at a low enough temperature. Basically, air temperature, air velocity over the product, and the interior dimensions and thermal characteristics of the product are the variables that determine the rate of cooling when using forced-air cooling. The temperature and velocity of the air are affected by the refrigeration system, and the interior characteristics of the product are outside the control of the plant operator.

Plant operators frequently want to calculate the cooling rate for a particular product to ensure the recommended temperature is achieved in a specified time. A number of technical papers have been written on this subject, but the process is so complex there may never be practical means for accurate predictions. Instead, a combination of analytical and experimental results for the specific product and package configuration may be the most practical approach. The plant operator is able to measure the air temperature and hopefully control its value. The operator also may be able to increase or decrease the air velocity, but he or

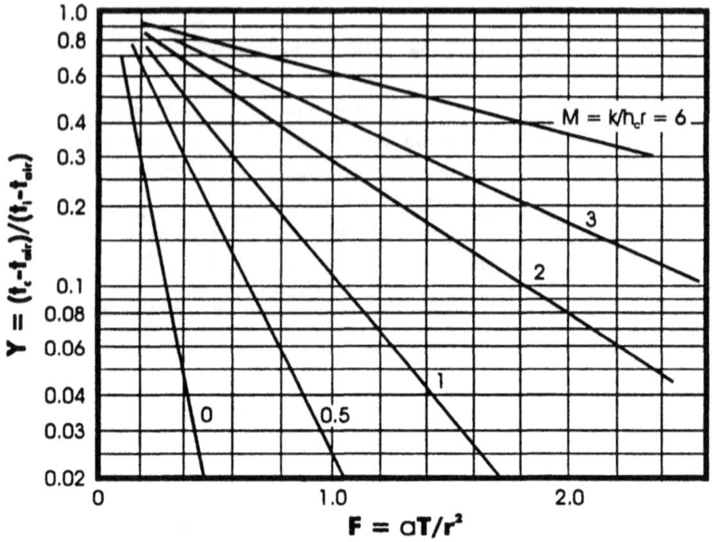

FIGURE 17.1
Cooling rate curve.

she might be hard-pressed to predict convection coefficients of the air over the surface of the product because of the complex geometry. With respect to the interior characteristics of the food product, these are functions of the thermal conductivity, specific heat, density, and shape of the product, as well as the manner in which the individual units are packed.

The complexity of the prediction may suggest throwing up one's hands and only measuring a few actual center temperatures of the product in the refrigerated storage room to acquire data for future operations. There could, however, be several benefits of examining idealized cooling. An example of a classical idealized cooling chart is the Gurney-Lurie graph[4], as shown in Fig. 17.1, which is applicable to spheres of homogeneous composition. The various factors affecting the calculation are incorporated into dimensionless groups Y, F, and M:

$$Y = (t_c - t_{air})/(t_i - t_{air})$$

$$F = \frac{kT}{\rho c_p r^2} = \frac{\alpha T}{r^2}$$

$$M = \frac{k}{h_c r}$$

where:

t_c	=	center temperature at time T, °C (°F)
t_{air}	=	air temperature, °C (°F)
t_i	=	initial center temperature, °C (°F)
k	=	thermal conductivity, W/m·K (Btu/hr·ft·°F)
T	=	time from start of cooldown, seconds
ρ	=	density of product, kg/m^3 (lb/ft^3)
c_p	=	specific heat of product, kJ/kg·K (Btu/lb·°F)
r	=	radius of sphere that idealizes the shape of the product, m (ft)
α	=	thermal diffusivity = $k/\rho c_p$, m^2/s (ft^2/hr)
h_c	=	convection coefficient, air to surface of the product, W/m^2·K (Btu/hr·ft^2·°F)

The shape of most food products is not precisely spherical, the convection coefficient, h_c, is not uniform over the entire sphere, and since the food product is probably not homogeneous, the internal properties, k, ρ, and c_p are difficult to determine. Figure 17.1 provides some useful insight, however, independently as well as in conjunction with a limited number of field measurements.

First consider the convection coefficient h_c. Doubling h_c reduces M by one half, which in turn changes F and the cooling time by a factor of about 0.6, regardless of the other conditions. An increase in h_c increases the cooling rate, but a cost penalty also results. Practically the only way to increase h_c is to increase the air velocity, which entails additional fan power. In fact, to double h_c the air velocity and flow rate must be increased by a factor of $2^{1.2 to 1.4}$. Since the fan power varies by a ratio of the flow rate cubed, doubling h_c may require the power to increase by a factor of $2^{3.6 to 4.2}$, as shown in Fig. 17.2. Increases in air flow rate should be made judiciously.

Another variable that has considerable influence on the cooling rate is the radius of the product. What, for example, is the effect on the cooling time by reducing the radius of the product, r, by one half? Call the F value in Fig. 17.1 F_{orig} for the original radius, r_{orig}. Thus, $F_{orig} = \alpha T_{orig}/(r_{orig})^2$. If the radius is reduced by one half to $r_{orig}/2$, M doubles, which according to Fig. 17.1 results in a new abscissa value $F_{new} = F_{orig}/0.6$. Therefore:

$$F_{new} = \frac{\alpha T_{new}}{(r_{orig}/2)^2} = \frac{F_{orig}}{0.6} = \frac{\alpha T_{orig}}{0.6 r_{orig}^2}$$

$$T_{new} = \frac{T_{orig}}{(4)(0.6)} = 0.42 T_{orig}$$

The above equations show an appreciable reduction in cooling time. The question is how to achieve something equivalent to this reduction in radius in practical situations. For packaged products, instead of stacking the packages together, as in Fig. 17.3a, much more rapid cooling results when air spaces are provided at as many surfaces as possible, as in Fig. 17.3b.

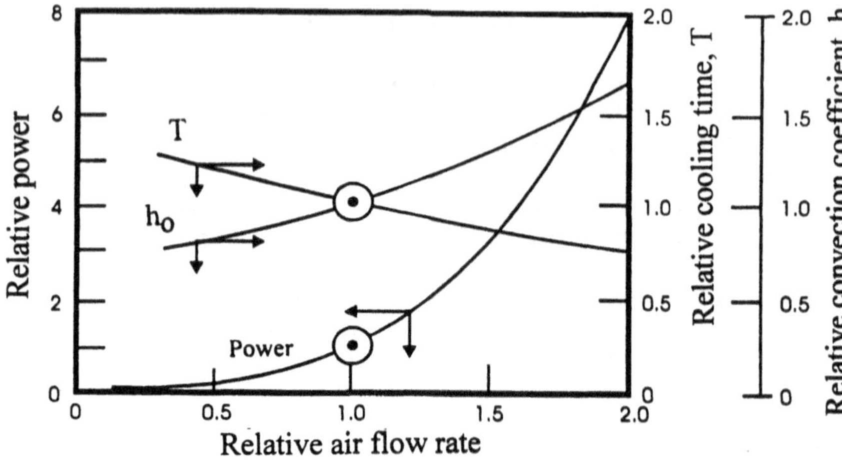

FIGURE 17.2
Influence of relative air flow rate on h_c, cooling time, and fan power.

A final cooling characteristic that can be derived from Fig. 17.1 is how to translate information from one data point on the cooling rate curve to another. If the center temperature at a given time after the start of cooling is known, the time required to achieve a lower center temperature can be predicted. The basis is a property of the shape of the lines on the semi-logarithmic plot of Fig. 17.1, in that they are almost straight lines radiating from the (Y=1, F=0) point. The lines are straight except for a slight curvature near the (1,0) point. Two points on any straight line emanating from (1,0) on a graph of these coordinates will show the same F_2/F_1 for a given Y_2/Y_1. The implication of this property of the graph is that since α/r^2 of a given product is constant, $F_2/F_1 = T_2/T_1$, regardless of the value of M.

A frequently used concept is the *half-cooling time*[5], which is the time required to drop the center temperature to midway between its original temperature and the air temperature. The half-cooling time is represented by the Y=0.5 horizontal line in Fig. 17.1, since along this line $t_c - t_{air} = 0.5(t_i - t_{air})$.

Example 17.1. A thermocouple is inserted into the interior of a crate of green beans to determine that the half-cooling time in a refrigerated storage room is 30 minutes. With the same product arrangement and same air flow rate and distribution, determine the time required to cool the center of the crate from 22°C (71.6°F) to 4°C (39.2°F) when the air temperature is 2°C (35.6°F).

Solution. The half-cooling time of the process where $F_1 = 0.5$ is 30 minutes. For the new situation, $F_2 = (4-2)/(22-2) = 0.1$. Choosing any of the lines on Fig. 17.1, for instance M = 1, $F_{Y=0.5} = 0.4$, and $F_{Y=0.1} = 1.05$, so the cooling time to drop the temperature to 4°C (39.2°F) is (30 min)(1.05/0.4) = 79 minutes.

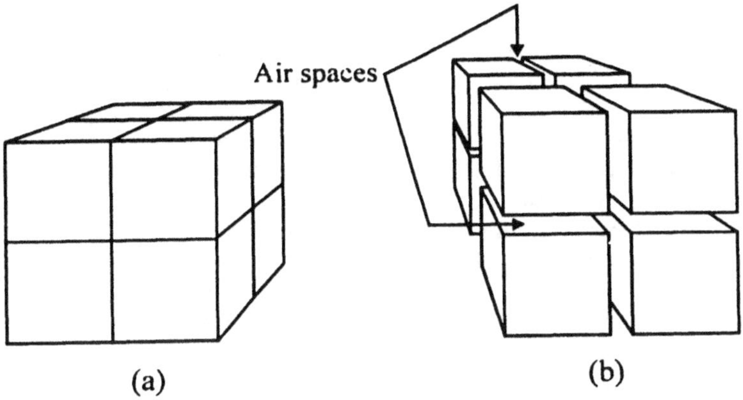

FIGURE 17.3
Shortening the cooling time for packages by changing from (a) a compacted arrangement to (b) providing air spaces between the packages.

17.5 COOLING AND STORAGE OF UNFROZEN MEAT AND POULTRY

Two dominant concerns occupy the operator of meat and poultry plants in the time interval from the receipt of the warm slaughtered product until it is processed and ready either for freezing or storage in the unfrozen state. Those concerns are: (1) proper chilling, including proper maintenance of temperatures and relative humidities, and (2) adhering to rigid sanitary standards. Lack of attention to either requirement can result in a product that is of poor quality or, worse yet, unsafe. We continue to emphasize the refrigeration side of the operation, but the strictest sanitary conditions are mandatory for the building, equipment, workers, water supplies, wash-up facilities, and the disposal of waste material.

Because of the differing sizes of the animals, the processes applicable to poultry are different from those for beef and pork. There are a variety of approaches to cooling beef and pork, but the basic procedure is to first cool the carcass, then cut it into smaller parts that are then enclosed in air-tight packages. These packages are then usually placed in boxes, perhaps for shipping and sale as unfrozen cuts or to be frozen. Beef carcasses are typically chilled overnight from the receiving temperature of about 40°C (104°F) to about 8°C (46°F). The carcasses then move to a holding chiller where the air temperature is about 0°C (32°F) for cooling the beef down to about 4°C (39°F). Hog carcasses, on the other hand, are smaller than beef and usually can be chilled to 4°C (39°F) overnight without the need for a holding chiller. Cutting and processing rooms are usually maintained at about 10°C (50°F), which is a compromise between the lower temperatures desirable from the standpoint of preserving the product and the higher temperatures that would be more comfortable for the workers.

An alternative to cooling the entire carcass is the practice of cutting, deboning, and packaging the smaller pieces while the meat is still warm. These smaller cuts can generally be cooled overnight without the need for the holding chiller.

The small size of birds relative to hog or beef carcasses permits a different strategy for cooling poultry compared to the larger animals. The regulation[6] of the United States Department of Agriculture requires that birds weighing 1.8 kg (4 lb) or less must be cooled to 4.4°C (40°F) in 4 hours or less. Poultry weighing between 1.8 and 3.6 kg (4 and 8 lb) must be cooled to 4.4°C (40°F) within 6 hours, and any products larger must be cooled within 8 hours. The most popular method of cooling poultry is by immersion in chilled water[7] or slush ice. The facility using chilled water can be configured into a continuous process where the birds are pushed counterflow to water that enters the trough at a temperature of about 1 to 2°C (33.8 to 35.6°F). Because of the direct contact of the chilled water with the product and the reuse of a certain fraction of the water, guarding against microbial contamination must be pursued vigilantly. An advantage of direct immersion cooling is that dehydration of the poultry is minimized. It is even possible for the poultry to absorb water, but the extent of this absorption is limited by law.

Returning to the cooling of beef and hog carcasses, the refrigeration plant operator attempts to cool the carcasses as rapidly as possible. At the same time, the operator must restrict the amount of dehydration to perhaps 1% of the product weight. Loss of a kg or pound of water is the same as an equivalent loss of meat. The rate of cooling of a carcass varies with air temperature, air velocity, configuration of the cooling room, and the size of the carcass. Some typical cooling rates[8] for beef are shown for two different carcass weights and several different air velocities in Figure 17.4. The data in Figure 17.4 apply to the deep leg temperature, which is a position very difficult to cool. The pattern of the straight lines on the semi-logarithmic plot resembles the classic Gurney-Lurie chart first shown in Figure 17.1, however, the dimensionless temperature ratio shown on the ordinate in Figure 17.4 is $(t_{product} - t_{air})/(t_{initial} - t_{air})$.

Along with the need to cool the carcass rapidly, a companion requirement is to limit the dehydration of the product. A frequent practice is to spray the carcasses with water once or twice during the cooling period. There are two beneficial purposes served by the water spray: inhibiting dehydration and aiding the cooling by the evaporation of the water.

17.6 REMOVAL OF FIELD HEAT FROM FRUITS AND VEGETABLES BY HYDROCOOLING

At harvest time an objective of the growers of fruits and vegetables is to remove field heat and ship the product as rapidly as possible. The three principal methods of post-harvest refrigeration are: (1) air cooling, (2) hydrocooling, and (3) vacuum cooling. Air cooling is similar to that used for other products, such as meats, as described in Section 17.5. The hydrocooling process brings chilled

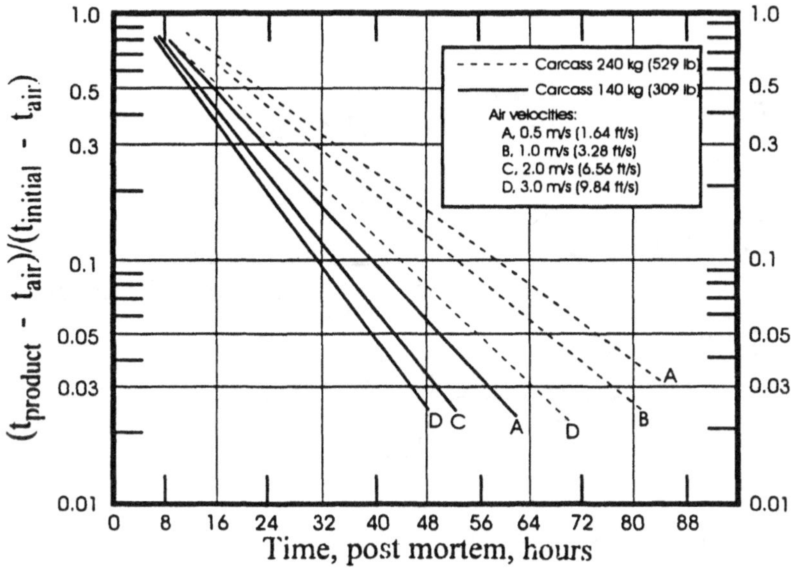

FIGURE 17.4
Cooling charts for two different beef carcass sizes with several different air velocities[8].

water into direct contact with the product either by spraying the water on the product or by immersing the product in the water. Products to which hydrocooling is usually applied include peaches, plums, nectarines, cherries, sweet corn, celery, radishes, and carrots[9].

The basic nature of the hydrocooling process has important implications for the refrigeration professional. A dominant fact is that rapid cooling is essential to preserve the quality of highly perishable fruits and vegetables. Slower cooling methods usually permit simpler and less costly facilities, but rapid cooling allows the shipper to cool and load the produce the day of the harvest. A comparison of the speed of cooling peaches by several different methods is shown in Fig. 17.5. Hydrocooling is generally able to cool the product within an hour.

What are the means available in the design and operation of the refrigeration plant to achieve rapid cooling? Only one variable can influence the cooling rate, and this variable is the water temperature. In air cooling, for example, the velocity of the air over the product influences the convection coefficient and thus the rate of cooling. In the case of hydrocooling, the convection coefficient between the water and the surface of the product is so high that the surface temperature and water temperature are essentially equal. With a given surface and water temperature, the thermal and geometric characteristics of the product regulate the internal cooling rate.

Addressing the one variable controllable by the refrigeration plant, both the entering and mean water temperature in contact with the product should be

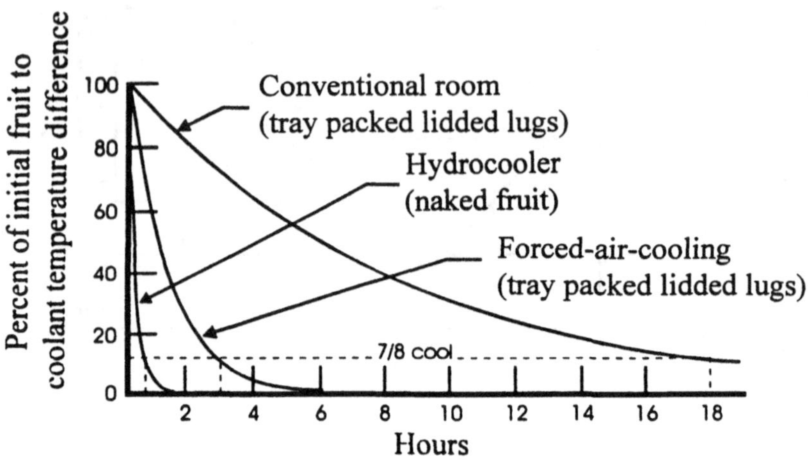

FIGURE 17.5
Comparative speed of cooling peaches by three different methods[10].

as low as possible. This means that a typical water chiller for air conditioning service operating with a 5°C (41°F) supply temperature and a 10°C (50°F) return temperature, for example, is not suitable for this assignment. Instead, a water chiller providing supply water at a temperature of 0.5°C (33°F) and a 1°C (34°F) return temperature is the target. A popular arrangement to achieve this objective is shown in Fig. 17.6.

From the standpoint of the refrigeration circuit, the water chiller shown in Fig. 17.6 is a flooded evaporator. A flooded evaporator is chosen in preference to a direct-expansion type, because the refrigerant should be at a constant low temperature throughout the evaporator. There will be no water warm enough to provide the superheat required by a superheat-controlled expansion valve to work properly. Since the water flows over the outside of the tubes instead of inside, there is no risk of bursting the tubes if some of the water freezes into ice. In fact, it is typical for a thin layer of ice to form on the exterior surface of the coil so that the water being chilled is in contact with a surface at 0°C (32°F). Temperature control is provided by an evaporator pressure regulator (EPR) that is set to provide an evaporating temperature several degrees below freezing. The thickness of the ice layer that may build up is self controlling, because with a given evaporating temperature, the ice surface temperature increases with the thickness of the ice. When the surface temperature reaches 0°C (32°F), the thickness of the ice layer remains constant.

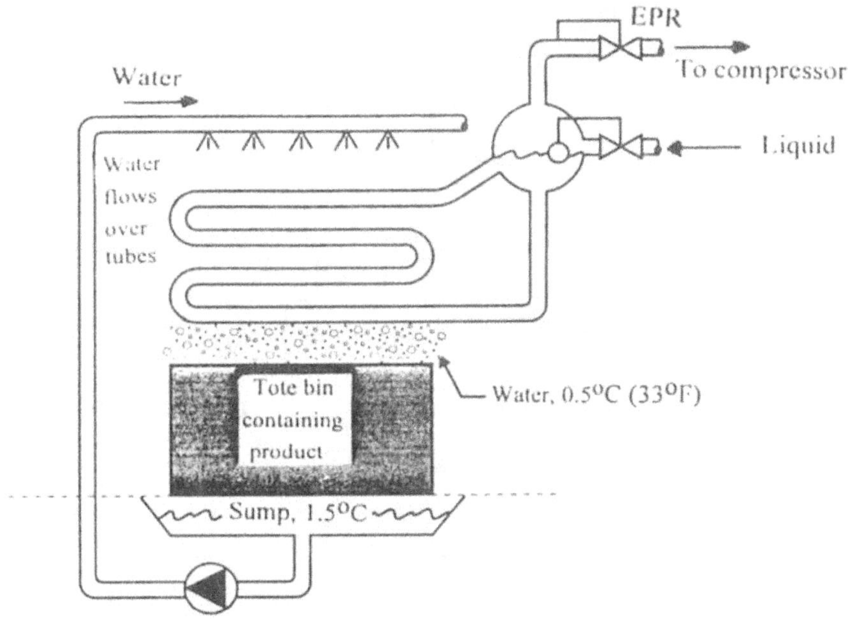

FIGURE 17.6
A chiller providing water temperatures approaching the freezing point to serve a hydrocooler.

17.7 VACUUM COOLING

Field heat can be removed from such leafy vegetables as lettuce, spinach, endive, parsley, and others using vacuum cooling. In vacuum cooling, the product is placed in an air-tight chamber which is then evacuated. When the partial pressure of the water vapor in the chamber drops below the vapor pressure of the water on the surface and in the tissue of the product, the liquid evaporates to cool the product. Vacuum chambers vary in size from small portable chambers that can accommodate one pallet to large chambers that can hold one or two truckloads. Moisture loss by the product during cooling may range from 0.5 to 5% without detriment to the product, but moisture losses higher than about 5% of the product weight are likely to result in some wilting. The time required for commercial vacuum cooling units to reduce the product temperature to the 0 to 1°C (32 to 34°F) range is about a half hour for lettuce to one or two hours for some other products.

The basic equipment for a vacuum cooling plant is a chamber and a vacuum pump. The vacuum pump removes the mixture of air and water vapor that initially is nearly 100% air, but as the cycle proceeds the mixture becomes primarily water vapor. In order to reach a saturation temperature of 0°C (32°F) of the moisture on the product, the water vapor pressure in the chamber must

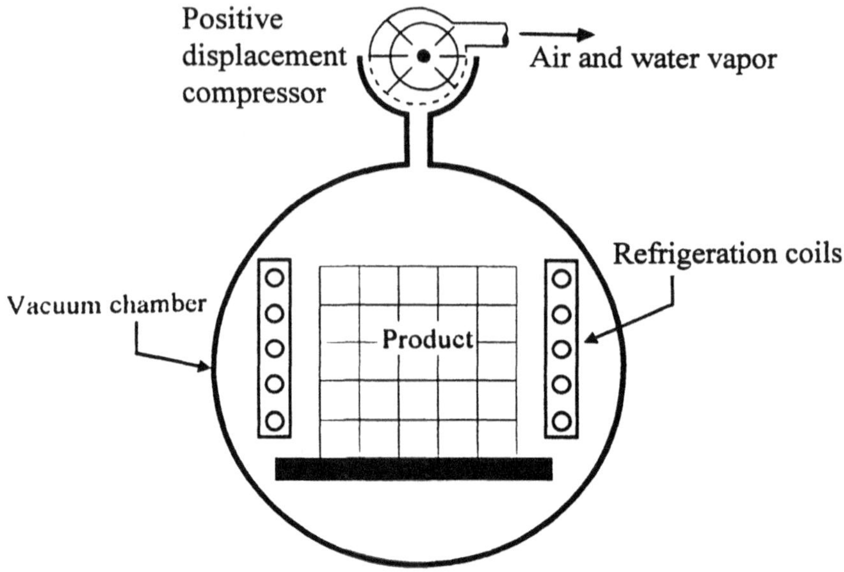

FIGURE 17.7
Vacuum cooling equipment combining refrigeration coils and a positive-displacement vacuum pump.

be reduced to 0.611 kPa (0.0886 psia or 0.18 inches of Hg), at which condition the specific volume of the water vapor is 206 m^3/kg (3,300 ft^3/lb). From a comparison of this specific volume with, for example, ammonia at −40°C (−40°F) that has a specific volume of 1.55 m^3/kg (25 ft^3/lb), it is obvious that the usual reciprocating and screw compressors used for refrigerants are not suitable for the vacuum pumping operation.

Three concepts of vacuum cooling are: (1) steam jet vacuum pump, (2) centrifugal compressor vacuum pump, and (3) positive displacement rotary or reciprocating vacuum pump in combination with refrigerated coils. The sequence of processes in the cycle when the steam jet or centrifugal vacuum pump are used is: (1) the reduction of total pressure, which reduces the partial pressures of both the air and the original water vapor in the chamber, (2) when the partial pressure of the water vapor drops to the saturation pressure of the water on the product, this water begins to vaporize, and (3) the final stage during which the partial pressure of the water vapor in the chamber continues to drop. The progressive drop in water vapor pressure permits the product moisture to continue to evaporate at progressively lower temperatures. The air in the chamber continues to be expelled, but the vaporization of water from the product continues to supply water vapor to the mixture. The steam jet and centrifugal compressor have passed out of favor because of their high cost and inconvenience.

Today, the most widely used concept for vacuum cooling is a system in which refrigeration coils are combined with a positive-displacement rotary or reciprocating pump, Fig. 17.7. If the temperature of the surface of the refrigeration coils is below the dew point of the air/water vapor mixture, which in turn is lower than the temperature of water on the product, water vaporizes off the product and then migrates to the cold refrigeration coils. In principle, the vacuum pump would not be needed at all, because if the coils were cold enough, the evaporative cooling process would take place even with the original mass of air present. Practically, however, the presence of air retards the rate of vapor migration from the product to the refrigeration coils and also forms a barrier to heat transfer at the surface of the refrigeration coils. The vacuum pump also removes some water vapor along with air, which assists the cooling process.

17.8 REFRIGERATION OF UNFROZEN DAIRY PRODUCTS

The major unfrozen dairy products include milk, cream, butter, and cheese. The special refrigeration concerns for ice cream and other frozen products is addressed in Section 17.12. The field of dairy product processing is a major topic in its own right, and the several paragraphs in this section concentrate only on the refrigeration needs and special requirements affecting the refrigeration plant. For a milk plant, the following refrigeration facilities would be expected:

- Coolers for incoming milk and raw milk storage
- Heat exchangers to chill the milk following pasteurization
- Milk storage room
- Heat removal from blow-molding machine if plastic containers are manufactured in the plant

Milk is cooled at the farm to about 5°C (41°F) and should arrive at the dairy no more than 1 or 2 degrees higher than that temperature. Because dairies normally cannot process the milk as rapidly as it is received, they are usually equipped with tall, insulated cylindrical tanks (silos) that are lined with stainless steel for storage of the raw milk. The incoming raw milk might be cooled before flowing to the silos, and in this case it may not be necessary for the silos to be refrigerated. When a silo is refrigerated, it is equipped with heat exchanger plates served with brine or evaporating refrigerant.

Another refrigeration requirement in a dairy is the final cooling of milk as it leaves the pasteurizer. Milk from the silo at 4 to 5°C (39 to 41°F) first flows through one side of a regenerative heat exchanger, into the pasteurizer where the milk is heated to at least 72°C (162°F) for 15 seconds, and finally back through the other side of the regenerative heat exchanger. Leaving the regenerative heat exchanger, the temperature of pasteurized milk will be approximately 32°C (90°F), and from this temperature the milk must be cooled with chilled water and possibly refrigerant. Chilled water is often called *sweet water* in a dairy, which distinguishes it from antifreeze, which is usually propylene glycol. Because the demand of chilled water is high for a short period of time, an ice thermal storage unit (Sections 16.11 and 16.12) is a standard piece of equipment in a dairy. The chilled water should be capable of lowering the milk temperature to about 4.5°C (40°F), and many dairies continue the cooling process to about 1°C (34°F) using propylene glycol or refrigerant at a temperature of −2°C (28°F). Care must be exercised not to operate the refrigerated surfaces at so low a temperature that milk freezes on them. The refrigeration capacity associated with the milk-cooling process is the product of the mass rate of flow of milk, the drop in temperature, and the specific heat. The specific heat of milk is 3.75 kJ/kg·K (0.90 Btu/lb·°F). The magnitude of the refrigeration load for cooling the milk leaving the pasteurizer is usually significant.

A popular type of heat exchanger used in dairies for cooling milk and other products is the *plate and frame* type. This heat exchanger consists of a series of vertical plates, spaced to provide channels for the flow of fluids. The channels for the flow of milk alternate with the channels for water or brine. The assembly of plates is bolted together so that it can be disassembled for cleaning after each production run.

Following pasteurization, the milk is packaged, which may increase its temperature by 0.5°C (1°F). At this point, the milk containers are stored in a refrigerated space before shipping. The packaged milk storage room is maintained between 0.6 and 4.5°C (33 and 40°F), and because of the low temperature of the entering milk, there is virtually no product load in this space. The refrigeration loads are those attributable to conduction through the enclosure, people, motors, and lights.

When the dairy packages milk in plastic bottles, these bottles are manufactured in the plant with blow-molding machines. The plastic must be heated to achieve the molding process, and this heat escapes into the space and must be removed by air-conditioning coils. The sizes of blow-molding machines vary, but an air-conditioning load imposed by each blow molder may be 100 to 150 kW (28 to 43 tons of refrigeration).

Refrigeration is an integral part of the cheese-making process. Different types of cheese undergo different processes, but for practically all cheeses the product is heated and must ultimately be cooled to temperatures in the 7 to 10°C (45 to 50°F) range. For example, after undergoing its chemical processes, mozzarella cheese[11] is at a temperature of 41°C (106°F). The cheese is then cooled by immersing molds containing the molten cheese in chilled water or brine. Further cooling takes place by floating the loaves of the semi-hardened product through chilled NaCl brine, a process that also introduces the required salt to the product.

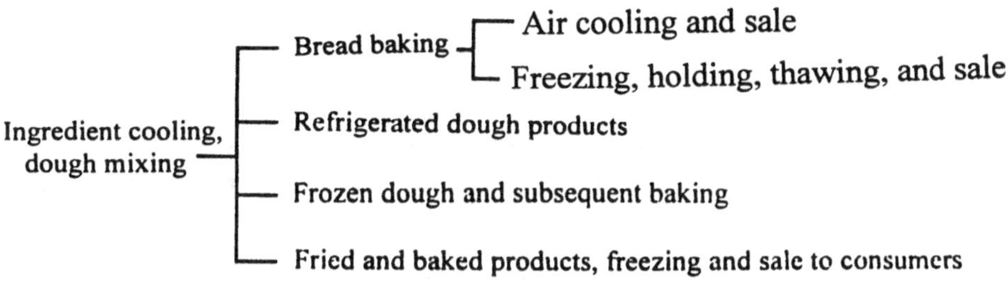

FIGURE 17.8
A sample of the variety of bakery products and processes, with those requiring refrigeration italicized.

17.9 REFRIGERATION OF BAKERY PRODUCTS

The variety of bakery products is extensive, and this number of items is multiplied by the variety of processes applicable to a given product. Many of these processes entail refrigeration at temperatures spanning the air-conditioning range to those air temperatures required to freeze a product. A sample of products and processes encompassed in the bakery industry is shown in Figure 6.8 (the item or process is italicized if it involves refrigeration).

Virtually all the products start as dough resulting from the mixing of the ingredients and fluids. To indicate that the temperature of the dough must be maintained at a temperature of about 27°C (80°F) may not sound demanding of a refrigeration system until it is realized that considerable heat is generated during the mixing process. In large commercial mixers, the paddles may be driven by a 50- to 100 kW motor, and all this heat must be removed. An even greater source of heat is attributable to the heat of hydration as the flour absorbs the water. The dough in a commercial bakery consists of a large mass with only a small area for heat transfer. Commercial dough mixers are equipped with a jacket cooled by antifreeze or occasionally by evaporating refrigerant. The required temperature for the antifreeze may be as low as −20°C (−4°F).

In conventional bread-making, the dough is baked and the bread from the oven is wrapped and air cooled, often requiring no refrigeration. To avoid paying overtime for labor, many establishments bake five days a week and freeze some of the product so that it can be sold seven days a week in the grocery store. The freezing process requires refrigeration, while the defrosting process needs low-humidity air at a temperature of 40 to 50°C (104 to 122°F). The freezing temperature of bread is lower than vegetables, for example, because bakery products have low moisture but high sugar contents. The freezing temperature may be between −15 to −10°C (5 to 14°F).

Another large sector in the bakery industry is the marketing of dough, either refrigerated unfrozen or frozen. The consumer is able to buy unfrozen dough at the supermarket to be baked later in the home. Unfrozen dough is packaged, then cooled sufficiently to arrest the fermentation process. The dough is then stored in a refrigerated space at a temperature of 3 to 5°C (37 to 41°F). Dough may also be frozen, and this large activity is usually confined to the food service sector. A popular development in recent years has been the bakery section of supermarkets that receive the frozen dough and bake it for quick sale. Many restaurants and institutional food establishments buy frozen dough and bake it as needed.

A fast-growing family of products, which includes pies, cakes, other pastries, waffles, and pizza, are the commodities that are baked or fried and then frozen. The consumer then only warms the product in a conventional oven, microwave oven, or toaster.

The applications of refrigeration in the bakery industry use evaporator coils that refrigerate air in the temperature range from comfort cooling down to the range needed to freeze products in tunnels or spiral freezers. In addition, refrigeration plants may produce ice and chill antifreezes to subfreezing temperatures.

17.10 THE FREEZING PROCESS

The biochemical and physical processes that occur during the freezing of food are complex, and the study of the freezing process can be a challenging career in itself. The refrigeration professional has many other assignments but would be well served by knowing some basic facts about the freezing process. A few of the key facts[12] applicable to food undergoing the freezing process are listed below:

- Freezing of many foods considerably lengthens the storage life of the product.
- Combining freezing with blanching, vacuum packing, or using additives will optimize the quality of certain frozen foods.
- Water is the major constituent of foods, comprising anywhere from 55% to 95% of the total mass.
- Formation of ice crystals as the water freezes is one of the major problems to be avoided.
- The formation of ice crystals breaks down the texture of the food and can rupture cell walls and possibly release degenerative enzymes.
- One of the most effective means of reducing ice crystal formation is to plunge the product through the freezing temperature range as rapidly as possible.
- The amount of energy that must be removed during the freezing process is normally greater than that required either above or below the freezing point.
- Foods do not freeze at a specific temperature, for example 0°C (32°F), but begin freezing slightly lower than that temperature and complete the freezing process several degrees below the temperature at the start of freezing.

The storage life of most foods can be lengthened from days or weeks to months by freezing the product. The emergence of the popularity of frozen food occurred almost simultaneously with the elevation of the quality of food in its frozen state. The greatest single factor in improving the quality was to discover that rapid freezing prevented the formation of ice crystals which resulted in inferior texture and taste when the food was thawed and cooked.

Even though water is the major constituent of food, this water does not occur in a free state that would freeze at 0°C (32°F). Instead, the water is in solution and forms a type of antifreeze. Not until the temperature has dropped somewhat below 0°C (32°F) does freezing begin. As the water freezes, it concentrates the remaining solution, which reduces the freezing temperature of the remaining liquid.

The undesirable formation of ice crystals occurs during slow freezing by allowing adequate time for water to migrate to a nucleation center where ice is forming. Rapid freezing keeps these nucelation centers dispersed so that large crystals cannot form.

17.11 FREEZING EQUIPMENT

The most popular types of equipment for freezing food products include blast freezers, contact plate freezers, individually quick-frozen (IQF) conveyor freezers, spiral freezers, and cryogenic freezers (Section 17.15).

One of the oldest and still widely used methods is blast freezing, wherein refrigerated air at perhaps −30 to −40°C (−22 to −40°F) is forced at high velocity over the product, as shown schematically in Fig. 17.9. The operation is a batch process in which pallets of product are loaded into the blast freezing cell and the process continued until the food is frozen. Depending on the product, and particularly on the form of packaging, the freezing process may require from perhaps 12 to 72 hours. A crucial requirement is that air be allowed access to all surfaces of the product or package, a requirement that can be aided by the installation of spacers vertically and horizontally between the packages.

A freezer that enjoys great popularity is the spiral freezer, as shown schematically in Fig. 17.10. Unwrapped product or individually wrapped packages pass in a long spiral path on a continuous metal conveyor belt. A chamber houses the spiral portion of the conveyor, and refrigerated air circulates at high velocity throughout the chamber. The spiral freezer arrangement permits a residence time of the product between one-half to two hours as necessary for freezing, and the facility requires only a moderate amount of space.

Another concept of freezing equipment is the plate freezer, as shown in Fig. 17.11. This freezer has a series of layers of packages held between multiple plates that are refrigerated by evaporating refrigerant. The plates are served by flexible refrigerant lines so that after the product is frozen, the plates may be lifted either manually or hydraulically so that the product can be pushed out of the freezer onto a conveyor belt. The direct contact of the cold plate with the package results in rapid freezing, but the disadvantage of this equipment is the number of mechanical devices that must be maintained.

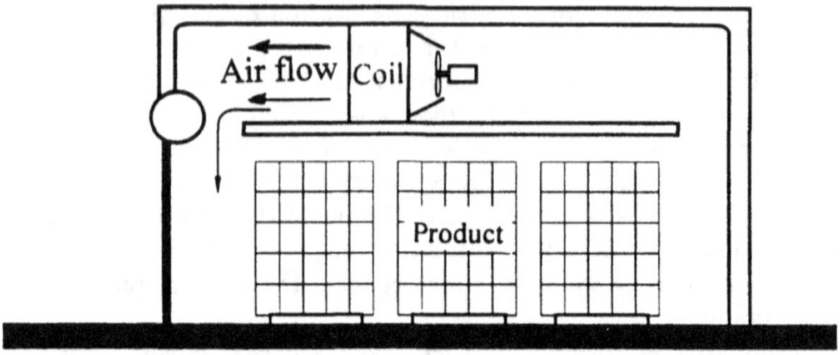

FIGURE 17.9
A blast freezer.

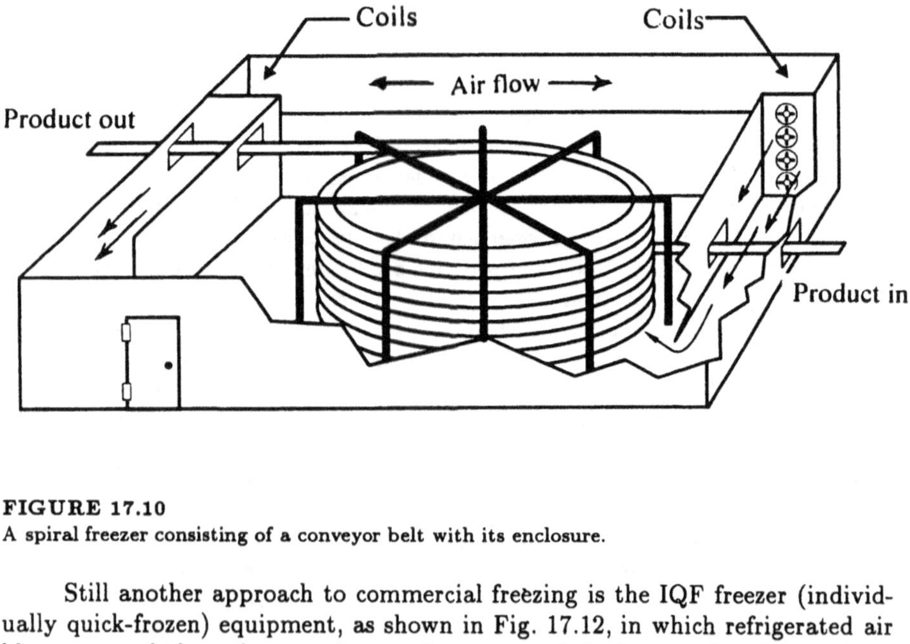

FIGURE 17.10
A spiral freezer consisting of a conveyor belt with its enclosure.

Still another approach to commercial freezing is the IQF freezer (individually quick-frozen) equipment, as shown in Fig. 17.12, in which refrigerated air blows upward through a moving mesh conveyor belt that carries the product. The air velocity is sufficiently high so that it levitates the product as the particles dance along the conveyor. The particles of the product are frozen individually and do not agglomerate. The residence time in the chamber is approximately several minutes. This IQF system is especially adaptable to small-size particles, such as peas and corn kernels, but also works well for larger pieces such as chips and french fries.

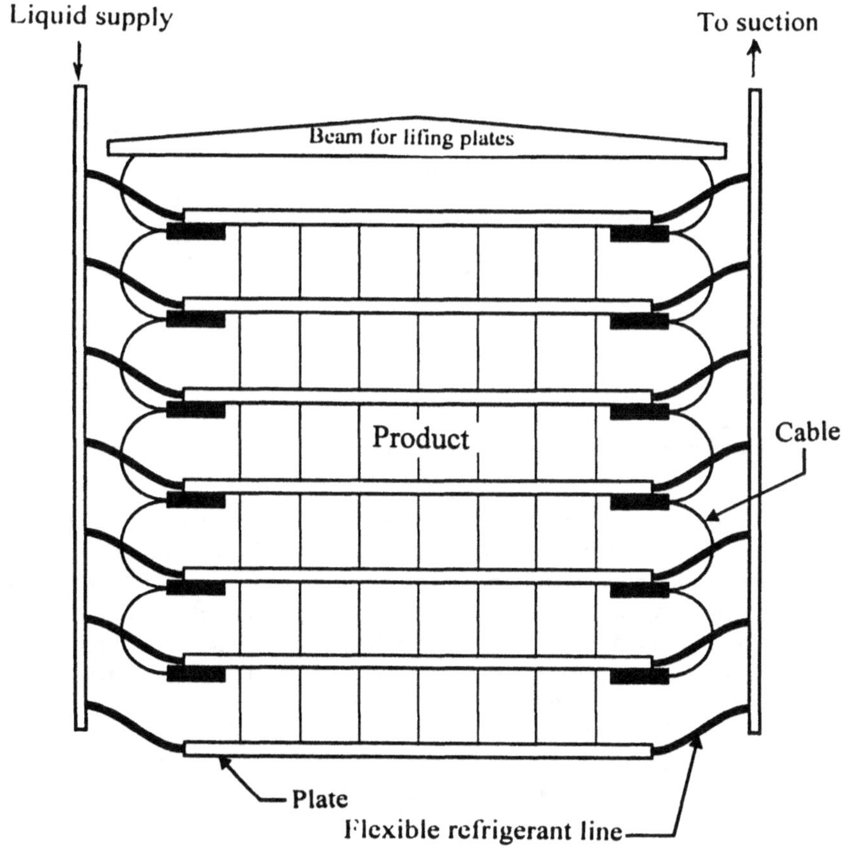

FIGURE 17.11
A plate-type freezer.

17.12 STORAGE TEMPERATURES FOR FROZEN FOOD

A broad statement, which must be qualified, is that the desirable temperature for the storage of frozen food is −23.3°C (−10°F). In the past, refrigerated warehouses in the United States were designed for and operated at a temperature of about −18°C (0°F), while in Europe the typical storage temperature was −23.3°C (−10°). From the standpoint of the product, the lower the temperature the better however, slightly higher operating costs may result. Practically all new U.S. warehouses are designed and built for operation at temperure of about −23.3°C (−10°F). Notable exceptions to this recommended storage temperature apply to ice cream and fish.

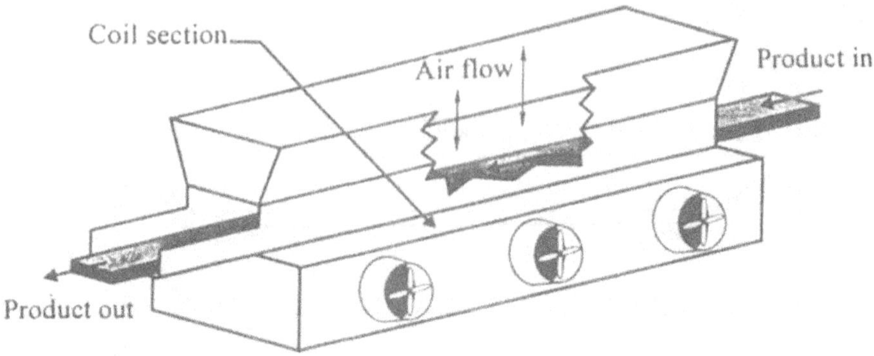

FIGURE 17.12
A fluidized-bed freezer for individually quick-frozen products.

The recommended storage temperature for ice cream is −29°C (−20°F) or below, thus lower than the temperature for most foods. Most of the water in ice cream is in a sugar solution, so the progressive freezing of water concentrates the remaining solution, lowering its freezing temperature. Depending on the butterfat content, freezing of ice cream begins at about −2°C (28.4°F), and the temperature may need to drop to about −55°C (−67°F) to freeze all of the water. At the recommended storage temperature of −29°C (−20°F) about 90% of the water is frozen, and in this state ice cream can be stored for several months. The temperature in the storage space should be held as constant as possible, because fluctuations in temperature result in objectionable growth in the size of ice crystals.

The quality of fish is particularly sensitive to its storage temperature, and, as Fig. 17.13 demonstrates, the required storage temperature is lower for fish than for a number of other products in order to achieve a comparable storage life. Cod is an example of lean fish, and salmon is typical of a fatty fish. The sensitivity of frozen fish to the storage temperature suggests maintaining the warehouse at −29°C (−20°F) or below.

The recommendations of storage temperatures given above are readily understandable, but a new concept in the freezing process should be monitored to be fully knowledgeable of the optimum storage temperature. This recently uncovered information is the existence of a *glass transition temperature* or the *mobility temperature*[13]. As a food product freezes, progressively more water freezes, leaving a high-concentration solute. The temperature at which the molecular mobility of this solute has disappeared is the glass transition temperature and appears to be the optimum storage temperature for that product. The commercial application of the knowledge of this phenomenon is complicated by the fact that each product seems to have a unique glass transition temperature. It is

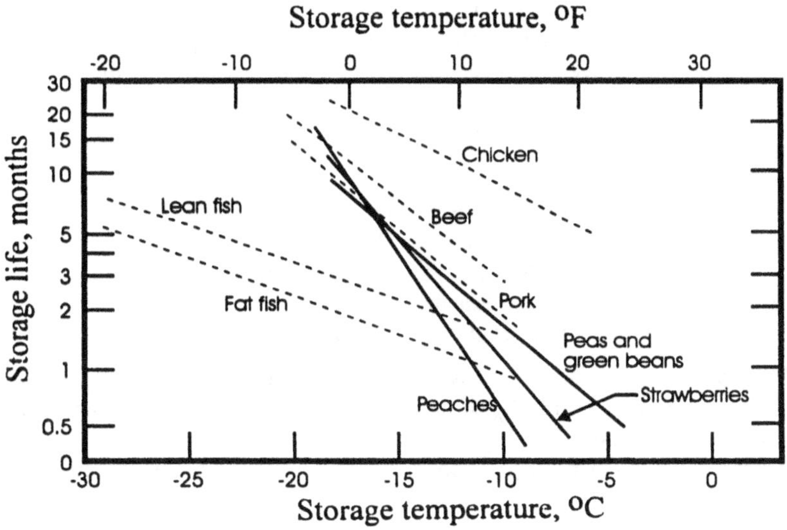

FIGURE 17.13
Typical storage life of some frozen food products showing lower recommended storage temperatures for fish. Fish, meats, and poultry are shown in dashed lines, while fruits and vegetables are represented by the solid lines.

impractical to have a large warehouse segregated into several regions of differing temperature. Compromises already exist in the choice of storage temperature, and the commercial application of knowledge of glass transition temperatures for specific products might influence in the future some minor shifts in storage temperature depending on the product mix stored.

17.13 PRODUCT REFRIGERATION LOADS

For unfrozen food, refrigeration loads in a facility that are attributable to the food product include the reduction in temperature and the removal of the heat of respiration. In a food freezing process, the refrigeration plant must first reduce the temperature of the product to the freezing temperature, next remove the heat of freezing, and finally refrigerate to the desired storage temperature. In a food freezing plant, the product load usually dominates, while for a refrigerated warehouse storing frozen food, the product load may be minor because of the expectation that the incoming product is already at the storage temperature. Despite this expectation, realistically, the incoming product is usually at a temperature somewhat higher than that at which it will be stored. Almost without exception, then, product refrigeration load exists wherever food is handled.

For completeness, the refrigeration load attributable to refrigerating the product will be listed with the other contributors to the refrigeration load in

Chapter 18, Refrigeration Load Calculations, but the tools to perform these calculations will be described in the next several paragraphs. The quantity of heat that must be removed in order to drop the temperature is expressed by the following equation:

$$Q = mc(t_{initial} - t_{final}) \tag{17.1}$$

where:

Q	=	heat removed from the product, kJ (Btu)
m	=	mass of product, kg (lb)
c	=	specific heat, kJ/kg·K (Btu/lb·°F)
$t_{initial}$	=	initial temperature of the product, °C (°F)
t_{final}	=	final temperature of the product, °C (°F)

The specific heat is defined as the amount of heat that must be added to or removed from 1 kg (1 lb) of substance in order to increase or decrease, respectively, the temperature by 1°C (1°F). Food products possess a different specific heat above freezing, c_{above}, in comparison to that below freezing, c_{below}. For unfrozen food, then:

$$Q_{above} = mc_{above}(t_{initial} - t_{final}) \tag{17.2}$$

Q_{above} divided by the time in seconds required for the cooling yields the refrigeration rate in kW (Btu/s). Equation 17.2 represents the only refrigeration load in such operations as post-harvest cooling where the product is harvested, cooled, and shipped immediately.

Another category of refrigeration load that must be accommodated when storing unfrozen produce is the heat of respiration. Unfrozen foods are living organisms and continue to give off heat. An excerpt of a much more complete table from the ASHRAE *Handbook—Fundamentals*[14] is shown in Table 17.3. The intensity of the heat of respiration increases with increasing temperature, which conforms to the expectation that lower temperatures slow the metabolism rate of the stored product.

In a freezing operation, there are three distinct components of the refrigeration process: (1) reducing the temperature to the freezing point, (2) freezing or removing the latent heat of freezing, and (3) reducing the temperature further to the storage temperature. The refrigeration loads associated with each of these operations are as follows:

$$Q_{total} = Q_{above} + Q_{freeze} + Q_{below} \tag{17.3}$$

or

$$Q_{total} = mc_{above}(t_{initial} - t_{freeze}) + mL + mc_{below}(t_{freeze} - t_{final}) \tag{17.4}$$

where:

t_{freeze}	=	freezing temperature of the product, °C (°F)
L	=	latent heat of freezing, kJ/kg (Btu/lb)

TABLE 17.3
Heat of respiration of several unfrozen fruits and vegetables[14].

	Heat of respiration, kJ/kg per day (Btu/lb per day)			
Commodity	0°C (32°F)	5°C (41°F)	15°C (59°F)	20°C (68°F)
Apples	.587—1.05	1.30—1.84	3.48—7.92	4.32—8.97
	(.253—.451)	(.559—.792)	(1.50—3.40)	(1.86—3.85)
Beans	4.52—8.97	7.46—15.6		54.2—69.2
	(1.95—3.85)	(3.21—6.72)		(23.3—29.8)
Celery	1.84	2.81	9.55	16.5
	(.793)	(1.21)	(4.11)	(7.11)
Lettuce	2.35—4.32	3.39—5.11	9.89—10.5	15.4
	(1.00—1.86)	(1.46—2.20)	(4.25—4.50)	(6.61)
Peaches	1.05—1.63	1.63—2.35	8.51—10.9	15.2—26.2
	(.451—.703)	(.702—1.01)	(3.66—4.67)	(6.52—11.3)
Pears	.795—1.76	1.30—2.56	3.85—15.4	7.71—17.9
	(.342—.757)	(.559—1.10)	(1.66—6.61)	(3.31—7.71)
Peas	12.1—19.4	20.3—24.9		89.4
	(5.21—8.32)	(8.72—10.7)		(38.4)
Strawberries	3.14—4.52	4.19—8.51	18.2—23.6	26.2—50.2
	(1.35—1.95)	(1.80—3.66)	(7.82—10.2)	(11.3—21.6)

TABLE 17.4
Specific heats and latent heats of several food products[14].

Product	Water content % mass	High freezing point °C (°F)	Specific heat kJ/kg·K (Btu/lb·°F) Above freezing	Below freezing	Latent heat of freezing kJ/kg (Btu/lb)
Apples	84	−1.1	3.78	1.90	281
		(30)	(0.902)	(0.453)	(121)
Chicken	74	−2.8	3.53	1.77	248
		(27)	(0.843)	(0.423)	(107)
Peas	74	−0.6	3.53	1.77	248
		(31)	(0.792)	(0.423)	(107)
Ham	56	−1.7	3.08	1.55	188
		(29)	(0.735)	(0.368)	(81)
Salmon	64	−2.2	3.28	1.65	214
		(28)	(0.783)	(0.392)	(92)
Sirloin beef	56		3.08	1.55	188
			(0.735)	(0.368)	(81)
Strawberries	90	−0.8	3.93	1.97	302
		(31)	(0.938)	(0.471)	(130)

Table 17.4 is an excerpt from a table of an extensive list of products tabulated in the ASHRAE *Handbook—Fundamentals*[14] providing specific heats above and below freezing as well latent heats of freezing. Several observations can be made concerning this information. First, the specific heat of a product below the freezing temperature is about half that above freezing. The reason is that water is the major constituent of food, and liquid water has about twice the specific heat of ice, 4.19 kJ/kg (1.0 Btu/lb) versus 1.94 kJ/kg (0.463 Btu/lb). A second observation is that the latent heat of freezing is greatest for products with high moisture contents.

Example 17.2. What is the refrigeration capacity needed to cool, freeze, and further refrigerate peas at a rate of 2.27 kg/s (300 lb/min) if the product enters the facility at 30°C (86°F) and after freezing is cooled to −20°C (−4°F)? Use Eq. 17.4.

Solution. From Table 17.4, the specific heat of peas above freezing is 3.53 kJ/kg·K (0.792 Btu/lb·°F) and below freezing, 1.77 kJ/kg·K (0.423 Btu/lb·°F). The high freezing point of peas is −0.6°C (31°F), and the latent heat of freezing is 248 kJ/kg (107 Btu/lb). To process 1 kg of peas requires the removal of:

$$Q_{total} = 3.53[30 - (-0.6)] + 248 + 1.77[(-0.6 - (-20)]$$

$$Q_{total} = 108 + 248 + 34.3 = 390.3 \text{ kJ/kg}$$

The rate of flow of product is 2.27 kg/s, so the required refrigerating capacity is (2.27)(390.3) = 886 kW (252 tons of refrigeration).

17.14 FREEZING RATES

The prediction of the time required for freezing a food product is both important and difficult. In the design of a freezing facility, it is necessary to know the freezing time in order to properly choose the refrigeration capacity. Suppose that a spiral freezer facility dedicated to freezing pizzas provides a residence time of 30 minutes with a product flow rate of 0.80 kg/s (6,360 lb/hr). The air temperature off the coils is −38°C (−36°F). Suppose also that the refrigeration capacity, as calculated using the procedures of Section 6.13, indicates a refrigeration load of 468 kW (133 tons of refrigeration). To select evaporator coils for the spiral freezer chamber that transfer a rate of 468 kW (133 tons) and provide an air temperature of −38°C (−36°F) with the available evaporating temperature of, for example, −42°C (−44°F) overlooks another important consideration. The −38°C (−36°F) supply air must be capable of freezing the pizzas in 30 minutes. If the rate of cooling/freezing/refrigerating is slower than expected, the return temperature of the air to the coil will be lower than design, resulting in the coil capacity and the demand on the refrigeration plant being less than 468 kW (133 tons). Design engineers can often make a reasonable estimate of the freezing time, because of their past experience with the same product and the same type of equipment.

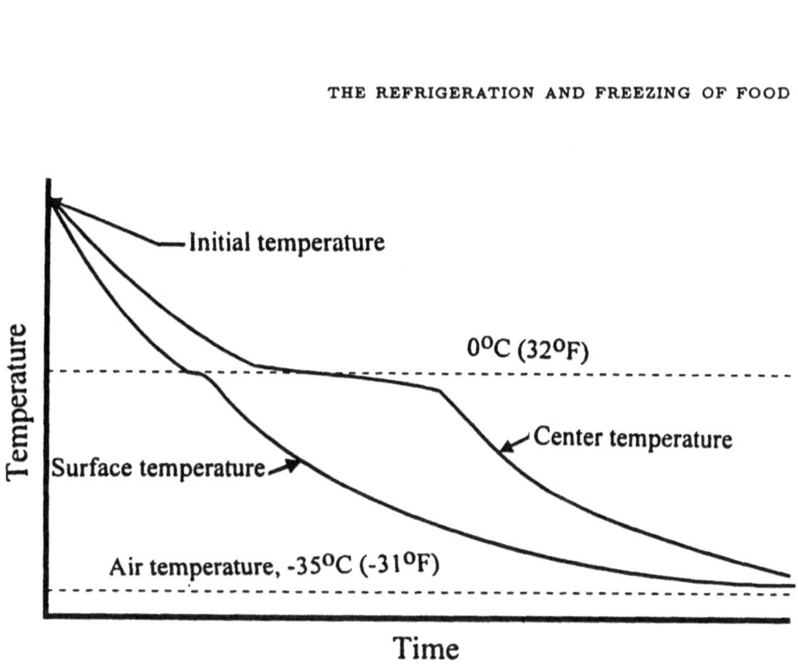

FIGURE 17.14
Surface and center temperatures of a product during the freezing process.

Analytical predictions of freezing times are mathematical challenges, because of the complex surface and internal heat-transfer processes. The approach taken in the past has been to approximate the shape of the product by a sphere, cube, or a cylinder and apply the necessary mathematical technique to the heat-transfer and solidification processes. In recent years, finite element analysis using computers has become popular. These activities have their value in two ways. The resulting equation or graphs indicate directions of trends and even quantify proposed changes in the process, such as in the air temperature or velocity. The other benefit of the analysis lies in the prediction applicable to new products, shapes, or freezing equipment.

The prediction of the freezing time using analytical methods alone can only be approximate, because of the unknowns and the variabilities in such factors as the convection coefficient at the surface, the non-uniformity of the product and/or package, and the deviations in the properties of the material such as thermal conductivity, density, and specific heat. When predicting cooling times for products above the freezing temperature (Section 17.4), the approach suggested was a combination of analytical techniques used to identify the trends and field tests to provide quantitative values. The same combination of tools will be proposed to predict freezing times.

The standard expectation of the temperature-time profiles of the surface and center of a product for given initial product and refrigerated air temperatures is shown in Fig. 17.14. When brought into contact with the low-temperature air, the surface temperature begins to drop rapidly. The temperature of the center of the product lags behind the change in surface temperature. When the surface temperature passes through the freezing temperature, a slight pause in the temperature drop occurs as the crust of the product freezes. From this point on, a freezing front moves steadily through the product toward the center. Some finite time is needed for the product to freeze at any location, during which period the heat extracted for freezing must pass by conduction to the surface. Following the complete freezing of the product, the center temperature continues its drop.

A rough estimate of the time required for freezing is provided by the Plank equation[15], one form[16] of which is as follows:

$$t_F = \frac{\rho L}{(t_f - t_a)}\frac{D}{N}\left(\frac{D}{4k} + \frac{1}{h_c}\right) \tag{17.5}$$

where:

t_F	=	freezing time, seconds
ρ	=	density, kg/m^3
L	=	latent heat of freezing, J/kg
t_f	=	freezing temperature, °C
t_a	=	air temperature, °C
D	=	distance to center parallel to main direction of flow, m
N	=	shape factor, 2, 4, and 6 for slab, cylinder, and sphere, respectively
k	=	thermal conductivity, W/m·K
h_c	=	convection coefficient at surface, W/m^2·K

Plank originally suggested that the initial temperature must be that of the freezing point, but the equation seems to be applicable to initial temperatures higher than freezing, so long as the enthalpy change in the total process replaces the latent heat, L.

In order to apply Eq. 17.5, the convection coefficient, h_c, and several additional properties of the food, namely density and thermal conductivity, must be known. The convection coefficient, h_c, can be approximated by either of the following equations:

$$h_c,\ \mathrm{W/m^2 \cdot K} = 25(V\ \mathrm{m/s})^{0.6} \tag{17.6}$$

$$\text{or}\quad h_c,\ \mathrm{Btu/hr \cdot ft^2 \cdot {}^\circ F} = 0.19(V\ \mathrm{fpm})^{0.6}$$

The thermal conductivity[17] of fruits and vegetables above the freezing temperature is primarily a function of the moisture content and can be represented by the following equations:

$$k,\ \mathrm{W/m \cdot K} = 0.148 + 0.493F_w \tag{17.7}$$

$$\text{or}\quad k,\ \mathrm{Btu/hr \cdot ft \cdot {}^\circ F} = 0.0855 + 0.285F_w$$

where F_w = fraction of water in the product. For meats, the corresponding equations for the thermal conductivities are as follows:

$$k,\ \mathrm{W/m\cdot K} = 0.08 + 0.52F_w \tag{17.8}$$

$$\text{or}\quad k,\ \mathrm{Btu/hr\cdot ft\cdot{}^\circ F} = 0.046 + 0.30F_w$$

The density of high-moisture content foods is about 10 percent higher than that of water, which is 1,000 kg/m^3 (62.4 lb/ft^3).

Example 17.3. An IQF freezer is freezing strawberries with −35°C (−31°F) air passing up through the product with a velocity of 5 m/s (984 fpm). The strawberries may be considered spheres with a depth to the center of 10 mm (0.39 inch). What is the estimated freezing time at the freezing temperature of strawberries?

Solution. Data needed to compute the freezing time from Eq. 17.5 are: the density of 1,100 kg/m^3 (68.7 lb/ft^3); the latent heat of freezing for strawberries from Table 17.4 of 302,000 J/kg (130 Btu/lb); strawberries begin freezing at −0.8°C (31°F). The fraction of water in strawberries is approximately 0.9, at which value the thermal conductivity from Equation 6-7 is 0.59 W/m·K (0.34 Btu/hr·ft·°F). From Eq. 17.6 at a velocity of 5 m/s, h_c = 65.7 W/m^2·K (11.6 Btu/hr·ft^2·°F); D = 0.01 m (0.39 in); and N for a sphere = 6. Therefore:

$$t_F = \frac{(1,100)(302,000)}{[-0.8-(-35)]}\frac{0.01}{6}\left[\frac{0.01}{(4)(0.59)} + \frac{1}{65.7}\right] = 315\,\mathrm{s}$$

$$t_F = 5.25\ \mathrm{min}$$

The Plank equation is most successful in predicting the freezing times of small individual, well-separated pieces of homogeneous product with good contact between the cooling medium and surface of the product. To predict the cooling/freezing time of a nonhomogeneous product in a complex packing arrangement, Section 17.4, which was directed toward predicting cooling times for products above freezing, applied the strategy of combining physical analysis with a few field measurements. The shape and position of the cooling curve is derived from the physics of the process, and the data from field tests fix the position of the curve. The same strategy[18] is sometimes applied to the cooling/freezing process, as shown in Figure 17.15. The process displayed in Figure 17.14 is divided into three distinct parts: cooling to the freezing point, freezing, and post-freezing reduction in temperature. At first it may seem that the curved lines of Figure 17.14 do not resemble the straight line cooling curves of Figure 17.1 until it is realized that the ordinate should be a dimensionless temperature, $\theta = (t_c - t_{air})/(t_i - t_{air})$, plotted on a logarithmic scale. The starting point for the prefreezing process is at zero time and $\theta = 1$. The line proceeds to a value of $\theta = (t_{freezing} - t_{air})/(t_i - t_{air})$ with a slope determined by the several field-test measurements.

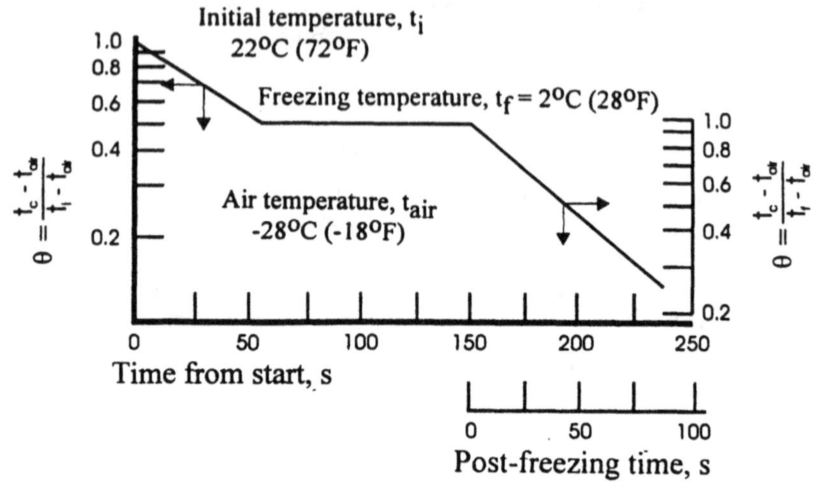

FIGURE 17.15
Prediction of the center temperature, t_c, of peas[19] during the cooling/freezing process by a combination of analytical tools and field tests.

The freezing time could also be determined by field measurements or estimated using the Plank equation. The line for the post-freezing reduction of temperature is another cooling curve, but in this case, the initial temperature is that of the frozen product and the dimensionless temperature, $\theta = (t_c - t_{air})/(t_f - t_{air})$, corresponding to the auxiliary ordinate of the curve. The data for the freezing of peas shown in Fig. 17.15 resembles the ideal model very closely, but attempts to use this model on a variety of other products does not always yield the ideal straight lines, but still provides an approximate estimate of the cooling/freezing times.

The slope of the cooling curve of the frozen product in Fig. 17.15 is steeper than that in the unfrozen phase. This more rapid rate of cooling in its frozen state can be explained, because the specific heat of the frozen product is about one-half that in the unfrozen state, and the thermal conductivity is about twice that of the unfrozen food. Both effects accelerate the cooling of the frozen product.

The final topic addressed in this section on freezing rates is the perennial question of the optimum air velocity for blast freezers. High air velocities increase the convection coefficient and thus reduce the cooling/freezing time, but they do so at the expense of additional fan power and heat introduced to the blast freezing cell, which must then be removed by the refrigeration plant. As Eq. 17.6 indicates, doubling the flow rate and velocity of the air increases the convection coefficient by a factor of 1.51. At the same time, doubling the air flow rate increases the fan power by a factor of 8. Another fact to keep in mind is that the refrigeration plant can only influence the surface temperature of the product, and what happens inside is strictly a function of the thermal properties and physical disposition of

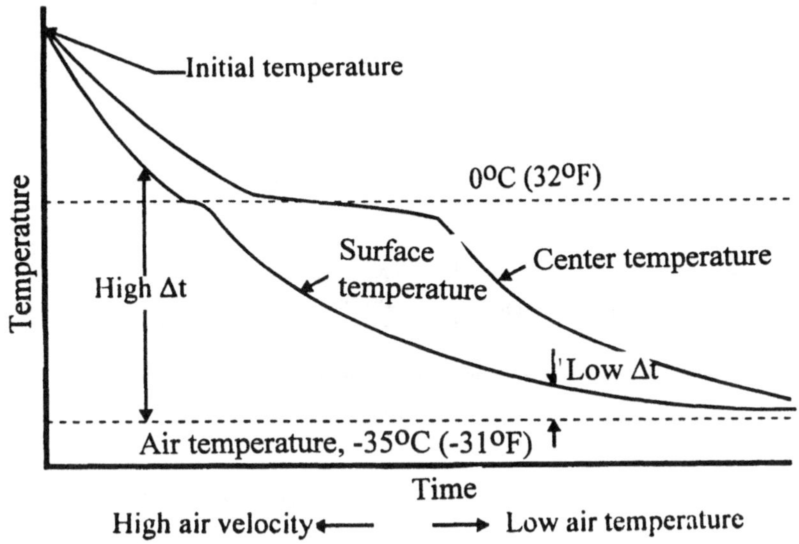

FIGURE 17.16
Optimum air flow rate in a blast freezer would be high in the early stages of the cycle and low near the end.

the product and package. It is understandable that no specific recommendations of optimum air flow rate can apply to all blast freezing systems, and even for a specific facility, computing the optimum combination of the profile of air velocity and temperature is extremely difficult.

There is some guidance that can be offered, however, on the optimum program, which can be visualized by reworking Fig. 17.14 into Fig. 17.16. The objective is to drop the surface temperature as rapidly as possible with the minimum combined power of the fan and compressor. In the prefreezing phase of the process, a large temperature difference prevails between the product surface and the air, so this is the portion of the process where improvements in the convection coefficient, h_c, are most effective. The temperature level of the refrigerated air is not as important. Near the end of the process, however, increasing h_c has only a modest influence on the surface temperature, while reducing the temperature of the refrigerated air is effective. This analysis suggests that the optimum program would be to use high air flow rates in the early part of the process, gradually decreasing the air flow rate and dropping the temperature of the refrigerated air in the latter stages. Such continuous changes during the cycle would have been impractical in the past but are beginning to appear now in plants under computer control.

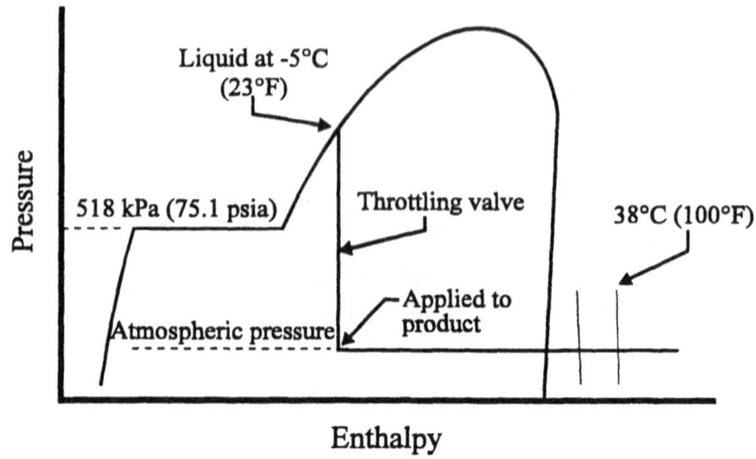

FIGURE 17.17
Pressure-enthalpy diagram of states of carbon dioxide in a cryogenic freezer.

17.15 CRYOGENIC FREEZING

Cold carbon dioxide (CO_2) or liquid nitrogen (N_2) may be sprayed directly on the product to achieve a rapid freezing. Sometimes these substances are used in combination with mechanical refrigeration, whereby CO_2 or N_2 freeze a crust on the surface to prevent loss of internal moisture. Following the application of CO_2 or N_2, the product is moved to one of the freezing devices using air.

Both CO_2 and N_2 are usually stored as high-pressure liquids and then passed through a throttling valve to expand to atmospheric pressure. In the case of N_2, the high-pressure liquid flashes into a mixture of liquid and vapor at a temperature of −195.8°C (−320°F). The vaporization of the liquid in the mixture provides most of the refrigeration capacity, and a small fraction is derived from the heating of the cold vapor.

After CO_2 passes through the throttling valve to be sprayed on the product, its state is a mixture of solid (dry ice) and vapor, as illustrated in the pressure-enthalpy diagram in Fig. 17.17. The sublimation of the solid CO_2 directly to vapor provides part of the refrigeration and the remainder derives from the cold vapor.

The cost of freezing per unit mass of product with a cryogenic system is likely to be from two to six times greater than that of a vapor-compression refrigeration system, but the investment cost will be considerably less. This cost comparison leads to the use of a cryogenic freezer when the annual usage is low or when the rapid freezing is required for valuable or delicate products.

REFERENCES

1. American Society of Heating, Refrigerating and Air-Conditioning Engineers, Inc., Atlanta, Georgia, from the 1994 ASHRAE *Handbook—Refrigeration Systems and Applications.*
2. Hardenburg, R. E., A. E. Watada, and C. Y. Wang, *The Commercial Storage of Fruits, Vegetables, and Florist and Nursery Stocks*, Agriculture Handbook No. 66, United States Department of Agriculture, 1986.
3. Anderson, R. W., "How to Eliminate Condensate and Save Energy in Meat and Poultry Plants," *ASHRAE Journal*, 24(9): 30, August 1982.
4. Gurney H. P. and J. Lurie, "Charts for Estimating Temperature Distributions in Heating and Cooling Solid Shapes," *Ind. Engr. Chem.*, 15: 1170-1172, 1923.
5. Stewart, J. K. and H. M. Couey, "Hydrocooling Vegetables—A Practical Guide to Predicting Final Temperatures and Cooling Times," Market Research Report 637, U. S. Dept. of Agr., 1963.
6. "Temperatures and Chilling and Freezing Procedures," *Code of Federal Regulations*, Title 9, Part 381.66, U. S. Department of Agriculture, May 16, 1972, Amended Sept. 12, 1975 and Jan. 30, 1984.
7. Hall, D. L., "Recirculated Water Chilling for Eviscerated Poultry," Proceedings of the Annual Meeting of the International Institute of Ammonia Refrigeration, Savannah, Georgia, March 1988.
8. James, S. J. and C. Bailey, "Temperature Changes, Weight Loss and Product Loads During Beef Chilling," *Recent Advances and Developments in the Refrigeration of Meat by Chilling*, pp 105-114, Proceedings of Commission C2 Meeting of the International Institute of Refrigeration, Bristol, United Kingdom, 1986.
9. Thompson, J. F. and Y. L. Chen, "Comparative Energy Use of Vacuum, Hydro, and Forced Air Coolers for Fruits and Vegetables," *ASHRAE Transactions*, 94(Part I): 1427-1433, 1988.
10. Mitchell, F. G., R. Guillou, and R. A. Parsons, "Commercial Cooling of Fruits and Vegetables," *California Agricultural Experiment Station*, Manual 43, 1972.
11. Burdick, R. S., "Salt Brine Cooling Systems in the Cheese Industry," Proceedings of the Annual Meeting of the International Institute of Ammonia Refrigeration, San Francisco, California, 1991.
12. "Physical and Chemical Action and Reactions of Freezing, Storage and Thawing Affecting Quality of Frozen Foods," *Information Bulletin*, The Refrigeration Research Foundation, Bethesda, Maryland, January, 1989.
13. D. S. Reid, "Glass Transition Temperature and Frozen Food Stability," *Research Summary*, No. 91-1, The Refrigeration Research Foundation, Bethesda, Maryland, July, 1991.
14. American Society of Heating, Refrigerating and Air-Conditioning Engineers, Inc., Atlanta, Georgia, from the 1993 ASHRAE *Hand- book—Fundamentals.*
15. R. Plank, "Messung und Berechnung der Gefrierzeit verpackter Lebensmittel, *Zeitschrift fuer der gesammte Kaelteindustrie*, p. 51, 1944.
16. "Recommendations for the Processing and Handling of Frozen Foods," 3rd ed., International Institute of Refrigeration, Paris, France, 1986.
17. Sweat, V. E., "Thermal Conductivity of Food: Present State of Data," *ASHRAE Transactions*, 91(Part 2B): 299, 1985.
18. Pham, Q. T., "Extension to Plank's Equation for Predicting Freezing Times of Foodstuffs of Simple Shapes," *International Journal of Refrigeration*, 7(6): 377-383, November, 1984.
19. Wilson, H. A. and R. P. Singh, "Numerical Simulation of Individual Quick Freezing of Spherical Foods," *International Journal of Refrigeration*, 10(3): 149, May, 1987.

CHAPTER 18

REFRIGERATION LOAD CALCULATIONS

18.1 MULTIPLE PURPOSES OF LOAD CALCULATIONS

It is mandatory that refrigeration load calculations be performed during the design stage in order to properly size the components. Load calculations are useful on other occasions as well, such as to evaluate the effectiveness in conserving energy of a proposed plant modification. Still another situation where a load calculation may be useful or necessary is in troubleshooting, when, for example, a refrigerated space cannot be maintained at the design temperature or a production rate is lower than design.

Many designers use computer programs and/or checklists as aids in performing load calculations so that all the major sources are included. The most significant error in load calculations probably results from the complete omission of a contributor, in contrast to a 20 to 30% error in any one calculation. Using a computer program for the load calculation will inherently incorporate all the items if the program has been written properly in the first place. The major contributors to refrigeration loads are: (1) the heat transmission through the roof, floor, and walls (2) infiltration through open doorways (3) internal loads from lights, people, motors, and lift trucks (4) defrost heat and (5) the product load—cooling, freezing, and maintaining the temperature of products. All of these categories will be addressed in this chapter.

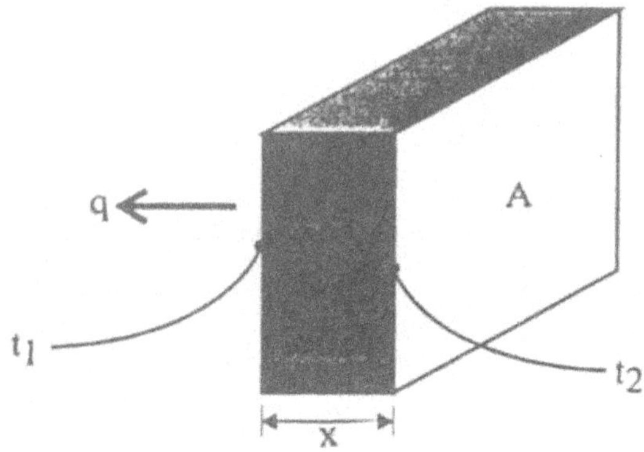

FIGURE 18.1
Heat transfer by conduction through a single wall.

18.2 HEAT TRANSFER THROUGH SINGLE MATERIALS

When heat flows by conduction through a wall, as in Fig. 18.1, the equation that expresses the rate of heat transfer is as follows:

$$q = k\frac{A}{x}(t_1 - t_2) \tag{18.1}$$

where:

q	=	rate of heat transfer, W (Btu/hr)
k	=	conductivity of material, W/m·K (Btu/hr·ft·°F)
A	=	cross-sectional area of wall, m^2 (ft^2)
x	=	thickness of wall, m (ft)
t_1 and t_2	=	temperatures on opposite surfaces of wall, °C (F)

Equation 18.1 applies when the temperatures are constant and the conductivity is uniform throughout the wall. The equation indicates that the rate of heat transfer is proportional to the conductivity, the cross-sectional area of the wall, and the temperature difference. The rate of heat transfer is inversely proportional to the thickness of the material. The thermal conductivities of some insulating materials used in low-temperature applications are shown in Table 18.1. Three columns of conductivities are shown—one in SI units and two in I-P units. Most heat-transfer books, when presenting conductivities in I-P units, use Btu/hr·ft·°F. In many catalogs, however, conductivity is expressed in Btu·in/hr·ft^2·°F. The user of conductivity values must express the thickness of the wall in feet when using k in Btu/hr·ft·°F and express the thickness in inches when using k in Btu·in/hr·ft^2·°F.

TABLE 18.1
Thermal conductivities of some insulating materials.

Material	Conductivity		
	W/m·K	Btu/hr·ft·°F	Btu·in/hr·ft²·°F
Cellular foamglass	0.050	0.029	0.35
Cellular polyurethane	0.023	0.013	0.16
Expanded polystyrene	0.035	0.020	0.24
Extruded polystyrene	0.027	0.015	0.18
Glass fiber	0.036	0.021	0.25
Polyisocyanurate	0.020	0.012	0.14

TABLE 18.2
Conductance and resistance of some materials.

Material	Conductance		Resistance	
	W/ m²·K	Btu/ hr·ft²·°F	m²·K/ W	hr·ft²·°F/ Btu
8-in Cinder concrete block	3.3	0.58	0.30	1.72
Hollow clay tile, 2 cells	3.75	0.66	0.27	1.51
Built-up roofing,				
... 10 mm (.394 in)	17.0	3.0	0.06	0.33
Asphalt roll roofing	36.9	6.50	0.027	0.154

18.3 CONDUCTANCE

The conductivity is one thermal characteristic of materials, but an alternate method of presenting thermal transfer data for materials is to use the *conductance*, C, which has units of W/m²·K or Btu/hr·ft²·°F. The conductance fits into the heat-transfer equation representing the process of Fig. 18.1 as follows:

$$q = CA(t_1 - t_2) \tag{18.2}$$

Comparison of Eqs. 18.1 and 18.2 shows that $C = k/x$ for slabs of uniform construction. Sometimes, it is more convenient to use the conductance in preference to the conductivity. A concrete block with hollow cores, for example, poses ambiguities as to what k values and x values to use, but can be expressed by the conductance such that the heat-transfer rate can be calculated by Eq. 18.2. Table 18.2 shows examples of several materials that are characterized better by conductance than by conductivity.

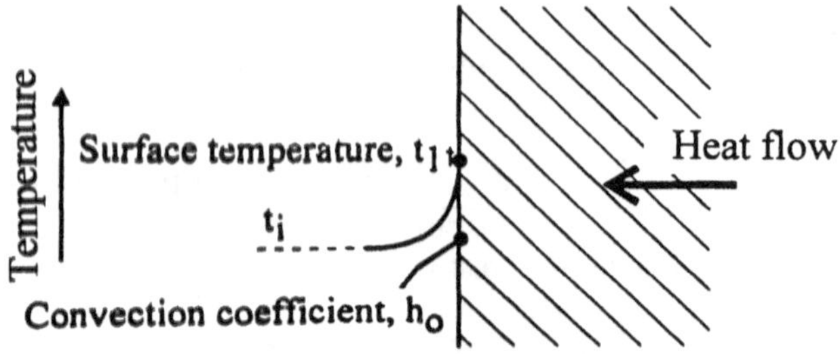

FIGURE 18.2
The convection coefficient.

18.4 CONVECTION AND THE CONVECTION COEFFICIENT

The calculation of the rate of heat transfer through a single material, as illustrated in Fig. 18.1, is a building block in the procedure for calculating the rate of heat transfer in walls and roofs of both refrigerated and nonrefrigerated structures. An additional concept that should be introduced now is that of convection. Equation 18.1 permits computation of the heat-transfer rate when the opposite surface temperatures are known. A more prevalent situation occurs when the air temperatures on either side of the wall are known, which then requires that the heat-transfer process between the air and surface be quantified. In Fig. 18.2, heat flows from the warm wall to the colder air, the surface temperature is t_1, and the air temperature is t_i. The rate of heat transfer is expressed by the following equation:

$$q = h_c A(t_1 - t_i) \tag{18.3}$$

where:

h_c = convection coefficient, W/m²·K (Btu/hr·ft²·°F)

Equation 18.3 can be thought of as the definition of the convection coefficient, which is designated by the symbol h_c. The letter h is also the symbol in refrigeration for enthalpy, but usually the context will provide the clue to the meaning of h. As a further means of avoiding confusion in this book, the following conventions will be employed:

- h with no subscript refers to an enthalpy.
- h_f or h_g refer to enthalpies of saturated liquid and vapor.
- h_a refers to an enthalpy of air.
- h_c, h_i, and h_o express convection coefficients, with the subscripts i and o indicating inside and outside, respectively.

The magnitude of the convection coefficient, h_c, is primarily a function of the air velocity along the surface. On outside surfaces of walls and roofs, the wind

TABLE 18.3
Convection coefficients at some surfaces.

Orientation and condition	Convection coefficient W/m²·K	Btu/hr·ft²·°F
6.7 m/s (15 mph) winter, any orientation	34	6.0
3.4 m/s (7.5 mph) summer, any orient.	23	4.0
Still air		
..vertical surface, heat flow horizontal	8.3	1.46
..horizontal surface, heat flow upward	9.3	1.63
..horizontal surface, heat flow downward	6.1	1.08

controls the velocity. On inside surfaces, even where there is no forced motion of the air, there is movement of the air due to natural convection. In this case, the orientation of the surface (whether the surface is horizontal or vertical) has a significant influence. Some typical values[1] of h_c are shown in Table 18.3.

18.5 THE RESISTANCE CONCEPT

The *resistance concept* transfers over from electrical terminology to heat transfer and is sometimes helpful in visualizing heat-transfer processes as well as to facilitate calculations. Ohms's law in electricity is as follows:

$$I = \frac{V_1 - V_2}{R} \quad \text{or Current, ampere} = \frac{\text{Voltage difference, volts}}{\text{Resistance, ohms}} \tag{18.4}$$

In general terms, the current, I, is the flow rate and the driving force is the voltage difference. The symbols applicable to electricity are shown in Fig. 18.3. Figure 18.3a represents the terms in Eq. 18.4. When several resistances are connected in series, as in Fig. 18.3b, the total resistance is the sum of the individual resistances, so that:

$$I = \frac{V_1 - V_2}{R_T} = \frac{V_1 - V_2}{R_1 + R_2 + R_3}$$

The use of the resistance concept to express rates of heat transfer has achieved wide acceptance. For the heat-transfer process of Fig. 18.1, the equation corresponding to Eq. 18.4 is as follows:

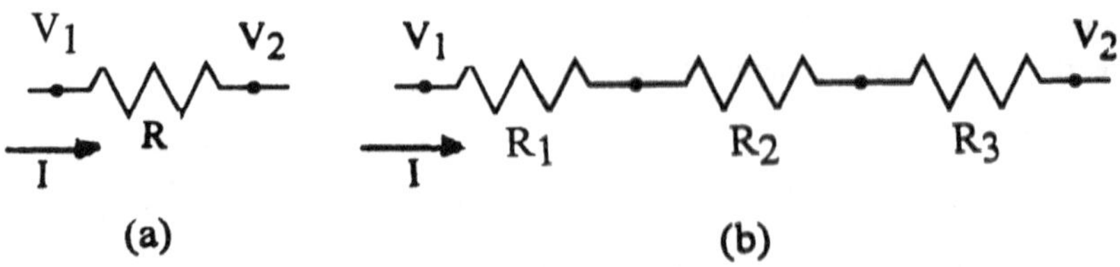

FIGURE 18.3
(a) The electrical analogy for a single resistance, and (b) the total resistance as the sum of individual resistances in series.

$$\frac{q}{A} = \frac{t_1 - t_2}{R} \tag{18.5}$$

where:
R in m^2·K/W (hr·ft^2·°F/Btu) = x/k

For the convective process in Fig. 18.2, the resistance is as follows:

$$R_c = 1/h_c$$

The resistance and conductance are reciprocals of one another, and Table 18.2 shows resistance values as well as the conductance values.

18.6 OVERALL HEAT-TRANSFER COEFFICIENTS FOR WALLS, ROOFS, AND FLOORS.

The tools are now available to compute the rate of heat transfer through the walls, roofs, and floors that are encountered in practice. Furthermore, the useful form of the equation expresses the rate of heat transfer in terms of the air temperatures on opposite sides of the solid sections, namely:

$$\frac{q}{A} = U(t_o - t_i) \tag{18.6}$$

where:
U = overall heat-transfer coefficient, which is also called the U-value, W/m^2·K (Btu/hr·ft^2·°F)

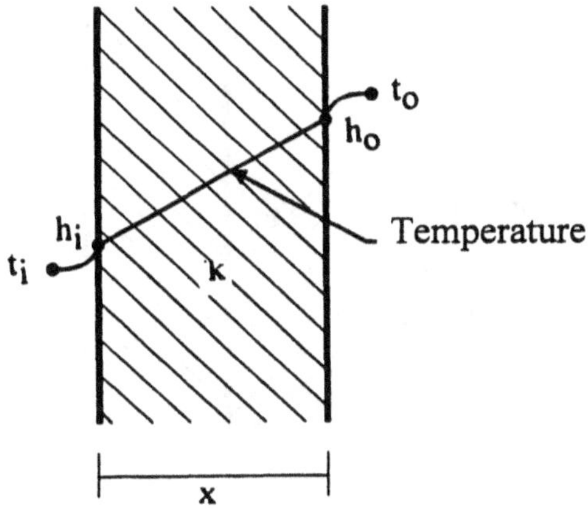

FIGURE 18.4
Heat transfer through a wall of single material with convection coefficients on both sides.

Comparison with Eq. 18.5 shows that the U-value is the reciprocal of the resistance:

$$U = \frac{1}{R}$$

For a single material with convection coefficients on each side, as shown in Fig. 18.4, the expressions for the rate of heat transfer and the U-value are as follows, respectively:

$$\frac{q}{A} = U(t_o - t_i)$$

where:

$$U = \frac{1}{R_1 + R_2 + R_3}$$

$$R_1 = \frac{1}{h_o}, \quad R_2 = \frac{x}{k}, \quad \text{and} \quad R_3 = \frac{1}{h_i}$$

The typical roof, floor, or wall consists of a combination of several materials, as illustrated in Fig. 18.5. The principal purpose of one of the materials might be to provide structural strength and that of another layer to provide the thermal insulation. There might even be another layer to provide a tough surface that is resistant to abrasion.

In the building member of Fig. 18.5, the total resistance is as follows:

$$R_T = \frac{1}{h_o} + \frac{x_1}{k_1} + \frac{x_2}{k_2} + \frac{1}{h_i}$$

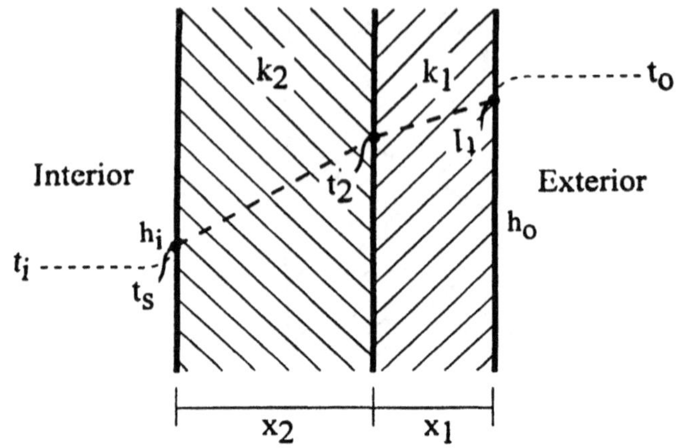

FIGURE 18.5
Heat transfer through a composite wall.

Example 18.1. The vertical wall shown in Fig. 18.5 is composed of 200-mm (8-in) cinder concrete block and 75-mm (3-in) polyisocyanurate insulation. In the summer condition, when the exterior temperature is 35°C (95°F) and the interior temperature is −18°C (0°F), what is the rate of heat transfer through an area of $6\,m^2$ ($64.6\,ft^2$)?

Solution. The following resistances apply:

exterior convection (from Table 18.3) $1/23 = 0.0435\,m^2{\cdot}K/W$ ($0.25\ hr{\cdot}ft^2{\cdot}°F/Btu$)

cinder concrete block (from Table 18.2) $0.3\,m^2{\cdot}K/W$ ($1.72\,hr{\cdot}ft^2{\cdot}°F/Btu$)

polyisocyanurate (from Table 18.1) $(0.075\,m)/0.020 = 3.75\,m^2{\cdot}K/W$ ($21.3\ hr{\cdot}ft^2{\cdot}°F/Btu$)

interior convection (from Table 18.3) $1/8.3 = 0.1205\ m^2{\cdot}K/W$ ($0.684\ hr{\cdot}ft^2{\cdot}°F/Btu$)

The total resistance of the wall is: $0.044 + 0.300 + 3.750 + 0.12 = 4.214\ m^2{\cdot}K/W$ ($23.94\ hr{\cdot}ft^2{\cdot}°F/Btu$)

The rate of heat transfer through the entire wall is as follows:

$$q = (6\,m^2)[35 - (-18)]/4.214 = 75.5\,W\ (257\,Btu/hr)$$

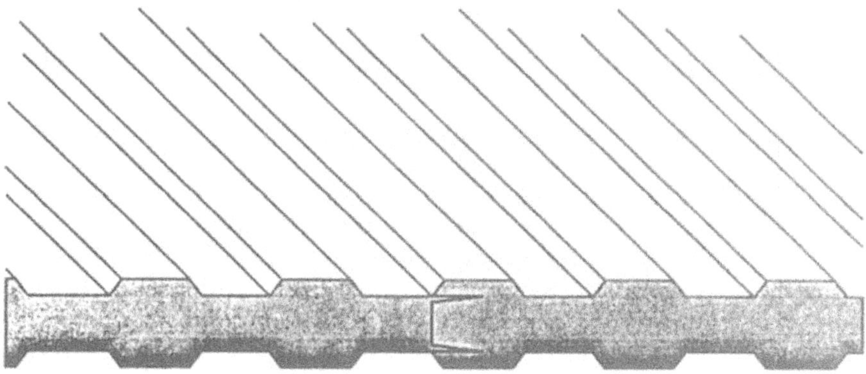

FIGURE 18.6
A wall or roof panel.

18.7 WALL AND ROOF PANELS

Combining a structural material with insulation, as in Fig. 18.5, and applying an abrasion-resistant coating to protect the insulation is a traditional method of constructing walls for refrigerated structures. Perhaps even surpassing the popularity of this method is now the use of wall and roof panels, as shown in Fig. 18.6. The panel is constructed as a sandwich of insulation foamed in place and tightly bonded to metal skins. Polyurethane, isocyanurate, or other insulating materials are protected by steel, aluminum, or plastic sheets. The edges of the panels are provided with some form of tongue-and-groove arrangement which facilitates sealing the joints.

To provide additional structural rigidity, the metal skin is usually corrugated in some fashion, which permits wall panels to be tilted in position during construction and ceiling panels to be laid on joists. Several panel thicknesses are available, providing a variety of thermal resistances, as shown in Table 18.4. Panel manufacturers normally provide a standard width of panel but produce lengths according to the customer's order. Thick panels are required for long spans, and these spans sometimes extend to lengths of perhaps 9 m (30 ft).

18.8 DESIGN TEMPERATURES

After the U-value or the R-value is determined, another set of data needed to calculate the rate of heat transfer from Eq. 18.6 are the interior and exterior air temperatures. The temperature inside the refrigerated space is determined by the activity within the space or the product being stored. For the refrigeration of food, the recommended storage temperatures for various products have been addressed in Chap. 17. The exterior temperature is a function of the weather in the particular locality, and since the maximum refrigeration load will usually occur at high outdoor temperatures, the summer design temperatures apply.

TABLE 18.4
Thermal resistances and recommended applications of prefabricated wall and ceiling panels.

Thickness mm (inch)	Thermal resistance $m^2 \cdot K/W$ ($hr \cdot ft^2 \cdot {}^\circ F/Btu$)	Recommended temperature application
50 (2)	2.92 (16.6)	Ambient
75 (3)	3.66 (20.8)	Down to 0°C (32°F)
100 (4)	5.87 (33.3)	Down to −29°C (−20°F)
125 (5)	7.33 (41.6)	Down to −45°C (−50°F)
150 (6)	8.81 (50.0)	Down to −57°C (−70°F)

TABLE 18.5
Design summer dry- and wet-bulb temperatures for five U.S. cities.

City	Dry-bulb temperature equaled or exceeded 1% of summer hours	Wet-bulb temperature coincident with 1% dry-bulb	Wet-bulb temperature equaled or exceeded 1% of summer hours
Atlanta	34°C (94°F)	23°C (74°F)	26°C (78°F)
Chicago	34°C (94°F)	24°C (75°F)	26°C (78°F)
Dallas	39°C (102°F)	24°C (75°F)	26°C (78°F)
Los Angeles	34°C (94°F)	21°C (70°F)	22°C (72°F)
New York	33°C (92°F)	23°C (74°F)	24°C (76°F)

Table 18.5 extracts temperatures for a few U.S. cities from more-extensive tables in the ASHRAE *Handbook—Fundamentals*[1].

The design temperature that is chosen is not the highest temperature likely to occur during the summer, because the peak temperatures occur for only a few hours at a time, and this peak is damped out because of the thermal capacity of the wall and roof structure. The first column of data in Table 18.5 indicates the outdoor dry-bulb temperature that is equaled or exceeded 1% of the hours during the summer.

The temperatures in Table 18.5 are used directly in design calculations for selecting condensers and computing the refrigeration load attributable to infiltration. For computing the rate of heat transfer by conduction through walls, roofs, and floors, the outdoor design temperatures are often not used directly. In the first place, the temperature applicable to the floor calculation is the soil temperature and not the air temperature. Chapter 19 explains the need for heating the soil beneath a low-temperature space, in order to maintain the soil temperature at about 10°C (50°F). For most refrigerated buildings, the temperature on the bottom side of the floor will be in the range of 5 to 10°C (41 to 50°F).

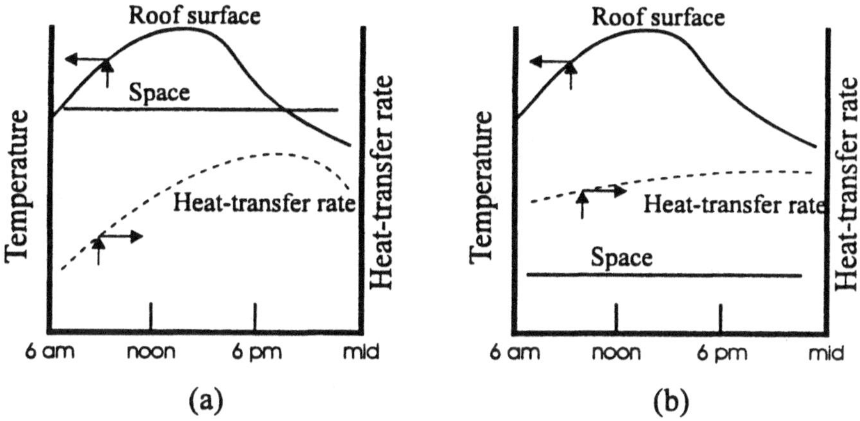

FIGURE 18.7
Roof surface temperatures and rate of heat entering the space in (a) an air-conditioning application, and (b) a refrigeration application.

In the case of walls and roofs, however, the outdoor air is indeed in contact with the outer surface of the building members, but another consideration arises, and that is the solar effect. The temperature of the outside surface of a roof may rise to 50 to 55°C (120 to 130°F) during the day, but even this temperature is not the outside temperature to be used in Eq. 18.6. Another factor that enters into the calculation is the thermal lag in the heat transfer through the wall or roof. Thus, if the peak temperature difference occurs at noon, for example, the peak rate of flow into the building might not occur until four or five hours later. Designers performing load calculations for air-conditioning systems must cope with more complications than do designers of refrigerated structures for the reason illustrated in Fig. 18.7. The designer calculating an air-conditioning load, in the situation reflected by Fig. 18.7a, must perform the calculations at several different times in order to identify the peak load. In the case of the large temperature differences encountered in refrigeration loads (Fig. 18.7b), the percentage change throughout the day is moderate. Many designers use a value of t_o in Eq. 18.6 higher than the design air temperature indicated by Table 18.5, but the influence of the adjustment of t_o on the total calculated load is small.

Table 18.5 also shows design wet-bulb temperatures that are used in computing the moisture load due to infiltration (Sec. 18.9) as well as in the selection of evaporative condensers or cooling towers. Column 3 shows the design wet-bulb temperature, but this value is not likely to occur simultaneously with the design dry-bulb temperature. Instead, column 2 shows the mean value of the wet-bulb temperature that coincides with the design dry-bulb temperature. It is this coincident design wet-bulb temperature that is normally used in conjunction with the design dry-bulb temperature.

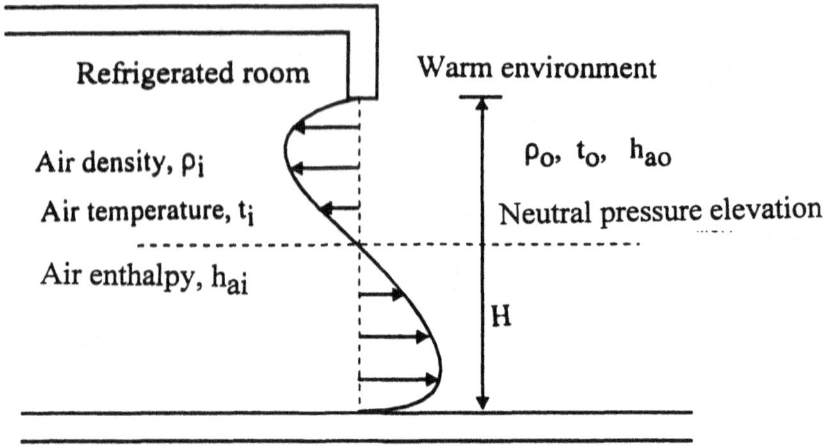

FIGURE 18.8
Infiltration of air due to exchange of air between a refrigerated room and the warm environment.

18.9 INFILTRATION THROUGH OPEN DOORWAYS

When an opening exists between a cold room and a warm exterior, there is a natural exchange of air between the two regions, as illustrated in Fig. 18.8. The air pressures of the cold and warm spaces equalize at the neutral pressure elevation. Above and below that elevation the air pressures differ inside and outside the room due to the static force of the column of air, which is a function of the air density. The air pressure increases at a greater rate with respect to elevation inside the cold room, because the air density is greater on the inside than on the outside. This means that the air pressure increases at a more rapid rate inside the room below the neutral elevation and decreases at a more rapid rate above the neutral elevation in comparison to pressure changes outside the room. These differences in air pressure cause the cold air to flow out from the cold room at the bottom and the warm air to flow into the room at the top of the opening. In the steady-state situation, the mass rate of flow into the room equals that flowing out.

Because of this air exchange, there is a combination of sensible load (due to the temperature differences of air) as well as a latent load, because the external air is likely to have a higher humidity ratio than that inside the cold room. The water vapor transfer has a double effect in that it not only constitutes a refrigeration load, but in the case of a room operating at subfreezing temperatures, the moisture will ultimately deposit on the coils, requiring defrost with the attendant load and loss in efficiency.

In the practical operation of a refrigerated facility, doors must be opened at times in order to move the product in and out. The infiltration load is one of the major loads in a refrigerated facility, varying from about one-fourth[2] to one-third of the total refrigeration load, depending on such conditions as the internal and external temperatures and humidities, the number and size of doors, and the amount of time the doors are open. There could be two different approaches to addressing the existence of infiltration. An operator of a plant may be satisfied to realize that infiltration constitutes a signficant load and try to reduce it by controlling door openings and/or installation of infiltration inhibitors (Sec. 18.10). The other approach is to quantify the load through calculations, which is certainly necessary for the designer of the facility, as well as for the operator in certain cases in order to economically evaluate methods of controlling infiltration.

The most widely used formula for computing infiltration is probably that of Gosney and Olama[3]. This equation is expressed as follows:

$$Q = C_{inf} A\sqrt{H} \left(\frac{\rho_i - \rho_o}{\rho_i}\right)^{1/2} \left[\frac{2}{1+(\rho_i/\rho_o)^{1/3}}\right]^{3/2} \tag{18.7}$$

where:

Q = volume rate of flow, m^3/s (ft^3/s)
C_{inf} = infiltration coefficient = 0.692 $\sqrt{m}/s$ (1.254 $\sqrt{ft}/s$)
A = area of doorway, m^2 (ft^2)
H = height of doorway, m (ft)
ρ_i and ρ_o = air densities, kg/m^3 (lb/ft^3) of the cold and warm air, respectively

The mass rate of flow, $\dot{m}$, is the product of the volume rate of flow and the mean density of air:

$$\dot{m} = Q\left(\frac{\rho_o + \rho_i}{2}\right)$$

where:

$\dot{m}$ = mass flow rate, kg/s (lb/s)

Finally, the refrigeration load (sensible plus latent) is the mass rate of flow multiplied by the difference in enthalpies:

$$q = \dot{m}(h_{a,o} - h_{a,i})$$

where:

q = refrigeration load, kW (Btu/s)

Example 18.2. The conditions in a refrigerated room are −18°C (−0.4 F) and 90% relative humidity, while the ambient conditions are 26°C (79°F) and 70% relative humidity. Compute the rate of heat transfer due to infiltration for 1 m^2 (ft^2) of door area if the open door is 3 m (9.84 ft) high.

Solution. Air properties may be obtained from the psychrometric charts:

$\rho_i = 1/(0.724 \text{ m}^3/\text{kg}) = 1.381 \text{ kg/m}^3$ (0.0864 lb/ft³)
$\rho_o = 1/(0.868 \text{ m}^3/\text{kg}) = 1.152 \text{ kg/m}^3$ (0.0719 lb/ft³)
$h_{a,i} = -16.6$ kJ/kg (0.7 Btu/lb)
$h_{a,o} = 63.8$ kJ/kg (35.2 Btu/lb)

Substituting into Eq. 18.7 gives

$$Q = 0.692(1 \text{ m}^2)\sqrt{3 \text{ m}}\left(\frac{1.381 - 1.152}{1.381}\right)^{1/2}\left[\frac{2}{1 + (1.381/1.152)^{1/3}}\right]^{3/2}$$

$Q = 0.466 \text{ m}^3/\text{s}$ through an area of 1 m² or (1.529 ft³/s through 1 ft²)

The mass rate of flow is

$$\dot{m} = 0.466\left(\frac{1.381 + 1.152}{2}\right)$$

$\dot{m} = 0.590$ kg/s through 1 m² (0.121 lb/s through 1 ft²)
The refrigeration load is:
$q = (0.590 \text{ kg/s})[63.8 - (-16.6) \text{ kJ/kg}]$
$q = 47.4$ kW per m² of door area (4.19 Btu/s per ft²)

The experimental facilities used to verify Eq. 18.7 were small scale, so there has been some concern about the applicability to commercial size doors and refrigerated warehouses. To check the validity for large-scale facilities, ASHRAE sponsored a research project in which Hendrix[7] acquired field data. The results showed that the Gosney-Olama equation, Eq. 18.7, indicated slightly greater infiltration rates (by 10%) than measured, but thus gave useful conservative predictions. Hendrix also explored two questions surrounding the transient behavior of the air flow through a doorway, and these were: how long does it require for the full velocities to develop after the door is opened, and does the flow rate subside during long durations of door openings? The field tests indicated that full velocities establish themselves within 2 or 3 seconds, so there is no basis to reduce the magnitudes of Eq. 18.7. With respect to potential reduction in the infiltration load after a long door opening, because of the air inside the room close to the door becoming warm, the 1989 ASHRAE study of large rooms showed no such phenomenon. These tests thus support the use of Eq. 18.7 in its basic form.

To facilitate the determination of the infiltration load of an open door, Figs. 18.9 and 18.10 present graphs of the infiltration load through open doorways for various temperatures of the infiltration air and the refrigerated space at four different ambient relative humidities. The relative humidity inside the space applicable to all the curves is 90%, which would be fairly typical of facilities storing both frozen and unfrozen foods. A deviation from the 90% relative humidity inside the space results in little error, because a change in percent relative humidity at low temperatures results in only a small change in humidity ratio. The curves in Figs. 18.9 and 18.10 are applicable to a door height of 3 m

(9.84 ft). For other door heights, the refrigeration load should be multiplied by a correction factor of $\sqrt{H, \text{m}/3}$ ($\sqrt{H, \text{ft}/9.84}$), as is apparent from Eq. 18.7. For comparison, Fig. 18.10 shows the same heat gain as calculated for the conditions applicable to Example 18.2.

A point to emphasize is that the refrigeration load contributed by an open door to the refrigerated space can be considerable. For example, suppose that a door is left standing open between a −20°C (−4°F) frozen food storage space and the dock, which is at a temperature of 10°C (50°F) and a humidity of 70%. If the door area is 9 m^2 (97 ft^2), for example, the rate of heat gain is 194 kW (55 tons of refrigeration). Since such a load can hardly be tolerated, the doors should be permitted to be open only when traffic is moving through them.

While Eq. 18.7 or the curves in Figs. 18.9 and 18.10 offer a reasonably accurate estimate of the refrigeration load due to infiltration, a much more illusive number is the fraction of time that a door is likely to be open. This quantity is primarily a function of the type of operation, and the selection of the fraction must be based heavily on past experience. A quantification of the doorway traffic can be derived from a standard industry term, *turnover*, which is the number of times the complete storage capacity is depleted and restocked. The number of turnovers per year is dependent upon the function of the space. For a refrigerated warehouse storing mostly seasonal products, the number may be 8 to 12, while for a grocery distribution warehouse it may be 30. With the latest concepts in material handling, annual turnovers may reach 100. Each turnover is associated with a lift truck moving in and out, causing an opening of the door of perhaps 10 to 15 seconds.

18.10 INHIBITING INFILTRATION LOADS

Because of the considerable magnitude of infiltration loads (perhaps 25 to 30% of the total refrigeration load), means of reducing these loads should be considered. Some possibilities are: (1) air curtains, (2) vestibules, (3) plastic strips, and (4) refrigerated loading docks.

The vertical type air curtain shown in Fig. 18.11 uses ambient air. The air curtain forms a two-dimensional free-stream jet, into which outside air is induced into the outside surface and cold air induced into the inside surface. When the jet reaches the floor, the induced air from the ambient spills back to the ambient, and the air induced from the refrigerated space spills back to that space. The air spilling back to the refrigerated space has mixed with warm humid air, so the infiltration load is that represented by the mass rate of flow of inside spill air multiplied by the difference in enthalpy of this spill air and the air in the refrigerated space.

Hayes and Stoecker[5,6] conducted a combined analytical and experimental study of air curtains and found that from 65 to 80% of the infiltration load can be avoided by using a properly applied air curtain in comparison to an open door. The tests verified the bending of the air curtain into the cold space at the

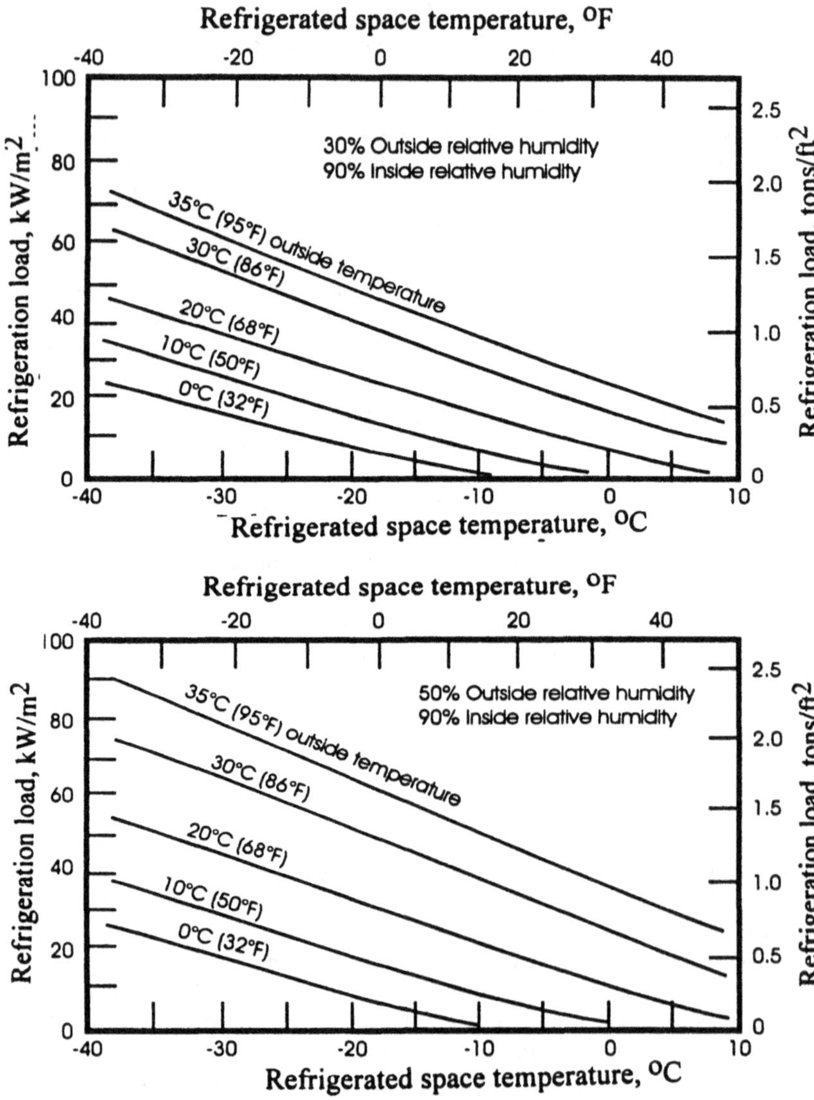

FIGURE 18.9
Refrigeration loads through open doorways for 30 and 50 percent outdoor relative humidities. For door heights other than 3 m (9.84 ft) multiply load by H m/3 (H ft/9.84).

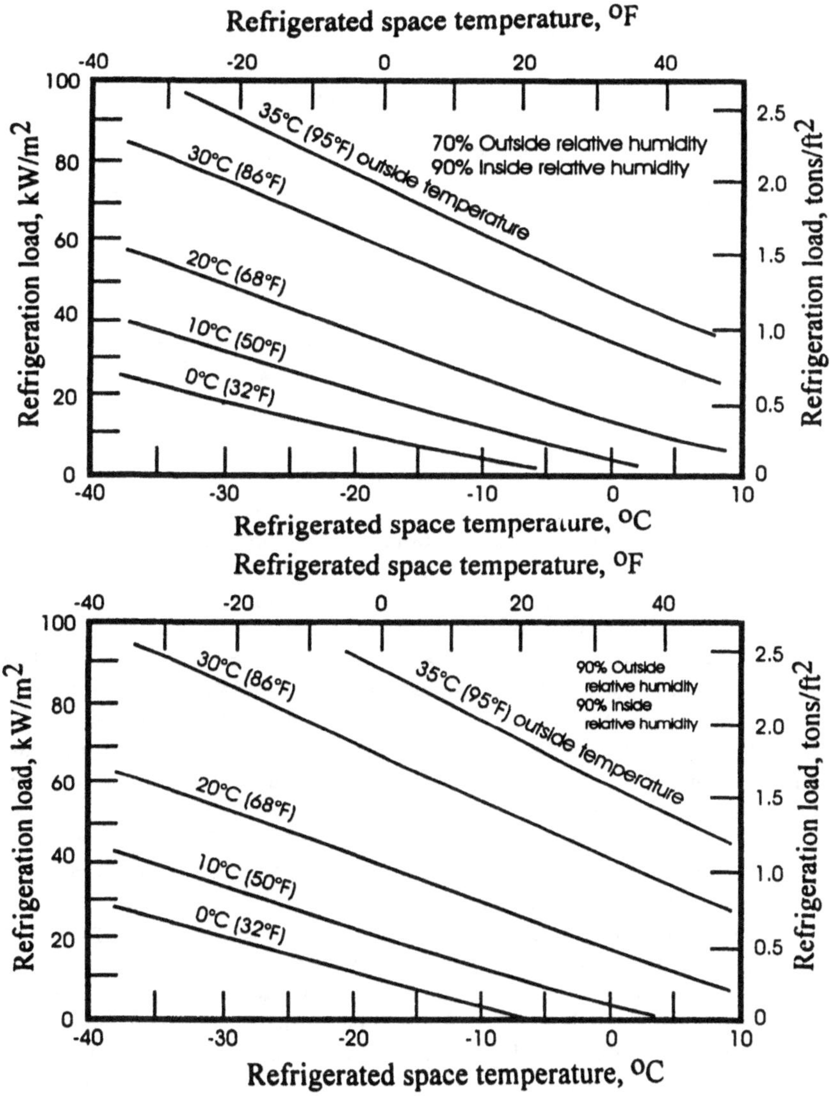

FIGURE 18.10
Refrigeration loads through open doorways for 70 and 90 percent outdoor relative humidities. For door heights other than 3 m (9.84 ft), multiply load by H m/3 (H ft/9.84).

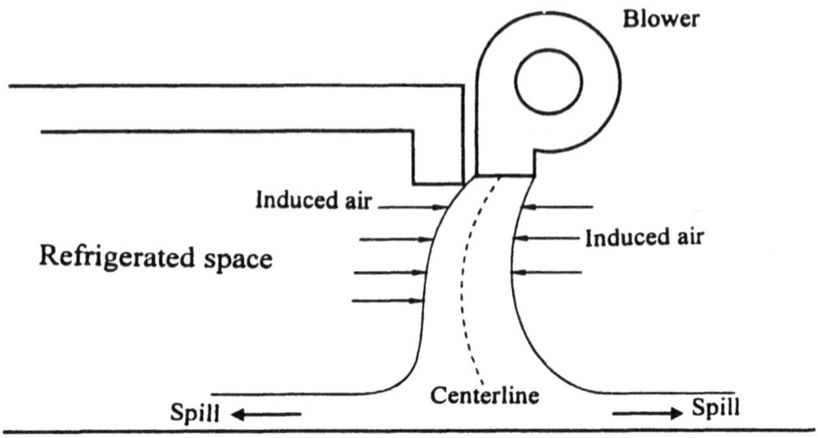

FIGURE 18.11
An air curtain resisting infiltration through an open door.

top and away from the space at the bottom. A failure of the air curtain occurs when the jet velocity is so low that the curtain *breaks* and the jet fails to reach the floor. In that case, cold air pours out at the floor level and is replaced by warm air flowing in at the top.

Some efforts in the Hayes/Stoecker tests to measure pressure differences between the cold space and the ambient suggest a fundamental rule for applying air curtains. Air curtains are very flexible therefore, they can easily deflect to change the rates of spill (as shown in Fig. 18.11) into the space and into the ambient. Thus, if a gust of wind strikes the outside surface of the curtain, the curtain deflects to allow net flow of air into the space to build up the pressure in the space, countering the overpressure on the outside. The air curtain breathes when one air curtain protects a door of a space that otherwise has few leaks. The installation of air curtains on opposite sides of a refrigerated space is to be avoided, because both air curtains will deflect to allow an unacceptable rate of infiltration due to the wind blowing directly through the space.

Another concept in air curtains is the horizontal curtain with recirculated air[7], shown schematically in Fig. 18.12a. The air is directed from the supply side on one side of the door, and most of the air is drawn into the return plenum on the opposite side of the door, as shown in Fig. 18.12b. This concept appears to provide a stiffer curtain than is possible in the vertical, nonrecirculated arrangement, and the conditions of the air in the active portion of the curtain are a blend of those of the ambient and the refrigerated space.

Vestibules are effective in reducing air infiltration but are expensive and also slow fork-lift traffic in and out of the space.

The third approach to reducing the infiltration load is to use plastic strips at the doorways in one of the commercially available forms. The simplest form

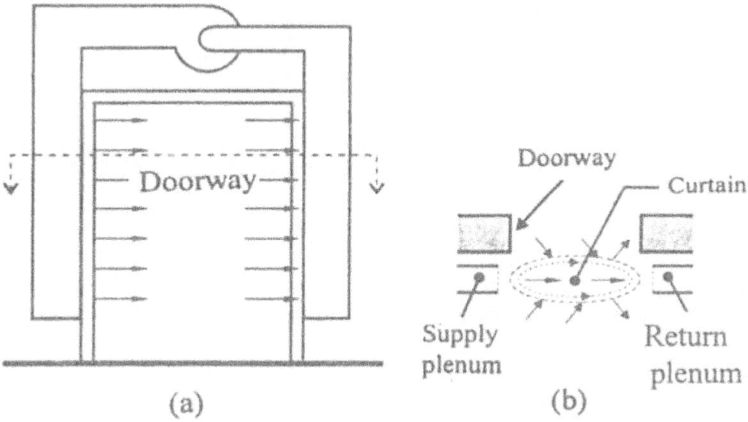

FIGURE 18.12
(a) A horizontal, recirculated air curtain, and (b) the plan view.

of plastic strip curtains is as a single fixed-in-place curtain immediately inside or outside the sliding door. Tests[8] conducted in a refrigerated warehouse indicate that these strips are about 90% effective in reducing infiltration in comparison to an open door when there is no traffic through the doorway. When there is fork-lift truck traffic through the doorway, the effectiveness is approximately 85%, depending on the intensity of the traffic. While fixed-in-place plastic strip curtains are effective in reducing infiltration, they can become cold and stiff and are subject to breakage. Also, they are not popular with fork-lift truck operators. To overcome these drawbacks and still achieve the benefits of plastic strips, they are available on hinged and sliding doors as well. The light weight of these doors facilitates easy movement of them.

18.11 CALCULATING THE REFRIGERATION LOAD FOR DOCKS

The fourth concept in reducing infiltration is to provide a refrigerated loading dock, as in Fig. 18.13, which is a design feature that is almost a requirement for a frozen food storage space. The first benefit, and perhaps the most important, is the protection of the frozen product. Rather than requiring any residence time on a non-refrigerated loading dock, the product has at least a cool ambient temperature during those situations when the product does not move directly between the truck and the storage space. The second benefit of refrigerated docks is in the reduction of infiltration load. During periods of heavy traffic, the infiltration air entering from the outdoors that is possibly warm and humid first enters the loading dock, and that infiltration load is absorbed by the evaporator coils lo-

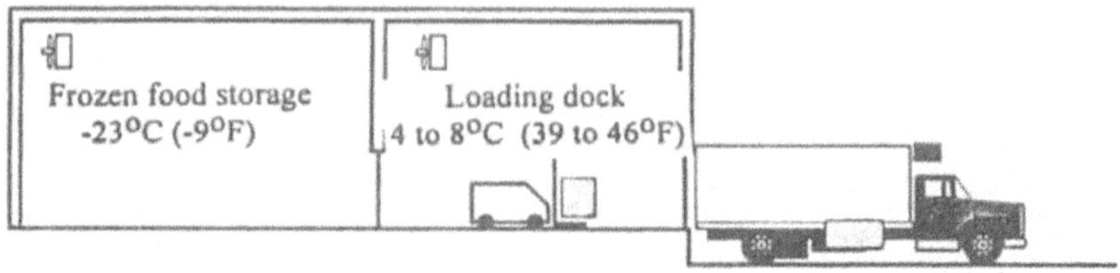

FIGURE 18.13
Refrigerated loading dock to reduce refrigeration load in the low-temperature space.

cated within the dock area. These coils can be served by refrigerant evaporating at an intermediate temperature level, which offers lower-cost refrigeration than in the frozen food space. The infiltration to the low-temperature space is of dock air, which is maintained in the 4-8°C (39-46°F) temperature range. Still another benefit of the refrigerated dock is that considerable moisture is condensed on the dock coils as liquid. Thus, the moisture is prevented from entering the low-temperature space, where it will deposit as frost on the coils and increase the frequency of defrost.

Example 18.3. A frozen food storage space operates at −20°C (−4°F) and 90% relative humidity. When the ambient conditions are 35°C (95°F) and 50% relative humidity, compute the rate of heat gain due to infiltration (a) if there is no refrigerated dock, and (b) if a refrigerated dock at 10°C (50°F) and 70% relative humidity protects the refrigerated space, as in Fig. 18.14.

Solution. (a) Figure 18.9 indicates a heat gain rate of 65 kW/m^2 (1.7 $tons/ft^2$) for the unprotected space.

(b) When the space is protected by a dock at 10°C (50°F) and 70% relative humidity, Fig. 18.10 shows that the rate of infiltration heat gain is only 21 kW/m^2 (0.55 $tons/ft^2$), which is about 32 percent of the load with no dock. Furthermore, each kg (lb) of air from the dock holds 0.0053 kg (lb) of water in comparison 0.0178 kg (lb) of ambient air.

Example 18.3 provides further insight into the influence of refrigerated docks. With the dock in Example 18.3b, the rate of heat gain to the space is 21 kW/m^2 (0.55 $tons/ft^2$), which is refrigeration load for the low-temperature space but *refrigerating capacity* for the dock. Furthermore, using Fig. 18.10, the infiltration load from the ambient to the dock is 26 kW/m^2 (0.68 $tons/ft^2$), which at least superficially means that hardly any refrigeration load is imposed on the dock cooling coils. The comparison between the two rates is actually not appropriate, because the area and degree of opening of the doors from the ambient to the dock are different than those between the dock and the refrigerated space. Also, the relative humidity of 70% was an assumption, and if it is lower then 70%, the net refrigeration load in the dock area increases.

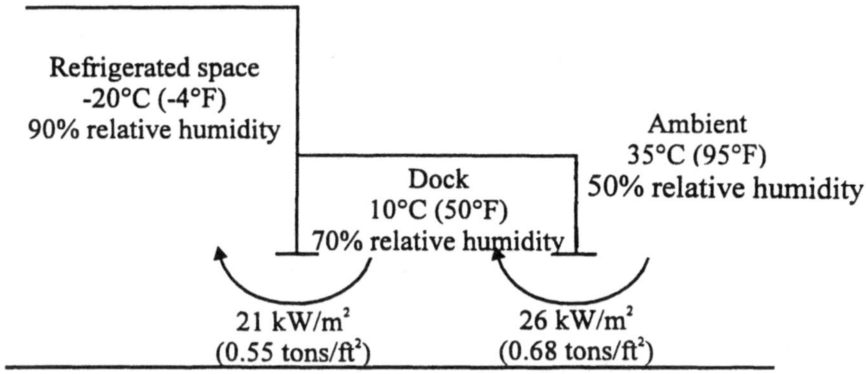

FIGURE 18.14
Heat flow rates in Example 18.3.

Even though the magnitudes of infiltration load given above may be subject to question, they do emphasize the fact that we really want the dock coils to be heavily loaded in order to remove water vapor from air before it reaches the refrigerated space. For this reason, some plants that experience heavy frosting conditions in the low-temperature space, even though equipped with refrigerated docks, may apply heat in the dock space to force the dock coils to operate more and thus perform dehumidification.

The choice of dock temperatures in the range of 2 to 6°C (35 to 43°F) conforms with tradition and represents a logical approach. One of the important reasons for a refrigerated dock serving a frozen food space is to reduce the refrigeration load, including the inflow of moisture to the low-temperature coils. If the dock temperature is too high, the infiltration load to the low-temperature space will be excessive. If the dock temperature is too low, some of the operating problems, such as the frosting of coils, will simply be shifted to the dock from the low-temperature space. Designers of new facilities should be alert to the potential innovations in the handling of frozen food that are appearing on the horizon. One possible trend is toward lower dock temperatures for protection of the product. One recently constructed frozen food storage facility[10] operates its dock at the same temperature as the storage space, −23°C (−10°F), so that the product is never subjected to temperatures higher than desired storage temperatures. The firm is convinced of the improvement in quality that results from this mode of operation. Several of the purposes relative to refrigeration loads of an intermediate-temperature loading dock are compromised, however. Also, the dock is less comfortable for the workers.

Another potential trend is toward larger-area docks. The storage time of products in refrigerated warehouses continues to shrink, and the ultimate extrapolation of this trend appears to be the situation now being experienced by one warehouse[11], which meets the need of their customers to move product from one vehicle to another. To repond to this need, the warehouse has constructed a large refrigerated trans-loading dock just for this purpose.

18.12 LOAD FROM LIGHTS, MOTORS, PEOPLE, AND OTHER INTERNAL LOADS

Heat liberated within the refrigerated space adds to the refrigeration load. In general, all the internal releases of heat convert directly and instantaneously into refrigeration loads. Some of the major contributors are lights, motors, people, fork-lift trucks, and processing equipment. The proportions of these loads are functions of the use of the refrigerated space. The temperature maintained in a meat-processing room, for example, may be 10 to 14°C (50 to 57°F). The lighting level would be high enough to illuminate detail work, and electrical consumption for lights could be in the range of 16 W/m^2 (1.5 W/ft^2). In a frozen food warehouse where the lights need only provide general illumination, the average refrigeration load attributable to lights could be 5 W/m^2 (0.48 W/ft^2). The concentration of people in the meat preparation room would be much higher than in the frozen food storage space.

In the meat-processing room, there will likely be some processing equipment that is electrically driven, such as slicers and grinders. Estimates of the contribution of such equipment to the refrigeration load may be computed by determining the electrical rating of the equipment. Since the motors are probably not being loaded to their full rating, a load factor of perhaps 0.6 to 0.8 can be applied. Except for stacker crane installations, fork-lift trucks operate in frozen food storage rooms. The power ratings of fork-lift trucks may vary from 5 to 10 kW, but once again the fork-lift truck does not operate at full power constantly.

People give off heat, and in heavily occupied work rooms the magnitude may be noticeable on the refrigeration plant. The rate of heat release from a person depends on the activity level and also on the temperature of the space. A comfort air-conditioning system may need to remove 100 W (350 Btu/hr) combined sensible and latent heat for each person seated quietly but twice that amount per person engaged in physical activity. In the case of refrigerated spaces, the rate of heat release increases as the temperature drops, as shown in Table 18.6. Also, when people go into refrigerated spaces for short durations of time, they carry in an appreciable amount of heat in their warm clothing.

The electric motors used to drive the fans on the cooling coils are in operation in virtually all refrigerated spaces. All of the electricity supplied to these motors eventually converts to heat and must be removed by the coil. Figure 18.15 shows fan power requirements as a function of coil capacity for a family of coils. The two different curves represent high and low fan speeds, which reflect themselves in different air flow rates and face velocities for a coil. Figure 18.15 applies to coils operating with a 5°C (9°F) temperature difference between entering air and refrigerant, typical of a frozen food warehouse or a blast freezer coil.

TABLE 18.6
Heat equivalence of occupants[9].

Space temperature, °C (°F)	Heat rate per person, W (Btu/hr)
10 (50)	210 (720)
0 (32)	270 (920)
−10 (14)	330 (1,130)
−20 (−4)	390 (1,330)

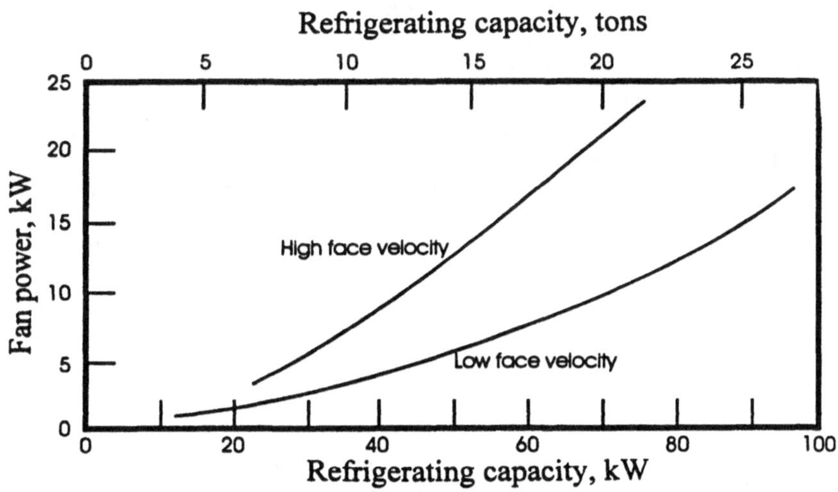

FIGURE 18.15
Fan power for a certain family of coils operating with two different face velocities. The temperature difference between the entering air and refrigerant is 5°C (9°F).

An important observation from Figure 18.15 is that, in the case of the low-velocity coil, approximately 8 to 15% of the refrigeration capacity is used simply to remove the heat caused by its own fan motor. For the high- velocity coil, the percentage of fan and motor heat is likely to be over 20% of the coil capacity. This contribution to the refrigeration load is critical, particularly in low-temperature applications where the temperature difference between the entering air and refrigerant is low. In the case of higher-temperature refrigerated spaces where the air-to-refrigerant temperature difference may be twice that chosen for Fig. 18.15, the percent of parasitic heat attributable to the fan and motor is half that shown.

A reasonable question is why choose the high face velocity coil if the fan power is relatively high? The answer is that a coil with a given surface area will transfer perhaps 30% more heat when operating at the high velocity for a given air-to-refrigerant temperature difference. Furthermore, the throw of the high-velocity coil will be longer than that of the low-velocity coil. While the first cost of the high-velocity coil will generally be less than that of the low-velocity coil, the designer should evaluate the lifetime costs of the selection.

18.13 DEFROST HEAT

Defrosting with hot gas is the dominant method applied in industrial refrigeration, and it cannot be denied that hot-gas defrost (also water defrost) adds heat to the refrigerated space and is thus a load. In addition, any refrigerant vapor that escapes through the pressure-regulating valve at the outlet of the evaporator during defrost increases the power requirement in the machine room to recompress this gas. Despite defrost constituting a refrigeration load, probably a minority of system designers incorporate it in their load calculation. The calculations are cumbersome if done for all coils in the facility, and the prediction is unreliable. The designers who do not include defrost load do not usually encounter trouble, because the calculations of so many other components of the refrigeration load are conservative.

The efficiency of the hot-gas defrost process may be defined as follows:

$$\text{Efficiency} = \left(\frac{\text{Heat to melt frost}}{\text{Heat provided by hot gas}}\right)100$$

The difference between the two thermal quantities in the efficiency equation are attributable to the energy used to raise the temperature of the coil and to the heat transferred to the refrigerated space during defrost. Some laboratory tests, field tests, and analyses[12,13,14] of the hot-gas defrost process suggest an efficiency approximately 20% and a hot-gas flow rate that is between one and two times the refrigerant flow rate during refrigeration. (Plant operators try to ensure that at least two or three evaporators are in their refrigerating mode while one evaporator is defrosting in order to provide the vapor that will be needed for defrost.)

Example 18.4. To check the order of magnitude of the recommended estimate of the defrost load, compute the average refrigeration load contributed by the defrost process if an ammonia coil with a nominal capacity of 106 kW (30 tons) defrosts twice a day for 20 min each.

Solution. If the refrigerant flow rate during refrigeration is 5.4 kg/min (12 lb/min), assume that the flow rate during defrost is 8.1 kg/min (18 lb/min) and that the latent heat of the ammonia is 1,240 kJ/kg (530 Btu/lb). The average rate of heat gain distributed over the 24 hour period is as follows:

Refrigeration load = (8.1)(1,240)(40 min)/(24 hrs) = 16,740 kJ/hr = 4.65 kW (15,900 Btu/hr or 1.3 tons)

The defrost load, in this case, is about 4% of the refrigeration load. The designer must decide whether this amount is significant.

When a pressure-regulating valve controls the release of condensed liquid refrigerant from the coil during hot-gas defrost, the escape of vapor through this valve adds to system load. This load can be reduced[15] through the use of a liquid trap at the coil outlet and/or by discharging the refrigerant to an intermediate pressure in contrast to the low evaporating pressure.

18.14 PRODUCT LOAD

The refrigeration load attributable to reducing and maintaining the temperature of products and/or freezing products may be the dominant contributor to the total load. Such is likely to be the case in a food processing and freezing plant where some loads, such as conduction through walls, roofs, and floors, and even infiltration might be insignificant. The calculation of the load due to the refrigerating and freezing of products, which was first introduced in Chapter 17, is in one sense straightforward but in another extremely difficult. It is simple to compute the quantity of heat that must be removed from the product in a refrigeration or freezing process. It is also easy to specify the desired time interval for the removal of this quantity of heat. This almost effortless calculation has failed to consider a key part of the process, however, which is the heat transfer between the cooling medium (refrigerant, air, water, etc.) and the product. The calculation of the refrigeration rate for the product load must ensure that, for example, air at a temperature of −34°C (−30°F) can actually freeze the desired number of boxes of meat in the blast freezer in the specified time interval.

The data on the product needed to make the load calculation are essentially specific heats, latent heat of freezing if freezing is part of the process, and heat of respiration for produce stored at a temperature above freezing. For cooling and immediate shipping of unfrozen products, only the information on the specific heat above freezing is needed. To calculate the refrigeration load for a freezing operation, the specific heats above and below freezing are needed as well as the latent heat of freezing of the product. When produce is stored in an unfrozen state, the heat of respiration must be considered. Such products are living organisms, and as such they give off heat (data for various products were presented in Table 17.3). Some magnitudes of properties were given in Chapter 17 on the refrigeration and freezing of foods, but more extensive data are available in Reference 1 for most food products. The refrigeration load encountered in freezing a product consists of three components: (1) reducing the temperature to the freezing point, (2) freezing, and (3) reducing the temperature further to the storage temperature. The refrigeration load in kJ (Btu) to convert a mass, m kg (lb), of warm product to a frozen food storage temperature is calculated by using the following equation:

$$Q_{total} = Q_{above} + Q_{freeze} + Q_{below}$$

The factors in the above equation are calculated as follows:

$$\begin{aligned} Q_{above} &= mc_{above}(t_{initial} - t_{freeze}) \\ Q_{freeze} &= mL \\ Q_{below} &= mc_{below}(t_{freeze} - t_{final}) \end{aligned}$$

where:

Q_{above} = heat removed above freezing, kJ (Btu)
m = mass of product, kg (lb)
Q_{freeze} = heat removed during freezing, kJ (Btu)
Q_{below} = heat removed below freezing, kJ (Btu)
c_{above} = specific heat above freezing, kJ/kg·K (Btu/lb·°F)
c_{below} = specific heat below freezing, kJ/kg·K (Btu/lb·°F)
L = latent heat of freezing, kJ/kg (Btu/lb)

Example 18.5. An IQF (individually-quick-frozen) freezer processing 0.8 kg/s (1.76 lb/s) of sweet corn kernels cools, freezes, and ultimately drops the temperature to −18°C (0°F). The product enters the process at a temperature of 40°C (104°F). What is the refrigeration capacity required for this IQF freezer?

Solution. The specific heats[1] of sweet corn are 3.53 kJ/kg·K (0.84 Btu/lb·°F) above freezing and 1.77 kJ/kg·K (0.42 Btu/lb·°F) below freezing, and the latent heat of freezing is 248 kJ/kg (106 Btu/lb). The freezing temperature is −0.6°C (31°F). The refrigeration rate is:
$q = (0.8)(3.53)[40 - (-0.6)] + (0.8)(248) + (0.8)(1.77)[-0.6 - (-18)]$
$q = 337$ kW (96 tons of refrigeration)

The product load in refrigerated storage facilities is often small, but an unexpected load might arise from the fact that the incoming product is not at the anticipated temperature. For example, product to be stored at 1°C (34°F) may come from a precooling facility also held at 1°C (34°F), so presumably there is no product load expected. It is possible, however, that the product was not held long enough at the initial site to bring the interior temperature of the product down to the steady-state temperature. Heat will continue to emerge from the product in its new location causing a product load. Another reason that the temperature of the incoming product may be higher than the ultimate storage temperature is that control of the temperature was lost in transit. Some public refrigerated warehouses routinely check the temperature of the interior of sample pallets of incoming product, because a high temperature may be the source of quality problems that become apparent later.

TABLE 18.7
Order of magnitude of refrigeration load for several applications.

Type of space	Refrigeration load	
	kW per 1,000 m^3	tons per 1,000 ft^3
Frozen food storage, single level	7.5 to 15	0.06 to 0.12
High-rise freezer storage	2.5 to 7.5	0.02 to 0.06
Produce storage	9.9 to 13.7	0.08 to 0.11
Shipping dock	13.7 to 25	0.11 to 0.2
Process area	20 to 60	0.16 to 0.5

18.15 LOAD CALCULATIONS IN PERSPECTIVE

Calculation of refrigeration loads is an inexact process, and because of this fact, most designers simplify some of the procedures outlined in this chapter. In making their short cuts, designers usually add generous safety factors resulting in extra refrigeration capacity for the plant. While this strategy results in a larger and more expensive refrigeration plant than initially required, this extra capacity is usually put to good use later as the facility expands or an unforseen load develops due to new products or processes.

This evaluation of load calculations should not suggest that sloppy calculations are not without their peril. An observation is that the total plant capacity provided by the load estimates of most competent designers is adequate, even generous. Where errors are more likely to occur is in the calculation of an individual load, resulting in the inability to maintain the desired temperature in a certain space or achieve the production rate of a certain freezing process. Experienced designers know that depending on the type of service (long-term food storage, short-term food storage, freezing operations, etc.), one of the components of the load will predominate. For example, the product load will be significant in a freezing operation, infiltration might dominate in short-term storage, while motors and lights are significant in processing areas.

Good designers intelligently use rules of thumb to check more-detailed calculations. Table 18.7[16] shows orders of magnitude of loads used by one designer, and if the calculation differs widely from these rules of thumb, further investigation of the calculation is warranted.

REFERENCES

1. American Society of Heating, Refrigerating and Air-Conditioning Engineers, Atlanta, Georgia, from the 1993 ASHRAE *Handbook— Fundamentals.*
2. Edmons, J., "Sizing and Application of Air Units for Low Temperature Storage Freezers," Seminar: A Rational Approach to Latent Heat in Low Temperature Rooms, Annual Meeting of ASHRAE, New York, New York, January 1987.

3. Gosney, W. B. and H. A. L. Olama, "Heat and Enthalpy Gains Through Cold Room Doorways," *Proceedings of the Institute of Refrigeration*, December 1975.
4. Hendrix, W. A., D. R. Henderson, and H. Z. Jackson, "Infiltration Heat Gains Through Cold Storage Room Doorways," *ASHRAE Transactions*, 95(Part 2): 1155-1168, 1989.
5. Hayes F. C. and W. F. Stoecker, "Heat Transfer Characteristics of Air Curtains," ASHRAE *Transactions*, 75(Part 2): 153-167, 1969.
6. Hayes F. C. and W. F. Stoecker, "Design Data for Air Curtains," ASHRAE *Transactions*, 75(Part 2): 168-180, 1969.
7. HCR, Inc., Lewistown, Montana, 1994.
8. Downing C. C. and W. A. Meffert, "Effectiveness of Cold-Storage Door Infiltration Protective Devices," ASHRAE *Transactions*, 99(Part 2): 356-366, 1993.
9. American Society of Heating, Refrigerating and Air-Conditioning Engineers, Atlanta, Georgia, from the 1994 ASHRAE *Handbook—Refrigeration.*
10. "Thrills and Chills," *Refrigerated and Frozen Foods*, p. 30, February 1994.
11. "Carving a Niche—State of Speciality Warehousing," *Warehouse Management*, p. 20, July/August 1994.
12. Niederer, D. H., "Frosting and Defrosting Effects on Coil Heat Transfer," ASHRAE *Transactions*, 82(Part 1): 467-473, 1976.
13. Stoecker, W. F., J. J. Lux, and R. J. Kooy, "Energy Considerations in Hot-Gas Defrosting of Industrial Refrigeration Coils," ASHRAE *Transactions*, 89(Part 2): 549-573, 1983.
14. Cole, R. A., "Refrigeration Loads in a Freezer due to Hot Gas Defrost and their Associated Costs," ASHRAE *Transactions*, 95(Part 2): 1149-1154, 1989.
15. Crombie, R. A., "Is Hot Gas Defrost Affordable?," *Meat Processing*, 1973.
16. Bonar, H. B. II, personal communication, 1994.

CHAPTER 19

REFRIGERATED STRUCTURES

19.1 REFRIGERATED ENCLOSURES

Refrigerated spaces might be needed for environment chambers to test products, to cold-treat metals, and for a variety of other special assignments. The refrigerated enclosures on which this chapter concentrates, however, are the spaces in which meat, fish, poultry, fruits, vegetables, dairy products, and other foods are cooled, maintained at a controlled temperature, and/or frozen. The design of refrigerated warehouses, spaces for cooling, and freezers incorporates such disciplines as structural design, materials-handling concepts, and thermal behavior. Because of the close connection of industrial refrigeration to the food industry, changes in food marketing and distribution impact quickly the design of refrigerated facilities. While this chapter emphasizes the thermal characteristics of refrigerated spaces, certain structural considerations must also be considered because of their interplay with the thermal requirements. Topics of special importance include types and characteristics of insulations, vapor barriers, underfloor heating, typical construction cross-sections of walls, roofs, and floors, and the junctions of walls with roofs and floors.

19.2 SPACE PLANNING

Even though some refrigerated storage warehouses and distribution centers are built without careful studies of effective solutions to the intended functions of the building, skillful analyses can usually result in a more efficient and profitable project. The owner should be pressed for data on the requirements of the building, and the designer should then translate these requirements into floor area and dimensions, building height, number and size of doors, and whether the product is stacked from the floor by fork-lift trucks or a high-rise automated stacker crane.

The plan for materials handling within the structure is becoming more important, because the typical number of turnarounds of the product per year is progressively increasing. The need for long-term storage applies to such situations as fruits and vegetables that are frozen during the harvest season and stored for steady use during the year for direct sale or for value-added packaging. Once a product is in a form for ultimate sale, the storage time may become very short, and the goal is to move the product as rapidly as possible to the supermarket or food service institution.

Methods for materials handling and record keeping of the location of stored products have become a sophisticated discipline. The traditional procedure for moving products completely in pallet loads with fork-lift trucks may in many cases be the most cost-effective method. Another approach involving greater mechanization is the application of automatic storage and retrieval systems (AS/RS). In the AS/RS concept, palletized unit loads are delivered to pickup and deposit stations at the end of crane aisles by fork-lift trucks. The stacker crane moves the product through the aisles and into the storage locations as well as recovers pallets that were previously stored. Several possible pallet storage configurations[1] are shown in Fig. 19.1. Because the AS/RS uses automated stacker cranes, the typical building height of perhaps 10 m (33 ft) no longer prevails. Instead, the building height may be 20 to 28 m (66 to 91 ft), which reduces the floor area for a given volume of storage.

Automatic storage/retrieval systems lend themselves to and, in fact, almost require computer control. Furthermore, modern methods of record keeping in storage and distribution warehouses employ electronic data interchange with the objective of eliminating the trail of paper that begins with a purchase order, continues with the record of receipt and the notation of storage location, indicates when the product is removed and shipped, and ultimately leads to the bill of lading and the invoice. The choice of concept for materials handling and method of record keeping are additional influences on the size and shape of the refrigerated structure.

19.3 INSULATION

Insulation is essential for low-temperature spaces in order to reduce the refrigeration load. The optimum thickness of insulation is based on such economic factors as the first cost of the insulation and the cost of refrigeration—both the first cost of the refrigeration plant per unit refrigeration capacity and the costs of energy and maintenance during the life of the facility. The classic expectation of the present worths of the contributing costs is shown in Fig. 19.2, where at

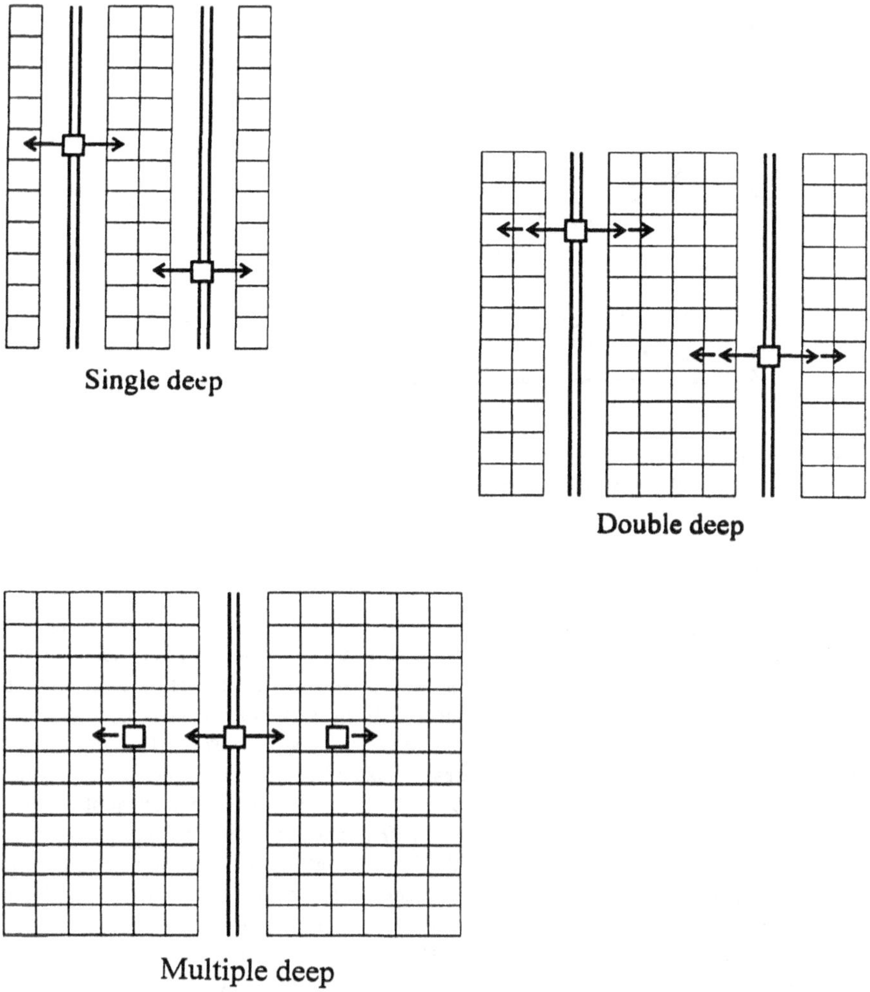

FIGURE 19.1
Several configurations for pallet storage when using stacker cranes.

low thicknesses of insulation, the operating cost is high and at large insulation thicknesses, the cost of the insulation is high. Since costs of insulation and energy change over time, it may be productive for the designer to periodically reevaluate optimum insulation thicknesses.

A variety of insulation materials is available, and most materials are available in either prefabricated panels or in board form. In addition, polyurethane insulation can be applied by foaming in place. Portable blending machines are available for foamed-in-place urethane, which incorporate a frothing nozzle to feed the partially-expanded mixture into wall, floor, or ceiling cavities. The in-

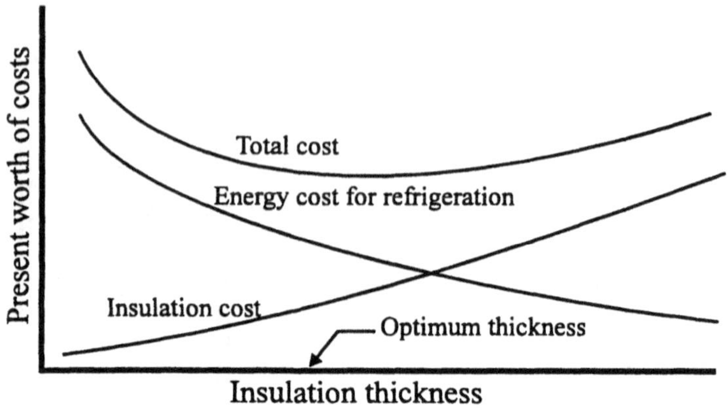

FIGURE 19.2
Influence of insulation thickness on costs.

sulation sets in a matter of seconds, and the result is a monolithic insulation construction without joints.

Some important characteristics of several different insulation materials in board form are given in Table 19.1. Cellular glass is dense, and it is used especially for floors where its weight is not a factor and its high compressive strength is an advantage. Glass fiber insulation is light but cannot support weight and offers negligible resistance to the permeability of water vapor. Polyurethane and polyisocyanurate are good insulators. Glass fiber and molded polystyrene are probably the most economical in first cost, while cellular glass is most expensive.

The need to replace CFCs as the refrigerant in many refrigeration systems has thrown the industry into turmoil. Not as visible has been the need to replace CFCs that have been used as blowing agents for many insulations. In the past, CFC-11 has been the most widely used blowing agent for polyurethanes and polyisocyanurates. CFC-12 has been the popular blowing agent for polystyrene. Because the use of CFCs is no longer permitted, alternative blowing agents are now in use and others are under development[2]. Possible blowing agents include air, CO_2, water, hydrocarbons (butane, pentane, and cyclopentane), HCFCs (R-22, R-123, R-124, R-141b, R-142b), and HFCs (R-134a and R-152a). At the present time, developments and applications of new blowing agents are occurring rapidly. Several manufacturers are currently providing commercial products using an HCFC-123/141b mixture, while another uses a water/CO_2 mixture. Table 19.2 provides a brief comparison of some of the candidates.

The major penalty attributable to the replacement of CFC blowing agents with HCFCs or natural substances shows up in increased cost, because not only is the cost of a section of insulation of a given area and thickness higher, the R-value is less. This means that either or both the thickness of insulation must be increased and/or the refrigeration load will be higher. The current state of development has, in general, limited the penalty of the substitution of other

TABLE 19.1
Some characteristics of popular board insulation.

Material	Density kg/m^3 (lb/ft^3)	R-value $m^2 \cdot K/W$ [$(hr \cdot ft^2 \cdot {}^\circ F)/Btu$]	Relative water-vapor perme-ability	Flame spread	Com-pressive strength kPa (psi)	Rel-ative cost
Molded polystyrene	16 (1.0)	0.7 (4)	Med.	Melted liquid burns	83-117 (12-17)	Low
Extruded polystyrene	86 (5.4)	0.95 (5.4)	Med.	"	124-276 (8-40)	Med.
Poly-urethane	32 (2.0)	1.14 (6.5)	Med.	30-75	172 (25)	Med.
Polyiso-cyanurate	32 (2.0)	1.16 (6.6)	Med.	25	138-207 (20-30)	Med.
Cellular glass	136 (8.5)	0.53 (3.0)	0	0	690 (100)	High
Glass fiber	173 (2.3)	0.74 (4.2)	High	15-20	N/A	Low

TABLE 19.2
Current and potential blowing agents for insulations.

Blowing agent	R-value of foam relative to CFC-11	Blowing agent flammable?
CFC-11	1.0	No
CFC-12	0.93	No
HCFC-22	0.89	No
HCFC-123	0.90	No
HCFC-141b	0.91	
HCFC-142b	0.87	Yes
HFC-134a	0.78	No
HFC-152a		Yes
Air	0.51	No
CO_2	0.72	No
Water/CO_2		No
Butane	0.75	Yes
Pentane	0.80	Yes

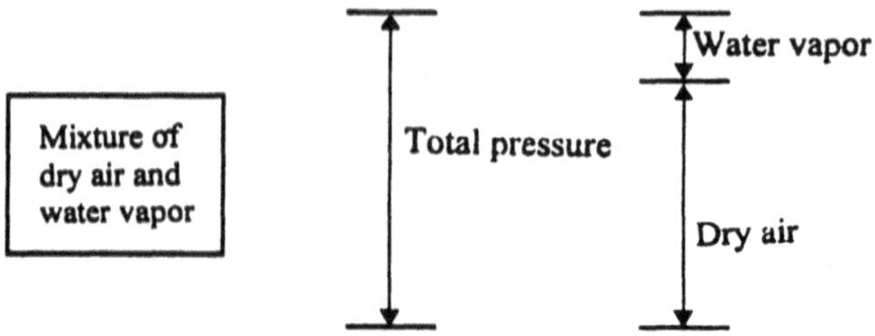

FIGURE 19.3
Atmospheric air as a mixture of air and water vapor.

blowing agents for the previously used CFCs to 10% or less. Possible reactions and degradations of the insulating value over time is still under study.

19.4 VAPOR BARRIERS

A frequent cause of structural damage in refrigerated structures is the formation of ice within walls, roofs, and under floors. The specific topic of prevention of ice formation in the earth below the floor will be treated in Section 19.6. The designers, builders, and operators of refrigerated facilities are constantly confronted with the consequences of water vapor in the ambient air. A few principles must first be explained and then applied to the walls and roofs of refrigerated enclosures to show how problems might arise and also how they can be prevented.

Atmospheric air always contains some water vapor. The total atmospheric pressure is the sum of the pressure exerted by the dry air and the pressure contributed by the water vapor, as shown in Figure 19.3. If progressively more water vapor were injected into the container in Figure 19.3, a state would be reached where some of the vapor would begin to condense into liquid. At that state the air is saturated, which is a condition shown by the saturation line on the psychrometric chart. The vapor pressure at which condensation begins is a function of temperature, and the saturation curve for water is shown in Fig. 19.4.

The next subject to consider is gradients—both of temperature and of water-vapor pressure—that exist in walls. Visualize a composite wall of two materials, as in Figure 19.5, with a high temperature outdoors and a low temperature indoors. Figure 19.5a shows the temperature gradient, with the steeper drop in temperatures occurring in the insulation. Figure 19.5b shows the gradient of the water-vapor pressure from the high vapor pressure of the warm outdoor air to the low vapor pressure inside the refrigerated space. Water vapor flows continuously through the wall from the ambient side of the wall to the interior. Figure 19.5c shows a composite graph of the water-vapor pressure distribution—the dotted line being the actual water-vapor pressure line from Figure 19.5b. The other curve in Figure 19.5c is the saturation pressure corresponding to the existing temperature at the various positions in the wall from the ambient to

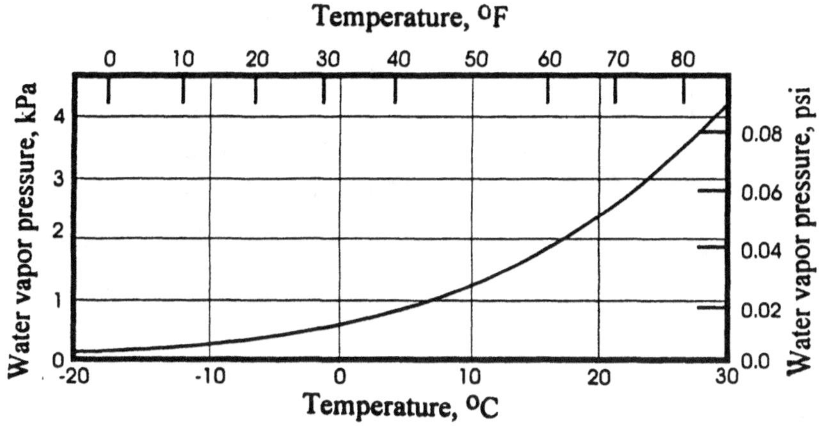

FIGURE 19.4
Saturation pressure of water vapor in air as a function of the air temperature.

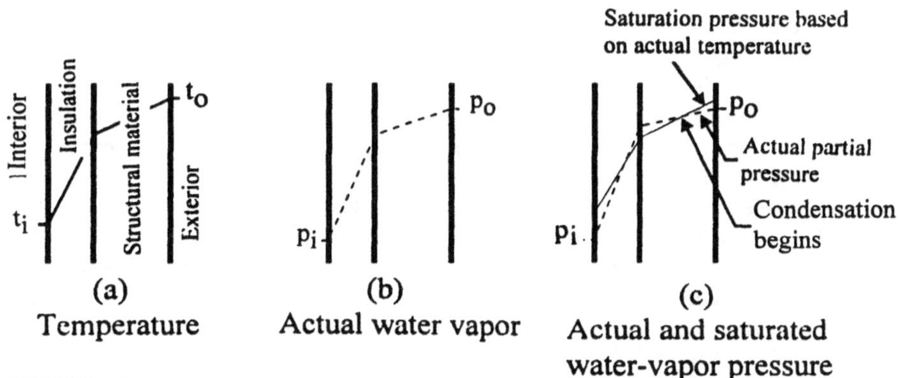

FIGURE 19.5
Gradients in a wall: (a) temperature, (b) actual water vapor, and (c) actual and saturated water-vapor pressure.

the interior. This distribution was obtained by translating the temperatures of Figure 19.5a through the saturation pressure curve of Figure 19.4.

One problem that Figure 19.5 illustrates is where the actual vapor pressure equals the saturation pressure, because here water will condense. The consequences of the liquid water are twofold: (1) the insulating value of most materials is diminished, and (2) if the temperature is below freezing, the resulting ice will expand and possibly damage the structure.

How can condensation of water within walls and roofs be prevented? The answer is by the careful installation of a vapor barrier. The vapor barrier, because of its extremely high resistance to the flow of vapor, causes an abruptly steep gradient, as shown in Fig. 19.6a. When applied to the *warm* side of a wall, as in

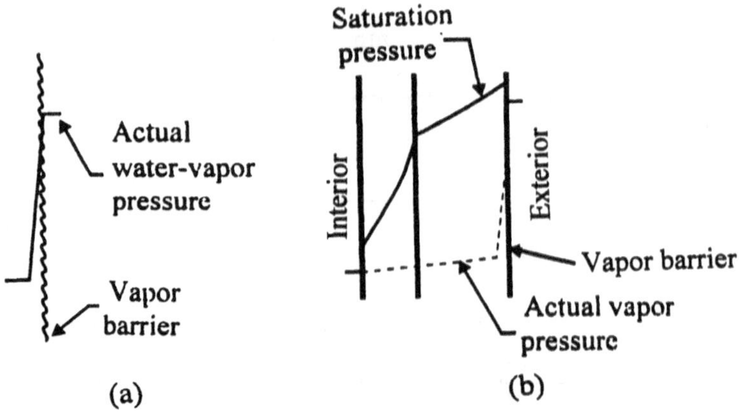

FIGURE 19.6
A vapor barrier: (a) providing a steep gradient of water-vapor pressure, and (b) maintaining the pressure below saturation throughout the wall.

Fig. 19.6b, the actual vapor pressure drops so low on the left of the vapor barrier that it remains below the saturation pressure throughout the wall.

The foregoing discussion pertained to walls and roofs in which the flow of water vapor is from the warm ambient to the cold refrigerated space. The vapor that does pass through the wall or roof section can ultimately enter the refrigerated space to be removed by the refrigerating coil. Another application in which moisture transmission must be considered is in the insulation covering pipes and vessels, as shown in Fig. 19.7. A troublesome fact is that the vapor that migrates from the ambient has no outlet on the cold side. In time, then, moisture will flow into the cold interior and condense. The only hope is to postpone the day the insulation needs to be replaced by devoting extreme care to moisture barriers on the exterior. Cellular insulation with its high impermeability to vapor flow is the recommended choice, as is the vapor-tight exterior layer.

19.5 CHARACTERIZING VAPOR BARRIERS.

Films made of such materials as polyethylene and Mylar are available as vapor barriers. Even thin films of most commercial vapor-barrier materials serve adequately to retard moisture flow, but a vapor barrier of adequate thickness should be chosen to prevent tearing and puncturing during installation. In the inch-pound system, the thickness of vapor barriers is usually given in *mils*, with one mil equaling 0.001 in. (0.025 mm). Most designers of refrigerated structures specify a minimum thickness of about 0.3 mm (12 mil) of polyethylene vapor barrier.

The measure of the ability of a material to transmit water vapor in the inch-pound system of units is called a *U.S. perm*[3]. A material exhibits one perm if it transmits 1 grain (1/7,000 lb) of water per hour through an area of 1 ft^2

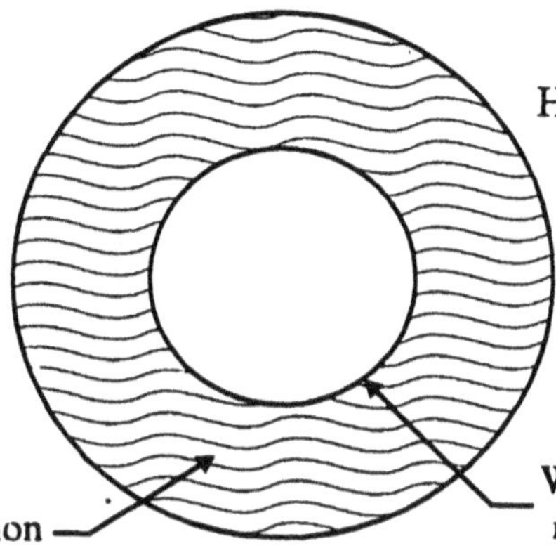

FIGURE 19.7
Moisture transmission through insulation on pipes and vessels.

when the vapor pressure difference is 1 inch of Hg. For other areas and pressure differences, multiply the number of perms by the prevailing values of the area and vapor pressure difference. The lower the perm value, the more effective the vapor barrier. A designer hardly ever computes the rate of vapor flow, and the merit of knowing typical values of perms is useful only for purposes of comparison. A 10-mil (0.25 mm) film of polyethylene might be expected to possess a perm value of about 0.03 grains per hour per ft^2 per inch of Hg vapor pressure difference. The most crucial concern is that the film is thick enough to maintain its integrity during installation.

19.6 PREVENTION OF FROST HEAVING BY UNDERFLOOR HEATING

Even though floor insulation is provided, as shown in Figure 19.8a, the soil beneath the floor of a low-temperature refrigerated space will become cold. If the temperature of the soil drops below freezing, the water it contains will freeze, expand, and could cause frost heaving, as shown in Figure 19.8b. The expansion that occurs when water freezes develops enormous pressures and is capable of buckling floors and even raising the columns that support the roof—all with disastrous results. Even if the soil is initially dry, because its temperature beneath the structure will be lower than the surroundings, moisture will migrate to that region. The fundamental method of preventing frost heaving, as shown in Figure 19.8c, is to provide enough heat under the floor to maintain soil temperatures above the freezing temperature.

Three different media[4] are used to provide heat beneath freezer floors—air, electric heat, or warm ethylene glycol or oil in tubes.

Air is a possible heating medium[5] under the floor in mild climates. Clearly no heat can be supplied by ambient air when the outdoor temperature is below

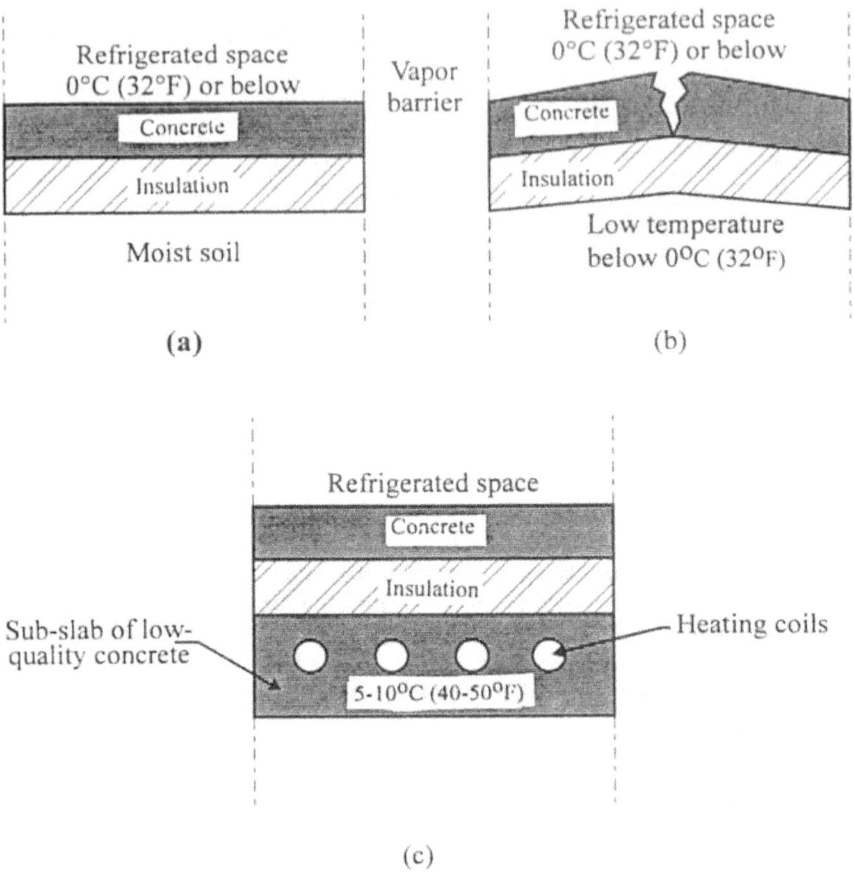

FIGURE 19.8
Sections of floor serving a refrigerated space operating at temperatures below 0°C (32°F). (a) Because the soil beneath the floor contains moisture, (b) it could freeze and expand, unless (c) the soil temperature is maintained high enough to prevent freezing.

freezing, so external means of heating the air must be provided in cold weather. When air is the heating medium, metal, plastic, or concrete ducts are encased in low-grade concrete, insulation is placed on the low-grade concrete, and then the concrete floor is poured over the insulated ducts. Two methods of air flow through the ducts are natural convection and forced convection. Natural convection ducts should be at least 200 mm (8 in) diameter, limited to 36 m (120 ft) in length, and spaced 1.5 m (5 ft) apart. These ducts must be sloped so that warm ambient air enters on the high side and the cooled air leaves at the low side. The ducts must have no traps in which condensed moisture could collect, and both ends should be kept free of obstructions but screened to prevent debris and rodents from

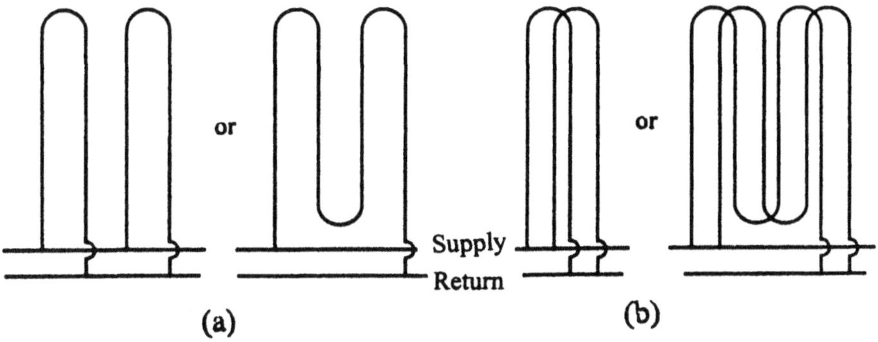

FIGURE 19.9
Placement of glycol heating tubes or electric cables to reduce the penalty of one circuit failing. (a) Elements of one circuit are adjacent to one another—not recommended, and (b) elements of two circuits are alternated—recommended.

entering. When using fans for forced convection, the diameter may be 100 mm (4 in) with the same spacing as used for natural convection ducts.

Electric resistance cables may be installed in loops with about 1 m (3 ft) spacing between the elements. By installing the cable in a conduit, a circuit may be replaced if it fails.

Ethylene glycol is a frequently used medium. The glycol passes through polybutylene tubes encased in the sub-slab, and often the source of heat for the glycol is rejected heat from refrigerant (Section 16.5). Tubing diameters of 25 mm (1 in) are typical, and the spacing between tubes is usually 1.2 to 1.5 m (4 to 5 ft). The supply temperature of the glycol need be no higher than 10 to 15°C (50 to 60°F).

Concern when using glycol-carrying tubes is potential blockage and a short or open circuit developing when using electric cables. Repair of such failures is usually costly. A practice that reduces the penalty of the loss of one circuit of glycol tubing or electric resistance cable is illustrated in Figure 19.9. If the lines of one circuit are adjacent to one another, as in Figure 19.9a, the failure of one circuit results in three spaces between heated elements (five spaces in the case of the double loop). On the other hand, if the lines from different circuits are crossed, as in Figure 19.9b, only two spaces exist between heated elements if one circuit fails.

The heat flow entering a refrigerated structure through the floor is a refrigeration load, so it is included in the load calculation. In designing an underfloor heating system, the heating capacity must also be calculated. These two calculations are identical in that all of the heat entering the underfloor system is assumed to flow into the space. The basic heat-transfer equation, Equation 18.6, is the relation used with the U-value equaling the reciprocal of the sum of the thermal resistances. The underfloor temperature, t_o, is usually assumed to be about 10°C (50°F), and t_i is the temperature of the refrigerated space.

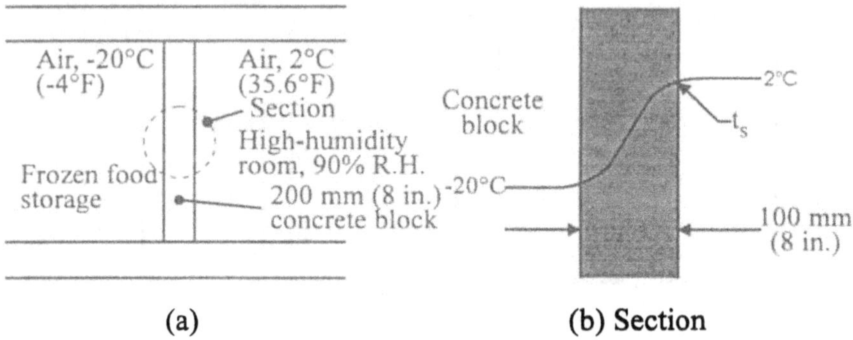

FIGURE 19.10
(a) A high-humidity room adjacent to a low-temperature room and separated by a concrete block wall, and (b) computing the surface temperature of the wall.

If frost heaving has taken place and deformed the floor and building, replacement of the damaged portions may not always be necessary. The first step in restoration is to inject heat beneath the floor by whatever means is possible. There are cases on record[6] where after the heat has been supplied under the floor, the ice melted and the entire distorted structure settled back closely enough to its original condition that normal operation of the facility could resume.

19.7 PREVENTING CONDENSATION IN HIGH-HUMIDITY ROOMS

For high-humidity rooms such as those storing unfrozen foods at 90–95% relative humidities, a check should be made to ensure prevention of condensation on the walls and ceiling. An example of a potential problem is illustrated in the food distribution center, Fig. 19.10a, where a high-humidity room operating at 2°C (35.6°F) and 90% relative humidity is separated by a concrete block wall 200 mm (8 in) thick from a room storing frozen food at −20°C (−4°F). The surface temperature on the high-humidity side of the wall should be checked to ensure that condensation of moisture will not occur.

The surface temperature, t_s, on the warm side in Fig. 19.10b can be determined by proportioning the 22°C (39.6°F) difference in air temperature. The convection coefficients at the vertical wall surfaces from Table 18.3 are 9.3 W/m^2·K (1.63 Btu/hr·ft^2·°F), so the R-values are 0.11 m^2·K/W. The resistance of the cinder block from Table 18.2 is 0.30 m^2·K/W (1.72 hr·ft^2· °F/Btu). Therefore:

$$t_s = 2 - 22\left(\frac{0.11}{0.11 + 0.30 + 0.11}\right) = -2.7^\circ\text{C} \quad (27.1^\circ\text{F})$$

This surface temperature is lower than the dewpoint temperature of the air, which can be determined from a psychrometric chart to be 0.5°C (32.9°F). Therefore, condensation will occur.

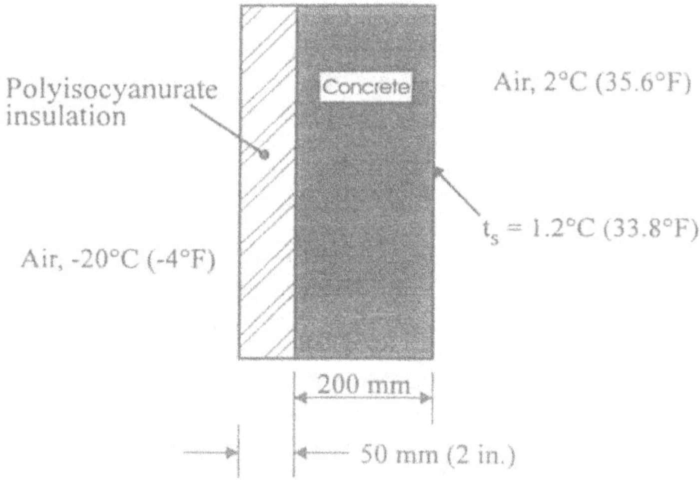

FIGURE 19.11
Preventing condensation on the wall surface by applying insulation.

Prevention of condensation can be accomplished by insulating the wall as shown in Fig. 19.11. A 50 m (2 in) layer of polyisocyanurate would add a resistance of 2.5 $m^2 \cdot K/W$ (0.44 $hr \cdot ft^2 \cdot {}^\circ F/Btu$) to the wall. This resistance is calculated from the conductivity in Table 18.1, $k = 0.02$ W/m·K, conductance $= 0.02/(0.050 \text{ m}) = 0.4 \text{ W/m}^2\cdot\text{K}$, and $R = 1/(\text{conductance}) = 2.5 \text{ m}^2\cdot\text{K/W}$. A new calculation of t_s yields the following:

$$t_s = 2 - 22\left(\frac{0.11}{0.11 + 0.30 + 2.5 + 0.11}\right) = 1.2^\circ\text{C} \quad (33.8^\circ\text{F})$$

The surface temperature is now higher than the dewpoint temperature of 0.5°C (32.9°F), so condensation is avoided.

19.8 DOORS

The cost of the doors that serve refrigerated spaces may constitute only several percent of the total cost of the facility, but that low cost does not truly reflect the influence on operating cost of the facility. Handling of the product in refrigerated facilities becomes progressively more important as a greater percentage of the product is in motion at any time in contrast to dormant storage. The choice of size, location, and type of door exerts a major influence on the avoidance of loss of time for the operating crew, as well as in controlling the amount of time the doors are open, which minimizes the refrigeration load.

The designer must decide[7] such questions with respect to doors as the size, number, type, opening/closing speed, and placement—both between the refrigerated space and the dock and between the dock and the truck or rail car openings.

Door size. Doors must be large enough for fork-lift trucks carrying a pallet of product to pass through unimpeded. Since the pallet is about 1.2 m (4 ft) wide, the width of the door should be at least 1.8 m (6 ft). The minimum size door available on the market is approximately 1.5 x 2.1 m (5 x 7 ft), and the largest is 4.3 x 4.3 m (14 x 14 ft). The door type and its opening/closing speed also influence the door size.

Number of doors. The number of doors between the frozen food space and the dock is influenced by the anticipated frequency of passage of fork-lift trucks, which in turn is controlled by the number of turnovers of product. A refrigerated space with the capacity of 10,000 pallets and 30 turnarounds a year must move 300,000 pallets in and out annually. Assuming 60 hours of operation a week, 144 truck movements in and out are required per hour. If 30 s is chosen as the minimum time per movement, each door can accommodate (288 movements)/(120 per hour) = 2.4 or three doors. The ability to maintain a uniform 30 s per truck movement would be difficult and result in delays at busy times. The number of doors in this situation might reasonably be increased to five. The number of dock doors to the outside must be chosen to adequately provide time to load and unload semi-trailers. Choosing a loading or unloading time of 20 min and a trailer capacity of 20 pallets requires handling 1 pallet per minute or 60 pallets per hour per door. Thus, for 432 pallets per hour, a minimum of seven doors is required.

After making the estimates of the required number of doors, the physical reality must be checked to be sure that the wall length is available for the desired number of doors. Also, additional dock doors are useful so that trailers can back in to the dock to be in position when the dock operation is ready.

Door type. Both vertically and horizontally operating doors are available, with horizontal doors much more common for the heavily used doors of the refrigerated space. For the loading docks, vertical doors are appropriate because they are usually left open while the trailer truck is loaded or unloaded and a seal provided between the dock space and the trailer. Sliding doors may be purchased in either single- or double-door models, and as expected, the double door is more expensive than the single panel. On the other hand, the time required to open the door of a given width is much less with the double door. Another design, shown in Fig. 19.12, is the bi-fold door, which is applicable to tunnels and corridors where mounting space is practically non-existent.

Opening/closing speed. Doors are available with movement speeds from about 0.6 m/s (25 in/s) to 1.2 m/s (50 in/s). Door are moved either directly by electric motors or by hydraulic motors using pressure provided by electric-driven hydraulic pumps. To withstand the repeated forces of acceleration of the starting and stopping of the door motion, doors must be firmly anchored structurally.

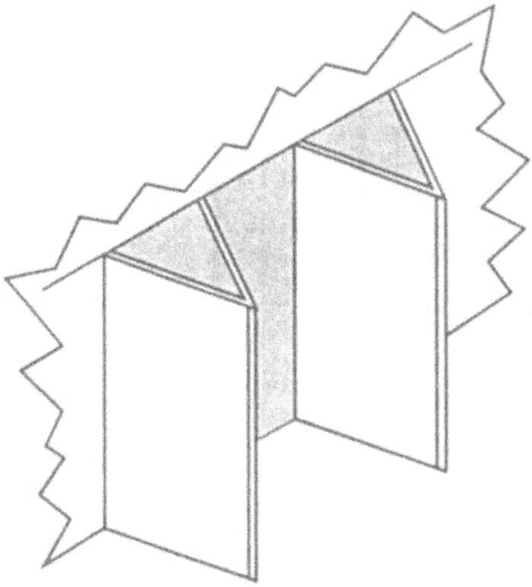

FIGURE 19.12
A bi-folding door.

Door placement. The placement of doors is often a compromise between the ideal location from the standpoint of traffic flow and the limits of wall space. Natural extension of the aisles are the preferred location for doors between the refrigerated space and the dock, and the space between doors must be adequate for the trailers to back into their stalls.

19.9 CONSTRUCTION METHODS

There are two basically different concepts[7] in the construction of refrigerated spaces, the structurally-interior and the structurally-exterior methods, as illustrated in Figure 19.13. In Figure 19.13a, the column, beam, and joist system is enclosed by the insulation and vapor barrier assembly with the walls supported by the vertical members and the roof supported by the beams and joist. In Figure 19.13b, the insulation and vapor barrier assembly is enclosed within the structural system.

The configuration of Figure 19.13a offers the opportunity of the fewest penetrations of the vapor barrier, and an important goal in the construction of any refrigerated space is to maintain the integrity of the vapor barrier. The interior placement of the insulation and vapor barrier system of Figure 19.13b is especially suitable for food processing areas that need to be washed periodically. In many installations, the interior finish must be chosen for sanitary washdowns. Insulated structural panels with metal exterior and metal or reinforced plastic interior faces may serve the need for the washdowns. Floor bumpers and curbs aid in protecting the finish from damage.

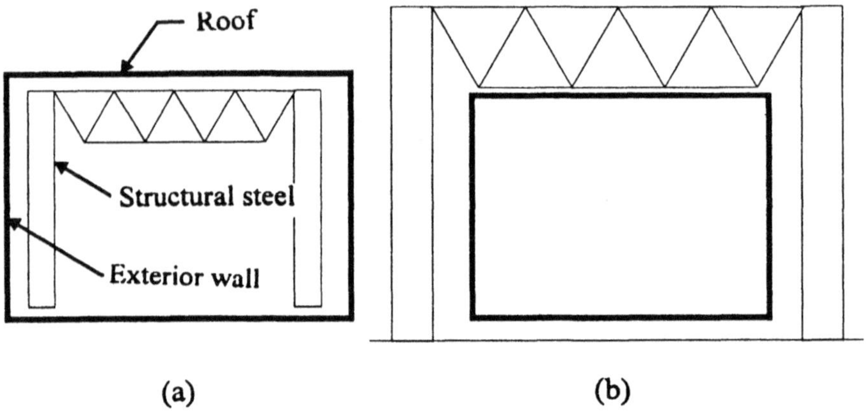

FIGURE 19.13
(a). Structural system interior to the insulation and vapor barrier assembly, and (b) exterior to the insulation and vapor barrier assmembly.

19.10 ROOF AND WALL/ROOF SECTIONS

Several important precautions in the construction of refrigerated spaces are: (1) to maintain structural integrity, (2) to carefully seal all joints to prevent access of air and water, and (3) to install the vapor barrier uninterrupted by penetrations. Since it is often impossible to avoid all penetrations of the vapor barrier, the points where the vapor barrier is pierced should be resealed with a vapor-proof compound. Of course, where a conduit must pass through the vapor barrier, the outside of the conduit must be sealed, but vapor can sometimes pass through the interior of the conduit as well. After the wiring is installed, the interior of the conduit should also be sealed.

The locations most vulnerable to openings are at the junctions—roof/wall and floor/wall. As is true in other types of buildings, the roof of refrigerated structures is frequently where water and vapor leaks occur. Several different roof structure concepts work successfully, and the construction details illustrated here are not exclusive. One possibility is to construct the roof with prefabricated panels, which provides a good roof that is usually slightly more expensive than a built-up roof. The construction details illustrated in Figure 19.14 apply to a refrigerated space. The structural system is enclosed in the insulation and vapor barrier assembly, as in Figure 19.13a.

Figure 19.14 shows a roof/wall corner section where the columns support the beams, which in turn support the bar joists. Insulation is supported by a metal deck which lays on the bar joists. Next comes the vapor barrier, for which it is crucial that it be brought over the wall panel and extended a short distance vertically down the panel and then sealed. The vapor barrier for the wall panel is its outside skin. If the membrane for the vapor barrier on the roof

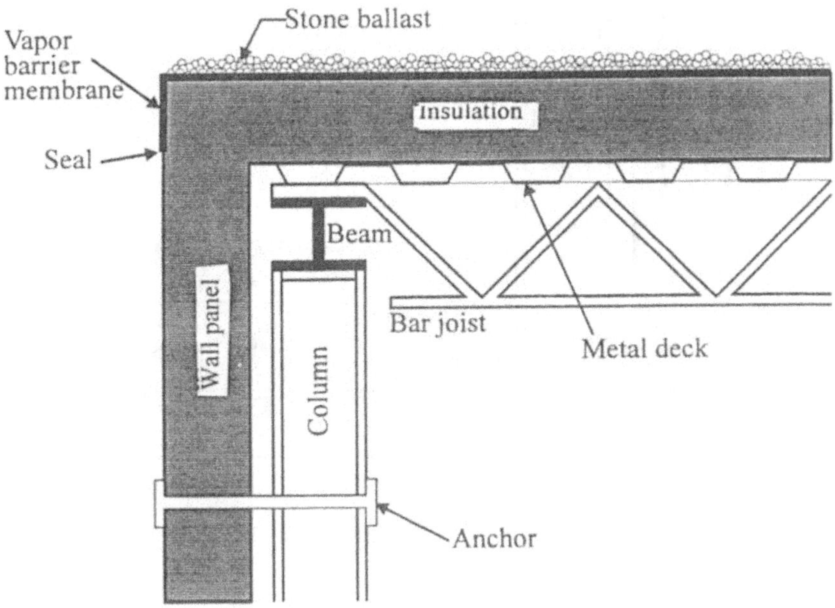

FIGURE 19.14
One concept of roof construction.

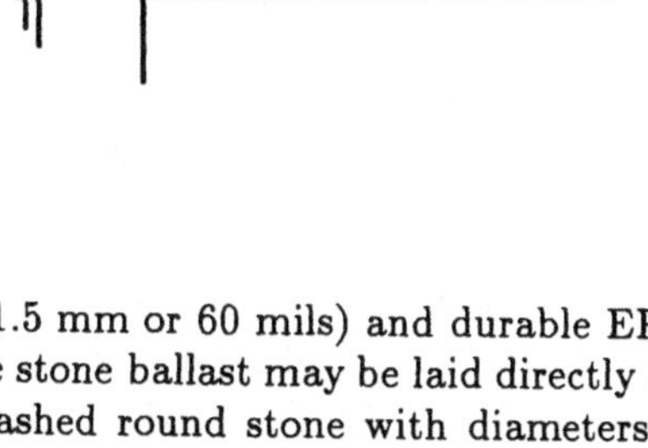

FIGURE 19.15
Roof curb.

is a thick (perhaps 1.5 mm or 60 mils) and durable EPDM (ethylene propylene diene monomer), the stone ballast may be laid directly atop the membrane. This ballast should be washed round stone with diameters between 18 and 38 mm (3/4 to 1-1/2 in). An alternative for the top cover is to install several layers of asphalt felt with asphalt applied between the layers. A roof curb of wood covered with a membrane, as shown in Fig. 19.15, can direct rain and melted snow to the drains.

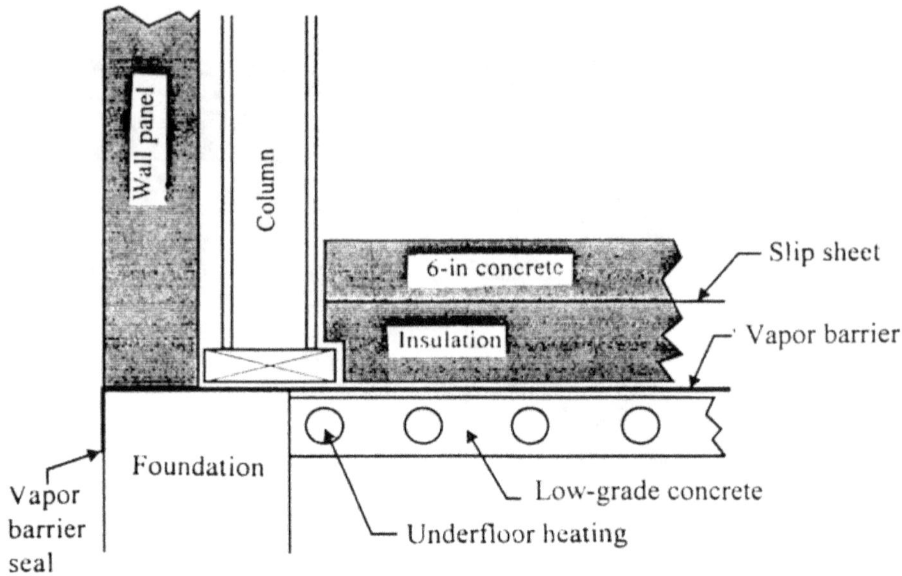

FIGURE 19.16
A floor/wall junction.

19.11 FLOOR AND WALL/FLOOR SECTIONS

Another critical location in the construction of a refrigerated building is the floor and the wall/floor junction. Here too there are alternate approaches to achieving satisfactory construction, but several of the basic considerations that must be addressed are: (1) continuity of the vapor barrier, (2) withstanding the compressive load on the floor, (3) underfloor heating when the storage temperature is below freezing, and (4) the transfer of the column load in such a way that there is a minimum conductivity of heat.

Figure 19.8c has already suggested the basic floor construction consisting of concrete on insulation over a low-grade concrete in which the underfloor heating pipes or ducts are placed. More details of the floor assembly and the floor/wall junction are shown in Figure 19.16.

Placing a slip sheet of film, perhaps 1.5 mm (6 mil) thick, between the concrete floor and insulation is advisable. This slip sheet, which during construction is placed on the insulation before pouring the concrete, serves two purposes. It permits differential expansion between the insulation and concrete, and it keeps the moisture of the concrete from contacting the insulation until the concrete dries.

Since the insulation must withstand the compressive stress of the concrete as well as the weight of pallets and fork-lift trucks, its inherent compressive strength is a factor in its selection. As Table 19.1 indicated, extruded polystyrene

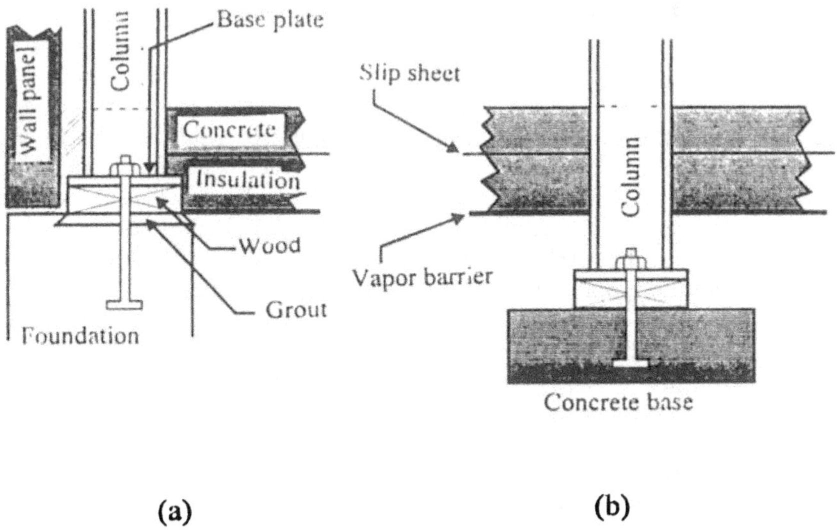

Column supports (a) for perimeter columns, and (b) for interior columns.

and cellular glass exhibit high compressive strengths, and one of these materials is usually chosen.

The vapor barrier is again placed on the warm side of the insulation, with the sections of the vapor barrier sheets sealed during installation to maintain the integrity of the barrier. Figure 19.16 shows the vapor barrier extending continuously to the exterior of the building where it is sealed. Alternatively, a vapor-proof sealant can be applied at the junctions of the column support and wall panel with the floor. The concrete beneath the insulation and vapor barrier holds the underfloor heating elements in position, gives a smooth support for the insulation, and prevents puncturing the vapor barrier.

19.12 COLUMNS

Because columns require adequate support at their base, they cannot normally rest on the insulation. This means that the columns extend past the insulation and become undesirable conductors of heat into the refrigerated space. An approach used frequently to reduce the conduction of heat is to set the column on a wood base that is in turn supported by the concrete foundation at the perimeter of the building, Figure 19.17a. In the case of an interior column, the wood rests on a concrete pad, as shown in Figure 19.17b. The wood chosen for these column pads is usually oak, perhaps 150 mm (6 in) thick and oriented so that the direction of the grain is vertical.

19.13 CONVERTIBLE SPACES

In order to provide greater flexibility in a refrigerated storage or distribution facility, some owners equip their space so that at least one room is convertible between the storage of frozen and unfrozen food products. Consideration of this possibility has become more important in recent years because of the increase in sales, and thus in the short-term storage as well, of unfrozen fruits and vegetables. From the standpoint of the facility, a convertible space is basically designed for frozen food storage. Convertible space must be insulated and provided with vapor barriers appropriate for low-temperature storage, and underfloor heating must be installed, even though it is shut off when the space operates at high temperature.

From the standpoint of the refrigeration system, the air-cooling coils chosen for low-temperature operation will normally be adequate for high-temperature operation, because the refrigeration load due to conduction and infiltration will be less. Also, if in the selection of the coils the temperature difference between the room air and refrigerant may be less for low-temperature operation, so the coils will have greater capacity when operating in the high-temperature space. Hot-gas defrost will be required for low-temperature operation and usually for high-temperature operation as well. The major provision on the refrigeration side is that the supply of liquid (assuming liquid recirculation) will be provided from an intermediate-pressure recirculator, and the suction gas and liquid from the coil must return to this intermediate-pressure recirculator. These requirements necessitate the installation of headers so that the source of the supply and the destination of the suction refrigerant can be interchanged.

REFERENCES

1. Cooksey, D. L., "Comparison of Conventional vs. Automated Refrigerated Storage Systems," Proceedings of Meeting of Commission D of the International Institute of Refrigeration, pp 77-86, Orlando, Florida, 1985.
2. "Energy Efficient Alternatives to Chlorofluorocarbons—A Research Needs Assessment," U.S. Department of Energy, Office of Energy Research, COE/ER/30115-H1, June, 1993.
3. American Society of Heating, Refrigerating and Air-Conditioning Engineers, Atlanta, Georgia, from the 1993 ASHRAE *Handbook— Fundamentals.*
4. Webber, J. F., "Preventing Floor Heavage in Refrigerated Warehouses," *ASHRAE Journal,* 23(12): 24-26, December, 1981.
5. Power, G. L., "Ambient Gravity Air Flow Prevents Freezer Floor Heaving," *ASHRAE Transactions,* 87(Part 2): 1107-1116, 1981.
6. Ballou, D. F., "A Case History of a Frost-Heaved Refrigerated Freezer Floor," *ASHRAE Transactions,* 87(Part 2): 1099-1105, 1981.
7. American Society of Heating, Refrigerating and Air-Conditioning Engineers, Atlanta, Georgia, from the 1994 ASHRAE *Handbook— Refrigeration.*

CHAPTER 20

SECONDARY COOLANTS—BRINES AND ANTIFREEZES

20.1 SECONDARY COOLANTS

A secondary coolant is the fluid chilled by the primary refrigeration system that then performs the refrigeration by absorbing heat through its rise in temperature. A secondary coolant could conceivably undergo a change in phase, condensing at the primary refrigeration system and boiling at the point of refrigeration load. In the pre-CFC era, R-11 was sometimes used for that purpose, and CO_2 continues to serve in this manner. The scope in this chapter will be restricted, however, to fluids that remain in the liquid phase and thus change temperature to accomplish heat transfer.

The terms secondary coolants, brines, and antifreezes are sometimes used interchangeably. This chapter will use *secondary coolant* as the general description of a fluid with freezing temperatures below those of water. Brines are one class of secondary coolants formed by the aqueous (with water) solution of a salt.

Secondary coolants have been used for years, and the number of applications seems to be growing rather than declining. Despite the cost penalties for pumping the secondary coolant and the lower evaporating temperature required to overcome the additional temperature differences due to the interposition of an extra fluid, some recent trends maintain the importance of secondary coolants. In the case of ammonia, the refrigerant charge can be contained in a compact chiller, and a leak in the distribution system will not cause damage to stored refrigerated products. The use of secondary coolants in ammonia systems supports the trend to low-refrigerant-charge systems. In the case of the halocarbon systems, the high cost of replacements for the CFCs and HCFCs motivates consideration of low-charge systems.

The number of candidates for secondary coolants is large, and this chapter will attempt to give key data on a wide selection. Of this large number of possibilities, certain choices are more popular than others, so additional characteristics will be provided. Key data include the following factors:

- Lowest solidification temperature
- Flammability
- Compatibility with food
- Corrosion tendency and inhibition possibilities

If a candidate possesses satisfactory key data to pass the first stage of the selection process, additional characteristics should be evaluated, such as the thermal and fluid properties, including the following:

- Viscosity
- Specific heat
- Specific gravity or density
- Thermal conductivity

Secondary coolants to be considered in this chapter include the following:

- Alcohol–ethyl alcohol
- Alcohol–methyl alcohol
- Chloride–calcium chloride
- Chloride–sodium chloride
- d-limonene
- Glycol–ethylene glycol
- Glycol–propylene glycol
- Halocarbons
- Polymers

20.2 PHASE DIAGRAM OF A SECONDARY COOLANT

A typical phase diagram for aqueous solutions is shown in Figure 20.1. The curves of the freezing points show that in many cases the solution of the two constituents has a lower freezing point than either substance individually. The chart shows the possible phases and mixtures that can exist at various concentrations and temperatures. Assume that the substance is a solution of salt and water. If the brine at temperature A has a concentration M, what will be the behavior of the

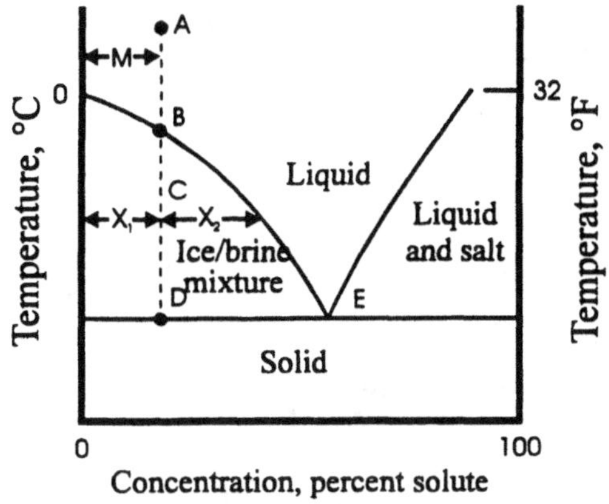

FIGURE 20.1
Phase diagram of an aqueous secondary coolant.

brine as it progressively cools? The brine remains a liquid until the temperature drops to B. Further cooling to C results in a slush, which is a mixture of ice and brine. The brine at C has concentrated itself by freezing out some of its water into ice. The percentage of ice and liquid, respectively, in the mixture at C is shown in the following equations:

$$\text{Percentage ice} = \left(\frac{x_2}{x_1 + x_2}\right) 100 \tag{20.1}$$

$$\text{Percentage liquid} = \left(\frac{x_1}{x_1 + x_2}\right) 100 \tag{20.2}$$

Cooling the solution below D solidifies the entire mixture. Point E is called the *eutectic point* and represents the concentration at which the lowest temperature can be reached with no solidification. Strengthening the solution beyond the eutectic concentration is fruitless, because the freezing temperature rises.

20.3 FREEZING TEMPERATURE

The freezing temperature of a secondary coolant must be lower than its expected temperature leaving the primary refrigeration system. In addition, there should be a safety factor to allow for occasional variations in control. An overview[1] of freezing temperatures for aqueous solutions is provided in Figure 20.2.

To supplement Figure 20.2, Table 20.1 shows the lowest freezing temperatures of some aqueous solutions and the concentration by mass of the solute at these minimum temperatures.

The secondary coolants listed in Table 20.1 fall roughly into two groups, based on their freezing temperature—the low temperature and the ultra-low temperature antifreezes. Those antifreezes applicable to the temperature range of −20 to −40°C (−4 to −40°F) are ethylene and propylene glycol and the calcium

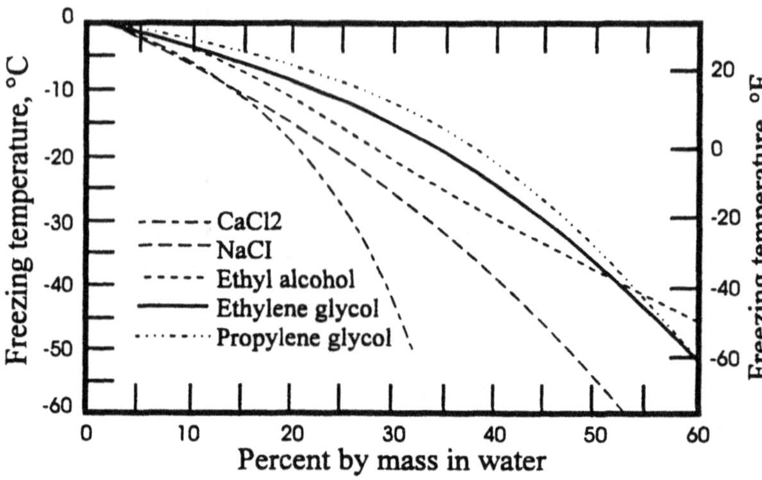

FIGURE 20.2
Freezing temperatures of some aqueous secondary coolants.

and sodium chloride brines. For temperatures below −40°C (−40°F), the choice can be made among the alcohols, acetone, d-limonene, or polydimethylsiloxane[3] types of secondary coolants.

Since the higher of the two temperature ranges is most often encountered, this chapter will present more engineering data on secondary coolants applicable to that range. The secondary coolants of the lower temperature range are particularly applicable to the low-temperature systems studied in Chapter 21, and selected data on these antifreezes will also be presented.

The next several sections will present data on the freezing temperatures, density, viscosity, specific heat, and thermal conductivity of various secondary coolants, along with comments on these properties as well as other characteristics.

20.4 SODIUM CHLORIDE

The two important brines for refrigeration service are aqueous solutions of either of two salts—sodium chloride (NaCl) or calcium chloride ($CaCl_2$). NaCl brine is probably the most economical of any available secondary coolant, and it can be applied in contact with food and in open systems because of low toxicity. Furthermore, NaCl brine is nonflammable and has favorable thermodynamic and transport properties. The freezing point, density, viscosity, specific heat, and thermal conductivity of NaCl brine are shown, respectively, in Figs. 20.3 to 20.7. These properties result in high heat-transfer coefficients, although not as good as $CaCl_2$ brine. Several disadvantages of NaCl brine are its relatively high freezing temperature and the fact that it is highly corrosive. NaCl brine is not nearly as popular as $CaCl_2$ brine, so the discussion of corrosion inhibition,

TABLE 20.1
Lowest freezing temperatures of some aqueous solutions and the concentration at which these minimum temperatures occur[2]. The practical operating temperature will be somewhat higher than these temperatures.

Solute	Minimum freezing temperature, °C (°F)	Concentration of solute, percent by mass
Acetone	−94.6 (−138)	100
Calcium chloride ($CaCl_2$)	−55 (−67)	30
d-limonene	−97 (−142)	100
Ethyl alcohol	−112 (−170)	100
Ethylene glycol	−48.3 (−55)	60
Methyl alcohol	−97.8 (−144)	100
Polydimethylsiloxane[3]	−111 (−168)	100
Propylene glycol	−51.1 (−60)	60
Sodium chloride (NaCl)	−20 (−4)	23

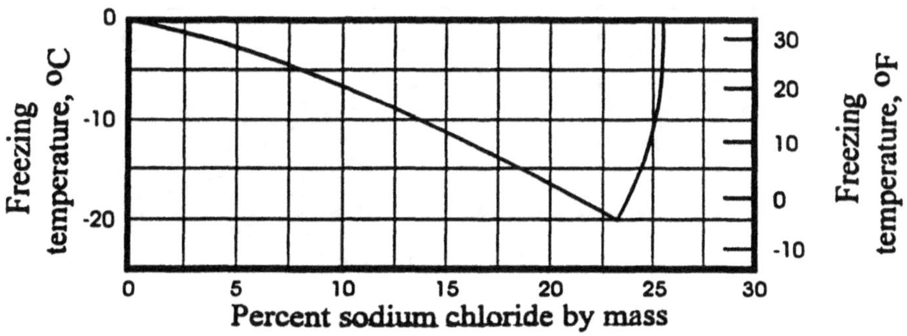

FIGURE 20.3
Freezing point of NaCl brine[2].

compatibility with materials of construction, and other factors that must be confronted when dealing with brines will be detailed in the next section.

20.5 CALCIUM CHLORIDE

Calcium chloride ($CaCl_2$) brine, certainly in the past and perhaps even now, is the most popular secondary coolant serving the industrial refrigeration field. Next to NaCl brine, its cost is the lowest of the conventional choices, and its freezing temperature is lower than NaCl brine. While NaCl brine is acceptable in the vicinity of foods, $CaCl_2$ brine, which is essentially non-toxic, should be kept from direct contact with foods. $CaCl_2$ brine enjoys lower viscosities in comparison to ethylene glycol and especially propylene glycol. $CaCl_2$ brine is nonflammable, and with the addition of suitable inhibitors, this brine is compatible with steel

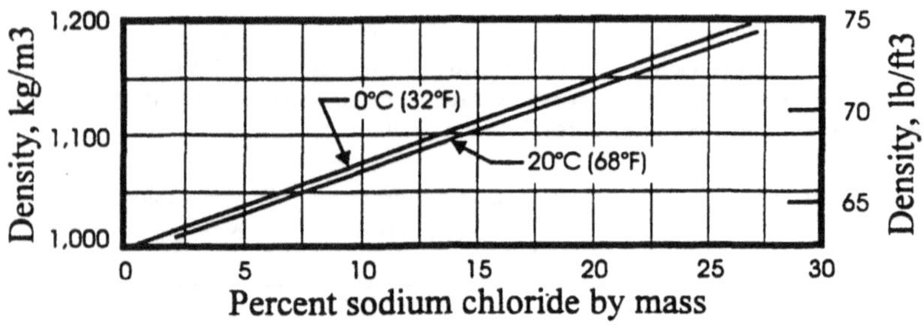

FIGURE 20.4
Density of NaCl brine[2].

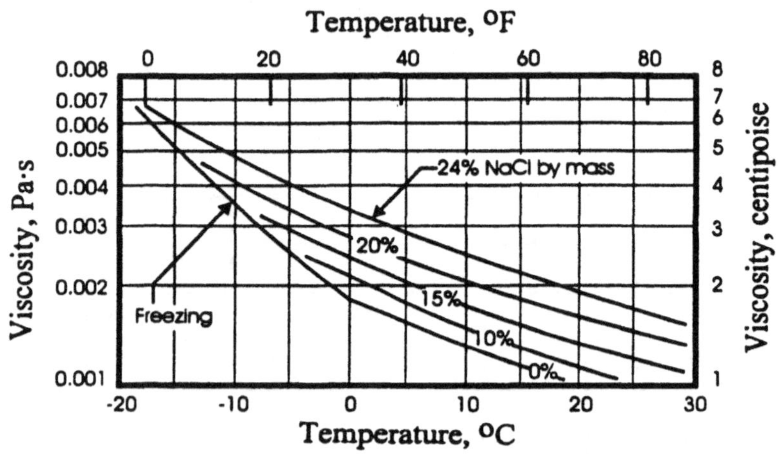

FIGURE 20.5
Viscosity of NaCl brine[2].

and aluminum. Neither NaCl nor $CaCl_2$ brines should be used with galvanized steel, because the brines attacks the zinc in the galvanized coating. As is true of many of the secondary coolants for industrial refrigeration, $CaCl_2$ brine is corrosive to the metals of its containment system. For this reason inhibitors must be added to the brine and a maintenance plan should include periodic checks of the brine. Oxygen facilitates corrosion, so the brine system should be closed and thus equipped with an expansion tank. However, oxygen seems to materialize even in a closed system. The freezing points, densities, viscosities, specific heats, and thermal conductivities of $CaCl_2$ brine are shown in Figs. 20.8 to 20.12, respectively.

A basic requirement is to maintain a slightly basic condition of the $CaCl_2$ brine with an ideal pH value of about 7.5 (a pH of 7 is neutral). Acidic solutions, indicated by low pH values, are particularly corrosive, but highly basic solutions are also corrosive.

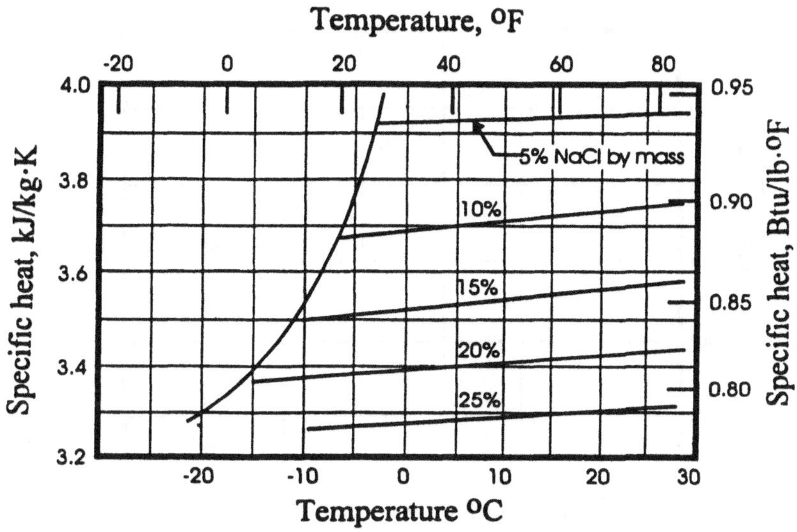

FIGURE 20.6
Specific heat of NaCl brine[2].

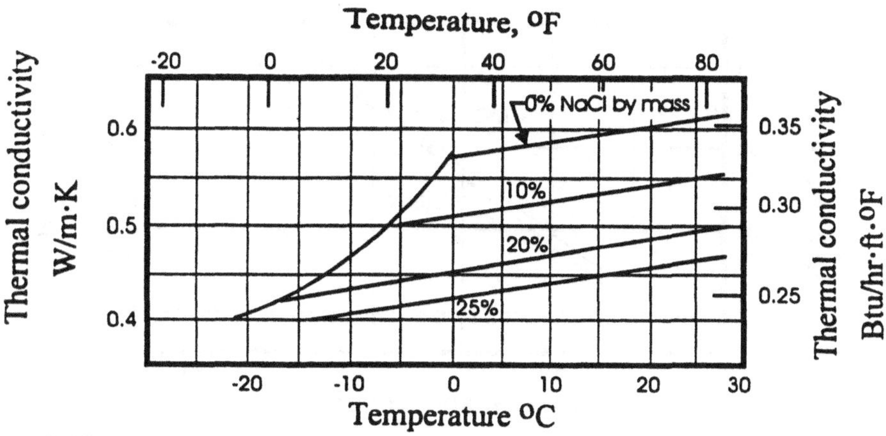

FIGURE 20.7
Thermal conductivity of NaCl brine[2].

The traditional inhibitor for brine is a chromate, in particular a solution of sodium dichromate[7]. The addition of sodium dichromate turns the solution acidic, therefore to bring the brine solution back to the target pH value, it is usually necessary as well to add caustic soda (sodium hydroxide, NaOH). Current environmental regulations are restricting the use of chromates, so before they are applied or disposed of from an existing system, federal, state, and local regulations should be consulted. New organic inhibitors are now available.

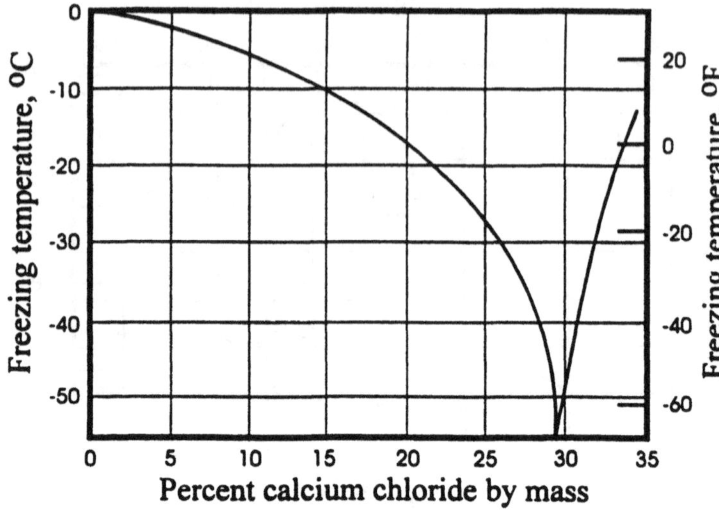

FIGURE 20.8
Freezing point of $CaCl_2$ brine[4].

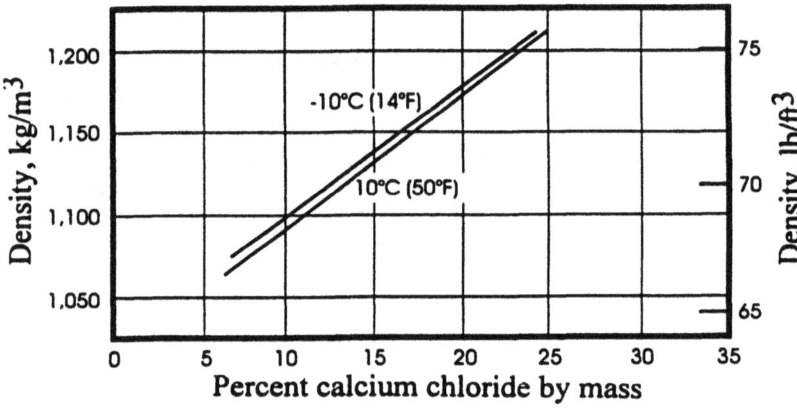

FIGURE 20.9
Density of $CaCl_2$ brine[4].

20.6 ETHYLENE GLYCOL.

After calcium chloride brine, aqueous solutions of ethylene glycol are probably the next most popular secondary coolant for industrial refrigeration systems. The freezing temperature of ethylene glycol is low enough to make it suitable for many industrial refrigeration applications. The solutions are nonflammable and can by used in piping systems of steel, aluminum, and copper. Just as is true of the brines, galvanized pipe and fittings should be avoided, and in addition, the temperature should be maintained below 60°C (140°F) in systems using

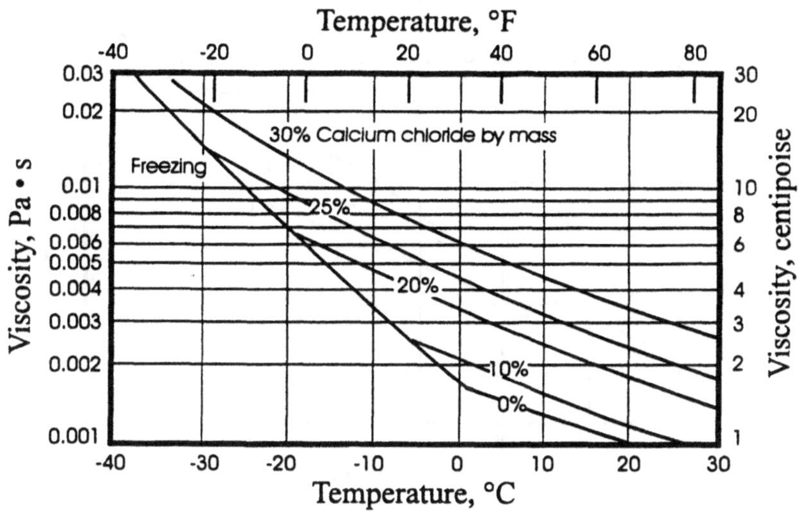

FIGURE 20.10
Viscosity of $CaCl_2$ brine[4].

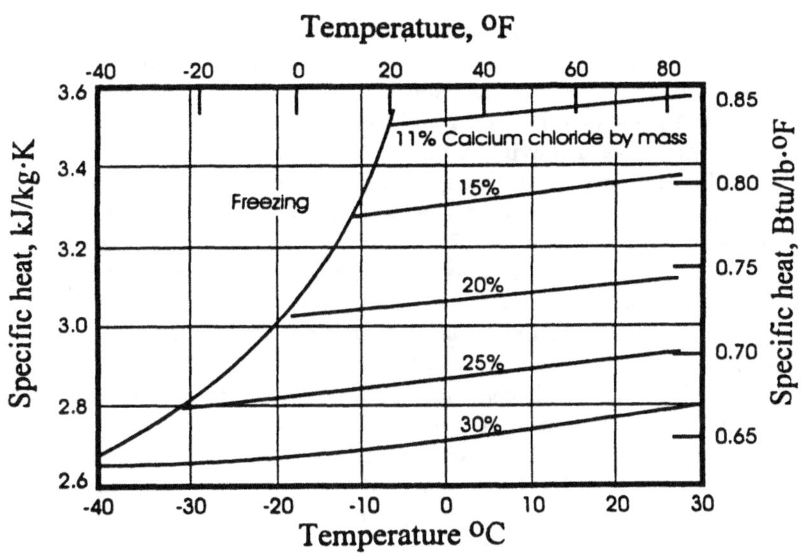

FIGURE 20.11
Specific heat of $CaCl_2$ brine[4].

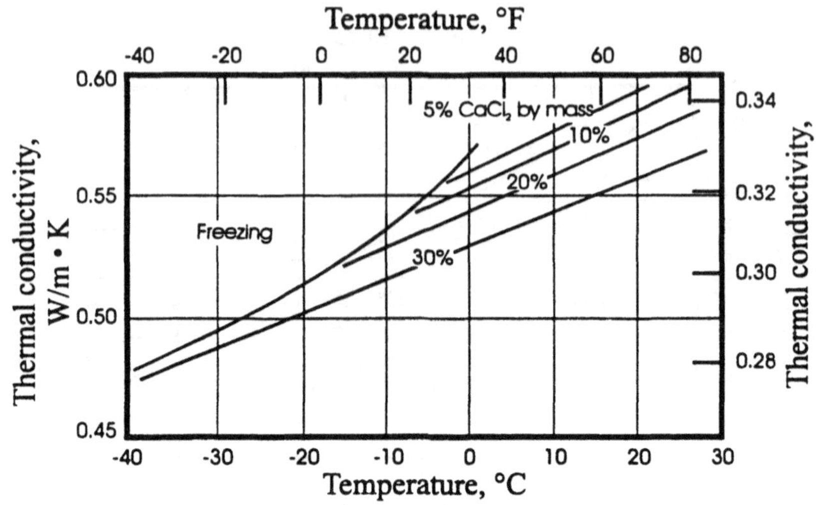

FIGURE 20.12
Thermal conductivity of $CaCl_2$ brine[4].

aluminum pipes. One of the important transport properties that affects both the pressure drop of the flowing coolant as well as the convection heat-transfer coefficient is the viscosity. A low viscosity is desirable from both standpoints, and the viscosity of ethylene glycol solutions lies between the lower values of calcium chloride brines and the higher values of propylene glycol solutions. The freezing points, densities, viscosities, specific heats, and thermal conductivities of ethylene glycol are shown in Figs. 20.13 to 20.17, respectively. Ethylene glycol is a moderate fire hazard, having a flash point in the neighborhood of 113°C (235°F). It is less corrosive than $CaCl_2$ brine, particularly when inhibited, which is the form in which ethylene glycol is usually sold. While it can be applied at temperatures down to about −40°C (−40°F), its high viscosity at those low temperatures is prohibitive, so it is usually considered to be a high-temperature coolant to be used at temperatures above −10°C (14°F). Ethylene glycol is somewhat toxic, so it should not be used in contact with food.

Two classes of inhibitors[2] are in common use with glycols, and these are: (1) corrosion inhibitors which coat the metal surfaces that come into contact with the glycol, and (2) environmental stabilizers which have the main purpose of regulating the pH to slightly above 7 in order to avoid an acidic condition.

The inclusion of an inhibitor affects the density and thus the ability to use the specific gravity to determine the solution strength. Accurate methods include gas chromatography or a refractive index measurement.

20.7 PROPYLENE GLYCOL

The freezing temperatures, densities, viscosities, specific heats, and thermal conductivities of propylene glycol solutions are illustrated in Figures 20.18 to 20.22,

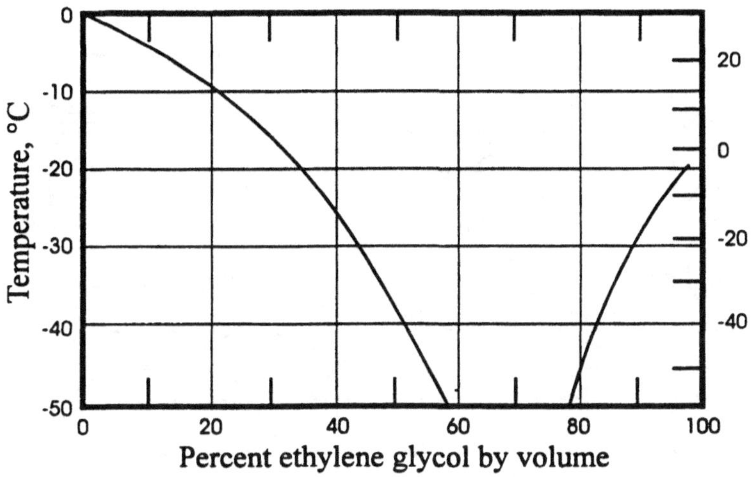

FIGURE 20.13
Freezing temperature of inhibited ethylene glycol solution[5].

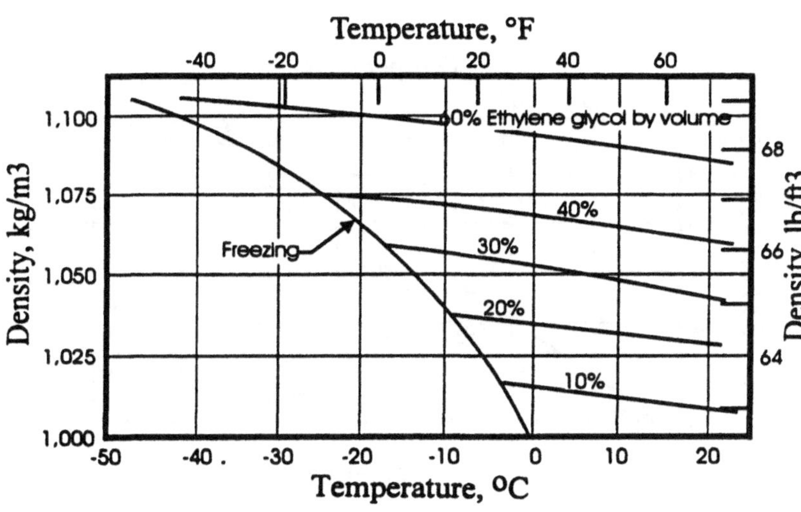

FIGURE 20.14
Density of inhibited ethylene glycol solution[5].

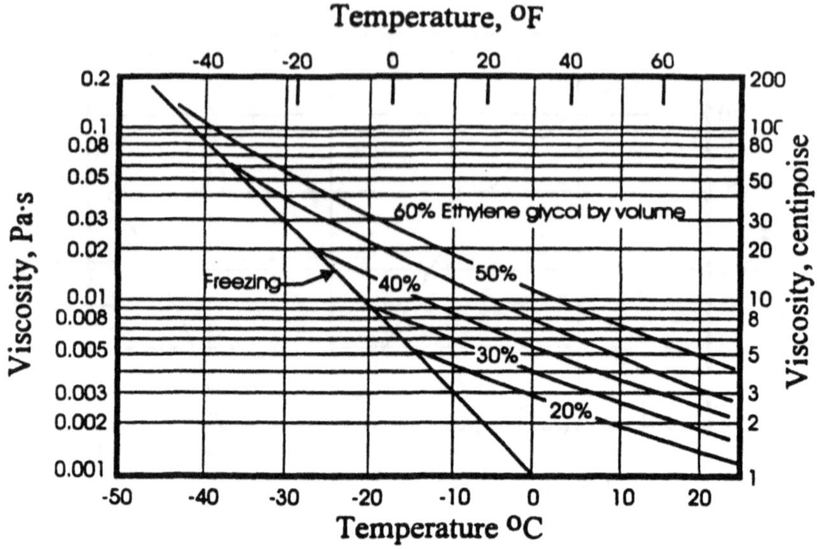

FIGURE 20.15
Viscosity of inhibited ethylene glycol solution[5].

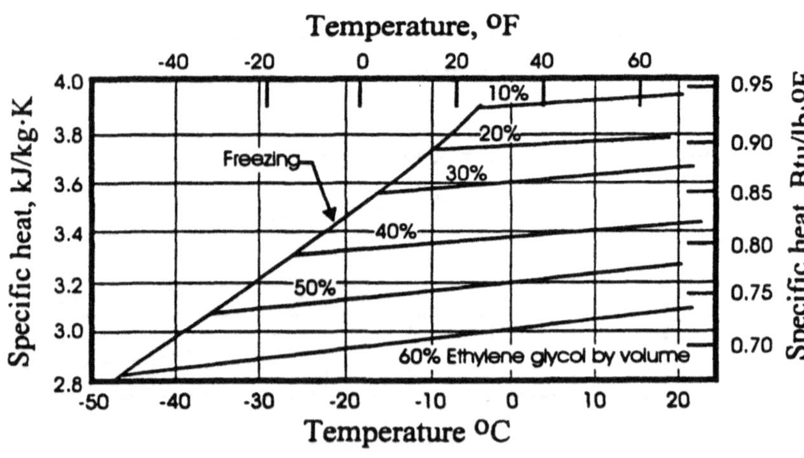

FIGURE 20.16
Specific heat of inhibited ethylene glycol solution[5].

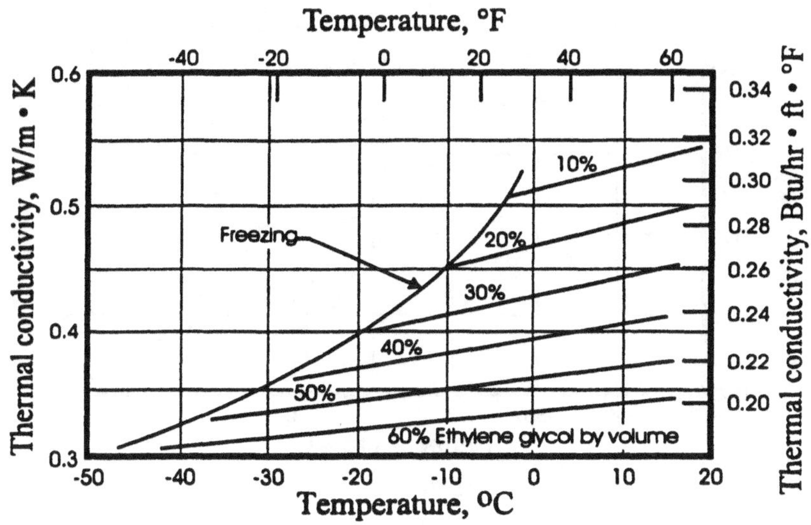

FIGURE 20.17
Thermal conductivity of inhibited ethylene glycol solution[5].

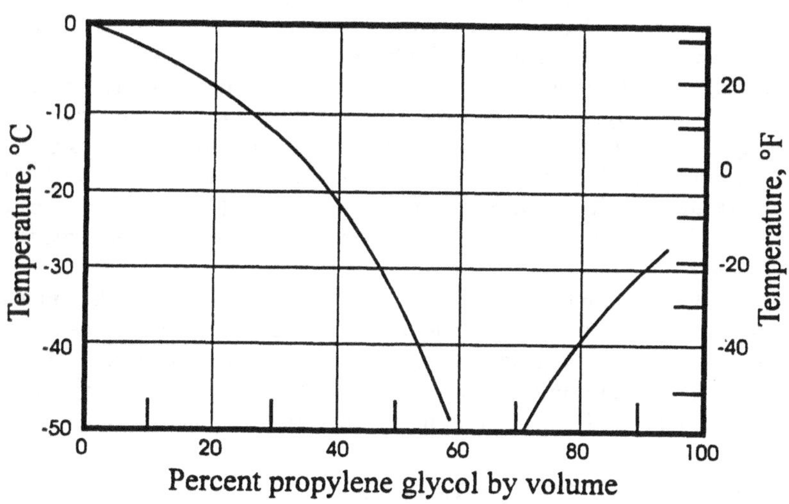

FIGURE 20.18
Freezing temperature of inhibited propylene glycol solution[6].

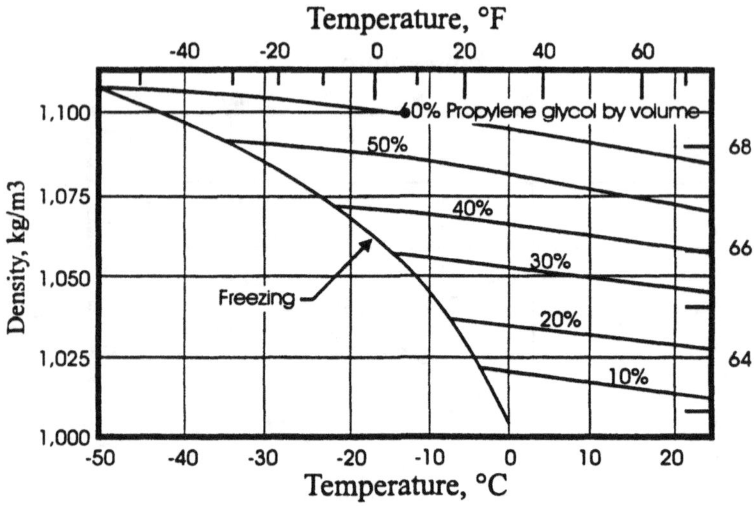

FIGURE 20.19
Density of inhibited propylene glycol solution[6].

respectively. A virtue of propylene glycol that is the chief contributor to its wide use is that it is nontoxic and suitable for direct contact with food. Its minimum freezing temperature is −51°C (−60°F), which is comparable to that of ethylene glycol and calcium chloride brine. It usually costs more than ethylene glycol. Propylene glycol would hardly ever be used close to the −51°C (−60°C) temperature, because of its high viscosity at high concentrations and low temperatures. In fact, its high viscosity for a given temperature service is one of the major drawbacks of propylene glycol. Its viscosity is two to four times greater than that of ethylene glycol, resulting in high pressure drops and low heat-transfer coefficients. When inhibited, it, like ethylene glycol, causes little corrosion. Practical low temperatures for its application are about −10°C (14°F). The flash point of propylene glycol is similar to that of ethylene glycol, so it is not considered a serious fire hazard.

Steel, aluminum, and copper are compatible with both glycols, but galvanized steel should be avoided as is true of calcium chloride brine. Some installers do not choose refrigeration grade copper but instead use the less-expensive standard grade. The removal of the active sludge that could absorb the inhibitors is critical, so the installation and periodic cleaning of filters is recommended.

20.8 POLYDIMETHYLSILOXANE

The four secondary coolants examined so far (two brines and two glycols) are aqueous solutions. The secondary coolant, polydimethylsiloxane[3] (PDS), is an example of a silcone-based polymer that is used full strength. A summary of the properties of PDS can therefore be presented in a straightforward table, as shown in Table 20.2.

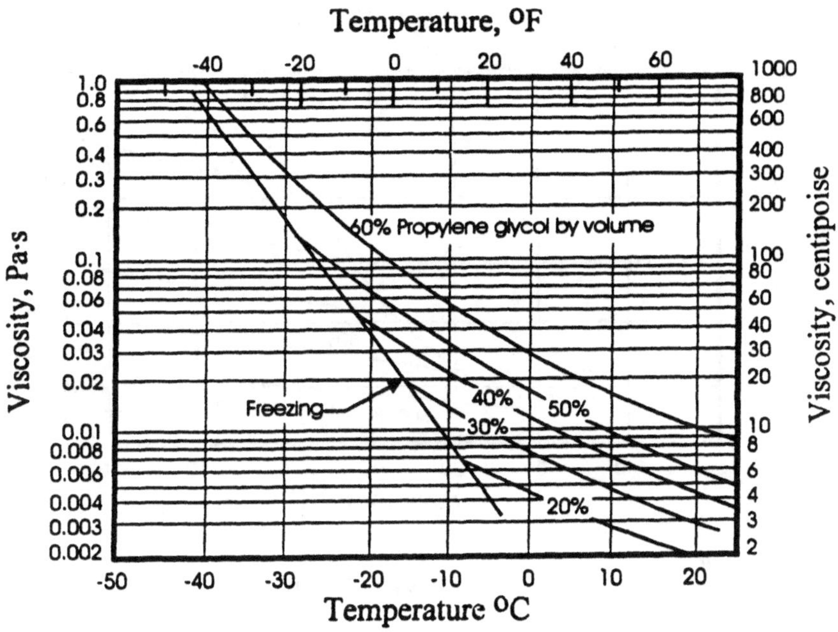

FIGURE 20.20
Viscosity of inhibited propylene glycol solution[6].

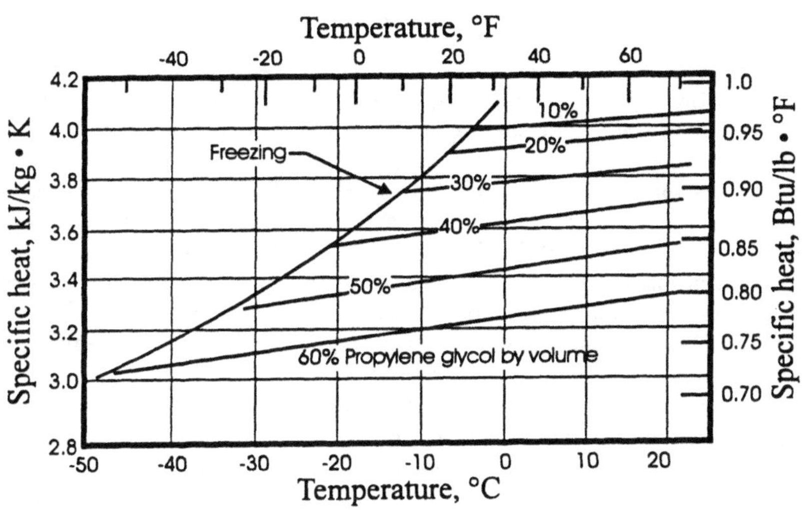

FIGURE 20.21
Specific heat of inhibited propylene glycol solution[6].

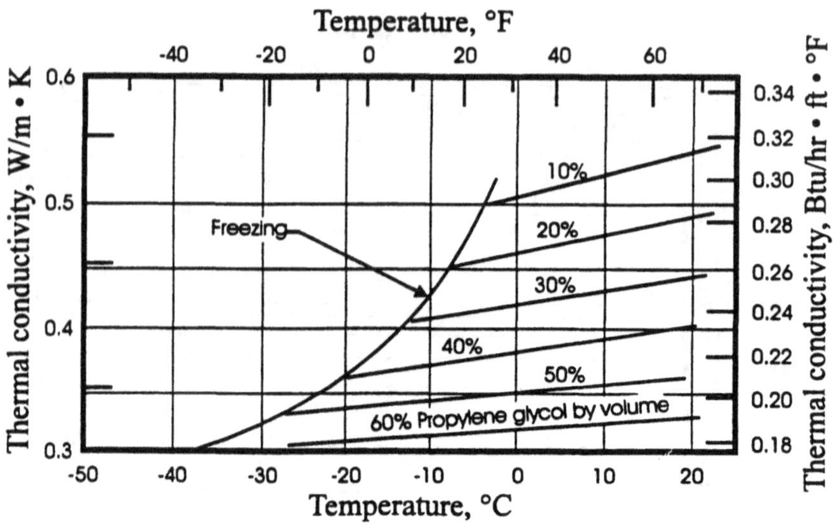

FIGURE 20.22
Thermal conductivity of inhibited propylene glycol solution[6].

The operational temperature range of PDS is −73 to 260°C (−100 to 500°F). Even though this chapter has confined its emphasis to low and extremely low temperatures, there are applications where, during an operating cycle, the heat-transfer fluid experiences a range of temperatures from very cold to very hot. Such an application[8] is the reaction that takes place in a jacketed vessel that is surrounded by fluid between −32 to 177°C (−25 to 350°F). The availability of one fluid capable of operating over this entire temperature range replaces the former procedure of passing steam, water, and glycol through the jacket at different times.

Some physical advantages advanced for PDS are that it is noncorrosive and essentially non-toxic. No worker exposure limits (TLVs or PELs) apply. The flash point is 47°C (116°F), the fire point is 60°C (140°F), and the auto-ignition temperature is 350°C (662°F). Because of its low vapor pressure, vapors of PDS do not pose a serious flammability hazard at room temperature. Carbon steel is the material predominantly used for piping and vessels, but low-alloy steels and stainless steel work quite satisfactorily.

An examination of several properties in Table 20.2, in comparison to previously presented data on brines and glycols, highlights an important advantage of PDS. The minimum freezing temperature as indicated by Table 20.1 is −111°C (−168°F), in comparison to the minimum freezing temperatures of calcium chloride brine and the glycols which are in the range of −50°C (−58°F). So while the lowest practical application temperature of $CaCl_2$ brine and the glycols may be about −30°C (−22°F), PDS would be practical down to temperatures in the neighborhood of −70°C (−94°F).

TABLE 20.2
Properties of polydimethylsiloxane[3] secondary coolant.

Temperature, °C	Viscosity, Pa·s	Density, kg/m^3	Specific heat, kJ/kg·K	Thermal conductivity, W/m·K
−73	0.01244	923.9	1.411	0.1294
−70	0.01126	921.4	1.418	0.1288
−60	0.00825	912.8	1.444	0.1269
−50	0.00621	904.3	1.470	0.1251
−40	0.00480	895.8	1.495	0.1231
−30	0.00379	887.3	1.521	0.1212
−20	0.00304	878.7	1.547	0.1192
−10	0.00249	870.1	1.572	0.1171
0	0.00207	861.5	1.598	0.1150
10	0.00174	852.7	1.624	0.1129
20	0.00148	843.9	1.649	0.1108
30	0.00127	835.0	1.675	0.1086
Temperature, °F	**Viscosity, centipoise**	**Density, lb/ft^3**	**Specific heat, Btu/lb·F**	**Thermal conductivity, Btu/hr·ft·°F**
−100	12.6	57.76	0.337	0.0748
−80	8.8	57.17	0.344	0.0736
−60	6.4	56.58	0.351	0.0724
−40	4.8	55.99	0.357	0.0711
−20	3.7	55.39	0.364	0.0699
0	2.9	54.80	0.371	0.0686
20	2.3	54.20	0.378	0.0673
40	1.9	53.60	0.385	0.0659
60	1.6	52.99	0.391	0.0646
80	1.3	52.37	0.398	0.0632

Other property comparisons of PDS with $CaCl_2$ brine and the glycols are for the viscosity, specific heat, and thermal conductivity, as shown in Table 20.3 for a temperature of −30°C (−22°F). The specific heat of PDS is about half that of the other coolants, therefore, it requires two times the flow rate. The thermal conductivity of PDS is about a third that of the others, which adversely affects the heat-transfer coefficient. The viscosity of PDS is, however, very favorable since it is perhaps one-tenth that of $CaCl_2$ brine and the glycols. This low viscosity reduces the pumping energy to help compensate for the greater flow rate of PDS. The low viscosity also exerts a positive influence on the heat-transfer coefficient.

A general conclusion that is subject to exceptions is to choose brine or the glycols for higher-temperature applications, but as the operating temperature of the coolant drops, consider the use of PDS.

TABLE 20.3
Comparison of several fluid properties at −30°C (−22°F).

Coolant	Viscosity, Pa·s (centipoise)	Specific heat, kJ/kg·K (Btu/lb·°F)	Thermal conductivity, W/m·K (Btu/hr·ft·°F)
Calcium chloride (28.2% by mass)	0.021 (21)	2.7 (0.645)	0.49 (0.283)
Ethylene glycol (52% by volume)	0.0482 (48.2)	3.04 (0.727)	0.329 (0.190)
Propylene glycol (54% by volume)	0.0270 (27)	3.24 (0.776)	0.316 (0.182)
PDS	0.00379 (3.79)	1.521 (0.363)	0.121 (0.0700)

20.9 D-LIMONENE

D-Limonene is the major component in citrus oils. It is obtained by distilling orange oil to produce food grade d-limonene and by pressing peelings to make the technical grade d-limonene used for cattle feed. The compound[9] is a terpene having a formula of $C_{10}H_{16}$, with the *d* designation indicating that the material is dextrorotatory, a fact that does not influence its use.

Major uses of D-Limonene are as a solvent and a cleaner, as well as a flavoring. The fact that it has the low freezing temperature of −97°C (−142°F), is available in food-grade form, and possesses a low viscosity has attracted interest as a possible secondary coolant. Because its use as a secondary coolant is not yet widespread, only preliminary measurements of the properties important to this use have been made so far. The data available[2] to date is shown in Table 20.4.

D-Limonene is compatible with steel, stainless steel, and aluminum. The major precaution in its use is the prevention of combustion. D-Limonene has a flash point of 46°C (115°F), and as a hydrocarbon, its vapor is combustible. The flammability limits at 150°C (302°F) are between 0.7% and 6.1%.

20.10 OTHER SECONDARY COOLANTS

In addition to the widely used secondary coolants that have been explored so far in this chapter, some additional fluids are available that may have favorable properties but also exhibit one or more important deficiencies. Also, the search continues for new substances as well as the development of new approaches to low-temperature fluids with desirable properties.

Methyl alcohol. Methyl alcohol or methanol is a widely used secondary coolant in the chemical and pharmaceutical industries. It is lower in cost and possesses superior heat-transfer characteristics in comparison to the glycols. The freezing temperature of pure methanol, for example, is −97.8°C (−144°F), which

TABLE 20.4
Thermal properties of d-limonene.

Temperature, °C	Viscosity, Pa·s	Density, kg/m³	Specific heat, kJ/kg·K	Thermal conductivity, W/m·K
−73	0.0038	914.3	1.27	0.137
−50	0.0030	897.1	1.39	0.133
−25	0.0023	878.3	1.51	0.128
0	0.0018	859.1	1.65	0.124
25	0.0014	839.8	1.78	0.119

Temperature, °F	Viscosity, centipoise	Density, lb/ft³	Specific heat, Btu/lb·°F	Thermal conductivity, Btu/hr·ft·°F
−100	3.8	57.1	0.30	0.0794
−50	2.8	55.8	0.34	0.0764
0	2.1	54.5	0.37	0.0734
50	1.6	53.2	0.41	0.0704

is lower than that of the glycols. So for less stringent temperature requirements, a solution of methanol in water is used. The major disadvantage of methanol is its toxicity, which makes it more harmful than ethylene glycol. Its use is usually limited to process applications where trained operators and technicians are available. Methanol is flammable, and thus introduces a fire hazard in its storage, handling, and use.

Ethyl alcohol. Pure ethyl alcohol or ethanol freezes at −112°C (−170°F), and in solutions with water, it is used as a secondary coolant. It is more expensive than methanol and has somewhat inferior heat-transfer coefficients. Ethanol is flammable and requires fire precautions.

Halocarbons. On occasion, some HCFC or HFC refrigerants are used as secondary coolants, because they usually have low freezing temperatures and are nontoxic and nonflammable. At the temperatures used, the refrigerant remains liquid throughout the circuit.

Ammonia-water. An aqua-ammonia solution performs effectively as a secondary coolant. A solution of 14% ammonia by mass exhibits a freezing temperature of −22°C (−7.0°F). Such a solution yields heat-transfer coefficients[10] of the same order of magnitude as $CaCl_2$ brine, which in turn is appreciably higher than the glycols, especially propylene glycol. The pumping energy required by an ammonia-water solution is very favorable and is the lowest of any popular secondary coolant. In a closed system with the assumed 14% ammonia solution, even at room temperature the pressure inside the system is below atmospheric, and at low operating temperatures the pressure is even lower. This means that in the event of a leak in the piping, air enters the system, but virtually no ammonia escapes.

Solid suspensions. A concept still in the development stage is to circulate a slurry, which consists of a liquid with solid particles that melt as they perform the refrigeration. Small particles of ice[11] frozen on a cold surface and removed constantly by scraping are introduced to a secondary coolant that is an aqueous solution. The pump in the circuit only handles liquid after the particles have melted. Another approach[12] has been to encapsulate minute quantities of a substance that is chosen so these particles freeze and thaw at the desired coolant temperature. The motivation behind these efforts is to exploit the latent heat of fusion of a substance to significantly reduce the flow rate of coolant.

20.11 INCENTIVES FOR NEW DEVELOPMENTS IN SECONDARY COOLANTS

We are now in a period of heightened interest in secondary coolants because of the appeal of indirect systems. Such systems decrease the amount of primary refrigerant required, which in turn reduces the cost of CFC replacements or provides a safer arrangement when ammonia is the refrigerant.

REFERENCES

1. Melinder A. and E. Granryd, "Secondary Refrigerants for Heat Pumps and Low Temperature Refrigeration," Report No. 92/6, Department of Applied Thermodynamics and Refrigeration, The Royal Institute of Technology, 1992.
2. American Society of Heating, Refrigerating and Air-Conditioning Engineers, Atlanta, Georgia, from the 1993 ASHRAE *Handbook— Fundamentals*.
3. "Syltherm XLT Heat Transfer Fluid," Product Technical Data, Dow Chemical Company, Midland, Michigan, 1993.
4. "Calcium Chloride for Refrigeration Brine," Calcium Chloride Institute, Washington, DC, 1956.
5. "Engineering Guide for Heat Transfer Fluids," UCARTHERM^R, Union Carbide Corporation, 1993.
6. "Engineering Manual for DOWFROST^R Heat Transfer Fluids," Dow Chemical USA, Midland, Michigan, 1990.
7. *Perry's Chemical Engineering Handbook*, 6th ed., editors R. H. Perry and D. Green, McGraw-Hill Book Company, New York, New York, 1984.
8. Lund, R. and D. Westbrook, "Low-Temperature Heat Transfer Fluid Improves Pharmaceutical Process," *Chemical Processing*, mid-April, 1990.
9. Florida Chemical Company, Inc., Lake Alfred, Florida, 1988.
10. American Society of Heating, Refrigerating and Air-Conditioning Engineers, Atlanta, Georgia, from the 1994 ASHRAE *Handbook— Refrigeration*.
11. Paul, J., "Binary Ice, an Alternative Cooling Method," Proceedings of Meeting of Commissions B1/2, International Institute of Refrigeration, Ghent, Belgium, 1993.
12. Pearson, S. F., "Development of Improved Secondary Refrigerants," Proceedings of the Institute of Refrigeration, London, England, March 1993.

CHAPTER 21

REFRIGERATION BELOW -40 DEGREES

21.1 REFRIGERATION AT TEMPERATURES BELOW −40 DEGREES

The term *low temperature* applies to the applications covered in this chapter and embraces the temperature range of approximately −40 to −80°C (−40 to −110°F). The description *low* is relative, because these temperatures are not low to the scientists and engineers who build and operate systems at temperatures approaching absolute zero. The band of temperatures addressed in this chapter is positioned below those temperatures normal for the freezing of food and well above cryogenic temperatures, which are usually defined as below −150°C (−240°F). Thus, the refrigeration temperatures explored in this chapter do not extend to the technologies of liquefaction of natural gas, nitrogen, oxygen, helium, and other industrial gases. Similarly, this chapter does not encompass the bulk of industrial refrigeration activities that span the range of evaporating temperatures from about 5°C down to −40°C (40°F down to −40°F).

Some justification should be advanced for treating refrigeration below −40° differently than systems operating above that temperature. Aren't the principles the same? Physical laws applicable above −40° do not suddenly lapse at temperatures below that value. However, there are some differences in the following areas:

- Refrigerants
- Systems (3 stage and cascade)
- Materials for pipe and vessels
- Insulation
- Secondary coolants
- Magnitude of boiling and condensing heat-transfer coefficients

Refrigeration technologies below 0°C (32°F) differ somewhat from those above that temperature, such that a competent designer and installer of air conditioning systems might not know all the requirements for systems operating below the freezing point of water. In a similar manner, designers and installers experienced with systems for freezing and preserving food may not possess all the required expertise to provide, for example, a successful liquid chiller for a pharmaceutical plant that refrigerates brine to a temperature of −70°C (−94°F).

Unfortunately, few designers, installers, and manufacturers of components enjoy a frequency of jobs high enough to permit a rapid learning curve in the low-temperature field. There is a special need, then, for the professionals active in this sphere of refrigeration to share experiences. In fact, this chapter has drawn on the accumulated experience of some of these designers, contractors, and operators.

21.2 SOME APPLICATIONS OF LOW-TEMPERATURE REFRIGERATION

A wide variety of applications call for evaporating temperatures below −40°, and while no statistical survey has been attempted, it is likely that environment chambers, fluid chillers in the chemical and process industries, freeze-drying chambers, and condensation for the recovery of vapors are dominant applications of low-temperature refrigeration. Some of the functions of low-temperature environment chambers[1] include determining the performance of mechanical and electrical equipment at low temperatures, studying the properties of materials at low temperatures, and testing humans and animals. Tests on personnel may be conducted to examine the performance of protective clothing or for altitude and space studies. Some environment chambers are small cabinets, while others are large enough to house an automobile under low-temperature operating conditions.

A sizeable business in the −50 to −70°C (−60 to −90°F) temperature range has developed for the condensation of volatile organic compounds (VOCs). These installations often incorporate a spare VOC condensation unit, because the entire plant operation may have to be shut down for environmental reasons when the condensation facility is out of operation for repair or maintenance. Other low-temperature applications in the chemical industry provide a low-temperature environment for chemical reactions that would race out of control at room temperatures.

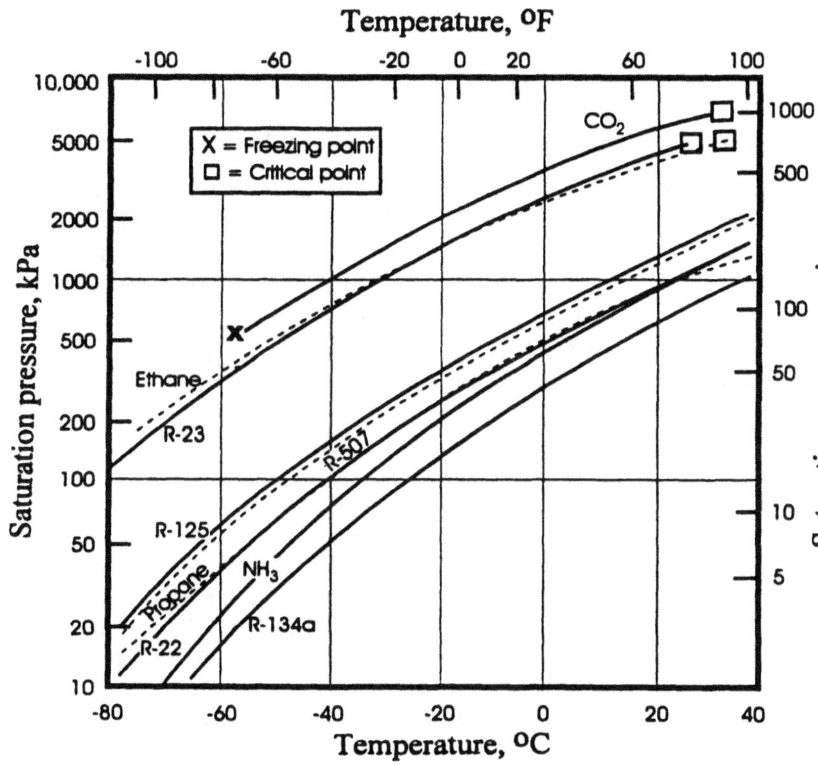

FIGURE 21.1
Saturation pressures of some low-temperature refrigerants as a function of temperature.

21.3 SYSTEMS AND REFRIGERANTS

The choices of system configurations and refrigerants are interrelated. The pressure should not be undesirably high at condensation nor undesirably low in low-temperature evaporators, but it is difficult to find a refrigerant that has suitable pressures at both ends of the temperature scale.

Figure 21.1 shows the saturation pressure curves for the family of refrigerants generally considered for low-temperature systems. The figure illustrates that the common refrigerants for industrial refrigeration systems (ammonia, R-22, R-507, and even R-134a) experience undesirably low evaporating pressures in the range of −40 to −80°C (−40 to −112°F). The refrigerants R-23, CO_2, and ethane, on the other hand, which have reasonable pressures in the evaporator, experience undesirably high pressures when condensing at typical ambient temperatures.

TABLE 21.1
Compressor displacement rates for several different refrigerants at three evaporating temperatures with 0°C (32°F) liquid entering the expansion valve.

Evaporating temperature	Compressor displacement, L/s per kW (cfm per ton)						
	R-22	R-507	Ammonia	R-23	CO_2	Propane	Ethane
−40°C (−40°F)	1.09 (8.13)	0.979 (7.30)	1.29 (9.62)	0.237 (1.77)	0.162 (1.208)	1.15 (8.58)	.244 (1.82)
−60°C (−76°F)	3.01 (22.4)	2.87 (21.4)	4.01 (29.9)	0.545 (4.06)	Freezes at −56.6°C	3.06 (22.8)	0.522 (3.89)
−80°C (−112°F)	10.56 (78.8)	11.12 (82.93)	Freezes at −77.7°C	1.50 (11.19)		10.13 (75.5)	1.28 (9.54)

A low evaporating pressure by itself may not preclude the use of a refrigerant if the circuit can be hermetically sealed, but the high specific volume that occurs at low temperatures may require the physical size and displacement rate of the compressor to be large. Table 21.1 shows compressor displacement rates of systems operating at several different evaporating temperatures with liquid entering the expansion device at a temperature of 0°C (32°F). This temperature of the supply liquid was arbitrarily chosen, but assumes that the refrigerant serving the evaporator is either on the low stage of a two-stage system or in a cascade system. The data in Table 21.1 shows that the displacement rate required at the compressor is very large when using the standard industrial refrigerants (R-22, ammonia, R-507, and propane) at low temperatures. On the other hand, the displacement rates are modest for R-23, CO_2, and ethane.

Also shown in Table 21.1 is the fact that ammonia freezes at −77.7°C (−108°F) and CO_2 freezes at −56.6°C (−69.8°F). The possibility of freezing in the case of ammonia simply means that the refrigerant cannot be used for evaporating temperatures lower than its freezing point. In the case of CO_2, however, the frozen form can be usefully applied. The skeleton pressure-enthalpy diagram of CO_2 is shown in Fig. 21.2. The lower region on the p-h diagram is especially interesting, because below a pressure of 518 kPa (75.1 psia), CO_2 cannot exist in liquid form—only as a solid or a vapor. The triple point (which is actually a line) is a special state at which vapor, liquid, and solid can coexist at saturated conditions. The triple point of CO_2 is at a pressure of 518 kPa (75.1 psia) and a temperature of −56.6°C (−69.8°F). What is called a triple point is actually a line at a pressure of 518 kPa (75.1 psia) extending from the saturated solid line to the saturated vapor line. It should not be implied that carbon dioxide is a rare substance possessing a triple point. Virtually all substances have triple points, but for CO_2, the property occurs at a temperature that is quite useful.

Carbon dioxide is sometimes used as an expendable refrigerant for freezing food. Carbon dioxide is supplied to the plant in liquid form, which means that

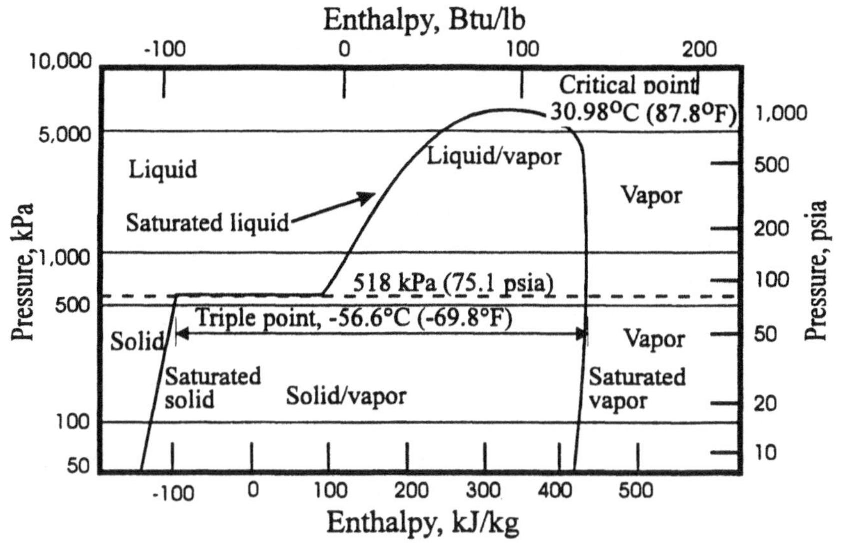

FIGURE 21.2
Skeleton pressure-enthalpy diagram of carbon dioxide (CO_2).

it must be under pressure. If the temperature of liquid CO_2 is 20°C (68°F), for example, the saturation pressure would be 5,728 kPa (831 psia). When storing large quantities of CO_2, the pressure of liquid in the vessel may be kept at a convenient value by maintaining it at a reduced temperature. Such a facility is equipped with a recompression unit that draws off vapor from the vessel, compresses it, then condenses it, and finally throttles and returns the mixture of liquid and vapor to the vessel. If CO_2 liquid expands to standard atmospheric pressure through a throttling valve, the resultant state is a mixture of CO_2 vapor and solid that appears as snow. The temperature after throttling is −79°C (−110°F). One commercial process presses the carbon dioxide snow into blocks of *dry ice*, which is a convenient, short-term, low-temperature cooling medium.

For large systems in industrial environments, hydrocarbons such as ethane and propane are used extensively. The requirements for dealing with flammable refrigerants, such as special electric gear, are at least partially offset by the relatively low cost of the refrigerant. Thermodynamic and fluid properties of hydrocarbon refrigerants are favorable at these low temperatures.

HFC-23 is one of the candidates to replace R-503. Its toxicity is low, its saturation pressure curve is close to that of R-503, and materials commonly used for R-503 are suitable for HFC-23. A system using HFC-23 can expect to use about 10% more energy than necessary with R-503. Early reports on the operation of HFC-23 systems are becoming available[2] and indicate that the mineral oils suitable for R-503 must be replaced with polyol ester oils. The expansion valve setting is expected to be different than with R-503.

21.4 MULTISTAGE SYSTEMS

Two-stage systems are very common in industrial refrigeration. The principal disadvantage of a two-stage plant in comparison to single stage is a higher first cost of the system. However, there are many advantages of the two-stage concept, including the following:

- Improved efficiency because of flash-gas removal at the intermediate pressure
- Improved efficiency attributable to desuperheating of discharge gas from the low-stage compressor
- Reduced pressure ratios imposed on the compressors which results in higher efficiency of each compressor
- Lower compressor discharge temperatures which is particularly important for reciprocating compressors
- Lower volume flow rate of refrigerants required for the low-stage compressor
- Smaller size of suction line from the low-temperature evaporator

A question arises as to whether to choose a three-stage system, of which there are many in operation, or a two-stage system. The need for three-stage compression is not as critical as it once was when reciprocating compressors were dominant. This situation has changed now with the availability of screw compressors, which can function with high pressure ratios, although not always so efficiently. Table 21.2 shows the ideal pressure ratios required for several different refrigerants serving an evaporator at −70°C (−94°F) and condensing at a temperature of 24°C (75°F) for single, double, and triple stage. Recall that reciprocating compressors are generally limited to pressure ratios of approximately 7 to 9. At the conditions of Table 21.2, the application of ammonia and R-134a in two-stage systems with reciprocating compressors would be questionable. It may appear that R-23 would be a good choice, but as Fig. 21.1 shows, it reaches a critical point at a temperature a few degrees higher than the 24°C (75°F) condensing temperature chosen in Table 21.2. The limitations of pressure ratios applicable to the reciprocating compressor do not apply to the screw compressor, which can operate with pressure ratios of 15 or higher. Thus, a system that is large enough to justify screw compressors would normally require no more than two stages of compression.

21.5 CASCADE SYSTEM

The difficulty of finding one refrigerant that is ideally suited to both the high and low temperatures and pressure ranges of a system operating with a large temperature lift leads to the concept of a cascade system. In the cascade system, as shown schematically in Fig. 21.3, two separate refrigerant circuits are connected thermally through a cascade condenser. The refrigerant chosen for the low-temperature circuit is a high-pressure refrigerant, such as R-23. For the high-

TABLE 21.2
Pressure ratios with single, double, and triple stage systems when the evaporating temperature is −70°C (−94°F) and the condensing temperature is 24°C (75°F).

Refrigerant	Single stage	Two stage	Three stage
R-22	49.7	7.05	3.68
R-23	23.8	4.88	2.88
R-125	43.0	6.56	3.50
R-134a	80.6	8.98	4.32
Ammonia	89.2	9.45	4.47

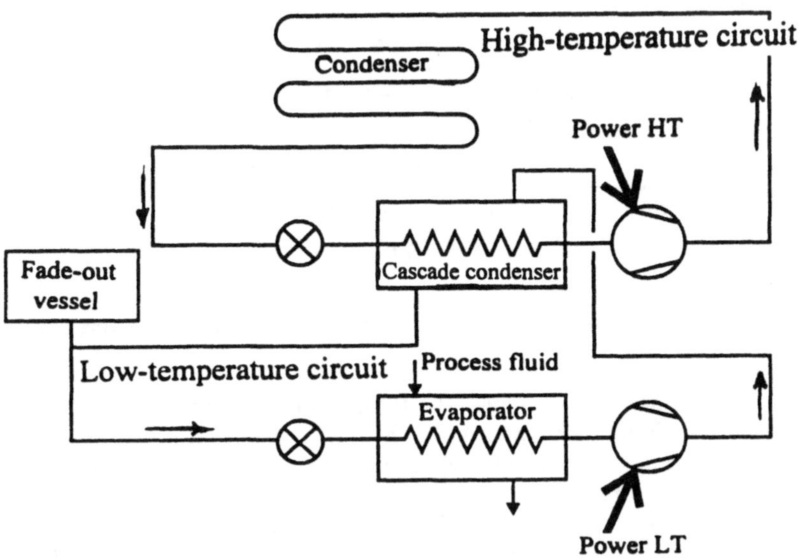

FIGURE 21.3
A cascade system.

temperature circuit, any of the typical industrial refrigerants, such as ammonia or R-22, may be selected.

The cascade condenser is the condenser of the low-temperature circuit (LTC) and the evaporator of the high-temperature circuit (HTC). Thus, the heat absorbed in the evaporator plus the energy introduced at the compressor of the LTC must be absorbed by the evaporator of the HTC.

A penalty suffered by the cascade system in comparison to a two-stage system using one refrigerant is that there must be an overlap of temperatures between the condensing temperature of the LTC and the evaporating temperature of the HTC. In systems that operate a large portion of the time, the energy requirements could be important. In this case, the designer should be able to

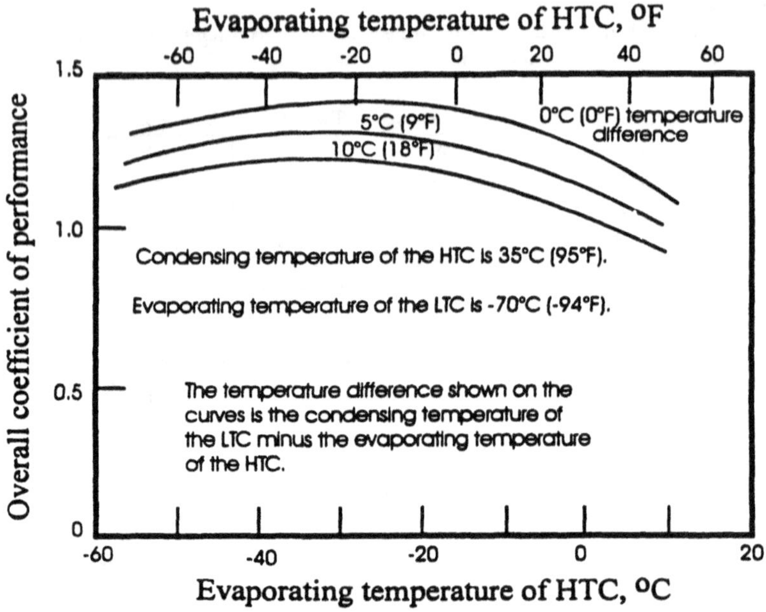

FIGURE 21.4
Overall coefficient of performance of a cascade system with ammonia serving the HTC and R-23 serving the LTC.

select components that minimize the total compressor power as well as provide the specified refrigeration capacity in the evaporator.

The overall coefficient of performance, $COP_{overall}$, is the refrigerating capacity in the evaporator divided by the total power required by the two circuits. The refrigeration load of the HTC is the heat gain in the evaporator plus the power of the LTC compressor. If the individual COPs are designated COP_{LTC} and COP_{HTC} of the low- and high-temperature circuits, respectively, the overall COP is expressed by the combination:

$$COP_{overall} = \frac{(COP_{LTC})(COP_{HTC})}{1 + COP_{LTC} + COP_{HTC}} \tag{21.1}$$

Figure 21.4 shows the $COP_{overall}$ of a cascade system, assuming isentropic compressions, in which R-23 serves the LTC and ammonia serves the HTC. The temperature difference between the condensing temperature of the LTC and the evaporating temperature of the HTC is strongly influenced by the size of the cascade condenser. As the heat-transfer capacity of the cascade condenser increases and the temperature difference in this heat exchanger decreases, the $COP_{overall}$ increases. Figure 21.4 also shows that there is an optimum intermediate temperature, just as is true of two-stage systems using a single refrigerant.

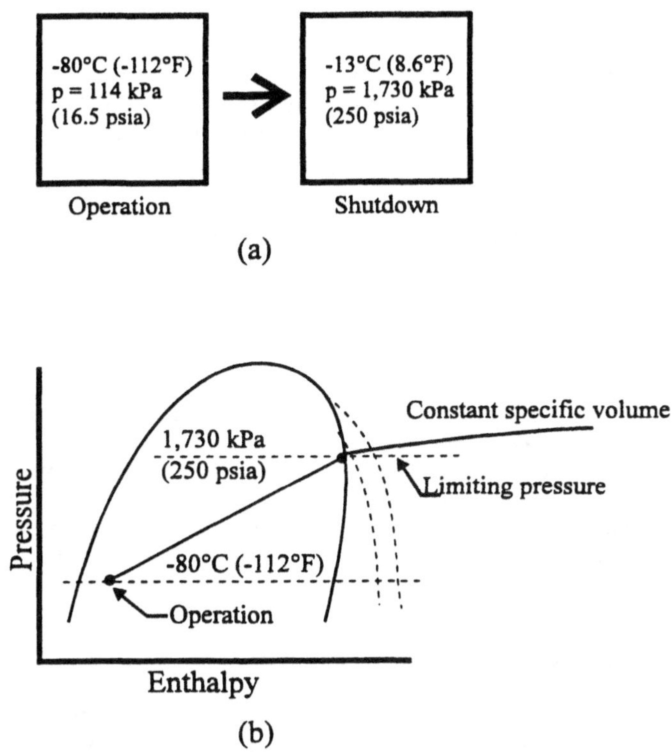

FIGURE 21.5
(a) Fade-out as a liquid/vapor mixture increases in temperature, and (b) the constant volume process shown in the pressure-enthalpy diagram.

21.6 FADE-OUT VESSEL

The diagram of the cascade system, Fig. 21.3, shows a fade-out vessel, which is sometimes called an expansion tank. Its purpose is to limit the pressure in the LTC during shutdown periods when the temperature of the liquid rises to room temperature. For example, the saturation pressure of R-23 at 20°C (68°F) is 4,182 kPa (606 psia). If liquid still exists in the system during shutdown, all vessels and pipes in the system must be able to contain a high pressure. A technique to limit the pressure is to install a fade-out vessel such that at the desired maximum pressure all of the liquid has vaporized. Beyond that point, the increase in pressure with temperature is modest and proportional only to the absolute temperature.

The fade-out process can be visualized, as in Fig. 21.5a, and represented by a constant-volume line on the pressure-enthalpy diagram of Fig. 21.5b. Assume that a vessel of R-23 containing 5% liquid by volume (which corresponds to 93.5%

liquid by mass) exists at a temperature of −80°C (−112°F). The pressure at this state is 113.8 kPa (16.5 psia). As the temperature of the refrigerant increases, the pressure increases according to the saturated temperature-pressure relationship. But another process that occurs is the progressive shift toward the saturated vapor line, which is reached at a temperature of −13°C (8.6°F) and a pressure of 1,730 kPa (250 psia). Once the saturated vapor line is reached, the increase in pressure with further increase in temperature is moderate.

Example 21.1. What is the size of the fade-out vessel in an R-23 LTC of a cascade system, if the pressure is to be limited to 2,000 kPa (290 psia) at the idle state when the temperature is 20°C (68°F)? The volume in the system apart from the fade-out vessel is 0.5 m^3 (17.7 ft^3), and the mass of R-23 in the system is 50 kg (110 lb).

Solution. At a pressure of 2,000 kPa (290 psia) and a temperature of 20°C (68°F), from the pressure-enthalpy diagram for R-23 the density of the superheated vapor is found to be 70 kg/m^3 (4.37 lb/ft^3). The total volume occupied by the charge of 50 kg (110 lb) would be (50 kg)/(70 kg/m^3) = 0.714 m^3 (25.2 ft^3). The available volume in the system is already 0.5 m^3 (17.7 ft^3), so the additional volume required in the fade-out vessel is 0.714 − 0.5 = 0.214 m^3 (7.5 ft^3).

21.7 CHOOSING THE TYPE OF EVAPORATOR

Assuming that the plant refrigerates a liquid, the shell-and-tube type of evaporator is the likely choice. Within that basic decision is the choice of whether to evaporate the refrigerant in the shell or in the tubes. Associated with the choice of evaporating in the tubes is whether to operate with liquid recirculation or direct expansion. Two of the many concerns associated with the choice of evaporator are maintaining high heat-transfer coefficients and achieving good oil handling. As will be pointed out in Section 21.10, the heat-transfer coefficients generally drop as the temperature drops, so care must be taken to avoid any unnecessary degradation of the refrigerant heat-transfer coefficient in the evaporator. Also, because of the low temperature in the evaporator, any oil that finds its way into the evaporator will be difficult to remove because of its high viscosity and low solubility in the refrigerant, as will be discussed in more detail in Section 21.11.

Usually the preference is to boil the refrigerant in the tubes rather than in the shell, because the return of oil from the evaporator is slightly easier with the refrigerant flowing in the tubes. The velocity of the suction gas in the tubes assists the return of oil to the compressor, in contrast to boiling in the shell where basically a separation of oil and refrigerant vapor occurs. Another choice is whether or not to install liquid recirculation. Even though liquid recirculation offers the possibility of assisting in oil return and helping to provide satisfactory refrigerant heat-transfer coefficients, the complications of the liquid recirculation equipment on what is inherently a rather sensitive plant usually deters its use.

By the process of elimination, the decision process leads to a shell-and-tube evaporator with refrigerant boiling in the tubes with some means of controlling the suction superheat. The amount of charge and the cost of the refrigerant also favor refrigerant boiling in the tubes. Conventional superheat control valves, or thermostatic expansion valves (TXVs), are used by some designers; however, on large systems, an industrial-grade controller incorporating pressure and temperature transmitters is used for the expansion valve. The amount of superheat leaving the evaporator should be maintained as low as possible to avoid the poor heat-transfer coefficients associated with the superheating section of the evaporator. Attempting to operate with only several degrees of superheat leaving the evaporator runs the risk of flooding out liquid from the evaporator during radical changes in load, so many plants are equipped with a liquid/suction heat exchanger to protect the compressor by boiling off any liquid that escapes the evaporator. This liquid/suction heat exchanger should not be so large that it chills the incoming liquid to such an extent that, even after passing the expansion valve, the refrigerant entering the evaporator is still subcooled. Some vapor in the refrigerant entering the evaporator helps maintain satisfactory heat-transfer coefficients.

21.8 AUTOCASCADE SYSTEMS

The cascade system just investigated consisted of two separate refrigeration systems, each with its own different refrigerant. Still another important type of cascade system is the *autocascade* cycle[3], which uses only one refrigerant, and all segments of the system are interconnected. Furthermore, the one refrigerant it uses is actually a blend of two refrigerants. This blend of refrigerants is called a nonazeotropic or a zeotropic mixture. Azeotropic and zeotropic mixtures have already been introduced in Section 12.3, but a more thorough explanation is warranted here in order to properly understand the operation of the autocascade system.

The existence and application of a mixture or blend of refrigerants is quite common. The now-supplanted R-502 and such new combinations as R-507 are mixtures of two refrigerants, but the choice and composition are such that the mixture is an azeotrope, which means that the combination behaves as a single refrigerant different from either of its constituents. For autocascade systems, the refrigerants are chosen such that the mixture is not azeotropic but zeotropic, as was illustrated in Fig. 11.2, a diagram reproduced in Fig. 21.6. Also labeled on Fig. 21.6 is the state of the refrigerant—superheated above the so-called dewpoint line and subcooled below the so-called bubblepoint line.

The behavior of a zeotrope during a condensation or evaporation process differs from that of a single refrigerant. The temperature of a single refrigerant remains constant during a constant-pressure condensation or evaporation, while the temperature of a zeotrope changes. Figure 21.7 shows a condensation process with the concentration remaining constant during the process. In Fig. 21.7a, Point 1 is superheated vapor flowing into the cool condenser tube. At Point

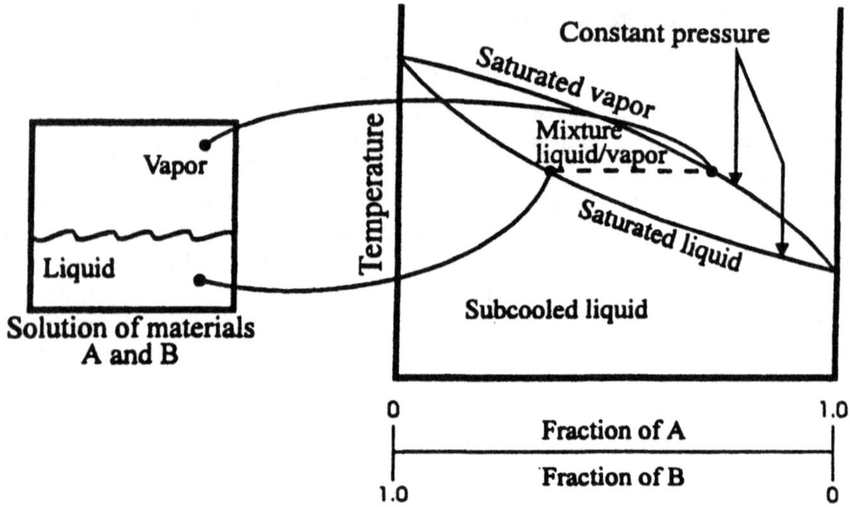

FIGURE 21.6
A zeotropic mixture of two refrigerants.

2 condensation begins, and as Fig. 21.7b shows, the temperature continues to drop as condensation proceeds to saturated liquid at Point 3. If cooling continues, Point 4 becomes subcooled liquid. The autocascade system capitalizes on this change in temperature, or temperature glide.

Figure 21.8 shows a schematic diagram of an autocascade system, and Fig. 21.9 identifies the state points on the temperature-concentration diagram for an R-23/R-134a zeotrope. The numbered points on the temperature-concentraton diagram correspond to those on the flow diagram. Begin tracing the processes at Point 1 entering the compressor, which is at the low pressure of 100 kPa (14.5 psia). In the common streams of the system, the concentration has been chosen to be 50% by mass. The compressor elevates the pressure to, in this example, 1,500 kPa (218 psia), for which the high-pressure dewpoint/bubblepoint curves apply. The refrigerant mixture leaves the compressor at Point 2.

The refrigerant flows from the compressor to the condenser, which only partially condenses the refrigerant to Point 3. The mixture of liquid and vapor then passes to the phase separator, which permits the liquid at Point 4 and the vapor at Point 7 to flow in different branches. The liquid has a high concentration of R-134a, while the vapor has a high concentration of R-23. The liquid stream moves from Point 4 to the expansion valve, where the pressure drops to 100 kPa (14.5 psia) at Point 5 and a small fraction of the liquid has converted to vapor. This stream flows through one side of the aftercondenser heat exchanger, vaporizing and leaving slightly superheated at Point 6.

The vapor that is rich in R-23 flows from the phase separator and condenses on the other side of the aftercondenser, leaving in a saturated state or as slightly subcooled liquid at Point 8. This fraction of the refrigerant mixture

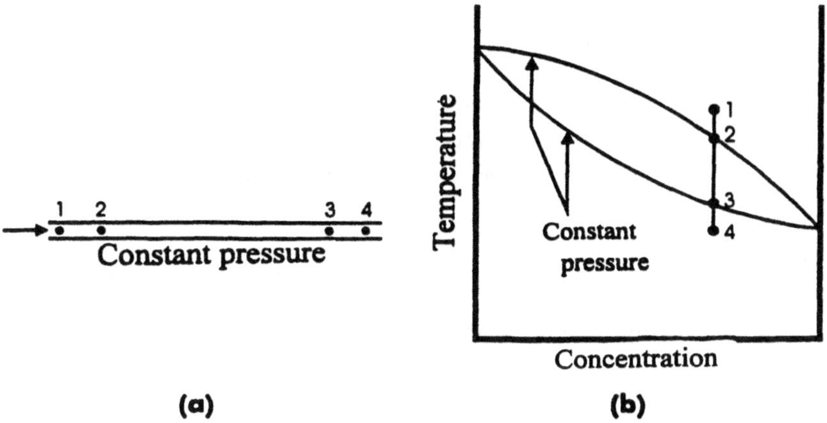

FIGURE 21.7
(a) A condensation process of a zeotrope, and (b) the change in temperature during the constant-pressure condensation.

is the ultimate refrigerant. The pressure drops to 100 kPa (14.5 psia) in the expansion valve and enters the evaporator (Point 9) at the lowest temperature that occurs in the system. During the evaporation process from Point 9 to Point 10, the refrigerant chills the process fluid. Finally, vapor at Point 10 mixes with vapor at Point 6 to restore the 50% proportions that will once again enter the compressor and repeat the cycle.

The autocascade system achieves a low temperature at Point 9 but does so with only one compressor in contrast to multistage or separate cascade circuits. In the example illustrated in Fig. 21.9, the condensing temperature is 20°C (68°F), and the lowest temperature in the evaporator is −73°C (−99.4°F). For a condensing temperature of 20°C (68°F) and an evaporating temperature of −73°C (−99.4°F), the pressure ratio if R-23 were used as a single refrigerant would be 25. If R-134a were a single refrigerant, the pressure ratio would be 85. The autocascade system provides the same temperature lift with a pressure ratio of 1,500/100 = 15. Admittedly, the autocascade system does not maintain −73°C (−99.4°F) throughout the entire evaporation process, so a slightly larger evaporator would have to be chosen for the autocascade system in order to provide the same cooling duty on the process fluid.

Analysis of the performance of the autocascade system is an intriguing exercize, because the condition leaving the partial condenser dictates the proportions of the low-boiling-temperature stream (7-8-9-10) and the high-boiling-temperature stream (4-5-6). But the energy balance at the aftercondenser requires that the product of the respective flow rate and change in enthalpy of one stream equals that of the other stream. The mean difference in temperatures of the condensing fluid in the aftercondenser (t_5/t_7) and those of the evaporating

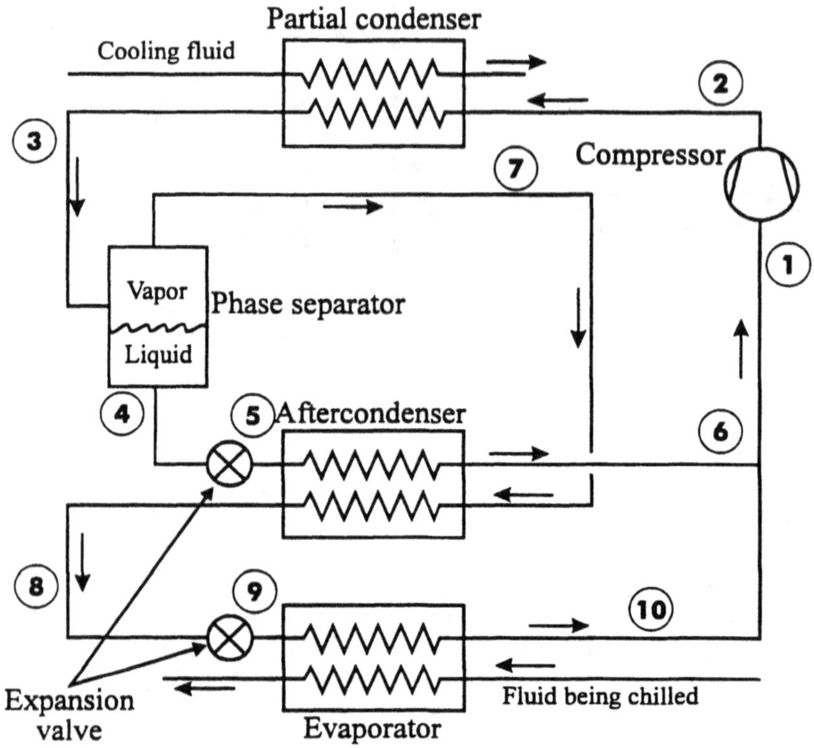

FIGURE 21.8
Flow diagram of an autocascade system.

fluid (t_5/t_6) must, in combination with the heat-transfer area and the overall heat-transfer coefficient, transfer the necessary rate of heat flow to boil the high-temperature fraction and condense the low-temperature fraction. The condensing temperature of 20°C (68°F) in the example would be difficult to achieve when using an air-cooled condenser. Extracting the mixture from the partial condenser with a temperature higher than Point 3, Fig. 21.9, shifts the fractions of the two streams, reducing the flow rate through the evaporator. The result of this shift is a lower flow rate through the evaporator, but there is at least a partial compensation, because the temperature leaving the aftercondenser is likely to drop, resulting in more subcooling at Point 8, Fig. 21.9. To restore a desirable proportioning of the two streams, the condensing pressure must increase, which could be achieved by adjusting the two expansion valves.

Autocascade systems are widely used for low-temperature applications. Their major application is for low- and moderate-capacity requirements for which

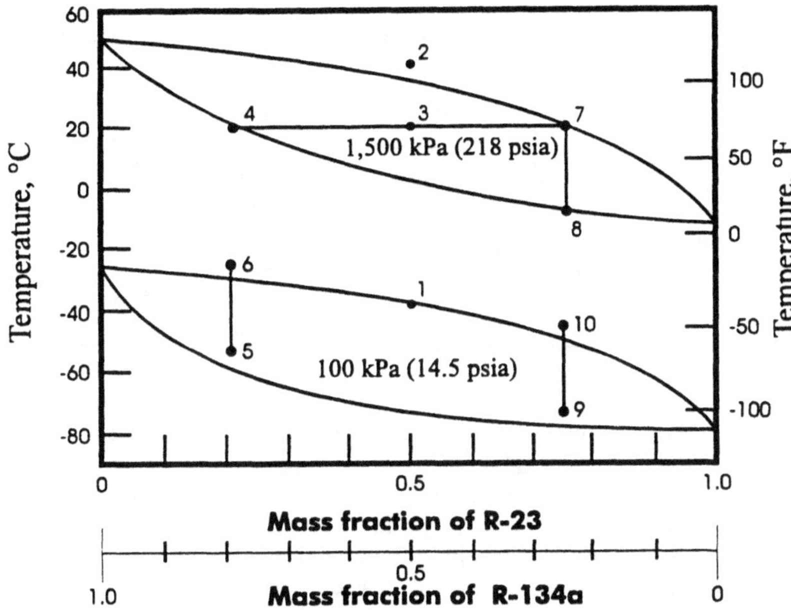

FIGURE 21.9
State points throughout the cycle of Figure 21-8 for a zeotrope of R-23/R-134a.

a hermetically sealed package can be assembled. Because the refrigerant charge is a zeotropic mixture, leaks of refrigerant are highly undesirable, because the proportions of the two refrigerants that have leaked will not be known. Field-assembled systems are more subject to leaks in contrast to the factory-assembled hermetic unit.

21.9 HEAT-TRANSFER FLUIDS

In many cases, the boiling refrigerant in the evaporator chills air in the low-temperature space or chills the product by direct contact with the metal of the heat exchanger. However, the application of a secondary coolant is also frequent, wherein the secondary coolant that is chilled in the evaporator then receives heat from the substance being refrigerated. The requirement of a low freezing temperature restricts the choice of the antifreeze. The popular choices in the temperature range above −40° such as calcium chloride brine, ethylene glycol, and propylene glycol either do not have adequately low freezing temperatures or become quite viscous, which in turn elevates the pumping costs. Instead, designers turn to such fluids as acetone, ethyl alcohol, methyl alcohol, and polydimethylsiloxane[4]. More complete information on the low-temperature properties of antifreezes is provided in Chapter 20, but two characteristics are especially important—the

TABLE 21.3
Effect of lowering the temperature on thermodynamic and transport properties and the resultant effect on heat-transfer coefficients.

	Change due to drop in temperature	
Property	Change in property	Influence on coefficient
Thermal conductivity	Decreases	Reduces
Viscosity of liquid	Increases	Reduces
Latent heat	Increases	Improves
Specific heat of liquid	Decreases	Reduces
Liquid density	Increases	Improves condensation Reduces boiling

specific heat and the viscosity. A high specific heat is preferred to reduce the rate of fluid flow of the antifreeze in transferring the specified rate of heat for a given change in temperature. A low viscosity is desired to minimize pumping costs. Water has a relatively high specific heat and low viscosity. So with antifreezes that are aqueous solutions, the lowest strength that provides adequate freeze protection results in the highest possible specific heat and lowest possible viscosity.

21.10 HEAT-TRANSFER COEFFICIENTS

Section 21.7 has already flagged the possibility of low boiling and condensing heat-transfer coefficients in low-temperature systems. The thermodynamic and transport properties that control the coefficients for a given geometric and flow condition almost all trend toward decreasing the coefficients as the temperature drops.

The thermodynamic and transport properties that exert the most influence on the boiling and condensing heat-transfer coefficients are thermal conductivity, viscosity, latent heat of vaporization, and specific heat. Liquid density has only a minor influence on heat-transfer coefficients. The changes in these properties with temperature and the effect on heat-transfer coefficients are shown in Table 21.3. The conclusion is that virtually all the properties that change with temperature result in lower heat-transfer coefficients at low temperatures.

An estimate of the effect of shifting to low temperatures on the heat-transfer coefficients calculated from correlations[5] for forced convection evaporation and film condensation is shown in Fig. 21.10. The coefficients for R-23 are shown as a ratio of values with respect to coefficients at 0°C (32°F). The point is that, for low-temperature systems, the designer must be keenly aware of the degradation of both boiling and condensing heat-transfer coefficients experienced at low temperatures.

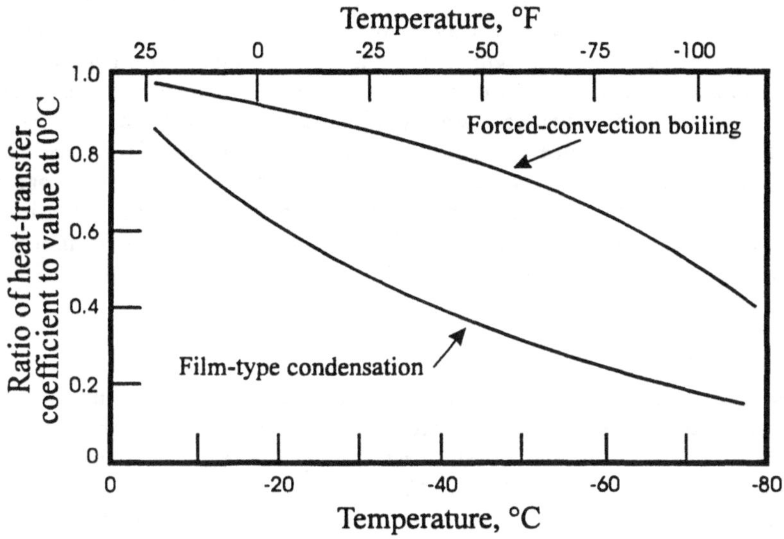

FIGURE 21.10
Trends in boiling and condensing heat-transfer coefficients of R-23 as a function of temperature.

21.11 OIL HANDLING

A refrigerant that is expected to be prominent in future low-temperature systems is HFC-23. For many years the azeotrope of R-503 (a blend of CFC-13 and HFC-23) was popular, but the CFC-13 content must now be eliminated. In all refrigeration systems, the oil that escapes to the condenser/evaporator accumulates in the evaporator. To remove this oil from the evaporator, the usual procedure in halocarbon systems is to dissolve the oil in the refrigerant and operate with suitably high vapor velocities in the suction line to carry the oil back to the compressor. The presence of CFC-13 in R-503 aided in this objective but is no longer available when HFC-23 alone is the refrigerant. Experience with the choice of oil and the evaporator configuration should be factored into future design.

An approach to oil return is to add a hydrocarbon to the refrigerant in the low-temperature circuits[6]. In a three-stage cascade system designed for temperatures down to about −130°C (−200°F), the designers added propane to the R-14 circuit, which was the lowest temperature of the three stages. The propane component was limited to a maximum of 7% in order to maintain a nonflammable mixture. For the R-503 circuit that served as the intermediate circuit, pentane in an amount of 5-10% by mass was added.

21.12 INSULATION

The choices of insulating material and the methods of application for the low-temperature systems dealt with in this chapter seem to lie somewhere between the approaches chosen for high-temperature industrial refrigeration plants and cryogenic facilities. As temperatures drop below −40°, the insulation thickness is likely to increase and additional care must be taken to provide effective vapor barriers. For the petrochemical industry, urethane is not acceptable, but cellular glass with an aluminum cover is frequently chosen. Studies are also in progress that adapt the approach of some cryogenic systems in the use of evacuated insulation panels. Such panels are formed by placing a low conductivity fill material into an enclosure, evacuating, and then sealing the panel.

21.13 MATERIALS

Materials suitable for pipe and vessels in low-temperature systems include carbon steel, stainless steel, and copper. For unitary systems using halocarbon refrigerants such as HFC-23, copper is widely used. Refrigeration-grade copper is used for temperatures even below −100°C (−150°F). For steel pipe, the ASME/ANSI B31.5 code for refrigeration piping[8] indicates that typical carbon steel pipe, such as ASTM A 333, when used at temperatures below −29°C (−20°F), may need to be impact tested. For this reason, materials not subject to impact tests at low temperatures are usually chosen, such as:

- Aluminum
- Types 304 and 321 austenitic stainless steel
- Copper

The piping code also states that impact tests are not required for ferrous materials used for temperatures between −29°C (−20°F) and −100°C (−150°F) "provided the maximum circumferential or longitudinal tensile stress resulting from coincident pressure, thermal contraction, or bending between supports does not exceed 40% of the allowable stress for materials." The option offered by this clause is that design calculations can be made which might permit a favorable choice, because normally at low temperatures, the pressures will also be low.

The most common choices of piping and vessel materials are copper for smaller diameter pipes with halocarbon systems and stainless steel for field-erected systems using any of the common choices of refrigerants.

21.14 DOCUMENTING THE DESIGN, CONSTRUCTING, AND OPERATING PROCEDURES

For most firms, low-temperature systems are not a day-to-day business, so the human memory should be bolstered with records. These records should show the supporting calculations and data that were a part of the design process. Included should be the bases for specifying the size of components, the arrangement, and the choice of materials. Procedures followed during fabrication and erection of field-assembled systems should also be recorded. Operating experiences should then be added to the records so that the next project can benefit from past experience.

REFERENCES

1. American Society of Heating, Refrigerating and Air-Conditioning Engineers, Atlanta, Georgia, from the 1994 ASHRAE *Handbook—Refrigeration*, Chapter 36, "Environment Test Facilities," Chapter 37, "Cryogenics," and Chapter 38, "Low Temperature Metallurgy."
2. "Using HFC-23 for Very Low Temperature (VLT) Refrigeration," Report ART-21, Du Pont Chemicals, Wilmington, Dela- ware, 1994.
3. Fulkerson, F., *Simplified Cascade System Servicing*, Business News Publishing Company, Troy, Michigan, 1988.
4. "Syltherm XLT," Dow Corning Corporation, Midland, Michigan, 1992.
5. American Society of Heating, Refrigerating and Air-Conditioning Engineers, Atlanta, Georgia, from the 1993 ASHRAE *Handbook— Refrigeration*, Chapter 4.
6. Downing, R. C. and W. D. Cooper, "Operation of a Three-Stage Cascade Refrigeration System," *ASHRAE Transactions*, 78(Part 2): 44-49, 1972.
7. "Energy Efficient Alternatives to Chlorofluorocarbons (CFCs)," Final Report of A Research Needs Assessment, U. S. Department of Energy, Office of Energy Research, DOE/ER/30115-H1, June 1993.
8. ASME/ANSI B31.5a-1992, "Refrigeration Piping," American Society of Mechanical Engineers, New York, New York, 1992.

APPENDIX

Conversions between SI and Inch-Pound Units

Properties of R-22

Properties of R-23

Conversions between SI and Inch-Pound Units.

Table A.1. Conversions of length, area, and volume.

To convert	From	To	Multiply by
length	inch	metre	0.0254
length	foot	metre	0.3048
length	metre	inch	39.3701
length	metre	foot	3.28084
area	square inch	square metre	0.00064516
area	square foot	square metre	0.092903
area	square metre	square inch	1,550.00
area	square metre	square foot	10.7639
volume	cubic inch	cubic metre	0.0000163871
volume	cubic foot	cubic metre	0.028317
volume	gallon	cubic metre	0.0037854
volume	cubic metre	litre	1,000.0
volume	cubic metre	cubic inch	61,023.74
volume	cubic metre	cubic foot	35.315
volume	cubic metre	gallon	264.173

Table A.2. Conversions of mass, density, and specific volume.

To convert	From	To	Multiply by
mass	lb	kg	0.453592
mass	kg	lb	2.20462
density	lb/ft^3	kg/m^3	16.0185
density	kg/m^3	lb/ft^3	0.062428
specific volume	ft^3/lb	m^3/kg	0.062428
specific volume	m^3/kg	ft^3/lb	16.0185

Table A.3. Conversions of temperature.

From	To	Operation
°F	°R	add 459.67
°R	°F	subtract 459.67
°C	K	add 273.15
K	°C	subtract 273.15
°C	°F	multiply by 9/5, then add 32
°F	°C	subtract 32, then multiply by 5/9

Table A.4. Conversions of velocity, acceleration, and force.

To convert	From	To	Multiply by
velocity	fps	m/s	0.304800
velocity	fpm	m/s	0.005080
velocity	mph	m/s	0.447040
velocity	m/s	fps	3.28084
velocity	m/s	fpm	196.850
velocity	m/s	mph	2.23694
acceleration	ft/s^2	m/s^2	0.304800
acceleration	m/s^2	ft/s^2	3.28084
force	lb	newton, N	4.44822
force	newton, N	lb	0.224809

Table A.5. Conversions of pressure.

To convert	From	To	Multiply by
pressure	psi	Pa	6,894.76
pressure	psi	kPa	6.89476
pressure	psf	Pa	47.8803
pressure	ft of water	Pa	2,988.98
pressure	in. of water	Pa	249.082
pressure	in. of Hg	Pa	3,376.85
pressure	Pa	psi	0.000145038
pressure	kPa	psi	0.145038
pressure	Pa	psf	0.0208854
pressure	Pa	ft of water	0.000334562
pressure	Pa	in. of water	0.00401474
pressure	Pa	in. of Hg	0.00029613

Table A.6. Conversions of flow rate.

To convert	From	To	Multiply by
volume flow rate	cfm	m^3/s	0.000471947
volume flow rate	cfs	m^3/s	0.02831685
volume flow rate	gpm	m^3/s	0.0000630902
volume flow rate	gpm	L/s	0.0630902
volume flow rate	m^3/s	cfm	2,118.88
volume flow rate	m^3/s	cfs	35.3147
volume flow rate	m^3/s	gpm	15,850.3
volume flow rate	L/s	gpm	15.8503
mass flow rate	lb/s	kg/s	0.453592
mass flow rate	lb/min	kg/s	0.00755987
mass flow rate	lb/hr	kg/s	0.000125998
mass flow rate	kg/s	lb/s	2.20462
mass flow rate	kg/s	lb/min	132.277
mass flow rate	kg/s	lb/hr	7,936.64

Table A.7. Conversions of energy and power.

To convert	From	To	Multiply by
energy	ft·lb	J	1.355818
energy	Btu	J	1,055.06
energy	J	ft·lb	0.73756212
energy	J	Btu	0.0009478133
power	Btu/hr	W	0.2930667
power	hp	W	745.6999
power	tons of refrig.	W	3,516.8
power	W	Btu/hr	3.41219
power	W	hp	0.001341022
power	W	tons of refrig.	0.0002843494

Table A.8. Conversions of enthalpy, specific heat, and entropy.

To convert	From	To	Multiply by
enthalpy	Btu/lb	kJ/kg	2.326009
enthalpy	kJ/kg	Btu/lb	0.42992095
specific heat	Btu/lb·°F	kJ/kg·K	4.186816
specific heat	kJ/kg·K	Btu/lb·°F	0.238845
entropy	Btu/lb·°R	kJ/kg·K	4.186816
entropy	kJ/kg·K	Btu/lb·°R	0.238845

Table A-9. Conversions of viscosities.

To convert	From	To	Multiply by
viscosity	lb/ft·s	Pa·s	1.48816
viscosity	lb/ft·hr	Pa·s	0.000413378
viscosity	centipoise	Pa·s	0.00100
viscosity	Pa·s	lb/ft·s	0.671971
viscosity	Pa·s	lb/ft·hr	2,419.09
viscosity	Pa·s	centipoise	1,000.0
kinematic viscosity	ft^2/s	m^2/s	0.092903
kinematic viscosity	m^2/s	ft^2/s	10.7639

Table A.10. Conversions of heat-transfer units.

To convert	From	To	Multiply by
conductivity	Btu/hr·ft·°F	W/m·K	1.730742
conductivity	W/m·K	Btu/hr·ft·°F	0.577787
U-value, conv. coeff.	Btu/hr·ft^2·°F	W/m^2·K	5.678286
U-value, conv. coeff.	W/m^2·K	Btu/hr·ft^2·°F	0.1761094

Table A.11. Some important SI constants.

standard gravitational acceleration, 9.80665 m/s^2
standard atmospheric pressure, 101.325 kPa
universal gas constant, $\mathcal{R}$, 8,314 J/mol$_{kg}$·K
gas constant for air, 287 J/kg·K

Table A.12. Nominal property values for air and water.

density of water, 1,000 kg/m^3
density of air, 1.2 kg/m^3
specific heat of liquid water, 4.19 kJ/kg·K
specific heat of air at constant pressure, 1.004 kJ/kg·K
specific heat of air at constant volume, 0.717 kJ/kg·K

Table A.13. Properties of R-22 saturated liquid and vapor in SI units*.

Temp., °C	Press., kPa	Enthalpy, kJ/kg h_f	h_g	Entropy, kJ/kg·K s_f	s_g	Liquid, ρ_f kg/m³	Vapor, v_g L/kg
−90	4.748	98.54	364.19	0.55017	2.0006	1542.8	3693.9
−85	7.084	104.51	366.62	0.58234	1.9754	1529.9	2539.4
−80	10.308	110.40	369.05	0.61320	1.9523	1516.8	1788.3
−75	14.662	116.10	371.48	0.64283	1.9312	1503.6	1287.0
−70	20.424	121.93	373.90	0.67134	1.9117	1490.3	944.77
−65	27.914	127.58	376.31	0.69882	1.8938	1476.7	706.09
−60	37.491	133.18	378.71	0.72537	1.8773	1463.1	536.41
−58	41.995	135.41	368.66	0.73575	1.8710	1457.6	482.61
−56	46.924	137.63	380.61	0.74601	1.8650	1452.0	435.20
−54	52.306	139.84	381.56	0.75614	1.8591	1446.4	393.32
−52	58.171	142.05	382.50	0.76616	1.8534	1440.8	356.23
−50	64.549	144.26	383.44	0.77607	1.8479	1435.2	323.30
−48	71.471	146.46	384.37	0.78587	1.8425	1429.5	294.00
−46	78.971	148.66	385.30	0.79557	1.8373	1423.8	267.86
−44	87.080	150.85	386.22	0.80518	1.8323	1418.1	244.50
−42	95.834	153.05	387.14	0.81469	1.8274	1412.3	223.58
−40	105.27	155.25	388.05	0.82412	1.8226	1406.5	204.80
−39	110.25	156.34	388.50	0.82880	1.8203	1403.6	196.13
−38	115.42	157.44	388.95	0.83347	1.8180	1400.7	187.90
−37	120.77	158.54	389.40	0.83811	1.8157	1397.7	180.10
−36	126.32	159.64	389.85	0.84273	1.8135	1394.8	172.68
−35	132.06	160.74	390.30	0.84734	1.8113	1391.8	165.64
−34	138.01	161.83	390.74	0.85193	1.8091	1388.9	158.94
−33	144.16	162.93	391.19	0.85650	1.8070	1385.9	152.57
−32	150.53	164.03	391.63	0.86105	1.8048	1382.9	146.51
−31	157.11	165.13	392.07	0.86559	1.8028	1379.9	140.74
−30	163.91	166.24	392.51	0.87011	1.8007	1376.9	135.24
−29	170.95	167.34	392.94	0.87461	1.7987	1373.9	130.01
−28	178.21	168.44	393.38	0.87910	1.7966	1370.9	125.02
−27	185.71	169.55	393.81	0.88357	1.7947	1367.9	120.26
−26	193.46	170.65	394.24	0.88803	1.7927	1364.8	115.73
−25	201.45	171.76	394.67	0.89248	1.7908	1361.8	111.40
−24	209.69	172.87	395.09	0.89691	1.7889	1358.7	107.26
−23	218.19	173.98	395.52	0.90132	1.7870	1355.7	103.31
−22	226.96	175.09	395.94	0.90573	1.7851	1352.6	99.540
−21	236.00	176.20	396.36	0.91012	1.7833	1349.5	95.935
−20	245.31	177.31	396.78	0.91450	1.7814	1346.4	92.487

*Thermophysical Properties of Refrigerants (R-22, Chlorodifluoromethane) Japanese Association of Refrigeration, 1986. Reference values of enthalpy and entropy are 200 kJ/kg and 1.0 kJ/kg·K, respectively, of saturated liquid at 0°C.

Table A.13. (cont.) Properties of R-22 saturated liquid and vapor in SI units.

Temp., °C	Press., kPa	Enthalpy, kJ/kg h_f	h_g	Entropy, kJ/kg·K s_f	s_g	Liquid, ρ_f kg/m^3	Vapor, v_g L/kg
−19	254.90	178.43	397.19	0.91887	1.7796	1343.2	89.190
−18	264.77	179.54	397.61	0.92322	1.7779	1340.1	86.034
−17	274.94	180.66	398.01	0.92757	1.7761	1337.0	83.014
−16	285.40	181.78	398.42	0.93190	1.7744	1333.8	80.122
−15	296.17	182.90	398.83	0.93622	1.7727	1330.6	77.352
−14	307.24	184.03	399.23	0.94054	1.7710	1327.5	74.698
−13	318.63	185.15	399.63	0.94484	1.7693	1324.3	72.153
−12	330.34	186.28	400.03	0.94913	1.7676	1321.1	69.714
−11	342.37	187.41	400.43	0.95341	1.7660	1317.8	67.374
−10	354.74	188.54	400.82	0.95769	1.7644	1314.6	65.128
−9	367.44	189.68	401.21	0.96195	1.7628	1311.4	62.972
−8	380.49	190.81	401.60	0.96621	1.7612	1308.1	60.904
−7	393.88	191.95	401.98	0.97046	1.7696	1304.8	58.917
−6	407.63	193.09	402.36	0.97470	1.7580	1301.5	57.008
−5	421.74	194.24	402.74	0.97898	1.7565	1298.2	55.173
−4	436.22	195.38	403.11	0.98316	1.7550	1294.9	53.409
−3	451.07	196.53	403.49	0.98738	1.7534	1291.6	51.713
−2	466.30	197.69	403.85	0.99159	1.7519	1288.2	50.082
−1	481.91	198.84	404.22	0.99580	1.7505	1284.9	48.512
0	497.92	200.00	404.58	1.00000	1.7490	1281.5	47.001
1	514.33	201.16	404.94	1.00419	1.7475	1278.1	45.547
2	531.13	202.32	405.30	1.00838	1.7461	1274.7	44.146
3	548.35	203.49	405.65	1.01256	1.7446	1271.2	42.797
4	565.99	204.66	406.00	1.01674	1.7432	1267.8	41.497
5	584.05	205.83	406.34	1.02091	1.7418	1264.3	40.244
6	602.54	207.01	406.69	1.02507	1.7404	1260.8	39.036
7	621.46	208.18	407.02	1.02924	1.7390	1257.3	37.871
8	640.83	209.37	407.36	1.03339	1.7376	1253.8	36.747
9	660.64	210.55	407.69	1.03755	1.7362	1250.3	35.663
10	680.91	211.74	408.02	1.04170	1.7349	1246.7	34.617
11	701.65	212.94	408.34	1.04584	1.7335	1243.1	33.607
12	722.85	214.13	408.66	1.04999	1.7322	1239.5	32.632
13	744.52	215.33	408.97	1.05413	1.7308	1235.9	31.690
14	766.68	216.54	409.28	1.05826	1.7295	1232.3	30.780
15	789.32	217.74	409.59	1.06240	1.7282	1228.6	29.900
16	812.46	218.95	409.89	1.06653	1.7269	1224.9	29.051
17	836.10	220.17	410.18	1.07066	1.7255	1221.1	28.229
18	860.25	221.39	410.48	1.07479	1.7242	1217.5	27.434
19	884.91	222.61	410.76	1.07891	1.7229	1213.8	26.666

Table A.13. (cont.) Properties of R-22 saturated liquid and vapor in SI units.

Temp., °C	Press., kPa	Enthalpy, kJ/kg h_f	h_g	Entropy, kJ/kg·K s_f	s_g	Liquid, $\rho_f$kg/m^3	Vapor, v_g L/kg
20	910.09	223.84	411.05	1.08304	1.7216	1210.0	25.922
21	935.81	225.07	411.32	1.08716	1.7203	1206.2	25.203
22	962.05	226.31	411.60	1.09128	1.7191	1202.4	24.506
23	988.84	227.55	411.86	1.09540	1.7178	1198.5	23.832
24	1016.2	228.79	412.13	1.09952	1.7165	1194.6	23.179
25	1044.1	230.04	412.38	1.10364	1.7152	1190.7	22.547
26	1072.5	231.30	412.64	1.10776	1.7139	1186.8	21.934
27	1101.6	232.56	412.88	1.11188	1.7127	1182.9	21.340
28	1131.2	233.82	413.12	1.11600	1.7114	1178.9	20.765
29	1161.3	235.09	413.36	1.12012	1.7101	1174.9	20.207
30	1192.1	236.36	413.59	1.12424	1.7089	1170.8	19.666
31	1223.5	237.64	413.81	1.12836	1.7076	1166.8	19.141
32	1255.5	238.92	414.03	1.13249	1.7063	1162.6	18.632
33	1288.1	240.21	414.24	1.13661	1.7050	1158.5	18.138
34	1321.3	241.51	414.44	1.14074	1.7038	1154.3	17.659
35	1355.1	242.80	414.64	1.14486	1.7025	1150.1	17.193
36	1389.6	244.11	414.82	1.14899	1.7012	1145.9	16.741
37	1424.7	245.42	415.01	1.15313	1.6999	1141.6	16.303
38	1460.5	246.74	415.18	1.15726	1.6986	1137.3	15.876
39	1496.9	248.06	415.35	1.16140	1.6973	1133.0	15.462
40	1534.0	249.38	415.51	1.16554	1.6960	1128.6	15.060
41	1571.8	250.72	415.66	1.16969	1.6947	1124.2	14.669
42	1610.2	252.06	415.81	1.17383	1.6934	1119.7	14.228
43	1649.4	253.40	415.94	1.17799	1.6921	1115.2	13.919
44	1689.2	254.76	416.07	1.18214	1.6908	1110.6	13.559
45	1729.7	256.12	416.19	1.18631	1.6894	1106.0	13.209
46	1771.0	257.48	416.30	1.19048	1.6881	1101.4	12.869
47	1812.9	258.85	416.40	1.19465	1.6867	1096.7	12.537
48	1855.6	260.23	416.49	1.19883	1.6854	1091.9	12.214
49	1899.1	261.62	416.57	1.20302	1.6840	1087.1	11.900
50	1943.2	263.01	416.64	1.20721	1.6826	1082.3	11.594
51	1988.2	264.41	416.69	1.21141	1.6812	1077.4	11.296
52	2033.9	265.82	416.74	1.21562	1.6798	1072.4	11.006
53	2080.3	267.24	416.78	1.21984	1.6783	1067.4	10.723
54	2127.6	268.67	416.80	1.22406	1.6769	1062.3	10.447
55	2175.6	270.10	416.82	1.22830	1.6754	1057.1	10.178
56	2224.4	271.54	416.82	1.23255	1.6739	1051.9	9.9153
57	2274.1	272.99	416.80	1.23681	1.6724	1046.6	9.6592
58	2324.5	274.45	416.77	1.24108	1.6709	1041.3	9.4093
59	2375.8	275.92	416.73	1.24536	1.6693	1035.8	9.1654

Table A.13. (concl.) Properties of R-22 saturated liquid and vapor in SI units.

Temp., °C	Press., kPa	Enthalpy, kJ/kg h_f	h_g	Entropy, kJ/kg·K s_f	s_g	Liquid, ρ_f kg/m^3	Vapor, v_g L/kg
60	2427.9	277.41	416.68	1.24966	1.6677	1030.3	8.9272
62	2534.7	280.40	416.52	1.25831	1.6644	1019.0	8.4673
64	2645.0	283.44	416.29	1.26702	1.6611	1007.3	8.0280
66	2758.8	286.53	415.98	1.27582	1.6575	995.25	7.6078
68	2876.3	289.67	415.60	1.28471	1.6539	982.72	7.2054
70	2997.5	292.87	415.12	1.29372	1.6500	969.68	6.8192
72	3122.5	296.15	414.55	1.30286	1.6459	956.06	6.4481
74	3251.5	299.50	413.85	1.31216	1.6416	941.79	6.0906
76	3384.6	302.95	413.02	1.32165	1.6369	926.77	5.7454
78	3521.8	306.50	412.04	1.33137	1.6319	910.86	5.4109
80	3663.3	310.18	410.88	1.34137	1.6265	893.89	5.0858
82	3809.3	314.02	409.49	1.35173	1.6206	875.65	4.7681
84	3959.9	318.04	407.84	1.36254	1.6140	855.82	4.4558
86	4115.3	322.31	405.84	1.37395	1.6065	833.93	4.1462
88	4275.7	326.89	403.38	1.38615	1.5979	809.27	3.8352
90	4441.3	331.94	400.26	1.39953	1.5877	780.60	3.5166
92	4612.5	337.72	396.12	1.41481	1.5747	745.43	3.1786
94	4789.7	344.97	389.88	1.43406	1.5564	697.15	2.7906
96.15	4973.8	361.48	361.48	1.47797	1.4780	512.00	1.95

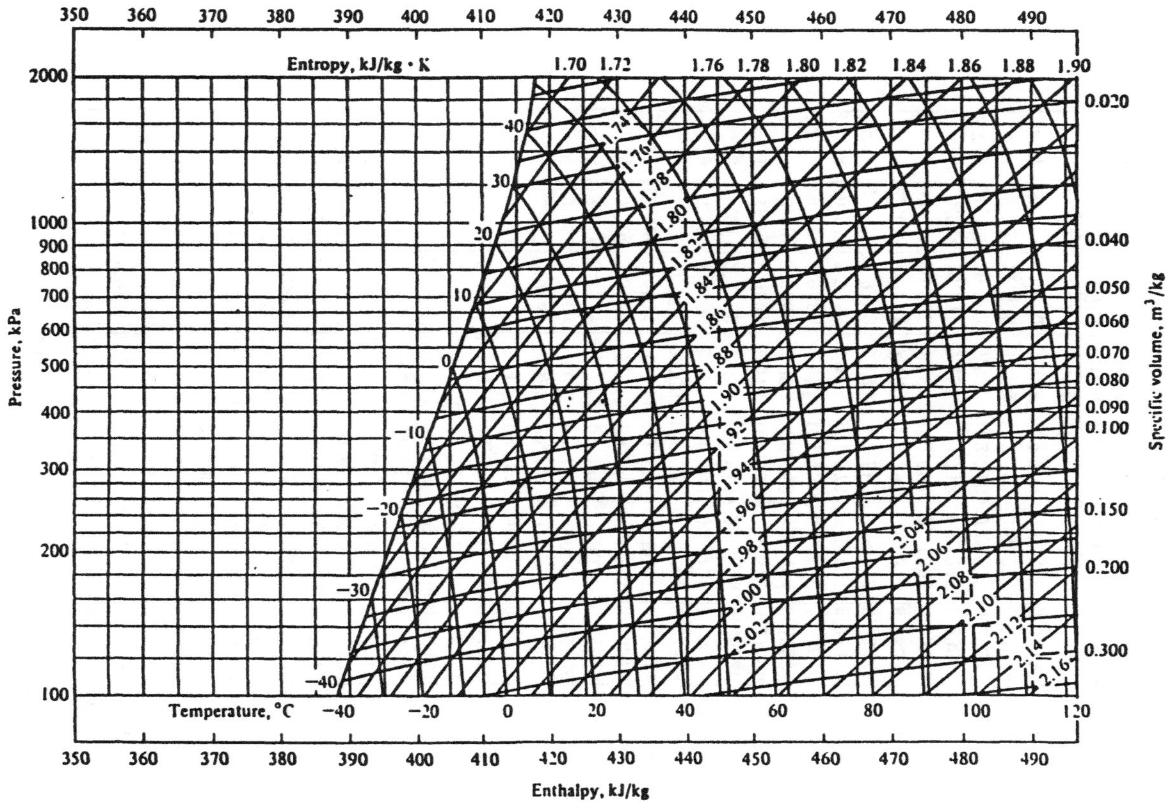

Fig. A.1. Pressure-enthalpy diagram for R-22 in SI units. (*Courtesy, Refrigeration Laboratory of the University of Denmark*).

Table A.14. Properties of R-22 saturated liquid and vapor in I-P units*.

Temp., °F	Pressure, psia	psig	Enthalpy, Btu/lb h_f	h_g	Entropy, Btu/lb·R s_f	s_g	Liquid, ρ_f lb/ft^3	Vapor, v_g ft^3/lb
−130	0.6892	28.599†	−24.38	89.831	−0.06543	0.28100	96.313	59.175
−120	1.0734	27.815†	−21.53	90.992	−0.05691	0.27434	95.416	39.081
−110	1.6211	26.696†	−18.72	92.153	−0.04878	0.26830	94.508	26.578
−100	2.3817	25.143†	−15.96	93.312	−0.04100	0.26282	93.590	18.558
−95	2.8596	24.168†	−14.60	93.890	−0.03723	0.26026	93.126	15.648
−90	3.4130	23.038†	−13.24	94.467	−0.03354	0.25782	92.660	13.268
−85	4.0506	21.736†	−11.89	95.043	−0.02993	0.25548	92.190	11.310
−80	4.7816	20.243†	−10.55	95.616	−0.02638	0.25325	91.717	9.6897
−75	5.6157	18.540†	−9.219	96.186	−0.02289	0.25112	91.241	8.3415
−70	6.5631	16.606†	−7.892	96.754	−0.01947	0.24908	90.761	7.2136
−65	7.6347	14.418†	−6.569	97.319	−0.01610	0.24713	90.278	6.2653
−60	8.8419	11.953†	−5.250	97.881	−0.01279	0.24525	89.791	5.4640
−58	9.3654	10.884†	−4.724	98.104	−0.01148	0.24453	89.595	5.1787
−56	9.9132	9.766†	−4.198	98.327	−0.01017	0.24381	89.399	4.9113
−54	10.486	8.596†	−3.672	98.550	−0.00888	0.24311	89.202	4.6605
−52	11.085	7.372†	−3.147	98.771	−0.00759	0.24241	89.005	4.4251
−50	11.711	6.095†	−2.622	98.993	−0.00631	0.24173	88.806	4.2040
−48	12.365	4.760†	−2.097	99.213	−0.00503	0.24106	88.608	3.9962
−46	13.046	3.368†	−1.573	99.432	−0.00376	0.24040	88.408	3.8008
−44	13.758	1.916†	−1.048	99.651	−0.00250	0.23976	88.208	3.6170
−42	14.499	0.402†	−0.524	99.870	−0.00125	0.23912	88.007	3.4438
−40	15.272	0.576	0.000	100.087	0.00000	0.23849	87.806	3.2806
−38	16.077	1.3806	0.525	100.303	0.00124	0.23787	87.603	3.1268
−36	18.914	2.2182	1.049	100.519	0.00248	0.23726	87.400	2.9817
−34	17.786	3.0897	1.574	100.734	0.00371	0.23666	87.197	2.8447
−32	18.692	3.9960	2.098	100.948	0.00494	0.23607	86.993	2.7154
−30	19.634	4.9380	2.623	101.161	0.00616	0.23549	86.788	2.5931
−28	20.613	5.9168	3.148	101.373	0.00737	0.23492	86.582	2.4775
−26	21.629	6.9332	3.674	101.585	0.00858	0.23435	86.375	2.3681
−24	22.684	7.9882	4.200	101.795	0.00978	0.23380	86.168	2.2646
−22	23.779	9.0829	4.726	102.004	0.01098	0.23325	85.960	2.1665
−20	24.914	10.218	5.253	102.213	0.01218	0.23271	85.751	2.0736
−19	25.498	10.802	5.516	102.316	0.01277	0.23244	85.646	2.0290
−18	26.091	11.395	5.780	102.420	0.01337	0.23218	85.541	1.9855
−17	26.696	12.000	6.043	102.523	0.01396	0.23191	85.436	1.9423
−16	27.311	12.615	6.307	102.626	0.01456	0.23165	85.331	1.9019

*Thermophysical Properties of Refrigerants (R-22 Chlorodifluoromethane) Japanese Association of Refrigeration, 1986.

† Inches of mercury below one standard atmosphere.

The zero values of enthalpy and entropy are at −40°F, saturated liquid.

Table A.14. (cont.) Properties of R-22 saturated liquid and vapor in I-P units.

Temp., °F	Pressure, psia	Pressure, psig	Enthalpy, Btu/lb h_f	Enthalpy, Btu/lb h_g	Entropy, Btu/lb·R s_f	Entropy, Btu/lb·R s_g	Liquid, ρ_f lb/ft^3	Vapor, v_g ft^3/lb
−15	27.937	13.241	6.571	102.729	0.01515	0.23139	85.225	1.8617
−14	28.575	13.878	6.836	102.831	0.01574	0.23113	85.120	1.8226
−13	29.223	14.527	7.100	102.934	0.01633	0.23088	85.014	1.7845
−12	29.883	15.187	7.365	103.035	0.01691	0.23062	84.908	1.7473
−11	30.554	15.858	7.629	103.137	0.01750	0.23037	84.801	1.7111
−10	31.237	16.541	7.894	103.238	0.01809	0.23012	84.695	1.6757
−9	31.931	17.235	8.159	103.339	0.01867	0.22987	84.588	1.6413
−8	32.638	17.942	8.424	103.440	0.01926	0.22962	84.481	1.6077
−7	33.356	18.660	8.690	103.541	0.01984	0.22938	84.374	1.5750
−6	34.087	19.391	8.955	103.641	0.02042	0.22913	84.266	1.5431
−5	34.829	20.133	9.221	103.741	0.02101	0.22889	84.158	1.5119
−4	35.585	20.889	9.487	103.840	0.02159	0.22865	84.051	1.4815
−3	36.352	21.656	9.753	103.940	0.02217	0.22841	83.942	1.4519
−2	37.133	22.437	10.020	104.039	0.02275	0.22818	83.834	1.4230
−1	37.926	23.230	10.286	104.137	0.02332	0.22794	83.726	1.3948
0	38.732	24.036	10.553	104.236	0.02390	0.22770	83.617	1.3672
1	39.552	24.855	10.820	104.334	0.02448	0.22747	83.508	1.3404
2	40.384	25.688	11.088	104.431	0.02505	0.22724	83.398	1.3141
3	41.230	26.534	11.355	104.529	0.02563	0.22701	83.289	1.2885
4	42.089	27.393	11.623	104.626	0.02620	0.22678	83.179	1.2635
5	42.962	28.266	11.891	104.722	0.02678	0.22655	83.069	1.2391
6	43.849	29.153	12.160	104.819	0.02735	0.22633	82.959	1.2152
7	44.749	30.053	12.428	104.915	0.02792	0.22610	82.848	1.1919
8	45.664	30.968	12.697	105.011	0.02849	0.22588	82.738	1.1692
9	46.593	31.897	12.966	105.106	0.02906	0.22566	82.627	1.1470
10	47.537	32.841	13.235	105.201	0.02963	0.22544	82.515	1.1253
11	48.495	33.799	13.505	105.296	0.03020	0.22522	82.404	1.1040
12	49.467	34.771	13.775	105.390	0.03077	0.22500	82.292	1.0833
13	50.455	35.759	14.045	105.484	0.03134	0.22479	82.180	1.0630
14	51.457	36.761	14.316	105.578	0.03190	0.22457	82.068	1.0432
15	52.474	37.778	14.587	105.671	0.03247	0.22436	81.955	1.0239
16	53.507	38.811	14.858	105.764	0.03303	0.22415	81.843	1.0050
17	54.555	39.859	15.129	105.856	0.03360	0.22393	81.729	0.9865
18	55.619	40.923	15.401	105.948	0.03416	0.22372	81.616	0.9684
19	56.698	42.002	15.673	106.040	0.03473	0.22352	81.503	0.9507
20	57.793	43.097	15.945	106.131	0.03529	0.22331	81.389	0.9334
21	58.904	44.208	16.218	106.222	0.03585	0.22310	81.274	0.9165
22	60.031	45.335	16.491	106.313	0.03642	0.22290	81.160	0.8999
23	61.175	46.479	16.764	106.403	0.03698	0.22269	81.045	0.8838
24	62.335	47.639	17.038	106.493	0.03754	0.22249	80.930	0.8679

Table A.14. (cont.) Properties of R-22 saturated liquid and vapor in I-P units.

Temp., °F	Pressure, psia	psig	Enthalpy, Btu/lb h_f	h_g	Entropy, Btu/lb·R s_f	s_g	Liquid, ρ_f lb/ft^3	Vapor, v_g ft^3/lb
25	63.511	48.815	17.312	106.582	0.03810	0.22229	80.815	0.8524
26	64.704	50.008	17.587	106.671	0.03866	0.22208	80.699	0.8373
27	65.915	51.219	17.861	106.759	0.03922	0.22188	80.583	0.8224
28	67.142	52.446	18.136	106.848	0.03978	0.22168	80.467	0.8079
29	68.386	53.690	18.412	106.935	0.04033	0.22149	80.351	0.7937
30	69.648	54.952	18.688	107.023	0.04089	0.22129	80.234	0.7798
31	70.927	56.231	18.964	107.109	0.04145	0.22109	80.117	0.7662
32	72.224	57.528	19.241	107.196	0.04201	0.22090	79.999	0.7529
33	73.539	58.843	19.518	107.282	0.04256	0.22070	79.882	0.7398
34	74.871	60.175	19.795	107.367	0.04312	0.22051	79.764	0.7270
35	76.222	61.526	20.073	107.452	0.04368	0.22032	79.645	0.7145
36	77.591	62.895	20.351	107.537	0.04423	0.22013	79.527	0.7023
37	78.979	64.283	20.630	107.621	0.04479	0.21994	79.408	0.6902
38	80.385	65.689	20.909	107.705	0.04534	0.21975	79.288	0.6785
39	81.810	67.114	21.188	107.788	0.04589	0.21956	79.169	0.6670
40	83.253	68.557	21.468	107.871	0.04645	0.21937	79.049	0.6557
41	84.716	70.020	21.748	107.953	0.04700	0.21918	78.928	0.6446
42	86.198	71.502	22.029	108.035	0.04755	0.21900	78.807	0.6338
43	87.699	73.003	22.310	108.116	0.04811	0.21881	78.686	0.6232
44	89.220	74.524	22.591	108.197	0.04866	0.21862	78.565	0.6128
45	90.761	76.065	22.873	108.277	0.04921	0.21844	78.443	0.6026
46	92.321	77.625	23.156	108.357	0.04976	0.21826	78.321	0.5926
47	93.902	79.206	23.438	108.437	0.05031	0.21807	78.199	0.5828
48	95.503	80.807	23.722	108.515	0.05087	0.21789	78.076	0.5731
49	97.124	82.428	24.005	108.594	0.05142	0.21771	77.953	0.5637
50	98.765	84.069	24.290	108.671	0.05197	0.21753	77.829	0.5545
51	100.43	85.731	24.574	108.749	0.05252	0.21735	77.705	0.5454
52	102.11	87.414	24.860	108.825	0.05307	0.21717	77.581	0.5365
53	103.81	89.118	25.145	108.901	0.05362	0.21699	77.456	0.5278
54	105.54	90.844	25.431	108.977	0.05417	0.21681	77.331	0.5193
55	107.29	92.590	25.718	109.052	0.05472	0.21663	77.206	0.5109
56	109.05	94.358	26.005	109.127	0.05527	0.21646	77.080	0.5027
57	110.84	96.148	26.293	109.201	0.05581	0.21628	76.953	0.4946
58	112.66	97.959	26.581	109.274	0.05636	0.21610	76.827	0.4867
59	114.49	99.792	26.870	109.347	0.05691	0.21593	76.700	0.4789
60	116.34	101.65	27.159	109.419	0.05746	0.21575	76.572	0.4713
61	118.22	103.53	27.448	109.419	0.05801	0.21558	76.444	0.4638
62	120.12	105.43	27.739	109.562	0.05856	0.21540	76.316	0.4565
63	122.05	107.35	28.029	109.632	0.05910	0.21523	76.187	0.4493
64	123.99	109.30	28.320	109.702	0.05965	0.21506	76.058	0.4422

Table A.14. (cont.) Properties of R-22 saturated liquid and vapor in I-P units.

Temp., °F	Pressure, psia	psig	Enthalpy, Btu/lb h_f	h_g	Entropy, Btu/lb·R s_f	s_g	Liquid, ρ_f lb/ft³	Vapor, v_g ft³/lb
65	125.96	111.26	28.612	109.771	0.06020	0.21488	75.928	0.4353
66	127.95	113.26	28.905	109.839	0.06075	0.21471	75.798	0.4285
67	129.97	115.27	29.197	109.907	0.06129	0.21454	75.667	0.4218
68	132.01	117.31	29.491	109.975	0.06184	0.21437	75.536	0.4152
69	134.07	119.37	29.785	110.041	0.06239	0.21420	75.405	0.4088
70	136.15	121.46	30.080	110.107	0.06294	0.21403	75.273	0.4024
71	138.26	123.57	30.375	110.173	0.06348	0.21385	75.140	0.3962
72	140.40	125.70	30.670	110.237	0.06403	0.21368	75.007	0.3901
73	142.56	127.86	30.967	110.301	0.06458	0.21351	74.874	0.3841
74	144.74	130.04	31.264	110.364	0.06512	0.21334	74.740	0.3782
75	146.95	132.25	31.561	110.427	0.06567	0.21317	74.606	0.3724
76	149.18	134.48	31.859	110.489	0.06622	0.21300	74.471	0.3668
77	151.44	136.74	32.158	110.550	0.06676	0.21283	74.335	0.3612
78	153.72	139.02	32.457	110.610	0.06731	0.21267	74.199	0.3557
79	156.03	141.33	32.757	110.670	0.06786	0.21250	74.063	0.3503
80	158.36	143.67	33.057	110.729	0.06840	0.21233	73.926	0.3450
81	160.72	146.03	33.359	110.787	0.06895	0.21216	73.788	0.3398
82	163.11	148.41	33.660	110.845	0.06950	0.21199	73.650	0.3346
83	165.52	150.82	33.963	110.901	0.07004	0.21182	73.511	0.3296
84	167.96	153.26	34.266	110.957	0.07059	0.21165	73.372	0.3247
85	170.42	155.72	34.570	111.012	0.07114	0.21148	73.232	0.3198
86	172.91	158.21	34.874	111.067	0.07168	0.21131	73.092	0.3150
87	175.43	160.73	35.179	111.120	0.07223	0.21115	72.951	0.3103
88	177.97	163.28	35.485	111.173	0.07278	0.21098	72.809	0.3057
89	180.54	165.85	35.791	111.225	0.07332	0.21081	72.667	0.3012
90	183.14	168.45	36.098	111.276	0.07387	0.21064	72.524	0.2967
91	185.77	171.07	36.406	111.326	0.07442	0.21047	72.381	0.2923
92	188.42	173.73	36.715	111.375	0.07497	0.21030	72.237	0.2880
93	191.10	176.41	37.024	111.423	0.07551	0.21013	72.092	0.2837
94	193.81	179.12	37.334	111.471	0.07606	0.20996	71.947	0.2795
95	196.55	181.86	37.644	111.517	0.07661	0.20979	71.801	0.2754
96	199.32	184.62	37.955	111.563	0.07716	0.20962	71.654	0.2714
97	202.11	187.42	38.268	111.608	0.07770	0.20945	71.507	0.2674
98	204.94	190.24	38.580	111.652	0.07825	0.20928	71.359	0.2635
99	207.79	193.10	38.894	111.694	0.07880	0.20911	71.210	0.2596
100	210.67	195.98	39.208	111.736	0.07935	0.20894	71.061	0.2558
101	213.58	198.89	39.523	111.777	0.07990	0.20877	70.910	0.2521
102	216.53	201.83	39.839	111.817	0.08045	0.20860	70.759	0.2484
103	219.50	204.80	40.156	111.856	0.08100	0.20843	70.608	0.2448
104	222.50	207.80	40.473	111.894	0.08155	0.20825	70.455	0.2412
105	225.53	210.83	40.791	111.930	0.08210	0.20808	70.302	0.2377
106	228.59	213.89	41.110	111.966	0.08265	0.20791	70.148	0.2343
107	231.68	216.98	41.430	112.001	0.08320	0.20773	69.993	0.2309

Table A.14. (concl.) Properties of R-22 saturated liquid and vapor in I-P units.

Temp., °F	Pressure, psia	Pressure, psig	Enthalpy, Btu/lb h_f	Enthalpy, Btu/lb h_g	Entropy, Btu/lb·R s_f	Entropy, Btu/lb·R s_g	Liquid, ρ_f lb/ft^3	Vapor, v_g ft^3/lb
108	234.80	220.11	41.751	112.034	0.08375	0.20756	69.838	0.2276
109	237.96	223.26	42.072	112.067	0.08430	0.20738	69.681	0.2243
110	241.14	226.44	42.395	112.098	0.08485	0.20721	69.524	0.2210
111	244.36	229.66	42.718	112.128	0.08540	0.20703	69.366	0.2178
112	247.60	232.91	43.042	112.157	0.08595	0.20686	69.207	0.2147
113	250.88	236.18	43.367	112.185	0.08651	0.20668	69.047	0.2116
114	254.19	239.50	43.692	112.211	0.08706	0.20650	68.886	0.2085
115	257.53	242.84	44.019	112.237	0.08761	0.20632	68.724	0.2055
116	260.91	246.21	44.347	112.261	0.08817	0.20614	68.561	0.2026
117	264.32	249.62	44.675	112.283	0.08872	0.20596	68.398	0.1997
118	267.76	253.06	45.005	112.305	0.08928	0.20578	68.233	0.1968
119	271.23	256.53	45.335	112.325	0.08983	0.20560	68.068	0.1940
120	274.74	260.04	45.666	112.344	0.09039	0.20541	67.901	0.1912
122	281.85	267.15	46.332	112.378	0.09150	0.20504	67.564	0.1857
124	289.10	274.40	47.002	112.405	0.09261	0.20467	67.224	0.1804
126	296.48	281.78	47.675	112.427	0.09373	0.20429	66.879	0.1753
128	304.00	289.31	48.353	112.443	0.09485	0.20391	66.529	0.1703
130	311.66	296.97	49.036	112.453	0.09597	0.20352	66.174	0.1654
132	319.47	304.77	49.723	112.455	0.09710	0.20313	65.814	0.1607
134	327.42	312.72	50.414	112.451	0.09823	0.20273	65.449	0.1561
136	335.51	320.82	51.111	112.440	0.09936	0.20232	65.078	0.1516
138	343.76	329.06	51.813	112.421	0.10050	0.20191	64.702	0.1472
140	352.15	337.45	52.520	112.395	0.10164	0.20149	64.319	0.1430
142	360.69	345.99	53.232	112.360	0.10278	0.20106	63.930	0.1389
144	369.39	354.69	53.951	112.316	0.10393	0.20062	63.534	0.1348
146	378.24	363.54	54.675	112.264	0.10509	0.20017	63.131	0.1309
148	387.25	372.56	55.406	112.202	0.10625	0.19972	62.720	0.1271
150	396.42	381.73	56.143	112.130	0.10742	0.19925	62.301	0.1233
152	405.76	391.06	56.888	112.047	0.10859	0.19877	61.874	0.1197
154	415.25	400.56	57.641	111.953	0.10977	0.19828	61.437	0.1161
156	424.92	410.22	58.401	111.846	0.11096	0.19777	60.991	0.1126
158	434.75	420.06	59.170	111.728	0.11216	0.19725	60.535	0.1092
160	444.58	430.06	59.948	111.595	0.11337	0.19672	60.067	0.1059
165	470.54	455.85	61.939	111.199	0.11644	0.19530	58.845	0.0979
170	497.46	482.76	64.005	110.692	0.11960	0.19374	57.532	0.0902
175	525.56	510.86	66.165	110.050	0.12287	0.19202	56.105	0.0829
180	554.89	540.19	68.447	109.235	0.12630	0.19006	54.533	0.0758
185	585.52	570.82	70.891	108.187	0.12994	0.18779	52.762	0.0689
190	617.52	602.82	73.568	106.807	0.13390	0.18507	50.703	0.0620
195	650.98	636.28	76.616	104.898	0.13839	0.18159	48.172	0.0549
200	686.02	671.32	80.416	101.929	0.14398	0.17659	44.671	0.0469
205.1	723.4	708.7	91.58	91.58	0.1605	0.1605	32.03	0.0312

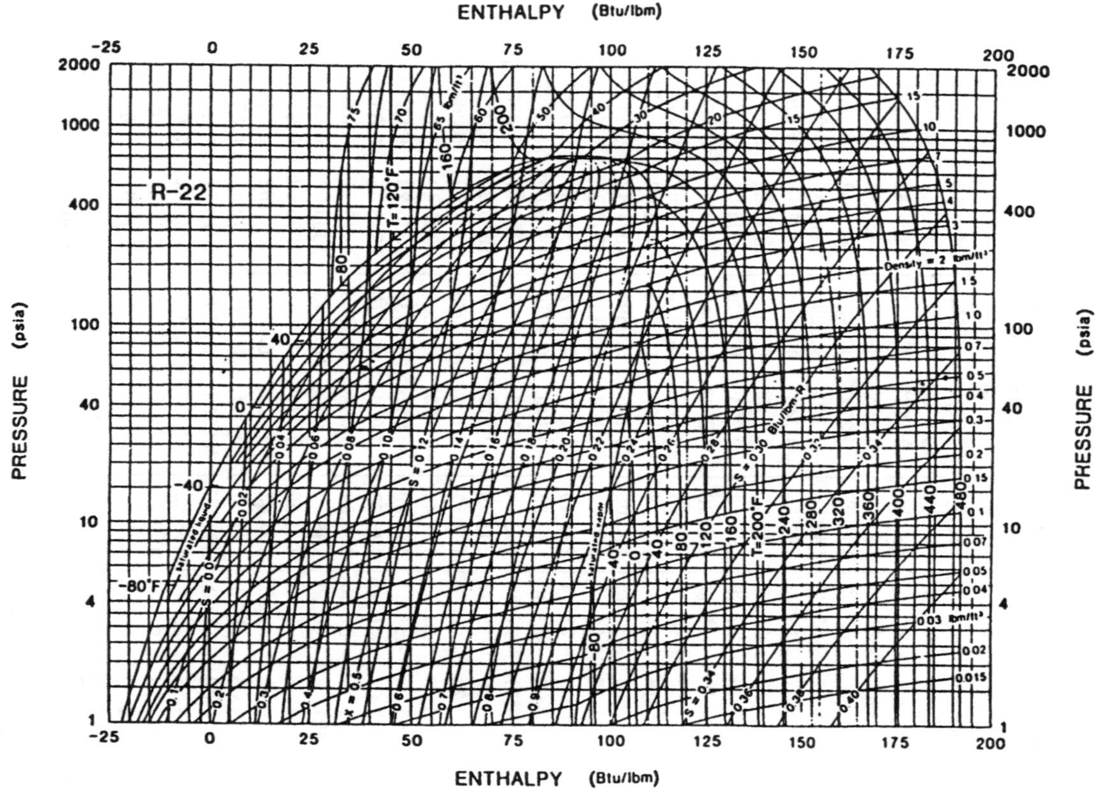

Fig. A.2. Pressure-enthalpy diagram for R-22 in I-P units. Reprinted with permission from 1997 ASHRAE *Handbook—Fundamentals*)

Table A.15. Properties of R-23 saturated liquid and vapor in SI units*.

Temp., °C	Press., kPa	Enthalpy, kJ/kg		Entropy, kJ/kg·K		Liquid,	Vapor,
		h_f	h_g	s_f	s_g	ρ_f kg/m^3	v_g L/kg
−100	31.68	−73.91	177.88	−0.3625	1.0917	1494.01	0.6355
−98	36.54	−71.59	178.79	−0.3491	1.0804	1488.26	0.5561
−96	42.00	−69.27	179.68	−0.3360	1.0693	1482.43	0.4883
−94	48.10	−66.96	180.57	−0.3231	1.0586	1476.50	0.4301
−92	54.89	−64.65	181.44	−0.3103	1.0482	1470.48	0.3801
−90	62.45	−62.35	182.30	−0.2977	1.0381	1464.36	0.3368
−88	70.82	−60.03	183.14	−0.2851	1.0283	1458.14	0.2994
−86	80.07	−57.72	183.97	−0.2727	1.0187	1451.82	0.2668
−84	90.26	−55.39	184.79	−0.2604	1.0094	1445.40	0.2385
−82	101.47	−53.06	185.59	−0.2481	1.0003	1438.88	0.2136
−80	113.76	−50.71	186.37	−0.2360	0.9915	1432.25	0.1919
−78	127.20	−48.35	187.14	−0.2239	0.9829	1425.51	0.1727
−76	141.86	−45.97	187.90	−0.2118	0.9745	1418.66	0.1558
−74	157.83	−43.58	188.63	−0.1998	0.9662	1411.69	0.1409
−72	175.18	−41.17	189.35	−0.1878	0.9582	1404.62	0.1277
−70	193.99	−38.75	190.05	−0.1759	0.9504	1397.42	0.1159
−68	214.35	−36.30	190.74	−0.1640	0.9427	1390.10	0.1054
−66	236.33	−33.84	191.40	−0.1521	0.9352	1382.66	0.0961
−64	260.02	−31.36	192.05	−0.1403	0.9279	1375.09	0.0877
−62	285.51	−28.85	192.68	−0.1284	0.9207	1367.40	0.0802
−60	312.88	−26.33	193.29	−0.1166	0.9137	1359.57	0.0735
−58	342.23	−23.78	193.88	−0.1049	0.9068	1351.60	0.0674
−56	373.64	−21.22	194.45	−0.0931	0.9001	1343.49	0.0620
−54	407.20	−18.63	195.00	−0.0814	0.8935	1335.23	0.0570
−52	443.01	−16.03	195.53	−0.0697	0.8870	1326.83	0.0525
−50	481.16	−13.40	196.04	−0.0580	0.8806	1318.27	0.0485
−48	521.75	−10.76	196.52	−0.0463	0.8743	1309.56	0.0448
−46	564.86	−8.10	196.99	−0.0347	0.8682	1300.68	0.0415
−44	610.59	−5.42	197.43	−0.0231	0.8622	1291.63	0.0384
−42	659.05	−2.72	197.85	−0.0115	0.8562	1282.41	0.0356
−40	710.33	−0.00	198.25	0.0000	0.8503	1273.01	0.0331
−38	764.53	2.73	198.63	0.0115	0.8446	1263.42	0.0307
−36	821.74	5.48	198.97	0.0230	0.8389	1253.63	0.0286
−34	882.08	8.25	199.30	0.0344	0.8332	1243.64	0.0266
−32	945.65	11.03	199.59	0.0457	0.8277	1233.44	0.0248

*From formulations in "Refrigerant Equations," by R. Downing, vol. 80, part 2, *ASHRAE Transactions*, pp. 158-169, 1974.

The enthalpy base is 200 kJ/kg of saturated liquid at −40°C, and the entropy base is 1.0 kJ/kg·K of saturated liquid at −40°C.

Table A.15. (concl.) Properties of R-23 saturated liquid and vapor in SI units.

Temp., °C	Press., kPa	Enthalpy, kJ/kg h_f	h_g	Entropy, kJ/kg·K s_f	s_g	Liquid, ρ_f kg/m^3	Vapor, v_g L/kg
−30	1012.6	13.83	199.86	0.0571	0.8222	1223.02	0.0231
−28	1082.9	16.64	200.10	0.0684	0.8167	1212.37	0.0216
−26	1156.8	19.47	200.31	0.0796	0.8113	1201.47	0.0202
−24	1234.4	22.32	200.48	0.0908	0.8059	1190.33	0.0189
−22	1315.7	25.18	200.63	0.1020	0.8006	1178.91	0.0176
−20	1401.0	28.07	200.73	0.1131	0.7952	1167.21	0.0165
−18	1490.2	30.97	200.79	0.1242	0.7899	1155.22	0.0154
−16	1583.7	33.89	200.82	0.1353	0.7845	1142.91	0.0145
−14	1681.5	36.84	200.79	0.1464	0.7791	1130.26	0.0136
−12	1783.7	39.81	200.72	0.1575	0.7737	1117.25	0.0127
−10	1890.5	42.82	200.59	0.1686	0.7682	1103.86	0.0119
−8	2002.0	45.86	200.41	0.1797	0.7626	1090.05	0.0112
−6	2118.5	48.94	200.15	0.1909	0.7569	1075.79	0.0104
−4	2240.1	52.07	199.82	0.2022	0.7511	1061.03	0.0098
−2	2367.0	55.26	199.41	0.2135	0.7452	1045.74	0.0092
−0	2499.4	58.52	198.91	0.2250	0.7390	1029.86	0.0086
2	2637.5	61.87	198.31	0.2367	0.7326	1013.31	0.0080
4	2781.5	65.31	197.58	0.2487	0.7259	996.01	0.0075
6	2931.7	68.87	196.71	0.2610	0.7189	977.87	0.0070
8	3088.4	72.58	195.68	0.2736	0.7115	958.74	0.0065
10	3251.8	76.47	194.45	0.2868	0.7035	938.46	0.0061
12	3422.3	80.59	192.98	0.3006	0.6948	916.80	0.0056
14	3600.2	84.98	191.22	0.3153	0.6853	893.42	0.0052
16	3785.8	89.74	189.07	0.3311	0.6746	867.87	0.0048
18	3979.6	94.99	186.40	0.3484	0.6624	839.38	0.0044
20	4181.9	100.9	182.98	0.3679	0.6478	806.63	0.0039
22	4393.2	108.0	178.33	0.3909	0.6293	766.91	0.0035
24	4614.0	117.1	171.12	0.4209	0.6025	712.57	0.0030

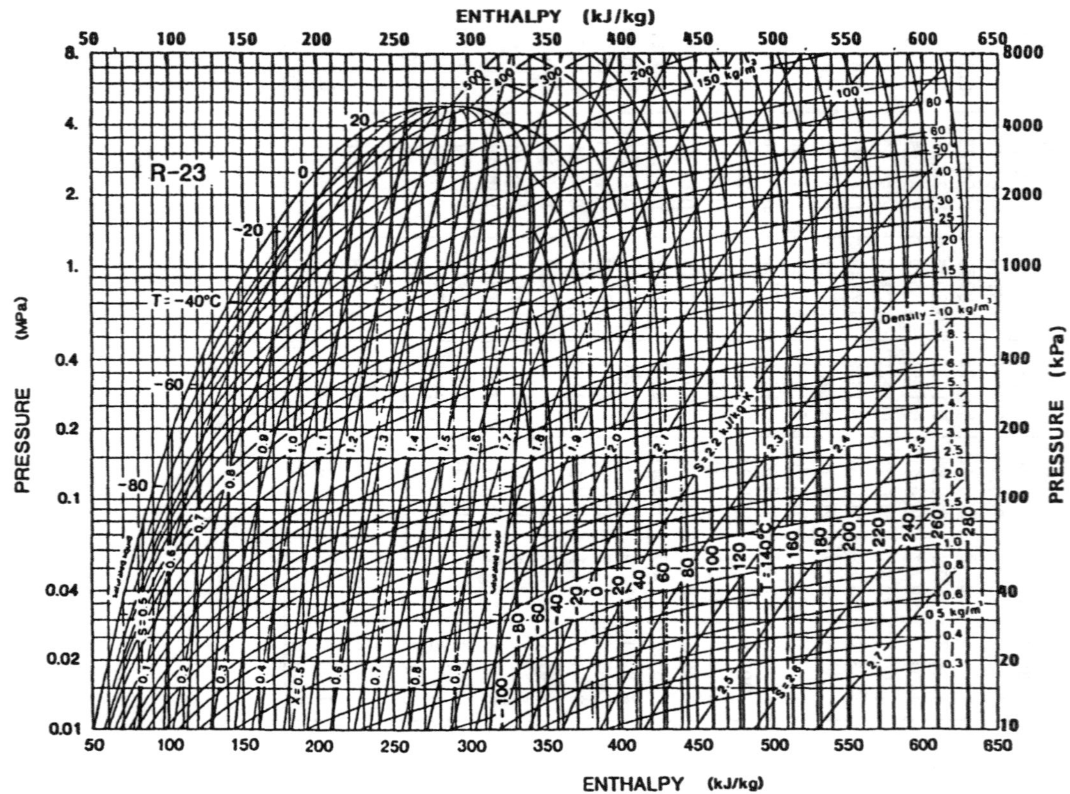

Fig. A.3. Pressure-enthalpy diagram for R-23 in SI units. (Reprinted with permission from 1997 ASHRAE *Handbook—Fundamentals*)

Table A.16. Properties of R-23 saturated liquid and vapor in I-P units*.

Temp., °F	Pressure, psia	Pressure, psig	Enthalpy, Btu/lb h_f	Enthalpy, Btu/lb h_g	Entropy, Btu/lb·R s_f	Entropy, Btu/lb·R s_g	Liquid, ρ_f lb/ft^3	Vapor, v_g ft^3/lb
−160	2.77	24.3†	−35.17	75.19	−0.0977	0.2706	94.42	16.3149
−155	3.44	22.9†	−33.76	75.76	−0.0930	0.2665	93.95	13.3345
−150	4.24	21.3†	−32.35	76.31	−0.0884	0.2625	93.46	10.9807
−145	5.18	19.4†	−30.96	76.85	−0.0840	0.2587	92.97	9.1063
−140	6.28	17.1†	−29.58	77.39	−0.0796	0.2550	92.46	7.6020
−135	7.57	14.5†	−28.20	77.91	−0.0753	0.2515	91.95	6.3858
−130	9.06	11.5†	−26.82	78.43	−0.0711	0.2481	91.42	5.3957
−125	10.78	8.0†	−25.44	78.93	−0.0670	0.2449	90.88	4.5844
−120	12.75	4.0†	−24.05	79.42	−0.0629	0.2417	90.32	3.9155
−115	15.00	0.3	−22.66	79.90	−0.0588	0.2387	89.76	3.3608
−110	17.56	2.9	−21.25	80.37	−0.0548	0.2358	89.18	2.8981
−105	20.45	5.8	−19.84	80.82	−0.0508	0.2330	88.59	2.5102
−100	23.71	9.0	−18.41	81.26	−0.0468	0.2303	87.98	2.1834
−95	27.36	12.7	−16.96	81.68	−0.0428	0.2277	87.36	1.9067
−90	31.43	16.7	−15.50	82.09	−0.0389	0.2251	86.73	1.6713
−85	35.96	21.3	−14.03	82.48	−0.0349	0.2226	86.08	1.4702
−80	40.99	26.3	−12.53	82.86	−0.0310	0.2202	85.42	1.2977
−75	46.53	31.8	−11.02	83.23	−0.0271	0.2179	84.74	1.1491
−70	52.64	37.9	−9.50	83.57	−0.0232	0.2157	84.04	1.0207
−65	59.34	44.6	−7.95	83.90	−0.0193	0.2135	83.33	0.9092
−60	66.67	52.0	−6.40	84.22	−0.0154	0.2113	82.60	0.8121
−55	74.67	60.0	−4.82	84.51	−0.0115	0.2092	81.84	0.7272
−50	83.37	68.7	−3.23	84.79	−0.0077	0.2072	81.07	0.6529
−45	92.81	78.1	−1.62	85.05	−0.0038	0.2052	80.28	0.5874
−40	103.02	88.3	0.00	85.29	0.0000	0.2032	79.47	0.5297
−35	114.06	99.4	1.64	85.51	0.0038	0.2013	78.64	0.4785
−30	125.95	111.3	3.28	85.71	0.0076	0.1994	77.78	0.4332
−25	138.74	124.0	4.95	85.89	0.0114	0.1976	76.89	0.3928
−20	152.46	137.8	6.62	86.04	0.0151	0.1958	75.98	0.3567
−15	167.17	152.5	8.31	86.17	0.0189	0.1940	75.04	0.3244
−10	182.90	168.2	10.01	86.27	0.0226	0.1922	74.07	0.2954
−5	199.70	185.0	11.73	86.35	0.0263	0.1904	73.07	0.2692

*From formulations in "Refrigerant Equations," by R. Downing, vol. 80, part 2, *ASHRAE Transactions*, pp. 158-169, 1974.
† indicates pressure in inches of mercury, vacuum.
The zero values of enthalpy and entropy are of saturated liquid at −40°F.

Table A.16. (concl.) Properties of R-23 saturated liquid and vapor in I-P units.

Temp., °F	Pressure, psia	psig	Enthalpy, Btu/lb h_f	h_g	Entropy, Btu/lb·R s_f	s_g	Liquid, ρ_f lb/ft^3	Vapor, v_g ft^3/lb
0	217.62	202.9	13.46	86.39	0.0300	0.1886	72.03	0.2457
5	236.70	222.0	15.21	86.39	0.0337	0.1869	70.96	0.2243
10	257.02	242.3	16.98	86.36	0.0373	0.1851	69.84	0.2049
15	278.61	263.9	18.78	86.28	0.0410	0.1832	68.67	0.1872
20	301.55	286.9	20.61	86.15	0.0447	0.1814	67.46	0.1710
25	325.90	311.2	22.48	85.96	0.0485	0.1794	66.19	0.1562
30	351.73	337.0	24.40	85.70	0.0523	0.1775	64.85	0.1426
35	379.13	364.4	26.37	85.36	0.0561	0.1754	63.43	0.1300
40	408.18	393.5	28.43	84.92	0.0601	0.1731	61.93	0.1183
45	438.98	424.3	30.60	84.37	0.0642	0.1707	60.32	0.1074
50	471.64	456.9	32.90	83.66	0.0685	0.1681	58.59	0.0971
55	506.27	491.6	35.39	82.75	0.0732	0.1652	56.68	0.0874
60	543.00	528.3	38.14	81.56	0.0783	0.1618	54.55	0.0780
65	581.99	567.3	41.27	79.97	0.0840	0.1578	52.08	0.0688
70	623.38	608.7	45.03	77.70	0.0908	0.1525	49.05	0.0594
75	667.38	652.7	50.13	73.85	0.1001	0.1445	44.72	0.0488

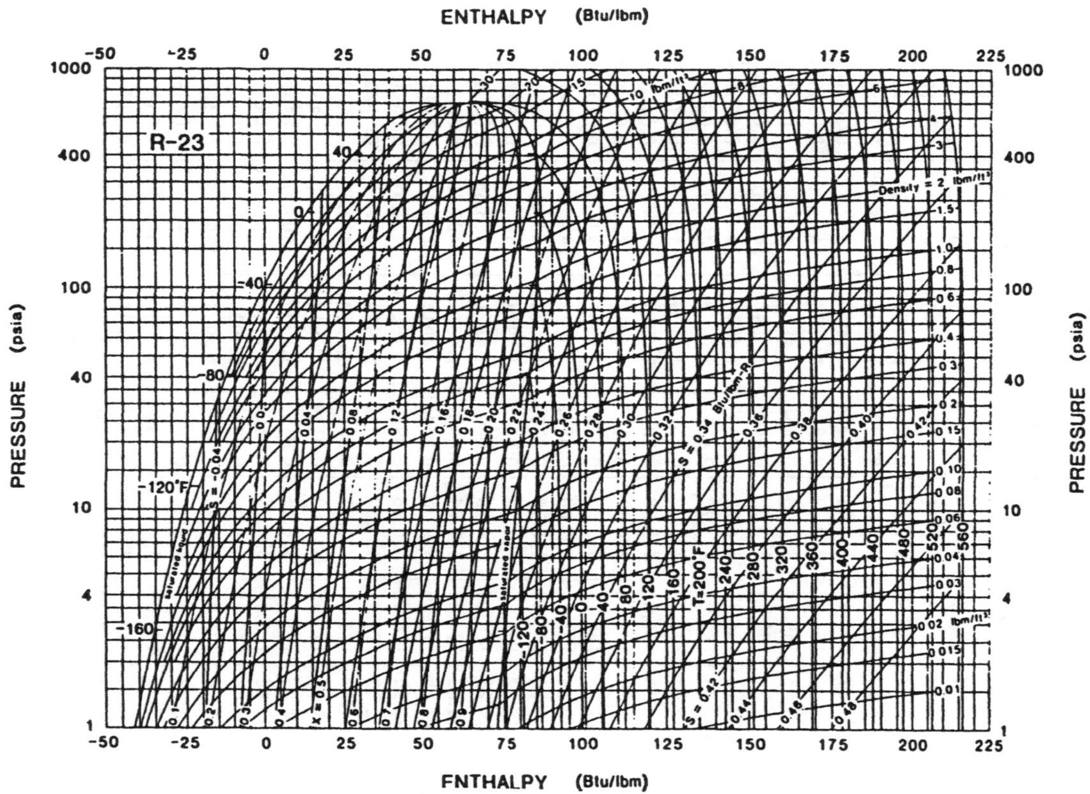

Fig. A.4. Pressure-enthalpy diagram for R-23 in SI units. (Reprinted with permission from 1997 ASHRAE *Handbook—Fundamentals*)

Table A.17. Properties of R-125 saturated liquid and vapor in SI units*.

Temp., °C	Press., kPa	Enthalpy, kJ/kg h_f	h_g	Entropy, kJ/kg·K s_f	s_g	Liquid, ρ_f kg/m³	Vapor, v_g L/kg
−45	120.89	152.72	309.03	0.8131	1.4982	1500.94	0.1243
−44	126.54	153.65	309.60	0.8172	1.4978	1497.32	0.1191
−43	132.40	154.58	310.18	0.8212	1.4973	1493.68	0.1141
−42	138.47	155.52	310.76	0.8253	1.4969	1490.03	0.1094
−41	144.76	156.47	311.33	0.8293	1.4964	1486.37	0.1049
−40	151.28	157.42	311.91	0.8334	1.4960	1482.69	0.1007
−39	158.02	158.37	312.48	0.8375	1.4956	1479.00	0.0966
−38	165.01	159.33	313.05	0.8415	1.4953	1475.29	0.0927
−37	172.24	160.30	313.62	0.8456	1.4949	1471.57	0.0891
−36	179.72	161.27	314.20	0.8497	1.4946	1467.83	0.0855
−35	187.45	162.24	314.77	0.8538	1.4942	1464.07	0.0822
−34	195.44	163.22	315.33	0.8579	1.4939	1460.29	0.0790
−33	203.70	164.21	315.90	0.8619	1.4936	1456.50	0.0760
−32	212.23	165.20	316.47	0.8660	1.4933	1452.70	0.0731
−31	221.05	166.20	317.03	0.8701	1.4931	1448.87	0.0703
−30	230.14	167.20	317.60	0.8742	1.4928	1445.03	0.0677
−29	239.54	168.20	318.16	0.8783	1.4926	1441.16	0.0651
−28	249.22	169.22	318.72	0.8825	1.4923	1437.28	0.0627
−27	259.22	170.23	319.29	0.8866	1.4921	1433.38	0.0604
−26	269.53	171.26	319.84	0.8907	1.4919	1429.46	0.0582
−25	280.15	172.29	320.40	0.8948	1.4917	1425.53	0.0561
−24	291.10	173.32	320.96	0.8989	1.4915	1421.57	0.0541
−23	302.38	174.36	321.52	0.9031	1.4914	1417.59	0.0521
−22	314.00	175.41	322.07	0.9072	1.4912	1413.59	0.0503
−21	325.97	176.46	322.62	0.9114	1.4910	1409.56	0.0485
−20	338.29	177.52	323.17	0.9155	1.4909	1405.52	0.0468
−19	350.96	178.58	323.72	0.9197	1.4908	1401.45	0.0452
−18	364.01	179.65	324.27	0.9239	1.4906	1397.36	0.0436
−17	377.43	180.73	324.81	0.9280	1.4905	1393.25	0.0421
−16	391.23	181.81	325.36	0.9322	1.4904	1389.12	0.0406
−15	405.41	182.90	325.90	0.9364	1.4903	1384.96	0.0393
−14	420.00	183.99	326.44	0.9406	1.4903	1380.77	0.0379
−13	434.98	185.09	326.98	0.9448	1.4902	1376.56	0.0367
−12	450.38	186.20	327.52	0.9490	1.4901	1372.32	0.0354
−11	466.19	187.31	328.05	0.9532	1.4900	1368.06	0.0343
−10	482.43	188.43	328.58	0.9574	1.4900	1363.77	0.0331
−9	499.10	189.56	329.11	0.9616	1.4899	1359.45	0.0320
−8	516.21	190.69	329.64	0.9659	1.4899	1355.10	0.0310
−7	533.77	191.83	330.17	0.9701	1.4899	1350.73	0.0300
−6	551.78	192.98	330.69	0.9744	1.4898	1346.32	0.0290

*From algorithms of *Genie*, Allied Signal, Inc. Morristown, NJ.
The enthalpy base is 200 kJ/kg of saturated liquid at −40°C,
and the entropy base is 1.0 kJ/kg·K of saturated liquid at −40°C.

Table A.17. (cont.) Properties of R-125 saturated liquid and vapor in SI units.

Temp., °C	Press., kPa	Enthalpy, kJ/kg h_f	h_g	Entropy, kJ/kg·K s_f	s_g	Liquid, ρ_f kg/m^3	Vapor, v_g L/kg
−5	570.26	194.13	331.21	0.9786	1.4898	1341.89	0.0281
−4	589.21	195.29	331.73	0.9829	1.4898	1337.42	0.0272
−3	608.64	196.46	332.24	0.9872	1.4898	1332.92	0.0263
−2	628.56	197.63	332.76	0.9914	1.4898	1328.38	0.0255
−1	648.97	198.82	333.27	0.9957	1.4898	1323.82	0.0247
0	669.88	200.00	333.78	1.0000	1.4898	1319.22	0.0239
1	691.30	201.20	334.28	1.0043	1.4898	1314.58	0.0232
2	713.25	202.40	334.78	1.0087	1.4898	1309.91	0.0224
3	735.71	203.62	335.28	1.0130	1.4898	1305.19	0.0218
4	758.72	204.83	335.78	1.0173	1.4898	1300.44	0.0211
5	782.27	206.06	336.27	1.0217	1.4898	1295.65	0.0204
6	806.37	207.30	336.76	1.0261	1.4899	1290.82	0.0198
7	831.03	208.54	337.25	1.0304	1.4899	1285.95	0.0192
8	856.26	209.79	337.73	1.0348	1.4899	1281.04	0.0186
9	882.06	211.05	338.21	1.0392	1.4899	1276.08	0.0181
10	908.45	212.32	338.69	1.0436	1.4900	1271.07	0.0175
11	935.44	213.59	339.16	1.0481	1.4900	1266.02	0.0170
12	963.02	214.88	339.63	1.0525	1.4900	1260.92	0.0165
13	991.22	216.18	340.10	1.0570	1.4900	1255.76	0.0160
14	1020.0	217.48	340.56	1.0614	1.4901	1250.56	0.0155
15	1049.5	218.79	341.02	1.0659	1.4901	1245.30	0.0151
16	1079.6	220.12	341.47	1.0704	1.4901	1239.99	0.0146
17	1110.3	221.45	341.92	1.0749	1.4901	1234.62	0.0142
18	1141.7	222.79	342.36	1.0795	1.4902	1229.19	0.0138
19	1173.7	224.14	342.80	1.0840	1.4902	1223.70	0.0134
20	1206.4	225.51	343.24	1.0886	1.4902	1218.14	0.0130
21	1239.8	226.88	343.67	1.0932	1.4902	1212.52	0.0126
22	1273.9	228.27	344.09	1.0978	1.4902	1206.83	0.0122
23	1308.7	229.66	344.51	1.1024	1.4902	1201.07	0.0119
24	1344.2	231.07	344.93	1.1070	1.4902	1195.24	0.0115
25	1380.4	232.49	345.34	1.1117	1.4902	1189.33	0.0112
26	1417.3	233.92	345.74	1.1164	1.4902	1183.34	0.0109
27	1455.0	235.37	346.14	1.1211	1.4902	1177.26	0.0106
28	1493.4	236.83	346.53	1.1259	1.4901	1171.10	0.0103
29	1532.5	238.30	346.92	1.1306	1.4901	1164.85	0.0100
30	1572.5	239.79	347.29	1.1354	1.4901	1158.51	0.0097
31	1613.2	241.29	347.67	1.1402	1.4900	1152.06	0.0094
32	1654.6	242.80	348.03	1.1451	1.4900	1145.51	0.0091
33	1696.9	244.33	348.39	1.1500	1.4899	1138.85	0.0089
34	1740.0	245.88	348.74	1.1549	1.4898	1132.08	0.0086

Table A.17. (concl.) Properties of R-125 saturated liquid and vapor in SI units.

Temp., °C	Press., kPa	Enthalpy, kJ/kg h_f	h_g	Entropy, kJ/kg·K s_f	s_g	Liquid, ρ_f kg/m^3	Vapor, v_g L/kg
35	1783.8	247.44	349.08	1.1599	1.4897	1125.19	0.0083
36	1828.5	249.03	349.42	1.1649	1.4896	1118.16	0.0081
37	1874.0	250.63	349.74	1.1699	1.4895	1111.01	0.0079
38	1920.4	252.25	350.06	1.1750	1.4893	1103.71	0.0076
39	1967.6	253.89	350.36	1.1801	1.4892	1096.27	0.0074
40	2015.6	255.55	350.66	1.1853	1.4890	1088.66	0.0072
41	2064.6	257.23	350.95	1.1905	1.4888	1080.89	0.0070
42	2114.4	258.94	351.22	1.1958	1.4886	1072.93	0.0068
43	2165.1	260.67	351.49	1.2011	1.4884	1064.78	0.0066
44	2216.6	262.43	351.74	1.2065	1.4881	1056.43	0.0064
45	2269.1	264.22	351.98	1.2120	1.4879	1047.85	0.0062
46	2322.5	266.04	352.21	1.2176	1.4876	1039.03	0.0060
47	2376.9	267.89	352.42	1.2232	1.4872	1029.96	0.0058
48	2432.1	269.78	352.62	1.2289	1.4869	1020.60	0.0056
49	2488.3	271.71	352.80	1.2347	1.4865	1010.93	0.0054
50	2545.5	273.68	352.96	1.2407	1.4860	1000.93	0.0053
51	2603.6	275.70	353.11	1.2467	1.4855	990.56	0.0051
52	2662.7	277.76	353.23	1.2529	1.4850	979.77	0.0049
53	2722.8	279.89	353.33	1.2592	1.4844	968.53	0.0048
54	2783.9	282.08	353.40	1.2658	1.4838	956.76	0.0046
55	2846.0	284.34	353.45	1.2725	1.4831	944.41	0.0044
56	2909.1	286.69	353.46	1.2794	1.4823	931.38	0.0043
57	2973.2	289.14	353.44	1.2866	1.4814	917.57	0.0041
58	3038.3	291.70	353.37	1.2941	1.4804	902.83	0.0039
59	3104.6	294.40	353.26	1.3021	1.4793	886.96	0.0038
60	3171.8	297.28	353.08	1.3105	1.4780	869.72	0.0036

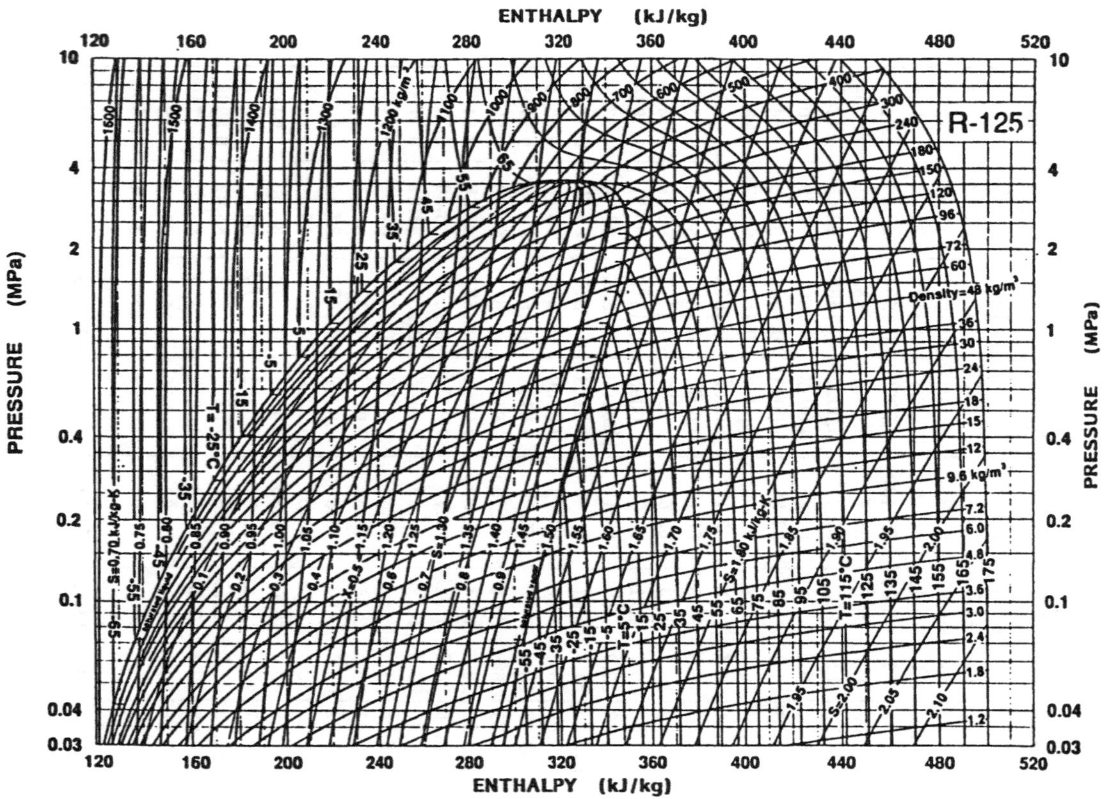

Fig. A.5. Pressure-enthalpy diagram for R-125 in SI units. (Reprinted with permission from 1997 ASHRAE *Handbook—Fundamentals*)

Table A.18. Properties of R-125 saturated liquid and vapor in I-P units.*

Temp., °F	Pressure, psia	Pressure, psig	Enthalpy, Btu/lb h_f	Enthalpy, Btu/lb h_g	Entropy, Btu/lb·R s_f	Entropy, Btu/lb·R s_g	Liquid, ρ_f lb/ft^3	Vapor, v_g ft^3/lb
−50	17.09	2.4	−2.24	65.09	−0.0054	0.1590	93.83	2.0394
−48	17.99	3.3	−1.80	65.36	−0.0043	0.1588	93.58	1.9441
−46	18.92	4.2	−1.35	65.64	−0.0032	0.1587	93.32	1.8541
−44	19.88	5.2	−0.91	65.91	−0.0022	0.1586	93.07	1.7690
−42	20.89	6.2	−0.45	66.19	−0.0011	0.1585	92.82	1.6885
−40	21.94	7.2	−0.00	66.46	0.0000	0.1584	92.56	1.6124
−38	23.03	8.3	0.46	66.74	0.0011	0.1583	92.31	1.5404
−36	24.16	9.5	0.91	67.01	0.0022	0.1582	92.05	1.4722
−34	25.34	10.6	1.38	67.28	0.0032	0.1581	91.79	1.4075
−32	26.56	11.9	1.84	67.56	0.0043	0.1580	91.53	1.3463
−30	27.83	13.1	2.31	67.83	0.0054	0.1579	91.27	1.2882
−28	29.14	14.4	2.78	68.10	0.0065	0.1578	91.01	1.2330
−26	30.50	15.8	3.25	68.37	0.0076	0.1577	90.74	1.1807
−24	31.92	17.2	3.73	68.64	0.0087	0.1577	90.48	1.1310
−22	33.38	18.7	4.21	68.91	0.0098	0.1576	90.21	1.0838
−20	34.90	20.2	4.69	69.18	0.0109	0.1575	89.94	1.0389
−18	36.46	21.8	5.17	69.45	0.0119	0.1575	89.67	0.9962
−16	38.09	23.4	5.66	69.72	0.0130	0.1574	89.40	0.9556
−14	39.77	25.1	6.15	69.99	0.0141	0.1574	89.13	0.9169
−12	41.51	26.8	6.64	70.25	0.0152	0.1573	88.86	0.8801
−10	43.31	28.6	7.14	70.52	0.0163	0.1573	88.58	0.8450
−8	45.16	30.5	7.64	70.78	0.0174	0.1572	88.30	0.8116
−6	47.08	32.4	8.14	71.05	0.0185	0.1572	88.02	0.7798
−4	49.06	34.4	8.65	71.31	0.0196	0.1571	87.74	0.7494
−2	51.11	36.4	9.16	71.57	0.0207	0.1571	87.46	0.7204
0	53.22	38.5	9.67	71.83	0.0218	0.1571	87.18	0.6927
2	55.40	40.7	10.18	72.09	0.0229	0.1571	86.89	0.6663
4	57.65	43.0	10.70	72.35	0.0241	0.1570	86.60	0.6411
6	59.97	45.3	11.22	72.61	0.0252	0.1570	86.32	0.6170
8	62.36	47.7	11.75	72.87	0.0263	0.1570	86.02	0.5939
10	64.82	50.1	12.28	73.13	0.0274	0.1570	85.73	0.5719
12	67.36	52.7	12.81	73.38	0.0285	0.1569	85.43	0.5508
14	69.97	55.3	13.34	73.64	0.0296	0.1569	85.14	0.5306
16	72.66	58.0	13.88	73.89	0.0308	0.1569	84.84	0.5113
18	75.43	60.7	14.42	74.14	0.0319	0.1569	84.54	0.4928
20	78.28	63.6	14.97	74.39	0.0330	0.1569	84.23	0.4751
22	81.21	66.5	15.52	74.64	0.0341	0.1569	83.93	0.4581
24	84.23	69.5	16.07	74.89	0.0353	0.1569	83.62	0.4418

*From algorithms of *Genie*, Allied Signal, Inc., Morristown, NJ. The zero values of enthalpy and entropy are of saturated liquid at −40°F.

Table A.18. (cont.) Properties of R-125 saturated liquid and vapor in I-P units.

Temp., °F	Pressure, psia	psig	Enthalpy, Btu/lb h_f	h_g	Entropy, Btu/lb·R s_f	s_g	Liquid, ρ_f lb/ft^3	Vapor, v_g ft^3/lb
26	87.33	72.6	16.63	75.14	0.0364	0.1569	83.31	0.4262
28	90.52	75.8	17.19	75.38	0.0375	0.1569	82.99	0.4112
30	93.79	79.1	17.75	75.63	0.0387	0.1569	82.68	0.3968
32	97.16	82.5	18.32	75.87	0.0398	0.1569	82.36	0.3830
34	100.61	85.9	18.89	76.11	0.0410	0.1569	82.03	0.3698
36	104.16	89.5	19.47	76.35	0.0421	0.1569	81.71	0.3570
38	107.81	93.1	20.05	76.59	0.0433	0.1569	81.38	0.3448
40	111.55	96.9	20.63	76.83	0.0444	0.1569	81.05	0.3331
42	115.39	100.7	21.22	77.06	0.0456	0.1569	80.72	0.3218
44	119.33	104.6	21.81	77.30	0.0467	0.1569	80.38	0.3109
46	123.37	108.7	22.41	77.53	0.0479	0.1569	80.04	0.3004
48	127.51	112.8	23.01	77.76	0.0491	0.1569	79.70	0.2904
50	131.76	117.1	23.62	77.99	0.0502	0.1569	79.35	0.2807
52	136.11	121.4	24.23	78.21	0.0514	0.1569	79.00	0.2714
54	140.57	125.9	24.84	78.44	0.0526	0.1569	78.65	0.2624
56	145.15	130.5	25.46	78.66	0.0538	0.1569	78.29	0.2537
58	149.83	135.1	26.09	78.88	0.0550	0.1569	77.92	0.2454
60	154.63	139.9	26.72	79.10	0.0562	0.1570	77.56	0.2373
62	159.54	144.8	27.35	79.31	0.0574	0.1570	77.19	0.2296
64	164.56	149.9	27.99	79.52	0.0586	0.1570	76.81	0.2221
66	169.71	155.0	28.64	79.73	0.0598	0.1570	76.43	0.2149
68	174.98	160.3	29.29	79.94	0.0610	0.1570	76.05	0.2079
70	180.36	165.7	29.95	80.15	0.0622	0.1570	75.66	0.2012
72	185.87	171.2	30.61	80.35	0.0634	0.1570	75.26	0.1947
74	191.51	176.8	31.28	80.55	0.0647	0.1570	74.86	0.1884
76	197.27	182.6	31.96	80.75	0.0659	0.1570	74.45	0.1823
78	203.17	188.5	32.64	80.94	0.0671	0.1570	74.04	0.1765
80	209.19	194.5	33.33	81.13	0.0684	0.1570	73.62	0.1708
82	215.35	200.7	34.02	81.32	0.0696	0.1570	73.20	0.1653
84	221.64	206.9	34.73	81.51	0.0709	0.1570	72.76	0.1600
86	228.06	213.4	35.44	81.69	0.0722	0.1569	72.32	0.1549
88	234.63	219.9	36.15	81.87	0.0735	0.1569	71.88	0.1499
90	241.33	226.6	36.88	82.04	0.0748	0.1569	71.42	0.1451
92	248.18	233.5	37.61	82.21	0.0761	0.1569	70.96	0.1404
94	255.17	240.5	38.36	82.37	0.0774	0.1569	70.48	0.1359
96	262.31	247.6	39.11	82.54	0.0787	0.1568	70.00	0.1315
98	269.59	254.9	39.87	82.69	0.0800	0.1568	69.51	0.1272

Table A.18. (concl.) Properties of R-125 saturated liquid and vapor in I-P units.

Temp., °F	Pressure, psia	psig	Enthalpy, Btu/lb h_f	h_g	Entropy, Btu/lb·R s_f	s_g	Liquid, ρ_f lb/ft³	Vapor, v_g ft³/lb
100	277.02	262.3	40.64	82.85	0.0814	0.1568	69.00	0.1230
102	284.61	269.9	41.42	82.99	0.0827	0.1567	68.49	0.1190
104	292.34	277.6	42.22	83.14	0.0841	0.1567	67.96	0.1151
106	300.24	285.5	43.02	83.27	0.0855	0.1566	67.42	0.1113
108	308.29	293.6	43.84	83.40	0.0869	0.1566	66.87	0.1076
110	316.49	301.8	44.67	83.53	0.0883	0.1565	66.30	0.1040
112	324.86	310.2	45.52	83.65	0.0898	0.1565	65.71	0.1005
114	333.39	318.7	46.38	83.76	0.0912	0.1564	65.11	0.0971
116	342.09	327.4	47.26	83.86	0.0927	0.1563	64.49	0.0938
118	350.95	336.3	48.16	83.96	0.0942	0.1562	63.85	0.0905
120	359.99	345.3	49.08	84.05	0.0958	0.1561	63.18	0.0873
122	369.19	354.5	50.02	84.13	0.0973	0.1560	62.49	0.0842
124	378.56	363.9	50.98	84.19	0.0989	0.1558	61.76	0.0812
126	388.11	373.4	51.98	84.25	0.1006	0.1557	61.01	0.0782
128	397.84	383.1	53.00	84.30	0.1023	0.1555	60.22	0.0752
130	407.75	393.1	54.06	84.33	0.1040	0.1554	59.39	0.0723
132	417.83	403.1	55.16	84.34	0.1059	0.1552	58.51	0.0695
134	428.10	413.4	56.31	84.34	0.1077	0.1549	57.58	0.0666
136	438.56	423.9	57.52	84.31	0.1097	0.1547	56.57	0.0638
138	449.20	434.5	58.80	84.26	0.1118	0.1544	55.49	0.0609
140	460.03	445.3	60.17	84.18	0.1140	0.1541	54.29	0.0580

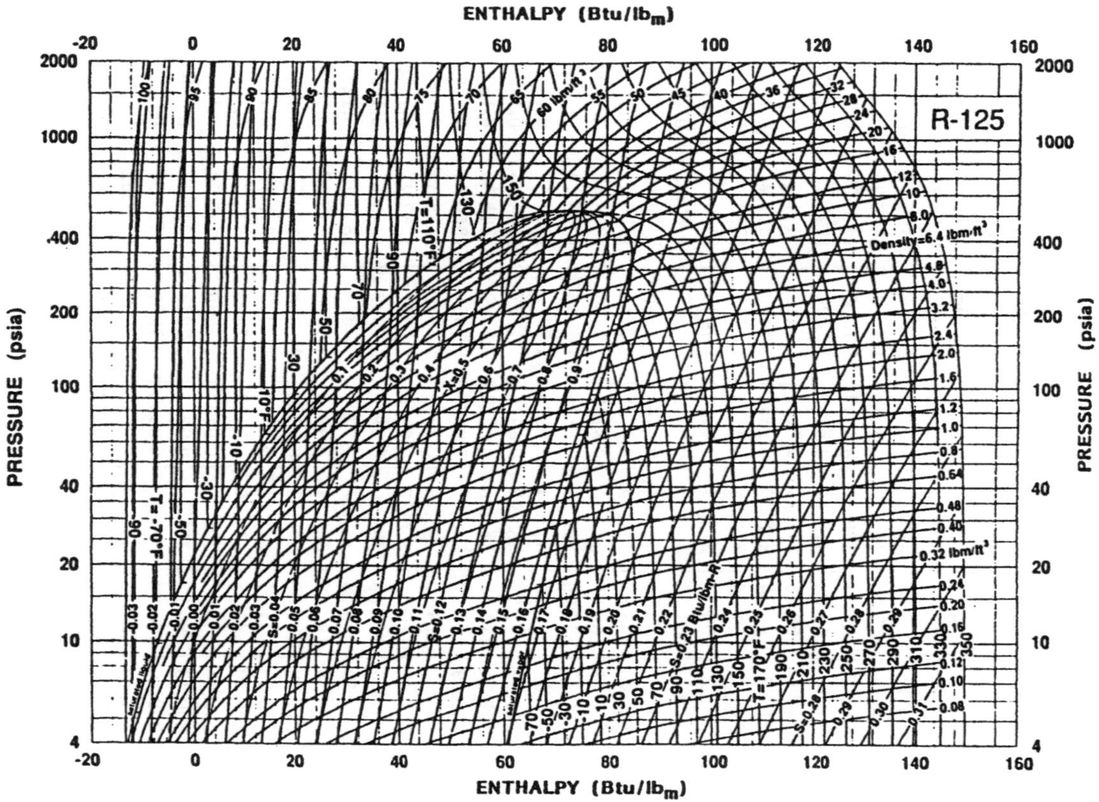

Fig. A.6. Pressure-enthalpy diagram for R-125 in IP units. (Reprinted with permission from 1997 ASHRAE *Handbook—Fundamentals*)

Table A.19. Properties of R-134a saturated liquid and vapor in SI units*.

Temp., °C	Press., kPa	Enthalpy, kJ/kg h_f	h_g	Entropy, kJ/kg·K s_f	s_g	Liquid, ρ_f kg/m^3	Vapor, v_g m^3/kg
−40	51.64	150.02	372.76	0.8032	1.7585	1417.38	0.3569
−39	54.38	151.20	373.38	0.8082	1.7571	1414.52	0.3400
−38	57.24	152.37	374.01	0.8132	1.7558	1411.64	0.3241
−37	60.22	153.56	374.64	0.8182	1.7544	1408.76	0.3090
−36	63.32	154.75	375.27	0.8232	1.7531	1405.86	0.2948
−35	66.55	155.94	375.89	0.8282	1.7519	1402.96	0.2813
−34	69.91	157.13	376.52	0.8332	1.7506	1400.05	0.2686
−33	73.40	158.33	377.14	0.8382	1.7494	1397.14	0.2565
−32	77.04	159.53	377.76	0.8432	1.7482	1394.21	0.2451
−31	80.82	160.74	378.39	0.8482	1.7470	1391.28	0.2343
−30	84.74	161.95	379.01	0.8532	1.7459	1388.33	0.2241
−29	88.82	163.17	379.63	0.8582	1.7448	1385.38	0.2144
−28	93.05	164.38	380.25	0.8631	1.7437	1382.42	0.2052
−27	97.44	165.61	380.87	0.8681	1.7426	1379.45	0.1965
−26	101.99	166.83	381.49	0.8730	1.7416	1376.47	0.1882
−25	106.71	168.06	382.10	0.8780	1.7406	1373.48	0.1803
−24	111.60	169.29	382.72	0.8829	1.7396	1370.48	0.1728
−23	116.67	170.53	383.34	0.8879	1.7386	1367.47	0.1657
−22	121.93	171.77	383.95	0.8928	1.7377	1364.45	0.1590
−21	127.36	173.02	384.56	0.8977	1.7367	1361.42	0.1525
−20	132.99	174.26	385.17	0.9027	1.7358	1358.38	0.1464
−19	138.81	175.52	385.78	0.9076	1.7349	1355.33	0.1406
−18	144.83	176.77	386.39	0.9125	1.7341	1352.27	0.1350
−17	151.05	178.03	387.00	0.9174	1.7332	1349.19	0.1298
−16	157.49	179.29	387.61	0.9223	1.7324	1346.11	0.1247
−15	164.13	180.56	388.21	0.9272	1.7316	1343.02	0.1199
−14	171.00	181.83	388.81	0.9321	1.7308	1339.91	0.1153
−13	178.08	183.11	389.41	0.9370	1.7300	1336.79	0.1110
−12	185.40	184.38	390.02	0.9419	1.7293	1333.66	0.1068
−11	192.95	185.67	390.61	0.9467	1.7286	1330.52	0.1028
−10	200.73	186.95	391.21	0.9516	1.7278	1327.37	0.0990
−9	208.76	188.24	391.81	0.9565	1.7271	1324.20	0.0953
−8	217.04	189.53	392.40	0.9613	1.7265	1321.02	0.0919
−7	225.57	190.83	392.99	0.9662	1.7258	1317.83	0.0885
−6	234.36	192.13	393.58	0.9710	1.7252	1314.63	0.0854
−5	243.42	193.43	394.17	0.9759	1.7245	1311.41	0.0823
−4	252.74	194.74	394.76	0.9807	1.7239	1308.18	0.0794
−3	262.33	196.05	395.35	0.9856	1.7233	1304.93	0.0766
−2	272.21	197.36	395.93	0.9904	1.7227	1301.67	0.0739
−1	282.37	198.68	396.51	0.9952	1.7221	1298.40	0.0714

*From algorithms of *Genie*, Allied Signal, Inc., Morristown, NJ, The enthalpy base is 200 kJ/kg of saturated liquid at −40°C, and the entropy base is 1.0 kJ/kg·K of saturated liquid at −40°C.

Table A.19. (cont.) Properties of R-134a saturated liquid and vapor in SI units.

Temp., °C	Press., kPa	Enthalpy, kJ/kg h_f	h_g	Entropy, kJ/kg·K s_f	s_g	Liquid, ρ_f kg/m^3	Vapor, v_g m^3/kg
0	292.82	200.00	397.09	1.0000	1.7216	1295.11	0.0689
1	303.57	201.33	397.67	1.0049	1.7210	1291.81	0.0665
2	314.62	202.66	398.24	1.0097	1.7205	1288.49	0.0643
3	325.98	203.99	398.82	1.0145	1.7200	1285.16	0.0621
4	337.65	205.33	399.39	1.0193	1.7195	1281.81	0.0600
5	349.64	206.67	399.96	1.0241	1.7190	1278.44	0.0580
6	361.95	208.01	400.52	1.0288	1.7185	1275.06	0.0561
7	374.59	209.36	401.09	1.0336	1.7180	1271.66	0.0543
8	387.57	210.71	401.65	1.0384	1.7176	1268.25	0.0525
9	400.89	212.07	402.21	1.0432	1.7171	1264.82	0.0508
10	414.55	213.43	402.77	1.0480	1.7167	1261.37	0.0491
11	428.57	214.79	403.33	1.0527	1.7162	1257.90	0.0476
12	442.95	216.16	403.88	1.0575	1.7158	1254.41	0.0460
13	457.69	217.53	404.43	1.0622	1.7154	1250.91	0.0446
14	472.80	218.90	404.98	1.0670	1.7150	1247.39	0.0432
15	488.29	220.28	405.52	1.0717	1.7146	1243.84	0.0418
16	504.17	221.66	406.07	1.0765	1.7142	1240.28	0.0405
17	520.43	223.05	406.61	1.0812	1.7139	1236.70	0.0393
18	537.08	224.44	407.14	1.0860	1.7135	1233.09	0.0381
19	554.14	225.83	407.68	1.0907	1.7131	1229.47	0.0369
20	571.61	227.23	408.21	1.0954	1.7128	1225.82	0.0358
21	589.49	228.63	408.74	1.1002	1.7124	1222.15	0.0347
22	607.79	230.04	409.26	1.1049	1.7121	1218.46	0.0336
23	626.51	231.45	409.78	1.1096	1.7118	1214.74	0.0326
24	645.67	232.87	410.30	1.1143	1.7114	1211.00	0.0317
25	665.27	234.29	410.82	1.1190	1.7111	1207.24	0.0307
26	685.31	235.71	411.33	1.1237	1.7108	1203.45	0.0298
27	705.80	237.14	411.84	1.1285	1.7105	1199.64	0.0289
28	726.76	238.57	412.34	1.1332	1.7102	1195.80	0.0281
29	748.17	240.01	412.84	1.1379	1.7099	1191.93	0.0273
30	770.07	241.45	413.34	1.1426	1.7096	1188.04	0.0265
31	792.43	242.90	413.83	1.1473	1.7093	1184.12	0.0257
32	815.29	244.35	414.32	1.1520	1.7090	1180.17	0.0250
33	838.63	245.80	414.81	1.1567	1.7087	1176.19	0.0243
34	862.48	247.26	415.29	1.1614	1.7084	1172.18	0.0236
35	886.83	248.73	415.77	1.1661	1.7081	1168.14	0.0229
36	911.69	250.20	416.24	1.1708	1.7079	1164.07	0.0223
37	937.08	251.68	416.71	1.1755	1.7076	1159.96	0.0216
38	962.99	253.16	417.17	1.1802	1.7073	1155.82	0.0210
39	989.43	254.65	417.63	1.1848	1.7070	1151.65	0.0204

Table A.19. (cont.) Properties of R-134a saturated liquid and vapor in SI units.

Temp., °C	Press., kPa	Enthalpy, kJ/kg h_f	h_g	Entropy, kJ/kg·K s_f	s_g	Liquid, ρ_f kg/m³	Vapor, v_g m³/kg
40	1016.41	256.14	418.09	1.1895	1.7067	1147.44	0.0199
41	1043.95	257.64	418.53	1.1942	1.7064	1143.20	0.0193
42	1072.04	259.14	418.98	1.1989	1.7061	1138.91	0.0188
43	1100.69	260.65	419.42	1.2036	1.7058	1134.59	0.0182
44	1129.91	262.16	419.85	1.2083	1.7056	1130.23	0.0177
45	1159.71	263.68	420.27	1.2131	1.7053	1125.83	0.0173
46	1190.09	265.21	420.70	1.2178	1.7050	1121.38	0.0168
47	1221.07	266.74	421.11	1.2225	1.7046	1116.90	0.0163
48	1252.64	268.28	421.52	1.2272	1.7043	1112.36	0.0159
49	1284.83	269.83	421.92	1.2319	1.7040	1107.78	0.0154
50	1317.63	271.39	422.31	1.2366	1.7037	1103.16	0.0150
51	1351.06	272.95	422.70	1.2414	1.7034	1098.48	0.0146
52	1385.12	274.51	423.08	1.2461	1.7030	1093.75	0.0142
53	1419.82	276.09	423.45	1.2508	1.7027	1088.97	0.0138
54	1455.17	277.67	423.82	1.2556	1.7023	1084.13	0.0135
55	1491.17	279.26	424.18	1.2603	1.7019	1079.24	0.0131
56	1527.84	280.86	424.52	1.2651	1.7016	1074.29	0.0127
57	1565.19	282.47	424.86	1.2699	1.7012	1069.27	0.0124
58	1603.22	284.08	425.19	1.2746	1.7008	1064.20	0.0121
59	1641.94	285.71	425.51	1.2794	1.7003	1059.05	0.0117
60	1681.36	287.34	425.82	1.2842	1.6999	1053.84	0.0114
61	1721.49	288.98	426.12	1.2890	1.6995	1048.56	0.0111
62	1762.34	290.64	426.41	1.2939	1.6990	1043.20	0.0108
63	1803.92	292.30	426.69	1.2987	1.6985	1037.76	0.0105
64	1846.24	293.97	426.95	1.3035	1.6980	1032.25	0.0102
65	1889.31	295.66	427.21	1.3084	1.6974	1026.65	0.0099
66	1933.13	297.35	427.45	1.3133	1.6969	1020.95	0.0097
67	1977.72	299.06	427.67	1.3182	1.6963	1015.17	0.0094
68	2023.09	300.78	427.89	1.3231	1.6957	1009.29	0.0091
69	2069.26	302.51	428.08	1.3280	1.6951	1003.30	0.0089
70	2116.22	304.26	428.26	1.3330	1.6944	997.21	0.0086
71	2163.99	306.02	428.43	1.3380	1.6937	991.00	0.0084
72	2212.58	307.79	428.57	1.3430	1.6929	984.68	0.0082
73	2262.01	309.58	428.70	1.3480	1.6922	978.22	0.0079
74	2312.29	311.39	428.81	1.3531	1.6913	971.63	0.0077
75	2363.42	313.21	428.90	1.3582	1.6905	964.90	0.0075
76	2415.43	315.05	428.96	1.3633	1.6895	958.02	0.0073
77	2468.32	316.91	429.00	1.3685	1.6886	950.97	0.0071
78	2522.11	318.79	429.01	1.3737	1.6875	943.76	0.0069
79	2576.81	320.69	429.00	1.3789	1.6865	936.35	0.0066

Table A.19. (concl.) Properties of R-134a saturated liquid and vapor in SI units.

Temp., °C	Press., kPa	Enthalpy, kJ/kg h_f	h_g	Entropy, kJ/kg·K s_f	s_g	Liquid, ρ_f kg/m^3	Vapor, v_g m^3/kg
80	2632.44	322.61	428.95	1.3842	1.6853	928.75	0.0064
81	2689.01	324.56	428.88	1.3895	1.6841	920.93	0.0063
82	2746.54	326.53	428.77	1.3949	1.6828	912.88	0.0061
83	2805.04	328.54	428.62	1.4003	1.6814	904.58	0.0059
84	2864.54	330.57	428.43	1.4059	1.6799	896.00	0.0057
85	2925.05	332.63	428.20	1.4114	1.6783	887.11	0.0055
86	2986.59	334.73	427.92	1.4171	1.6766	877.89	0.0053
87	3049.18	336.87	427.58	1.4229	1.6747	868.31	0.0051
88	3112.85	339.06	427.18	1.4287	1.6727	858.31	0.0050
89	3177.61	341.29	426.71	1.4347	1.6705	847.84	0.0048
90	3243.50	343.58	426.16	1.4408	1.6682	836.84	0.0046

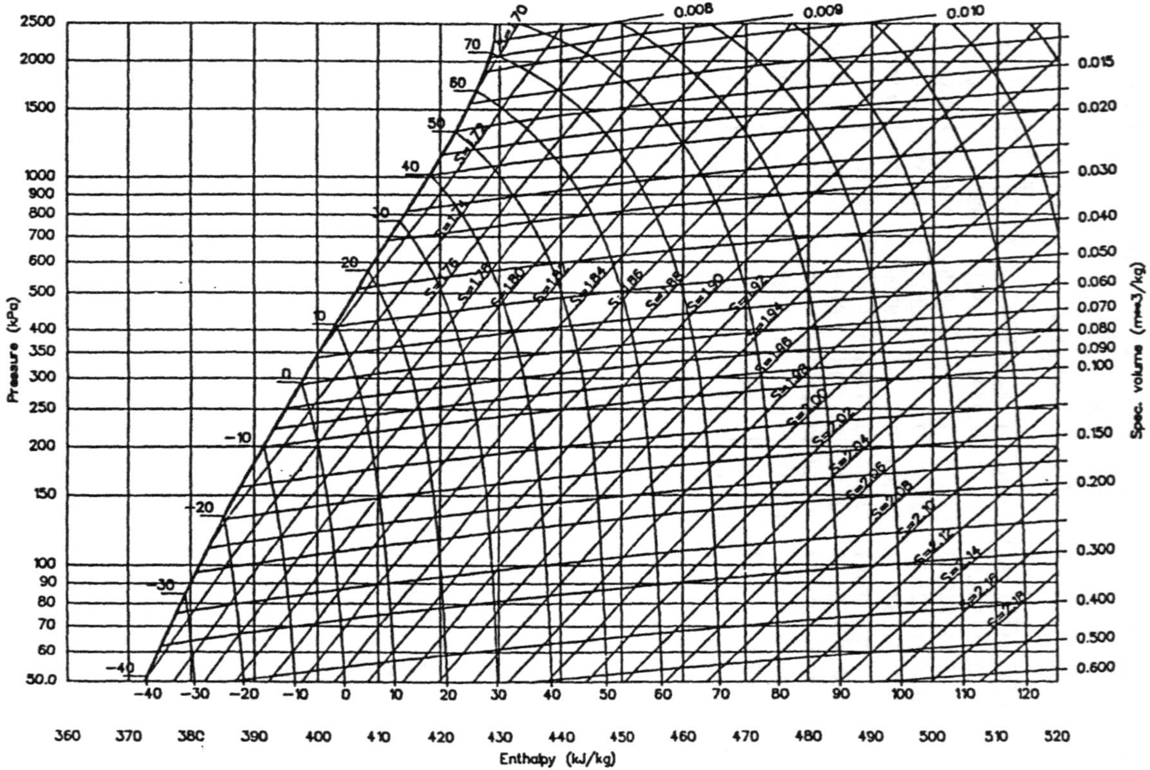

Fig. A.7. Pressure-enthalpy diagram for R-134a in SI units. (Courtesy Refrigeration Laboratory of the University of Denmark, from Wilson and Basu, *ASHRAE Transactions*, 1988)

Table A.20. Properties of R-134a saturated liquid and vapor in I-P units*.

Temp., °F	Pressure, psia	psig	Enthalpy, Btu/lb h_f	h_g	Entropy, Btu/lb·R s_f	s_g	Liquid, ρ_f lb/ft^3	Vapor, v_g ft^3/lb
−40	7.49	14.7†	−0.00	95.83	−0.0000	0.2283	88.48	5.7174
−38	7.93	13.8†	0.56	96.13	0.0013	0.2280	88.29	5.4174
−36	8.40	12.8†	1.13	96.43	0.0027	0.2276	88.09	5.1361
−34	8.88	11.8†	1.69	96.73	0.0040	0.2272	87.89	4.8721
−32	9.39	10.8†	2.26	97.02	0.0053	0.2269	87.68	4.6241
−30	9.92	9.7†	2.83	97.32	0.0066	0.2266	87.48	4.3912
−28	10.48	8.6†	3.40	97.62	0.0080	0.2262	87.28	4.1722
−26	11.05	7.4†	3.98	97.92	0.0093	0.2259	87.08	3.9661
−24	11.66	6.2†	4.55	98.22	0.0106	0.2256	86.87	3.7722
−22	12.29	4.9†	5.13	98.52	0.0119	0.2253	86.67	3.5895
−20	12.95	3.6†	5.71	98.81	0.0133	0.2250	86.47	3.4174
−18	13.63	2.2†	6.30	99.11	0.0146	0.2247	86.26	3.2551
−16	14.35	0.7†	6.88	99.40	0.0159	0.2244	86.05	3.1019
−14	15.09	0.4	7.47	99.70	0.0172	0.2242	85.85	2.9574
−12	15.87	1.2	8.06	99.99	0.0185	0.2239	85.64	2.8209
−10	16.67	2.0	8.65	100.29	0.0198	0.2236	85.43	2.6919
−8	17.51	2.8	9.24	100.58	0.0212	0.2234	85.22	2.5699
−6	18.38	3.7	9.83	100.87	0.0225	0.2231	85.01	2.4546
−4	19.29	4.6	10.43	101.17	0.0238	0.2229	84.80	2.3454
−2	20.23	5.5	11.03	101.46	0.0251	0.2227	84.59	2.2420
0	21.20	6.5	11.63	101.75	0.0264	0.2224	84.38	2.1440
1	21.71	7.0	11.93	101.89	0.0270	0.2223	84.27	2.0970
2	22.22	7.5	12.23	102.04	0.0277	0.2222	84.16	2.0512
3	22.74	8.0	12.53	102.18	0.0283	0.2221	84.06	2.0066
4	23.27	8.6	12.84	102.33	0.0290	0.2220	83.95	1.9632
5	23.81	9.1	13.14	102.47	0.0296	0.2219	83.84	1.9209
6	24.35	9.7	13.44	102.62	0.0303	0.2218	83.73	1.8797
7	24.91	10.2	13.75	102.76	0.0309	0.2217	83.63	1.8395
8	25.48	10.8	14.05	102.91	0.0316	0.2216	83.52	1.8004
9	26.06	11.4	14.36	103.05	0.0322	0.2215	83.41	1.7623
10	26.65	12.0	14.66	103.19	0.0329	0.2214	83.30	1.7251
11	27.25	12.6	14.97	103.34	0.0335	0.2213	83.19	1.6889
12	27.86	13.2	15.27	103.48	0.0342	0.2212	83.08	1.6536
13	28.48	13.8	15.58	103.62	0.0348	0.2211	82.97	1.6192
14	29.11	14.4	15.89	103.76	0.0355	0.2210	82.86	1.5856
15	29.76	15.1	16.20	103.91	0.0361	0.2209	82.76	1.5529
16	30.41	15.7	16.50	104.05	0.0368	0.2208	82.65	1.5210
17	31.07	16.4	16.81	104.19	0.0374	0.2207	82.54	1.4898

*From algorithms of *Genie*, AlliedSignal, Inç., Morrisotwn, NJ.

† Inches of mercury below one standard atmosphere.

The zero values of enthalpy and entropy are at −40°F, saturated liquid.

Table A.20. (cont.) Properties of R-134a saturated liquid and vapor in I-P units.

Temp., °F	Pressure, psia	psig	Enthalpy, Btu/lb h_f	h_g	Entropy, Btu/lb·R s_f	s_g	Liquid, ρ_f lb/ft^3	Vapor, v_g ft^3/lb
18	31.75	17.1	17.12	104.33	0.0381	0.2206	82.42	1.4595
19	32.44	17.7	17.43	104.47	0.0387	0.2205	82.31	1.4299
20	33.14	18.4	17.74	104.62	0.0393	0.2205	82.20	1.4010
21	33.85	19.2	18.05	104.76	0.0400	0.2204	82.09	1.3728
22	34.57	19.9	18.36	104.90	0.0406	0.2203	81.98	1.3452
23	35.30	20.6	18.68	105.04	0.0413	0.2202	81.87	1.3184
24	36.05	21.4	18.99	105.18	0.0419	0.2201	81.76	1.2922
25	36.81	22.1	19.30	105.32	0.0426	0.2200	81.64	1.2666
26	37.58	22.9	19.61	105.46	0.0432	0.2200	81.53	1.2416
27	38.36	23.7	19.93	105.60	0.0438	0.2199	81.42	1.2172
28	39.16	24.5	20.24	105.74	0.0445	0.2198	81.31	1.1934
29	39.97	25.3	20.56	105.88	0.0451	0.2197	81.19	1.1701
30	40.79	26.1	20.87	106.02	0.0458	0.2196	81.08	1.1474
31	41.62	26.9	21.19	106.16	0.0464	0.2196	80.97	1.1252
32	42.47	27.8	21.50	106.29	0.0470	0.2195	80.85	1.1035
33	43.33	28.6	21.82	106.43	0.0477	0.2194	80.74	1.0824
34	44.21	29.5	22.14	106.57	0.0483	0.2194	80.62	1.0617
35	45.09	30.4	22.46	106.71	0.0490	0.2193	80.51	1.0415
36	45.99	31.3	22.77	106.85	0.0496	0.2192	80.39	1.0217
37	46.91	32.2	23.09	106.98	0.0502	0.2191	80.28	1.0024
38	47.84	33.1	23.41	107.12	0.0509	0.2191	80.16	0.9835
39	48.78	34.1	23.73	107.26	0.0515	0.2190	80.04	0.9651
40	49.74	35.0	24.05	107.39	0.0521	0.2189	79.93	0.9470
41	50.71	36.0	24.37	107.53	0.0528	0.2189	79.81	0.9294
42	51.70	37.0	24.69	107.66	0.0534	0.2188	79.69	0.9122
43	52.70	38.0	25.01	107.80	0.0541	0.2187	79.58	0.8953
44	53.71	39.0	25.34	107.93	0.0547	0.2187	79.46	0.8788
45	54.74	40.0	25.66	108.07	0.0553	0.2186	79.34	0.8627
46	55.79	41.1	25.98	108.20	0.0560	0.2186	79.22	0.8469
47	56.85	42.2	26.31	108.34	0.0566	0.2185	79.10	0.8315
48	57.93	43.2	26.63	108.47	0.0572	0.2184	78.98	0.8164
49	59.02	44.3	26.95	108.60	0.0579	0.2184	78.86	0.8016
50	60.13	45.4	27.28	108.74	0.0585	0.2183	78.74	0.7871
51	61.25	46.6	27.60	108.87	0.0591	0.2183	78.62	0.7730
52	62.39	47.7	27.93	109.00	0.0598	0.2182	78.50	0.7591
53	63.54	48.8	28.26	109.14	0.0604	0.2182	78.38	0.7456
54	64.71	50.0	28.58	109.27	0.0610	0.2181	78.26	0.7323
55	65.90	51.2	28.91	109.40	0.0617	0.2180	78.14	0.7193
56	67.11	52.4	29.24	109.53	0.0623	0.2180	78.02	0.7066
57	68.33	53.6	29.57	109.66	0.0629	0.2179	77.90	0.6942
58	69.57	54.9	29.90	109.79	0.0636	0.2179	77.77	0.6820
59	70.82	56.1	30.23	109.92	0.0642	0.2178	77.65	0.6701

Table A.20. (cont.) Properties of R-134a saturated liquid and vapor in I-P units.

Temp., °F	Pressure, psia	psig	Enthalpy, Btu/lb h_f	h_g	Entropy, Btu/lb·R s_f	s_g	Liquid, ρ_f lb/ft^3	Vapor, v_g ft^3/lb
60	72.09	57.4	30.56	110.05	0.0648	0.2178	77.53	0.6584
61	73.38	58.7	30.89	110.18	0.0654	0.2177	77.40	0.6469
62	74.69	60.0	31.22	110.31	0.0661	0.2177	77.28	0.6357
63	76.01	61.3	31.55	110.44	0.0667	0.2176	77.15	0.6247
64	77.36	62.7	31.88	110.57	0.0673	0.2176	77.03	0.6140
65	78.72	64.0	32.22	110.70	0.0680	0.2175	76.90	0.6034
66	80.09	65.4	32.55	110.82	0.0686	0.2175	76.78	0.5931
67	81.49	66.8	32.88	110.95	0.0692	0.2174	76.65	0.5830
68	82.90	68.2	33.22	111.08	0.0698	0.2174	76.53	0.5731
69	84.34	69.6	33.55	111.20	0.0705	0.2174	76.40	0.5633
70	85.79	71.1	33.89	111.33	0.0711	0.2173	76.27	0.5538
71	87.26	72.6	34.22	111.46	0.0717	0.2173	76.14	0.5445
72	88.75	74.1	34.56	111.58	0.0724	0.2172	76.01	0.5353
73	90.26	75.6	34.90	111.71	0.0730	0.2172	75.89	0.5263
74	91.79	77.1	35.24	111.83	0.0736	0.2171	75.76	0.5175
75	93.33	78.6	35.57	111.95	0.0742	0.2171	75.63	0.5089
76	94.90	80.2	35.91	112.08	0.0749	0.2170	75.50	0.5004
77	96.49	81.8	36.25	112.20	0.0755	0.2170	75.37	0.4921
78	98.09	83.4	36.59	112.32	0.0761	0.2170	75.23	0.4840
79	99.72	85.0	36.93	112.44	0.0767	0.2169	75.10	0.4760
80	101.37	86.7	37.27	112.57	0.0774	0.2169	74.97	0.4682
81	103.04	88.3	37.62	112.69	0.0780	0.2168	74.84	0.4605
82	104.73	90.0	37.96	112.81	0.0786	0.2168	74.70	0.4530
83	106.43	91.7	38.30	112.93	0.0792	0.2168	74.57	0.4456
84	108.16	93.5	38.65	113.05	0.0799	0.2167	74.44	0.4383
85	109.92	95.2	38.99	113.17	0.0805	0.2167	74.30	0.4312
86	111.69	97.0	39.33	113.29	0.0811	0.2166	74.17	0.4242
87	113.48	98.8	39.68	113.40	0.0817	0.2166	74.03	0.4174
88	115.30	100.6	40.03	113.52	0.0824	0.2166	73.89	0.4106
89	117.13	102.4	40.37	113.64	0.0830	0.2165	73.76	0.4040
90	118.99	104.3	40.72	113.76	0.0836	0.2165	73.62	0.3975
91	120.87	106.2	41.07	113.87	0.0842	0.2164	73.48	0.3912
92	122.78	108.1	41.42	113.99	0.0849	0.2164	73.34	0.3849
93	124.70	110.0	41.77	114.10	0.0855	0.2164	73.20	0.3788
94	126.65	112.0	42.12	114.22	0.0861	0.2163	73.07	0.3728
95	128.62	113.9	42.47	114.33	0.0867	0.2163	72.92	0.3668
96	130.62	115.9	42.82	114.44	0.0873	0.2162	72.78	0.3610
97	132.63	117.9	43.17	114.56	0.0880	0.2162	72.64	0.3553
98	134.68	120.0	43.52	114.67	0.0886	0.2162	72.50	0.3497
99	136.74	122.0	43.88	114.78	0.0892	0.2161	72.36	0.3442

Table A.20. (cont.) Properties of R-134a saturated liquid and vapor in I-P units.

Temp., °F	Pressure, psia	psig	Enthalpy, Btu/lb h_f	h_g	Entropy, Btu/lb·R s_f	s_g	Liquid, ρ_f lb/ft^3	Vapor, v_g ft^3/lb
100	138.83	124.1	44.23	114.89	0.0898	0.2161	72.21	0.3388
101	140.94	126.2	44.59	115.00	0.0905	0.2161	72.07	0.3335
102	143.07	128.4	44.94	115.11	0.0911	0.2160	71.92	0.3283
103	145.23	130.5	45.30	115.22	0.0917	0.2160	71.78	0.3231
104	147.42	132.7	45.65	115.33	0.0923	0.2159	71.63	0.3181
105	149.63	134.9	46.01	115.43	0.0930	0.2159	71.49	0.3131
106	151.86	137.2	46.37	115.54	0.0936	0.2159	71.34	0.3083
107	154.12	139.4	46.73	115.65	0.0942	0.2158	71.19	0.3035
108	156.40	141.7	47.09	115.75	0.0948	0.2158	71.04	0.2988
109	158.71	144.0	47.45	115.86	0.0955	0.2158	70.89	0.2942
110	161.04	146.3	47.81	115.96	0.0961	0.2157	70.74	0.2896
111	163.40	148.7	48.17	116.06	0.0967	0.2157	70.59	0.2851
112	165.79	151.1	48.54	116.17	0.0973	0.2156	70.44	0.2807
113	168.20	153.5	48.90	116.27	0.0980	0.2156	70.28	0.2764
114	170.64	155.9	49.26	116.37	0.0986	0.2156	70.13	0.2722
115	173.10	158.4	49.63	116.47	0.0992	0.2155	69.97	0.2680
116	175.59	160.9	50.00	116.57	0.0998	0.2155	69.82	0.2639
117	178.11	163.4	50.36	116.67	0.1005	0.2154	69.66	0.2599
118	180.65	166.0	50.73	116.76	0.1011	0.2154	69.51	0.2559
119	183.23	168.5	51.10	116.86	0.1017	0.2154	69.35	0.2520
120	185.82	171.1	51.47	116.96	0.1023	0.2153	69.19	0.2481
121	188.45	173.8	51.84	117.05	0.1030	0.2153	69.03	0.2443
122	191.11	176.4	52.21	117.15	0.1036	0.2152	68.87	0.2406
123	193.79	179.1	52.59	117.24	0.1042	0.2152	68.71	0.2369
124	196.50	181.8	52.96	117.33	0.1048	0.2151	68.54	0.2333
125	199.24	184.5	53.33	117.42	0.1055	0.2151	68.38	0.2298
126	202.00	187.3	53.71	117.51	0.1061	0.2150	68.21	0.2263
127	204.80	190.1	54.09	117.60	0.1067	0.2150	68.05	0.2229
128	207.62	192.9	54.46	117.69	0.1074	0.2150	67.88	0.2195
129	210.48	195.8	54.84	117.78	0.1080	0.2149	67.71	0.2162
130	213.36	198.7	55.22	117.86	0.1086	0.2149	67.54	0.2129
131	216.27	201.6	55.60	117.95	0.1093	0.2148	67.37	0.2096
132	219.22	204.5	55.98	118.03	0.1099	0.2148	67.20	0.2065
133	222.19	207.5	56.37	118.11	0.1105	0.2147	67.03	0.2033
134	225.19	210.5	56.75	118.19	0.1112	0.2147	66.86	0.2003
135	228.23	213.5	57.14	118.27	0.1118	0.2146	66.68	0.1972
136	231.29	216.6	57.52	118.35	0.1124	0.2145	66.51	0.1942
137	234.39	219.7	57.91	118.43	0.1131	0.2145	66.33	0.1913
138	237.51	222.8	58.30	118.51	0.1137	0.2144	66.15	0.1884
139	240.67	226.0	58.69	118.58	0.1143	0.2144	65.97	0.1855

Table A.20. (concl.) Properties of R-134a saturated liquid and vapor in I-P units.

Temp., °F	Pressure, psia	psig	Enthalpy, Btu/lb h_f	h_g	Entropy, Btu/lb·R s_f	s_g	Liquid, ρ_f lb/ft^3	Vapor, v_g ft^3/lb
140	243.86	229.2	59.08	118.66	0.1150	0.2143	65.79	0.1827
142	250.33	235.6	59.86	118.80	0.1162	0.2142	65.42	0.1772
144	256.94	242.2	60.65	118.94	0.1175	0.2141	65.05	0.1719
146	263.67	249.0	61.45	119.07	0.1188	0.2139	64.67	0.1667
148	270.54	255.8	62.25	119.19	0.1201	0.2138	64.29	0.1616
150	277.54	262.8	63.06	119.31	0.1214	0.2137	63.89	0.1567
152	284.67	270.0	63.87	119.42	0.1227	0.2135	63.50	0.1519
154	291.95	277.3	64.70	119.52	0.1240	0.2133	63.09	0.1473
156	299.37	284.7	65.52	119.62	0.1253	0.2132	62.68	0.1428
158	306.93	292.2	66.36	119.71	0.1266	0.2130	62.25	0.1384
160	314.64	299.9	67.20	119.78	0.1279	0.2128	61.82	0.1341
162	322.49	307.8	68.05	119.85	0.1293	0.2126	61.38	0.1299
164	330.49	315.8	68.90	119.91	0.1306	0.2124	60.93	0.1258
166	338.65	324.0	69.77	119.96	0.1320	0.2122	60.47	0.1218
168	346.96	332.3	70.65	119.99	0.1333	0.2119	60.00	0.1180
170	355.43	340.7	71.53	120.02	0.1347	0.2117	59.52	0.1142
172	364.05	349.4	72.43	120.03	0.1361	0.2114	59.02	0.1105
174	372.84	358.1	73.33	120.02	0.1375	0.2111	58.51	0.1068
176	381.80	367.1	74.25	120.00	0.1389	0.2108	57.98	0.1033
178	390.93	376.2	75.18	119.97	0.1403	0.2105	57.44	0.0998
180	400.22	385.5	76.13	119.91	0.1417	0.2101	56.88	0.0964
182	409.70	395.0	77.09	119.83	0.1432	0.2098	56.29	0.0931
184	419.35	404.7	78.07	119.74	0.1446	0.2094	55.69	0.0898
186	429.18	414.5	79.06	119.61	0.1461	0.2089	55.06	0.0865
188	439.20	424.5	80.08	119.46	0.1476	0.2084	54.41	0.0833
190	449.41	434.7	81.12	119.28	0.1492	0.2079	53.72	0.0802
192	459.82	445.1	82.18	119.06	0.1508	0.2074	53.00	0.0771
194	470.43	455.7	83.27	118.80	0.1524	0.2067	52.24	0.0740
196	481.24	466.5	84.40	118.49	0.1540	0.2060	51.43	0.0709
198	492.27	477.6	85.56	118.13	0.1558	0.2053	50.57	0.0678
200	503.52	488.8	86.78	117.70	0.1575	0.2044	49.64	0.0647

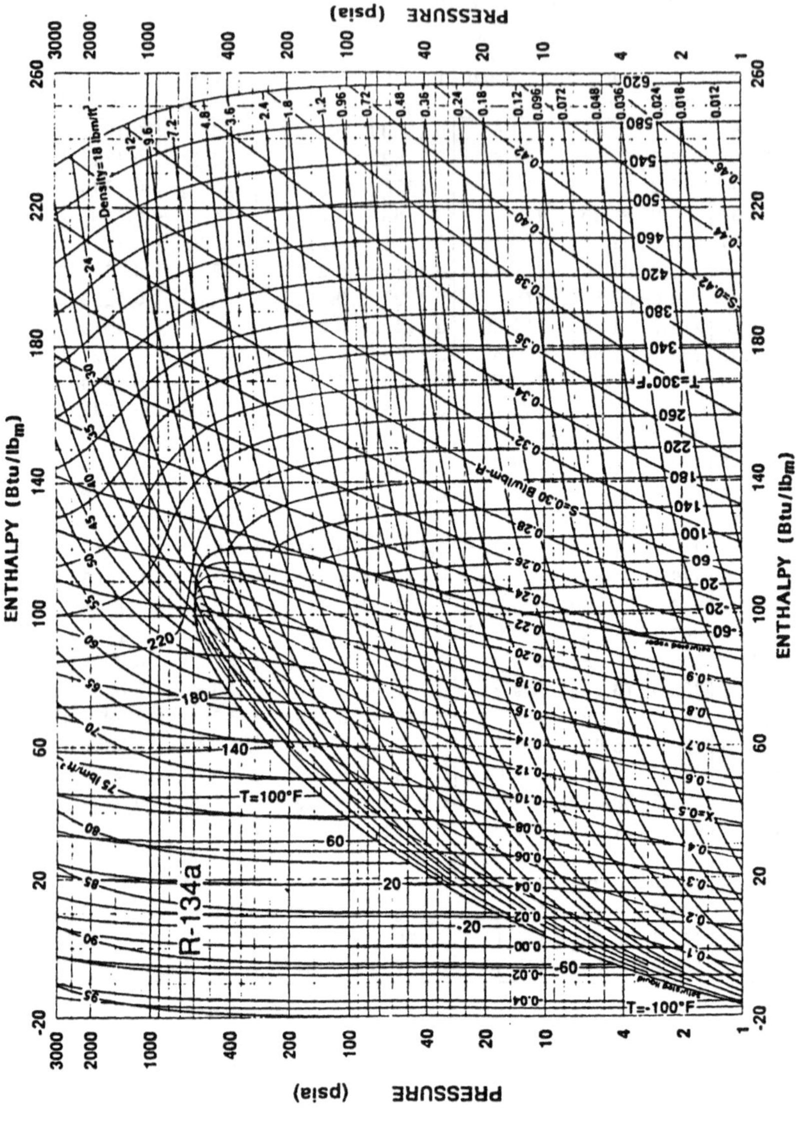

Fig. A.8. Pressure-enthalpy diagram for R-134a in IP units. (Reprinted with permission from 1997 ASHRAE *Handbook—Fundamentals*).

Table A.21. Properties of R-404a saturated liquid and vapor in SI units*.

Temp., °C	Pressure, kPa Liquid	Pressure, kPa Vapor	Enthalpy, kJ/kg h_f	Enthalpy, kJ/kg h_g	Entropy, kJ/kg·K s_f	Entropy, kJ/kg·K s_g	Liquid, ρ_f kg/m^3	Vapor, v_g m^3/kg
−60	50.5	48.3	121.0	330.8	0.6764	1.6628	1343.8	0.3683
−55	66.1	63.4	127.1	334.0	0.7042	1.6552	1329.1	0.2856
−50	85.2	82.1	133.1	337.3	0.7318	1.6487	1314.1	0.2244
−48	94.0	90.7	135.6	338.6	0.7427	1.6463	1308.0	0.2045
−46	103.5	100.0	138.1	339.9	0.7537	1.6440	1301.8	0.1866
−44	113.8	110.1	140.6	341.2	0.7645	1.6419	1295.7	0.1706
−42	124.8	120.9	143.1	342.5	0.7754	1.6399	1289.5	0.1563
−40	136.7	132.5	145.6	343.8	0.7862	1.6380	1283.2	0.1434
−39	142.9	138.6	146.9	344.5	0.7916	1.6371	1280.1	0.1374
−38	149.4	144.9	148.1	345.1	0.7970	1.6362	1277.0	0.1317
−37	156.1	151.1	149.4	345.8	0.8024	1.6353	1273.8	0.1263
−36	163.0	158.3	150.7	346.4	0.8077	1.6345	1270.7	0.1212
−35	170.1	165.3	152.4	347.0	0.8150	1.6337	1267.5	0.1163
−34	177.5	172.6	153.7	347.7	0.8203	1.6630	1264.4	0.1117
−33	185.2	179.1	155.0	348.3	0.8256	1.6322	1261.2	0.1073
−32	193.0	187.8	156.3	349.0	0.8309	1.6315	1258.0	0.1031
−31	201.2	195.8	157.5	349.6	0.8362	1.6308	1254.9	0.0991
−30	209.5	204.1	159.9	350.3	0.8460	1.6301	1251.7	0.0953
−29	218.2	212.5	161.2	350.9	0.8512	1.6295	1248.5	0.0917
−28	227.0	221.3	162.5	351.5	0.8563	1.6289	1245.3	0.0882
−27	236.2	230.3	163.7	352.2	0.8615	1.6283	1242.2	0.0849
−26	245.7	239.7	165.0	352.8	0.8667	1.6277	1239.0	0.0818
−25	255.4	249.3	166.3	353.4	0.8718	1.6271	1235.8	0.0788
−24	265.4	259.2	167.6	354.0	0.8769	1.6265	1232.5	0.0759
−23	275.8	269.4	168.9	354.7	0.8821	1.6260	1229.3	0.0732
−22	286.4	279.9	170.2	355.3	0.8872	1.6255	1226.1	0.0705
−21	297.4	290.7	171.5	355.9	0.8924	1.6250	1222.8	0.0680
−20	308.7	301.8	172.8	356.5	0.8975	1.6245	1219.6	0.0656
−19	320.3	313.3	174.1	357.1	0.9026	1.6240	1216.3	0.0633
−18	332.2	325.1	175.4	357.8	0.9078	1.6235	1213.1	0.0611
−17	344.5	337.2	176.7	358.4	0.9129	1.6231	1209.8	0.0589
−16	357.1	349.7	178.0	359.0	0.9180	1.6228	1206.5	0.0569
−15	370.1	362.5	179.4	359.6	0.9231	1.6222	1203.2	0.0550
−14	383.4	375.7	180.7	360.2	0.9282	1.6218	1199.8	0.0531
−13	397.1	389.2	182.1	360.8	0.9334	1.6214	1196.5	0.0513
−12	411.1	403.1	183.4	361.4	0.9385	1.6210	1193.2	0.0496
−11	425.6	417.4	184.7	362.0	0.9436	1.6206	1189.8	0.0479
−10	440.4	432.1	186.1	362.6	0.9487	1.6202	1186.4	0.0463
−9	455.6	447.2	187.5	363.2	0.9538	1.6198	1183.0	0.0448
−8	471.2	462.6	188.8	363.8	0.9589	1.6195	1179.6	0.0433
−7	487.2	478.5	190.2	364.3	0.9641	1.6191	1176.1	0.0419
−6	503.6	494.7	191.6	364.9	0.9692	1.6188	1172.7	0.0405

*From "Thermodynamic Properties of SUVA^R HP62 Refrigerant," T-HP62-SI, DuPont Fluorochemicals, Wilmington, Delaware, 1995.

Table A.21. (cont.) Properties of R-404a saturated liquid and vapor in SI units.

Temp., °C	Pressure, kPa Liquid	Pressure, kPa Vapor	Enthalpy, kJ/kg h_f	Enthalpy, kJ/kg h_g	Entropy, kJ/kg·K s_f	Entropy, kJ/kg·K s_g	Liquid, ρ_f kg/m^3	Vapor, v_g m^3/kg
−5	520.5	511.4	193.0	365.5	0.9743	1.6184	1169.2	0.0392
−4	537.7	528.5	194.4	366.1	0.9794	1.6181	1165.7	0.0380
−3	555.4	546.1	195.8	366.6	0.9846	1.6178	1162.2	0.0368
−2	573.5	564.1	197.2	367.2	0.9897	1.6175	1158.6	0.0356
−1	592.1	582.5	198.6	367.7	0.9948	1.6171	1155.1	0.0345
0	611.1	601.3	200.0	368.3	1.0000	1.6168	1151.5	0.0334
1	630.6	620.7	201.4	368.9	1.0051	1.6165	1147.8	0.0323
2	650.6	640.5	202.9	369.4	1.0102	1.6162	1144.2	0.0313
3	671.0	660.7	204.3	369.9	1.0154	1.6159	1140.5	0.0304
4	691.9	681.5	205.7	370.5	1.0205	1.6156	1136.8	0.0294
5	713.3	702.7	207.2	371.0	1.0257	1.6153	1133.0	0.0285
6	735.1	724.5	208.6	371.5	1.0308	1.6150	1129.1	0.0276
7	757.5	746.7	210.1	372.1	1.0360	1.6147	1125.4	0.0268
8	780.4	769.5	211.6	372.6	1.0412	1.6144	1121.6	0.0260
9	803.8	792.7	213.1	373.1	1.0464	1.6142	1117.7	0.0252
10	827.8	816.5	214.5	363.6	1.0515	1.6138	1113.7	0.0244
11	852.2	840.8	216.0	374.1	1.0567	1.6135	1109.8	0.0237
12	877.2	865.7	217.5	374.6	1.0619	1.6132	1105.7	0.0230
13	902.8	891.1	219.1	375.1	1.0671	1.6129	1101.7	0.0223
14	928.9	917.1	220.6	375.6	1.0723	1.6126	1097.5	0.0216
15	955.6	943.6	222.1	376.0	1.0776	1.6123	1093.4	0.0210
16	982.8	970.7	223.6	376.5	1.0828	1.6120	1089.1	0.0204
17	1010.6	998.4	225.2	377.0	1.0880	1.6116	1084.9	0.0198
18	1039.0	1026.7	226.7	377.4	1.0933	1.6113	1080.5	0.0192
19	1068.1	1055.6	228.3	377.9	1.0985	1.6110	1076.1	0.0186
20	1097.7	1085.1	229.9	378.3	1.1038	1.6106	1071.7	0.0181
21	1127.9	1115.2	231.5	378.7	1.1091	1.6103	1067.2	0.0176
22	1158.7	1145.9	233.0	379.2	1.1144	1.6099	1062.6	0.0171
23	1190.2	1177.2	234.6	379.6	1.1197	1.6095	1057.9	0.0166
24	1222.3	1209.2	236.3	380.0	1.1250	1.6091	1053.2	0.0161
25	1255.0	1241.8	237.9	380.4	1.1304	1.6087	1048.4	0.0156
26	1288.4	1275.1	239.5	380.8	1.1356	1.6083	1043.5	0.0152
27	1322.5	1309.1	241.1	381.1	1.1411	1.6079	1038.5	0.0147
28	1357.2	1343.7	242.8	381.5	1.1465	1.6075	1033.5	0.0143
29	1392.6	1379.0	244.5	381.9	1.1519	1.6070	1028.3	0.0139
30	1428.7	1415.0	246.1	382.2	1.1574	1.6065	1023.1	0.0135
31	1465.4	1451.7	247.9	382.6	1.1628	1.6060	1017.8	0.0131
32	1502.9	1489.1	249.6	382.9	1.1683	1.6055	1012.3	0.0127
33	1541.1	1527.2	251.3	383.2	1.1738	1.6050	1006.8	0.0123
34	1580.0	1566.0	253.0	383.5	1.1793	1.6044	1001.1	0.0120

Table A.21. (concl.) Properties of R-404a saturated liquid and vapor in SI units.

Temp., °C	Pressure, kPa Liquid	Pressure, kPa Vapor	Enthalpy, kJ/kg h_f	Enthalpy, kJ/kg h_g	Entropy, kJ/kg·K s_f	Entropy, kJ/kg·K s_g	Liquid, ρ_f kg/m^3	Vapor, v_g m^3/kg
35	1619.7	1605.6	254.8	383.8	1.1848	1.6038	995.4	0.0116
36	1660.1	1645.9	256.5	384.1	1.1904	1.6032	989.5	0.0113
37	1701.2	1687.0	258.3	384.3	1.1960	1.6026	983.5	0.0110
38	1743.1	1728.8	260.1	384.6	1.2016	1.6019	977.4	0.0106
39	1785.8	1771.4	361.9	384.8	1.2073	1.6012	971.1	0.0103
40	1829.2	1814.8	263.8	385.0	1.2130	1.6005	964.7	0.0110
41	1873.4	1859.0	265.6	385.2	1.2187	1.5998	958.2	0.0097
42	1918.4	1904.0	267.5	385.4	1.2245	1.5990	951.5	0.0094
43	1964.2	1949.8	269.4	385.6	1.2303	1.5981	944.6	0.0092
44	2010.8	1996.4	271.3	385.7	1.2362	1.5973	937.6	0.0089
46	2106.6	2092.2	275.1	385.9	1.2480	1.5954	923.0	0.0084
48	2205.6	2191.3	279.1	386.1	1.2600	1.5933	907.7	0.0079
50	2308.2	2294.0	283.2	386.2	1.2723	1.5910	891.5	0.0074
52	2414.2	2400.2	287.4	386.0	1.2849	1.5884	874.3	0.0069
54	2523.8	2510.0	291.7	385.8	1.2977	1.5855	856.0	0.0065
56	2637.1	2623.7	296.2	385.5	1.3110	1.5822	836.5	0.0061
58	2754.2	2741.1	300.9	384.9	1.3247	1.5784	815.6	0.0057
60	2875.1	2862.6	305.8	384.2	1.3389	1.5742	792.9	0.0053
62	2999.9	2988.0	311.0	383.1	1.3539	1.5692	768.3	0.0049
64	3128.7	3117.6	316.5	381.8	1.3697	1.5633	741.0	0.0045

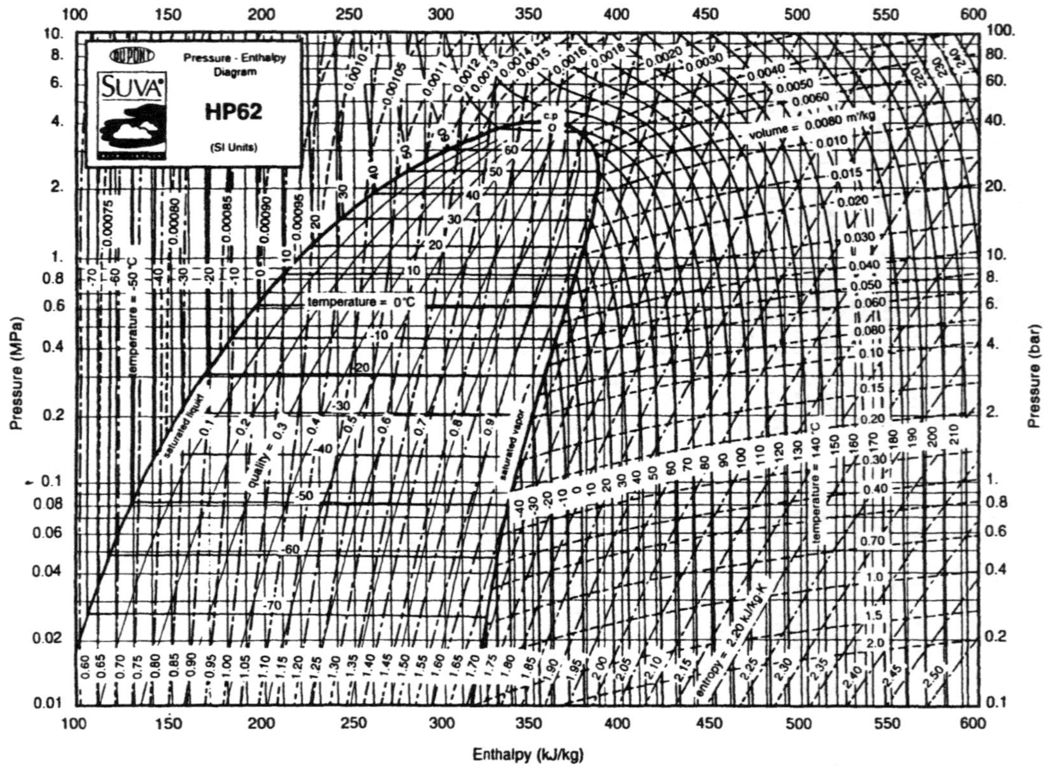

Fig. A.9. Pressure-enthalpy diagram for R-404a in SI units. (*Courtesy DuPont Fluorochemicals.*)

Table A.22. Properties of R-404a saturated liquid and vapor in I-P units*.

Temp., °F	Pressure, psia Liquid	Pressure, psia Vapor	Enthalpy, Btu/lb h_f	Enthalpy, Btu/lb h_g	Entropy, Btu/lb·R s_f	Entropy, Btu/lb·R s_g	Liquid, ρ_f lb/ft³	Vapor, v_g ft³/lb
−150	0.39	0.36	−30.6	68.4	−0.0842	0.2369	89.78	94.3396
−145	0.51	0.46	−29.3	69.1	−0.0800	0.2341	89.53	74.6269
−140	0.64	0.59	−28.0	69.9	−0.0759	0.2315	89.25	59.5238
−135	0.81	0.75	−26.7	70.6	−0.0719	0.2290	88.95	47.6190
−130	1.01	0.94	−25.4	71.3	−0.0678	0.2267	88.62	38.4615
−125	1.25	1.17	−24.1	72.1	−0.0638	0.2245	88.26	31.3480
−120	1.54	1.44	−22,7	72.9	−0.0599	0.2225	87.88	25.7732
−115	1.89	1.77	−21.4	73.6	−0.0559	0.2206	87.49	21.3220
−110	2.29	2.15	−20.0	74.4	−0.0520	0.2188	87.07	17.7305
−105	2.76	2.60	−18.7	75.2	−0.0482	0.2172	86.64	14.8368
−100	3.31	3.13	−17.3	75.9	−0.0443	0.2156	86.19	12.5000
−95	3.94	3.74	−15,9	76.7	−9.0405	0.2142	85.74	10.5920
−90	4.68	4.45	−14.5	77.5	−0.0367	0.2128	85.26	9.0171
−85	5.51	5.25	−13.1	78.3	−0.0330	0.2116	84.78	7.7160
−80	6.47	6.18	−11.7	79.0	−0.0292	0.2104	84.29	6.6357
−75	7.55	7.23	−10.3	79.8	−0.0255	0.2093	83.79	5.7307
−70	8.77	8.42	−8.8	80.6	−0.0218	0.2083	83.28	4.9727
−65	10.15	9.76	−7.4	81.4	−0.0181	0.2073	82.77	4.3328
−60	11.69	11.26	−5.9	82.2	−0.0145	0.02065	82.24	3.7893
−55	13.42	12.95	−4.5	82.9	−0.0108	0.2057	81.72	3.3256
−50	15.34	14.82	−3.0	83.7	−0.0072	0.2049	81.19	2.9300
−48	16.16	15.63	−2.4	84.0	−0.0058	0.2046	80.87	2.7196
−46	17.02	16.47	−1.8	84.3	−0.0043	0.2044	80.76	2.6539
−44	17.92	17.35	−1.2	84.7	−0.0029	0.2041	80.54	2.5278
−42	18.85	18.26	−0.6	85.0	−0.0014	0.2038	80.33	2.4085
−40	19.82	19.21	0.0	85.3	0.0000	0.2036	80.11	2.2962
−38	20.83	20.20	0.6	85.6	0.0014	0.2033	79.89	2.1906
−36	21.88	21.23	1.2	85.9	0.0029	0.2031	79.68	2.0603
−34	22.97	22.30	1.8	86.2	0.0043	0.2029	79.46	1.9960
−32	24.10	23.41	2.6	86.5	0.0062	0.2027	79.24	1.9066
−30	25.27	24.56	3.2	86.8	0.0076	0.2025	79.02	1.8222
−28	26.48	25.75	3.8	87.1	0.0090	0.2023	78.80	1.7422
−26	27.74	26.99	4.5	87.4	0.0104	0.2021	78.58	1.6665
−24	29.04	28.27	5.1	87.7	0.0118	0.2019	78.36	1.5946
−22	30.39	29.60	6.2	88.0	0.0143	0.2017	78.14	1.5267
−20	31.78	30.97	6.8	88.4	0.0157	0.2015	77.92	1.4626
−18	33.22	32.39	7.4	88.7	0.0170	0.2014	77.70	1.4025
−16	34.71	33.85	8.0	89.0	0.0184	0.2012	77.48	1.3436
−14	36.25	35.37	8.6	89.3	0.0198	0.2011	77.26	1.2885
−12	37.85	36.95	9.2	89.6	0.0211	0.2009	77.03	1.2362
−10	39.49	38.57	9.8	89.9	0.0225	0.2008	76.81	1.1864

*From "Thermodynamic Properties of SUVA^R HP62 Refrigerant," T-HP62-ENG, DuPont Fluorochemicals, Wilmington Delaware, 1995.

Table A.22 (cont.) Properties of R-404a saturated liquid and vapor in I-P units.

Temp., °F	Pressure, psia Liquid	Pressure, psia Vapor	Enthalpy, Btu/lb h_f	Enthalpy, Btu/lb h_g	Entropy, Btu/lb·R s_f	Entropy, Btu/lb·R s_g	Liquid, ρ_f lb/ft^3	Vapor, v_g ft^3/lb
−8	41.20	40.25	10.4	90.2	0.0239	0.2006	76.59	1.1390
−6	42.95	41.98	11.1	90.5	0.0252	0.2005	76.36	1.0937
−4	44.77	43.78	11.7	90.7	0.0266	0.2004	76.14	1.0506
−2	46.46	45.63	12.3	91.0	0.0280	0.2002	75.91	1.0096
0	48.57	47.54	12.9	91.3	0.0293	0.2001	75.68	0.9705
1	49.56	48.51	13.3	91.5	0.0300	0.2000	75.57	0.9516
2	50.57	49.51	13.6	91.6	0.0307	0.2000	75.46	0.9332
3	51.59	50.51	13.9	91.8	0.0314	0.1999	75.34	0.9152
4	52,62	51.54	14.2	91.9	0.0320	0.1999	75.23	0.8975
5	53.67	52.58	14.5	92.1	0.0327	0.1998	75.11	0.8803
6	54.74	53.63	14.9	92.2	0.0334	0.1998	75.00	0.8635
7	55.82	54.70	15.2	92.3	0.0341	0.1997	74.88	0.8470
8	56.92	55.79	15.5	92.5	0.0348	0.1996	74.76	0.8310
9	58.04	56.90	15.8	92.6	0.0354	0.1996	74.65	0.8153
10	59.17	58.02	16.1	92.8	0.0361	0.1995	74.53	0.7999
11	60.32	59.16	16.5	92.9	0.0368	0.1995	74.42	0.7849
12	61.49	60.31	16.8	93.1	0.0375	0.1994	74.30	0.7702
13	62.67	61.48	17.1	93.2	0.0382	0.1994	74.18	0.7559
14	63.87	62.67	17.4	93.3	0.0388	0.1993	74.06	0.7418
15	65.09	63.88	17.8	93.5	0.0395	0.1993	73.95	0.7281
16	66.33	65.10	18.1	93.6	0.0402	0.1992	73.83	0.7146
17	67.58	66.34	18.4	93.8	0.0409	0.1992	73.71	0.7015
18	68.85	67.60	18.7	93.9	0.0416	0.1991	73.59	0.6887
19	70.14	68.88	19.1	94.0	0.0422	0.1991	73.47	0.6761
20	71.45	70.18	19.4	94.2	0.0429	0.1990	73.35	0.6637
21	72.78	71.49	19.7	94.3	0.0436	0.1990	73.23	0.6517
22	74.12	72.82	20.1	94.5	0.0443	0.1990	73.11	0.6399
23	75.49	74.18	20.4	94.6	0.0450	0.1989	72.99	0.6284
24	76.87	75.55	20.7	94.7	0.0456	0.1989	72.87	0.6171
25	78.27	76.94	21.0	94.9	0.0463	0.1988	72.75	0.6060
26	79.69	78.35	21.4	95.0	0.0470	0.1988	72.63	0.5952
27	81.13	79.78	21.7	95.1	0.0477	0.1987	72.50	0.5845
28	82.60	81.22	22.1	95.3	0.0484	0.1987	72.38	0.5742
29	84.08	82.69	22.4	95.4	0.0490	0.1987	72.26	0.5640
30	85.58	84.18	22.7	95.5	0.0497	0.1986	72.13	0.5540
31	87.10	85.69	23.1	95.7	0.0504	0.1986	72.01	0.5442
32	88.64	87.22	23.4	95.8	0.0511	0.1985	71.88	0.5347
33	90.20	88.77	23.7	95.9	0.0518	0.1985	71.76	0.5253
34	91.78	90.34	24.1	96.1	0.0525	0.1984	71.63	0.5161
35	93.38	91.93	24.4	96.2	0.0531	0.1984	71.52	0.5071
36	95.01	93.54	24.8	96.3	0.0538	0.1984	71.38	0.4983
37	96.65	95.17	25.1	96.5	0.0545	0.1983	71.25	0.4896
38	98.32	96.83	25.5	96.6	0.0552	0.1983	71.22	0.4812
39	100.01	98.51	25.8	96.7	0.0559	0.1982	70.99	0.4729

Table A.22 (cont.) Properties of R-404a saturated liquid and vapor in I-P units.

Temp., °F	Pressure, psia Liquid	Pressure, psia Vapor	Enthalpy, Btu/lb h_f	Enthalpy, Btu/lb h_g	Entropy, Btu/lb·R s_f	Entropy, Btu/lb·R s_g	Liquid, ρ_f lb/ft^3	Vapor, v_g ft^3/lb
40	101.72	100.20	26.1	96.8	0.0566	0.1982	70.86	0.4647
41	103.45	101.92	26.5	97.0	0.0572	0.1982	70.73	0.4567
42	105.20	103.67	26.8	97.1	0.0579	0.1981	70.60	0.4489
43	106.98	105.43	27.2	97.2	0.0586	0.1981	70.47	0.4412
44	108.78	107.22	27.5	97.4	0.0593	0.1980	70.34	0.4337
45	110.60	109.03	27.9	97.5	0.0600	0.1980	70.20	0.4263
46	112.45	110.86	28.2	97.6	0.0607	0.1980	70.07	0.4191
47	114.31	112.72	28.6	97.7	0.0614	0.1979	69.94	0.4120
48	116.20	114.60	29.0	97.8	0.0620	0.1979	69.80	0.4050
49	118.12	116.50	29.3	98.0	0.0627	0.1978	69.67	0.3982
50	120.06	118.43	29.7	98.1	0.0634	0.1978	69.53	0.3915
51	122.02	120.38	30.0	98.2	0.0641	0.1978	69.39	0.3849
52	124.00	122.35	30.4	98.3	0.0648	0.1977	69.25	0.3784
53	126.02	124.35	30.7	98.4	0.0655	0.1977	69.11	0.3721
54	128.05	126.37	31.1	98.6	0.0662	0.1976	68.97	0.3659
55	130.11	128.42	31.5	98.7	0.0669	0.1976	68.83	0.3598
56	132.19	130.49	31.8	98.8	0.0676	0.1976	68.69	0.3538
57	134.30	132.59	32.2	98.9	0.0683	0.1975	68.55	0.3479
58	136.44	134.70	32.5	99.0	0.0689	0.1975	68.40	0.3421
59	138.59	136.86	32.9	99.1	0.0696	0.1974	68.26	0.3365
60	140.78	139.04	33.3	99.2	0.0703	0.1974	68.11	0.3309
61	142.99	141.24	33.6	99.4	0.0710	0.1974	67.96	0.3255
62	145.23	143.46	34.0	99.5	0.0717	0.1973	67.82	0.3201
63	147.49	145.71	34.4	99.6	0.0724	0.1973	67.67	0.3148
64	149.78	147.99	34.8	99.7	0.0731	0.1972	67.52	0.3096
65	152.09	150.30	35.1	99.8	0.0738	0.1972	67.36	0.3046
66	154.44	152.63	35.5	99.9	0.0745	0.1971	67.21	0.2996
67	156.81	154.99	35.9	100.0	0.0752	0.1971	67.06	0.2947
68	159.20	157.38	36.3	100.1	0.0759	0.1970	66.90	0.2898
69	161.63	159.79	36.6	100.2	0.0766	0.1970	66.75	0.2851
70	164.08	162.23	37.0	100.3	0.0773	0.1969	66.58	0.2804
71	166.56	164.70	37.4	100.4	0.0780	0.1969	66.43	0.2759
72	169.06	167.20	37.8	100.5	0.0787	0.1969	66.27	0.2714
73	171.60	169.72	38.2	100.6	0.0794	0.1968	66.11	0.2670
74	174.16	172.28	38.5	100.7	0.0801	0.1968	65.95	0.2626
75	176.75	174.86	38.9	100.8	0.0808	0.1967	65.78	0.2584
76	179.37	177.47	39.3	100.9	0.0816	0.1966	65.61	0.2542
77	182.02	180.11	39.7	101.0	0.0823	0.1966	65.45	0.2500
78	184.70	182.78	40.1	101.1	0.0830	0.1965	65.28	0.2460
79	187.41	185.48	40.5	101.2	0.0837	0.1965	65.11	0.2420
80	190.15	188.21	40.9	101.3	0.0844	0.1964	64.94	0.2381
81	192.92	190.97	41.3	101.4	0.0851	0.1964	64.76	0.2342

Table A.22 (cont.) Properties of R-404a saturated liquid and vapor in I-P units.

Temp., °F	Pressure, psia Liquid	Pressure, psia Vapor	Enthalpy, Btu/lb h_f	Enthalpy, Btu/lb h_g	Entropy, Btu/lb·R s_f	Entropy, Btu/lb·R s_g	Liquid, ρ_f lb/ft^3	Vapor, v_g ft^3/lb
82	195.72	193.76	41.7	101.5	0.0858	0.1963	64.59	0.2304
83	198.54	196.58	42.1	101.5	0.0866	0.1963	64.41	0.2267
84	201.40	199.43	42.5	101.6	0.0873	0.1962	64.23	0.2230
85	204.29	202.31	42.9	101.7	0.0880	0.1961	64.05	0.2194
86	207.21	205.23	43.4	101.8	0.0887	0.1961	63.87	0.2159
87	210.16	208.17	43.7	101.9	0.0894	0.1960	63.69	0.2124
88	213.15	211.15	44.1	102.0	0.0902	0.1959	63.50	0.2089
89	216.16	214.15	44.5	102.0	0.0909	0.1959	63.31	0.2055
90	219.21	217.19	44.9	102.1	0.0916	0.1958	63.12	0.2022
91	222.28	220.26	45.3	102.2	0.0923	0.1957	62.93	0.1989
92	225.39	223.37	45.7	102.3	0.0931	0.1957	62.74	0.1957
93	228.53	226.50	46.1	102.3	0.0938	0.1956	62.54	0.1925
94	231.71	229.67	46.6	102.4	0.0945	0.1955	62.34	0.1894
95	234.92	232.87	47.0	102.5	0.0953	0.1954	62.14	0.1863
96	238.16	236.11	47.4	102.5	0.0960	0.1953	61.94	0.1833
97	241.43	239.38	47.8	102.6	0.0968	0.1953	61.73	0.1803
98	244.74	242.68	48.2	102.7	0.0975	0.1952	61.53	0.1774
99	248.08	246.02	48.7	102.7	0.0982	0.1951	61.32	0.1745
100	251.46	249.39	49.1	102.8	0.0990	0.1950	61.10	0.1716
101	254.87	252.80	49.5	102.8	0.0997	0.1949	60.89	0.1688
102	258.31	256.24	50.0	102.9	0.1055	0.1948	60.67	0.1661
103	261.79	259.71	50.4	103.0	0.1013	0.1947	60.45	0.1633
104	265.30	263.22	50.8	103.0	0.1020	0.1946	60.23	0.1607
105	268.85	266.77	51.3	103.1	0.1028	0.1945	60.00	0.1580
106	272.43	270.35	51.7	103.1	0.1035	0.1944	59.77	0.1554
107	276.05	273.97	52.2	103.1	0.1043	0.1943	59.54	0.1528
108	279.71	277.62	52.8	103.2	0.1051	0.1942	59.31	0.1503
109	283.40	281.31	53.1	103.2	0.1058	0.1942	59.07	0.1478
110	287.13	285.04	53.5	103.3	0.1066	0.1940	58.83	0.1454
111	290.89	288.80	54.0	103.3	0.1074	0.1939	58.58	0.1429
112	294.69	292.60	54.4	103.3	0.1082	0.1938	58.34	0.1405
113	298.53	296.44	54.9	103.4	0.1090	0.1936	58.08	0.1382
114	302.40	300.32	55.4	103.4	0.1097	0.1935	57.83	0.1359
116	310.27	308.18	56.3	103.4	0.1113	0.1932	57.31	0.1313
118	318.28	316.20	57.2	103.5	0.1129	0.1930	56.77	0.1269
120	326.45	324.38	58.2	103.5	0.1146	0.1927	56.22	0.1225
122	334.77	332.71	59.2	103.5	0.1162	0.1924	55.65	0.1183
124	343.25	341.20	60.2	103.5	0.1178	0.1920	55.06	0.1142
126	351.89	349.86	61.2	103.4	0.1195	0.1917	54.46	0.1102
130	369.66	367.67	63.3	103.3	0.1230	0.1909	53.18	0.1025
135	392.80	290.88	66.0	103.1	0.1274	0.1897	51.43	0.0932
140	417.00	415.18	68.9	102.6	0.1321	0.1883	49.50	0.0844
145	442,39	440.62	72.1	102.0	0.1371	0.1877	47.32	0.0579
150	468.72	467.23	75.5	101.0	0.1426	0.1844	44.79	0.0676

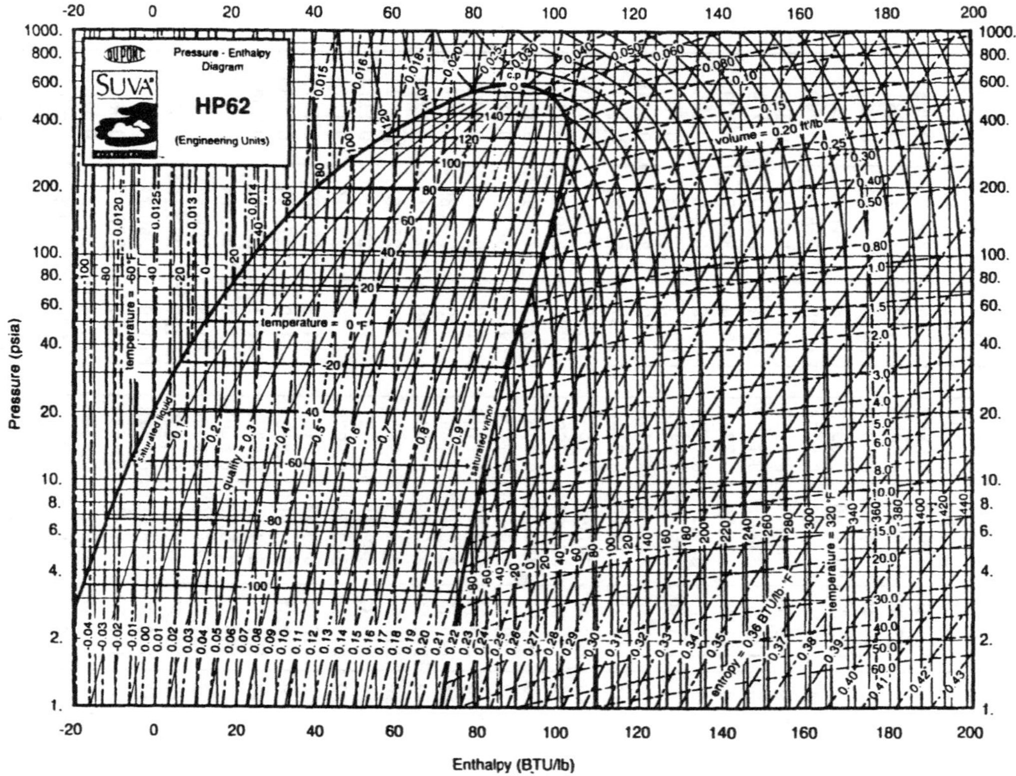

Fig. A.10. Pressure-enthalpy diagram for R-404a in IP units. (*Courtesy DuPont Fluorochemicals.*)

Table A.23. Properties of R-507 saturated liquid and vapor in SI units*.

Temp., °C	Press., kPa	Enthalpy, kJ/kg h_f	h_g	Entropy, kJ/kg·K s_f	s_g	Liquid, ρ_f kg/m³	Vapor, v_g m³/kg
−60	50.31	117.56	328.30	0.6610	1.6497	1360.32	0.3465
−59	53.24	118.94	328.91	0.6674	1.6479	1357.19	0.3286
−58	56.30	120.32	329.52	0.6738	1.6462	1354.05	0.3119
−57	59.49	121.69	330.13	0.6802	1.6445	1350.91	0.2961
−56	62.83	123.07	330.74	0.6865	1.6429	1347.76	0.2813
−55	66.31	124.45	331.35	0.6929	1.6413	1344.61	0.2674
−54	69.95	125.83	331.96	0.6992	1.6398	1341.45	0.2544
−53	73.74	127.21	332.57	0.7054	1.6383	1338.28	0.2421
−52	77.69	128.59	333.17	0.7117	1.6368	1335.10	0.2305
−51	81.80	129.97	333.78	0.7179	1.6354	1331.92	0.2196
−50	86.08	131.35	334.39	0.7241	1.6340	1328.73	0.2093
−49	90.53	132.73	334.99	0.7303	1.6326	1325.53	0.1996
−48	95.16	134.12	335.60	0.7364	1.6313	1322.33	0.1905
−47	99.97	135.50	336.20	0.7425	1.6300	1319.11	0.1819
−46	104.97	136.88	336.80	0.7486	1.6287	1315.89	0.1737
−45	110.16	138.26	337.40	0.7546	1.6275	1312.66	0.1660
−44	115.54	139.65	338.01	0.7607	1.6263	1309.43	0.1587
−43	121.13	141.03	338.61	0.7667	1.6252	1306.18	0.1518
−42	126.92	142.41	339.20	0.7726	1.6240	1302.92	0.1452
−41	132.93	143.79	339.80	0.7786	1.6229	1299.66	0.1390
−40	139.15	145.17	340.40	0.7845	1.6219	1296.39	0.1331
−39	145.59	146.56	340.99	0.7904	1.6208	1293.10	0.1276
−38	152.26	147.94	341.59	0.7963	1.6198	1289.81	0.1223
−37	159.16	149.32	342.18	0.8021	1.6188	1286.51	0.1173
−36	166.29	150.70	342.78	0.8079	1.6179	1283.19	0.1125
−35	173.67	152.08	343.37	0.8137	1.6169	1279.87	0.1079
−34	181.30	153.46	343.96	0.8194	1.6160	1276.54	0.1036
−33	189.17	154.84	344.54	0.8252	1.6151	1273.19	0.0995
−32	197.31	156.22	345.13	0.8309	1.6143	1269.84	0.0956
−31	205.71	157.59	345.72	0.8366	1.6135	1266.47	0.0919
−30	214.37	158.97	346.30	0.8422	1.6127	1263.10	0.0884
−29	223.31	160.35	346.89	0.8478	1.6119	1259.71	0.0850
−28	232.54	161.72	347.47	0.8534	1.6111	1256.30	0.0818
−27	242.04	163.10	348.05	0.8590	1.6104	1252.89	0.0787
−26	251.84	164.47	348.63	0.8645	1.6097	1249.46	0.0758
−25	261.93	165.85	349.21	0.8700	1.6090	1246.02	0.0730
−24	272.33	167.22	349.78	0.8755	1.6083	1242.57	0.0703
−23	283.03	168.60	350.36	0.8810	1.6076	1239.10	0.0678
−22	294.04	169.97	350.93	0.8864	1.6070	1235.62	0.0653
−21	305.38	171.34	351.50	0.8919	1.6064	1232.13	0.0630

*From algorithms of *Genie*, Allied Signal, Inc., Morristown, NJ. The enthalpy base is 200 kJ/kg of saturated liquid at 0°C; and the entropy base is 1.0 kJ/kg·K of saturated liquid at 0°C.

Table A.23. (cont.) Properties of R-507 saturated liquid and vapor in SI units.

Temp., °C	Press., kPa	Enthalpy, kJ/kg h_f	h_g	Entropy, kJ/kg·K s_f	s_g	Liquid, ρ_f kg/m^3	Vapor, v_g m^3/kg
−20	317.04	172.71	352.07	0.8972	1.6058	1228.62	0.0608
−19	329.03	174.08	352.64	0.9026	1.6052	1225.10	0.0586
−18	341.37	175.45	353.21	0.9079	1.6046	1221.56	0.0566
−17	354.04	176.82	353.77	0.9133	1.6041	1218.00	0.0546
−16	367.06	178.18	354.33	0.9185	1.6036	1214.43	0.0528
−15	380.44	179.55	354.89	0.9238	1.6030	1210.85	0.0510
−14	394.18	180.92	355.45	0.9290	1.6025	1207.24	0.0492
−13	408.29	182.28	356.01	0.9343	1.6021	1203.62	0.0476
−12	422.77	183.65	356.57	0.9394	1.6016	1199.98	0.0460
−11	437.64	185.01	357.12	0.9446	1.6011	1196.32	0.0445
−10	452.89	186.38	357.67	0.9498	1.6007	1192.65	0.0430
−9	468.53	187.74	358.22	0.9549	1.6003	1188.95	0.0416
−8	484.58	189.10	358.77	0.9600	1.5999	1185.24	0.0403
−7	501.03	190.47	359.31	0.9651	1.5995	1181.50	0.0390
−6	517.90	191.83	359.85	0.9701	1.5991	1177.75	0.0377
−5	535.18	193.19	360.39	0.9751	1.5987	1173.97	0.0365
−4	552.90	194.55	360.93	0.9802	1.5983	1170.17	0.0354
−3	571.05	195.92	361.47	0.9852	1.5980	1166.35	0.0342
−2	589.64	197.28	362.00	0.9901	1.5976	1162.51	0.0332
−1	608.68	198.64	362.53	0.9951	1.5973	1158.64	0.0321
0	628.17	200.00	363.06	1.0000	1.5970	1154.75	0.0312
1	648.13	201.37	363.58	1.0049	1.5967	1150.83	0.0302
2	668.56	202.73	364.11	1.0098	1.5964	1146.89	0.0293
3	689.47	204.09	364.63	1.0147	1.5961	1142.92	0.0284
4	710.86	205.46	365.14	1.0196	1.5958	1138.93	0.0275
5	732.74	206.83	365.66	1.0244	1.5955	1134.90	0.0267
6	755.13	208.19	366.17	1.0293	1.5952	1130.85	0.0259
7	778.03	209.56	366.68	1.0341	1.5949	1126.77	0.0251
8	801.44	210.93	367.18	1.0389	1.5947	1122.65	0.0244
9	825.38	212.30	367.68	1.0437	1.5944	1118.51	0.0237
10	849.85	213.67	368.18	1.0485	1.5942	1114.33	0.0230
11	874.86	215.05	368.67	1.0533	1.5939	1110.12	0.0223
12	900.43	216.43	369.16	1.0580	1.5937	1105.87	0.0216
13	926.55	217.81	369.65	1.0628	1.5934	1101.59	0.0210
14	953.24	219.19	370.13	1.0675	1.5932	1097.27	0.0204
15	980.51	220.58	370.61	1.0722	1.5929	1092.91	0.0198
16	1008.37	221.96	371.09	1.0770	1.5927	1088.52	0.0193
17	1036.81	223.36	371.56	1.0817	1.5925	1084.08	0.0187
18	1065.87	224.75	372.03	1.0864	1.5922	1079.60	0.0182
19	1095.54	226.15	372.49	1.0911	1.5920	1075.08	0.0176

Table A.23. (cont.) Properties of R-507 saturated liquid and vapor in SI units.

Temp., °C	Press., kPa	Enthalpy, kJ/kg h_f	h_g	Entropy, kJ/kg·K s_f	s_g	Liquid, ρ_f kg/m^3	Vapor, v_g m^3/kg
20	1125.83	227.56	372.94	1.0958	1.5917	1070.51	0.0171
21	1156.76	228.97	373.40	1.1005	1.5915	1065.89	0.0167
22	1188.33	230.38	373.84	1.1052	1.5913	1061.23	0.0162
23	1220.55	231.81	374.29	1.1099	1.5910	1056.52	0.0157
24	1253.44	233.23	374.72	1.1146	1.5908	1051.75	0.0153
25	1287.01	234.67	375.15	1.1193	1.5905	1046.93	0.0149
26	1321.26	236.11	375.58	1.1240	1.5903	1042.05	0.0144
27	1356.21	237.56	376.00	1.1288	1.5900	1037.11	0.0140
28	1391.87	239.01	376.41	1.1335	1.5897	1032.11	0.0136
29	1428.25	240.48	376.82	1.1382	1.5895	1027.05	0.0133
30	1465.36	241.95	377.22	1.1430	1.5892	1021.92	0.0129
31	1503.21	243.44	377.61	1.1477	1.5889	1016.73	0.0125
32	1541.82	244.93	378.00	1.1525	1.5886	1011.46	0.0122
33	1581.20	246.44	378.38	1.1573	1.5883	1006.11	0.0118
34	1621.36	247.96	378.75	1.1622	1.5880	1000.68	0.0115
35	1662.32	249.49	379.11	1.1670	1.5876	995.17	0.0112
36	1704.08	251.04	379.46	1.1719	1.5873	989.58	0.0109
37	1746.66	252.60	379.81	1.1768	1.5869	983.89	0.0106
38	1790.08	254.18	380.14	1.1817	1.5866	978.11	0.0103
39	1834.35	255.77	380.47	1.1867	1.5862	972.22	0.0100
40	1879.47	257.38	380.79	1.1917	1.5858	966.23	0.0097
41	1925.48	259.02	381.09	1.1967	1.5853	960.12	0.0094
42	1972.38	260.67	381.39	1.2018	1.5849	953.90	0.0091
43	2020.19	262.34	381.67	1.2070	1.5844	947.55	0.0089
44	2068.92	264.04	381.94	1.2122	1.5839	941.06	0.0086
45	2118.59	265.77	382.20	1.2175	1.5834	934.43	0.0084
46	2169.22	267.53	382.44	1.2228	1.5829	927.65	0.0081
47	2220.82	269.31	382.67	1.2282	1.5823	920.71	0.0079
48	2273.41	271.13	382.88	1.2337	1.5817	913.60	0.0077
49	2327.01	272.98	383.08	1.2393	1.5811	906.30	0.0074
50	2381.63	274.87	383.26	1.2450	1.5804	898.79	0.0072
51	2437.29	276.80	383.42	1.2507	1.5797	891.07	0.0070
52	2494.02	278.78	383.56	1.2566	1.5789	883.12	0.0068
53	2551.83	280.81	383.68	1.2627	1.5781	874.91	0.0066
54	2610.74	282.90	383.78	1.2688	1.5772	866.42	0.0063
55	2670.77	285.04	383.85	1.2752	1.5763	857.62	0.0061
56	2731.94	287.25	383.90	1.2817	1.5753	848.48	0.0059
57	2794.27	289.54	383.92	1.2884	1.5743	838.97	0.0057
58	2857.78	291.91	383.90	1.2953	1.5731	829.04	0.0055
59	2922.50	294.37	383.85	1.3025	1.5719	818.62	0.0053

Table A.23. (concl.) Properties of R-507 saturated liquid and vapor in in SI units.

Temp., °C	Press., kPa	Enthalpy, kJ/kg h_f	Enthalpy, kJ/kg h_g	Entropy, kJ/kg·K s_f	Entropy, kJ/kg·K s_g	Liquid, ρ_f kg/m^3	Vapor, v_g m^3/kg
60	2988.45	296.95	383.75	1.3100	1.5706	807.67	0.0052
61	3055.64	299.65	383.62	1.3179	1.5692	796.09	0.0050
62	3124.11	302.50	383.42	1.3261	1.5676	783.79	0.0048
63	3193.88	305.52	383.17	1.3348	1.5659	770.62	0.0046
64	3264.97	308.75	382.84	1.3442	1.5639	756.41	0.0044
65	3337.40	312.25	382.43	1.3543	1.5618	740.89	0.0042
66	3411.21	316.09	381.90	1.3653	1.5593	723.71	0.0040
67	3486.42	320.40	381.22	1.3777	1.5565	704.28	0.0038
68	3563.05	325.39	380.32	1.3920	1.5530	681.59	0.0036
69	3641.13	331.51	379.08	1.4096	1.5486	653.59	0.0033
70	3720.70	340.11	377.15	1.4343	1.5422	614.54	0.0030

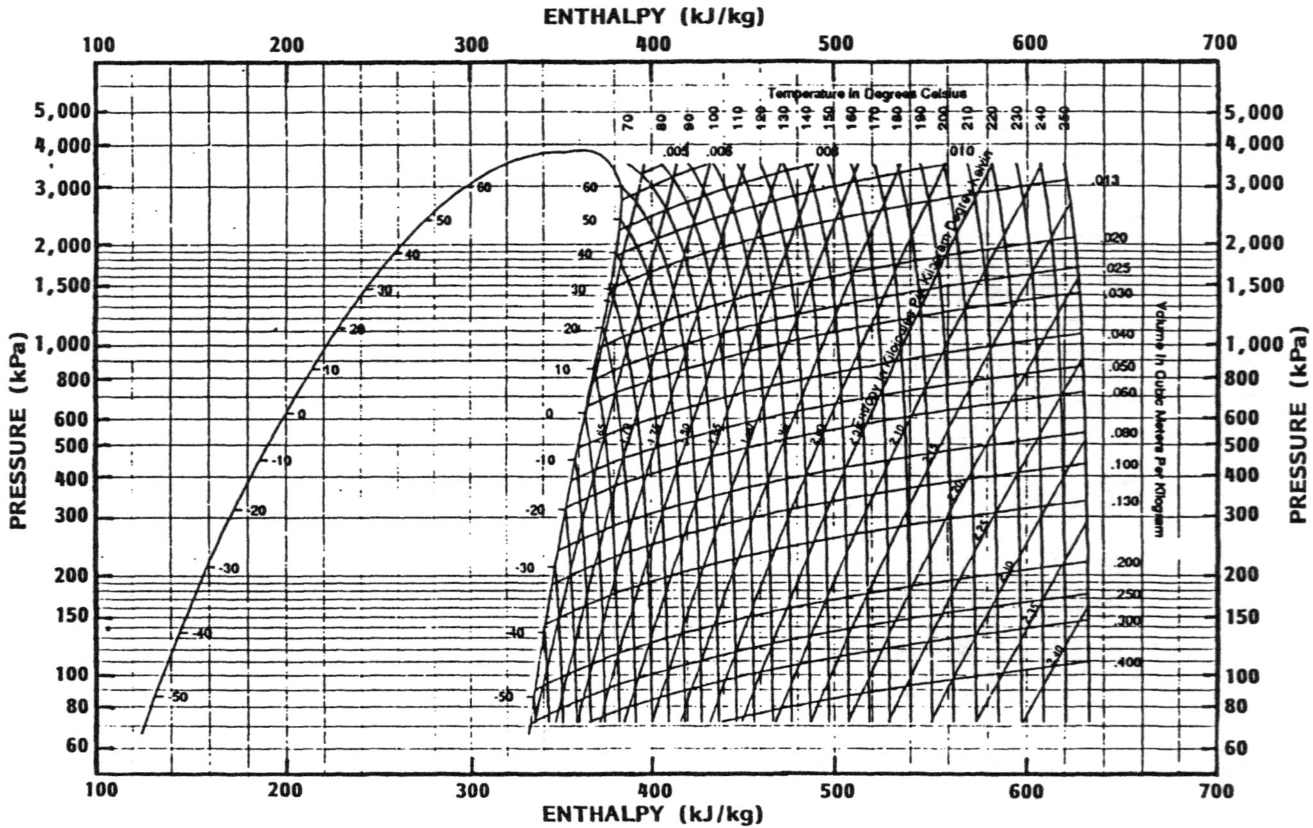

Fig. A.11. Pressure-enthalpy diagram for R-507 in SI units. (*Courtesy*

Table A.24. Properties of R-507 saturated liquid and vapor in I-P units*.

Temp., °F	Pressure, psia	psig	Enthalpy, Btu/lb h_f	h_g	Entropy, Btu/lb·R s_f	s_g	Liquid, ρ_f lb/ft^3	Vapor, v_g ft^3/lb
−70	8.79	12.0†	−9.91	79.66	−0.0245	0.2054	84.27	4.6627
−68	9.33	10.9†	−9.25	79.95	−0.0228	0.2049	84.05	4.4059
−66	9.91	9.7†	−8.59	80.24	−0.0211	0.2045	83.83	4.1662
−64	10.51	8.5†	−7.93	80.53	−0.0194	0.2041	83.61	3.9422
−62	11.14	7.2†	−7.27	80.82	−0.0178	0.2037	83.39	3.7327
−60	11.80	5.9†	−6.61	81.11	−0.0161	0.2034	83.17	3.5367
−58	12.48	4.5†	−5.95	81.40	−0.0145	0.2030	82.95	3.3532
−56	13.20	3.0†	−5.29	81.69	−0.0128	0.2026	82.73	3.1812
−54	13.95	1.5†	−4.63	81.98	−0.0112	0.2023	82.51	3.0200
−52	14.74	0.0	−3.97	82.27	−0.0096	0.2019	82.28	2.8686
−50	15.56	0.9	−3.31	82.55	−0.0080	0.2016	82.06	2.7264
−49	15.98	1.3	−2.98	82.70	−0.0072	0.2014	81.95	2.6586
−48	16.41	1.7	−2.65	82.84	−0.0064	0.2013	81.83	2.5928
−47	16.85	2.2	−2.32	82.99	−0.0056	0.2011	81.72	2.5290
−46	17.30	2.6	−1.98	83.13	−0.0048	0.2010	81.61	2.4672
−45	17.75	3.1	−1.65	83.27	−0.0040	0.2008	81.50	2.4072
−44	18.22	3.5	−1.32	83.42	−0.0032	0.2007	81.38	2.3489
−43	18.70	4.0	−0.99	83.56	−0.0024	0.2005	81.27	2.2924
−42	19.18	4.5	−0.66	83.70	−0.0016	0.2004	81.16	2.2376
−41	19.68	5.0	−0.33	83.84	−0.0008	0.2002	81.04	2.1844
−40	20.18	5.5	0.00	83.99	0.0000	0.2001	80.93	2.1327
−39	20.70	6.0	0.33	84.13	0.0007	0.1999	80.82	2.0825
−38	21.22	6.5	0.66	84.27	0.0015	0.1998	80.70	2.0337
−37	21.76	7.1	0.99	84.41	0.0023	0.1997	80.59	1.9864
−36	22.30	7.6	1.32	84.56	0.0031	0.1995	80.47	1.9404
−35	22.86	8.2	1.65	84.70	0.0038	0.1994	80.36	1.8957
−34	23.42	8.7	1.98	84.84	0.0046	0.1993	80.25	1.8523
−33	24.00	9.3	2.31	84.98	0.0054	0.1992	80.13	1.8101
−32	24.59	9.9	2.64	85.12	0.0062	0.1990	80.02	1.7690
−31	25.19	10.5	2.97	85.26	0.0069	0.1989	79.90	1.7291
−30	25.80	11.1	3.30	85.40	0.0077	0.1988	79.78	1.6903
−29	26.42	11.7	3.63	85.55	0.0085	0.1987	79.67	1.6526
−28	27.05	12.4	3.96	85.69	0.0092	0.1986	79.55	1.6159
−27	27.70	13.0	4.29	85.83	0.0100	0.1984	79.44	1.5802
−26	28.35	13.7	4.62	85.97	0.0107	0.1983	79.32	1.5455
−25	29.02	14.3	4.95	86.11	0.0115	0.1982	79.20	1.5117
−24	29.70	15.0	5.27	86.25	0.0122	0.1981	79.09	1.4788
−23	30.39	15.7	5.60	86.39	0.0130	0.1980	78.97	1.4468

*From algorithms of *Genie*, Allied Signal, Inc., Morristown, NJ.
† indicates inches of mercury, vacuum.
The zero values of enthalpy and entropy are of saturated liquid at −40°C.

Table A.24. (cont.) Properties of R-507 saturated liquid and vapor in I-P units.

Temp., °F	Pressure, psia	Pressure, psig	Enthalpy, Btu/lb h_f	Enthalpy, Btu/lb h_g	Entropy, Btu/lb·R s_f	Entropy, Btu/lb·R s_g	Liquid, ρ_f lb/ft^3	Vapor, v_g ft^3/lb
−22	31.09	16.4	5.93	86.53	0.0137	0.1979	78.85	1.4156
−21	31.81	17.1	6.26	86.67	0.0145	0.1978	78.74	1.3853
−20	32.54	17.8	6.59	86.81	0.0152	0.1977	78.62	1.3557
−19	33.28	18.6	6.92	86.94	0.0160	0.1976	78.50	1.3269
−18	34.03	19.3	7.25	87.08	0.0167	0.1975	78.38	1.2989
−17	34.79	20.1	7.58	87.22	0.0175	0.1974	78.26	1.2716
−16	35.57	20.9	7.91	87.36	0.0182	0.1973	78.14	1.2450
−15	36.37	21.7	8.24	87.50	0.0189	0.1972	78.03	1.2191
−14	37.17	22.5	8.56	87.64	0.0197	0.1971	77.91	1.1938
−13	37.99	23.3	8.89	87.78	0.0204	0.1970	77.79	1.1692
−12	38.82	24.1	9.22	87.91	0.0211	0.1969	77.67	1.1452
−11	39.67	25.0	9.55	88.05	0.0219	0.1968	77.55	1.1218
−10	40.53	25.8	9.88	88.19	0.0226	0.1967	77.43	1.0990
−9	41.40	26.7	10.20	88.33	0.0233	0.1966	77.31	1.0767
−8	42.29	27.6	10.53	88.46	0.0240	0.1966	77.19	1.0550
−7	43.19	28.5	10.86	88.60	0.0247	0.1965	77.06	1.0338
−6	44.11	29.4	11.19	88.74	0.0255	0.1964	76.94	1.0132
−5	45.04	30.3	11.52	88.87	0.0262	0.1963	76.82	0.9930
−4	45.98	31.3	11.84	89.01	0.0269	0.1962	76.70	0.9734
−3	46.94	32.2	12.17	89.14	0.0276	0.1962	76.58	0.9542
−2	47.92	33.2	12.50	89.28	0.0283	0.1961	76.46	0.9355
−1	48.91	34.2	12.83	89.42	0.0290	0.1960	76.33	0.9172
0	49.91	35.2	13.15	89.55	0.0297	0.1959	76.21	0.8994
1	50.94	36.2	13.48	89.69	0.0304	0.1959	76.09	0.8819
2	51.97	37.3	13.81	89.82	0.0311	0.1958	75.96	0.8649
3	53.03	38.3	14.13	89.95	0.0318	0.1957	75.84	0.8483
4	54.09	39.4	14.46	90.09	0.0325	0.1957	75.72	0.8321
5	55.18	40.5	14.79	90.22	0.0332	0.1956	75.59	0.8162
6	56.28	41.6	15.11	90.36	0.0339	0.1955	75.47	0.8008
7	57.40	42.7	15.44	90.49	0.0346	0.1955	75.34	0.7856
8	58.53	43.8	15.77	90.62	0.0353	0.1954	75.22	0.7708
9	59.68	45.0	16.09	90.76	0.0360	0.1953	75.09	0.7564
10	60.85	46.2	16.42	90.89	0.0367	0.1953	74.96	0.7423
11	62.03	47.3	16.75	91.02	0.0374	0.1952	74.84	0.7285
12	63.23	48.5	17.07	91.15	0.0381	0.1951	74.71	0.7150
13	64.45	49.8	17.40	91.29	0.0388	0.1951	74.58	0.7018
14	65.69	51.0	17.72	91.42	0.0394	0.1950	74.45	0.6889
15	66.94	52.2	18.05	91.55	0.0401	0.1950	74.33	0.6763
16	68.21	53.5	18.38	91.68	0.0408	0.1949	74.20	0.6640
17	69.50	54.8	18.70	91.81	0.0415	0.1949	74.07	0.6519
18	70.81	56.1	19.03	91.94	0.0422	0.1948	73.94	0.6401
19	72.13	57.4	19.35	92.07	0.0428	0.1948	73.81	0.6286

Table A.24. (cont.) Properties of R-507 saturated liquid and vapor in I-P units.

Temp., °F	Pressure, psia	Pressure, psig	Enthalpy, Btu/lb h_f	Enthalpy, Btu/lb h_g	Entropy, Btu/lb·R s_f	Entropy, Btu/lb·R s_g	Liquid, ρ_f lb/ft^3	Vapor, v_g ft^3/lb
20	73.48	58.8	19.68	92.20	0.0435	0.1947	73.68	0.6173
21	74.84	60.1	20.00	92.33	0.0442	0.1947	73.55	0.6062
22	76.22	61.5	20.33	92.46	0.0449	0.1946	73.42	0.5954
23	77.62	62.9	20.66	92.59	0.0455	0.1946	73.29	0.5848
24	79.04	64.3	20.98	92.72	0.0462	0.1945	73.16	0.5744
25	80.48	65.8	21.31	92.85	0.0468	0.1945	73.03	0.5643
26	81.94	67.2	21.63	92.97	0.0475	0.1944	72.89	0.5544
27	83.42	68.7	21.96	93.10	0.0482	0.1944	72.76	0.5447
28	84.91	70.2	22.28	93.23	0.0488	0.1943	72.63	0.5351
29	86.43	71.7	22.61	93.36	0.0495	0.1943	72.49	0.5258
30	87.97	73.3	22.93	93.48	0.0502	0.1942	72.36	0.5167
31	89.53	74.8	23.26	93.61	0.0508	0.1942	72.22	0.5077
32	91.11	76.4	23.59	93.74	0.0515	0.1941	72.09	0.4990
33	92.71	78.0	23.91	93.86	0.0521	0.1941	71.95	0.4904
34	94.33	79.6	24.24	93.99	0.0528	0.1941	71.82	0.4820
35	95.97	81.3	24.56	94.11	0.0534	0.1940	71.68	0.4737
36	97.63	82.9	24.89	94.24	0.0541	0.1940	71.54	0.4657
37	99.32	84.6	25.22	94.36	0.0547	0.1939	71.41	0.4578
38	101.02	86.3	25.54	94.48	0.0554	0.1939	71.27	0.4500
39	102.75	88.1	25.87	94.61	0.0560	0.1939	71.13	0.4424
40	104.50	89.8	26.19	94.73	0.0567	0.1938	70.99	0.4349
41	106.27	91.6	26.52	94.85	0.0573	0.1938	70.85	0.4276
42	108.07	93.4	26.85	94.98	0.0579	0.1937	70.71	0.4205
43	109.89	95.2	27.17	95.10	0.0586	0.1937	70.57	0.4135
44	111.73	97.0	27.50	95.22	0.0592	0.1937	70.43	0.4066
45	113.59	98.9	27.83	95.34	0.0599	0.1936	70.28	0.3998
46	115.48	100.8	28.16	95.46	0.0605	0.1936	70.14	0.3932
47	117.39	102.7	28.48	95.58	0.0611	0.1936	70.00	0.3867
48	119.32	104.6	28.81	95.70	0.0618	0.1935	69.85	0.3803
49	121.28	106.6	29.14	95.82	0.0624	0.1935	69.71	0.3740
50	123.26	108.6	29.47	95.94	0.0630	0.1935	69.57	0.3679
51	125.27	110.6	29.80	96.06	0.0637	0.1934	69.42	0.3619
52	127.30	112.6	30.13	96.17	0.0643	0.1934	69.27	0.3560
53	129.35	114.7	30.45	96.29	0.0649	0.1934	69.13	0.3502
54	131.43	116.7	30.78	96.41	0.0656	0.1933	68.98	0.3445
55	133.53	118.8	31.11	96.53	0.0662	0.1933	68.83	0.3389
56	135.67	121.0	31.44	96.64	0.0668	0.1933	68.68	0.3334
57	137.82	123.1	31.77	96.76	0.0675	0.1932	68.53	0.3280
58	140.00	125.3	32.11	96.87	0.0681	0.1932	68.38	0.3227
59	142.21	127.5	32.44	96.99	0.0687	0.1932	68.23	0.3175

Table A.24. (cont.) Properties of R-507 saturated liquid and vapor in I-P units.

Temp., °F	Pressure, psia	psig	Enthalpy, Btu/lb h_f	h_g	Entropy, Btu/lb·R s_f	s_g	Liquid, ρ_f lb/ft³	Vapor, v_g ft³/lb
60	144.44	129.7	32.77	97.10	0.0693	0.1931	68.08	0.3124
61	146.70	132.0	33.10	97.21	0.0700	0.1931	67.92	0.3074
62	148.99	134.3	33.43	97.33	0.0706	0.1931	67.77	0.3025
63	151.31	136.6	33.77	97.44	0.0712	0.1930	67.62	0.2976
64	153.65	139.0	34.10	97.55	0.0719	0.1930	67.46	0.2929
65	156.01	141.3	34.43	97.66	0.0725	0.1930	67.30	0.2882
66	158.41	143.7	34.77	97.77	0.0731	0.1930	67.15	0.2836
67	160.83	146.1	35.11	97.88	0.0737	0.1929	66.99	0.2791
68	163.29	148.6	35.44	97.99	0.0744	0.1929	66.83	0.2747
69	165.77	151.1	35.78	98.10	0.0750	0.1929	66.67	0.2703
70	168.28	153.6	36.12	98.20	0.0756	0.1928	66.51	0.2661
71	170.81	156.1	36.45	98.31	0.0762	0.1928	66.35	0.2618
72	173.38	158.7	36.79	98.42	0.0769	0.1928	66.19	0.2577
73	175.98	161.3	37.13	98.52	0.0775	0.1927	66.02	0.2536
74	178.60	163.9	37.47	98.63	0.0781	0.1927	65.86	0.2497
75	181.26	166.6	37.81	98.73	0.0787	0.1927	65.69	0.2457
76	183.95	169.3	38.16	98.84	0.0794	0.1926	65.53	0.2419
77	186.66	172.0	38.50	98.94	0.0800	0.1926	65.36	0.2381
78	189.41	174.7	38.84	99.04	0.0806	0.1926	65.19	0.2343
79	192.19	177.5	39.19	99.14	0.0812	0.1925	65.02	0.2306
80	195.00	180.3	39.53	99.24	0.0819	0.1925	64.85	0.2270
81	197.84	183.1	39.88	99.34	0.0825	0.1925	64.68	0.2235
82	200.71	186.0	40.23	99.44	0.0831	0.1924	64.50	0.2200
83	203.62	188.9	40.58	99.54	0.0837	0.1924	64.33	0.2165
84	206.56	191.9	40.93	99.64	0.0844	0.1924	64.15	0.2131
85	209.53	194.8	41.28	99.73	0.0850	0.1923	63.98	0.2098
86	212.53	197.8	41.63	99.83	0.0856	0.1923	63.80	0.2065
87	215.57	200.9	41.99	99.92	0.0863	0.1922	63.62	0.2033
88	218.64	203.9	42.34	100.01	0.0869	0.1922	63.44	0.2001
89	221.74	207.0	42.70	100.11	0.0875	0.1922	63.25	0.1970
90	224.88	210.2	43.06	100.20	0.0882	0.1921	63.07	0.1939
91	228.05	213.4	43.42	100.29	0.0888	0.1921	62.88	0.1908
92	231.26	216.6	43.78	100.38	0.0894	0.1920	62.70	0.1878
93	234.50	219.8	44.14	100.47	0.0901	0.1920	62.51	0.1849
94	237.78	223.1	44.51	100.55	0.0907	0.1920	62.32	0.1820
95	241.10	226.4	44.88	100.64	0.0914	0.1919	62.13	0.1791
96	244.45	229.8	45.25	100.73	0.0920	0.1919	61.93	0.1763
97	247.83	233.1	45.62	100.81	0.0927	0.1918	61.74	0.1735
98	251.26	236.6	45.99	100.89	0.0933	0.1918	61.54	0.1708
99	254.72	240.0	46.36	100.97	0.0940	0.1917	61.34	0.1681

Table A.24. (cont.) Properties of R-507 saturated liquid and vapor in I-P units.

Temp., °F	Pressure, psia	Pressure, psig	Enthalpy, Btu/lb h_f	Enthalpy, Btu/lb h_g	Entropy, Btu/lb·R s_f	Entropy, Btu/lb·R s_g	Liquid, ρ_f lb/ft^3	Vapor, v_g ft^3/lb
100	258.22	243.5	46.74	101.05	0.0946	0.1917	61.14	0.1655
101	261.75	247.1	47.12	101.13	0.0953	0.1916	60.94	0.1628
102	265.33	250.6	47.50	101.21	0.0959	0.1916	60.73	0.1603
103	268.94	254.2	47.89	101.29	0.0966	0.1915	60.53	0.1577
104	272.59	257.9	48.27	101.36	0.0973	0.1915	60.32	0.1552
105	276.28	261.6	48.66	101.44	0.0979	0.1914	60.11	0.1527
106	280.02	265.3	49.05	101.51	0.0986	0.1913	59.90	0.1503
107	283.79	269.1	49.45	101.58	0.0993	0.1913	59.68	0.1479
108	287.60	272.9	49.84	101.65	0.1000	0.1912	59.46	0.1455
109	291.45	276.8	50.25	101.71	0.1007	0.1912	59.24	0.1432
110	295.34	280.6	50.65	101.78	0.1013	0.1911	59.02	0.1409
111	299.28	284.6	51.06	101.85	0.1020	0.1910	58.79	0.1386
112	303.25	288.6	51.47	101.91	0.1027	0.1910	58.57	0.1363
113	307.27	292.6	51.88	101.97	0.1034	0.1909	58.33	0.1341
114	311.34	296.6	52.30	102.03	0.1041	0.1908	58.10	0.1319
115	315.44	300.7	52.72	102.08	0.1049	0.1908	57.86	0.1298
116	319.59	304.9	53.15	102.14	0.1056	0.1907	57.62	0.1276
117	323.78	309.1	53.58	102.19	0.1063	0.1906	57.38	0.1255
118	328.02	313.3	54.01	102.24	0.1070	0.1905	57.13	0.1234
119	332.30	317.6	54.45	102.29	0.1078	0.1904	56.88	0.1214
120	336.63	321.9	54.89	102.34	0.1085	0.1904	56.63	0.1193
121	341.00	326.3	55.34	102.38	0.1092	0.1903	56.37	0.1173
122	345.42	330.7	55.80	102.43	0.1100	0.1902	56.11	0.1153
123	349.89	335.2	56.26	102.47	0.1108	0.1901	55.84	0.1134
124	354.40	339.7	56.72	102.50	0.1115	0.1900	55.57	0.1114
125	358.97	344.3	57.19	102.54	0.1123	0.1899	55.30	0.1095
126	363.57	348.9	57.67	102.57	0.1131	0.1898	55.02	0.1076
127	368.23	353.5	58.16	102.60	0.1139	0.1897	54.73	0.1057
128	372.94	358.2	58.65	102.62	0.1147	0.1896	54.44	0.1038
129	377.70	363.0	59.15	102.65	0.1155	0.1894	54.15	0.1020
130	382.50	367.8	59.65	102.67	0.1164	0.1893	53.85	0.1001
131	387.36	372.7	60.17	102.68	0.1172	0.1892	53.54	0.0983
132	392.27	377.6	60.70	102.69	0.1181	0.1891	53.23	0.0965
133	397.23	382.5	61.23	102.70	0.1190	0.1889	52.90	0.0947
134	402.24	387.5	61.77	102.71	0.1198	0.1888	52.58	0.0930
135	407.30	392.6	62.33	102.71	0.1207	0.1887	52.24	0.0912
136	412.42	397.7	62.90	102.70	0.1217	0.1885	51.90	0.0895
137	417.59	402.9	63.47	102.70	0.1226	0.1883	51.54	0.0877
138	422.82	408.1	64.07	102.68	0.1236	0.1882	51.18	0.0860
139	428.10	413.4	64.67	102.66	0.1246	0.1880	50.81	0.0843

Table A.24. (concl.) Properties of R-507 saturated liquid and vapor in I-P units.

Temp., °F	Pressure, psia	Pressure, psig	Enthalpy, Btu/lb h_f	Enthalpy, Btu/lb h_g	Entropy, Btu/lb·R s_f	Entropy, Btu/lb·R s_g	Liquid, ρ_f lb/ft^3	Vapor, v_g ft^3/lb
140	433.44	418.7	65.29	102.64	0.1256	0.1878	50.42	0.0826
142	444.28	429.6	66.59	102.57	0.1276	0.1874	49.62	0.0792
144	455.34	440.6	67.96	102.47	0.1299	0.1870	48.75	0.0758
146	466.65	451.9	69.43	102.34	0.1322	0.1866	47.82	0.0723
148	478.19	463.5	71.02	102.17	0.1348	0.1860	46.80	0.0689
150	489.97	475.3	72.77	101.95	0.1376	0.1854	45.67	0.0653
152	502.00	487.3	74.74	101.65	0.1407	0.1847	44.39	0.0616
154	514.29	499.6	77.02	101.26	0.1443	0.1838	42.89	0.0577
156	526.83	512.1	79.84	100.70	0.1488	0.1827	41.02	0.0533
158	539.64	524.9	83.86	99.80	0.1553	0.1811	38.36	0.0476

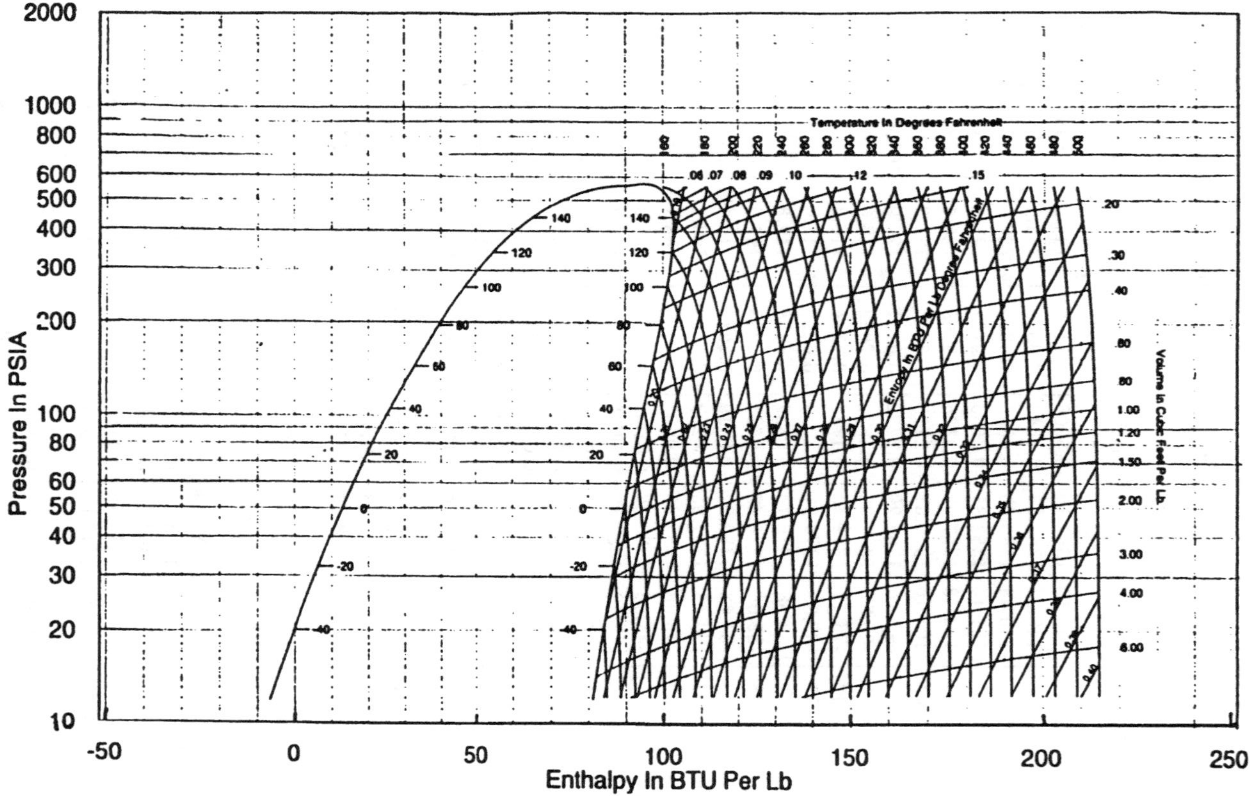

Fig. A.12. Pressure-enthalpy diagram for R-507 in IP units. (*Courtesy AlliedSignal, Inc.*)

Table A.25. Properties of ammonia saturated liquid and vapor in SI units*

Temp., °C	Press., kPa	Enthalpy, kJ/kg h_f	h_g	Entropy, kJ/kg·K s_f	s_g	Liquid, ρ_f kg/m^3	Vapor, v_g L/kg
-78	5.90	-149.12	1342.19	-0.5008	7.1410	734.3	16084.
-76	6.92	-140.37	1345.93	-0.4549	7.0828	732.1	13853.
-74	8.08	-130.44	1349.65	-0.4055	7.0260	729.8	11974.
-72	9.41	-121.44	1353.35	-0.3603	6.9708	727.5	10385.
-70	10.91	-112.18	1357.02	-0.3156	6.9169	725.4	9036.2
-68	12.61	-103.16	1360.67	-0.2719	6.8644	723.2	7887.6
-66	14.54	-94.37	1364.29	-0.2282	6.8131	720.9	6906.2
-64	16.70	-85.57	1367.88	-0.1870	6.7632	718.6	6064.7
-62	19.13	-76.40	1371.45	-0.1435	6.7144	716.3	5341.1
-60	21.85	-67.95	1374.98	-0.1036	6.6668	714.0	4716.9
-59	23.32	-63.58	1376.74	-0.0832	6.6434	712.8	4437.2
-58	24.88	-59.06	1378.48	-0.0624	6.6203	711.5	4176.8
-57	26.53	-54.68	1380.22	-0.0419	6.5975	710.4	3934.3
-56	28.26	-50.46	1381.95	-0.0222	6.5749	709.3	3708.2
-55	30.09	-46.08	1383.68	-0.0021	6.5526	708.1	3497.3
-54	32.01	-41.96	1385.39	0.0165	6.5306	706.9	3300.4
-53	34.03	-37.49	1387.10	0.0372	6.5088	705.7	3116.5
-52	36.16	-32.95	1388.79	0.0578	6.4872	704.5	2944.6
-51	38.40	-28.68	1390.48	0.0772	6.4659	703.3	2783.9
-50	40.76	-24.54	1392.16	0.0955	6.4449	702.1	2633.4
-49	43.23	-20.19	1393.83	0.1150	6.4240	700.8	2492.5
-48	45.82	-15.67	1395.49	0.1348	6.4034	699.6	2360.4
-47	48.54	-11.25	1397.14	0.1541	6.3830	698.4	2236.6
-46	51.39	-6.94	1398.78	0.1736	6.3629	697.2	2120.5
-45	54.38	-2.42	1400.41	0.1936	6.3429	696.0	2011.4
-44	57.51	1.85	1402.03	0.2117	6.3232	694.8	1909.0
-43	60.78	6.13	1403.64	0.2304	6.3037	693.6	1812.7
-42	64.21	10.56	1405.24	0.2500	6.2844	692.4	1722.1
-41	67.79	15.14	1406.83	0.2695	6.2653	691.1	1636.9
-40	71.54	19.50	1408.41	0.2881	6.2464	689.9	1556.7
-39	75.45	23.92	1409.98	0.3072	6.2277	688.7	1481.1
-38	79.54	28.43	1411.54	0.3262	6.2092	687.4	1409.9
-37	83.80	32.59	1413.09	0.3442	6.1908	686.2	1342.7
-36	88.25	37.13	1414.62	0.3629	6.1727	685.0	1279.4
-35	92.89	41.42	1416.15	0.3818	6.1547	683.7	1219.5
-34	97.72	46.02	1417.67	0.4005	6.1370	682.5	1163.0
-33	102.76	50.34	1419.17	0.4184	6.1194	681.2	1109.7
-32	108.01	54.85	1420.66	0.4370	6.1019	680.0	1059.2
-31	113.47	59.25	1422.15	0.4553	6.0847	678.8	1011.4

*From equations of Haar and Gallagher, "Thermodynamic Properties of Ammonia," *Journal of Phys. and Chem. Reference Data*, v. 7, no. 3, 1978.

Table A.25. (cont.) Properties of ammonia saturated liquid and vapor in SI units.

Temp., °C	Press., kPa	Enthalpy, kJ/kg h_f	h_g	Entropy, kJ/kg·K s_f	s_g	Liquid, ρ_f kg/m^3	Vapor, v_g L/kg
-30	119.15	63.95	1423.62	0.4745	6.0676	677.5	966.25
29	125.06	68.13	1425.08	0.4920	6.0507	676.2	923.47
-28	131.20	72.83	1426.52	0.5108	6.0339	675.0	882.94
-27	137.58	77.09	1427.96	0.5285	6.0174	673.7	844.54
-26	144.21	81.64	1429.38	0.5464	6.0009	672.5	808.12
-25	151.10	86.19	1430.79	0.5648	5.9846	671.2	773.58
-24	158.25	90.60	1432.20	0.5825	5.9685	670.0	740.81
-23	165.66	95.15	1433.58	0.6009	5.9525	668.7	709.69
-22	173.35	99.51	1434.96	0.6181	5.9367	667.4	680.13
-21	181.33	104.12	1436.32	0.6363	5.9210	666.2	652.05
-20	189.59	108.61	1437.67	0.6541	5.9055	664.9	625.35
-19	198.16	113.11	1439.01	0.6717	5.8901	663.6	599.96
-18	207.02	117.58	1440.34	0.6892	5.8748	662.3	575.80
-17	216.21	122.20	1441.65	0.7073	5.8597	661.0	552.81
-16	225.71	126.73	1442.95	0.7248	5.8447	659.8	530.92
-15	235.54	131.16	1444.24	0.7421	5.8298	658.5	510.06
-14	245.71	135.78	1445.51	0.7597	5.8151	657.2	490.19
-13	256.23	140.28	1446.77	0.7770	5.8005	655.9	471.24
-12	267.10	144.85	1448.02	0.7945	5.7860	654.6	453.18
-11	278.33	149.46	1449.25	0.8120	5.7716	653.3	435.94
-10	289.93	153.98	1450.47	0.8292	5.7574	652.0	419.49
-9	301.90	158.57	1451.68	0.8466	5.7433	650.7	403.79
-8	314.27	163.16	1452.87	0.8638	5.7293	649.3	388.79
-7	327.03	167.77	1454.05	0.8811	5.7154	648.0	374.45
-6	340.20	172.37	1455.21	0.8983	5.7016	646.7	360.76
-5	353.77	176.98	1456.36	0.9153	5.6879	645.4	347.66
-4	367.77	181.54	1457.50	0.9323	5.6744	644.0	335.14
-3	382.21	186.18	1458.62	0.9494	5.6609	642.7	323.15
-2	397.08	190.70	1459.72	0.9660	5.6476	641.3	311.68
-1	412.40	195.30	1460.81	0.9828	5.6343	640.0	300.70
0	428.18	200.00	1461.89	1.0000	5.6212	638.6	290.19
1	444.43	204.60	1462.95	1.0167	5.6081	637.3	280.11
2	461.15	209.26	1463.99	1.0336	5.5951	635.9	270.46
3	478.36	213.88	1465.02	1.0503	5.5823	634.5	261.20
4	496.07	218.57	1466.03	1.0671	5.5695	633.2	252.33
5	514.29	223.19	1467.03	1.0837	5.5568	631.8	243.81
6	533.02	227.90	1468.01	1.1004	5.5442	630.4	235.64
7	552.27	232.56	1468.98	1.1170	5.5317	629.0	227.80
8	572.06	237.20	1469.92	1.1334	5.5193	627.6	220.27
9	592.40	241.88	1470.85	1.1500	5.5070	626.2	213.04

Table A.25. (cont.) Properties of ammonia saturated liquid and vapor in SI units.

Temp., °C	Press., kPa	Enthalpy, kJ/kg h_f	h_g	Entropy, kJ/kg·K s_f	s_g	Liquid, ρ_f kg/m^3	Vapor, v_g L/kg
10	613.29	246.56	1471.77	1.1663	5.4947	624.8	206.10
11	634.75	251.26	1472.67	1.1828	5.4826	623.4	199.42
12	656.78	256.00	1473.54	1.1993	5.4705	622.0	193.01
13	679.40	260.64	1474.41	1.2155	5.4584	620.6	186.84
14	702.61	265.43	1475.25	1.2320	5.4465	619.1	180.91
15	726.43	270.11	1476.08	1.2482	5.4346	617.7	175.20
16	750.87	274.89	1476.88	1.2646	5.4228	616.3	169.71
17	775.94	279.64	1477.67	1.2809	5.4111	614.8	164.42
18	801.64	284.39	1478.44	1.2971	5.3994	613.3	159.33
19	828.00	289.15	1479.19	1.3132	5.3878	611.9	154.43
20	855.01	293.91	1479.93	1.3293	5.3763	610.4	149.71
21	882.69	298.68	1480.64	1.3454	5.3648	608.9	145.16
22	911.06	303.47	1481.33	1.3615	5.3534	607.5	140.77
23	940.11	308.26	1482.00	1.3776	5.3421	606.0	136.54
24	969.87	313.05	1482.65	1.3935	5.3308	604.5	132.47
25	1000.35	317.87	1483.29	1.4095	5.3195	603.0	128.53
26	1031.55	322.68	1483.90	1.4255	5.3083	601.5	124.74
27	1063.48	327.51	1484.48	1.4414	5.2972	600.0	121.08
28	1096.17	332.33	1485.05	1.4573	5.2861	598.4	117.55
29	1129.61	337.18	1485.60	1.4732	5.2751	596.9	114.13
30	1163.83	342.04	1486.12	1.4890	5.2641	595.4	110.84
31	1198.82	346.90	1486.62	1.5048	5.2532	593.8	107.66
32	1234.61	351.77	1487.10	1.5206	5.2423	592.3	104.58
33	1271.21	356.63	1487.56	1.5363	5.2314	590.7	101.61
34	1308.62	361.52	1487.99	1.5521	5.2206	589.1	98.74
35	1346.86	366.42	1488.40	1.5678	5.2099	587.6	95.96
36	1385.94	371.34	1488.78	1.5835	5.1991	586.0	93.28
37	1425.88	376.26	1489.14	1.5992	5.1884	584.4	90.68
38	1466.67	381.17	1489.48	1.6148	5.1778	582.8	88.17
39	1508.35	386.11	1489.79	1.6304	5.1671	581.2	85.74
40	1550.91	391.06	1490.07	1.6460	5.1565	579.6	83.39
41	1594.37	396.01	1490.33	1.6615	5.1459	577.9	81.11
42	1638.75	400.97	1490.56	1.6770	5.1354	576.3	78.91
43	1684.05	405.95	1490.77	1.6926	5.1249	574.7	76.77
44	1730.28	410.93	1490.94	1.7081	5.1144	573.0	74.70
45	1777.47	415.93	1491.10	1.7235	5.1039	571.3	72.70
46	1825.62	420.94	1491.22	1.7390	5.0935	569.7	70.76
47	1874.74	425.97	1491.31	1.7545	5.0830	568.0	68.88
48	1924.85	431.00	1491.38	1.7698	5.0726	566.3	67.05
49	1975.96	436.04	1491.42	1.7852	5.0622	564.6	65.28

Table A.25. (concl.) Properties of ammonia saturated liquid and vapor in SI units.

Temp., °C	Press., kPa	Enthalpy, kJ/kg h_f	h_g	Entropy, kJ/kg·K s_f	s_g	Liquid, ρ_f kg/m^3	Vapor, v_g L/kg
50	2028.08	441.11	1491.42	1.8007	5.0518	562.9	63.57
51	2081.22	446.17	1491.40	1.8160	5.0415	561.2	61.90
52	2135.41	451.26	1491.35	1.8314	5.0311	559.4	60.29
53	2190.65	456.35	1491.26	1.8467	5.0207	557.7	58.72
54	2246.95	461.46	1491.14	1.8620	5.0104	555.9	57.20
55	2304.32	466.58	1491.00	1.8774	5.0001	554.2	55.73
56	2362.79	471.72	1490.81	1.8927	4.9897	552.4	54.29
57	2422.36	476.86	1490.60	1.9079	4.9794	550.6	52.90
58	2483.05	482.03	1490.35	1.9232	4.9691	548.8	51.55
59	2544.87	487.21	1490.07	1.9385	4.9587	547.0	50.24
60	2607.83	492.41	1489.75	1.9538	4.9484	545.2	48.96
62	2737.25	502.85	1489.01	1.9843	4.9277	541.5	46.52
64	2871.40	513.35	1488.12	2.0148	4.9069	537.8	44.21
66	3010.41	523.93	1487.08	2.0453	4.8861	534.0	42.03
68	3154.39	534.58	1485.88	2.0759	4.8652	530.1	39.96
70	3303.44	545.31	1484.51	2.1064	4.8443	526.2	38.01
72	3457.70	556.13	1482.98	2.1370	4.8232	522.2	36.15
74	3617.28	567.02	1481.26	2.1676	4.8020	518.2	34.39
76	3782.30	578.02	1479.36	2.1982	4.7806	514.1	32.72
78	3952.88	589.12	1477.27	2.2290	4.7591	509.9	31.14
80	4129.15	600.32	1474.97	2.2598	4.7374	505.6	29.63
82	4311.25	611.64	1472.46	2.2907	4.7154	501.3	28.20
84	4499.29	623.08	1469.73	2.3218	4.6932	496.8	26.84
86	4693.43	634.65	1466.76	2.3530	4.6708	492.3	25.54
88	4893.79	646.37	1463.54	2.3844	4.6480	487.6	24.30
90	5100.52	658.23	1460.06	2.4160	4.6248	482.8	23.12
92	5313.77	670.26	1456.30	2.4478	4.6013	477.9	21.99
94	5533.69	682.47	1452.23	2.4799	4.5773	472.9	20.91
96	5760.43	694.87	1447.84	2.5123	4.5528	467.7	19.88
98	5994.16	707.48	1443.11	2.5450	4.5278	462.4	18.89
100	6235.05	720.32	1438.00	2.5781	4.5022	456.9	17.95
105	6869.77	753.59	1423.37	2.6629	4.4348	442.2	15.74
110	7553.34	788.91	1405.48	2.7515	4.3614	425.9	13.73
115	8289.24	827.06	1383.24	2.8459	4.2793	407.5	11.88
120	9081.61	869.39	1354.73	2.9491	4.1841	385.8	10.14
125	9935.83	918.91	1316.13	3.0684	4.0665	358.5	8.456
130	10860.18	986.36	1257.57	3.2299	3.9029	317.4	6.701

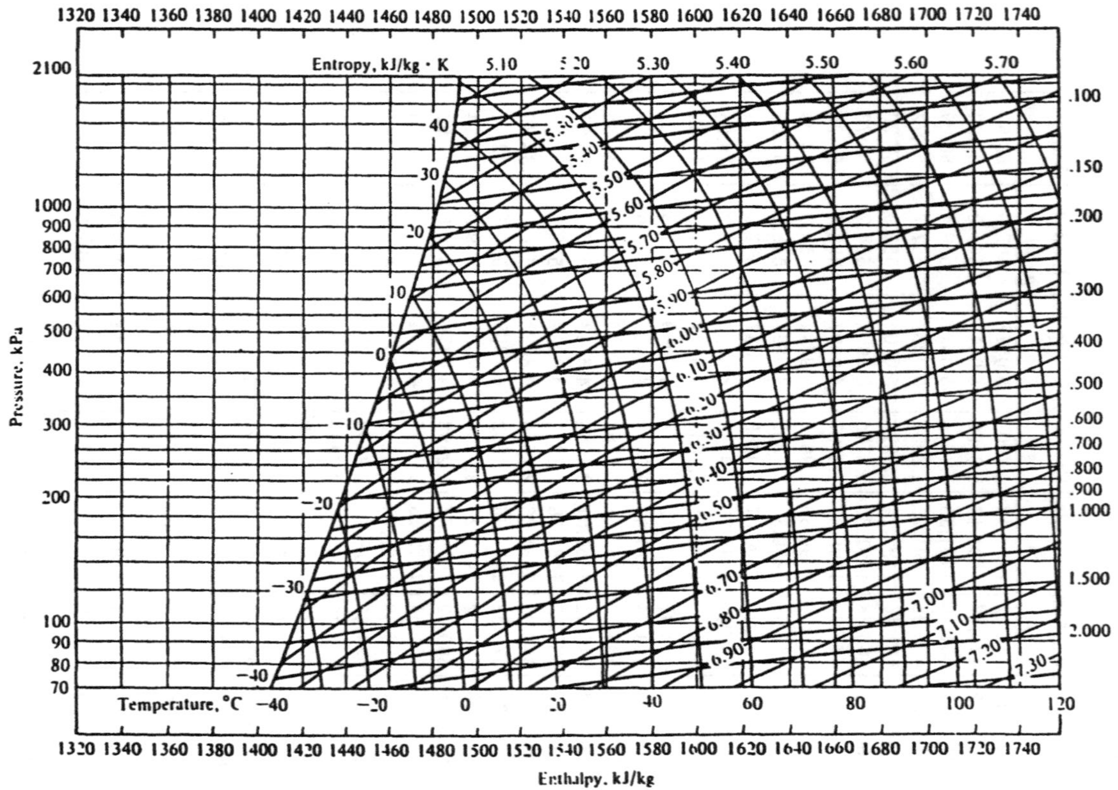

Fig. A.13. Pressure-enthalpy diagram for ammonia in SI units. (*Courtesy Refrigeration Laboratory of the University of Denmark.*)

Table A.26. Properties of superheated ammonia in SI units.

Saturation, t=−50°C p=40.76 kPa

Temp. °C	Enthalpy kJ/kg	Entropy kJ/(kg·K)	Sp. Vol. L/kg
−50	1392.2	6.4449	2633.4
−45	1402.7	6.4914	2696.2
−40	1413.2	6.5370	2758.8
−35	1423.7	6.5816	2821.1
−30	1434.2	6.6252	2883.2
−25	1444.7	6.6680	2945.1
−20	1455.2	6.7099	3006.8
−15	1465.7	6.7509	3068.4
−10	1476.2	6.7912	3129.7
−5	1486.7	6.8308	3191.0
0	1497.2	6.8697	3252.1

Saturation, t=−45°C, p=54.38 kPa

Temp. °C	Enthalpy kJ/kg	Entropy kJ/(kg·K)	Sp. Vol. L/kg
−45	1400.4	6.3429	2011.4
−40	1411.1	6.3892	2059.0
−35	1421.7	6.4345	2106.4
−30	1432.4	6.4787	2153.5
−25	1443.0	6.5220	2200.4
−20	1453.6	6.5644	2247.1
−15	1464.3	6.6059	2293.6
−10	1474.9	6.6466	2340.0
−5	1485.5	6.6865	2386.3
0	1496.1	6.7257	2432.4
5	1506.7	6.7642	2478.4

Saturation, t=−40°C, p=71.54 kPa

Temp. °C	Enthalpy kJ/kg	Entropy kJ/(kg·K)	Sp. Vol. L/kg
−40	1408.4	6.2464	1556.7
−35	1419.3	6.2925	1593.3
−30	1430.1	6.3375	1629.7
−25	1440.9	6.3815	1665.9
−20	1451.7	6.4245	1701.8
−15	1462.4	6.4665	1737.6
−10	1473.2	6.5077	1773.3
−5	1483.9	6.5481	1808.8
0	1494.6	6.5877	1844.1
5	1505.3	6.6265	1879.4
10	1516.0	6.6647	1914.5

Saturation, t=−35°C, p=92.89 kPa

Temp. °C	Enthalpy kJ/kg	Entropy kJ/(kg·K)	Sp. Vol. L/kg
−35	1416.1	6.1547	1219.5
−30	1427.2	6.2007	1248.1
−25	1438.2	6.2456	1276.5
−20	1449.2	6.2894	1304.6
−15	1460.2	6.3322	1332.6
−10	1471.1	6.3740	1360.4
−5	1481.9	6.4150	1388.1
0	1492.8	6.4551	1415.6
5	1503.6	6.4944	1443.1
10	1514.4	6.5329	1470.4
15	1525.3	6.5708	1497.6

Saturation, t=−30°C, p=119.15 kPa

Temp. °C	Enthalpy kJ/kg	Entropy kJ/(kg·K)	Sp. Vol. L/kg
−30	1423.6	6.0676	966.2
−25	1434.9	6.1136	988.9
−20	1446.1	6.1584	1011.2
−15	1457.3	6.2021	1033.5
−10	1468.4	6.2448	1055.5
−5	1479.5	6.2864	1077.4
0	1490.5	6.3272	1099.2
5	1501.5	6.3670	1120.8
10	1512.5	6.4061	1142.4
15	1523.4	6.4444	1163.8
20	1534.4	6.4820	1185.2

Saturation, t=−25°C, p=151.10 kPa

Temp. °C	Enthalpy kJ/kg	Entropy kJ/(kg·K)	Sp. Vol. L/kg
−25	1430.8	5.9846	773.6
−20	1442.3	6.0307	791.7
−15	1453.8	6.0756	809.6
−10	1465.2	6.1192	827.4
−5	1476.5	6.1618	844.9
0	1487.7	6.2033	862.4
5	1498.9	6.2439	879.7
10	1510.1	6.2836	897.0
15	1521.2	6.3225	914.1
20	1532.2	6.3606	931.2
25	1543.3	6.3980	948.1

Table A.26. (cont.) Properties of superheated ammonia in SI units.

Saturation, t=−20°C, p=189.59 kPa

Temp. °C	Enthalpy kJ/kg	Entropy kJ/(kg·K)	Sp. Vol. L/kg
−20	1437.7	5.9055	625.3
−15	1449.5	5.9517	640.0
−10	1461.2	5.9967	654.5
−5	1472.8	6.0404	668.9
0	1484.3	6.0829	683.1
5	1495.8	6.1244	697.2
10	1507.1	6.1649	711.1
15	1518.4	6.2044	725.0
20	1529.7	6.2432	738.8
25	1540.9	6.2811	752.48
30	1552.1	6.3183	766.11

Saturation, t=−15°C, p=235.54 kPa

Temp. °C	Enthalpy kJ/kg	Entropy kJ/(kg·K)	Sp. Vol. L/kg
−15	1444.2	5.8298	510.1
−10	1456.4	5.8764	522.1
−5	1468.3	5.9214	534.0
0	1480.2	5.9652	545.7
5	1491.9	6.0078	557.3
10	1503.6	6.0493	568.8
15	1515.1	6.0897	580.1
20	1526.6	6.1292	591.4
25	1538.0	6.1678	602.6
30	1549.4	6.2056	613.8
35	1560.7	6.2427	624.8

Saturation, t=−10°C, p=289.93 kPa

Temp. °C	Enthalpy kJ/kg	Entropy kJ/(kg·K)	Sp. Vol. L/kg
−10	1450.5	5.7574	419.5
−5	1462.9	5.8043	429.5
0	1475.2	5.8496	439.3
5	1487.3	5.8935	449.0
10	1499.3	5.9362	458.5
15	1511.1	5.9777	468.0
20	1522.9	6.0181	477.3
25	1534.5	6.0575	486.6
30	1546.1	6.0961	495.8
35	1557.6	6.1338	505.0
40	1569.1	6.1707	514.1

Saturation, t=−5°C, p=353.77 kPa

Temp. °C	Enthalpy kJ/kg	Entropy kJ/(kg·K)	Sp. Vol. L/kg
−5	1456.4	5.6879	347.7
0	1469.1	5.7352	356.0
5	1481.7	5.7808	364.2
10	1494.1	5.8249	372.3
15	1506.3	5.8677	380.3
20	1518.4	5.9093	388.1
25	1530.4	5.9497	395.9
30	1542.2	5.9892	403.6
35	1554.0	6.0277	411.2
40	1565.7	6.0653	418.8
45	1577.3	6.1022	426.332

Saturation, t = 0°C, p=428.18 kPa

Temp. °C	Enthalpy kJ/kg	Entropy kJ/(kg·K)	Sp. Vol. L/kg
0	1461.9	5.6212	290.2
5	1475.0	5.6689	297.2
10	1487.9	5.7148	304.2
15	1500.6	5.7592	311.0
20	1513.1	5.8022	317.7
25	1525.4	5.8439	324.3
30	1537.6	5.8844	330.8
35	1549.7	5.9239	337.2
40	1561.6	5.9624	343.6
45	1573.5	6.0000	350.0
50	1585.3	6.0368	356.2

Saturation, t = 5°C, p=514.29 kPa

Temp. °C	Enthalpy kJ/kg	Entropy kJ/(kg·K)	Sp. Vol. L/kg
5	1467.0	5.5568	243.8
10	1480.6	5.6051	249.8
15	1493.8	5.6515	255.7
20	1506.8	5.6962	261.5
25	1519.6	5.7394	267.2
30	1532.2	5.7812	272.8
35	1544.6	5.8219	278.3
40	1556.9	5.8614	283.7
45	1569.0	5.9000	289.1
50	1581.1	5.9376	294.5
55	1593.1	5.9744	300.0

Table A.26. (cont.) Properties of superheated ammonia in SI units.

Saturation, $t = 10°C$, $p=613.29$ kPa

Temp. °C	Enthalpy kJ/kg	Entropy kJ/(kg·K)	Sp. Vol. L/kg
10	1471.8	5.4947	206.1
15	1485.7	5.5436	211.3
20	1499.3	5.5904	216.3
25	1512.7	5.6355	221.3
30	1525.7	5.6790	226.1
35	1538.6	5.7211	230.9
40	1551.3	5.7619	235.6
45	1563.8	5.8015	240.3
50	1576.2	5.8401	244.9
55	1588.4	5.8778	249.4
60	1600.6	5.9145	253.9

Saturation, $t = 15°C$, $p=726.43$ kPa

Temp. °C	Enthalpy kJ/kg	Entropy kJ/(kg·K)	Sp. Vol. L/kg
15	1476.1	5.4346	175.2
20	1490.5	5.4842	179.7
25	1504.5	5.5316	184.1
30	1518.2	5.5771	188.3
35	1531.6	5.6209	192.5
40	1544.7	5.6632	196.6
45	1557.6	5.7042	200.7
50	1570.4	5.7440	204.7
55	1583.0	5.7827	208.6
60	1595.5	5.8204	212.5
65	1607.8	5.8572	216.4

Saturation, $t = 20°C$, $p=855.01$ kPa

Temp. °C	Enthalpy kJ/kg	Entropy kJ/(kg·K)	Sp. Vol. L/kg
20	1479.9	5.3763	149.7
25	1494.8	5.4266	153.6
30	1509.2	5.4746	157.5
35	1523.3	5.5206	161.2
40	1537.0	5.5648	164.8
45	1550.5	5.6074	168.4
50	1563.7	5.6486	171.9
55	1576.7	5.6886	175.3
60	1589.5	5.7274	178.7
65	1602.2	5.7652	182.1
70	1614.8	5.8021	185.4

Saturation, $t = 25°C$, $p=1000.3$ kPa

Temp. °C	Enthalpy kJ/kg	Entropy kJ/(kg·K)	Sp. Vol. L/kg
25	1483.3	5.3195	128.5
30	1498.7	5.3707	132.0
35	1513.5	5.4193	135.4
40	1528.0	5.4659	138.6
45	1542.1	5.5105	141.8
50	1555.8	5.5535	144.9
55	1569.4	5.5949	148.0
60	1582.6	5.6351	151.0
65	1595.7	5.6741	154.0
70	1608.7	5.7120	156.9
75	1621.4	5.7490	159.8

Saturation, $t = 30°C$, $p=1163.8$ kPa

Temp. °C	Enthalpy kJ/kg	Entropy kJ/(kg·K)	Sp. Vol. L/kg
30	1486.1	5.2641	110.8
35	1502.0	5.3162	113.9
40	1517.4	5.3656	116.9
45	1532.3	5.4127	119.8
50	1546.7	5.4579	122.6
55	1560.8	5.5012	125.3
60	1574.7	5.5430	128.0
65	1588.2	5.5834	130.7
70	1601.6	5.6227	133.3
75	1614.8	5.6607	135.8
80	1627.8	5.6979	138.3

Saturation, $t = 35°C$, $p=1346.9$ kPa

Temp. °C	Enthalpy kJ/kg	Entropy kJ/(kg·K)	Sp. Vol. L/kg
35	1488.4	5.2099	95.96
40	1504.9	5.2630	98.72
45	1520.8	5.3132	101.37
50	1536.1	5.3610	103.93
55	1551.0	5.4067	106.43
60	1565.4	5.4505	108.86
65	1579.6	5.4927	111.24
70	1593.5	5.5334	113.57
75	1607.1	5.5728	115.86
80	1620.5	5.6111	118.12
85	1633.8	5.6484	120.34

Table A.26. (cont.) Properties of superheated ammonia in SI units.

Saturation t = 40° C, p=1550.9 kPa

Temp. °C	Enthalpy kJ/kg	Entropy kJ/(kg·K)	Sp. Vol. L/kg
40	1490.1	5.1565	83.389
45	1507.2	5.2108	85.867
50	1523.6	5.2620	88.247
55	1539.4	5.3106	90.545
60	1554.7	5.3568	92.774
65	1569.6	5.4011	94.943
70	1584.1	5.4437	97.061
75	1598.3	5.4848	99.135
80	1612.2	5.5245	101.17
85	1625.9	5.5630	103.17
90	1639.4	5.6005	105.14

Saturation t = 45° C, p=1777.5 kPa

Temp. °C	Enthalpy kJ/kg	Entropy kJ/(kg·K)	Sp. Vol. L/kg
45	1491.1	5.1039	72.699
50	1508.9	5.1595	74.948
55	1525.9	5.2118	77.100
60	1542.3	5.2611	79.172
65	1558.0	5.3081	81.176
70	1573.3	5.3530	83.122
75	1588.2	5.3960	85.019
80	1602.7	5.4375	86.873
85	1617.0	5.4775	88.689
90	1631.0	5.5163	90.473
95	1644.7	5.5540	92.227

Saturation t = 50° C, p=2028.1 kPa

Temp. °C	Enthalpy kJ/kg	Entropy kJ/(kg·K)	Sp. Vol. L/kg
50	1491.4	5.0518	63.567
55	1510.0	5.1089	65.622
60	1527.7	5.1623	67.581
65	1544.6	5.2126	69.460
70	1560.8	5.2604	71.273
75	1576.5	5.3059	73.030
80	1591.8	5.3494	74.738
85	1606.7	5.3913	76.405
90	1621.3	5.4318	78.036
95	1635.6	5.4709	79.635
100	1649.7	5.5088	81.206

Saturation t = 55° C, p=2304.3 kPa

Temp. °C	Enthalpy kJ/kg	Entropy kJ/(kg·K)	Sp. Vol. L/kg
55	1491.0	5.0001	55.727
60	1510.4	5.0588	57.618
65	1528.8	5.1135	59.412
70	1546.3	5.1649	61.127
75	1563.1	5.2135	62.777
80	1579.3	5.2597	64.372
85	1595.0	5.3038	65.919
90	1610.3	5.3462	67.427
95	1625.2	5.3871	68.899
100	1639.8	5.4266	70.340
105	1654.2	5.4648	71.754

Saturation t = 60° C, p=2607.8 kPa

Temp. °C	Enthalpy kJ/kg	Entropy kJ/(kg·K)	Sp. Vol. L/kg
60	1489.8	4.9484	48.965
65	1510.1	5.0090	50.716
70	1529.2	5.0652	52.370
75	1547.4	5.1177	53.945
80	1564.8	5.1672	55.455
85	1581.5	5.2142	56.911
90	1597.6	5.2591	58.320
95	1613.3	5.3020	59.690
100	1628.7	5.3433	61.025
105	1643.6	5.3832	62.331
110	1658.3	5.4218	63.611

Saturation t = 65° C, p=2849.3 kPa

Temp. °C	Enthalpy kJ/kg	Entropy kJ/(kg·K)	Sp. Vol. L/kg
65	1487.6	4.8965	43.105
70	1509.0	4.9593	44.739
75	1529.0	5.0171	46.273
80	1547.9	5.0710	47.728
85	1565.9	5.1216	49.117
90	1583.1	5.1695	50.453
95	1599.8	5.2150	51.743
100	1615.9	5.2586	52.994
105	1631.6	5.3004	54.212
110	1647.0	5.3408	55.401
115	1662.0	5.3798	56.564

Table A.26. (concl.) Properties of superheated ammonia in SI units.

Saturation t = 70° C, p=3303.4 kPa

Temp. °C	Enthalpy kJ/kg	Entropy kJ/(kg·K)	Sp. Vol. L/kg
70	1484.5	4.8443	38.006
75	1507.1	4.9095	39.540
80	1528.0	4.9692	40.972
85	1547.7	5.0245	42.323
90	1566.3	5.0763	43.609
95	1584.2	5.1252	44.841
100	1601.4	5.1716	46.028
105	1618.0	5.2158	47.176
110	1634.2	5.2582	48.292
115	1649.9	5.2991	49.379
120	1665.3	5.3385	50.441

Saturation t = 75° C, p=3699.1 kPa

Temp. °C	Enthalpy kJ/kg	Entropy kJ/(kg·K)	Sp. Vol. L/kg
75	1480.3	4.7913	33.547
80	1504.2	4.8594	35.000
85	1526.2	4.9213	36.344
90	1546.8	4.9783	37.606
95	1566.2	5.0314	38.802
100	1584.7	5.0813	39.944
105	1602.4	5.1286	41.041
110	1619.6	5.1736	42.100
115	1636.2	5.2167	43.127
120	1652.4	5.2581	44.125
125	1668.2	5.2980	45.100

Saturation t = 80° C, p=4129.2 kPa

Temp. °C	Enthalpy kJ/kg	Entropy kJ/(kg·K)	Sp. Vol. L/kg
80	1475.0	4.7374	29.632
85	1500.4	4.8088	31.017
90	1523.5	4.8731	32.289
95	1545.1	4.9320	33.474
100	1565.3	4.9866	34.592
105	1584.5	5.0377	35.656
110	1602.9	5.0860	36.674
115	1620.6	5.1318	37.655
120	1637.7	5.1756	38.604
125	1654.3	5.2176	39.525
130	1670.5	5.2581	40.422

Saturation t = 85° C, p=4595.6 kPa

Temp. °C	Enthalpy kJ/kg	Entropy kJ/(kg·K)	Sp. Vol. L/kg
85	1468.3	4.6820	26.179
90	1495.4	4.7574	27.510
95	1520.0	4.8245	28.720
100	1542.6	4.8855	29.840
105	1563.7	4.9418	30.891
110	1583.7	4.9942	31.886
.115	1602.7	5.0436	32.836
120	1621.0	5.0904	33.749
125	1638.6	5.1350	34.629
130	1655.8	5.1777	35.482
135	1672.4	5.2187	36.312

Saturation t = 90° C, p=5100.5 kPa

Temp. °C	Enthalpy kJ/kg	Entropy kJ/(kg·K)	Sp. Vol. L/kg
90	1460.1	4.6248	23.118
95	1489.3	4.7048	24.410
100	1515.4	4.7752	25.569
105	1539.2	4.8387	26.633
110	1561.4	4.8968	27.626
115	1582.2	4.9508	28.561
120	1601.9	5.0014	29.451
125	1620.8	5.0492	30.303
130	1639.1	5.0946	31.123
135	1656.7	5.1381	31.917
140	1673.8	5.1798	32.686

Saturation t = 95° C, p=5646.2 kPa

Temp. °C	Enthalpy kJ/kg	Entropy kJ/(kg·K)	Sp. Vol. L/kg
95	1450.1	4.5651	20.392
100	1481.9	4.6509	21.658
105	1509.7	4.7251	22.776
110	1534.9	4.7913	23.793
115	1558.2	4.8516	24.735
120	1579.9	4.9072	25.619
125	1600.5	4.9592	26.456
130	1620.1	5.0081	27.255
135	1638.9	5.0545	28.022
140	1657.1	5.0987	28.762
145	1674.7	5.1412	29.479
150	1691.9	5.1820	30.175

Table A.27. Properties of ammonia saturated liquid and vapor in I-P units*.

Temp., °F	Pressure, psia	Enthalpy, Btu/lb h_f	Enthalpy, Btu/lb h_g	Entropy, Btu/lb·R s_f	Entropy, Btu/lb·R s_g	Liquid, ρ_f lb/ft^3	Vapor, v_g ft^3/lb
-105	0.9947	-68.756	570.18	-0.1776	1.6237	45.709	223.74
-100	1.2334	-63.128	572.40	-0.1623	1.6049	45.517	182.85
-95	1.5191	-57.424	574.60	-0.1465	1.5868	45.322	150.40
-90	1.8589	-52.295	576.78	-0.1327	1.5694	45.130	124.48
-85	2.2608	-47.067	578.94	-0.1182	1.5525	44.931	103.63
-80	2.7335	-41.679	581.07	-0.1041	1.5362	44.727	86.761
-78	2.9444	-39.673	581.92	-0.0989	1.5299	44.649	80.932
-76	3.1688	-37.496	582.76	-0.0933	1.5236	44.570	75.557
-74	3.4073	-35.495	583.60	-0.0881	1.5174	44.491	70.599
-72	3.6607	-33.415	584.44	-0.0827	1.5113	44.408	66.019
-70	3.9295	-31.341	585.27	-0.0773	1.5052	44.327	61.785
-68	4.2147	-29.173	586.09	-0.0718	1.4993	44.247	57.869
-66	4.5168	-27.091	586.91	-0.0665	1.4934	44.161	54.242
-64	4.8367	-25.123	587.73	-0.0613	1.4876	44.079	50.881
-62	5.1752	-22.958	588.54	-0.0561	1.4818	43.997	47.764
-60	5.5331	-20.809	589.35	-0.0507	1.4762	43.911	44.871
-59	5.7195	-19.856	589.75	-0.0484	1.4734	43.869	43.502
-58	5.9112	-18.887	590.15	-0.0459	1.4706	43.829	42.183
-57	6.1081	-17.835	590.55	-0.0431	1.4678	43.787	40.911
-56	6.3104	-16.796	590.95	-0.0407	1.4650	43.745	39.684
-55	6.5182	-15.694	591.34	-0.0379	1.4623	43.702	38.501
-54	6.7317	-14.683	591.74	-0.0354	1.4596	43.659	37.359
-53	6.9508	-13.568	592.13	-0.0328	1.4569	43.621	36.257
-52	7.1758	-12.520	592.52	-0.0302	1.4542	43.577	35.194
-51	7.4068	-11.603	592.92	-0.0280	1.4515	43.538	34.167
-50	7.6438	-10.431	593.31	-0.0251	1.4489	43.493	33.177
-49	7.8870	-9.357	593.69	-0.0225	1.4462	43.450	32.220
-48	8.1366	-8.356	594.08	-0.0201	1.4436	43.408	31.295
-47	8.3926	-7.338	594.47	-0.0176	1.4410	43.364	30.402
-46	8.6552	-6.255	594.85	-0.0149	1.4384	43.323	29.540
-45	8.9245	-5.274	595.24	-0.0126	1.4358	43.281	28.706
-44	9.2006	-4.290	595.62	-0.0100	1.4333	43.241	27.900
-43	9.4837	-3.033	596.00	-0.0072	1.4307	43.196	27.122
-42	9.7739	-2.088	596.38	-0.0049	1.4282	43.154	26.368
-41	10.071	-1.009	596.76	-0.0023	1.4257	43.111	25.640
-40	10.376	0.000	597.13	0.0000	1.4232	43.069	24.936
-39	10.688	1.030	597.51	0.0025	1.4207	43.026	24.255
-38	11.008	2.111	597.88	0.0051	1.4182	42.983	23.595
-37	11.336	3.234	598.26	0.0077	1.4157	42.941	22.957
-36	11.671	4.231	598.63	0.0101	1.4133	42.896	22.340
-35	12.015	5.201	599.00	0.0124	1.4109	42.855	21.742

*From equations of Haar and Gallagher, "Thermodynamic Properties of Ammonia," *Journal of Phys. and Chem. Reference Data*, vol. 7, no. 3, 1978.

Table A.27. (cont.) Properties of ammonia saturated liquid and vapor in I-P units.

Temp., °F	Pressure, psia	Enthalpy, Btu/lb		Entropy, Btu/lb·R		Liquid,	Vapor,
		h_f	h_g	s_f	s_g	ρ_f lb/ft^3	v_g ft^3/lb
-34	12.367	6.381	599.37	0.0151	1.4084	42.813	21.164
-33	12.727	7.413	599.73	0.0175	1.4060	42.769	20.603
-32	13.095	8.479	600.10	0.0201	1.4036	42.726	20.061
-31	13.473	9.572	600.46	0.0226	1.4013	42.682	19.535
-30	13.859	10.552	600.83	0.0248	1.3989	42.641	19.026
-29	14.254	11.636	601.19	0.0275	1.3966	42.597	18.533
-28	14.658	12.635	601.55	0.0297	1.3942	42.554	18.055
-27	15.071	13.763	601.90	0.0324	1.3919	42.511	17.591
-26	15.493	14.819	602.26	0.0347	1.3896	42.469	17.142
-25	15.926	15.892	602.62	0.0371	1.3873	42.425	16.707
-24	16.367	16.922	602.97	0.0395	1.3850	42.384	16.285
-23	16.819	17.956	603.32	0.0419	1.3827	42.339	15.875
-22	17.281	19.057	603.67	0.0444	1.3805	42.297	15.478
-21	17.753	20.116	604.02	0.0468	1.3782	42.252	15.093
-20	18.235	21.237	604.37	0.0494	1.3760	42.209	14.719
-19	18.728	22.251	604.72	0.0517	1.3737	42.165	14.356
-18	19.232	23.308	605.06	0.0541	1.3715	42.123	14.004
-17	19.746	24.398	605.40	0.0566	1.3693	42.079	13.662
-16	20.271	25.451	605.74	0.0589	1.3671	42.034	13.330
-15	20.808	26.511	606.08	0.0613	1.3650	41.992	13.008
-14	21.356	27.583	606.42	0.0637	1.3628	41.948	12.695
-13	21.915	28.681	606.76	0.0661	1.3606	41.904	12.392
-12	22.486	29.752	607.09	0.0685	1.3585	41.860	12.097
-11	23.069	30.759	607.43	0.0708	1.3564	41.815	11.810
-10	23.664	31.824	607.76	0.0732	1.3542	41.772	11.531
-9	24.272	32.986	608.09	0.0757	1.3521	41.728	11.261
-8	24.891	34.019	608.42	0.0781	1.3500	41.683	10.998
-7	25.524	35.085	608.75	0.0803	1.3479	41.640	10.742
-6	26.169	36.176	609.07	0.0828	1.3459	41.596	10.494
-5	26.827	37.263	609.39	0.0851	1.3438	41.552	10.252
-4	27.498	38.346	609.72	0.0875	1.3417	41.508	10.017
-3	28.183	39.385	610.04	0.0898	1.3397	41.465	9.7887
-2	28.881	40.460	610.35	0.0921	1.3376	41.420	9.5665
-1	29.593	41.544	610.67	0.0945	1.3356	41.374	9.3503
0	30.318	42.619	610.99	0.0969	1.3336	41.330	9.1401
1	31.058	43.712	611.30	0.0992	1.3316	41.286	8.9355
2	31.812	44.839	611.61	0.1016	1.3296	41.241	8.7364
3	32.581	45.905	611.92	0.1039	1.3276	41.197	8.5426
4	33.364	46.960	612.23	0.1062	1.3256	41.152	8.3540
5	34.163	48.045	612.54	0.1085	1.3237	41.107	8.1704
6	34.976	49.143	612.84	0.1109	1.3217	41.062	7.9917
7	35.805	50.218	613.15	0.1131	1.3198	41.017	7.8177

Table A.27. (cont.) Properties of ammonia saturated liquid and vapor in I-P units.

Temp., °F	Pressure, psia	Enthalpy, Btu/lb h_f	h_g	Entropy, Btu/lb·R s_f	s_g	Liquid, ρ_f lb/ft^3	Vapor, v_g ft^3/lb
8	36.649	51.321	613.45	0.1155	1.3178	40.973	7.6482
9	37.509	52.376	613.75	0.1177	1.3159	40.928	7.4831
10	38.384	53.456	614.04	0.1200	1.3140	40.882	7.3224
11	39.276	54.575	614.34	0.1224	1.3121	40.837	7.1657
12	40.184	55.633	614.64	0.1247	1.3101	40.791	7.0132
13	41.109	56.707	614.93	0.1269	1.3083	40.747	6.8645
14	42.050	57.814	615.22	0.1292	1.3064	40.701	6.7196
15	43.008	58.923	615.51	0.1316	1.3045	40.655	6.5784
16	43.984	60.025	615.79	0.1339	1.3026	40.610	6.4408
17	44.977	61.099	616.08	0.1362	1.3008	40.564	6.3066
18	45.987	62.207	616.36	0.1385	1.2989	40.518	6.1758
19	47.015	63.288	616.64	0.1407	1.2971	40.473	6.0483
20	48.062	64.400	616.92	0.1430	1.2952	40.426	5.9240
21	49.126	65.472	617.20	0.1452	1.2934	40.380	5.8027
22	50.209	66.603	617.48	0.1476	1.2916	40.334	5.6844
23	51.311	67.669	617.75	0.1498	1.2898	40.288	5.5690
24	52.431	68.780	618.02	0.1521	1.2880	40.242	5.4565
25	53.571	69.898	618.29	0.1544	1.2862	40.195	5.3466
26	54.730	70.982	618.56	0.1566	1.2844	40.150	5.2395
27	55.908	72.090	618.83	0.1588	1.2826	40.103	5.1349
28	57.106	73.196	619.09	0.1611	1.2808	40.056	5.0328
29	58.325	74.304	619.35	0.1634	1.2791	40.009	4.9332
30	59.563	75.389	619.61	0.1656	1.2773	39.963	4.8360
31	60.822	76.493	619.87	0.1678	1.2756	39.915	4.7410
32	62.102	77.602	620.13	0.1701	1.2738	39.869	4.6484
33	63.403	78.711	620.38	0.1723	1.2721	39.821	4.5578
34	64.725	79.803	620.63	0.1745	1.2704	39.774	4.4695
35	66.068	80.937	620.88	0.1768	1.2686	39.727	4.3831
36	67.433	82.025	621.13	0.1790	1.2669	39.680	4.2988
37	68.820	83.134	621.38	0.1812	1.2652	39.632	4.2165
38	70.229	84.270	621.62	0.1835	1.2635	39.585	4.1360
39	71.660	85.357	621.86	0.1856	1.2618	39.537	4.0574
40	73.114	86.470	622.10	0.1879	1.2601	39.489	3.9806
41	74.591	87.580	622.34	0.1900	1.2585	39.441	3.9055
42	76.091	88.697	622.57	0.1923	1.2568	39.393	3.8321
43	77.614	89.817	622.81	0.1945	1.2551	39.345	3.7604
44	79.161	90.931	623.04	0.1967	1.2535	39.297	3.6903
45	80.731	92.054	623.26	0.1989	1.2518	39.249	3.6218
46	82.326	93.173	623.49	0.2011	1.2502	39.201	3.5548
47	83.945	94.283	623.72	0.2033	1.2485	39.152	3.4893
48	85.589	95.393	623.94	0.2054	1.2469	39.104	3.4253
49	87.257	96.516	624.16	0.2076	1.2453	39.055	3.3626

Table A.27. (cont.) Properties of ammonia saturated liquid and vapor in I-P units.

Temp., °F	Pressure, psia	Enthalpy, Btu/lb		Entropy, Btu/lb·R		Liquid,	Vapor,
		h_f	h_g	s_f	s_g	ρ_f lb/ft^3	v_g ft^3/lb
50	88.950	97.648	624.37	0.2099	1.2436	39.006	3.3014
51	90.669	98.756	624.59	0.2120	1.2420	38.957	3.2415
52	92.413	99.892	624.80	0.2142	1.2404	38.908	3.1828
53	94.183	101.01	625.01	0.2164	1.2388	38.859	3.1255
54	95.980	102.14	625.22	0.2186	1.2372	38.810	3.0694
55	97.802	103.26	625.43	0.2207	1.2356	38.760	3.0145
56	99.651	104.40	625.63	0.2229	1.2340	38.711	2.9608
57	101.53	105.52	625.83	0.2251	1.2324	38.661	2.9082
58	103.43	106.65	626.03	0.2272	1.2308	38.611	2.8568
59	105.36	107.77	626.23	0.2294	1.2293	38.562	2.8064
60	107.32	108.90	626.42	0.2316	1.2277	38.511	2.7571
61	109.30	110.05	626.61	0.2337	1.2261	38.462	2.7089
62	111.32	111.17	626.80	0.2359	1.2246	38.411	2.6616
63	113.36	112.30	626.99	0.2380	1.2230	38.361	2.6154
64	115.43	113.43	627.17	0.2402	1.2215	38.311	2.5701
65	117.53	114.57	627.35	0.2423	1.2199	38.260	2.5257
66	119.66	115.72	627.53	0.2445	1.2184	38.209	2.4823
67	121.82	116.85	627.71	0.2466	1.2169	38.158	2.4397
68	124.01	117.99	627.88	0.2487	1.2153	38.107	2.3981
69	126.23	119.13	628.05	0.2509	1.2138	38.056	2.3572
70	128.48	120.28	628.22	0.2530	1.2123	38.005	2.3173
71	130.76	121.42	628.39	0.2552	1.2108	37.954	2.2781
72	133.07	122.55	628.55	0.2573	1.2093	37.902	2.2397
73	135.41	123.70	628.71	0.2594	1.2078	37.851	2.2021
74	137.78	124.86	628.87	0.2616	1.2063	37.799	2.1652
75	140.18	126.01	629.02	0.2637	1.2048	37.747	2.1291
76	142.62	127.15	629.18	0.2658	1.2033	37.695	2.0936
77	145.09	128.29	629.33	0.2679	1.2018	37.643	2.0589
78	147.59	129.43	629.47	0.2700	1.2003	37.591	2.0249
79	150.12	130.59	629.62	0.2721	1.1988	37.538	1.9915
80	152.69	131.75	629.76	0.2743	1.1973	37.486	1.9588
81	155.29	132.90	629.90	0.2764	1.1959	37.433	1.9268
82	157.92	134.05	630.03	0.2785	1.1944	37.380	1.8953
83	160.59	135.21	630.16	0.2806	1.1929	37.327	1.8645
84	163.29	136.36	630.29	0.2827	1.1915	37.274	1.8342
85	166.03	137.53	630.42	0.2848	1.1900	37.221	1.8046
86	168.80	138.67	630.55	0.2869	1.1885	37.167	1.7755
87	171.60	139.84	630.67	0.2890	1.1871	37.114	1.7470
88	174.45	141.00	630.78	0.2911	1.1856	37.060	1.7190
89	177.32	142.17	630.90	0.2932	1.1842	37.006	1.6915

Table A.27. (cont.) Properties of ammonia saturated liquid and vapor in I-P units.

Temp., °F	Pressure, psia	Enthalpy, Btu/lb h_f	h_g	Entropy, Btu/lb·R s_f	s_g	Liquid, ρ_f lb/ft³	Vapor, v_g ft³/lb
90	180.24	143.32	631.01	0.2953	1.1828	36.952	1.6646
91	183.18	144.49	631.12	0.2974	1.1813	36.898	1.6381
92	186.17	145.66	631.23	0.2994	1.1799	36.844	1.6122
93	189.19	146.82	631.33	0.3015	1.1784	36.790	1.5867
94	192.25	148.00	631.43	0.3036	1.1770	36.735	1.5617
95	195.35	149.16	631.52	0.3057	1.1756	36.680	1.5372
96	198.48	150.34	631.62	0.3078	1.1742	36.626	1.5131
97	201.65	151.51	631.71	0.3099	1.1727	36.570	1.4895
98	204.86	152.68	631.79	0.3119	1.1713	36.515	1.4663
99	208.11	153.86	631.88	0.3140	1.1699	36.460	1.4436
100	211.40	155.03	631.96	0.3161	1.1685	36.404	1.4212
101	214.72	156.21	632.03	0.3182	1.1671	36.349	1.3993
102	218.09	157.39	632.11	0.3202	1.1657	36.293	1.3777
103	221.50	158.57	632.18	0.3223	1.1643	36.237	1.3565
104	224.94	159.75	632.24	0.3244	1.1628	36.181	1.3358
105	228.43	160.93	632.31	0.3264	1.1614	36.124	1.3153
106	231.95	162.11	632.37	0.3285	1.1600	36.068	1.2953
107	235.52	163.30	632.42	0.3306	1.1586	36.011	1.2756
108	239.13	164.49	632.47	0.3326	1.1572	35.954	1.2563
109	242.78	165.68	632.52	0.3347	1.1559	35.897	1.2373
110	246.47	166.87	632.57	0.3367	1.1545	35.840	1.2186
111	250.20	168.07	632.61	0.3388	1.1531	35.783	1.2003
112	253.98	169.25	632.65	0.3408	1.1517	35.725	1.1822
113	257.80	170.45	632.68	0.3429	1.1503	35.667	1.1645
114	261.66	171.65	632.71	0.3450	1.1489	35.609	1.1471
115	265.57	172.84	632.74	0.3470	1.1475	35.551	1.1300
116	269.52	174.04	632.76	0.3491	1.1461	35.493	1.1132
117	273.51	175.24	632.78	0.3511	1.1447	35.435	1.0967
118	277.55	176.44	632.80	0.3531	1.1434	35.376	1.0805
119	281.63	177.65	632.81	0.3552	1.1420	35.317	1.0645
120	285.76	178.85	632.82	0.3572	1.1406	35.258	1.0488
121	289.93	180.06	632.82	0.3593	1.1392	35.199	1.0334
122	294.15	181.27	632.82	0.3613	1.1378	35.140	1.0183
123	298.41	182.48	632.82	0.3634	1.1365	35.080	1.0033
124	302.72	183.69	632.81	0.3654	1.1351	35.020	0.98869
125	307.08	184.91	632.80	0.3674	1.1337	34.960	0.97428
126	311.48	186.12	632.78	0.3695	1.1323	34.900	0.96010
127	315.93	187.34	632.76	0.3715	1.1310	34.840	0.94616
128	320.43	188.56	632.74	0.3735	1.1296	34.779	0.93246
129	324.98	189.77	632.71	0.3756	1.1282	34.718	0.91897

Table A.27. (cont.) Properties of ammonia saturated liquid and vapor in I-P units.

Temp., °F	Pressure, psia	Enthalpy, Btu/lb h_f	h_g	Entropy, Btu/lb·R s_f	s_g	Liquid, ρ_f lb/ft³	Vapor, v_g ft³/lb
130	329.57	191.00	632.68	0.3776	1.1269	34.657	0.90571
131	334.21	192.22	632.64	0.3796	1.1255	34.596	0.89267
132	338.91	193.45	632.60	0.3817	1.1241	34.534	0.87984
133	343.65	194.68	632.55	0.3837	1.1227	34.473	0.86721
134	348.44	195.91	632.50	0.3857	1.1214	34.411	0.85479
135	353.28	197.14	632.45	0.3878	1.1200	34.349	0.84257
136	358.17	198.37	632.39	0.3898	1.1186	34.286	0.83055
137	363.11	199.60	632.32	0.3918	1.1173	34.224	0.81872
138	368.10	200.84	632.26	0.3938	1.1159	34.161	0.80708
139	373.14	202.09	632.18	0.3959	1.1145	34.098	0.79562
140	378.23	203.33	632.11	0.3979	1.1131	34.034	0.78435
141	383.38	204.57	632.02	0.3999	1.1118	33.971	0.77325
142	388.58	205.82	631.94	0.4019	1.1104	33.907	0.76233
143	393.83	207.07	631.84	0.4040	1.1090	33.843	0.75157
144	399.13	208.31	631.75	0.4060	1.1076	33.779	0.74099
145	404.49	209.57	631.64	0.4080	1.1063	33.714	0.73057
146	409.90	210.82	631.54	0.4100	1.1049	33.649	0.72032
147	415.36	212.08	631.43	0.4121	1.1035	33.584	0.71022
148	420.88	213.34	631.31	0.4141	1.1021	33.519	0.70027
149	426.45	214.60	631.19	0.4161	1.1008	33.453	0.69048
150	432.08	215.87	631.06	0.4181	1.0994	33.388	0.68084
151	437.76	217.13	630.93	0.4202	1.0980	33.321	0.67135
152	443.50	218.40	630.79	0.4222	1.0966	33.255	0.66200
153	449.30	219.67	630.65	0.4242	1.0952	33.188	0.65279
154	455.15	220.95	630.50	0.4262	1.0938	33.121	0.64373
155	461.06	222.22	630.35	0.4283	1.0924	33.054	0.63479
156	467.02	223.50	630.19	0.4303	1.0911	32.986	0.62600
157	473.04	224.78	630.02	0.4323	1.0897	32.919	0.61733
158	479.12	226.07	629.85	0.4343	1.0883	32.850	0.60879
159	485.26	227.36	629.68	0.4364	1.0869	32.782	0.60038
160	491.46	228.65	629.50	0.4384	1.0855	32.713	0.59209
162	504.03	231.24	629.11	0.4425	1.0827	32.575	0.57588
164	516.84	233.84	628.71	0.4465	1.0799	32.435	0.56014
166	529.89	236.45	628.28	0.4506	1.0770	32.293	0.54485
168	543.19	239.08	627.83	0.4546	1.0742	32.151	0.53000
170	556.73	241.72	627.35	0.4587	1.0714	32.007	0.51557
172	570.53	244.37	626.84	0.4628	1.0685	31.861	0.50155
174	584.58	247.04	626.31	0.4669	1.0656	31.714	0.48792
176	598.88	249.72	625.75	0.4710	1.0627	31.565	0.47466
178	613.45	252.42	625.16	0.4751	1.0598	31.414	0.46177
180	628.28	255.13	624.55	0.4792	1.0569	31.262	0.44924
182	643.38	257.86	623.90	0.4833	1.0540	31.108	0.43704
184	658.75	260.61	623.22	0.4874	1.0510	30.952	0.42517

Table A.27. (concl.) Properties of ammonia saturated liquid and vapor in I-P units.

Temp., °F	Pressure, psia	Enthalpy, Btu/lb h_f	Enthalpy, Btu/lb h_g	Entropy, Btu/lb·R s_f	Entropy, Btu/lb·R s_g	Liquid, ρ_f lb/ft^3	Vapor, v_g ft^3/lb
186	674.39	263.37	622.51	0.4916	1.0480	30.794	0.41361
188	690.31	266.15	621.77	0.4957	1.0450	30.635	0.40236
190	706.51	268.95	621.00	0.4999	1.0420	30.473	0.39140
192	723.00	271.78	620.19	0.5041	1.0389	30.308	0.38072
194	739.77	274.62	619.34	0.5083	1.0359	30.142	0.37032
196	756.83	277.48	618.46	0.5125	1.0327	29.973	0.36018
198	774.19	280.37	617.53	0.5167	1.0296	29.802	0.35030
200	791.85	283.28	616.57	0.5210	1.0264	29.628	0.34066
205	837.34	290.67	613.98	0.5317	1.0183	29.180	0.31758
210	884.78	298.23	611.10	0.5426	1.0100	28.714	0.29585
215	934.25	305.99	607.90	0.5537	1.0013	28.225	0.27534
220	985.81	313.98	604.33	0.5650	0.9923	27.711	0.25593
225	1039.5	322.24	600.34	0.5765	0.9829	27.168	0.23753
230	1095.5	330.80	595.87	0.5884	0.9729	26.590	0.22001
235	1153.8	339.73	590.83	0.6007	0.9624	25.971	0.20326
240	1214.6	349.12	585.10	0.6136	0.9510	25.300	0.18718
245	1277.9	359.08	578.52	0.6271	0.9386	24.565	0.17163
250	1343.9	369.79	570.83	0.6415	0.9249	23.743	0.15645
255	1412.7	381.57	561.67	0.6573	0.9094	22.802	0.14146
260	1484.5	395.01	550.39	0.6752	0.8912	21.673	0.12635
265	1559.7	411.63	535.80	0.6973	0.8687	20.191	0.11063

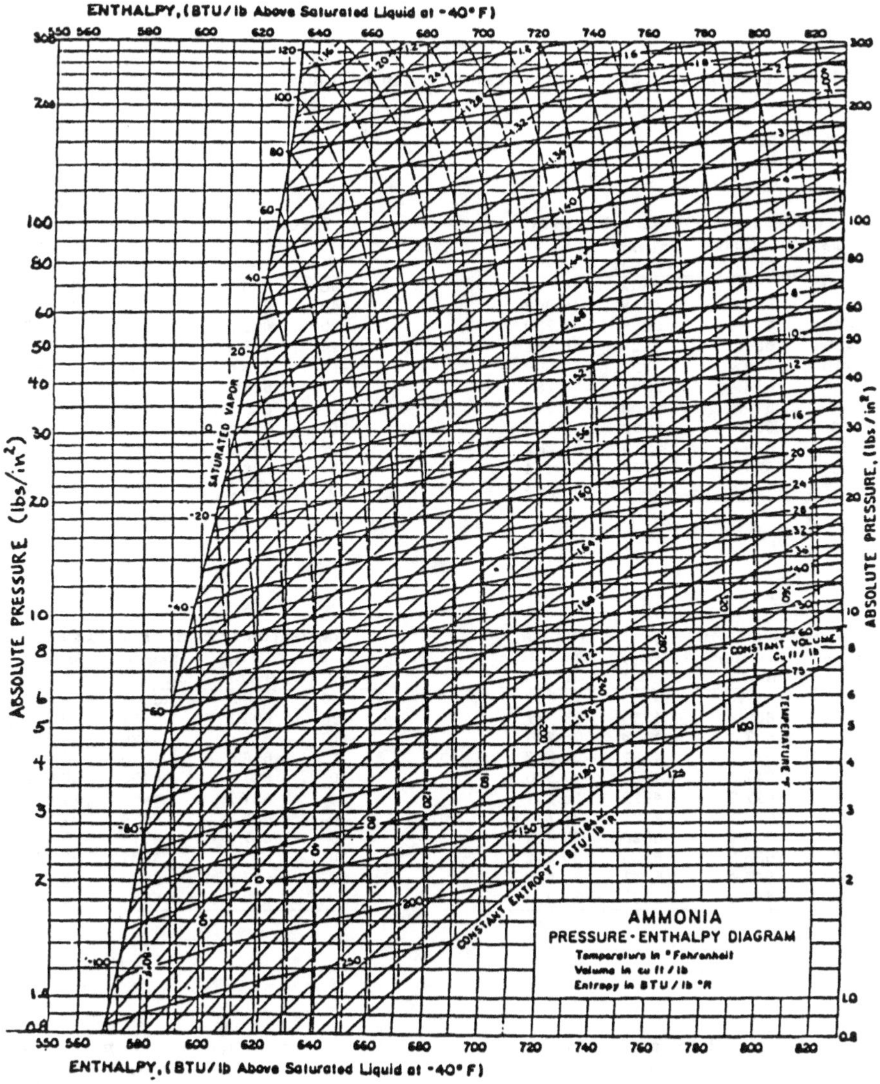

Fig. A.14. Pressure-enthalpy diagram for ammonia in IP units.

Table A.28. Properties of superheated ammonia in I-P units.

Saturation, t=−60° F, p=5.533 psia

Temp. °F	Enthalpy Btu/lb	Entropy Btu/lb·R	Sp. Vol. ft^3/lb
−60	589.3	1.4762	44.871
−50	594.4	1.4885	46.063
−40	599.4	1.5006	47.250
−30	604.4	1.5124	48.431
−20	609.4	1.5239	49.609
−10	614.4	1.5351	50.782
0	619.4	1.5462	51.952
10	624.4	1.5569	53.119
20	629.4	1.5675	54.282
30	634.4	1.5778	55.444
40	639.4	1.5880	56.603

Saturation, t=−50° F, p=7.644 psia

Temp. °F	Enthalpy Btu/lb	Entropy Btu/lb·R	Sp. Vol. ft^3/lb
−50	593.3	1.4489	33.177
−40	598.4	1.4611	34.049
−30	603.5	1.4731	34.917
−20	608.6	1.4848	35.781
−10	613.6	1.4962	36.640
0	618.7	1.5073	37.495
10	623.7	1.5182	38.348
20	628.8	1.5289	39.197
30	633.9	1.5393	40.044
40	638.9	1.5496	40.889
50	644.0	1.5596	41.732

Saturation, t=−40° F, p=10.378 psia

Temp. °F	Enthalpy Btu/lb	Entropy Btu/lb·R	Sp. Vol. ft^3/lb
−40	597.1	1.4232	24.936
−30	602.3	1.4354	25.587
−20	607.5	1.4473	26.234
−10	612.7	1.4589	26.877
0	617.8	1.4702	27.516
10	622.9	1.4812	28.151
20	628.1	1.4920	28.784
30	633.2	1.5026	29.414
40	638.3	1.5129	30.042
50	643.4	1.5231	30.668
60	648.5	1.5330	31.292

Saturation, t=−30° F, p=13.859 psia

Temp. °F	Enthalpy Btu/lb	Entropy Btu/lb·R	Sp. Vol. ft^3/lb
−30	600.8	1.3989	19.026
−20	606.1	1.4111	19.521
−10	611.4	1.4229	20.012
0	616.6	1.4345	20.499
10	621.9	1.4457	20.982
20	627.1	1.4567	21.463
30	632.3	1.4674	21.940
40	637.5	1.4779	22.416
50	642.6	1.4882	22.889
60	647.8	1.4982	23.361
70	653.0	1.5081	23.831

Saturation, t=−20° F, p=18.24 psia

Temp. °F	Enthalpy Btu/lb	Entropy Btu/lb·R	Sp. Vol. ft^3/lb
−20	604.4	1.3760	14.719
−10	609.8	1.3882	15.101
0	615.2	1.4000	15.479
10	620.5	1.4115	15.854
20	625.8	1.4227	16.226
30	631.1	1.4336	16.595
40	636.4	1.4443	16.962
50	641.7	1.4547	17.326
60	646.9	1.4649	17.689
70	652.1	1.4749	18.050
80	657.4	1.4847	18.410

Saturation, t=−10° F, p=23.66 psia

Temp. °F	Enthalpy Btu/lb	Entropy Btu/lb·R	Sp. Vol. ft^3/lb
−10	607.8	1.3542	11.531
0	613.3	1.3665	11.831
10	618.8	1.3783	12.128
20	624.3	1.3898	12.421
30	629.7	1.4010	12.711
40	635.1	1.4119	12.999
50	640.4	1.4225	13.284
60	645.8	1.4329	13.568
70	651.1	1.4430	13.850
80	656.4	1.4529	14.131
90	661.7	1.4626	14.410

Table A.28. (cont.) Properties of superheated ammonia in I-P units.

Saturation, t=0° F, p= 30.32 psia

Temp. °F	Enthalpy Btu/lb	Entropy Btu/lb·R	Sp. Vol. ft^3/lb
0	611.0	1.3336	9.1401
10	616.7	1.3459	9.3788
20	622.3	1.3578	9.6141
30	627.9	1.3693	9.8464
40	633.4	1.3805	10.0762
50	638.9	1.3913	10.3038
60	644.4	1.4019	10.5294
70	649.8	1.4122	10.7532
80	655.2	1.4223	10.9756
90	660.6	1.4322	11.1967
100	665.9	1.4419	11.4165

Saturation, t=10° F, p= 38.38 psia

Temp. °F	Enthalpy Btu/lb	Entropy Btu/lb·R	Sp. Vol. ft^3/lb
10	614.0	1.3140	7.3224
20	619.9	1.3263	7.5149
30	625.7	1.3383	7.7044
40	631.4	1.3498	7.8911
50	637.1	1.3610	8.0754
60	642.6	1.3719	8.2578
70	648.2	1.3824	8.4383
80	653.7	1.3927	8.6172
90	659.2	1.4028	8.7948
100	664.6	1.4126	8.9712
110	670.1	1.4222	9.1466

Saturation, t = 20° F, p= 48.06 psia

Temp. °F	Enthalpy Btu/lb	Entropy Btu/lb·R	Sp. Vol. ft^3/lb
20	616.9	1.2952	5.9240
30	623.0	1.3077	6.0813
40	628.9	1.3197	6.2357
50	634.8	1.3313	6.3876
60	640.5	1.3425	6.5374
70	646.2	1.3534	6.6853
80	651.9	1.3640	6.8315
90	657.5	1.3743	6.9764
100	663.1	1.3843	7.1199
110	668.6	1.3941	7.2624
120	674.1	1.4037	7.4039

Saturation, t = 30° F, p= 59.56 psia

Temp. °F	Enthalpy Btu/lb	Entropy Btu/lb·R	Sp. Vol. ft^3/lb
30	619.6	1.2773	4.8360
40	625.8	1.2899	4.9661
50	632.0	1.3020	5.0935
60	638.0	1.3137	5.2186
70	643.9	1.3249	5.3416
80	649.7	1.3358	5.4630
90	655.5	1.3464	5.5829
100	661.2	1.3567	5.7014
110	666.8	1.3667	5.8188
120	672.4	1.3765	5.9352
130	678.0	1.3861	6.0507

Saturation, t = 40° F, p= 73.11 psia

Temp. °F	Enthalpy Btu/lb	Entropy Btu/lb·R	Sp. Vol. ft^3/lb
40	622.1	1.2601	3.9806
50	628.5	1.2729	4.0894
60	634.8	1.2851	4.1957
70	641.0	1.2969	4.2998
80	647.0	1.3082	4.4021
90	653.0	1.3191	4.5028
100	658.9	1.3297	4.6021
110	664.7	1.3400	4.7002
120	670.4	1.3500	4.7972
130	676.2	1.3598	4.8933
140	681.8	1.3694	4.9887

Saturation, t = 50° F, p= 88.95 psia

Temp. °F	Enthalpy Btu/lb	Entropy Btu/lb·R	Sp. Vol. ft^3/lb
50	624.4	1.2436	3.3014
60	631.0	1.2566	3.3934
70	637.5	1.2689	3.4830
80	643.8	1.2808	3.5706
90	650.0	1.2921	3.6565
100	656.1	1.3031	3.7409
110	662.1	1.3138	3.8240
120	668.1	1.3241	3.9061
130	673.9	1.3341	3.9871
140	679.8	1.3439	4.0673
150	685.5	1.3534	4.1467

Table A.28. (cont.) Properties of superheated ammonia in I-P units.

Saturation, t = 60°F, p = 107.32 psia

Temp. °F	Enthalpy Btu/lb	Entropy Btu/lb·R	Sp. Vol. ft^3/lb
60	626.4	1.2277	2.7571
70	633.3	1.2408	2.8357
80	640.0	1.2533	2.9121
90	646.5	1.2653	2.9865
100	652.9	1.2768	3.0594
110	659.1	1.2878	3.1308
120	665.3	1.2985	3.2011
130	671.3	1.3089	3.2703
140	677.3	1.3189	3.3386
150	683.2	1.3287	3.4062
160	689.1	1.3383	3.4730

Saturation, t = 70°F, p = 128.47 psia

Temp. °F	Enthalpy Btu/lb	Entropy Btu/lb·R	Sp. Vol. ft^3/lb
70	628.2	1.2123	2.3173
80	635.4	1.2257	2.3851
90	642.3	1.2383	2.4507
100	649.0	1.2504	2.5146
110	655.5	1.2620	2.5770
120	661.9	1.2732	2.6380
130	668.2	1.2839	2.6980
140	674.4	1.2943	2.7570
150	680.5	1.3044	2.8151
160	686.5	1.3142	2.8725
170	692.5	1.3237	2.9292

Saturation, t = 80°F, p = 152.7 psia

Temp. °F	Enthalpy Btu/lb	Entropy Btu/lb·R	Sp. Vol. ft^3/lb
80	629.8	1.1973	1.9588
90	637.2	1.2110	2.0179
100	644.3	1.2238	2.0749
110	651.2	1.2361	2.1302
120	658.0	1.2478	2.1841
130	664.5	1.2590	2.2367
140	671.0	1.2698	2.2883
150	677.3	1.2803	2.3390
160	683.5	1.2904	2.3889
170	689.6	1.3002	2.4381
180	695.7	1.3098	2.4867

Saturation, t = 90°F, p = 180.2 psia

Temp. °F	Enthalpy Btu/lb	Entropy Btu/lb·R	Sp. Vol. ft^3/lb
90	631.0	1.1828	1.6646
100	638.7	1.1967	1.7165
110	646.1	1.2098	1.7664
120	653.3	1.2222	1.8147
130	660.2	1.2340	1.8617
140	666.9	1.2454	1.9075
150	673.5	1.2563	1.9522
160	680.0	1.2668	1.9961
170	686.3	1.2770	2.0393
180	692.6	1.2868	2.0818
190	698.8	1.2965	2.1237

Saturation, t = 100°F, p = 211.4 psia

Temp. °F	Enthalpy Btu/lb	Entropy Btu/lb·R	Sp. Vol. ft^3/lb
100	632.0	1.1685	1.4212
110	640.0	1.1827	1.4673
120	647.7	1.1961	1.5114
130	655.0	1.2087	1.5540
140	662.2	1.2207	1.5952
150	669.1	1.2321	1.6353
160	675.9	1.2432	1.6745
170	682.5	1.2538	1.7128
180	689.0	1.2640	1.7504
190	695.4	1.2739	1.7874
200	701.7	1.2836	1.8239

Saturation, t = 110°F, p = 246.5 psia

Temp. °F	Enthalpy Btu/lb	Entropy Btu/lb·R	Sp. Vol. ft^3/lb
110	632.6	1.1545	1.2186
120	640.9	1.1690	1.2598
130	648.9	1.1827	1.2991
140	656.6	1.1955	1.3369
150	663.9	1.2077	1.3734
160	671.1	1.2193	1.4089
170	678.0	1.2304	1.4434
180	684.8	1.2411	1.4771
190	691.5	1.2515	1.5102
200	698.0	1.2615	1.5426
210	704.4	1.2712	1.5746

Table A.28. (cont.) Properties of superheated ammonia in I-P units.

Saturation, t = 120°F, p = 285.8 psia

Temp. °F	Enthalpy Btu/lb	Entropy Btu/lb·R	Sp. Vol. ft^3/lb
120	632.8	1.1406	1.0488
130	641.6	1.1556	1.0860
140	649.9	1.1696	1.1214
150	657.8	1.1827	1.1552
160	665.4	1.1950	1.1877
170	672.8	1.2068	1.2193
180	679.9	1.2181	1.2499
190	686.9	1.2289	1.2798
200	693.7	1.2393	1.3090
210	700.4	1.2494	1.3377
220	707.0	1.2591	1.3659

Saturation, t = 130°F, p = 329.6 psia

Temp. °F	Enthalpy Btu/lb	Entropy Btu/lb·R	Sp. Vol. ft^3/lb
130	632.7	1.1269	0.9057
140	641.9	1.1423	0.9396
150	650.5	1.1566	0.9715
160	658.8	1.1700	1.0020
170	666.6	1.1826	1.0313
180	674.2	1.1946	1.0595
190	681.6	1.2060	1.0869
200	688.7	1.2170	1.1136
210	695.7	1.2275	1.1396
220	702.6	1.2376	1.1651
230	709.3	1.2475	1.1901

Saturation, t = 140°F, p = 378.2 psia

Temp. °F	Enthalpy Btu/lb	Entropy Btu/lb·R	Sp. Vol. ft^3/lb
140	632.1	1.1131	0.7843
150	641.8	1.1291	0.8154
160	650.8	1.1439	0.8446
170	659.4	1.1576	0.8723
180	667.6	1.1705	0.8988
190	675.4	1.1826	0.9242
200	683.0	1.1942	0.9489
210	690.4	1.2053	0.9728
220	697.6	1.2160	0.9962
230	704.6	1.2262	1.0189
240	711.5	1.2362	1.0413

Saturation, t = 150°F, p = 432.1 psia

Temp. °F	Enthalpy Btu/lb	Entropy Btu/lb·R	Sp. Vol. ft^3/lb
150	631.1	1.0994	0.6808
160	641.3	1.1160	0.7096
170	650.8	1.1312	0.7364
180	659.7	1.1453	0.7618
190	668.2	1.1585	0.7859
200	676.3	1.1709	0.8090
210	684.2	1.1827	0.8313
220	691.7	1.1939	0.8529
230	699.1	1.2047	0.8740
240	706.3	1.2151	0.8945
250	713.4	1.2251	0.9146

Saturation, t = 160°F, p = 491.5 psia

Temp. °F	Enthalpy Btu/lb	Entropy Btu/lb·R	Sp. Vol. ft^3/lb
160	629.5	1.0855	0.5921
170	640.3	1.1029	0.6189
180	650.3	1.1186	0.6438
190	659.7	1.1331	0.6671
200	668.5	1.1466	0.6893
210	676.9	1.1593	0.7104
220	685.0	1.1713	0.7307
230	692.9	1.1827	0.7503
240	700.5	1.1936	0.7694
250	707.9	1.2042	0.7880
260	715.1	1.2143	0.8061

Saturation, t = 170°F, p = 556.7 psia

Temp. °F	Enthalpy Btu/lb	Entropy Btu/lb·R	Sp. Vol. ft^3/lb
170	627.3	1.0714	0.5156
180	638.9	1.0896	0.5409
190	649.5	1.1060	0.5461
200	659.3	1.1210	0.5857
210	668.5	1.1348	0.6061
220	677.3	1.1478	0.6255
230	685.6	1.1600	0.6442
240	693.7	1.1717	0.6621
250	701.5	1.1828	0.6795
260	709.2	1.1935	0.6964
270	716.6	1.2036	0.7128

Table A.28. (concl.) Properties of superheated ammonia in I-P units.

Saturation, t = 180°F, p = 628.3 psia

Temp. °F	Enthalpy Btu/lb	Entropy Btu/lb·R	Sp. Vol. ft^3/lb
180	624.5	1.0569	0.4492
190	637.0	1.0762	0.4733
200	648.2	1.0933	0.4951
210	658.5	1.1088	0.5153
220	668.1	1.1231	0.5343
230	677.2	1.1364	0.5522
240	685.9	1.1489	0.5694
250	694.3	1.1608	0.5858
260	702.4	1.1721	0.6018
270	710.2	1.1829	0.6172
280	717.9	1.1933	0.6322
290	725.3	1.2033	0.6469

Saturation, t = 190°F, p = 706.5 psia

Temp. °F	Enthalpy Btu/lb	Entropy Btu/lb·R	Sp. Vol. ft^3/lb
190	621.0	1.0420	0.3914
200	634.4	1.0625	0.4145
210	646.4	1.0805	0.4352
220	657.2	1.0966	0.4542
230	667.3	1.1114	0.4719
240	676.9	1.1251	0.4886
250	685.9	1.1379	0.5045
260	694.6	1.1500	0.5197
270	702.9	1.1615	0.5343
280	711.0	1.1725	0.5485
290	718.9	1.1831	0.5623
300	726.5	1.1932	0.5757

Saturation, t = 200°F, p = 791.9 psia

Temp. °F	Enthalpy Btu/lb	Entropy Btu/lb·R	Sp. Vol. ft^3/lb
200	616.6	1.0264	0.3407
210	631.2	1.0485	0.3632
220	644.0	1.0674	0.3830
230	655.5	1.0843	0.4009
240	666.2	1.0996	0.4175
250	676.1	1.1137	0.4331
260	685.5	1.1269	0.4479
270	694.5	1.1393	0.4620
280	703.2	1.1510	0.4756
290	711.5	1.1623	0.4886
300	719.6	1.1730	0.5013
310	727.5	1.1833	0.5137

Saturation, t = 210°F, p = 884.8 psia

Temp. °F	Enthalpy Btu/lb	Entropy Btu/lb·R	Sp. Vol. ft^3/lb
210	611.1	1.0100	0.2958
220	627.3	1.0340	0.3180
230	641.1	1.0541	0.3371
240	653.4	1.0718	0.3543
250	664.6	1.0877	0.3700
260	675.0	1.1023	0.3846
270	684.8	1.1159	0.3984
280	694.2	1.1286	0.4116
290	703.1	1.1406	0.4241
300	711.8	1.1521	0.4363
310	720.1	1.1630	0.4480
320	728.2	1.1734	0.4594

INDEX

www.ingramcontent.com/pod-product-compliance
Lightning Source LLC
LaVergne TN
LVHW021001300926
849160LV00007B/211
* 9 7 8 1 2 6 5 8 3 0 9 9 1 *